A CONCISE COURSE IN A-LEVEL STATISTICS

Second Edition

A CONCISE COURSE IN A-LEVEL STATISTICS

With Worked Examples

Second Edition

J CRAWSHAW BSc

Head of Mathematics Department
Clifton High School, Bristol

J CHAMBERS MA

Head of Mathematics Department
Sutton High School GPDST, Surrey

STANLEY THORNES (PUBLISHERS) LTD

First published in 1984 by

Stanley Thornes (Publishers) Ltd
Ellenborough House
Wellington Street
CHELTENHAM GL50 1YD
UK

First Edition 1984
Reprinted 1985 (twice)
Reprinted 1986
Reprinted 1987
Reprinted 1988 (twice)
Reprinted 1989
Second Edition 1990
Reprinted 1991
Reprinted 1992
Reprinted 1993

British Library Cataloguing in Publication Data

Crawshaw, J.
A concise course in A-level statistics. With worked examples
2nd ed.
1. Statistics
I. Title II. Chambers, J.
519.5

ISBN 0-7487-0455-8

Typeset by Tech-Set, Gateshead, Tyne & Wear.
Printed and bound in Great Britain at The Bath Press, Avon.

CONTENTS

PREFACE

This text is intended primarily for use by students and teachers of the statistics section of A-level Pure Mathematics with Statistics, an increasingly popular course.

Points of theory are presented concisely and illustrated by suitable worked examples, many taken from previous A-level papers. These are then supported by very carefully graded exercises which serve to consolidate the theory, link it with previous work and build up the confidence of the reader. There are frequent summaries of main points and miscellaneous exercises containing mainly A-level questions.

Throughout the text we have aimed to provide the reader with a mathematical structure and a logical framework within which to work. We have given special attention to topics which, in our experience, cause great difficulty. These include probability theory, the theory of continuous random variables and significance testing.

The text covers the main theory required by all the major examining boards. We are very grateful to the following for permission to reproduce questions:

University of Cambridge Local Examinations Syndicate (C)
The Southern Universities' Joint Board (SUJB)
Joint Matriculation Board (JMB)
University of London (L)
Oxford and Cambridge School Examinations Board (O & C)
incorporating School Mathematics Project (SMP)
Mathematics in Education and Industry (MEI)
The Associated Examining Board (AEB)
Oxford Delegacy of Local Examinations (O)

A-level questions are followed by the name of the board. Questions from Additional Mathematics papers are indicated by the word Additional, and (P) indicates a part-question.

We are particularly indebted to The Associated Examining Board and The Southern Universities' Joint Board for allowing us to use some of their questions as worked examples, and would stress that they are in no way involved in, or responsible for, this working.

We extend our thanks to our families, colleagues and students for all their encouragement and support, in particular to Audrey Shepherd and Jane Ziesler.

J Crawshaw
J Chambers

PREFACE TO THE SECOND EDITION

In order to give a fully comprehensive coverage of the present A-level syllabuses the following material has been added:

Chapter 4	— The use of binomial and Poisson cumulative probability tables. The geometric distribution
Chapter 5	— The negative exponential distribution
Chapter 6	— The use of the standard normal cumulative tables $\Phi(z)$ (with the use of tables giving $Q(z)$ retained in the Appendix)
Chapter 7	— Random sampling and the use of random number tables
Chapter 9	— Significance testing relating to the binomial and Poisson distributions
Chapter 11	— A fuller treatment of correlation and linear regression, including significance testing relating to Spearman's and Kendall's coefficients of correlation.

Numerous recent A-level questions taken from all the major examining boards have been added, together with worked examples from the University of London Schools Examination Board which we would stress is in no way responsible for these solutions.

J Crawshaw
J Chambers
1990

1

DESCRIPTIVE STATISTICS

DISCRETE DATA

These are the marks obtained by 30 pupils in a test:

6	3	5	9	0	1	8	5	6	7	4	4	3	1	0
2	2	7	10	9	7	5	4	6	6	2	1	0	8	8

This is an example of **discrete raw data.**

Discrete data can assume only exact values, for example

the number of cars passing a checkpoint in a certain time,

the shoe sizes of children in a class,

the number of tomatoes on each of the plants in a greenhouse.

The data is 'raw' because it has not been ordered in any way.

To illustrate the data more concisely, a **frequency distribution** can be formed. We count the number of 0's, 1's, 2's, . . . , and form a table:

Mark	0	1	2	3	4	5	6	7	8	9	10	
Frequency	3	3	3	2	3	3	4	3	3	2	1	Total 30

Discrete data can be grouped into 'classes', but once this has been done some of the original information is lost:

Mark	0–1	2–3	4–5	6–7	8 and over	
Frequency	6	5	6	7	6	Total 30

CONTINUOUS DATA

These are the heights of 20 children in a school. The heights have been measured correct to the nearest cm.

133	136	120	138	133
131	127	141	127	143
130	131	125	144	128
134	135	137	133	129

This is an example of **continuous raw data.**

Continuous data cannot assume exact values, but can be given only within a certain range or measured to a certain degree of accuracy, for example

144 cm (correct to the nearest cm) could have arisen from any value in the range 143.5 cm $\leqslant h <$ 144.5 cm.

Other examples of continuous data are

the speeds of vehicles passing a particular point,

the masses of cooking apples from a tree,

the time taken by each of a class of children to perform a task.

FREQUENCY DISTRIBUTIONS

To form a frequency distribution for the heights of the 20 children we group the information into 'classes' or 'intervals':

	(Alternative ways of writing the interval)	
$119.5 \leqslant h < 124.5$	119.5–124.5	120–124
$124.5 \leqslant h < 129.5$	124.5–129.5	125–129
$129.5 \leqslant h < 134.5$	129.5–134.5	130–134
$134.5 \leqslant h < 139.5$	134.5–139.5	135–139
$139.5 \leqslant h < 144.5$	139.5–144.5	140–144

The values 119.5, 124.5, 129.5, . . . , are called the **class boundaries.**

NOTE: the upper class boundary (u.c.b.) of one interval is the lower class boundary (l.c.b.) of the next interval.

The width of an interval = u.c.b. − l.c.b.

Therefore the width of the first interval $= 124.5 - 119.5$

$= 5$

In fact, in this example, each of the classes has been chosen so that the width is 5.

To group the heights into the following classes it helps to use a 'tally' column, entering the numbers in the first row, then the second row, and so on.

~~133~~	~~136~~	~~120~~	~~138~~
131	127	141	127
130	131	125	144
134	135	137	133
133	143	128	129

Height (cm)	Tally
$119.5 \leqslant h < 124.5$	I
$124.5 \leqslant h < 129.5$	
$129.5 \leqslant h < 134.5$	I
$134.5 \leqslant h < 139.5$	II
$139.5 \leqslant h < 144.5$	

The final frequency distribution should read:

Height (cm)	Tally	Frequency
$119.5 \leqslant h < 124.5$	I	1
$124.5 \leqslant h < 129.5$	~~IIII~~	5
$129.5 \leqslant h < 134.5$	~~IIII~~ II	7
$134.5 \leqslant h < 139.5$	IIII	4
$139.5 \leqslant h < 144.5$	III	3
		Total 20

Example 1.1 The following table gives the diameters of 40 ball-bearings, each measured in cm correct to 2 decimal places (d.p.). Form a frequency distribution by taking classes of width 0.02 cm.

3.98	3.94	3.96	3.97	4.02	3.96	3.97	3.98
3.94	3.97	3.96	3.97	4.00	4.00	3.98	3.97
3.99	3.99	4.00	3.95	4.03	3.95	4.00	4.01
3.99	3.99	3.98	4.01	3.98	4.00	4.04	4.03
3.99	4.02	4.03	4.00	3.93	4.01	4.00	3.93

Solution 1.1 The smallest value in the table is 3.93 and the largest value is 4.04. As measurements have been taken in cm correct to 2 d.p., the lowest class boundary is 3.925 cm. As the class width is 0.02 cm, the first interval must have an upper class boundary of 3.945 cm.

So we take as class boundaries 3.925, 3.945, 3.965, . . . , 4.045.

The frequency distribution is as follows:

Diameter (cm)	Tally	Frequency
$3.925 \leqslant d < 3.945$	IIII	4
$3.945 \leqslant d < 3.965$	~~IIII~~	5
$3.965 \leqslant d < 3.985$	~~IIII~~ ~~IIII~~	10
$3.985 \leqslant d < 4.005$	~~IIII~~ ~~IIII~~ II	12
$4.005 \leqslant d < 4.025$	~~IIII~~	5
$4.025 \leqslant d < 4.045$	IIII	4
		Total 40

NOTE: The intervals are often written

Diameter (cm)
3.93–3.94
3.95–3.96
3.97–3.98
and so on

Remember to work out the class boundaries.

The following frequency distributions show some of the ways in which data may be grouped.

(i) Frequency distribution to show the lengths of 30 rods. Lengths have been measured to the nearest mm.

Length (mm)	27–31	32–36	37–46	47–51
Frequency	4	11	12	3

The interval '27–31' means 26.5 mm $\leqslant$ length $<$ 31.5 mm.

The class boundaries are 26.5, 31.5, 36.5, 46.5, 51.5
The class widths are 5, 5, 10, 5

(ii) Frequency distribution to show the marks in a test of 100 students

Mark	30–39	40–49	50–59	60–69	70–79	80–89
Frequency	10	14	26	20	18	12

The class boundaries are 29.5, 39.5, 49.5, 59.5, 69.5, 79.5, 89.5
The class widths are 10, 10, 10, 10, 10, 10,

(iii) Frequency distribution to show the lengths of 50 telephone calls

Length of call (min)	0–	3–	6–	9–	12–	18–
Frequency	9	12	15	10	4	0

The interval '3–' means 3 minutes $\leqslant$ time $<$ 6 minutes, so any time including 3 minutes and up to (but not including) 6 minutes comes into this interval.

The class boundaries are 0, 3, 6, 9, 12, 18
The class widths are 3, 3, 3, 3, 6

(iv) Frequency distribution to show the masses of 40 packages brought to a particular counter at a post office

Mass (g)	-100	-250	-500	-800
Frequency	8	10	16	6

The interval '-250' means 100 g $<$ mass $\leqslant$ 250 g; so any mass over 100 grams up to and including 250 grams comes into this interval.

The class boundaries are 0, 100, 250, 500, 800
The class widths are 100, 150, 250, 300

(v) Frequency distribution to show the speeds of 50 cars passing a checkpoint

Speed (km/h)	20–30	30–40	40–60	60–80	80–100
Frequency	2	7	20	16	5

The class '30–40' means 30 km/h $\leqslant$ speed $<$ 40 km/h.

The class boundaries are 20, 30, 40, 60, 80, 100
The class widths are 10, 10, 20, 20, 20

(vi) Frequency distribution to show ages (in completed years) of applicants for a teaching post

Age (years)	21–24	25–28	29–32	33–40	41–52
Frequency	4	2	2	1	1

As the ages are in completed years (not to the nearest year) then '21–24' means 21 $\leqslant$ age $<$ 25. Someone who is 24 years and 11 months would come into this category. Sometimes this interval is written '21–' and the next is '25–', etc.

The class boundaries are 21, 25, 29, 33, 41, 53
The class widths are 4, 4, 4, 8, 12

HISTOGRAMS

Grouped data can be displayed in a histogram.

In a histogram rectangles are drawn so that the area of each rectangle is proportional to the frequency in the range covered by it.

We have area $\propto$ frequency

(a) Histograms with equal class widths

Example 1.2 The lengths of 30 Swiss cheese plant leaves were measured and the information grouped as shown. Measurements were taken correct to the nearest cm. Draw a histogram to illustrate the data.

Length of leaf (cm)	10–14	15–19	20–24	25–29
Frequency	3	8	12	7

Solution 1.2 The class boundaries are 9.5, 14.5, 19.5, 24.5, 29.5
The class widths are 5, 5, 5, 5

Now, area of rectangle = class width $\times$ height of rectangle

As the class width is 5 for each interval,

$$\text{area of rectangle} = 5 \times \text{height of rectangle}$$

So $\quad \text{area} \propto \text{height of rectangle}$

Now, if we make the height of each rectangle the same as the frequency,

we have $\quad \text{area} \propto \text{frequency}, \quad$ as required.

When all the class intervals are of equal width the frequency can be used for the height of each rectangle.

Histogram to show the lengths of 30 leaves

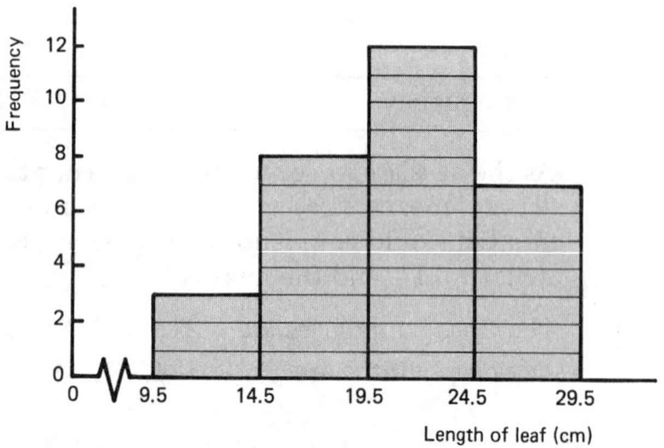

(b) Histograms with unequal class widths

Example 1.3 The frequency distribution gives the masses of 35 objects, measured to the nearest kg. Draw a histogram to illustrate the data.

Mass (kg)	6–8	9–11	12–17	18–20	21–29
Frequency	4	6	10	3	12

Solution 1.3 The class boundaries are 5.5, 8.5, 11.5, 17.5, 20.5, 29.5
The class widths are 3, 3, 6, 3, 9

As the class widths are not equal we cannot make the height of each rectangle equal to the frequency.

So we choose a convenient width as a 'standard' and adjust the heights of the rectangles accordingly, as follows.

If we choose a class width of 3 as standard, then the first two rectangles can be 4 and 6 units high respectively. However, as the third interval is twice the standard width we must make the height of the rectangle equal to half the frequency.

Similarly, as the last interval is $3 \times$ standard we must make the height of the rectangle equal to one-third of the frequency.

As the heights of the rectangles have been adjusted, we are considering frequency per standard width. We will write this as 'standard frequency'.

Mass (kg)	Class width		Frequency	Height of rectangle (standard frequency)
6–8	3	standard	4	4
9–11	3	standard	6	6
12–17	6	2 × standard	10	$\frac{1}{2} \times 10 = 5$
18–20	3	standard	3	3
21–29	9	3 × standard	12	$\frac{1}{3} \times 12 = 4$

We have now ensured that the area of each rectangle is proportional to the frequency, and the histogram is drawn as shown.

Histogram to show the masses of 35 objects

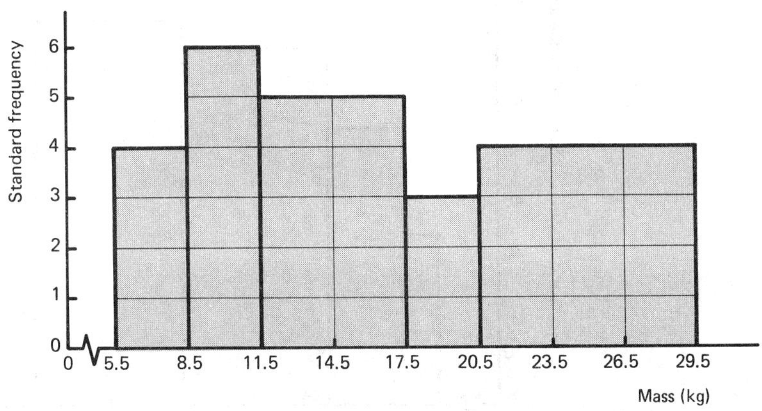

In general, choose a 'standard' width.

If $\quad$ class width $= n \times$ standard width

then $\quad$ height of rectangle $= \dfrac{1}{n} \times$ corresponding frequency

Example 1.4 The following table gives the distribution of the interest paid to 460 investors in a particular year.

Interest (£)	25–	30–	40–	60–	80–	110–
Frequency	17	55	142	153	93	0

Draw a histogram to illustrate this information.

Solution 1.4 The class boundaries are 25, 30, 40, 60, 80, 110

The class widths are 5, 10, 20, 20, 30

We will choose a class width of 10 as the standard width.

Interest (£)	Class width		Frequency	Standard frequency
25–	5	$\frac{1}{2}\times$ standard	17	$\frac{1}{1/2}\times 17 = 34$
30–	10	standard	55	55
40–	20	$2\times$ standard	142	$\frac{1}{2}\times 142 = 71$
60–	20	$2\times$ standard	153	$\frac{1}{2}\times 153 = 76.5$
80–	30	$3\times$ standard	93	$\frac{1}{3}\times\ 93 = 31$

Histogram to show the interest paid to 460 investors

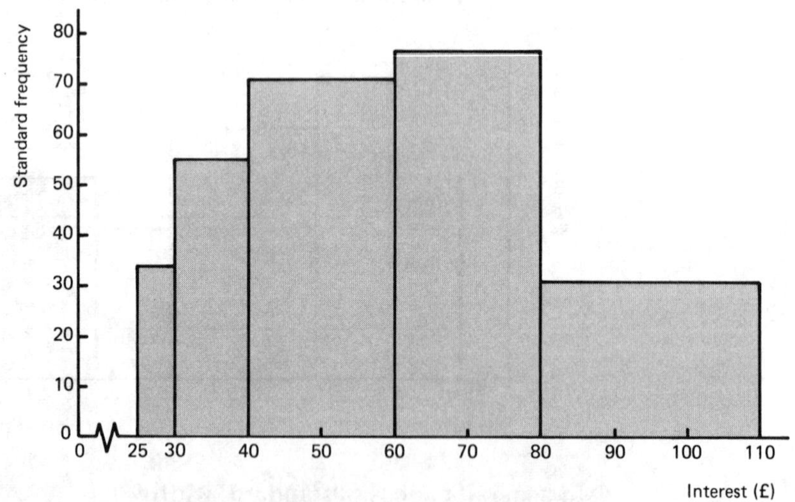

Example 1.5 The following table gives the distribution of marks of 60 pupils in a test. Draw a histogram to illustrate the data.

Mark	0–9	10–14	15–19	20–24	25–34
Frequency	13	19	12	7	9

Solution 1.5 The class boundaries are 0, 9.5, 14.5, 19.5, 24.5, 34.5

The class widths are 9.5 5, 5, 5, 10

We will choose a class width of 5 as the standard width.

Marks	Class width		Frequency	Standard frequency
0–9	9.5	$\frac{9.5}{5}\times$ standard	13	$\frac{5}{9.5}\times 13 = 6.8$
10–14	5	standard	19	19
15–19	5	standard	12	12
20–24	5	standard	7	7
25–34	10	$2\times$ standard	9	$\frac{1}{2}\times 9 = 4.5$

Histogram to show the marks of 60 pupils

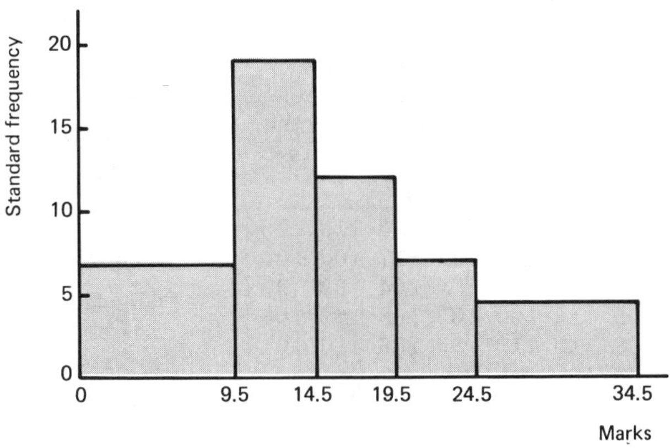

Alternative Approach The first interval can be regarded as having a lower class boundary of -0.5, in which case the width of the first interval is 10. Therefore the height of the first rectangle is 6.5 and the histogram would look like this:

Histogram to show the marks of 60 pupils

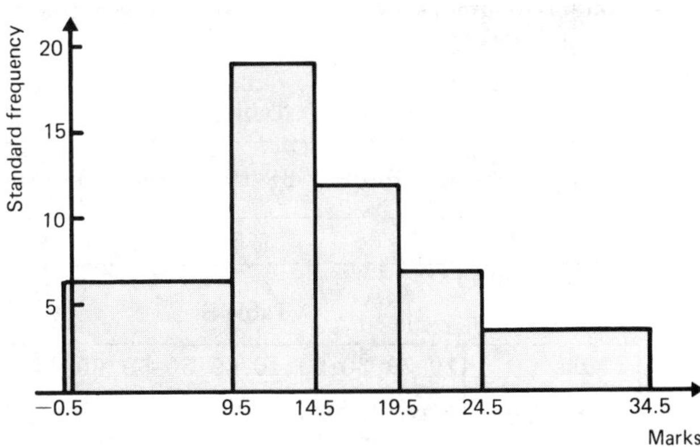

Exercise 1a

1. The following table gives the distribution of marks obtained by 101 pupils in a test.

47	62	61	60	39	34	37	46	81	72	74	62
65	53	47	52	38	25	42	40	70	63	62	83
19	26	59	49	53	52	88	91	51	52	24	80
69	59	32	33	27	18	22	73	59	36	34	31
21	63	72	54	18	71	44	71	43	44	18	93
12	15	60	71	82	13	61	64	25	63	82	71
11	58	64	39	16	23	10	92	83	75	36	74
43	29	85	65	42	57	70	63	54	55	49	81
47	72	65	63	60							

Construct a frequency distribution, taking equal class intervals 10–19, 20–29, 30–39, . . . , 90–99.

Draw a histogram to illustrate the data.

2. The masses of 50 apples (measured to the nearest g) were noted and shown in the table.

86	101	114	118	87	92	93	116	105
102	97	93	101	111	96	117	100	106
118	101	107	96	101	102	104	92	99
107	98	105	113	100	103	108	92	109
95	100	103	110	113	99	106	116	101
105	86	88	108	92				

Construct a frequency distribution, using equal class intervals of width 5 g, and taking the lower class boundary of the first interval as 84.5 g.

Draw a histogram to illustrate the data.

3. The masses (measured to the nearest g) of washers are recorded in the table. Draw a histogram to illustrate the data.

Mass (g)	0–2	3–5	6–11	12–14	15–17
Frequency	5	6	14	4	3

4. 100 people were asked to record how many television programmes they watched in a week. The results were as follows:

Number of programmes	0–	10–	18–	30–	35–	45–	50–	60–
Number of viewers	3	16	36	21	12	9	3	0

Draw a histogram to illustrate the data.

5. 68 smokers were asked to record their consumption of cigarettes each day for several weeks. The table shown is based on the information obtained.

Average number of cigarettes smoked per day	0–	8–	12–	16–	24–	28–	34–50
Number of smokers	4	6	12	28	8	6	4

Illustrate these data by means of a histogram. (C Additional) P

6. On a particular day, the length of stay of each car at a city car park was recorded.

The length of stay was measured to the nearest minute. The results were as shown in in Table A below.

Taking 20 minutes as 'standard', draw a histogram to illustrate this information.

7. The marks awarded to 136 pupils in an examination are summarised in Table B below.

Draw a histogram to illustrate the data.

8. 38 children solved a simple problem and the time taken by each was noted.

Time (seconds)	5–	10–	20–	25–	40–	45–
Frequency	2	12	7	15	2	0

Draw a histogram to illustrate this information.

Table A

Length of stay (min)	6–25	26–60	61–80	81–105	106–115	116–150	151–200	201–300
Frequency	62	70	88	125	56	105	30	15

Table B

Marks	10–29	30–39	40–49	50–59	60–64	65–69	70–84
Frequency	22	18	22	24	14	12	24

9. Table C below shows the number of pupils gaining marks within various groups in an examination.

Table C

Marks	5–29	30–39	40–49	50–59	60–79	80–99
Number of pupils	30	30	65	48	40	20

Draw a histogram to illustrate these data.
(C Additional) P

CIRCULAR DIAGRAMS OR PIE DIAGRAMS

Another useful way of displaying data is to draw a pie diagram, sometimes called a pie chart. Here again, area is proportional to frequency.

The sales (in thousands of litres) of petrol from four petrol stations A, B, C and D are noted for the first week of March, and are shown in the table:

Petrol station	A	B	C	D
Sales (thousands of litres)	90	140	30	20

Construct a pie diagram to illustrate this information.

Solution 1.6 The total angle of 360° at the centre of a circle is divided according to the sales at each of the stations.

The total sales (thousands of litres) $= 90+140+30+20 = 280$

The angle representing the sales of petrol at station A is given by

$$\left(\frac{90}{280}\right)(360) = 115.7° \quad (1\,\text{d.p.})$$

and so, for each of the petrol stations we have

Petrol station	Sales (thousands of litres)	Sector angle
A	90	$\left(\frac{90}{280}\right)(360) = 115.7°$
B	140	$\left(\frac{140}{280}\right)(360) = 180°$
C	30	$\left(\frac{30}{280}\right)(360) = 38.6°$
D	20	$\left(\frac{20}{280}\right)(360) = 25.7°$
	280	Total = 360°

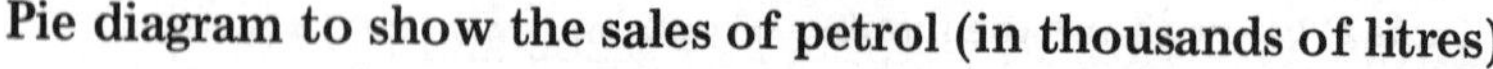

Pie diagram to show the sales of petrol (in thousands of litres)

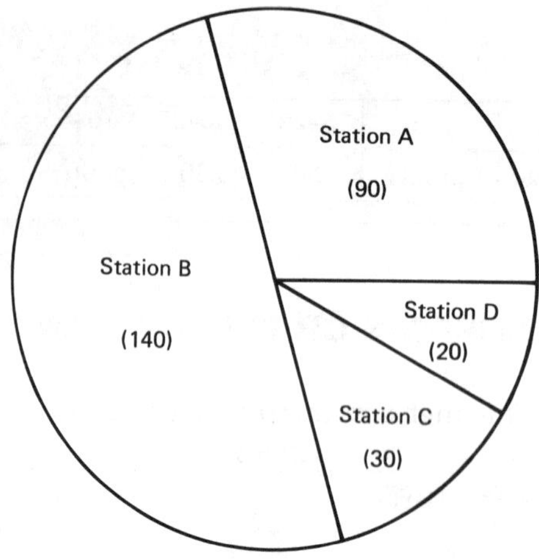

COMPARISON OF DATA USING PIE DIAGRAMS

Pie diagrams are particularly useful when we wish to **compare two or more sets of similar data.**

Suppose that we are given information about the land use (for barley, wheat and woodland) in three parishes. We can draw three pie diagrams to illustrate the land use in each parish. However, if we wish to compare the sets of data we must make the size (or area) of each circle proportional to the total land for each parish. In this example we will refer to the total amount of land as the 'frequency' F.

The area of a circle is πr^2, so we will require, with obvious notation,

$$\pi r_1^2 : \pi r_2^2 : \pi r_3^2 = F_1 : F_2 : F_3$$

so $$r_1^2 : r_2^2 : r_3^2 = F_1 : F_2 : F_3 \quad \text{(cancelling } \pi\text{)}$$

i.e. $$r_1 : r_2 : r_3 = \sqrt{F_1} : \sqrt{F_2} : \sqrt{F_3} \quad \text{(taking square roots)}$$

So, the radii of the circles are proportional to the square roots of the frequencies.

We then choose a convenient scale and draw the circles.

Example 1.7 The following agricultural statistics refer to the land use, in hectares, of three parishes. Draw three pie diagrams to compare these data.

Parish	Barley	Wheat	Woodland	Total land, F
Appleford	1830	1640	550	$F_1 = 4020$
Burnford	645	435	120	$F_2 = 1200$
Carnford	320	160	150	$F_3 = 630$

Solution 1.7 Now $F_1 = 4020$, $F_2 = 1200$ and $F_3 = 630$.

So
$$\begin{aligned} r_1 : r_2 : r_3 &= \sqrt{F_1} : \sqrt{F_2} : \sqrt{F_3} \\ &= \sqrt{4020} : \sqrt{1200} : \sqrt{630} \\ &= 63.40 : 34.64 : 25.10 \\ &= 3.17 : 1.732 : 1.255 \end{aligned}$$

For convenience, we take $r_1 = 3.2$ cm, $r_2 = 1.7$ cm and $r_3 = 1.3$ cm.

The angles in the pie diagrams are calculated as shown in the table:

Parish	Barley	Wheat	Woodland
Appleford	$\left(\frac{1830}{4020}\right)(360) = 163.9°$	$\left(\frac{1640}{4020}\right)(360) = 146.9°$	$\left(\frac{550}{4020}\right)(360) = 49.2°$
Burnford	$\left(\frac{645}{1200}\right)(360) = 193.5°$	$\left(\frac{435}{1200}\right)(360) = 130.5°$	$\left(\frac{120}{1200}\right)(360) = 36°$
Carnford	$\left(\frac{320}{630}\right)(360) = 182.9°$	$\left(\frac{160}{630}\right)(360) = 91.4°$	$\left(\frac{150}{630}\right)(360) = 85.7°$

Pie diagrams to show land use (in hectares) in three parishes

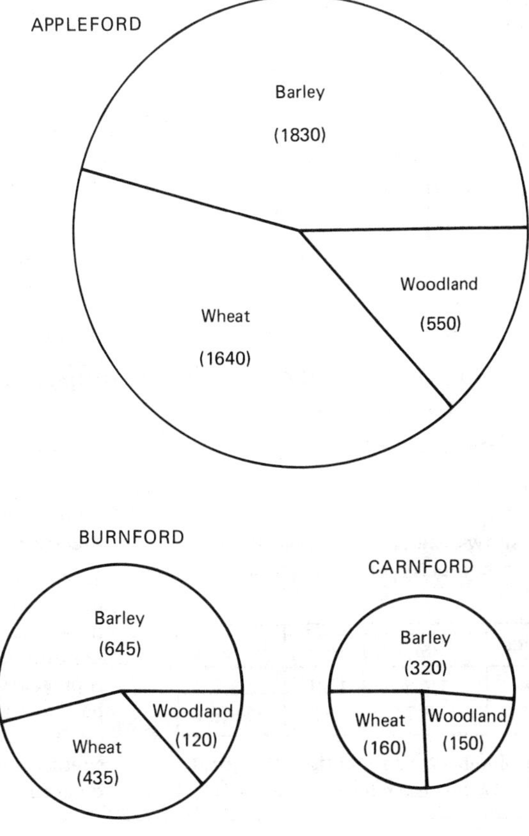

Exercise 1b

1. Construct a pie diagram to illustrate the scores obtained when a die is thrown 120 times.

Number on die	1	2	3	4	5	6
Frequency	22	20	15	25	10	28

2. The results of the voting in an election were as follows:

Mr P	2045 votes
Mr Q	4238 votes
Mrs R	8605 votes
Miss S	12012 votes

Represent this information on a pie diagram.

3. The pie chart, which is not drawn to scale, shows the distribution of various types of land and water in a certain county. Calculate
   (i) the area of woodland,
   (ii) the angle of the urban sector,
   (iii) the total area of the county.

(C Additional) P

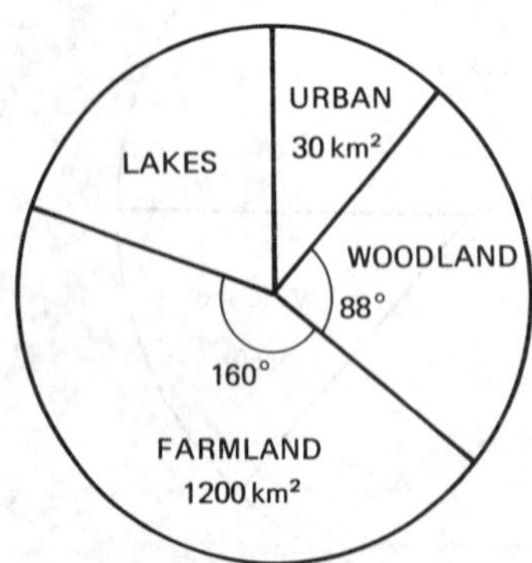

4. The table shows the sales, in millions of dollars, of a company in two successive years.

Year	Africa	America	Asia	Europe
1972	8.4	12.2	15.6	23.8
1973	5.5	6.7	13.2	19.6

Draw two pie charts which allow the total annual sales to be compared. (C Additional)

5. Five companies form a group. The sales of each company during the year ending 5th April, 1978, are shown in the table below.

Company	*A*	*B*	*C*	*D*	*E*
Sales (in £1000's)	55	130	20	35	60

Draw a pie chart of radius 5 cm to illustrate this information.

For the year ending 5th April, 1979, the total sales of the group increased by 20%, and this growth was maintained for the year ending 5th April, 1980.

If pie charts were drawn to compare the total sales for each of these years with the total sales for the year ending 5th April, 1978, what would be the radius of each of these pie charts?

If the sales of company *E* for the year ending 5th April, 1980, were again £60 000, what would be the angle of the sector representing them? (C Additional)

6. Mr Williams worked out how much it had cost him to run his car for each of 3 consecutive years. The results were as follows:

	Tax and insurance	Maintenance	Petrol
Year 1	£150.00	£72.50	£190.00
Year 2	£187.00	£116.00	£205.00
Year 3	£175.00	£289.90	£253.10

Draw three pie diagrams to compare this information.

7. Housewives were asked how much they spent last week on various items. Mrs M replied as follows:

	Item *A*	Item *B*	Item *C*
Mrs M	£1.50	£3.50	£3.00

Draw a pie diagram with radius 4 cm to illustrate this information.

A comparison was then made with the pie diagram drawn to illustrate Mrs N's replies in which the circle representing the total amount had a radius of 5 cm, the sector representing the amount spent on item *A* had an angle of 72° and the amount spent on item *B* was £4.00. Find the amount spent on item *C* by Mrs N and draw a pie diagram to illustrate her expenditure.

FREQUENCY POLYGONS

A frequency distribution may be displayed as a frequency polygon.

(a) Ungrouped data

Example 1.8 Twenty pieces of material, each of length 10 m, were examined for flaws and the number of flaws in each length was noted. Draw a frequency polygon to illustrate this information.

Number of flaws	0	1	2	3	4	5
Frequency	2	4	6	5	2	1

Solution 1.8 Points are plotted, with the number of flaws on the horizontal axis and the frequency on the vertical axis.

A frequency polygon to show the number of flaws in lengths of material

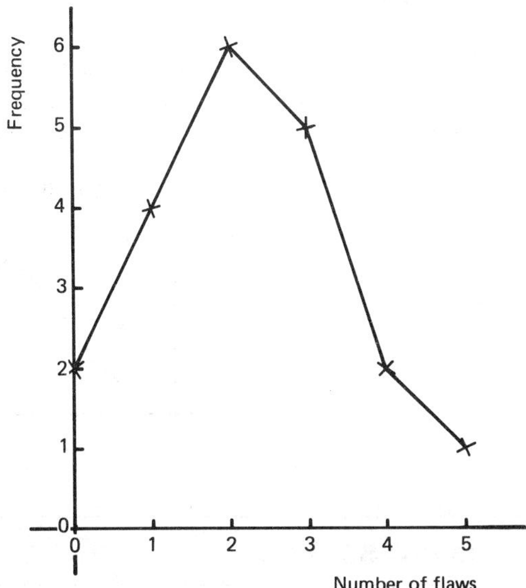

(b) Grouped data

A frequency polygon may be superimposed on a histogram by joining the mid-points of the tops of the rectangles.

Example 1.9 Construct a frequency polygon for the data given in Example 1.3.

Solution 1.9

Frequency polygon to show the masses of 35 objects

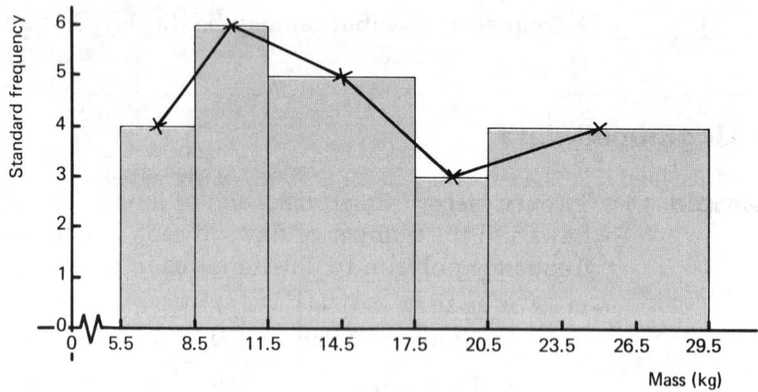

The frequency polygon can be constructed without drawing the histogram first. To do this, plot standard frequency against the mid-point of the interval.

> *NOTE*: the mid-point of the interval (a, b) is $\frac{1}{2}(a+b)$.

FREQUENCY CURVES

If the number of intervals is large, then the frequency polygon will consist of a large number of line segments. The frequency polygon approaches a smooth curve, known as a frequency curve.

Frequency curve

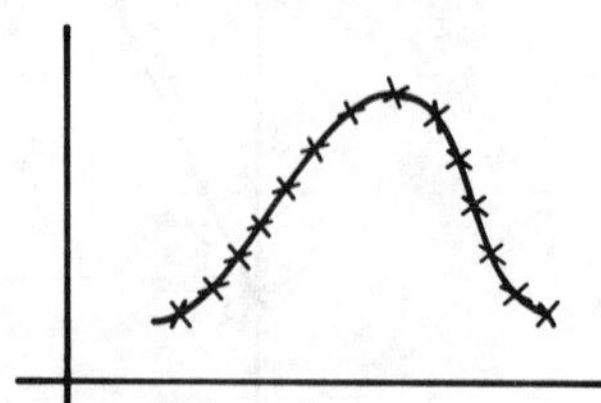

Exercise 1c

1. In a competition to grow the tallest hollyhock, the heights recorded by 50 competitors were as follows. Heights were measured to the nearest cm (see Table A below).

   Draw a histogram and superimpose the frequency polygon.

Table A

Height (cm)	177–186	187–191	192–196	197–201	202–206	207–216
Frequency	12	8	8	9	7	6

2. (*a*) The following table shows the weekly sales of television sets in a department store in one year.

Number of sets sold/week	5-13	14-22	23-31	32-40	41-49
Number of weeks	1	19	23	7	2

Draw a frequency polygon to illustrate this information.

(*b*) The following year the sales were as follows:

Number of sets sold/week	5-13	14-22	23-31	32-40	41-49
Number of weeks	3	16	20	12	1

Draw a frequency polygon to show the sales in the second year, on the same grid as part (*a*).

3. The table shows the duration, in minutes, of 64 telephone calls made from a high street call box in one day.

Length of call (min)	0-	$1\frac{1}{2}$-	3-	6-	12-	15-	21-
Frequency	3	7	22	20	6	6	0

Draw a frequency polygon to illustrate the information.

4. The table shows the ages (in completed years) of women who gave birth to a child at Anytown Maternity Hospital during a particular year.

Age (years)	16-	20-	25-	30-	35-	45-
Number of births	70	470	535	280	118	0

Draw a frequency polygon to illustrate this information. Do not draw a histogram first.

CUMULATIVE FREQUENCY

The cumulative frequency is the total frequency up to a particular item or class boundary. Sometimes this is thought of as a 'running total'.

(a) Ungrouped data

Example 1.10 The marks of 40 pupils in a test are shown in the table. Construct a cumulative frequency distribution.

Mark	4	5	6	7	8	9	10
Frequency	2	5	8	10	7	5	3

Solution 1.10 The cumulative frequency distribution for the marks is as follows.

Mark	Cumulative frequency
Up to and including 4	2
Up to and including 5	$2+5 = 7$
Up to and including 6	$2+5+8 = 15$
Up to and including 7	$2+5+8+10 = 25$
Up to and including 8	$2+5+8+10+7 = 32$
Up to and including 9	$2+5+8+10+7+5 = 37$
Up to and including 10	$2+5+8+10+7+5+3 = 40$

NOTE: the final value in the cumulative frequency column must be 40, as *all* the pupils obtained 10 marks or less.

(b) Grouped data

When data is grouped we consider the total frequency up to the *upper* class boundary of each interval.

Example 1.11 The heights of 30 broad bean plants were measured, correct to the nearest cm, 6 weeks after planting. The frequency distribution is given below. Construct the cumulative frequency table.

Height (cm)	3–5	6–8	9–11	12–14	15–17	18–20
Frequency	1	2	11	10	5	1

Solution 1.11 The upper class boundaries are 5.5, 8.5, 11.5, 14.5, 17.5, 20.5.

The lower boundary of the first class is 2.5.

Cumulative frequency table to show heights of plants

Height (cm)	Cumulative frequency
<2.5	0
<5.5	1
<8.5	3
<11.5	14
<14.5	24
<17.5	29
<20.5	30

CUMULATIVE FREQUENCY CURVE

The information in a cumulative frequency table can be shown on a graph, called a cumulative frequency curve, or ogive.

The cumulative frequencies are plotted against the upper class boundaries.

Example 1.12 (a) Construct a cumulative frequency curve for the data in Example 1.11.

(b) Estimate from the curve

(*i*) the number of plants that were less than 10 cm tall;

(*ii*) the value of x, if 10% of the plants were of height x cm or more.

Solution 1.12 (a) **Cumulative frequency curve to show the heights of 30 broad bean plants**

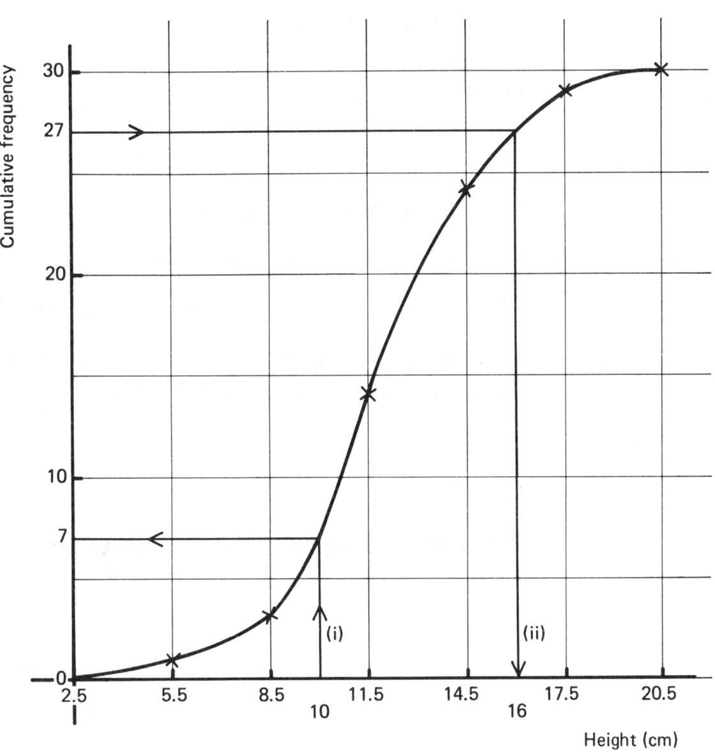

(b)

(i) To find how many plants were less than 10 cm tall, find the height, 10 cm on the horizontal axis. Draw a vertical line to meet the curve and then draw a horizontal line to meet the cumulative frequency axis.

From the graph, 7 plants were less than 10 cm tall.

(ii) 10% of the plants were of height x cm or more,
so 90% of the plants were less than x cm tall,
i.e. 27 plants were less than x cm tall.

Find 27 on the cumulative frequency axis and draw a horizontal line to meet the curve. Then draw a vertical line to meet the height axis.

From the graph, 27 plants were less than 16 cm tall.

Therefore 10% of the plants were of height 16 cm or more, and the value of x is 16.

Example 1.13 Pupils were asked how long it took them to walk to school on a particular morning. A cumulative frequency distribution was formed:

Time taken (minutes)	<5	<10	<15	<20	<25	<30	<35	<40	<45
Cumulative frequency	28	45	81	143	280	349	374	395	400

(a) Draw a cumulative frequency curve.

(b) Estimate how many pupils took less than 18 minutes.

(c) 6% of the pupils took x minutes or longer. Find x.

(d) Taking equal class intervals of 0–, 5–, 10–, . . . , construct a frequency distribution and draw a histogram.

Solution 1.13 (a) Cumulative frequency curve to show the times taken to walk to school

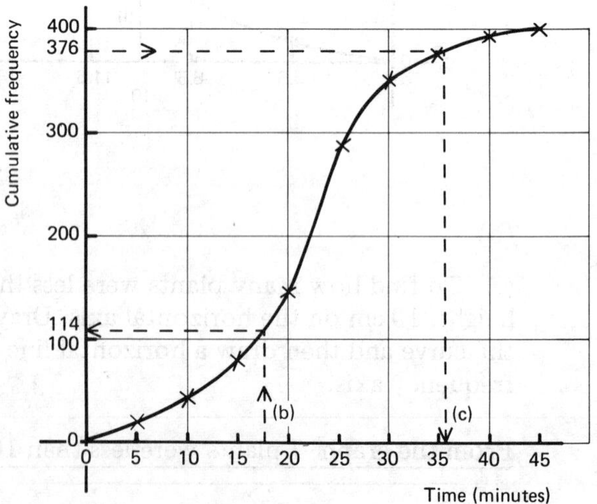

(b) From the graph we estimate that 114 pupils took less than 18 minutes.

(c) 6% of the pupils took x minutes or longer,
so 24 pupils took x minutes or longer,
and 376 pupils took less than x minutes.

From the graph, $x = 36$

Therefore, 6% of the pupils took 36 minutes or longer.

(d) We form the frequency distribution as follows:

Upper class boundary	Cumulative frequency	Time (min)	Frequency
5	28	0-	28
10	45	5-	45 − 28 = 17
15	81	10-	81 − 45 = 36
20	143	15-	143 − 81 = 62
25	280	20-	280 − 143 = 137
30	349	25-	349 − 280 = 69
35	374	30-	374 − 349 = 25
40	395	35-	395 − 374 = 21
45	400	40-(45)	400 − 395 = 5
			Total = 400

Histogram to show times taken by 400 pupils to walk to school

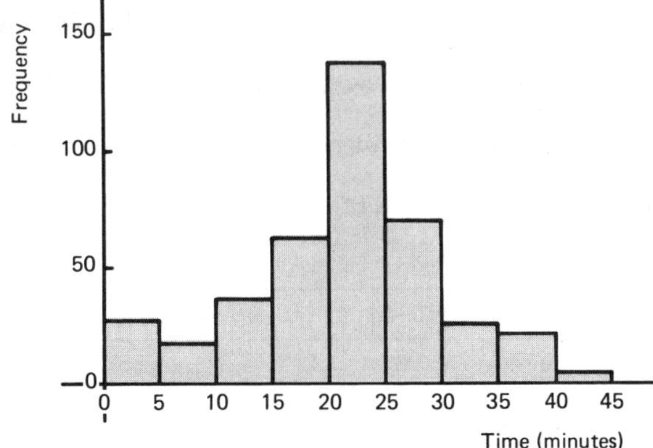

Exercise 1d

1. Table A below gives the distribution of marks of candidates in an examination.
   (*a*) Construct a cumulative frequency distribution and draw a cumulative frequency curve.
   (*b*) Use your curve to estimate
   (i) the percentage of candidates who passed, if the pass mark was 45;
   (ii) the range of marks gained by all candidates except the top 10% and the bottom 10%.

2. The cumulative frequency curve overleaf has been drawn from information about the amount of time spent by 50 people in a supermarket on a particular day.

Table A

Mark	0-9	10-19	20-29	30-39	40-49	50-59	60-69	70-79	80-89	90-99
Frequency	20	15	35	55	65	105	155	100	35	15

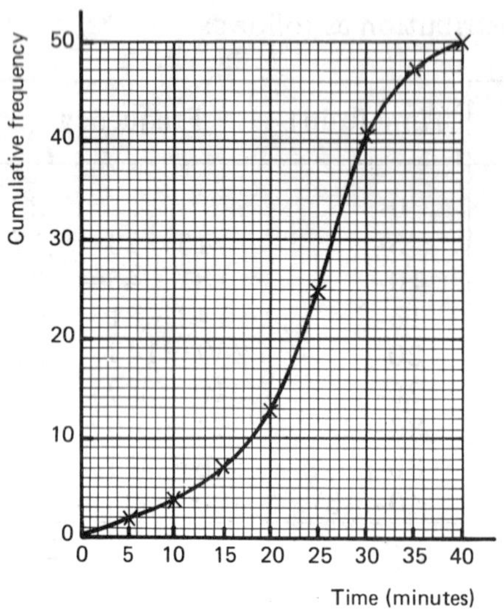

(*a*) Construct the cumulative frequency table, taking boundaries ⩽ 5, ⩽ 10, ...
(*b*) How many people spent between 17 and 27 minutes in the supermarket?
(*c*) 60% of the people spent less than or equal to t minutes. Find t.
(*d*) 60% of the people spent longer than s minutes. Find s.
(*e*) Taking equal class intervals of -5, -10, -15, . . . , construct the frequency distribution and draw a histogram.

3. Table B below shows the frequency distribution of the masses of 52 women students at a college. Measurements have been recorded to the nearest kg.

(*a*) Construct a cumulative frequency table and draw a cumulative frequency curve.
(*b*) How many students were of mass less than 57 kg?
(*c*) How many students were of mass greater than 61 kg?
(*d*) What was the mass exceeded by 20% of the students?

4. 49 soil samples were collected in an area of woodland, and the pH value for each sample was found. The cumulative frequency distribution was constructed as shown in Table C below.

(*a*) Draw a cumulative frequency curve.
(*b*) What percentage of the samples had a pH value less than 7?
(*c*) 50% of the samples had a pH value greater than x. Find x.
(*d*) Taking equal class intervals of 4.4-, 4.8-, 5.2-, . . . , construct the frequency distribution and draw a histogram.

Table B

Mass (kg)	40-44	45-49	50-54	55-59	60-64	65-69	70-74
Frequency	3	2	7	18	18	3	1

Table C

pH value	<4.8	<5.2	<5.6	<6.0	<6.4	<6.8	<7.2	<7.6	<8.0	<8.4
Cumulative frequency	1	2	5	10	19	38	43	46	48	49

THE MEDIAN

The median is the middle value of a set of numbers arranged in order of magnitude.

If there are n numbers, the median is the $\frac{1}{2}(n+1)$th value.

(a) Raw data

Example 1.14 Find the median of each of the sets

(a) 7, 7, 2, 3, 4, 2, 7, 9, 31
(b) 36, 41, 27, 32, 29, 38, 39, 43.

Solution 1.14 (a) Arranging in order of magnitude

$$2, 2, 3, 4, \boxed{7}, 7, 7, 9, 31$$

$n = 9$, and the median is the $\frac{1}{2}(9+1)$th value, i.e. the 5th value

So median = 7.

(b) Arranging in order of magnitude

$$27, 29, 32, \boxed{36, 38}, 39, 41, 43$$

$n = 8$, and the median is the $\frac{1}{2}(8+1)$th value, i.e. the $4\frac{1}{2}$th value.

This does not exist, so we consider the 4th and 5th values

$$\begin{aligned} \text{median} &= \tfrac{1}{2}(36+38) \\ &= 37 \end{aligned}$$

So median = 37.

In general,

If n is odd then there is a middle value, and this is the median.

If n is even and the two middle values are c and d, then the median is $\frac{1}{2}(c+d)$.

(b) Ungrouped frequency distribution

The median can be found directly from the cumulative frequency distribution.

Example 1.15 The table shows the number of children in the family for 35 families in a certain area. Find the median number of children per family.

Number of children	0	1	2	3	4	5
Frequency	3	5	12	9	4	2

Solution 1.15 Form a cumulative frequency distribution:

Number of children	Cumulative frequency
0	3
Up to and including 1	8
Up to and including 2	20
Up to and including 3	29
Up to and including 4	33
Up to and including 5	35

There are 35 values, so the median is the $\frac{1}{2}(35+1)$th value, i.e. the 18th value.

We could have written out all the values in order from the frequency table, thus 0, 0, 0, 1, 1, 1, 1, 1, 2, 2, ... However, we can see from the cumulative frequency table that the 18th value will be 2, as the first 8 values are 0 or 1 and the first 20 values are 0 or 1 or 2.

Therefore the median number of children per family is 2.

Exercise 1e

1. Find the median of each of the following sets of numbers:
(*a*) 4, 6, 18, 25, 9, 16, 22, 5, 20, 4, 8
(*b*) 192, 217, 189, 210, 214, 204
(*c*) 1267, 1896, 895, 3457, 2164
(*d*) 0.7, 0.4, 0.65, 0.78, 0.45, 0.32, 1.9, 0.0078

2. The table shows the scores obtained when a die is thrown 60 times. Form a cumulative frequency table and use it to find the median score.

Score, x	1	2	3	4	5	6
Frequency, f	12	9	8	13	9	9

3. Find the median of each of the following frequency distributions:

(*a*)

x	5	6	7	8	9	10
f	6	11	15	18	6	5

(*b*)

x	12	13	14	15	16
f	3	9	11	15	7

(*c*)

x	5	9	13	17	21
f	3	7	32	29	16

(*d*)

x	2.4	5.4	8.4	11.4	14.4
f	16	31	18	17	6

(c) Grouped frequency distribution

Once the information has been grouped and the raw data lost we can only estimate a value for the median.

This can be done by one of the following methods:

(a) by calculation,

(b) from a cumulative frequency curve,

(c) from a histogram.

Example 1.16 below is done in the three different ways to illustrate the methods.

Example 1.16 The masses, measured to the nearest kg, of 49 boys are noted and the distribution formed. Estimate the median mass.

Mass (kg)	60–64	65–69	70–74	75–79	80–84	85–89
Frequency	2	6	12	14	10	5

Solution 1.16 First form a cumulative frequency distribution.

The upper class boundaries are 64.5, 69.5, 74.5, 79.5, 84.5, 89.5.

The lower class boundary of the first class is 59.5.

Mass (kg)	Frequency	Mass (kg)	Cumulative frequency
-59	0	< 59.5	0
60–64	2	< 64.5	2
65–69	6	< 69.5	8
70–74	12	< 74.5	20
75–79	14	< 79.5	34
80–84	10	< 84.5	44
85–89	5	< 89.5	49

The median is the $\frac{1}{2}(49+1)$th value, i.e. the 25th value.

Method (a) — By calculation The 25th value lies in the class 74.5–79.5.

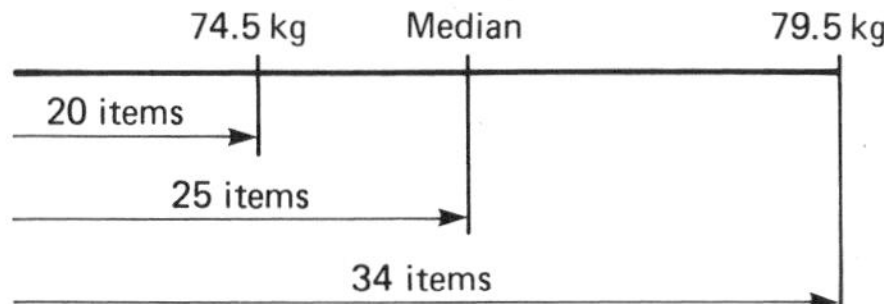

There are 14 items in the class 74.5–79.5 and from the diagram the median is $\frac{5}{14}$ of the interval of 5 kg from 74.5 to 79.5.

$$\begin{aligned}\text{Estimate of the median mass} &= 74.5 + \left(\frac{5}{14}\right)(5)\\ &= 76.3\text{ kg} \quad (1\text{ d.p.})\end{aligned}$$

Therefore we estimate the median to be 76.3 kg (1 d.p.)

Method (b) — From the cumulative frequency curve Draw the cumulative frequency curve and read off the value corresponding to a cumulative frequency of 25.

Cumulative frequency curve to show the masses of 49 boys

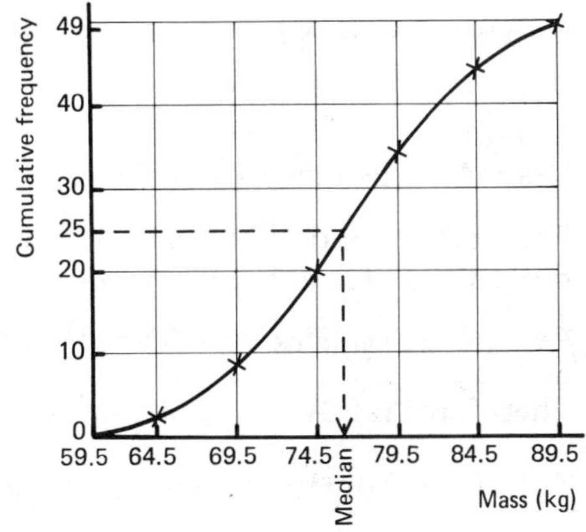

From the graph, the value corresponding to the cumulative frequency of 25 is 76.3 kg.

Therefore an estimate of the median is 76.3 kg.

Method (c) — From a histogram First, draw the histogram. As the classes are of equal width the vertical axis can be labelled 'frequency'.

Histogram to show the masses of 49 boys

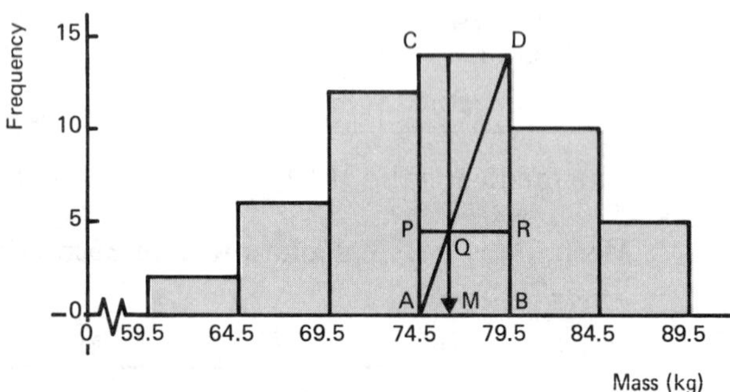

In a histogram, the area is proportional to the frequency. The median is the middle value, so it will divide the area under the histogram into two equal parts. Altogether, contained in the histogram, there are 49 'units' of area. We want to draw a line to the left of which there are $24\frac{1}{2}$ units of area, and to the right of which there are $24\frac{1}{2}$ units of area.

Consider the class 74.5–79.5; it contains 14 units. To the left of it there is a total of 20 units of area. So we need to divide the area in this class in the ratio $4\frac{1}{2}:(14-4\frac{1}{2})$, i.e. $4\frac{1}{2}:9\frac{1}{2}$. This will give $24\frac{1}{2}$ units of area to the left of the median.

Now, to divide AB in the ratio $4\frac{1}{2}:9\frac{1}{2}$, first find the point P which is at $4\frac{1}{2}$ on the vertical axis.

So $AP:PC = 4\frac{1}{2}:9\frac{1}{2}$ and $AP:AC = 4\frac{1}{2}:14$.

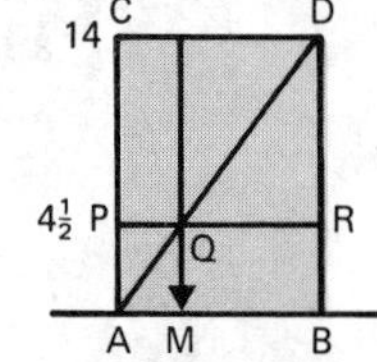

Draw AD, the diagonal of the rectangle.

Draw a horizontal line PR and drop the vertical line QM from the point where PR cuts the diagonal AD.

By similar triangles $PQ:CD = 4\frac{1}{2}:14$.

Therefore $AM:AB = 4\frac{1}{2}:14$ and so $AM:MB = 4\frac{1}{2}:9\frac{1}{2}$.

Therefore, from the histogram, an estimate of the median is 76.3 kg.

Example 1.17 The haemoglobin levels were measured in a sample of 50 people and the results were as follows, each being correct to 1 d.p.:

13.5	15.6	16.3	12.3	13.1	14.2	12.4	11.3	14.0	14.6
13.6	14.8	12.7	10.9	11.0	11.4	15.0	10.1	15.4	11.3
10.7	14.6	13.5	15.1	12.1	12.0	14.2	11.4	15.0	13.3
13.2	9.1	16.9	14.2	15.0	13.6	14.8	11.4	14.8	15.7
13.5	13.5	12.9	13.8	13.7	16.2	11.6	13.8	14.2	10.7

(a) Group the data into eight classes, 9.0–9.9, 10.0–10.9, . . . , 16.0–16.9.

(b) What are the smallest and largest possible measurements which could be included in the class 9.0–9.9?

(c) Draw a histogram of the grouped data and use it to estimate the median value of the sample, showing your working.

(d) Find the true median of the sample. (SUJB)

Solution 1.17 (a)

Haemoglobin level	Tally	Frequency
9.0–9.9	|	1
10.0–10.9	||||	4
11.0–11.9	~~||||~~ ||	7
12.0–12.9	~~||||~~ |	6
13.0–13.9	~~||||~~ ~~||||~~ ||	12
14.0–14.9	~~||||~~ ~~||||~~	10
15.0–15.9	~~||||~~ ||	7
16.0–16.9	|||	3
		$\Sigma f = 50$

(b) If the haemoglobin level y is in the class 9.0–9.9 then, as levels have been measured correct to 1 d.p., we have $8.95 \leqslant y < 9.95$.

The smallest measurement is 8.95 and the largest measurement is a, such that $a < 9.95$. The upper class boundary of the interval is 9.95.

(c) The class boundaries are 8.95, 9.95, 10.95, 11.95, . . . , 16.95.

The class widths are each equal to 1. As the class widths are equal we can label the vertical axis ‘frequency’.

In a histogram the area is proportional to frequency. So the median divides the histogram into two equal parts. We need to draw on the histogram a line which will have 25 ‘units’ of area to the left of it and 25 ‘units’ to the right of it.

Consider the class 13.0–13.9; there are 18 units of area to the left of the lower class boundary point of 12.95, so we need another 7 units. If we find P such that AP = 7 then AP : AC = 7 : 12 and, by similar triangles, PQ : CD = 7 : 12. Hence AM : AB = 7 : 12.

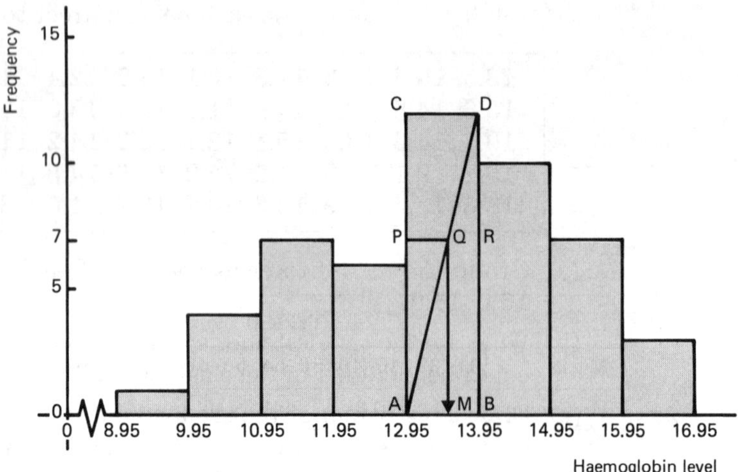

So there are 7 units of area in the class 13.0–13.9 to the left of the line QM.

From the histogram, an estimate of the median is 13.45.

(d) In the sample the median is the $\frac{1}{2}(50+1)$th value, i.e. the $25\frac{1}{2}$th value.

Now there are 18 readings as far as a haemoglobin level of 12.95. Arranging the items in the class 13.0–13.9 gives

19th	20th	21st	22nd	23rd	24th	25th		26th	...
13.1	13.2	13.3	13.5	13.5	13.5	13.5	↑ median	13.6	...

So the true median is $\frac{1}{2}(13.5+13.6) = 13.55$.

Exercise 1f

1. Estimate the median of the following frequency distribution
   (*a*) by calculation,
   (*b*) from a cumulative frequency curve,
   (*c*) from a histogram.

   The frequency distribution shows the times taken by 55 pupils to do their mathematics homework. Times have been measured to the nearest minute.

Time (min)	5-14	15-24	25-34	35-44	45-54
Frequency	5	7	19	17	7

2. Eggs laid at Hill Farm are weighed and the results grouped as shown:

Mass (gm)	-50	-54	-58	-62	-66	-70	-74
Frequency	3	2	5	12	10	6	2

   Construct a cumulative frequency table and draw a cumulative frequency curve. Use the curve to estimate the median mass.

3. The table shows the frequency distribution of the speeds of cars passing along a marked stretch of road of length 1 kilometre. Estimate the median speed.

Speed (km/h)	0-	20-	40-	60-	80-	100-
Frequency	2	8	22	9	2	0

4. Estimate the median diameter of rods produced by a particular machine by drawing a histogram of the data given in Table A below. Explain your method.

5. The distribution of marks obtained by 199 students in a mathematics examination is shown in Table B below.

Construct a cumulative frequency table and use it to estimate the median mark.

6. The length of life (to the nearest hour) of each of 50 electric light bulbs is noted and the results shown in Table C below. Calculate the median length of life.

Table A

Diameter (cm)	0.49–0.51	0.52–0.54	0.55–0.57	0.58–0.60	0.61–0.63
Frequency	12	23	32	18	15

Table B

Marks	10–39	40–49	50–54	55–59	60–69	70–79	80–89
Frequency	14	18	29	45	54	22	17

Table C

Length of life (h)	650–669	670–679	680–689	690–699	700–719
Frequency	3	7	20	17	3

QUARTILES, PERCENTILES, INTERQUARTILE RANGE

The three values which split a distribution into four equal portions are known as the quartiles.

The 99 values which split a distribution into 100 equal portions are the percentiles.

Consider n items, arranged in ascending order:

Lower quartile	Q_1	$\frac{1}{4}(n+1)$th value
Median	Q_2	$\frac{1}{2}(n+1)$th value
Upper quartile	Q_3	$\frac{3}{4}(n+1)$th value
10th percentile	P_{10}	$\frac{10}{100}(n+1)$th value
90th percentile	P_{90}	$\frac{90}{100}(n+1)$th value

and so on.

The interquartile range $=$ upper quartile $-$ lower quartile

$$= Q_3 - Q_1$$

The semi-interquartile range $= \frac{1}{2}(Q_3 - Q_1)$

NOTE: the advantage of these ranges is that they depend entirely on the middle half of the readings and they are not affected by extreme values.

The 10 to 90 percentile range $= P_{90} - P_{10}$.

Example 1.18 Find the semi-interquartile range of the following set of numbers:

$$2, 3, 3, 9, 6, 6, 12, 11, 8, 2, 3, 5, 7, 5, 4, 4, 5, 12, 9$$

Solution 1.18 First, arrange the numbers in ascending order:

$$2, 2, 3, 3, \textcircled{3}, 4, 4, 5, 5, 5, 6, 6, 7, 8, \textcircled{9}, 9, 11, 12, 12$$

There are 19 numbers.

Q_1 is the $\frac{1}{4}(19+1)$th value, i.e. the 5th value

$$Q_1 = 3$$

Q_3 is the $\frac{3}{4}(19+1)$th value, i.e. the 15th value

$$Q_3 = 9$$

Therefore semi-interquartile range $= \frac{1}{2}(Q_3 - Q_1)$

$$= \frac{1}{2}(9-3)$$

$$= 3$$

The semi-interquartile range of the set of numbers is 3.

When data has been grouped into classes, the values of the quartiles and percentiles may be obtained in the same way as the median.

Example 1.19 The table gives the cumulative distribution of the heights (in cm) of 400 children in a certain school:

Height (cm)	<100	<110	<120	<130	<140	<150	<160	<170
Cumulative frequency	0	27	85	215	320	370	395	400

(a) Draw a cumulative frequency curve.
(b) Find an estimate of the median.
(c) Determine the interquartile range.
(d) Determine the 10 to 90 percentile range.

Solution 1.19 Consider the 400 values arranged in ascending order:

Median	Q_2 or M	$\frac{1}{2}(401)$th value $\approx$ 200th value
Lower quartile	Q_1	$\frac{1}{4}(401)$th value $\approx$ 100th value
Upper quartile	Q_3	$\frac{3}{4}(401)$th value $\approx$ 300th value
10th percentile	P_{10}	$\frac{10}{100}(401)$th value $\approx$ 40th value
90th percentile	P_{90}	$\frac{90}{100}(401)$th value $\approx$ 360th value

(a) **Cumulative frequency curve to show the heights of 400 children**

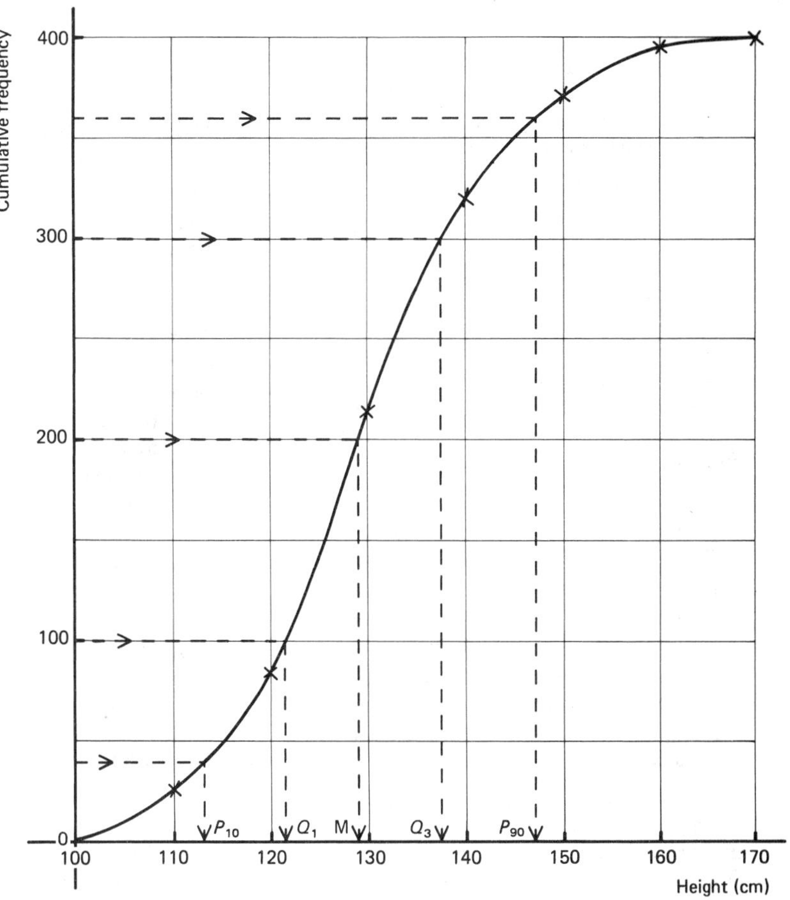

From the curve,

(b) An estimate of the median is 129 cm.

(c) $Q_3 = 137.5$ cm, $Q_1 = 121.5$ cm.

$$\text{The interquartile range} = Q_3 - Q_1 = 137.5 - 121.5 = 16 \text{ cm}$$

The middle half of the readings, that is the interquartile range, has a range of 16 cm.

(d) $P_{90} = 147$ cm, $P_{10} = 113$ cm.

$$\text{The 10 to 90 percentile range} = P_{90} - P_{10} = 147 - 113 = 34 \text{ cm}$$

Therefore the middle 80% of the readings have a range of 34 cm.

Exercise 1g

1. Find (*a*) the median, (*b*) the lower quartile Q_1, (*c*) the upper quartile Q_3 for each of the following sets of data:
(i) Test marks of 11 students:

52, 61, 78, 49, 47, 79, 54, 58, 62, 73, 72

(ii)

Number of peas per pod	5	6	7	8	9	10	11	12
Frequency	10	13	18	24	22	19	8	5

2. The marks scored by 63 pupils in a test are shown in the frequency distribution. Calculate (*a*) the median, (*b*) the interquartile range for the set of marks.

Mark	0	1	2	3	4	5	6	7	8	9	10
Frequency	2	2	3	4	6	11	15	10	6	3	1

3. Table A below shows the marks, collected into groups, of 400 candidates in an examination. The maximum mark was 99.

Compile the cumulative frequency table and draw the cumulative frequency curve.

Use your curve to estimate (i) the median, (ii) the 20th percentile.

If the minimum mark for Grade A was fixed at 74, estimate from your curve the percentage of candidates obtaining Grade A. (C Additional)

4. An inspection of 34 aircraft assemblies revealed a number of missing rivets as shown in Table B below.

Draw a cumulative frequency curve. Use this curve to estimate the median and the quartiles of the distribution. (O & C)

5. From the soil of an English garden 100 earthworms were collected. Their lengths were recorded to the nearest millimetre and grouped as shown in Table C below.

Write down the cumulative frequency table and draw a cumulative frequency curve to illustrate this information.

Estimate
(i) the median length of worm,
(ii) the semi-interquartile range,
(iii) the percentage of worms which are over 180 mm in length. (C Additional)

6. Every day at 08 28 a train departs from one city and travels to a second city. The times taken for the journey were recorded in minutes over a certain period and were grouped as shown in Table D below.

(The interval '-90' indicates all times greater than 85 minutes up to and including 90 minutes.)

From these figures draw a cumulative frequency curve and from this curve estimate
(i) the median time for the journey,
(ii) the semi-interquartile range,
(iii) the number of trains which arrived at the second city between 10 00 and 10 15. (C Additional)

Table A

Marks	0–9	10–19	20–29	30–39	40–49	50–59	60–69	70–79	80–89	90–99
No. of candidates	10	26	42	66	83	71	52	30	14	6

Table B

Number of rivets missing	0–2	3–5	6–8	9–11	12–14	15–17	18–20	21–23
Frequency	4	9	11	6	2	1	0	1

Table C

Length (mm)	95–109	110–124	125–139	140–154	155–169	170–184	185–199	200–214
Number of worms	2	8	17	26	24	16	6	1

Table D

Time	–80	–85	–90	–95	–100	–105	–110	–115	–120	–125	over 125
Frequency	0	6	12	22	31	15	7	4	2	1	0

7. 30 specimens of sheet steel are tested for tensile strength, measured in kN m^{-2}. Table E below gives the distribution of the measurements.

Draw a cumulative frequency diagram of this distribution.

Estimate the median and the 10th and 90th percentiles. (O & C)

8. The figure below shows the cumulative frequency diagram for the distribution of the number of marks, N, in the range 0 to 99 inclusive, obtained by 120 candidates in an examination. From the diagram, estimate
(a) the median mark,
(b) the inter-quartile range,
(c) the number of candidates who scored more than 59 marks.

State why the diagram has to be read at $N = 9.5, 19.5, \ldots$ if a grouped frequency table showing how many candidates are in the class intervals 0–9, 10–19, ... is to be found.

Draw up such a table and illustrate it by drawing a histogram. Mark on your diagram the median mark. (L Additional)

9. The distribution of the times taken when a certain task was performed by each of a large number of people was such that its twentieth percentile was 25 minutes, its fortieth percentile was 50 minutes, its sixtieth percentile was 64 minutes and its eightieth percentile was 74 minutes. Use linear interpolation to estimate (i) the median of the distribution, (ii) the upper quartile of the distribution, (iii) the percentage of persons who performed the task in forty minutes or less. (JMB)

10. The frequency distribution, given in the table, refers to the heights, in cm, of 50 men, corrected to the nearest 10 cm.

Height (cm)	140	150	160	170	180	190
Frequency	1	6	8	21	10	4

(a) State the least possible height of the one man whose height is recorded in the table as 140 cm.

(b) Draw on graph paper a histogram to illustrate the data of the table, drawing five columns, with the first column representing the seven shortest men. Label the axes carefully and explain clearly how frequency has been represented on your histogram.

(c) Draw a cumulative frequency diagram on graph paper for the data given in the table. From your diagram, estimate the upper and lower quartiles, the median height and the interquartile range. (L Additional)

11. The following data concern a random sample of 1000 men with heights in the given ranges.

Height (cm)	No. of men	Height (cm)	No. of men
168–	8	180–	130
170–	31	182–	131
172–	84	184–	94
174–	108	186–	64
176–	142	188–	42
178–	141	190-192	25

Table E

Tensile strength	405–415	415–425	425–435	435–445	445–455	455–465
Number of specimens	4	3	6	10	5	2

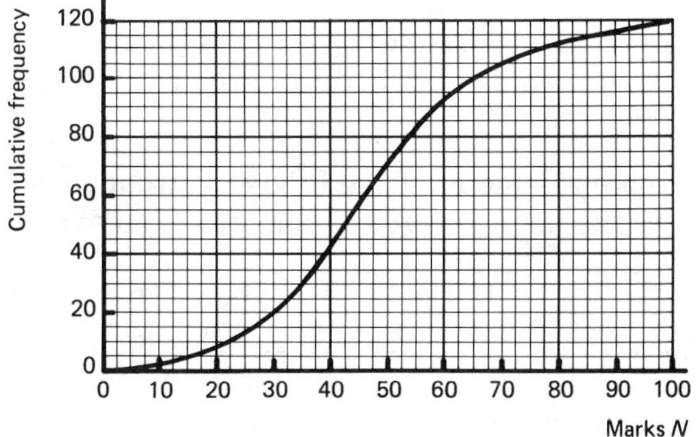

Draw a cumulative frequency diagram to illustrate these data. Use your diagram to estimate
(*a*) the median height,
(*b*) the range of heights for men who are between the fortieth and seventieth percentiles,
(*c*) the number of men in the sample with heights of at least 183 cm.
(L Additional)

MEASURES OF LOCATION

There are three main statistical measures which attempt to locate a 'typical' value. These are

the median (which we have already investigated, p. 22),
the mode,
the arithmetic mean.

THE MODE

The mode is the value that occurs most often.

The mode has the advantage that it is easy to calculate and it eliminates the effects of extreme values, but it is generally unsuitable for further calculation and it is not used widely.

(a) Raw data

Example 1.20 Find the mode of each of the following sets:

(a) 4, 5, 5, 1, 2, 9, 5, 6, 4, 5, 7, 5, 5 (b) 1, 8, 19, 12, 3, 4, 6, 9
(c) 2, 2, 3, 5, 8, 2, 5, 6, 6, 5

Solution 1.20 (a) The mode is 5, as it occurs most often.

(b) The mode does not exist.

(c) The modes are 2 and 5. The distribution is said to be bimodal.

(b) Grouped data

When data has been grouped into classes, the class which has the largest standard frequency is called the **modal class**. An estimate of the mode can be obtained from the modal class.

Example 1.21 Estimate the mode of the following frequency distribution which shows the marks of 330 candidates in an examination.

Marks	11-20	21-30	31-40	41-50	51-60	61-70	71-80	81-90	91-100
Frequency	20	40	80	100	50	20	10	10	0

Solution 1.21 First, a histogram is constructed.

Histogram to show examination marks

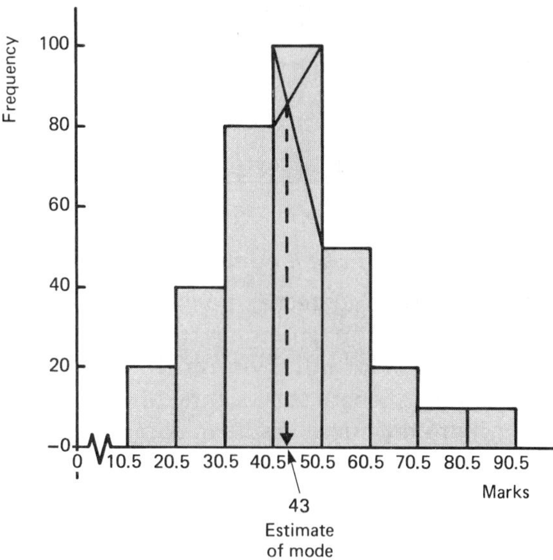

The modal class is 41–50.

Now the modal class contains 20 more than the class below and 50 more than the class above. So the mode is likely to divide the modal class in the ratio $20:50 = 2:5$.

An estimate of the mode can be found from the histogram by drawing lines as shown in the diagram. This gives a value of 43 marks.

Estimate of the mode by calculation:

An estimate of the mode is $\dfrac{20}{20+50}$ of the interval of 10 marks from 40.5 to 50.5. So

$$\begin{aligned}\text{estimate of mode} &= 40.5 + \left(\frac{20}{20+50}\right)(10)\\ &= 40.5 + \left(\frac{2}{7}\right)(10)\\ &= 43.4\end{aligned}$$

An estimate of the mode is 43.4 marks.

From the example it can be seen that the relevant information concerns the modal class and the class on either side of it. Hence we can save time if we extract this information instead of drawing the complete histogram. For example,

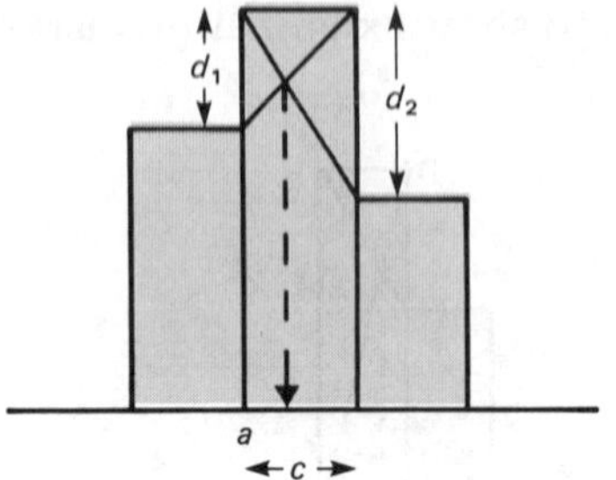

a is the l.c.b. of the modal class

c is the width of the modal class

Estimate of the mode

$$= a + \left(\frac{d_1}{d_1 + d_2}\right)(c)$$

Estimate of the mode from the cumulative frequency curve

The rate of increase of a cumulative frequency curve is greatest at the point corresponding to the mode. Therefore, at the mode there is a point of inflexion on the cumulative frequency curve. To estimate where this occurs, place a ruler along the curve and find where the curve has its maximum gradient.

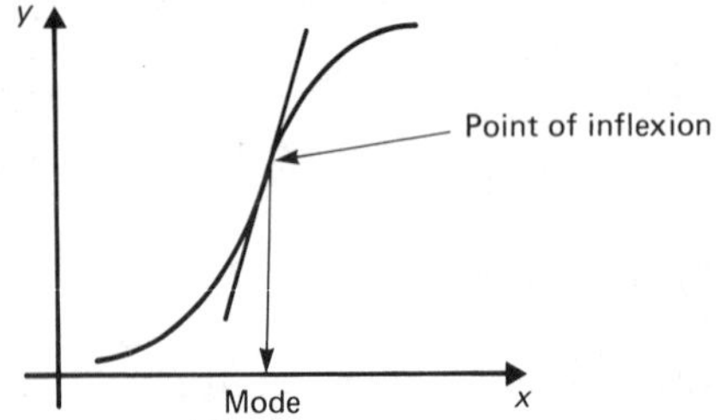

Example 1.22 Draw a cumulative frequency curve for the data in Example 1.21 and use it to estimate the mode.

Solution 1.22 The upper class boundaries are 20.5, 30.5, 40.5, ..., 90.5. The lower boundary of the first class is 10.5.

The cumulative frequency distribution is as follows:

Marks	<10.5	<20.5	<30.5	<40.5	<50.5	<60.5	<70.5	<80.5	<90.5
Cumulative frequency	0	20	60	140	240	290	310	320	330

Cumulative frequency curve to show examination marks

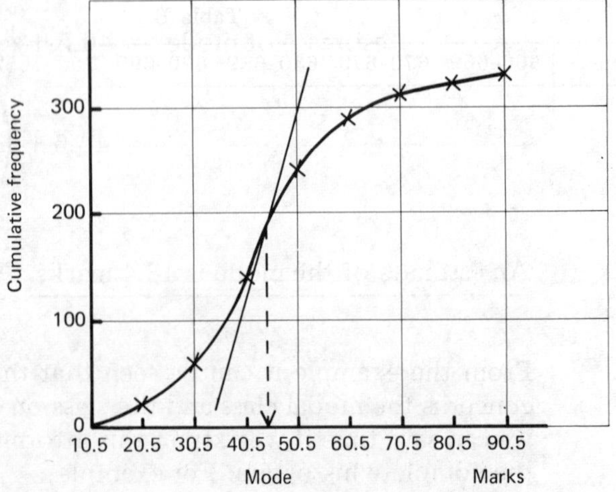

We see from the curve that the point of inflexion occurs when the mark is 44.5 (approximately).

Therefore an estimate of the mode is 44.5 marks.

Exercise 1h

1. Find the mode of each of the following sets of numbers:
   (*a*) 27, 29, 27, 25, 24, 27, 25, 29
   (*b*) 412, 426, 435, 412, 427, 428, 435
   (*c*) 4, 6, 4, 8, 9, 2, 4, 2, 6, 7, 8, 6, 5, 5, 4, 6
   (*d*) 101, 106, 99, 108, 76, 87, 102, 93

2. Find the mode of the following frequency distribution:

Shoe size	2	3	4	5	6
Frequency	8	15	23	20	14

3. State the modal class and using the histograms which you drew previously estimate the mode:
   (*a*) for the data of Question 2, Exercise 1a,
   (*b*) for the data of Question 3, Exercise 1a,
   (*c*) for the data of Question 4, Exercise 1a.
   For part (*a*) refer to the original data and state the true mode.

4. Table A below shows the number of men in various age groups with some form of paid employment in the village of West Morton. The age recorded for each man is the number of completed years lived.
   (*a*) Construct the cumulative frequency table and draw the cumulative frequency curve.
   (*b*) From the cumulative frequency curve, estimate the mode.
   (*c*) Draw a histogram, and use it to estimate the mode.

5. The lives of 80 electric light bulbs were recorded in hours to the nearest hour and grouped as shown in Table B below.
   (*a*) State the limits between which the actual life of each bulb in the first group must lie.
   (*b*) Construct the cumulative frequency table and draw the cumulative frequency curve.
   (*c*) Use your curve to estimate
   (i) the median,
   (ii) the 90th percentile.
   (*d*) Explain how the curve may be used to estimate the mode of the distribution.
   (C Additional)

Table A

Age (years)	14–20	21–30	31–40	41–50	51–60	61–70	71–90
Frequency	12	14	26	35	23	5	1

Table B

Life (in hours)	660–669	670–679	680–689	690–699	700–709	710–719	720–729	730–739
No. of bulbs	4	5	12	24	15	10	7	3

THE ARITHMETIC MEAN

The mean of the n numbers $x_1, x_2, \ldots, x_n$ is $\bar{x}$ where

$$\bar{x} = \frac{x_1 + x_2 + \ldots + x_n}{n} = \frac{\Sigma x_i}{n} \quad \text{for} \quad i = 1, 2, \ldots, n$$

NOTE: for simplicity we often drop the subscript i and write $\bar{x} = \dfrac{\Sigma x}{n}$.

(a) Raw data

Example 1.23 Find the mean of the set of numbers

$$63, 65, 67, 68, 69, 70, 71, 72, 74, 75$$

Solution 1.23 $n = 10$

$$\Sigma x = 63+65+67+68+69+70+71+72+74+75 = 694$$

Therefore
$$\bar{x} = \frac{\Sigma x}{n} = \frac{694}{10} = 69.4$$

The mean of the set of numbers is 69.4.

The calculations can be made easier by using an *assumed mean*, written $\bar{x}_a$.

We then consider the deviation, y, of each reading from $\bar{x}_a$. So, for $x_1, x_2, \ldots, x_n$ we have

$$\begin{aligned} y_1 &= x_1 - \bar{x}_a \\ y_2 &= x_2 - \bar{x}_a \\ &\vdots \\ y_i &= x_i - \bar{x}_a \\ &\vdots \\ y_n &= x_n - \bar{x}_a \end{aligned}$$

Summing $\quad \Sigma y_i = \Sigma x_i - n\bar{x}_a \quad$ for $\quad i = 1, 2, \ldots, n$

So $\quad \dfrac{\Sigma y_i}{n} = \dfrac{\Sigma x_i}{n} - \bar{x}_a$

Therefore $\quad \bar{y} = \bar{x} - \bar{x}_a$

Rearranging $\quad \bar{x} = \bar{x}_a + \bar{y}$

In general, if $y = x - \bar{x}_a$, then $\bar{y} = \bar{x} - \bar{x}_a$

and so $\quad \bar{x} = \bar{x}_a + \bar{y}$

Example 1.24 Find the mean of the set of numbers given in Example 1.23, using an assumed mean $\bar{x}_a$ of 70.

Solution 1.24 $\quad \bar{x}_a = 70, \quad$ so $\quad y = x - 70$

Mark, x	$y = x - 70$
63	-7
65	-5
67	-3
68	-2
69	-1
70	0
71	1
72	2
74	4
75	5
	$\Sigma y = -18 + 12 = -6$

Now $y = x - 70$

Therefore

$$\bar{y} = \bar{x} - 70$$

so

$$\bar{x} = 70 + \bar{y} = 70 + \frac{\Sigma y}{n} = 70 + \frac{(-6)}{10} = 69.4$$

Therefore the mean, $\bar{x} = 69.4$, as before.

Exercise 1i

1. Find the mean for each of the following sets of numbers (*a*) without using an assumed mean, (*b*) using an assumed mean.
   (i) 5, 6, 6, 8, 8, 9, 11, 13, 14, 17
   (ii) 148, 153, 156, 157, 160
   (iii) $44\frac{1}{2}, 47\frac{1}{2}, 48\frac{1}{2}, 51\frac{1}{2}, 52\frac{1}{2}, 54\frac{1}{2}, 55\frac{1}{2}, 56\frac{1}{2}$
   (iv) 1769, 1771, 1772, 1775, 1778, 1781, 1784
   (v) 0.85, 0.88, 0.89, 0.93, 0.94, 0.96

2. If the mean of the following numbers is 17, find the value of c:

   12, 18, 21, c, 13

3. The mean of 10 numbers is 8. If an eleventh number is now included in the results, the mean becomes 9. What is the value of the eleventh number?

4. The mean of 4 numbers is 5, and the mean of 3 different numbers is 12. What is the mean of the 7 numbers together?

5. The mean of n numbers is 5. If the number 13 is now included with the n numbers, the new mean is 6. Find the value of n.

(b) Ungrouped frequency distribution

For a frequency distribution

$$\bar{x} = \frac{\Sigma fx}{\Sigma f}$$

Example 1.25 The 30 members of an orchestra were asked how many instruments each could play. The results are set out in the frequency distribution. Calculate the mean number of instruments played:

Number of instruments, x	1	2	3	4	5
Frequency, f	11	10	5	3	1

Solution 1.25

x	f	fx
1	11	11
2	10	20
3	5	15
4	3	12
5	1	5
	$\Sigma f = 30$	$\Sigma fx = 63$

$$\bar{x} = \frac{\Sigma fx}{\Sigma f} = \frac{63}{30} = 2.1$$

The mean number of instruments played is 2.1.

Again, an assumed mean may be used, and we have

$$\bar{x} = \bar{x}_a + \bar{y} \quad \text{where} \quad \bar{y} = \frac{\Sigma fy}{\Sigma f} \quad \text{and} \quad y = x - \bar{x}_a$$

Example 1.26 For the data of Example 1.25, find the mean using an assumed mean of 2.

Solution 1.26

$$\bar{x}_a = 2, \quad \text{so} \quad y = x - 2 \quad \text{and} \quad \bar{y} = \bar{x} - 2$$

x	f	$y = x - 2$	fy
1	11	-1	-11
2	10	0	0
3	5	1	5
4	3	2	6
5	1	3	3
	$\Sigma f = 30$		$\Sigma fy = -11 + 14 = 3$

Now $\qquad \bar{x} = 2 + \bar{y} \quad \text{and} \quad \bar{y} = \dfrac{\Sigma fy}{\Sigma f}$

so

$$\bar{x} = 2 + \frac{3}{30} = 2.1$$

The mean number of instruments played is 2.1, as before.

It is particularly useful to use an assumed mean when dealing with large numbers or ones involving fractions or decimals.

(c) Grouped frequency distribution

When data has been grouped into intervals, the mid-point, x, of the interval is taken to represent the interval.

Example 1.27 The lengths of 40 bean pods were measured to the nearest cm and grouped as shown. Find the mean length, giving the answer to 1 d.p.

Length (cm)	4–8	9–13	14–18	19–23	24–28	29–33
Frequency, f	2	4	7	14	8	5

Solution 1.27 **Method 1** Without using an assumed mean

Length (cm)	Mid-point, x	f	fx
4–8	6	2	12
9–13	11	4	44
14–18	16	7	112
19–23	21	14	294
24–28	26	8	208
29–33	31	5	155
		$\Sigma f = 40$	$\Sigma fx = 825$

$$\bar{x} = \frac{\Sigma fx}{\Sigma f} = \frac{825}{40} = 20.6 \quad (1\text{ d.p.})$$

Therefore the mean length of the bean pods is 20.6 cm (1 d.p.).

Method 2 Using an assumed mean

Let $\bar{x}_a = 21$, then $y = x - 21$, where x is the mid-point of an interval, and $\bar{y} = \bar{x} - 21$.

Length (cm)	Mid-point, x	f	$y = x - 21$	fy
4–8	6	2	−15	−30
9–13	11	4	−10	−40
14–18	16	7	−5	−35
19–23	21	14	0	0
24–28	26	8	5	40
29–33	31	5	10	50
		$\Sigma f = 40$		$\Sigma fy = -105 + 90 = -15$

$$\bar{x} = 21 + \bar{y} \quad \text{where} \quad \bar{y} = \frac{\Sigma fy}{\Sigma f}$$

$$= 21 + \frac{-15}{40}$$

$$= 20.6 \quad (1\text{ d.p.})$$

Therefore the mean length of the bean pods is 20.6 cm (1 d.p.), as before.

Exercise 1j

1. Find the mean for each of the following frequency distributions (*a*) without using an assumed mean, (*b*) using an assumed mean.

(i)

x	1	2	3	4	5	6	7
f	4	5	8	10	17	5	1

(ii)

x	27	28	29	30	31	32
f	30	43	51	49	42	35

(iii)

x	121	122	123	124	125
f	14	25	32	23	6

(iv)

Interval	5-9	10-14	15-19	20-24	25-29	30-34
f	4	6	12	10	7	1

(v)

Interval	101-104	105-108	109-112	113-116	117-120
f	13	18	21	12	6

2. If the mean of the following frequency distribution is 3.66, find the value of a, and find the median and the mode of the distribution.

x	1	2	3	4	5	6
f	3	9	a	11	8	7

3. (*a*) State briefly the meaning of the terms mode, median and mean. Give an example of a situation where the most appropriate average would be (i) the mode, (ii) the median.
(*b*) A bag contained five balls each bearing one of the numbers 1, 2, 3, 4, 5. A ball was drawn from the bag, its number noted, and then replaced. This was repeated 50 times and the table below shows the resulting frequency distribution.

Number	1	2	3	4	5
Frequency	x	11	y	8	9

If the mean is 2.7, (i) determine the value of x and y, (ii) state the mode and median of this distribution. (C Additional)

4. The maximum daytime temperature was recorded for each day in February and the results noted as shown in Table A below.

Table A

Temperature (°C)	1	2	3	4	5	6	7	8	9	10	11	12
Frequency	1	0	3	1	4	1	3	5	3	5	1	1

(*a*) Calculate the mean maximum day time temperature for February. (*b*) Find the mode. (*c*) Find the median.

5. A sample of 100 boxes of matches was taken and a record made of the number of matches per box. The results were as follows:

Number of matches per box	47	48	49	50	51
Frequency	4	20	35	24	17

Calculate the mean number of matches per box, using an assumed mean of 49.

6. The table shows the speeds of 200 vehicles passing a particular point:

Speed (km/h)	30-	40-	50-	60-	70-	80-
Frequency	14	30	52	71	33	0

Using an assumed mean of 55 km/h, calculate the mean speed.

7. After an exciting friendly rugby match between the A and B teams at a certain school the matron had to treat all 30 boys for bruises. She recorded the number of bruises sustained by each boy and submitted her report to the headmaster. However, she gave it to a very grubby boy to deliver and what the headmaster finally saw looked something like this:

Number of bruises	1	2	3	4	5	6	7	8
Number of boys	3	1	•	3	•	6	3	6

The headmaster overheard one boy saying that if everyone had received the same number of bruises he would have 5, instead of 8.
(*a*) How many boys received (i) 3 bruises, (ii) 5 bruises?
(*b*) How many boys were fortunate enough to sustain fewer than the median number of bruises?

8. On a certain day the number of books on 40 shelves in a library was noted and grouped as shown. Find the mean number of books on a shelf, using an assumed mean. Give your answer to 2 significant figures.

Number of books	31-35	36-40	41-45	46-50	51-55	56-60
Number of shelves	4	6	10	13	5	2

Using a method of 'coding' or 'scaling' to find the mean

The idea of using an assumed mean may be extended to make calculations easier. We now illustrate this method of 'coding' or 'scaling'.

Example 1.28 Find the mean of the set of numbers 5693, 5700, 5714, 5721, 5735.

Solution 1.28 Choose as assumed mean $\bar{x}_a = 5714$. However, when we write out the values of $x - \bar{x}_a$ we note that all the numbers are multiples of 7.

So, we introduce a further column $y = \dfrac{x - 5714}{7}$.

x	$x - 5714$	$y = \dfrac{x-5714}{7}$
5693	-21	-3
5700	-14	-2
5714	0	0
5721	7	1
5735	21	3
		$\Sigma y = -5 + 4 = -1$

We have used the 'coding'

$$y = \frac{x - 5714}{7}$$

Rearranging $\quad x = 5714 + 7y$

Summing, for all 5 values

$$\Sigma x = (5)(5714) + \Sigma 7y$$

$$= (5)(5714) + 7\Sigma y$$

So $\quad \dfrac{\Sigma x}{5} = 5714 + 7\dfrac{\Sigma y}{5}$

and $\quad \bar{x} = 5714 + 7\bar{y}$

Therefore $\quad \bar{x} = 5714 + 7\dfrac{(-1)}{5}$

$$= 5712.6$$

The mean of the set of numbers is 5712.6.

In general, if the set of numbers $x_1, x_2, \ldots, x_n$ is transformed to the set of numbers $y_1, y_2, \ldots, y_n$ by means of the coding

$$y = \frac{x - \bar{x}_a}{b}$$

then $$x = \bar{x}_a + by$$

and $$\bar{x} = \bar{x}_a + b\bar{y}$$

Proof Let

$$y_i = \frac{x_i - \bar{x}_a}{b} \qquad i = 1, 2, \ldots, n$$

Rearranging $$x_i = \bar{x}_a + by_i$$

Summing $$\Sigma x = n\bar{x}_a + b\Sigma y$$

so $$\frac{\Sigma x}{n} = \bar{x}_a + b\frac{\Sigma y}{n}$$

i.e. $$\bar{x} = \bar{x}_a + b\bar{y}$$

This method is particularly useful when the data is in the form of a frequency distribution and the intervals are of equal width.

In this case $$\bar{y} = \frac{\Sigma f y_i}{\Sigma f_i}$$

For data grouped into classes of equal width:

(a) use the mid-point of each interval to represent the class,

(b) choose a central value as assumed mean, $\bar{x}_a$,

(c) divide by the class width, b.

Example 1.29 A girl measured her waiting time (in minutes, to the nearest minute) for the school bus on 30 mornings, and obtained the following results:

Waiting time (min)	1–4	5–8	9–12	13–16	17–20
Frequency	3	6	10	7	4

Use a method of coding to find her mean waiting time.

Solution 1.29 The mid-points are 2.5, 6.5, 10.5, 14.5, 18.5. Choose a central value of 10.5, say, as assumed mean, so $\bar{x}_a = 10.5$.

The class widths are each equal to 4 so take $b = 4$.

So we use

$$y = \frac{x - \bar{x}_a}{b} \quad \text{where} \quad \bar{x}_a = 10.5 \quad \text{and} \quad b = 4$$

i.e. $$y = \frac{x - 10.5}{4} \quad \text{and} \quad x = 10.5 + 4y$$

Waiting time (min)	Mid-point, x	f	$y = \frac{x-10.5}{4}$	fy
1-4	2.5	3	−2	−6
5-8	6.5	6	−1	−6
9-12	10.5	10	0	0
13-16	14.5	7	1	7
17-20	18.5	4	2	8
		$\Sigma f = 30$		$\Sigma fy = -12 + 15 = 3$

Now $$\bar{x} = \bar{x}_a + b\bar{y} \quad \text{where} \quad \bar{y} = \frac{\Sigma fy}{\Sigma f}$$

So $$\bar{x} = 10.5 + 4\left(\frac{3}{30}\right)$$
$$= 10.5 + 0.4$$
$$= 10.9$$

The mean waiting time is 10.9 minutes.

Exercise 1k

1. Find the mean, $\bar{x}$, of the set of numbers 10, 20, 30, 40, 50, 60 using the coding $y = \frac{x - 40}{10}$.

2. Find the mean, $\bar{x}$, of the numbers 217, 222, 227, 237, 242, 252 using the coding $y = \frac{x - 237}{5}$.

3. The table shows the masses of a group of male students at a college. Measurements have been taken to the nearest kg.

Mass (kg)	60-64	65-69	70-74	75-79	80-84	85-89
Frequency	4	27	42	60	35	12

Find the mean, using a method of coding.

4. Find the mean, using a method of coding, for each of the following frequency distributions:

(a)

Interval	15-21	22-28	29-35	36-42	43-49
Frequency	2	18	23	17	9

(b)

Interval	0-	10-	20-	30-	40-	50-	60-
Frequency	10	15	23	32	18	2	0

(c)

Interval	1-2	3-4	5-6	7-8	9-10	11-12	13-14
Frequency	5	8	12	14	13	6	2

5. In a practical class students timed how long it took for a sample of their saliva to break down a 2% starch solution. The times, to the nearest second, are shown in Table A below. Find the mean time, using a method of coding.

6. Each morning for a month the owner of a smallholding timed how long it took to feed the animals. The results were as shown:

Time (min)	-15	-20	-25	-30	-35	-40	-45	-50
Frequency	0	3	2	6	10	7	2	1

Calculate the mean time taken to feed the animals, using a method of coding.

Table A

Time (seconds)	11-20	21-30	31-40	41-50	51-60	61-70	71-90
Frequency	1	2	5	11	8	2	1

MEASURES OF DISPERSION

There are several ways of obtaining a measure of the 'spread' of a set of observations.

THE RANGE

The range is the difference between the highest and the lowest value. It is based entirely on the extreme values.

RANGES BASED ON QUARTILE AND PERCENTILE OBSERVATIONS

Interquartile range = $Q_3 - Q_1$ where Q_3 is the upper quartile
Q_1 is the lower quartile

Semi-interquartile range = $\frac{1}{2}(Q_3 - Q_1)$

NOTE: these ranges depend entirely on the middle half of the observations.

The 10 to 90 percentile range = $P_{90} - P_{10}$ where P_{90} is the 90th percentile
P_{10} is the 10th percentile

NOTE: this range depends on the middle 80% of the observations.

THE MEAN DEVIATION FROM THE MEAN

The mean deviation from the mean makes use of all the observations.

The mean deviation from the mean of a set of n numbers, $(x_1, x_2, \ldots, x_n)$, is given by

$$\frac{\Sigma|x_i - \bar{x}|}{n} \qquad i = 1, 2, \ldots, n$$

where $\bar{x}$ is the mean of the set of numbers.

NOTE: $|x_i - \bar{x}|$ is the positive difference between x_i and $\bar{x}$ and is called the modulus of $(x_i - \bar{x})$.

Example 1.30 Two machines, A and B, are used to pack biscuits. A sample of 10 packets was taken from each machine and the mass of each packet, measured to the nearest gram, was noted. Find the mean deviation from the mean of the masses of the packets taken in the sample for each machine. Comment on your answer.

Machine A (mass in g)	196, 198, 198, 199, 200, 200, 201, 201, 202, 205
Machine B (mass in g)	192, 194, 195, 198, 200, 201, 203, 204, 206, 207

Solution 1.30 Machine A

$$\bar{x} = \frac{\Sigma x}{n} = \frac{2000}{10} = 200$$

x	$x-200$	$\lvert x-200\rvert$
196	−4	4
198	−2	2
198	−2	2
199	−1	1
200	0	0
200	0	0
201	1	1
201	1	1
202	2	2
205	5	5
		18

$$\text{Mean deviation} = \frac{\Sigma|x-200|}{10} = 1.8$$

Machine B

$$\bar{x} = \frac{\Sigma x}{n} = \frac{2000}{10} = 200$$

x	$x-200$	$\lvert x-200\rvert$
192	−8	8
194	−6	6
195	−5	5
198	−2	2
200	0	0
201	1	1
203	3	3
204	4	4
206	6	6
207	7	7
		42

$$\text{Mean deviation} = \frac{\Sigma|x-200|}{10} = 4.2$$

The larger number for machine B indicates that the masses are more widely spread than those from machine A.

Therefore machine A is more reliable.

For a frequency distribution

$$\text{mean deviation from the mean} = \frac{\Sigma|f_i(x_i-\bar{x})|}{\Sigma f_i} \qquad i = 1, 2, \ldots, n$$

It is possible to find the mean deviation from the median and the mean deviation from the mode. However, the mean deviation is not widely used.

THE STANDARD DEVIATION, *s*

The square of the deviation from the mean is considered for each value of x.

The standard deviation of a set of n numbers, $x_1, x_2, \ldots, x_n$, with mean $\bar{x}$ is given by s, where

$$s = \sqrt{\frac{\Sigma(x_i - \bar{x})^2}{n}} \qquad i = 1, 2, \ldots, n$$

The standard deviation is the most useful measure of spread. For most distributions the bulk of the readings lie within ± 2 standard deviations of the mean, i.e. within the interval $(\bar{x} \pm 2s)$.

The units of standard deviation are the same as the units of the original data.

NOTE: sometimes the abbreviation s.d. is used for standard deviation.

THE VARIANCE

The **variance** of the set of numbers is given by s^2 where

$$s^2 = \frac{\Sigma(x_i - \bar{x})^2}{n}$$

We have standard deviation $= \sqrt{\text{variance}}$

Example 1.31 For the data given in Example 1.30, calculate the standard deviation of each machine, given $\bar{x} = 200$ g in each case.

Solution 1.31

Machine *A*

x	$x-200$	$(x-200)^2$
196	−4	16
198	−2	4
198	−2	4
199	−1	1
200	0	0
200	0	0
201	1	1
201	1	1
202	2	4
205	5	25
		56

Machine *B*

x	$x-200$	$(x-200)^2$
192	−8	64
194	−6	36
195	−5	25
198	−2	4
200	0	0
201	1	1
203	3	9
204	4	16
206	6	36
207	7	49
		240

$$s^2 = \frac{\Sigma(x-200)^2}{10} = 5.6 \qquad s = \sqrt{5.6} = 2.37 \quad (2\text{ d.p.})$$

$$s^2 = \frac{\Sigma(x-200)^2}{10} = 24 \qquad s = \sqrt{24} = 4.90 \quad (2\text{ d.p.})$$

The standard deviation for machine A is 2.37 g and the standard deviation for machine B is 4.90 g, once again indicating that machine A is more reliable.

Alternative form of the formula for standard deviation The formula $s = \sqrt{\dfrac{\Sigma(x-\bar{x})^2}{n}}$ is sometimes difficult to use, especially when x is not an integer, so an alternative form is often used.

Now

$$s^2 = \frac{1}{n}\Sigma(x_i-\bar{x})^2 \qquad i = 1, 2, \ldots, n$$

$$= \frac{1}{n}\Sigma(x_i^2 - 2\bar{x}x_i + \bar{x}^2)$$

$$= \frac{1}{n}(\Sigma x_i^2 - 2\bar{x}\,\Sigma x_i + \Sigma\bar{x}^2)$$

$$= \frac{\Sigma x_i^2}{n} - 2\bar{x}\frac{\Sigma x_i}{n} + \frac{n\bar{x}^2}{n}$$

$$= \frac{\Sigma x_i^2}{n} - 2\bar{x}(\bar{x}) + \bar{x}^2$$

$$= \frac{\Sigma x_i^2}{n} - \bar{x}^2$$

So we have

$$s = \sqrt{\frac{\Sigma x_i^2}{n} - \bar{x}^2} \qquad i = 1, 2, \ldots, n$$

Example 1.32 Find the mean and the standard deviation of the set of numbers

2, 3, 5, 6, 8

Solution 1.32

$$\bar{x} = \frac{\Sigma x}{n} = \frac{24}{5} = 4.8$$

Standard deviation:

Method 1—using $s = \sqrt{\frac{\Sigma(x-\bar{x})^2}{n}}$

x	$x-\bar{x}$	$(x-\bar{x})^2$
2	−2.8	7.84
3	−1.8	3.24
5	0.2	0.04
6	1.2	1.44
8	3.2	10.24
		22.80

$$s^2 = \frac{22.80}{5} = 4.56$$

$$s = \sqrt{4.56} = 2.14 \quad (2\,\text{d.p.})$$

Method 2—using $s = \sqrt{\frac{\Sigma x^2}{n} - \bar{x}^2}$

x	x^2
2	4
3	9
5	25
6	36
8	64
	138

$$s^2 = \frac{138}{5} - (4.8)^2 = 4.56$$

$$s = \sqrt{4.56} = 2.14 \quad (2\,\text{d.p.})$$

Therefore the standard deviation of the set of numbers is 2.14 (2 d.p.).

NOTE: in this case there is far less working involved in method 2.

Exercise 1I

1. Find the mean and the standard deviation of the following sets of numbers. For questions (*a*), (*b*) and (*c*) try using both forms of the formula for the standard deviation. Use whichever you wish for parts (*d*), (*e*) and (*f*). *Do not use the programmed functions on your calculator.*
   (*a*) 2, 4, 5, 6, 8
   (*b*) 6, 8, 9, 11
   (*c*) 11, 14, 17, 23, 29
   (*d*) 5, 13, 7, 9, 16, 15
   (*e*) 4.6, 2.7, 3.1, 0.5, 6.2
   (*f*) 200, 203, 206, 207, 209

2. The mean of the numbers 3, 6, 7, a, 14 is 8. Find the standard deviation of the set of numbers.

3. For a set of 10 numbers $\Sigma x = 290$ and $\Sigma x^2 = 8469$. Find the mean and the variance.

4. For a set of 9 numbers $\Sigma(x-\bar{x})^2 = 234$. Find the standard deviation of the numbers.

5. For a set of 9 numbers $\Sigma(x-\bar{x})^2 = 60$ and $\Sigma x^2 = 285$. Find the mean of the numbers.

6. The numbers $a, b, 8, 5, 7$ have a mean of 6 and a variance of 2. Find the values of a and b, if $a > b$.

7. Find the mean and the standard deviation of the set of integers $1, 2, 3, \ldots, 20$.

8. Find the mean and the standard deviation of the first n integers.

   You may use

   $$\sum_{r=1}^{n} r = \tfrac{1}{2}n(n+1);$$

   $$\sum_{r=1}^{n} r^2 = \tfrac{1}{6}n(n+1)(2n+1)$$

9. From the information given about each of the following sets of data, work out the missing values in the table:

	n	Σx	Σx^2	$\bar{x}$	s
(*a*)	63	7623	924 800		
(*b*)		152.6		10.9	1.7
(*c*)	52		57 300	33	
(*d*)	18			57	4

10. Calculate the mean and the standard deviation of the four numbers

2, 3, 6, 9

Two numbers, a and b, are to be added to this set of four numbers, such that the mean is increased by 1 and the variance is increased by 2.5. Find a and b.

(L Additional)

THE USE OF CALCULATORS

If your calculator has SD (standard deviation) mode then it can be used to calculate standard deviations, and you will have access to the following information $\bar{x}$, s, n, Σx, Σx^2.

The following example has been done using two types of calculator, and you should consult your calculator instructions if yours does not appear to follow one of the patterns.

Example 1.33 Find the mean and standard deviation of the numbers

33, 28, 26, 35, 38

Solution 1.33

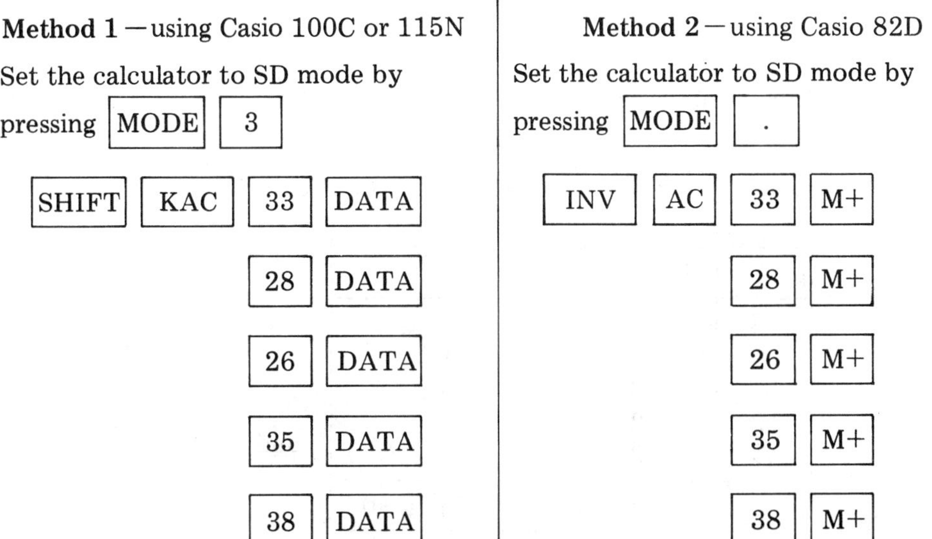

(Try to use both hands, left hand for the numbers and right hand for DATA or M+.)

Key		Result	Key		Result
SHIFT	1	gives $\bar{x} = 32$	INV	7	gives $\bar{x} = 32$
SHIFT	2	gives $s = 4.427\ldots$	INV	8	gives $s = 4.427\ldots$
K out	3	gives $n = 5$	INV	6	gives $n = 5$
K out	2	gives $\Sigma x = 160$	INV	5	gives $\Sigma x = 160$
K out	1	gives $\Sigma x^2 = 5218$	INV	4	gives $\Sigma x^2 = 5218$

Therefore the mean is 32 and the standard deviation is 4.43 (2 d.p.)

NOTE: To clear the SD mode, press MODE 0.

Exercise 1m

Do Exercise 1l question 1 using your calculator in SD mode.

Calculations involving the mean and the standard deviation

Example 1.34 For the set of numbers 3, 6, 7, 9, 10 the mean is 7 and the standard deviation is $\sqrt{6}$. If each number in the set is increased by 3, find the new mean and standard deviation. Comment on your answers.

Solution 1.34 The new set of numbers is 6, 9, 10, 12, 13.

$$\text{The mean} = \frac{\Sigma x}{n} = \frac{(6+9+10+12+13)}{5} = 10$$

$$\text{Now} \quad s^2 = \frac{\Sigma(x-\bar{x})^2}{n} \quad \text{where } \bar{x} = 10$$

x	$(x-\bar{x})$	$(x-\bar{x})^2$
6	-4	16
9	-1	1
10	0	0
12	2	4
13	3	9
		$\Sigma(x-\bar{x})^2 = 30$

So $s^2 = \frac{30}{5}$

$= 6$

and $s = \sqrt{6}$

Therefore, if each member of the set of numbers is increased by 3, then the mean is increased by 3 but the standard deviation remains unaltered.

In general, consider the set of n numbers $x_1, x_2, \ldots, x_n$ with mean $\bar{x}$ and standard deviation s_1.

(i) Increase each number by a constant, c Then

$$y_i = x_i + c \quad \text{for} \quad i = 1, 2, \ldots, n$$

Summing
$$\Sigma y_i = \Sigma x_i + nc$$

and
$$\frac{\Sigma y_i}{n} = \frac{\Sigma x_i}{n} + c$$

So
$$\bar{y} = \bar{x} + c$$

If each number is increased by a constant c, the mean is increased by c.

For the new set of numbers

$$s_2^2 = \frac{\Sigma (y_i - \bar{y})^2}{n}$$

$$= \frac{\Sigma [x_i + c - (\bar{x} + c)]^2}{n}$$

$$= \frac{\Sigma (x_i - \bar{x})^2}{n}$$

$$= s_1^2$$

Therefore
$$s_2 = s_1$$

If each number is increased by a constant c, the standard deviation remains unaltered.

(ii) Multiply each number by a constant k It can be shown that

$$\text{new mean} = k\bar{x}$$

$$\text{new standard deviation} = ks_1$$

If each number is multiplied by a constant k, both the mean and the standard deviation are multiplied by k.

Exercise 1n

1. By considering the set of numbers 3, 6, 7, 9, 10, with mean 3 and standard deviation $\sqrt{6}$, investigate the effect on the mean and on the standard deviation of multiplying each term by 3.

2. The set of numbers $x_1, x_2, \ldots, x_n$ has mean $\bar{x}$ and standard deviation s_1. Each of the numbers is multiplied by a constant term k. Show that the new mean is $k\bar{x}$ and the new standard deviation $s_2 = ks_1$.

3. (a) Find the mean and the standard deviation of the set of numbers 4, 6, 9, 3, 5, 6, 9.
(b) Deduce the mean and the standard deviation of the set of numbers 514, 516, 519, 513, 515, 516, 519.
(c) Deduce the mean and the standard deviation of the set of numbers 52, 78, 117, 39, 65, 78, 117.

4. (a) Find the mean and the variance of the ordered set of numbers

$A = \{1, 2, 3, 4, 5, 6, 7\}$.

Hence find the mean and the variance of the following ordered sets

$B = \{4, 5, 6, 7, 8, 9, 10\}$

$C = \{10, 20, 30, 40, 50, 60, 70\}$

$D = \{13, 23, 33, 43, 53, 63, 73\}$

(b) If a_i is the ith member of A and d_i is the ith member of D, find a relationship between a_i and d_i in the form $d_i = la_i + m$ where l and m are constants.
(c) The two ordered sets X and Y each have n elements and $y_i = px_i + q$ where p and q are constants. If the mean and the variance of X are $\bar{x}$ and s_x^2, show that the mean $\bar{y}$ and the variance s_y^2 of Y are $\bar{y} = p\bar{x} + q$ and $s_y^2 = p^2 s_x^2$.
(L Additional)

5. A set of values of a variable X has mean 5 and standard deviation 2. Values of a new variable are obtained by using the formula $Y = 4X - 3$. Find the mean and the standard deviation of the set of values of Y.

Scaling similar sets of data for comparison

If we wish to compare two sets of data, e.g. examination marks in two papers, we 'scale' one of the sets of data so that the two means are the same and the two standard deviations are the same.

Example 1.35 A set of marks has a mean of 40 and a standard deviation of 5. The marks are to be scaled so that the mean becomes 50 and the standard deviation becomes 8. If the equation of the transformation is $y = ax + b$, find the values of the constants a and b. Find also the scaled mark which corresponds to a mark of 45 in the original set.

Solution 1.35 If there are n marks, then $y_i = ax_i + b$ for each $i = 1, 2, \ldots, n$.

Summing $$\Sigma y_i = a\Sigma x_i + nb$$

$$\frac{\Sigma y_i}{n} = a\frac{\Sigma x_i}{n} + b$$

So $$\bar{y} = a\bar{x} + b$$

Hence $$50 = a(40) + b \qquad 40a + b = 50 \quad \text{(i)}$$

Let s_1 be the original standard deviation, and s_2 the new standard deviation

Then $$s_2^2 = \frac{\Sigma (y_i - \bar{y})^2}{n} \qquad i = 1, 2, \ldots, n$$

$$= \frac{\Sigma\,[ax_i + b - (a\bar{x} + b)]^2}{n}$$

$$= a^2 \frac{\Sigma (x_i - \bar{x})^2}{n}$$

$$= a^2 s_1^2$$

or $$s_2 = as_1$$

So $$8 = 5a \qquad a = \tfrac{8}{5} \quad \text{(ii)}$$

Substituting for a from (ii) into (i),

$$40(\tfrac{8}{5}) + b = 50$$
$$b = -14$$

Therefore the equation of the transformation is $y = \frac{8}{5}x - 14$.

If $x = 45$, then $y = \frac{8}{5}(45) - 14 = 58$.

Therefore a mark of 45 in the original set becomes a mark of 58 when scaled.

Exercise 1o

1. It is proposed to convert a set of marks whose mean is 52 and standard deviation is 4 to a set of marks with mean 61 and standard deviation 3. The equation for the transformation necessary to convert the marks is $y = ax + b$. Find (i) the values of a and b, (ii) the value of the scaled mark which corresponds to a mark of 64 in the original data, (iii) the value in the original data if the scaled mark is 79.

2. The marks of 5 students in a mathematics test were 27, 31, 35, 47, 50.
(i) Calculate the mean mark and the standard deviation.
(ii) The marks are scaled so that the mean and standard deviation become 50 and 20 respectively. Calculate, to the nearest whole number, the new marks corresponding to the original marks of 31 and 50.
(C Additional)

3. In order to compare the performances of candidates in two schools a test was given. The mean mark at school A was 45, and the mean mark at school B was 31 with a standard deviation of 5. The marks of school A are scaled so that the mean and standard deviation are the same as school B and a mark of 85 at school A becomes 63. Find the values of a and b if the transformation used is $y = ax + b$. Find also the original standard deviation of the marks from school A.

4. Given that the mean and standard deviation of a set of figures are μ and σ respecttively, write down the new values of the mean and standard deviation when
(i) each figure is increased by a constant c,
(ii) each figure is multiplied by a constant k.

A group of students sat two examinations, one in algebra and one in biology. In order to compare the results the algebra marks were scaled linearly (that is, a mark of x became a mark of $ax + b$ where a and b are constants) so that the means and standard deviations of the marks in both examinations became the same. The original means and standard deviations are shown in the table.

	Algebra	Biology
Mean mark	48	62
Standard deviation	12	10

Find a and b.

The original marks of a particular student are 36 in algebra, 48 in biology. In what sense, if any, has he done better in algebra than in biology? (C Additional)

5. A linear function $f(x) = ax + b$ transforms
$$X = \{1, 2, 3, 5, 8, 11\}$$
into a set Y, so that $f(5) = 13$ and $f(1) = 5$.
(*a*) Find f.
(*b*) Calculate the mean and the variance of X.
(*c*) Hence calculate the mean and the variance of Y.

An element, k, is added to X forming a set Z. Given that the mean of Z is three greater than the mean of X, find
(*d*) the value of k,
(*e*) the variance of Z. (L Additional)

6. Show that the standard deviation of the integers
$$1, 2, 3, 4, 5, 6, 7$$
is 2.

Using this result find the standard deviation of the numbers
(*a*) 101, 102, 103, 104, 105, 106, 107.
(*b*) 100, 200, 300, 400, 500, 600, 700.
(*c*) 2.01, 3.02, 4.03, 5.04, 6.05, 7.06, 8.07.
(*d*) Write down seven integers which have mean 5 and standard deviation 6.
(L Additional)

Combining sets of numbers

Example 1.36 A set of 12 numbers has a mean of 4 and a standard deviation of 2.

A second set of 20 numbers has a mean of 5 and a standard deviation of 3.

Find the mean and the standard deviation of the combined set of 32 numbers.

Solution 1.36 *For the first set of numbers:* $n_1 = 12, \bar{x}_1 = 4, s_1 = 2$

Therefore

$$\sum_{i=1}^{n_1} x_i = n_1\bar{x}_1 = (12)(4) = 48$$

and

$$s_1^2 = \frac{\sum_{i=1}^{n_1} x_i^2}{n_1} - \bar{x}_1^2$$

so

$$\sum_{i=1}^{n_1} x_i^2 = n_1(\bar{x}_1^2 + s_1^2) = (12)(4^2 + 2^2) = 240$$

For the second set of numbers: $n_2 = 20, \bar{x}_2 = 5, s_2 = 3$

Therefore

$$\sum_{j=1}^{n_2} x_j = n_2\bar{x}_2 = (20)(5) = 100$$

and

$$\sum_{j=1}^{n_2} x_j^2 = n_2(\bar{x}_2^2 + s_2^2) = 20(5^2 + 3^2) = 680$$

To find the mean of the combined set of numbers:

$$\bar{x} = \frac{\sum_{i=1}^{n_1} x_i + \sum_{j=1}^{n_2} x_j}{n_1 + n_2} = \frac{48 + 100}{32} = 4.625$$

To find the standard deviation of the combined set of numbers:

$$\sum_{\text{all } x} x^2 = \sum_{i=1}^{n_1} x_i^2 + \sum_{j=1}^{n_2} x_j^2 = 240 + 680 = 920$$

So
$$s^2 = \frac{\sum_{\text{all } x} x^2}{n} - \bar{x}^2$$

$$= \frac{\sum_{i=1}^{n_1} x_i^2 + \sum_{j=1}^{n_2} x_j^2}{n_1 + n_2} - \bar{x}^2$$

$$= \frac{920}{32} - (4.625)^2$$

$$= 7.359$$

So
$$s = \sqrt{7.359}$$

$$= 2.71 \quad (2\,\text{d.p.})$$

Therefore the mean of the combined set of numbers is 4.625 and the standard deviation is 2.71 (2 d.p.).

In general, the mean of the combined set of numbers is given by

$$\bar{x} = \frac{n_1\bar{x}_1 + n_2\bar{x}_2}{(n_1 + n_2)}$$

The formula for the standard deviation of the combined set of numbers is very complicated to write out. It is better to work out $\sum_{i=1}^{n_1} x_i^2$ and $\sum_{j=1}^{n_2} x_j^2$ as in Example 1.36 and proceed from there.

Example 1.37 Suppose that the values of a random sample taken from some population are $x_1, x_2, \ldots, x_n$. Prove the formula

$$\sum_{i=1}^{n}(x_i - \bar{x})^2 = \sum_{i=1}^{n} x_i^2 - n\bar{x}^2$$

Prior to the start of delicate wage negotiations in a large company, the unions and the management take independent samples of the work force and ask them at what percentage level they believe a settlement should be made. The results are as follows:

Sample	Size	Mean	Standard deviation
'management'	350	12.4%	2.1%
'union'	237	10.7%	1.8%

Assuming that no individual was consulted by both sides, calculate the mean and standard deviation for these 587 workers.

(AEB 1979)

Solution 1.37 For first part, see p. 49.

Management $n_1 = 350, \quad \bar{x}_1 = 12.4, \quad s_1 = 2.1$

Union $n_2 = 237, \quad \bar{x}_2 = 10.7, \quad s_2 = 1.8$

For the combined set of 587 workers

$$\bar{x} = \frac{n_1\bar{x}_1 + n_2\bar{x}_2}{n_1 + n_2}$$

$$= \frac{(350)(12.4) + (237)(10.7)}{587}$$

$$= 11.7 \quad \text{(1 d.p.)}$$

Therefore the mean percentage level for the 587 workers is 11.7% (1 d.p.).

For the combined variance $s^2 = \dfrac{\Sigma x^2}{n} - \bar{x}^2$

We need to find Σx^2 for 'management' and 'union'.

Management

$$\sum_{i=1}^{n_1} x_i^2 = n_1(\bar{x}_1{}^2 + s_1{}^2) = 350(12.4^2 + 2.1^2) = 55\,359.5$$

Union

$$\sum_{j=1}^{n_2} x_j^2 = n_2(\bar{x}_2{}^2 + s_2{}^2) = 237(10.7^2 + 1.8^2) = 27\,902.0$$

So, for all 587 workers:

$$s^2 = \frac{\sum_{i=1}^{n_1} x_i^2 + \sum_{j=1}^{n_2} x_j^2}{n_1 + n_2} - \bar{x}^2$$

$$= \frac{55\,359.5 + 27\,902.0}{587} - (11.7)^2$$

$$= 4.95$$

Hence $s = 2.2 \quad \text{(1 d.p.)}$

The standard deviation for the 587 workers is 2.2% (1 d.p.).

Exercise 1p

1. For each of the following sets of data, find the mean and the standard deviation of the combined set.

   (*a*) $n_1 = 12, \quad \bar{x}_1 = 6, \quad s_1 = 2$
   $n_2 = 8, \quad \bar{x}_2 = 10, \quad s_2 = 3$

   (*b*) $n_1 = 30, \quad \bar{x}_1 = 27, \quad s_1 = 5.6$
   $n_2 = 40, \quad \bar{x}_2 = 33, \quad s_2 = 6.4$

   (*c*) $n_1 = 12, \quad \bar{x}_1 = 15, \quad s_1 = 2.7$
   $n_2 = 15, \quad \bar{x}_2 = 14, \quad s_2 = 3.1$
   $n_3 = 13, \quad \bar{x}_3 = 12, \quad s_3 = 2.4$

2. For a set of 20 numbers $\Sigma x = 300$ and $\Sigma x^2 = 5500$. For a second set of 30 numbers $\Sigma x = 480$ and $\Sigma x^2 = 9600$. Find the mean and the standard deviation of the combined set of 50 numbers.

3. Prove the formula $\Sigma (x-\bar{x})^2 = \Sigma x^2 - n\bar{x}^2$.
   In a Middle School there are 253 girls whose ages have a mean 11.8 yr and a standard deviation 1.7 yr. There are also 312 boys whose ages have a mean 12.3 yr and a standard deviation 1.9 yr. Calculate the mean and standard deviation of the ages of all the 565 pupils. (AEB 1976)

4. Suppose that the values of a random sample taken from some population are $x_1, x_2, \ldots, x_n$. Prove the formula

   $$\sum_{i=1}^{n} (x_i - \bar{x})^2 = \sum_{i=1}^{n} x_i^2 - n\bar{x}^2$$

   Parplan Opinion Polls Ltd. conducted a nationwide survey into the attitudes of teenage girls. One of the questions asked was 'What is the ideal age for a girl to have her first baby?' In reply, the sample of 165 girls from the Northern zone gave a mean of 23.4 years and a standard deviation of 1.6 years. Subsequently, the overall sample of 384 girls (Northern plus Southern zones) gave a mean of 24.8 years and a standard deviation of 2.2 years.
   Assuming that no girl was consulted twice, calculate the mean and standard deviation for the 219 girls from the Southern zone. (AEB 1981)

Standard deviation —data in the form of a frequency distribution

If $(x_1, x_2, \ldots, x_n)$ occur with frequencies $(f_1, f_2, \ldots, f_n)$ then the standard deviation s is given by:

$$s = \sqrt{\frac{\Sigma f_i (x_i - \bar{x})^2}{\Sigma f_i}} \qquad i = 1, 2, \ldots, n$$

The alternative form of the formula is

$$s = \sqrt{\frac{\Sigma f_i x_i^2}{\Sigma f_i} - \bar{x}^2} \qquad i = 1, 2, \ldots, n$$

When class intervals are given, the mid-point of an interval is taken to represent the interval.

Example 1.38 The table shows the number of children per family for a group of 20 families. The mean number of children per family is 2.9. Find the standard deviation.

Number of children per family, x	1	2	3	4	5
Frequency, f	3	4	8	2	3

Solution **1.38** Method 1 — using $s = \sqrt{\dfrac{\Sigma f(x-\bar{x})^2}{\Sigma f}}$

x	$x-2.9$	$(x-2.9)^2$	f	$f(x-2.9)^2$
1	−1.9	3.61	3	10.83
2	−0.9	0.81	4	3.24
3	0.1	0.01	8	0.08
4	1.1	1.21	2	2.42
5	2.1	4.41	3	13.23
			$\Sigma f = 20$	$\Sigma f(x-\bar{x})^2 = 29.80$

So
$$s^2 = \frac{\Sigma f(x-2.9)^2}{\Sigma f} = \frac{29.80}{20} = 1.49$$
$$s = \sqrt{1.49} = 1.22 \quad (2\,\text{d.p.})$$

The standard deviation of the number of children per family is 1.22 (2 d.p.).

Method 2 — using $s = \sqrt{\dfrac{\Sigma fx^2}{\Sigma f} - \bar{x}^2}$

x	f	x^2	fx^2
1	3	1	3
2	4	4	16
3	8	9	72
4	2	16	32
5	3	25	75
	$\Sigma f = 20$		$\Sigma fx^2 = 198$

So
$$s^2 = \frac{\Sigma fx^2}{\Sigma f} - (2.9)^2 = \frac{198}{20} - (2.9)^2 = 1.49$$
$$s = \sqrt{1.49} = 1.22 \quad (2\,\text{d.p.})$$

The standard deviation is 1.22 (2 d.p.), as before.

Method 3 — using the calculator in SD mode.

This time we need to take account of the frequencies, and this is done as follows:

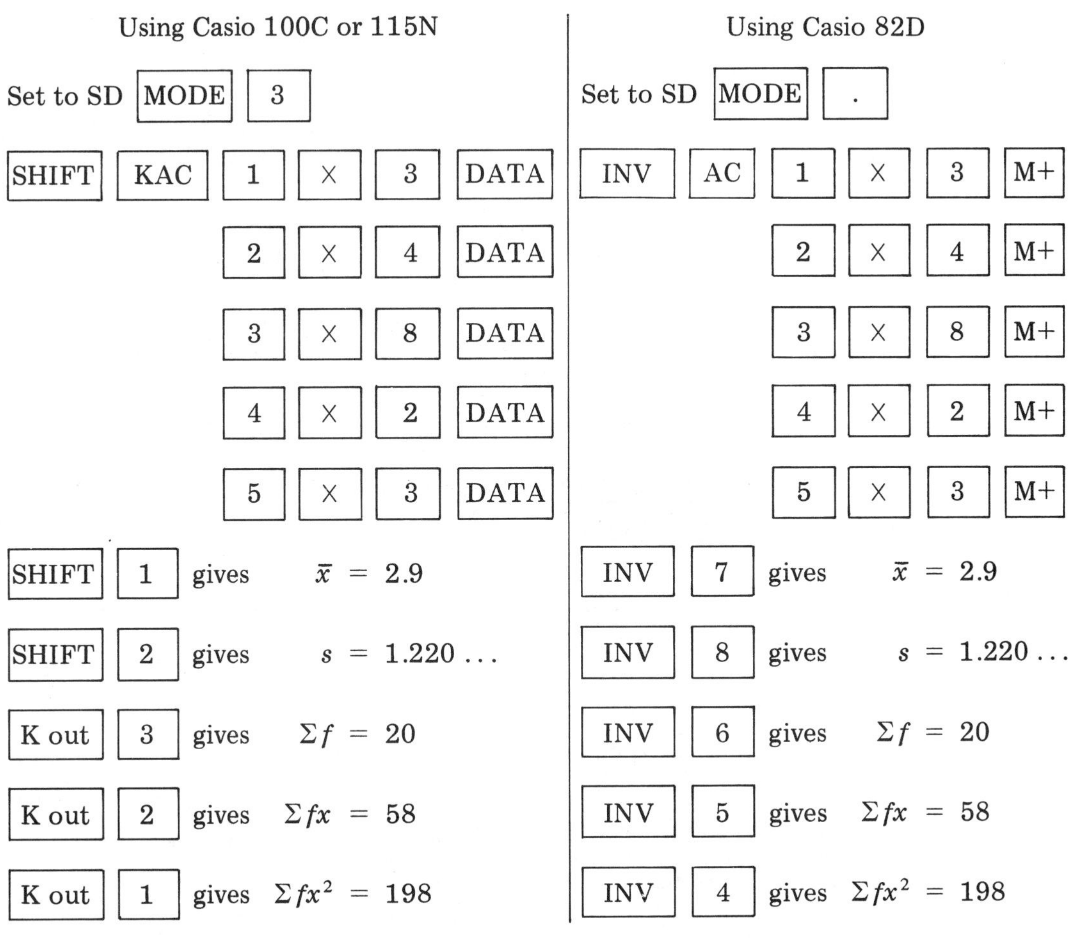

Therefore the standard deviation is 1.22 (2 d.p.), as before.

Example 1.39 The lengths of 32 leaves were measured correct to the nearest mm. Find the mean length and the standard deviation.

Length (mm)	20–22	23–25	26–28	29–31	32–34
Frequency	3	6	12	9	2

Solution 1.39 The mid-points, x, of each interval are considered:

Length (mm)	Mid-point, x	x^2	f	fx	fx^2
20–22	21	441	3	63	1323
23–25	24	576	6	144	3456
26–28	27	729	12	324	8748
29–31	30	900	9	270	8100
32–34	33	1089	2	66	2178
			$\Sigma f = 32$	$\Sigma fx = 867$	$\Sigma fx^2 = 23\,805$

Now $$\bar{x} = \frac{\Sigma fx}{\Sigma f} = \frac{867}{32} = 27.1 \quad (1\text{ d.p.})$$

and $$s^2 = \frac{\Sigma fx^2}{\Sigma f} - \bar{x}^2 = \frac{23\,805}{32} - \left(\frac{867}{32}\right)^2 = 9.835$$

$$s = \sqrt{9.835} = 3.14 \quad (2\text{ d.p.})$$

The mean length of the leaves is 27.1 mm and the standard deviation is 3.14 mm (2 d.p.).

Exercise 1q

Do questions 1, 2 and 3 without using the calculator in SD mode, and then check them using SD mode.

1. **The score for a round of golf for each of 50 club members was noted. Find the mean score for a round and the standard deviation.**

Score, x	66	67	68	69	70	71	72	73
Frequency, f	2	5	10	12	9	6	4	2

2. **The scores in an IQ test for 60 candidates are shown in the table. Find the mean score and the standard deviation.**

Score	100–106	107–113	114–120	121–127	128–134
Frequency	8	13	24	11	4

3. **Find the mean and the standard deviation for each of the following sets of data:**

(a)

x	1	2	3	4	5	6
f	2	6	11	15	8	3

(b)

Interval	1–3	4–6	7–9	10–12	13–15
Frequency	2	4	8	5	1

(c)

Interval	20–24	25–29	30–34	35–39	40–44
Frequency	1	6	10	2	1

(d)

x	10	20	30	40	50	60
f	3	9	14	10	6	4

(e)

Interval	1–7	8–14	15–21	22–28
Frequency	4	5	10	6

(f)

x	7	8	9	10	11	12	13
f	1	3	5	7	5	3	1

4. **For a particular set of observations $\Sigma f = 20$, $\Sigma fx^2 = 16\,143$, $\Sigma fx = 563$. Find the values of the mean and the standard deviation.**

5. **For a given frequency distribution $\Sigma f(x-\bar{x})^2 = 182.3$, $\Sigma fx^2 = 1025$, $\Sigma f = 30$. Find Σfx.**

6. **From the information given about each of the following frequency distributions, work out the missing values in the table:**

	Σf	Σfx	Σfx^2	$\Sigma f(x-\bar{x})^2$	$\bar{x}$	s
(a)	20	563	16 143			
(b)		270		160	27	
(c)	50				10	3
(d)	30		1025	182.3		
(e)		240	5100		20	

Using the method of coding to find the standard deviation

Consider the set of numbers $x_1, x_2, \ldots, x_n$. To find the standard deviation we can use the coding

$$y = \frac{x-\bar{x}_a}{b}$$ where $\bar{x}_a$ is the assumed mean and b is a suitable constant

Now $$x = \bar{x}_a + by$$

and $$\bar{x} = \bar{x}_a + b\bar{y} \quad \text{(see p. 44)}$$

Also
$$s^2 = \frac{\Sigma(x_i-\bar{x})^2}{n} \quad \text{for} \quad i = 1, 2, \ldots, n$$
$$= \frac{\Sigma(\bar{x}_a + by_i - (\bar{x}_a + b\bar{y}))^2}{n}$$
$$= \frac{\Sigma b^2(y_i-\bar{y})^2}{n}$$

and so $$s = b\sqrt{\frac{\Sigma(y_i-\bar{y})^2}{n}} \quad \text{for} \quad i = 1, 2, \ldots, n$$

Using the alternative form of the formula, we have

$$s = b\sqrt{\frac{\Sigma y_i^2}{n} - \bar{y}^2}$$

If the numbers occur with frequencies $f_1, f_2, \ldots, f_n$, then the corresponding formulae for the standard deviation are:

$$s = b\sqrt{\frac{\Sigma f_i(y_i-\bar{y})^2}{\Sigma f_i}} \quad \text{or} \quad s = b\sqrt{\frac{\Sigma f_i y_i^2}{\Sigma f_i} - \bar{y}^2}$$

These formulae look very complicated, but in fact they are very easy to use.

Example 1.40 Find the standard deviation of the set of numbers

$$327, 332, 342, 347, 352$$

Solution 1.40 We will use the coding $y = \dfrac{x-\bar{x}_a}{b}$ where $\bar{x}_a = 342$, $b = 5$.

so $$y = \frac{x-342}{5}$$

x	$y = \dfrac{x-342}{5}$	y^2
327	−3	9
332	−2	4
342	0	0
347	1	1
352	2	4
	$\Sigma y = -2$	$\Sigma y^2 = 18$

Now $$s^2 = b^2\left(\frac{\Sigma y^2}{n} - \bar{y}^2\right) \quad \text{where} \quad \bar{y} = \frac{\Sigma y}{n}$$

$$= 25\left[\frac{18}{5} - \left(\frac{-2}{5}\right)^2\right]$$

$$= 25[3.6-(-0.4)^2]$$

$$= 86$$

So $$s = \sqrt{86}$$

$$= 9.27 \quad (2\text{ d.p.})$$

The standard deviation of the set of numbers is 9.27 (2 d.p.).

Example 1.41 For the data given in Example 1.39, find the standard deviation, using a method of coding.

Solution 1.41 We note that the class widths are each equal to 3 and a central value is 27. Therefore, we choose the coding $y = \dfrac{x-27}{3}$.

Length (mm)	Mid-point, x	f	$y = \dfrac{x-27}{3}$	y^2	fy	fy^2
20–22	21	3	−2	4	−6	12
23–25	24	6	−1	1	−6	6
26–28	27	12	0	0	0	0
29–31	30	9	1	1	9	9
32–34	33	2	2	4	4	8
		$\Sigma f = 32$			$\Sigma fy = 1$	$\Sigma fy^2 = 35$

Now $$\bar{y} = \frac{\Sigma fy}{\Sigma f} = \frac{1}{32}$$

We have $$s^2 = b^2\left(\frac{\Sigma fy^2}{\Sigma f} - \bar{y}^2\right) \quad \text{where} \quad b = 3$$

$$= 9\left[\frac{35}{32} - \left(\frac{1}{32}\right)^2\right]$$

$$= 9.835$$

So $$s = \sqrt{9.835}$$

$$= 3.14 \quad (2\text{ d.p.})$$

The standard deviation of the length of the leaves is 3.14 mm, as before.

Exercise 1r

1. Find the mean and the standard deviation of the following sets of data, using a method of coding:

(*a*)

x	304	308	312	316	320	324
f	1	5	9	4	4	2

(*b*)

Interval	10–19	20–29	30–39	40–49	50–59	60–69
f	3	7	12	18	12	6

(*c*)

x	1250	1500	1750	2000	2250	2500	2750
f	5	19	27	35	24	12	3

(*d*)

Interval	0–	5–	10–	15–	20–	25–	30–	35–
f	5	10	13	18	12	6	6	0

(*e*)

x	0.1	0.4	0.7	1.0	1.3	1.6	1.9	2.2
f	4	6	12	15	9	6	5	1

(*f*)

Interval	-200	-250	-300	-350	-400	-450	-600
f	0	20	33	35	25	14	3

2. The marks obtained in an examination by 190 students are recorded in Table A below. Find the mean mark and the standard deviation by using an assumed mean between 50 and 60 and a coding factor of 10.

3. A farmer grows two different varieties of potatoes, Desirée and Pentland Squire. A sample of 50 potatoes of each variety is taken and the potatoes are weighed. The results are shown in Table B below. Find the mean and the standard deviation for each sample. Use a method of coding.

4. The table shows the times taken on 30 consecutive days for a coach to complete one journey on a particular route. Times have been given to the nearest minute. Find the mean time for the journey and the standard deviation.

Time (min)	60–63	64–67	68–71	72–75	76–79
Frequency	1	3	12	10	4

Table A

Mark	0–	10–	20–	30–	40–	50–	60–	70–	80–	90–	100–
Frequency	3	5	5	6	25	33	49	40	15	9	0

Table B

Mass (g)	0–60	60–120	120–180	180–240	240–300	300–360	360–420	420–480
Desirée frequency	1	3	4	7	12	15	5	3
Pentland Squire frequency	3	2	4	8	17	12	3	1

MISCELLANEOUS WORKED EXAMPLES

Example 1.42 In order to estimate the mean length of leaves from a certain tree a sample of 100 leaves was chosen and their lengths measured correct to the nearest mm. A grouped frequency table was set up and the results were as follows:

Mid-interval value (cm)	2.2	2.7	3.2	3.7	4.2	4.7	5.2	5.7	6.2
Frequency	3	5	8	12	18	24	20	8	2

(a) Display the table in the form of a frequency polygon and describe the distribution exhibited by this polygon.

(b) Calculate estimates for the mean and standard deviation of the leaf lengths using an assumed mean of 4.7 cm.

(c) What are the boundaries of the interval whose mid-point is 3.7 cm?

(d) Construct a cumulative frequency table and use it to estimate the sample median. (SUJB)

Solution 1.42 (a) **Frequency polygon to show lengths of leaves**

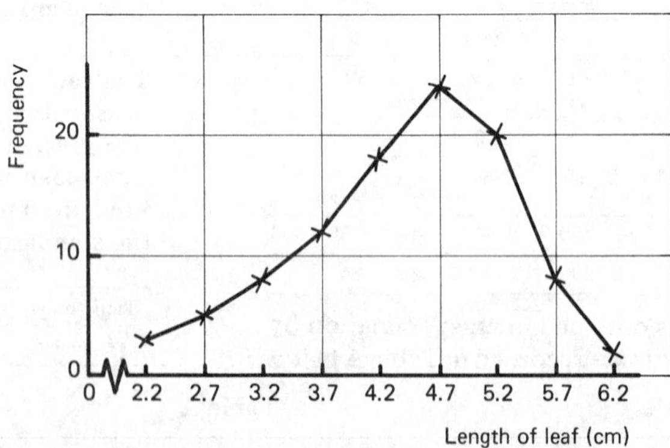

This is a frequency distribution which is skewed to the left; an estimate of the modal length of leaf is 4.7 cm.

(b) As the intervals each have a width of 0.5, and we are told to take 4.7 as assumed mean, we use the code $y = \dfrac{x-4.7}{0.5}$ to make the working easier when finding the mean and the standard deviation.

Mid-point, x	f	$y = \dfrac{x - 4.7}{0.5}$	y^2	fy	fy^2
2.2	3	-5	25	-15	75
2.7	5	-4	16	-20	80
3.2	8	-3	9	-24	72
3.7	12	-2	4	-24	48
4.2	18	-1	1	-18	18
4.7	24	0	0	0	0
5.2	20	1	1	20	20
5.7	8	2	4	16	32
6.2	2	3	9	6	18
	$\Sigma f = 100$			$\Sigma fy = -59$	$\Sigma fy^2 = 363$

Now $\qquad y = \dfrac{x - 4.7}{0.5}$

so $\qquad x = 4.7 + 0.5y$

and $\qquad \bar{x} = 4.7 + 0.5\bar{y} \quad$ where $\quad \bar{y} = \dfrac{\Sigma fy}{\Sigma f}$

Therefore

$$\bar{x} = 4.7 + 0.5\left(\frac{-59}{100}\right)$$

$$= 4.405$$

Now

$$s^2 = (0.5)^2\left(\frac{\Sigma fy^2}{\Sigma f} - \bar{y}^2\right)$$

$$= (0.25)\left[\frac{363}{100} - \left(\frac{-59}{100}\right)^2\right]$$

$$= 0.8205$$

$$s = \sqrt{0.8205}$$

$$= 0.91 \quad (2\text{ d.p.})$$

Therefore the mean length of the leaves is 4.405 cm and the standard deviation is 0.91 cm.

(c) Each class width is 0.5 cm, so the class with mid-point 3.7 has

l.c.b. $= 3.7 - 0.25 = 3.45$

u.c.b. $= 3.7 + 0.25 = 3.95$

Therefore the interval whose mid-point is 3.7 cm is defined by

$$3.45 \text{ cm} \leqslant \text{length of leaf} < 3.95 \text{ cm}$$

(d) **Cumulative frequency table to show lengths of leaves**

Length (cm) (mid-point)	Frequency	Length (cm)	Cumulative frequency
1.7	0	<1.95	0
2.2	3	<2.45	3
2.7	5	<2.95	8
3.2	8	<3.45	16
3.7	12	<3.95	28
4.2	18	<4.45	46
4.7	24	<4.95	70
5.2	20	<5.45	90
5.7	8	<5.95	98
6.2	2	<6.45	100

The median is the $\frac{1}{2}(\Sigma f+1)$th item, i.e. the 50.5th item.

From the table, we see that this lies in the interval 4.45–4.95.

There are 24 items in the interval 4.45–4.95 and the median is $\frac{4.5}{24}$ of the interval of 0.5 cm from 4.45 to 4.95.

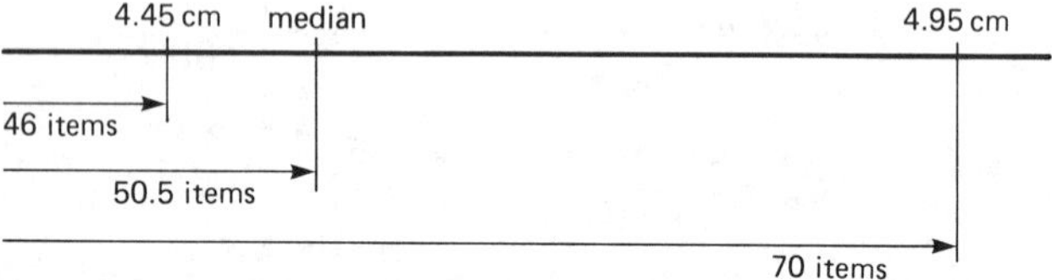

An estimate of the median is $4.45+\left(\frac{4.5}{24}\right)(0.5)=4.54$ cm (2 d.p.).

Example 1.43 In a certain village of 400 inhabitants the distribution of ages is as follows:

Age (years)	0–9	10–19	20–29	30–39	40–49	50–59	60–69	70–79	80–99
Frequency	44	56	64	78	60	40	36	18	4

Explain why the mid-interval value of the 80–99 group is 90 years.

Use the data to estimate the mean age and the standard deviation, in years, each correct to 1 d.p.

Draw a cumulative frequency graph and estimate the percentage of the population that have ages within 1 s.d. of the mean. (SUJB)

Solution 1.43 Assuming that ages have been given in completed years then the interval 80–99 means $80 \leqslant \text{age} < 100$; someone who is 99 years and 11 months, for example, would come into this interval.

$$\text{The mid-point of the interval} = \tfrac{1}{2}(\text{u.c.b.} + \text{l.c.b.})$$
$$= \tfrac{1}{2}(100 + 80)$$
$$= 90$$

To find the mean and the standard deviation we use a method of 'coding'. The class widths (with the exception of the last one) are each equal to 10 and a 'central' value, suitable to be chosen as an assumed mean, is the mid-point of the interval '40–49' which is 45.

We use the code $y = \dfrac{x-45}{10}$

Age (years)	Mid-points	f	y	y^2	fy	fy^2
0-9	5	44	−4	16	−176	704
10-19	15	56	−3	9	−168	504
20-29	25	64	−2	4	−128	256
30-39	35	78	−1	1	−78	78
40-49	45	60	0	0	0	0
50-59	55	40	1	1	40	40
60-69	65	36	2	4	72	144
70-79	75	18	3	9	54	162
80-99	90	4	4.5	20.25	18	81
		$\Sigma f = 400$			$\Sigma fy = -366$	$\Sigma fy^2 = 1969$

Now $y = \dfrac{x-45}{10}$

So $x = 45 + 10y$

and $\bar{x} = 45 + 10\bar{y}$ where $\bar{y} = \dfrac{\Sigma fy}{\Sigma f}$

So $\bar{x} = 45 + 10\left(\dfrac{-366}{400}\right)$

$= 35.9$ (1 d.p.)

We have $s^2 = b^2\left(\dfrac{\Sigma fy^2}{\Sigma f} - \bar{y}^2\right)$ where $b = 10$

$= 100\left[\dfrac{1969}{400} - \left(\dfrac{-366}{400}\right)^2\right]$

$= 408.53$

So $s = 20.2$ (1 d.p.)

Therefore the mean age of the population is 35.9 years and the standard deviation is 20.2 years.

The cumulative frequency distribution is as shown:

Age (years)	Frequency	Age (years)	Cumulative frequency
0–9	44	<10	44
10–19	56	<20	100
20–29	64	<30	164
30–39	78	<40	242
40–49	60	<50	302
50–59	40	<60	342
60–69	36	<70	378
70–79	18	<80	396
80–99	4	<100	400

Cumulative frequency curve to show ages of population in a village

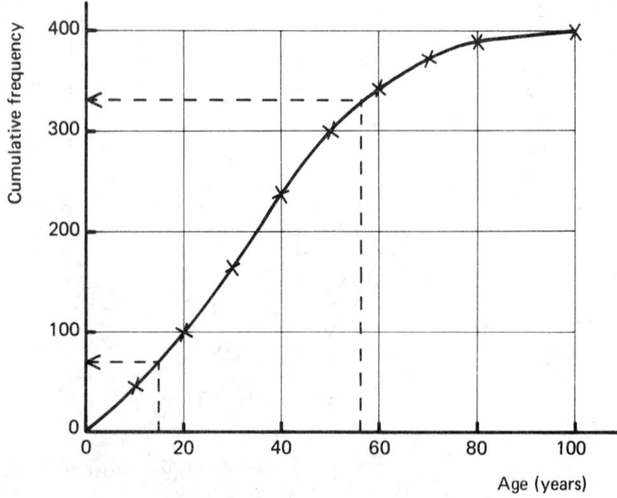

We want to find the number of people with ages within one standard deviation of the mean, i.e. the number of people with ages in the interval $35.9 \pm 20.2 = (15.7, 56.1)$.

From the graph, we estimate that

72 people have ages below 15.7 years,
330 people have ages below 56.1 years.

So 258 people have ages in the interval (15.7, 56.1),

i.e. the percentage of the population $= \left(\frac{258}{400}\right)(100)\%$

$= 64.5\%$

Therefore 64.5% of the population of the village have ages which are within 1 s.d. of the mean.

Example 1.44 A set of n values has mean μ and variance s_1^2. A second set of values has mean $a\mu$ and variance s_2^2. Given that s is the standard deviation of the combined set of $2n$ values, show that

$$s^2 = \tfrac{1}{2}(s_1^2 + s_2^2) + \tfrac{1}{4}\mu^2(a-1)^2$$

Solution 1.44

	Number	Mean	Variance	Σx
First set	n	μ	s_1^2	$n\mu$
Second set	n	$a\mu$	s_2^2	$an\mu$

Let $\bar{x}$ be the mean of the combined set.

$$\begin{aligned}\bar{x} &= \frac{\Sigma x_1 + \Sigma x_2}{2n}\\ &= \frac{n\mu + an\mu}{2n}\\ &= \frac{n\mu(a+1)}{2n}\\ &= \tfrac{1}{2}\mu(a+1)\\ s_1^2 &= \frac{\Sigma x_1^2}{n} - \mu^2\\ \Sigma x_1^2 &= n(s_1^2 + \mu^2)\end{aligned}$$

Similarly $\quad \Sigma x_2^2 = n(s_2^2 + a^2\mu^2)$

For the combined set

$$\begin{aligned}\Sigma x^2 &= n(s_1^2 + \mu^2 + s_2^2 + a^2\mu^2)\\ &= n(s_1^2 + s_2^2 + \mu^2(1+a^2))\\ s^2 &= \frac{\Sigma x^2}{2n} - \bar{x}^2\\ &= \frac{n(s_1^2 + s_2^2 + (1+a^2)\mu^2)}{2n} - \frac{(1+a)^2\mu^2}{4}\\ &= \tfrac{1}{2}(s_1^2 + s_2^2) + \tfrac{1}{2}(1+a^2)\mu^2 - \tfrac{1}{4}(1+a)^2\mu^2\\ &= \tfrac{1}{2}(s_1^2 + s_2^2) + \tfrac{1}{4}\mu^2(2(1+a^2) - (1+2a+a^2))\\ &= \tfrac{1}{2}(s_1^2 + s_2^2) + \tfrac{1}{4}\mu^2(2 + 2a^2 - 1 - 2a - a^2)\\ &= \tfrac{1}{2}(s_1^2 + s_2^2) + \tfrac{1}{4}\mu^2(a^2 - 2a + 1)\\ &= \underline{\tfrac{1}{2}(s_1^2 + s_2^2) + \tfrac{1}{4}\mu^2(a-1)^2 \quad \text{as required}}\end{aligned}$$

Miscellaneous Exercise 1s

1. Two hundred and fifty Army recruits have the following heights.

Height (cm)	165-	170-	175-	180-	185-	190-195
No. of recruits	18	37	60	65	48	22

Plot the data in the form of a cumulative frequency curve. Use the curve to estimate (*a*) the median height, (*b*) the lower quartile height.

The tallest 40% of the recruits are to be formed into a special squad. Estimate (*a*) the median, (*b*) the upper quartile of the heights of the members of this squad. (SUJB Additional)

2. Below are given the number n of hours worked in a week by 64 men.

36.5	15.6	30.8	27.6	33.6	39.4	39.7	39.1
33.0	40.1	21.8	40.6	33.9	36.9	39.1	27.4
29.8	37.0	45.4	42.5	9.6	26.3	36.1	35.2
38.4	28.5	30.5	44.4	38.4	40.6	26.5	38.7
24.4	41.9	52.7	35.7	28.9	38.2	30.4	41.7
43.2	24.5	34.8	37.8	38.0	43.7	40.8	11.7
31.5	37.6	40.1	23.7	31.8	42.0	29.1	41.5
36.3	29.7	37.3	28.4	39.6	22.9	35.2	42.4

(i) Group the numbers into intervals of width 3 hours defined by $9.5 \leqslant n < 12.5$, $12.5 \leqslant n < 15.5, \ldots$.
(ii) Use the grouped data to calculate estimates of the mean and standard deviation of n.
(iii) Estimate the percentage of workmen for whom n is within one standard deviation of the mean. (MEI)

3. The following table shows the durations of 40 telephone calls from an office via the office switchboard.

Duration in minutes	$\leqslant 1$	1-2	2-3	3-5	5-10	$\geqslant 10$
Number of calls	6	10	15	5	4	0

Obtain an estimate of the mean and standard deviation of the data. Estimate the median, and the lower and upper quartiles. (O & C)

4. Table A below gives an analysis by numbers of employees of the size of UK factories of less than 1000 employees manufacturing clothing and footwear.

Calculate as accurately as the data allow the mean and the median of this distribution, showing your working.

If 90% of the factories have less than N employees, estimate N. (O & C)

5. The numbers $4, 6, 12, 4, 10, 12, 3, x, y$ have a mean of 7 and a mode of 4. Find (i) the values of the two numbers x and y, (ii) the median of this set of nine numbers.

When two additional numbers $7+n$ and $7-n$ are included the standard deviation of all eleven numbers is found to be 4. Write down the mean of these eleven numbers and calculate the value of n. (C Additional)

6. The sum of 20 numbers is 320 and the sum of their squares is 5840. Calculate the mean of the 20 numbers and the standard deviation.
(i) Another number is added to these 20 so that the mean is unchanged. Show that the standard deviation is decreased.
(ii) Another set of 10 numbers is such that their sum is 130 and the sum of their squares is 2380. This set is combined with the original 20 numbers. Calculate the mean and standard deviation of all 30 numbers. (C Additional)

7. A weather station recorded the number of hours of sunshine each day for 80 days, with the results as shown in Table B below.

Table A

Number of employees	11-19	20-24	25-99	100-199	200-499	500-999	Total
Number of factories	1500	800	2300	700	400	100	5800

Table B

Hours of sunshine	0	0-1	1-2	2-3	3-4	4-5	5-6	6-7	7-8	8-12	over 12
Number of days	10	2	6	17	22	11	5	3	2	2	0

[The grouping symbol 2–3, for example, denotes greater than 2 hours and less than or equal to 3 hours.]

State which is the modal group.

Construct a cumulative frequency table and draw the cumulative frequency curve. Use your curve to estimate (i) the median, (ii) the inter-quartile range, (iii) the percentage of days for which more than $3\frac{1}{2}$ hours of sunshine were recorded.

(C Additional)

8. (*a*) Sketch the expected frequency curves for each of the following distributions:
   (i) the number of light bulbs broken in boxes containing 125 bulbs, assuming that the modal number of breakages is 0,
   (ii) the age at marriage of females.

   (*b*) State the assumption that is made in obtaining measures of average and dispersion from grouped frequency tables.

   The table below shows the ages, at last birthday, of the employees of a certain firm.

Age (last birthday)	Less than 20	20–	25–	30–	40–	50 and over
No. of employees	32	35	39	47	45	42

Without drawing a cumulative frequency curve, estimate (i) the semi-interquartile range, (ii) the number of employees aged 37 and over. (C Additional)

9. Table C below shows the durations of 60 journeys on the same route by a lorry, the variations in journey times being caused by varying traffic conditions.

   Calculate, to the nearest minute, estimates of the mean and standard deviation for the duration of the journeys.

   When the times for 40 other journeys were taken, it was found that the mean and standard deviation for the times of these 40 journeys were 6 h 24 min and 18 min, respectively. Find, also to the nearest minute, the estimated mean and standard deviation for the duration of all 100 journeys. (C)

10. Table D below gives the cumulative frequency distribution of the masses x in kilogrammes of a group of 200 eighteen-year-old boys.

    Draw a cumulative frequency graph and from this estimate the median.

    Compile a frequency distribution from the data and hence estimate the mean and standard deviation of the sample. State a well known probability distribution which you would expect to fit such data.

    (JMB)

11. 100 pupils were tested to determine their intelligence quotient (I.Q.), and the results were as follows:

I.Q.	45–	55–	65–	75–	85–	95–	105–	115–	125–134
No. of pupils	1	1	2	6	21	29	24	12	4

All I.Q.'s are given to the nearest integer.
(i) Calculate the mean, and the standard deviation.
(ii) Draw a cumulative frequency graph, and estimate how many pupils have I.Q.'s within 1 s.d. on either side of the mean.

(SUJB)

12. (*a*) Find the median, mean, and standard deviation of the set of numbers 3, 5, 12, 1, 6, 3, 12.
    (*b*) A set of digits consists of m zeros and n ones. Find the mean of this set and show that the standard deviation is

$$\frac{\sqrt{(mn)}}{(m+n)}$$

(C Additional)

13. (*a*) A set of values of a variable X has a mean μ and a standard deviation σ. State the new value of the mean and of the standard deviation when each of the variables is (i) increased by k, (ii) multiplied by p.

Table C

Time of journey in hours	5.6–5.8	5.8–6.0	6.0–6.2	6.2–6.4	6.4–6.6	6.6–6.8
Number of journeys	2	7	16	21	12	2

Table D

x	30	35	40	45	50	55	60	65	70	75	80	85	90	95
Number with mass less than x	0	1	4	11	25	47	79	114	146	171	187	195	198	200

Values of a new variable Y are obtained by using the formula $Y = 3X + 5$. Find the mean and the standard deviation of the set of values of Y.

(*b*) It is proposed to convert a set of values of a variable X, whose mean and standard deviation are 20 and 5 respectively, to a set of values of a variable Y whose mean and standard deviation are 42 and 8 respectively. If the conversion formula is $Y = aX + b$, calculate the value of a and of b. (C Additional)

14. A set of numbers has mean μ and standard deviation σ. A new set of numbers is obtained by subtracting μ from each number and dividing the result by σ. Write down the mean and standard deviation of the new set of numbers.

In an examination in Statistics the mean mark of a group of 120 students was 68 and the standard deviation was 6. In Algebra the mean mark of the group was 62 and the standard deviation was 5. One student scored 76 in Statistics and 70 in Algebra. By scaling the marks for each subject so that each set of marks has the same mean and standard deviation compare the performances of this student in the two subjects. (C Additional)

15. 200 candidates sat an examination and the distribution was obtained as shown in Table E below.

If the limits of class 40–49 are 39.5 to 49.5, what is the mid-interval value of this class?

Calculate the mean of the marks explaining any limitations of your calculation.

Plot a cumulative frequency curve and use it to estimate the upper and lower quartiles.

Assuming that your estimates are exact, find values for a and b correct to 2 significant figures, in order that the above marks can be scaled by the equation $y = ax + b$, where y is the new mark, so that the mean becomes 45 and the lower quartile becomes 35.

State, with reason, whether the quartiles of the original marks will scale into the quartiles of the scaled marks. (SUJB)

16. The table shows the yield, in litres, of milk produced by 131 cows at a certain farm on a given day.

Yield (litres)	5-10	11-16	17-22	23-28	29-34	35-40
Frequency	15	28	37	26	18	7

(*a*) State the modal class and estimate the mode. (*b*) By calculation, estimate the median yield. (*c*) Draw a cumulative frequency curve, and from it estimate the semi-interquartile range. (*d*) Calculate the mean and the standard deviation of the distribution, using a method of coding.

17. In a certain industry, the numbers of thousands of employees in 1970 were as shown in Table F below, by age groups.

Calculate the arithmetic mean, median, variance and standard deviation of the ages of employees in the industry.

Estimate the percentage of the employees whose ages lie within one standard deviation of the arithmetic mean. (AEB 1975)

18. Twelve members of a class of children each estimate the height of the top of a church tower. Their estimates, in metres, are

47, 52, 52, 54, 52, 50, 51, 50, 48, 53, 54, 49

(*a*) Calculate the median of these estimates.
(*b*) Calculate the mode, m, that is the number which has the highest frequency.
(*c*) Two extra children join the class and each makes an estimate. The mode for the set of 14 estimates is different from m and unique. Suggest what the two new estimates could be.
(*d*) Calculate the arithmetic mean, $\bar{x}$, of the original 12 estimates.

Table E

Marks (x)	10-19	20-29	30-39	40-49	50-59	60-69	70-79	80-89	90-99
Frequency	10	18	20	30	49	46	20	5	2

Table F

Age last birthday	15-19	20-24	25-29	30-34	35-39	40-44	45-49	50-54	55-59	60-64
Number of thousands	66	65	56	50	42	37	35	30	24	22

(*e*) One member of the original class of 12 revises his estimate and the new mean for the 12 estimates is $\bar{x}+0.5$. Find the increase in the estimate of this member.
(*f*) The teacher of the class makes an estimate of the height of the church tower and when her estimate is taken with the original 12, the mean of all 13 estimates is $\bar{x}+0.5$. Find the teacher's estimate.
(*g*) Two extra children, different from those mentioned in (*c*), join the class and each make an estimate so that the mean of their two estimates and the original 12 estimates is $\bar{x}+0.5$.

Find the sum of their two estimates. (L Additional)

19. Ten values of a variable x are

8.2, 8.0, 8.1, 8.2, 8.4, 7.9, 8.0, 8.3, 7.8, 8.1

Express each of these values in the form $8+0.1y$. Calculate the arithmetic mean and the variance of the ten values of y and hence, or otherwise, deduce the mean and the variance of the ten values of x.

Hence find the mean and the variance of the set of ten numbers

824, 804, 814, 824, 844,
794, 804, 834, 784, 814

A transformation of the form $z=a+bx$, where $b>0$, is applied to the first set of ten values of x so that the mean is increased by 0.9 and the standard deviation is doubled. Find the values of the constants a and b. (L Additional)

20. Show, from the basic definition, why the standard deviation of a set of observations $x_1, x_2, x_3, \ldots, x_n$ with mean $\bar{x}$ may be found by evaluating

$$\sqrt{\frac{\Sigma x_i^2}{n}-\bar{x}^2}.$$

(*a*) Find, showing your working clearly and not using any pre-programmed function on your calculator, the standard deviation of the following frequency distribution:

x	25	26	27	28
f	2	0	15	11

(*b*) The average height of 20 boys is 160 cm, with a standard deviation of 4 cm. The average height of 30 girls is 155 cm, with a standard deviation of 3.5 cm. Find the standard deviation of the whole group of 50 children. (SUJB)

21. (*a*) Sketch frequency curves for distributions which have one mode and for which (i) the mode, median and mean coincide, (ii) the mode is less than the median, indicating on each sketch the positions of these measures.
(*b*) The mean of the set of numbers 3, 1, 7, 2, 1, 1, 7, x, y, where x and y are single digit positive whole numbers, is known to be 4. Show that $x+y=14$.

Hence, or otherwise, find the mode of this set of numbers when (i) $x=y$, (ii) $x \neq y$.

If the standard deviation is $\frac{1}{3}\sqrt{76}$ find x and y, assuming that $x \leqslant y$. (C Additional)

22. A random sample of 1000 surnames is drawn from a local telephone directory. The distribution of the lengths of the names is as shown in Table G below.

Calculate the sample mean and sample standard deviation. Obtain the upper quartile.

Represent graphically the data in the table.

Give a reason why the sample of names obtained in this way may not be truly representative of the population of Great Britain. (JMB)

23. In an agricultural experiment the gains in mass, in kilograms, of 100 pigs during a certain period were recorded as follows:

Gain in mass (kilograms)	5-9	10-14	15-19	20-24	25-29	30-34
Frequency	2	29	37	16	14	2

Construct a histogram and a relative cumulative frequency polygon of these data. Obtain (i) the median and the semi-interquartile range, (ii) the mean and the standard deviation.

Which of these pairs of statistics do you consider more appropriate in this case, and why? (AEB 1977)

Table G

Number of letters in surname	3	4	5	6	7	8	9	10	11	12
Frequency	13	102	186	237	215	113	83	32	13	6

24. Table H below gives the ages in completed years of the 113 persons convicted of shop-lifting in a British town in 1986. Working in years and giving answers correct to 1 place of decimals, calculate

(*a*) the mean age and standard deviation,

(*b*) the coefficient of skewness given by

(mean − mode)/standard deviation,

(*c*) the median age.

Which do you consider to be best as a representative average of the distribution —the mean, median or mode? Give reasons for your choice.

Draw a histogram of the data with a class interval of 2 years. (SUJB)

25. A grouped frequency distribution of the ages of 358 employees in a factory is shown in Table I below. Estimate, to the nearest month, the mean and the standard deviation of the ages of these employees.

Graphically, or otherwise, estimate

(*a*) the median and the interquartile range of the ages, each to the nearest month,

(*b*) the percentage, to one decimal place, of the employees who are over 27 years old and under 55 years old. (L)

26. The following is a set of 109 examination marks ordered for convenience.

6	11	11	12	13	14	16	17	18	20
21	21	23	24	25	25	25	25	26	26
27	27	28	28	28	29	29	29	30	31
31	32	32	32	33	33	34	34	35	36
36	37	37	37	37	38	38	38	39	39
39	39	39	39	39	39	40	40	40	40
40	40	41	41	41	42	42	42	42	43
43	43	44	45	46	46	47	47	47	47
48	50	50	51	51	52	52	52	53	53
54	54	55	57	58	58	59	59	61	62
63	64	66	66	67	70	76	77	82	

(*a*) Construct a grouped frequency distribution using a class width of 10 and starting with 0-9.

(*b*) Draw a histogram and comment on the shape of the distribution.

(*c*) Using the frequency table estimate the mean and standard deviation of the marks.

(*d*) The marks are to be scaled linearly by the relation $Y = a + bX$ where X is the old mark and Y the new mark. The new mean and standard deviation are to be 50 and 10 respectively. Using your estimates in (*c*) calculate suitable values for a and b. (SUJB)

27. A travel agency has two shops, *R* and *S*. The number of holidays purchased in a particular week and the mean and standard deviation of the costs of these holidays at each shop are shown in the following table.

	Number of holidays	Mean cost (£)	S.D. (£)
Shop *R*	32	190.35	10.4
Shop *S*	24	202.25	15.5

Calculate the mean, and, to the nearest penny, the standard deviation of the costs of all the 56 holidays purchased. (L)P

28. The following are the ignition times in seconds (correct to the nearest 100th of a second) of samples of 80 upholstery materials They are arranged in numerical order by columns.

1.20	2.11	3.10	4.11	5.09	5.92	6.85	8.80
1.38	2.20	3.24	4.19	5.11	5.92	6.90	9.20
1.42	2.32	3.49	4.32	5.12	6.20	7.35	9.45
1.47	2.46	3.62	4.50	5.15	6.25	7.40	9.65
1.52	2.46	3.75	4.54	5.21	6.37	7.41	9.70
1.58	2.50	3.78	4.56	5.33	6.40	7.60	9.79
1.70	2.51	3.81	4.71	5.40	6.43	7.86	10.60
1.76	2.58	3.87	4.72	5.50	6.43	7.95	11.25
1.79	2.65	3.90	4.79	5.62	6.75	8.64	11.75
1.92	2.80	4.04	4.90	5.84	6.77	8.75	12.80

(*a*) Group these data into 8 equal classes commencing 1.00–2.49, 2.50–3.99, ... and arrange them in a frequency table.

Table H

Age	12	13	14	15	16	17	18	19	20	21	22	23	24	25	26	27	28	29	30–49
f	1	1	5	8	17	19	15	9	8	6	4	3	3	2	2	2	1	1	6

Table I

Age (last birthday)	16-20	21-25	26-30	31-35	36-40	41-45	46-50	51-60	61-
Number of employees	36	56	58	52	46	38	36	36	0

(*b*) Using the frequency table obtain estimates for the mean time and standard deviation.

(*c*) Construct a frequency polygon for the distribution and comment on its shape.

(*d*) Chebychev's Theorem states that, for any distribution, the proportion of the population that lies outside k standard deviations from the mean is less than $1/k^2$. Verify this for the above distribution when $k = 1.5$. (SUJB)

29. In a borehole the thickness, in mm, of the 25 strata are shown in the table.

Thickness (mm)	0–	20–	30–	40–	50–	60–
Number of strata	2	5	9	8	1	0

Draw a histogram to illustrate these data. Construct a cumulative frequency table and draw a cumulative frequency polygon. Hence, or otherwise, estimate the median and the interquartile range for these data.

Find the proportion of the strata that are less than 28 mm thick. (L)P

30. Three random samples of 50, 30 and 20 bags respectively are taken from the production line of '12 kg bags' of cat litter. The contents of each bag are then weighed. A summary of the results is shown in the table.

Sample	Size	Mean wt. (kg)	S.D. (kg)
1	50	11.8	0.5
2	30	12.1	0.9
3	20	11.7	1.1

Find, in kg to 2 decimal places, the mean weight per bag and the standard deviation for the 100 bags. (L)P

31. Referring to your projects if possible, give an example of a graphical representation of

(*a*) a discrete frequency distribution,

(*b*) a grouped frequency distribution.

Given the frequency distribution

x	1	2	3	4	5	6	7	8	9
f	1	3	7	9	13	9	5	2	1

find the median and the semi-interquartile range when

(*c*) x is a discrete variable,

(*d*) x is a continuous variable whose values were recorded to the nearest integer.

Calculate also, to 2 decimal places, the mean and the variance of the above distribution. (L)

Table J

Lifetime (to nearest hour)	690-709	710-719	720-729	730-739	740-744	745-749	750-754	755-759	760-769	770-789
Number of discs	3	7	15	38	41	35	21	16	14	10

2

PROBABILITY

An experiment can result in several possible outcomes. For example

(a) One toss of a coin results in the outcomes (H, T). If the coin is fair, then each outcome is equally likely.

(b) Two tosses of a coin result in the outcomes (HH, HT, TH, TT). Again, if the coin is fair, then each outcome is equally likely.

(c) If a machine produces articles, some of which are defective, the outcomes are (defective, not defective). In this case the outcomes should not be equally likely.

(d) If a coin is tossed repeatedly until a head is obtained, the outcomes are (H, TH, TTH, TTTH, TTTTH, . . .).

(e) The outcomes of a race being run by A and B could be (A wins, B wins, there is a dead heat). These outcomes may not be equally likely.

Each possible outcome is called a sample point and the set of all possible outcomes is the possibility space S.

If the possibility space has a finite number of sample points then we denote the number of points in S by $n(S)$.

Consider an event E which is a subset of S, then $n(E) \leqslant n(S)$.

For example, for one throw of an ordinary die the possibility space $S = (1, 2, 3, 4, 5, 6)$ and $n(S) = 6$.

Let E_1 be the event 'the number is odd', then $E_1 = (1, 3, 5)$ and $n(E_1) = 3$.

Let E_2 be the event 'the number is less than 3' then $E_2 = (1, 2)$ and $n(E_2) = 2$.

CLASSICAL DEFINITION OF PROBABILITY

If the possibility space S consists of a finite number of equally likely outcomes, then the probability of an event E, written $P(E)$ is defined as

$$P(E) = \frac{n(E)}{n(S)}$$

So, in the example on the previous page

$$P(E_1) = \frac{n(E_1)}{n(S)} = \frac{3}{6} = \frac{1}{2}$$

$$P(E_2) = \frac{n(E_2)}{n(S)} = \frac{2}{6} = \frac{1}{3}$$

In order to investigate the rules which apply when considering probabilities, we will consider the situation in the classical definition—that of a finite possibility space with *equally likely outcomes*. However, the results apply in general and will be used in other situations in the problems.

IMPORTANT RESULTS

Let the number of sample points in the possibility space be n, so that $n(S) = n$.

Let the event A have r sample points, so that $n(A) = r$.

Result 1 We have

$$P(A) = \frac{n(A)}{n(S)} = \frac{r}{n}$$

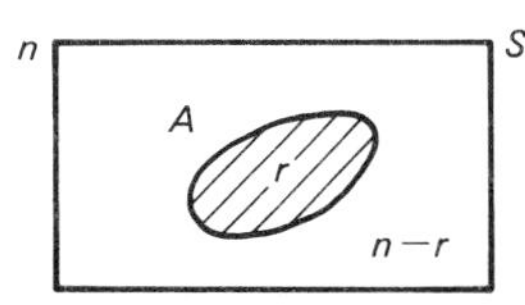

Now, since A is a subset of S

$$0 \leqslant r \leqslant n$$

i.e.

$$0 \leqslant \frac{r}{n} \leqslant 1$$

Hence

$$0 \leqslant P(A) \leqslant 1$$

NOTE: the probability of an event A is a number between 0 and 1 inclusive.

If $P(A) = 0$ then the event cannot possibly occur.

If $P(A) = 1$ then the event is certain to occur.

For example, if a card is drawn from the clubs suit of a pack of cards, then

$$P(\text{card is red}) = 0$$

$$P(\text{card is black}) = 1$$

Result 2 Let $\bar{A}$ denote the event 'A does not occur'.

Now
$$P(\bar{A}) = \frac{n(\bar{A})}{n(S)} = \frac{n-r}{n} = 1-\frac{r}{n} = 1-P(A)$$

Therefore
$$P(\bar{A}) = 1-P(A)$$

or
$$P(A)+P(\bar{A}) = 1$$

Example 2.1 A card is drawn at random from an ordinary pack of 52 playing cards. Find the probability that the card **(a)** is a seven, **(b)** is not a seven.

Solution 2.1 The possibility space $S =$ (the pack of 52 cards) and $n(S) = 52$. Let A be the event 'the card is a seven', then $n(A) = 4$.

(a) Now
$$P(A) = \frac{n(A)}{n(S)} = \frac{4}{52} = \frac{1}{13}$$

Therefore the probability that the card drawn is a seven is $\frac{1}{13}$.

(b) Let $\bar{A}$ be the event 'the card is not a seven'.

Now
$$P(\bar{A}) = 1-P(A) = 1-\frac{1}{13} = \frac{12}{13}$$

Therefore the probability that the card drawn is not a seven is $\frac{12}{13}$.

Example 2.2 Compare the probabilities of scoring a 4 with one die and a total of 8 with two dice.

Solution 2.2 *With one die*

The possibility space $S = (1, 2, 3, 4, 5, 6)$ and $n(S) = 6$.

Let A be the event 'a 4 occurs', then $n(A) = 1$.

So
$$P(A) = \frac{n(A)}{n(S)} = \frac{1}{6}$$

The probability of scoring 4 with one die is $\frac{1}{6}$.

With two dice

The possibility space S has 36 sample points, each of which is equally likely to occur. These can be represented on a diagram as shown. The dots in the first column represent the outcomes (1, 1), (1, 2), (1, 3), (1, 4), (1, 5), (1, 6), . . . , and so on for the other columns.

Let B be the event 'the sum on the two dice is 8'.

The sample points which give a sum of 8 are ringed on the diagram.

We see that $n(B) = 5$.

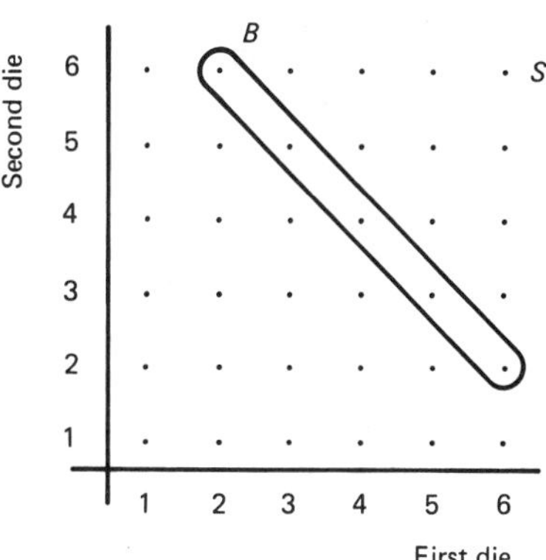

So
$$P(B) = \frac{n(B)}{n(S)} = \frac{5}{36}$$

The probability of obtaining a total of 8 with two dice is $\frac{5}{36}$.

So $P(\text{scoring an 8 with two dice}) < P(\text{scoring a 4 with one die})$

Example 2.3 Two fair coins are tossed. Illustrate the possible outcomes on a possibility space diagram and find the probability that two heads are obtained.

Solution 2.3 Each coin is equally likely to show a head or a tail. The possibility space for the outcomes when two coins are tossed is as shown.

$$n(S) = 4$$

Let A be the event 'two heads are obtained'. From the diagram $n(A) = 1$.

A
S
Second coin
H HH TH
T HT TT
H T
First coin

Therefore
$$P(A) = \frac{n(A)}{n(S)}$$
$$= \frac{1}{4}$$

The probability that two heads are obtained when two fair coins are tossed is $\frac{1}{4}$.

Exercise 2a

1. An ordinary die is thrown. Find the probability that the number obtained (*a*) is a multiple of 3, (*b*) is less than 7, (*c*) is a factor of 6.

2. A card is drawn at random from an ordinary pack containing 52 playing cards. Find the probability that the card drawn (*a*) is the four of spades, (*b*) is the four of spades or any diamond, (*c*) is not a picture card (Jack, Queen, King) of any suit.

3. From a set of cards numbered 1 to 20 a card is drawn at random. Find the probability that the number (*a*) is divisible by 4, (*b*) is greater than 15, (*c*) is divisible by 4 and greater than 15.
   If the card is divisible by 4 and it is not replaced, find the probability that (*d*) the second card drawn is even.

4. A counter is drawn from a box containing 10 red, 15 black, 5 green and 10 yellow counters. Find the probability that the counter is (*a*) black, (*b*) not green or yellow, (*c*) not yellow, (*d*) red or black or green, (*e*) not blue.

5. Two ordinary dice are thrown. Find the probability that (*a*) the sum on the two dice is 3, (*b*) the sum on the two dice exceeds 9, (*c*) the two dice show the same number, (*d*) the numbers on the two dice differ by more than 2, (*e*) the product of the two numbers is even.

6. The pupils in a class were asked how many brothers and sisters they had. Their answers are shown in the table:

Number of brothers and sisters	0	1	2	3	4	5
Number of pupils	4	12	8	3	2	1

   If a child is chosen at random, find the probability that there are three children in his or her family.

7. If
$$\mathscr{E} = \{x : x \text{ is an integer and } 1 \leqslant x \leqslant 20\}$$
$$A = \{x : x \text{ is a multiple of } 3\}$$
$$B = \{x : x \text{ is a multiple of } 4\}$$
   and an integer is picked at random from $\mathscr{E}$, find the probability that (*a*) it is in A, (*b*) it is not in B.

8. A die is in the form of a tetrahedron and its faces are marked 1, 2, 3 and 4. The 'score' is the number on which the die lands. Find the probability that when a tetrahedral die is thrown the score is (*a*) an even number, (*b*) a prime number. (*NOTE*: 1 is not a prime number.)
   If two tetrahedral dice are thrown find the probability that (*c*) the sum of the two scores is 5, (*d*) the difference of the two scores is 1, (*e*) the product of the two scores is a multiple of 4.

9. An ordinary die and a fair coin are thrown together. Show the possible outcomes on a possibility space diagram and find the probability that (*a*) a head and a 2 is obtained, (*b*) a tail and a 7 is obtained, (*c*) a head and an even number is obtained.

10. An ordinary die and two coins are thrown together. Show the possible outcomes on a possibility space diagram and find the probability that (*a*) two heads and a number less than 3 is obtained, (*b*) the coins show different faces and a 4 is shown on the die, (*c*) the die shows an odd number and the coins show the same face, (*d*) a 6 and at least one head is obtained.

11. Two dice are thrown simultaneously. The scores are to be multiplied. Denoting by $P(n)$ the probability that the number n will be obtained, calculate (*a*) $P(9)$, (*b*) $P(4)$, (*c*) $P(14)$, (*d*) $\sum_{m=15}^{30} P(m)$.

Given that $P(t) = \frac{1}{9}$, find the possible values of t. (L Additional)

Result 3

If A and B are any two events of the same experiment such that $P(A) \neq 0$ and $P(B) \neq 0$ then

$$P(A \text{ or } B) = P(A) + P(B) - P(A \text{ and } B)$$

Note that 'A or B' means 'A occurs, or B occurs, or both A and B occur'.

Writing the result in set notation we have

$$P(A \cup B) = P(A) + P(B) - P(A \cap B)$$

To illustrate this result let $n(S) = n$ where S is the possibility space,

$$n(A) = r$$
$$n(B) = s$$
$$n(A \cap B) = t$$

The Venn diagram is as shown. The shaded area represents $A \cup B$.

Now

$$\begin{aligned} P(A \cup B) &= \frac{n(A \cup B)}{n(S)} \\ &= \frac{(r-t)+t+(s-t)}{n} \\ &= \frac{r+s-t}{n} \\ &= \frac{r}{n} + \frac{s}{n} - \frac{t}{n} \\ &= P(A) + P(B) - P(A \cap B) \end{aligned}$$

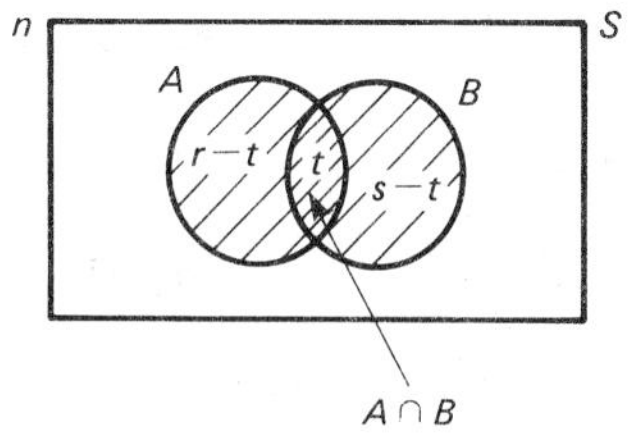

Example 2.4 A coin and a die are thrown together. Draw a possibility space diagram and find the probability of obtaining **(a)** a head, **(b)** a number greater than 4, **(c)** a head and a number greater than 4, **(d)** a head or a number greater than 4.

Solution 2.4 The possibility space S is as shown.

Let A be the event 'a head is obtained', so $n(A) = 6$.

Let B be the event 'a number greater than 4 is obtained', so $n(B) = 4$.

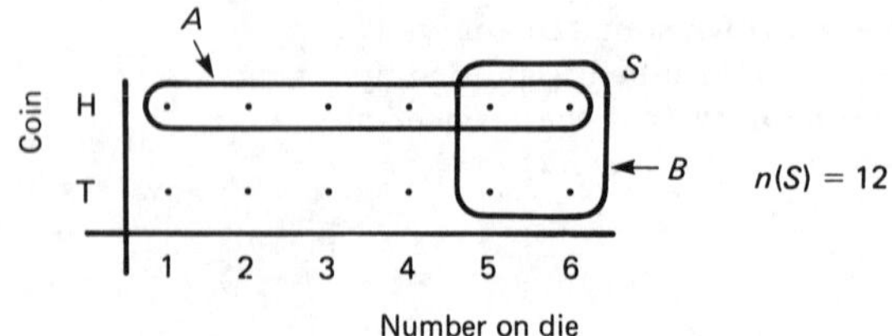

(a) $$P(A) = \frac{n(A)}{n(S)} = \frac{6}{12} = \frac{1}{2}$$

The probability of obtaining a head is $\frac{1}{2}$.

(b) $$P(B) = \frac{n(B)}{n(S)} = \frac{4}{12} = \frac{1}{3}$$

The probability of obtaining a number greater than 4 is $\frac{1}{3}$.

(c) $$\begin{aligned} P(\text{head and a number greater than 4}) &= P(A \cap B) \\ &= \frac{n(A \cap B)}{n(S)} \\ &= \frac{2}{12} \\ &= \frac{1}{6} \end{aligned}$$

The probability of obtaining a head and a number greater than 4 is $\frac{1}{6}$.

(d) $$\begin{aligned} P(\text{a head or a number greater than 4}) &= P(A \cup B) \\ &= \frac{n(A \cup B)}{n(S)} \\ &= \frac{8}{12} \\ &= \frac{2}{3} \end{aligned}$$

The probability of obtaining a head or a number greater than 4 is $\frac{2}{3}$.

We now check that this satisfies $P(A \cup B) = P(A) + P(B) - P(A \cap B)$.

$$\text{left hand side} = P(A \cup B) = \frac{2}{3}$$

$$\text{right hand side} = P(A)+P(B)-P(A\cap B)$$
$$= \frac{1}{2}+\frac{1}{3}-\frac{1}{6}$$
$$= \frac{8}{12}$$
$$= \frac{2}{3}$$

Therefore left hand side = right hand side and
$P(A\cup B) = P(A)+P(B)-P(A\cap B)$.

Example 2.5 Events A and B are such that $P(A) = \frac{19}{30}$, $P(B) = \frac{2}{5}$ and $P(A\cup B) = \frac{4}{5}$. Find $P(A\cap B)$.

Solution 2.5 Now

$$P(A\cup B) = P(A)+P(B)-P(A\cap B)$$

so

$$\frac{4}{5} = \frac{19}{30}+\frac{2}{5}-P(A\cap B)$$
$$P(A\cap B) = \frac{19}{30}+\frac{12}{30}-\frac{24}{30}$$
$$= \frac{7}{30}$$

Therefore $P(A\cap B) = \frac{7}{30}$.

Example 2.6 In a group of 20 adults, 4 out of the 7 women and 2 out of the 13 men wear glasses. What is the probability that a person chosen at random from the group is a woman or someone who wears glasses?

Solution 2.6 Let W be the event 'the person chosen is a woman' and G be the event 'the person chosen wears glasses'.

Now

$$P(W) = \frac{7}{20},\quad P(G) = \frac{6}{20},\quad P(W \text{ and } G) = P(W\cap G) = \frac{4}{20}$$
$$P(W \text{ or } G) = P(W\cup G) = P(W)+P(G)-P(W\cap G)$$
$$= \frac{7}{20}+\frac{6}{20}-\frac{4}{20}$$
$$= \frac{9}{20}$$

Therefore the probability that the person is a woman or someone who wears glasses is $\frac{9}{20}$.

MUTUALLY EXCLUSIVE EVENTS

Result 4

If an event A can occur *or* an event B can occur but *not both* A and B can occur, then the two events A and B are said to be mutually exclusive.

In this case $n(A \cap B) = 0$ and $A \cap B = \emptyset$.

When A and B are mutually exclusive events

$$P(A \cup B) = P(A) + P(B)$$

and $\quad P(A \cap B) = 0$

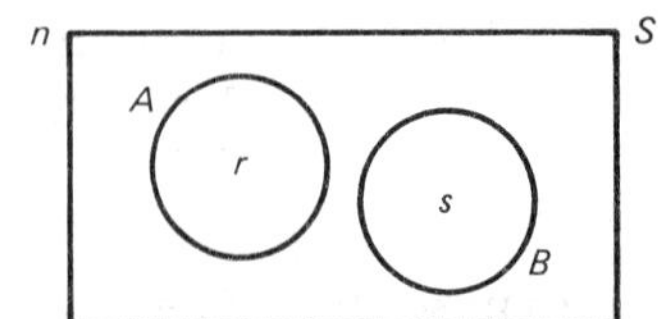

This is known as the **addition law for mutually exclusive events.**

Examples of mutually exclusive events:

(i) A number is chosen from the set of integers from 1 to 10 inclusive. If A is the event 'the number is odd' and B is the event 'the number is a multiple of 4' then A and B are mutually exclusive events, as an event cannot be both odd and a multiple of 4.

(ii) Two men are standing for election as chairman of a committee. Let A be the event 'Mr Smith is elected' and Y be the event 'Mr Jones is elected'. Then A and Y are mutually exclusive events as both cannot be elected as chairman.

Example 2.7 In a race the probability that John wins is $\frac{1}{3}$, the probability that Paul wins is $\frac{1}{4}$ and the probability that Mark wins is $\frac{1}{5}$. Find the probability that **(a)** John or Mark wins, **(b)** neither John nor Paul wins. Assume that there are no dead heats.

Solution 2.7 We assume that only one person can win, so the events are mutually exclusive.

(a) $$P(\text{John or Mark wins}) = P(\text{John wins}) + P(\text{Mark wins}) = \frac{1}{3} + \frac{1}{5} = \frac{8}{15}$$

$\underline{P(\text{John or Mark wins}) = \frac{8}{15}}$

(b) $P(\text{neither John nor Paul wins}) = 1 - P(\text{John or Paul wins})$

$$= 1 - \left(\frac{1}{3} + \frac{1}{4}\right)$$

$$= 1 - \frac{7}{12}$$

$$= \frac{5}{12}$$

$P(\text{neither John nor Paul wins}) = \frac{5}{12}$.

Example 2.8 A card is drawn at random from an ordinary pack of 52 playing cards. Find the probability that the card is (a) a club or a diamond, (b) a club or a king.

Solution 2.8 The possibility space S = (the pack of 52 cards) so $n(S) = 52$.

Let C be the event 'a club is drawn', D be the event 'a diamond is drawn', K be the event 'a king is drawn'.

(a)
$$P(\text{club}) = \frac{n(C)}{n(S)} = \frac{13}{52} = \frac{1}{4}$$

$$P(\text{diamond}) = \frac{n(D)}{n(S)} = \frac{13}{52} = \frac{1}{4}$$

Now the events C and D are mutually exclusive since they cannot occur together; a card cannot be both a club and a diamond.

Therefore

$$P(\text{club} \cup \text{diamond}) = P(\text{club}) + P(\text{diamond})$$

$$= \frac{1}{4} + \frac{1}{4}$$

$$= \frac{1}{2}$$

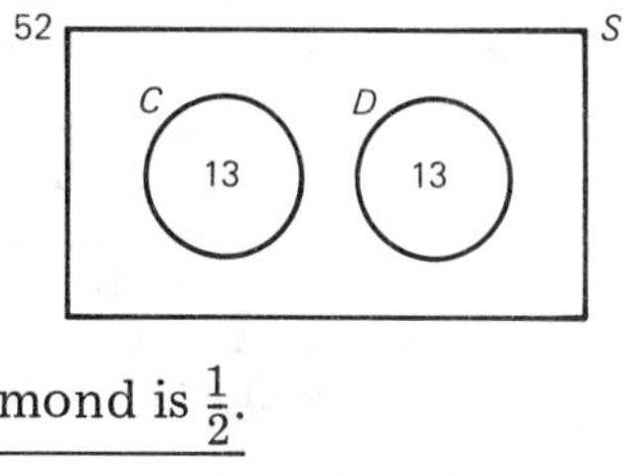

52 S
C D
13 13

The probability of drawing a club or a diamond is $\frac{1}{2}$.

(b)
$$P(\text{club}) = \frac{13}{52}$$

$$P(\text{king}) = \frac{n(K)}{n(S)} = \frac{4}{52}$$

Now
$$P(\text{king} \cap \text{club}) = P(\text{king of clubs})$$

$$= \frac{1}{52}$$

The events C and K are not mutually exclusive as a card can be both a king and a club.

Therefore

$$\begin{aligned} P(\text{club} \cup \text{king}) &= P(\text{club}) + P(\text{king}) - P(\text{club} \cap \text{king}) \\ &= \frac{13}{52} + \frac{4}{52} - \frac{1}{52} \\ &= \frac{16}{52} \\ &= \frac{4}{13} \end{aligned}$$

The probability of drawing a club or a king is $\frac{4}{13}$.

In this example we could have noted straight away that the event 'a club or a king is drawn' has 16 sample points:

(A♣, 2♣, 3♣, 4♣, 5♣, 6♣, 7♣, 8♣, 9♣, 10♣,
J♣, Q♣, K♣, K♦, K♠, K♥)

and the possibility space has 52 sample points so $P(\text{clubs} \cup \text{king}) = \frac{16}{52} = \frac{4}{13}$ as before.

Exercise 2b

1. An ordinary die is thrown. Find the probability that the number obtained is (*a*) even, (*b*) prime, (*c*) even or prime.

2. In a group of 30 students all study at least one of the subjects physics and biology. 20 attend the physics class and 21 attend the biology class. Find the probability that a student chosen at random studies both physics and biology.

3. From an ordinary pack of 52 playing cards the seven of diamonds has been lost. A card is dealt from the well-shuffled pack. Find the probability that it is (*a*) a diamond, (*b*) a queen, (*c*) a diamond or a queen, (*d*) a diamond or a seven.

4. For events A and B it is known that $P(A) = \frac{2}{3}$, $P(A \cup B) = \frac{3}{4}$ and $P(A \cap B) = \frac{5}{12}$. Find $P(B)$.

5. In a street containing 20 houses, 3 households do not own a television set; 12 households have a black and white set and 7 households have a colour and a black and white set. Find the probability that a household chosen at random owns a colour television set.

6. For events A and B it is known that $P(A) = P(B)$ and $P(A \cap B) = 0.1$ and $P(A \cup B) = 0.7$. Find $P(A)$.

7. The probability that a boy in class 2 is in the football team is 0.4 and the probability that he is in the chess team is 0.5. If the probability that a boy in the class is in both teams is 0.2, find the probability that a boy chosen at random is in the football or the chess team.

8. Two ordinary dice are thrown. Find the probability that the sum of the scores obtained (*a*) is a multiple of 5, (*b*) is greater than 9, (*c*) is a multiple of 5 or is greater than 9, (*d*) is a multiple of 5 and is greater than 9.

9. Given that $P(\bar{A}) = \frac{2}{3}$, $P(B) = \frac{1}{2}$ and $P(A \cap B) = \frac{1}{12}$, find $P(A \cup B)$.

10. Two ordinary dice are thrown. Find the probability that (*a*) at least one 6 is thrown, (*b*) at least one 3 is thrown, (*c*) at least one 6 or at least one 3 is thrown.

EXHAUSTIVE EVENTS

Result 5

If two events A and B are such that $A \cup B = S$ then $P(A \cup B) = 1$ and the events A and B are said to be exhaustive.

For example

(i) Let $S = (1, 2, 3, 4, 5, 6, 7, 8, 9, 10)$.

If $A = (1, 2, 3, 4, 5, 6)$ and $B = (5, 6, 7, 8, 9, 10)$ then $A \cup B = S$ and A and B are exhaustive events.

(ii) Let S be the possibility space when an ordinary die is thrown.

If A is the event 'the number is less than 5' and B is the event 'the number is greater than 3' then the events A and B are exhaustive as $A \cup B = S$.

Example 2.9 Events A and B are such that they are both mutually exclusive and exhaustive. Find a relationship between A and B. Give an example of such events.

Solution 2.9 If A and B are mutually exclusive then $P(A \cup B) = P(A) + P(B)$

If A and B are exhaustive then $P(A \cup B) = 1$

Therefore

$$P(A) + P(B) = 1$$

so $P(B) = 1 - P(A)$

But $P(\bar{A}) = 1 - P(A)$

Therefore $P(B) = P(\bar{A})$

i.e. $\underline{B = \bar{A}}$

Similarly $\underline{A = \bar{B}}$

Toss a coin. Let A be the event 'a head is obtained', B be the event 'a tail is obtained'.

Now A and B are mutually exclusive, as the coin cannot show both a head and a tail.

A and B are exhaustive as the probability that the outcome is a head or a tail is 1.

Therefore A and B are both mutually exclusive and exhaustive.

CONDITIONAL PROBABILITY

Result 6

If A and B are two events and $P(A) \neq 0$ and $P(B) \neq 0$, then the probability of A, given that B has already occurred is written $P(A|B)$

and $$P(A|B) = \frac{P(A \cap B)}{P(B)}$$

Illustrating this by means of the Venn diagram, the possibility space is B, since we know that B has already occurred.

$$\begin{aligned} P(A|B) &= \frac{n(A \cap B)}{n(B)} \\ &= \frac{t}{s} \\ &= \frac{t/n}{s/n} \\ &= \frac{P(A \cap B)}{P(B)} \end{aligned}$$

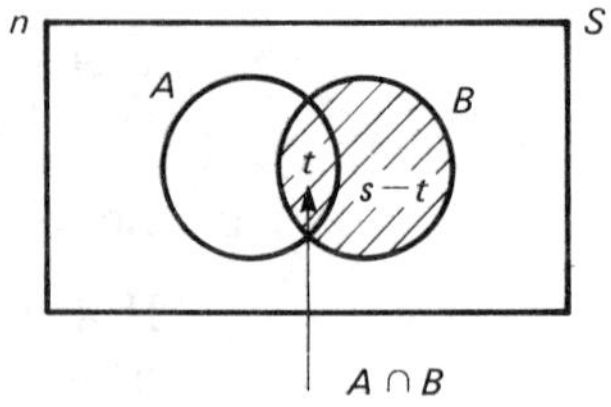

This result is often written

$$P(A \cap B) = P(A|B) \cdot P(B)$$

NOTE: if A and B are mutually exclusive events then, as $P(A \cap B) = 0$ and $P(B) \neq 0$, it follows that $P(A|B) = 0$.

Example 2.10 Given that a heart is picked at random from a pack of 52 playing cards, find the probability that it is a picture card.

Solution 2.10 We require

$$\begin{aligned} P(\text{picture card}\,|\,\text{heart}) &= \frac{P(\text{picture card} \cap \text{heart})}{P(\text{heart})} \\ &= \frac{3/52}{13/52} \\ &= \frac{3}{13} \end{aligned}$$

The probability that it is a picture card, given that it is a heart, is $\frac{3}{13}$.

Example 2.11 When a die is thrown, an odd number occurs. What is the probability that the number is prime?

Solution 2.11

$$P(\text{prime} \mid \text{odd}) = \frac{P(\text{prime} \cap \text{odd})}{P(\text{odd})}$$

$$= \frac{2/6}{3/6} \quad \text{(The odd prime numbers are 3 and 5)}$$

$$= \frac{2}{3}$$

The probability that the number is prime, given that it is odd, is $\frac{2}{3}$.

Result 7

As $P(A \mid B) = \dfrac{P(A \cap B)}{P(B)}$ we have $P(A \cap B) = P(A \mid B) \cdot P(B)$

It follows that

$P(B \mid A) = \dfrac{P(B \cap A)}{P(A)}$ and $P(B \cap A) = P(B \mid A) \cdot P(A)$

Now

$$P(A \cap B) = P(B \cap A)$$

Therefore $\quad P(A \mid B) \cdot P(B) = P(B \mid A) \cdot P(A)$

Example 2.12 Two tetrahedral dice, with faces labelled 1, 2, 3 and 4, are thrown and the number on which each lands is noted. The 'score' is the sum of these two numbers. Find the probability that **(a)** the score is even, given that at least one die lands on a 3, **(b)** at least one die lands on a 3, given that the score is even.

Solution 2.12 There are 16 sample points in the possibility space S, as shown in the diagram, so $n(S) = 16$.

Let A be the event 'at least one die lands on a 3' and let B be the event 'the score is even'.

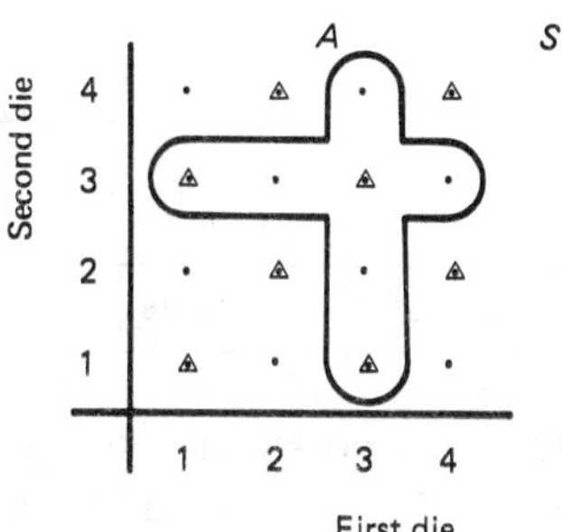

The sample space A is $((1,3),(2,3),(3,3),(4,3),(3,1),(3,2),(3,4))$, so

$$n(A) = 7 \quad \text{and} \quad P(A) = \frac{n(A)}{n(S)} = \frac{7}{16}$$

Sample space B is $((1,1),(1,3),(2,2),(2,4),(3,1),(3,3),(4,2),(4,4))$.

B has been marked Δ on the diagram.

We have $$n(B) = 8 \quad \text{and} \quad P(B) = \frac{n(B)}{n(S)} = \frac{8}{16}$$

There are 3 sample points which are in both A and B, so

$$P(A \cap B) = \frac{n(A \cap B)}{n(S)} = \frac{3}{16}$$

(a) $P(\text{score is even} \mid \text{at least one die lands on a 3}) = P(B \mid A)$

$$= \frac{P(B \cap A)}{P(A)} = \frac{3/16}{7/16} = \frac{3}{7}$$

Therefore the probability that the score is even, given that at least one die lands on a 3, is $\frac{3}{7}$.

NOTE: this result could have been obtained directly from the diagram. The possibility space has been reduced to the 7 sample points in A. For 3 of these the event B occurs, so $P(B \mid A) = \frac{3}{7}$.

(b) $P(\text{at least one die lands on a 3} \mid \text{score is even}) = P(A \mid B)$.

Using $$P(A \mid B) \cdot P(B) = P(B \mid A) \cdot P(A)$$

we have $$P(A \mid B) \cdot \frac{8}{16} = \left(\frac{3}{7}\right)\left(\frac{7}{16}\right)$$

$$P(A \mid B) = \frac{3}{8}$$

Therefore the probability that at least one die lands on a 3, given that the score is even, is $\frac{3}{8}$.

NOTE: the possibility space has been reduced to the 8 sample points in B. For three of these the event A occurs, so $P(A \mid B) = \frac{3}{8}$.

Example 2.13 A bag contains 10 counters, of which 7 are green and 3 are white. A counter is picked at random from the bag and its colour is noted. The counter is not replaced. A second counter is then picked out.

Find the probability that (a) the first counter is green, (b) the first counter is green and the second counter is white, (c) the counters are of different colours.

Solution 2.13 (a) Let G_1 be the event 'the first counter is green'.

$$P(G_1) = \frac{7}{10} \quad \text{(as there are 10 counters, of which 7 are green)}$$

The probability that the first counter is green is $\frac{7}{10}$.

(b) Let W_2 be the event 'the second counter picked is white'. Now

$$P(W_2 | G_1) = \frac{3}{9} = \frac{1}{3} \quad \text{(as there are 9 counters in the bag, of which 3 are white)}$$

We require

$$\begin{aligned} P(W_2 \cap G_1) &= P(W_2 | G_1) \cdot P(G_1) \\ &= \left(\frac{1}{3}\right)\left(\frac{7}{10}\right) \\ &= \frac{7}{30} \end{aligned}$$

The probability that the first counter is green and the second counter is white is $\frac{7}{30}$.

(c) With obvious notation, we require $P(W_2 \cap G_1) + P(G_2 \cap W_1)$.
Now $P(W_1) = \frac{3}{10}$ and $P(G_2 | W_1) = \frac{7}{9}$. Therefore

$$\begin{aligned} P(G_2 \cap W_1) &= P(G_2 | W_1) \cdot P(W_1) \\ &= \left(\frac{7}{9}\right)\left(\frac{3}{10}\right) \\ &= \frac{7}{30} \end{aligned}$$

So

$$\begin{aligned} P(W_2 \cap G_1) + P(G_2 \cap W_1) &= \frac{7}{30} + \frac{7}{30} \\ &= \frac{7}{15} \end{aligned}$$

Therefore the probability that the counters are different colours is $\frac{7}{15}$.

Exercise 2c

1. A card is picked at random from a pack of 20 cards numbered 1, 2, 3, . . . , 20. Given that the card shows an even number, find the probability that it is a multiple of 4.

2. If $P(A|B) = \frac{2}{5}$, $P(B) = \frac{1}{4}$, $P(A) = \frac{1}{3}$, find (*a*) $P(B|A)$, (*b*) $P(A \cap B)$.

3. Two digits are chosen at random from a table of random numbers containing the digits 0, 1, 2, . . . , 9. Find the probability that (*a*) the sum of the two numbers is greater than 9, given that the first number is 3, (*b*) the second number is 2, given that the sum of the two numbers is greater than 7, (*c*) the first number is 4, given that the difference between the two numbers is 4.

4. A bag contains 4 red counters and 6 black counters. A counter is picked at random from the bag and not replaced. A second counter is then picked. Find the probability that (*a*) the second counter is red, given that the first counter is red, (*b*) both counters are red, (*c*) the counters are of different colours.

5. Two cards are drawn successively from an ordinary pack of 52 playing cards and kept out of the pack. Find the probability that (*a*) both cards are hearts, (*b*) the first card is a heart and the second card is a spade, (*c*) the second card is a diamond, given that the first card is a club.

6. X and Y are two events such that $P(X) = \frac{2}{5}$, $P(X|Y) = \frac{1}{2}$ and $P(Y|X) = \frac{2}{3}$. Find (*a*) $P(X \cap Y)$, (*b*) $P(Y)$, (*c*) $P(X \cup Y)$.

7. A box contains two yellow and two black tickets numbered 1 and 2. Two tickets are drawn from the box. Indicate the sample space by listing all possible pairs of results.

   What is the probability that both tickets drawn will be yellow, (*a*) if nothing is known about either of them, (*b*) if one is known to be yellow, (*c*) if one is known to be yellow ticket numbered 1?

   (SUJB Additional)

8. A number is picked at random from the digits 1, 2, . . . , 9. Given that the number is a multiple of 3, find the probability that the number is (*a*) even, (*b*) a multiple of 4.

9. Two tetrahedral dice are thrown; one is red and the other is blue. The number on which each lands is noted, the faces being marked 1, 2, 3 and 4. Find the probability that (*a*) the sum of the numbers on which the dice land is 6 given that the red die lands on an odd number, (*b*) the blue die lands on a 2 or a 3, given that the red die lands on a 2.

10. If events A and B are such that $P(A) = \frac{1}{2}$, $P(B) = \frac{1}{5}$ and $P(A|B) = 0$. (*a*) Find $P(A \cup B)$. (*b*) Are events A and B exhaustive? (Give a reason.)

11. A and B are two events such that $P(A) = \frac{8}{15}$, $P(B) = \frac{2}{3}$ and $P(A \cap B) = \frac{1}{5}$. Are A and B exhaustive events?

12. A and B are exhaustive events and it is known that $P(A|B) = \frac{1}{4}$ and $P(B) = \frac{2}{3}$. Find $P(A)$.

13. Give two examples of events which are both mutually exclusive and exhaustive.

14. Two coins are tossed. A is the event 'at least one head is obtained'. Describe an event B such that A and B are exhaustive events.

15. If

$$\mathscr{E} = \{(x,y): x \text{ and } y \text{ are positive integers}\}$$
$$A = \{(x,y): 2 \leqslant x \leqslant 5 \text{ and } 1 \leqslant y \leqslant 4\}$$
$$B = \{(x,y): x+y=5\}$$
$$C = \{(x,y): y=2\}$$

   Find the probability that a member of A chosen at random will also be a member of (*a*) B, (*b*) C, (*c*) $B \cap C$, (*d*) $B \cup C$.

INDEPENDENT EVENTS

Result 8 If the occurrence or non-occurrence of an event A does not influence in any way the probability of an event B, then event B is independent of event A and $P(B|A) = P(B)$.

If events A and B are independent, then $P(A|B) = P(A)$
and $P(B|A) = P(B)$

Now $$P(A \cap B) = P(A|B) \cdot P(B)$$

Therefore $$P(A \cap B) = P(A) \cdot P(B)$$

This is known as the **multiplication law for independent events.**

NOTE: If two events are mutually exclusive, $P(A \cap B) = 0$. So for two events to be both independent and mutually exclusive we must have $P(A) \cdot P(B) = 0$. This is possible only if either $P(A) = 0$ or $P(B) = 0$.

Example 2.14 A die is thrown twice. Find the probability of obtaining a 4 on the first throw and an odd number on the second throw.

Solution 2.14 Let A be the event 'a 4 is obtained on the first throw', then $P(A) = \frac{1}{6}$.

Let B be the event 'an odd number is obtained on the second throw'.

Now the result on the second throw is not affected in any way by the result on the first throw. Therefore A and B are independent events and $P(B) = \frac{3}{6} = \frac{1}{2}$.

As A and B are independent events

$$\begin{aligned} P(A \cap B) &= P(A) \cdot P(B) \\ &= (\tfrac{1}{6})(\tfrac{1}{2}) \\ &= \frac{1}{12} \end{aligned}$$

The probability that the first throw results in a 4 and the second throw results in an odd number is $\frac{1}{12}$.

Example 2.15 A bag contains 5 red counters and 7 black counters. A counter is drawn from the bag, the colour is noted and the counter is replaced. A second counter is then drawn. Find the probability that the first counter is red and the second counter is black.

Solution 2.15 Let R_1 be the event 'the first counter is red'.

Then $$P(R_1) = \frac{5}{12}$$

Let B_2 be the event 'the second counter is black'.

Now, as the first counter is replaced before the second draw is made R_1 and B_2 are independent events.

Now $$P(B_2) = \frac{7}{12}$$

and
$$\begin{aligned} P(R_1 \cap B_2) &= P(R_1) \cdot P(B_2) \\ &= \left(\frac{5}{12}\right)\left(\frac{7}{12}\right) \\ &= \frac{35}{144} \end{aligned}$$

The probability that the first counter is red and the second counter is black is $\frac{35}{144}$.

Example 2.16 A fair die is thrown twice. Find the probability that **(a)** neither throw results in a 4, **(b)** at least one throw results in a 4.

Solution 2.16 Let A be the event 'the number on the first throw is 4'.

Let B be the event 'the number on the second throw is 4'.

Now $P(A) = \frac{1}{6}$, so $P(\bar{A}) = \frac{5}{6}$ where $\bar{A}$ is the event 'the number on the first throw is not a 4'.

Similarly $P(\bar{B}) = \frac{5}{6}$.

NOTE: $\bar{A}$ and $\bar{B}$ are independent events.

(a)
$$\begin{aligned} P(\text{neither throw results in a 4}) &= P(\bar{A} \cap \bar{B}) \\ &= P(\bar{A}) \cdot P(\bar{B}) \\ &= \left(\frac{5}{6}\right)\left(\frac{5}{6}\right) \\ &= \frac{25}{36} \end{aligned}$$

The probability that neither throw results in a 4 is $\frac{25}{36}$.

(b)
$$\begin{aligned} P(\text{at least one throw results in a 4}) &= 1 - P(\text{neither results in a 4}) \\ &= 1 - \frac{25}{36} \\ &= \frac{11}{36} \end{aligned}$$

The probability that at least one throw results in a 4 is $\frac{11}{36}$.

Example 2.17 Events A and B are such that $P(A) = \frac{1}{3}$ and $P(A \cap B) = \frac{1}{12}$. If A and B are independent events, find **(a)** $P(B)$, **(b)** $P(A \cup B)$.

Solution 2.17 (a) As A and B are independent events $P(A \cap B) = P(A) \cdot P(B)$

so
$$\frac{1}{12} = \frac{1}{3}P(B)$$

Hence
$$P(B) = \frac{1}{4}$$

(b) Now
$$P(A \cup B) = P(A) + P(B) - P(A \cap B)$$

so
$$P(A \cup B) = \frac{1}{3} + \frac{1}{4} - \frac{1}{12}$$
$$= \frac{1}{2}$$

Therefore $P(B) = \frac{1}{4}$ and $P(A \cup B) = \frac{1}{2}$.

Example 2.18 Two events A and B are such that $P(A) = \frac{1}{4}$, $P(A | B) = \frac{1}{2}$ and $P(B | A) = \frac{2}{3}$.

(**a**) Are A and B independent events? (**b**) Are A and B mutually exclusive events? (**c**) Find $P(A \cap B)$. (**d**) Find $P(B)$.

Solution 2.18 (a) If A and B are independent events then $P(A | B) = P(A)$.

Now $P(A | B) = \frac{1}{2}$ and $P(A) = \frac{1}{4}$.

Therefore $P(A | B) \neq P(A)$ and A and B are not independent events.

(b) If A and B are mutually exclusive events then $P(A | B) = 0$.

But we are given that $P(A | B) = \frac{1}{2}$.

Therefore A and B are not mutually exclusive events.

(c) Now
$$P(A \cap B) = P(B | A) \cdot P(A)$$
$$= \left(\frac{2}{3}\right)\left(\frac{1}{4}\right)$$
$$= \frac{1}{6}$$

Therefore $P(A \cap B) = \frac{1}{6}$.

(d) Now
$$P(A | B) \cdot P(B) = P(B | A) \cdot P(A)$$

so
$$\tfrac{1}{2}P(B) = \left(\frac{2}{3}\right)\left(\frac{1}{4}\right)$$
$$P(B) = \frac{1}{3}$$

Example 2.19 The probability that a certain type of machine will break down in the first month of operation is 0.1. If a firm has two such machines which are installed at the same time, find the probability that, at the end of the first month, just one has broken down.

Solution 2.19 We assume that the performances of the two machines are independent.

Let A be the event 'machine 1 breaks down' and let B be the event 'machine 2 breaks down'.

Then, if just one machine breaks down, either machine 1 breaks down and machine 2 is still working, or machine 2 breaks down and machine 1 is still working. Therefore we require

$$\begin{aligned} P(A \cap \bar{B}) + P(\bar{A} \cap B) &= P(A) \cdot P(\bar{B}) + P(\bar{A}) \cdot P(B) \\ &= (0.1)(0.9) + (0.9)(0.1) \\ &= 0.18 \end{aligned}$$

NOTE: A and $\bar{B}$ are independent events, as are $\bar{A}$ and B.

Therefore the probability that after 1 month just one machine has broken down is 0.18.

Exercise 2d

1. A die is thrown twice. Find the probability of obtaining a number less than 3 on both throws.

2. A card is picked from a pack containing 52 playing cards. It is then replaced and a second card is picked. Find the probability that (*a*) both cards are the seven of diamonds, (*b*) the first card is a heart and the second card is a spade, (*c*) one card is from a black suit and the other is from a red suit, (*d*) at least one card is a queen.

3. A coin is tossed and a die is thrown. What is the probability of obtaining a head on the coin and an even number on the die?

4. Two men fire at a target. The probability that Alan hits the target is $\frac{1}{2}$ and the probability that Bob does not hit the target is $\frac{1}{3}$. Alan fires at the target first, then Bob fires at the target. Find the probability that (*a*) both Alan and Bob hit the target, (*b*) only one hits the target, (*c*) neither hits the target.

5. The probability that I am late for work is 0.05. Find the probability that, on two consecutive mornings, (*a*) I am late for work twice, (*b*) I am late for work once.

6. Events A and B are such that $P(A) = \frac{2}{5}$ and $P(B) = \frac{1}{4}$. If A and B are independent events, find (*a*) $P(A \cap B)$, (*b*) $P(A \cap \bar{B})$, (*c*) $P(\bar{A} \cap \bar{B})$.

7. If events A and B are such that they are independent and $P(A) = 0.3$, $P(B) = 0.5$, find (*a*) $P(A \cap B)$, (*b*) $P(A \cup B)$.
Are events A and B mutually exclusive?

8. Events A and B are such that $P(A) = \frac{2}{3}$, $P(A \mid B) = \frac{2}{3}$, $P(B) = \frac{1}{4}$. Find (*a*) $P(B \mid A)$, (*b*) $P(A \cap B)$.

9. In a group of 120 girls, each is either freckled or blonde or both; 80 are freckled and 60 are blonde. A girl is to be chosen at random from the group. A is the event 'a freckled girl is chosen' and B is the event 'a blonde girl is chosen'. (*a*) Calculate $P(A \cap B)$. (*b*) State, giving a reason, if you think A and B are independent events. (L Additional)

10. A and B are independent events and $P(A) = \frac{1}{3}$, $P(B) = \frac{3}{4}$. Find the probability that (*a*) both A and B occur, (*b*) only one occurs.

11. The probability that I have to wait at the traffic lights on my way to school is $\frac{1}{4}$. Find the probability that, on two consecutive mornings, I have to wait on at least one morning.

Result 9

For events A and B we have $P(B) = P(B \cap A) + P(B \cap \bar{A})$.

Illustrating this by means of the Venn diagram:

$$P(B) = \frac{s}{n}$$
$$= \frac{t + (s-t)}{n}$$
$$= \frac{t}{n} + \frac{(s-t)}{n}$$
$$= P(B \cap A) + P(B \cap \bar{A})$$

n S A B t s − t
$B \cap A$ $B \cap \bar{A}$

This result is often written

$$P(B) = P(B \mid A) \cdot P(A) + P(B \mid \bar{A}) \cdot P(\bar{A})$$

Example 2.20 The probability that it will be sunny tomorrow is $\frac{1}{3}$. If it is sunny, the probability that Susan plays tennis is $\frac{4}{5}$. If it is not sunny, the probability that Susan plays tennis is $\frac{2}{5}$. Find the probability that Susan plays tennis tomorrow.

Solution 2.20 Let A be the event 'it is sunny tomorrow' and let B be the event 'Susan plays tennis tomorrow'.

Then $\bar{A}$ is the event 'it is not sunny tomorrow'.

$P(A) = \frac{1}{3}$ and $P(\bar{A}) = \frac{2}{3}$; also $P(B \mid A) = \frac{4}{5}$ and $P(B \mid \bar{A}) = \frac{2}{5}$.

We require
$$P(B) = P(B \mid A) \cdot P(A) + P(B \mid \bar{A}) \cdot P(\bar{A})$$
$$= \left(\frac{4}{5}\right)\left(\frac{1}{3}\right) + \left(\frac{2}{5}\right)\left(\frac{2}{3}\right)$$
$$= \frac{8}{15}$$

Therefore the probability that Susan plays tennis tomorrow is $\frac{8}{15}$.

Example 2.21 If events A and B are independent, show that events $\bar{A}$ and B are independent.

Solution 2.21 Now

$$P(B) = P(B \cap A) + P(B \cap \bar{A})$$

so

$$\begin{aligned} P(B \cap \bar{A}) &= P(B) - P(B \cap A) \\ &= P(B) - P(B)\cdot P(A) \quad \text{as } A \text{ and } B \text{ are independent} \\ &= P(B)[1 - P(A)] \\ &= P(B)\cdot P(\bar{A}) \end{aligned}$$

Therefore $P(B \cap \bar{A}) = P(B)\cdot P(\bar{A})$ and so $\bar{A}$ and B are independent.

Exercise 2e

1. A bag contains 6 white counters and 4 blue counters. A counter is drawn, its colour is noted and it is not put back into the bag. A second counter is then drawn. Find the probability that the second counter drawn is blue.

2. In a restaurant 40% of the customers choose steak for their main course. If a customer chooses steak, the probability that he will choose ice cream to follow is 0.6. If he does not have steak, the probability that he will choose ice cream is 0.3. Find the probability that a customer picked at random will choose (*a*) steak and ice cream, (*b*) ice cream.

3. Events C and D are such that $P(C) = \frac{4}{7}$, $P(C \cap \bar{D}) = \frac{1}{3}$, $P(C \mid D) = \frac{5}{14}$. Find (*a*) $P(C \cap D)$, (*b*) $P(D)$, (*c*) $P(D \mid C)$.

4. Exactly 60% of the members of a form are boys, and 90% of these boys and 75% of the girls each buy one raffle ticket, the rest buying none. Calculate the probability that the winning ticket will be bought by a boy. (L Additional)

5. $P(X) = \frac{1}{2}$ and $P(Y) = \frac{1}{4}$. Given that X and Y are mutually exclusive, find (*a*) $P(X \cup Y)$, (*b*) $P(Y \cap \bar{X})$.

6. It is estimated that one-quarter of the drivers on the road between 11 p.m. and midnight have been drinking during the evening. If a driver has not been drinking, the probability that he will have an accident at that time of night is 0.004%; if he has been drinking, the probability of an accident goes up to 0.02%. What is the probability that a car selected at random at that time of night will have an accident?

   A policeman on the beat at 11.30 p.m. sees a car run into a lamp-post, and jumps to the conclusion that the driver has been drinking. What is the probability that he is right? (SMP)

SUMMARY — PROBABILITY LAWS

For a finite possibility space S with equally likely outcomes, and a subset E of S,

$$P(E) = \frac{n(E)}{n(S)}$$

$0 \leqslant P(E) \leqslant 1$

$P(S) = 1$

$P(E) + P(\bar{E}) = 1$

$P(A \cup B) = P(A) + P(B) - P(A \cap B)$

If A and B are exhaustive, then $P(A \cup B) = 1$

If A and B are mutually exclusive, then $P(A \cap B) = 0$

and $P(A \cup B) = P(A) + P(B)$ Addition law for mutually exclusive events

$P(A \mid B) = \dfrac{P(A \cap B)}{P(B)}$ i.e. $P(A \cap B) = P(A \mid B) \cdot P(B)$

$P(B \mid A) = \dfrac{P(B \cap A)}{P(A)}$ i.e. $P(B \cap A) = P(B \mid A) \cdot P(A)$

so that $P(A \mid B)P(B) = P(B \mid A)P(A)$

If A and B are independent, $P(A \mid B) = P(A)$

and $P(A \cap B) = P(A) \cdot P(B)$ Multiplication law for independent events

If A and B are mutually exclusive, $P(A \cap B) = 0$

so that $P(A \mid B) = 0$

If $P(A) \neq 0$ and $P(B) \neq 0$ then events A and B cannot be both independent and mutually exclusive, as $P(A \cap B) = P(A) \cdot P(B) \neq 0$

$P(A) = P(A \cap B) + P(A \cap \overline{B})$

or $P(A) = P(A \mid B) \cdot P(B) + P(A \mid \overline{B}) \cdot P(\overline{B})$

Miscellaneous Exercise 2f

1. Bag A contains 5 red and 4 white counters. Bag B contains 6 red and 3 white counters. A counter is picked at random from bag A and placed in bag B. A counter is now picked from bag B. Find the probability that this counter is white.

2. In a set of 28 dominoes each domino has from 0 to 6 spots at each end. Each domino is different from every other and the ends are indistinguishable so that, for example, the two diagrams in the figure represent the same domino.

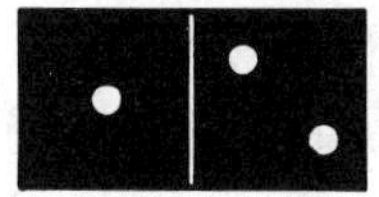
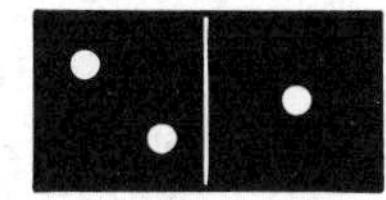

A domino which has no spots at all or the same number of spots at each end is called a 'double'.

A domino is drawn from the set. Let the event A be 'The domino is a double', event B be 'The sum of the spots is 6' and event C be 'The number of spots at each end differ by more than 3'. On graph paper draw a diagram to represent the possibility space with, for example, the point $(1, 2)$ representing the selection of the domino shown in the figure. On your diagram mark clearly the set of elements associated with each of the events A, B and C. Using your diagram find the probability that (a) both A and B occur, (b) both A and C occur, (c) both B and C occur.

State a pair of events which are independent and also a pair which are mutually exclusive. Find the probability that A occurs and B does not occur.

(L Additional)

3. The probability that a person in a particular evening class is left-handed is $\frac{1}{6}$. From the class of 15 women and 5 men a person is chosen at random. Assuming that 'left-handedness' is independent of the sex of a person, find the probability that the person chosen is a man or is left handed.

4. Two events A and B are such that $P(A) = 0.2$, $P(A' \cap B) = 0.22$, $P(A \cap B) = 0.18$.
Evaluate (*a*) $P(A \cap B')$, (*b*) $P(A|B)$. (JMB)
(*NOTE*: B' is the event 'B does not occur'.)

5. In a group of 100 people, 40 own a cat, 25 own a dog and 15 own a cat and a dog. Find the probability that a person chosen at random (*a*) owns a dog or a cat, (*b*) owns a dog or a cat, but not both, (*c*) owns a dog, given that he owns a cat, (*d*) does not own a cat, given that he owns a dog.

6. The two events A and B are such that $P(A) = 0.6$, $P(B) = 0.2$, $P(A|B) = 0.1$. Calculate the probabilities that (i) both of the events occur, (ii) at least one of the events occur, (iii) exactly one of the events occurs, (iv) B occurs, given that A has occurred. (JMB)

7. Of a group of pupils studying at A-level in schools in a certain area, 56% are boys and 44% are girls. The probability that a boy of this group is studying Chemistry is $\frac{1}{5}$ and the probability that a girl of this group is studying Chemistry is $\frac{1}{11}$.
(*a*) Find the probability that a pupil selected at random from this group is a girl studying Chemistry.
(*b*) Find the probability that a pupil selected at random from this group is not studying Chemistry.
(*c*) Find the probability that a Chemistry pupil selected at random from this group is male.
(You may leave your answers as fractions in their lowest terms.) (O & C)

8. At a fête the vicar has a board in the shape of a circle, having sectors coloured red and green, with an arrow which can be spun above it: you have to try to guess the colour on which the arrow will come to rest when it is next spun. It is made so that the results of successive spins are independent, and

$$P(\text{the arrow rests on red}) = 0.6$$

Find the probability of guessing correctly
(i) if you always guess 'green';
(ii) if you toss a fair coin and guess 'green' if it comes down 'head' and 'red' otherwise;
(iii) if your guess is always the colour the arrow is resting on before the spin. (SMP)

9. Two soldiers, Alan and Bill, are shooting at a target with independent probabilities of $\frac{4}{5}$ and $\frac{3}{4}$ respectively of hitting the bull with a single shot. If they each fire two shots, copy and complete the tables which show the possible outcomes, together with their associated probabilities.

Alan's two shots at the target

Number of bulls	0	1	2
Probability		$\frac{8}{25}$	

Bill's two shots at the target

Number of bulls	0	1	2
Probability			$\frac{9}{16}$

Calculate the probability that
(*a*) Alan records two bulls *and* Bill records two misses,
(*b*) *either* Alan records two bulls *or* Bill records two misses with the two events not occurring simultaneously,
(*c*) the soldiers record two bulls and two misses between them. (L Additional)

EXTENSION OF RESULTS TO MORE THAN TWO EVENTS

The result $P(A \cup B) = P(A) + P(B) - P(A \cap B)$ can be extended for three events A, B and C:

$$P(A \cup B \cup C) = P(A) + P(B) + P(C) - P(A \cap B) - P(B \cap C) - P(C \cap A) + P(A \cap B \cap C)$$

To illustrate this, consider the Venn diagram with the number of elements in each part as shown:

The shaded area shows

$A \cup B \cup C$

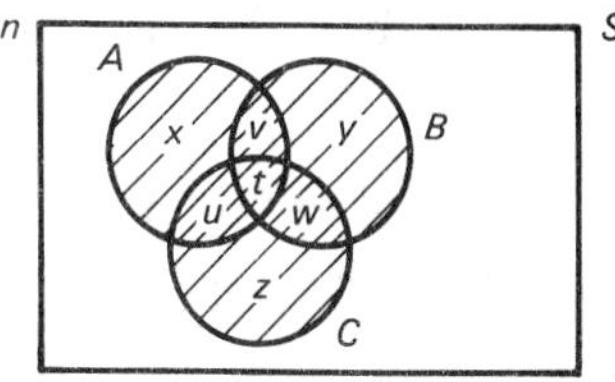

$$\begin{aligned}
\text{right hand side} &= P(A)+P(B)+P(C)-P(A\cap B)-P(B\cap C) \\
&\quad -P(C\cap A)+P(A\cap B\cap C) \\
&= \frac{(t+u+v+x)}{n}+\frac{(t+v+w+y)}{n}+\frac{(t+u+w+z)}{n} \\
&\quad -\frac{(t+v)}{n}-\frac{(t+w)}{n}-\frac{(t+u)}{n}+\frac{t}{n} \\
&= \frac{t+u+v+w+x+y+z}{n} \\
&= P(A\cup B\cup C) \\
&= \text{left hand side}
\end{aligned}$$

Example 2.22 In the Good Grub Restaurant customers may (if they wish) order any combination of chips, peas and salad to accompany the main course. The probability that a customer chooses salad is 0.45, peas and chips 0.19, salad and peas 0.15, salad and chips 0.25, salad or peas 0.6, salad or chips 0.84, salad or chips or peas 0.9. Find the probability that a customer chooses **(a)** peas, **(b)** chips, **(c)** all three, **(d)** none of these.

Solution 2.22 Let A be the event 'salad is chosen', E the event 'peas are chosen' and C the event 'chips are chosen'.

Then

$$P(A) = 0.45, \quad P(E\cap C) = 0.19, \quad P(A\cap E) = 0.15,$$
$$P(A\cap C) = 0.25, \quad P(A\cup E) = 0.6, \quad P(A\cup C) = 0.84,$$
$$P(A\cup E\cup C) = 0.9$$

(a) We require $P(\text{peas are chosen}) = P(E)$.

Now $\quad P(A\cup E) = P(A)+P(E)-P(A\cap E)$

so $\quad 0.6 = 0.45+P(E)-0.15$

$$P(E) = 0.3$$

The probability that a customer chooses peas is 0.3.

(**b**) We require $P(C)$.

$$\begin{aligned} \text{Now} \quad P(A \cup C) &= P(A) + P(C) - P(A \cap C) \\ \text{so} \quad 0.84 &= 0.45 + P(C) - 0.25 \\ P(C) &= 0.64 \end{aligned}$$

The probability that a customer chooses chips is 0.64.

(**c**) We require $P(A \cap E \cap C)$.

Now

$$\begin{aligned} P(A \cup E \cup C) &= P(A) + P(E) + P(C) - P(A \cap E) - P(E \cap C) \\ &\quad - P(C \cap A) + P(A \cap E \cap C) \\ \text{so} \quad 0.9 &= 0.45 + 0.3 + 0.64 - 0.15 - 0.19 - 0.25 \\ &\quad + P(A \cap E \cap C) \\ P(A \cap E \cap C) &= 0.1 \end{aligned}$$

The probability that a customer chooses all three is 0.1.

(**d**)
$$\begin{aligned} P(\text{customer chooses none}) &= P(\bar{A} \cap \bar{E} \cap \bar{C}) \\ &= 1 - P(A \cup E \cup C) \\ &= 1 - 0.9 \\ &= 0.1 \end{aligned}$$

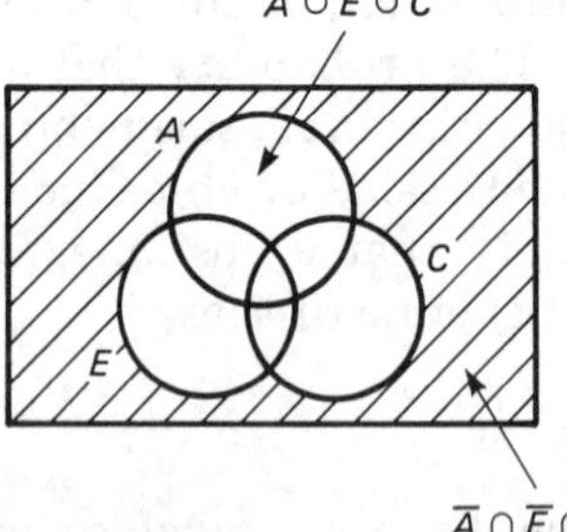

The probability that a customer chooses none of these is 0.1.

Mutually exclusive events

If events A, B and C are mutually exclusive, then

$$P(A \cup B \cup C) = P(A) + P(B) + P(C)$$

This can be extended to any number of mutually exclusive events:

$$P(A_1 \cup A_2 \cup \ldots \cup A_n) = P(A_1) + P(A_2) + \ldots + P(A_n)$$

Example 2.23 Records in a music shop are classed in the following sections: classical, popular, rock, folk and jazz. The respective probabilities that a customer buying a record will choose from each section are 0.3, 0.4, 0.2, 0.05 and 0.05. Find the probability that a person **(a)** will choose a record from the classical or the folk or the jazz sections, **(b)** will not choose a record from the rock or folk or classical sections.

Solution 2.23 A record cannot be classed in more than one section, so the events are mutually exclusive.

(a)
$$\begin{aligned} P(\text{classical or folk or jazz}) &= P(\text{classical}) + P(\text{folk}) + P(\text{jazz}) \\ &= 0.3 + 0.05 + 0.05 \\ &= 0.4 \end{aligned}$$

The probability that the record will be classical or folk or jazz is 0.4.

(b)
$$\begin{aligned} P(\text{rock or folk or classical}) &= P(\text{rock}) + P(\text{folk}) + P(\text{classical}) \\ &= 0.2 + 0.05 + 0.3 \\ &= 0.55 \end{aligned}$$

$$\begin{aligned} P(\text{not rock nor folk nor classical}) &= 1 - 0.55 \\ &= 0.45 \end{aligned}$$

Therefore the probability that the record is not rock nor folk nor classical is 0.45.

The result $P(A \cap B) = P(A) \cdot P(B | A)$ can be extended for three events A, B and C as follows:

$$\begin{aligned} P(A \cap B \cap C) &= P[(A \cap B) \cap C] \\ &= P(A \cap B) \cdot P[C | (A \cap B)] \\ &= P(A) \cdot P(B | A) \cdot P(C | A \cap B) \end{aligned}$$

So
$$P(A \cap B \cap C) = P(A) \cdot P(B | A) \cdot P(C | A \cap B)$$

Example 2.24 A bag of sweets contains 4 red 'fruities' and 5 green ones. A child picks out 3 fruities one after the other and eats them. Find the probability that the first is red, the second is green and the third is red.

Solution 2.24 Let R_1 be the event 'the first fruitie is red', G_2 be the event 'the second fruitie is green', R_3 be the event 'the third fruitie is red'.

We require $P(R_1 \cap G_2 \cap R_3) = P(R_1) \cdot P(G_2 | R_1) \cdot P(R_3 | G_2 \cap R_1)$.

Now $P(R_1) = \dfrac{4}{9}$

and

$P(G_2 | R_1) = \dfrac{5}{8}$ as there are now 8 fruities, of which 5 are green

$$P(R_3|G_2 \cap R_1) = \frac{3}{7} \quad \text{as there are now 7 fruities of which 3 are red}$$

So
$$P(R_1 \cap G_2 \cap R_3) = \left(\frac{4}{9}\right)\left(\frac{5}{8}\right)\left(\frac{3}{7}\right) = \frac{5}{42}$$

Therefore the probability that the first is red, the second is green and the third is red is $\frac{5}{42}$.

Independent events

If A, B and C are independent events, then

$$P(A \cap B \cap C) = P(A) \cdot P(B) \cdot P(C).$$

This result can be extended to n independent events so that

$$P(A_1 \cap A_2 \cap \ldots \cap A_n) = P(A_1) \cdot P(A_2) \ldots P(A_n)$$

Example 2.25 A die is thrown four times. Find the probability that a 5 is obtained each time.

Solution 2.25 Let 5_1 be the event 'a 5 is obtained on the first throw', 5_2 be the event 'a 5 is obtained on the second throw' and so on.

The events are independent, so

$$P(5_1 \cap 5_2 \cap 5_3 \cap 5_4) = P(5_1) \cdot P(5_2) \cdot P(5_3) \cdot P(5_4) = \left(\frac{1}{6}\right)\left(\frac{1}{6}\right)\left(\frac{1}{6}\right)\left(\frac{1}{6}\right) = \frac{1}{1296}$$

Therefore the probability that a 5 is obtained each time is $\frac{1}{1296}$.

Example 2.26 Three men in an office decide to enter a marathon race. The respective probabilities that they will complete the marathon are 0.9, 0.7 and 0.6. Find the probability that at least two will complete the marathon. Assume that the performance of each is independent of the performances of the others.

Solution 2.26 Let A be the event 'the first man completes the marathon', then $P(A) = 0.9$.

Let B be the event 'the second man completes the marathon', then $P(B) = 0.7$.

Let C be the event 'the third man completes the marathon', then $P(C) = 0.6$.

$P(\text{all complete the marathon}) = P(A \cap B \cap C)$. We will abbreviate $P(A \cap B \cap C)$ as $P(ABC)$. Then

$$\begin{aligned} P(\text{all complete the marathon}) &= P(ABC) \\ &= P(A)\cdot P(B)\cdot P(C) \quad \text{(independent events)} \\ &= (0.9)(0.7)(0.6) \\ &= 0.378 \end{aligned}$$

$$\begin{aligned} P(\text{two out of the three complete the marathon}) &= P(AB\bar{C}) + P(A\bar{B}C) + P(\bar{A}BC) \\ &= (0.9)(0.7)(0.4) + (0.9)(0.3)(0.6) + (0.1)(0.7)(0.6) \\ &= 0.456 \end{aligned}$$

$$\begin{aligned} P(\text{at least two complete the marathon}) &= 0.378 + 0.456 \\ &= 0.834 \end{aligned}$$

Therefore the probability that at least two complete the marathon is 0.834.

Exercise 2g

1. Three cards are drawn from a pack containing 52 playing cards. Find the probability that they are a heart, club and spade, in that order, if (*a*) the card is looked at and then replaced after each draw, (*b*) the card is not replaced after each draw.

2. A die is thrown three times. What is the probability of scoring a 2 on just one occasion?

3. A coin is tossed four times. Find the probability of obtaining less than two heads.

4. A box contains 4 black, 6 white and 2 red balls. Balls are picked out of the box without replacement. With obvious notation, find (*a*) $P(B_1 \cap W_2)$, (*b*) $P(W_2)$, (*c*) $P(B_1 \cup W_2)$, (*d*) $P(B_1 \cap W_2 \cap R_3)$, (*e*) P(the first three are different colours).

5. Of 24 boys in a class, 8 play rugby, 6 play hockey and 13 play soccer. One boy plays both soccer and rugby. Every boy plays at least one game but not one plays all three games. Two boys play both hockey and rugby. A boy is to be picked at random from the group.
(*a*) Draw a possibility space diagram to illustrate the situation.
(*b*) Calculate the probability of a boy being selected who (i) only plays hockey, (ii) plays both hockey and soccer.
(*c*) If S is the event 'a boy is chosen who plays soccer', H is the event 'a boy is chosen who plays hockey' and R is the event 'a boy is chosen who plays rugby', state, giving reasons for your answers, (i) two events which are independent, (ii) two events which are mutually exclusive. (L Additional)

6. In a lucky draw a first prize is given, then a second, then a third prize. 8 boys and 4 girls each buy one ticket. (*a*) Find the probability that (i) a girl has the first prize, a boy the second and a girl the third, (ii) the prizes go to 3 boys. (*b*) If the first and third prizes go to members of one sex and the second to a member of the opposite sex, find the probability that the second prize goes to a boy.

7. Three fair cubical dice are thrown. Find the probability that (i) the sum of the scores is 18, (ii) the sum of the scores is 5, (iii) none of the three dice shows a 6, (iv) the product of the scores is 90. (AEB 1974)

8. (*a*) A, B and C represent 3 events. If $A \cap B$ is the event that both A and B occur and $P(B|A)$ is the probability that B occurs given that A has already occurred, show that

$$P(A \cap B) = P(A) \cdot P(B|A)$$

Deduce, or show otherwise, that

$$P(A \cap B \cap C) = P(A) \cdot P(B|A) \cdot P(C|A \cap B)$$

(*b*) An athlete aims to measure his fitness by subjecting himself to a sequence of 3 physical tests, the completion of each test in a specified time being classed by him as a 'pass'. The probability that he passes the first test in the sequence is p, but the probability of passing any subsequent test is half the probability of passing the immediately preceding test. Show that, if the probability of passing all 3 tests is $\frac{1}{216}$ the value of p is $\frac{1}{3}$. Hence find the probabilities (i) that he fails all the tests, (ii) that he passes exactly 2 of the 3 tests. (MEI)

9. A, B and C are three events and, for example, $A \cup B$ denotes the event that either A or B or both A and B occur, $A \cap B$ denotes the event that both A and B occur, $\bar{A}$ is the event complementary to the event A. Find $Pr(\bar{A} \cap \bar{B} \cap \bar{C})$ given that

$$Pr(A) = \frac{11}{16}, \quad Pr(B) = \frac{9}{16}, \quad Pr(C) = \frac{19}{32},$$

$$Pr(A \cap B) = \frac{5}{16}, \quad Pr(A \cap C) = \frac{7}{16},$$

$$Pr(B \cap C) = \frac{3}{8}, \quad Pr(A|B \cap C) = \frac{5}{8}$$

(MEI)

PROBABILITY TREES

A useful way of tackling many probability problems is to draw a 'probability tree'. The method is illustrated in the following examples.

Example 2.27 A bag contains 8 white counters and 3 black counters. Two counters are drawn, one after the other. Find the probability of drawing one white and one black counter, in any order, (**a**) if the first counter is replaced, (**b**) if the first counter is not replaced.

Solution 2.27 (a) *With replacement* Let

W_1 be the event 'a white counter is drawn first',

W_2 be the event 'a white counter is drawn second',

B_1 be the event 'a black counter is drawn first',

B_2 be the event 'a black counter is drawn second'.

The results of the first draw and the second draw are shown on the 'tree' opposite. As the counter is replaced after the first draw the events along any one 'branch' of the tree are independent.

We have $$P(W_1 \cap W_2) = P(W_1) \cdot P(W_2)$$

We multiply terms as we go along the branch.

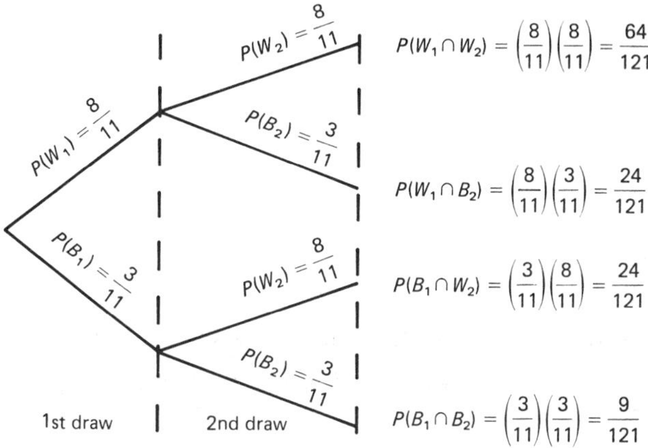

NOTE: these events are mutually exclusive, so check that the sum of the probabilities is 1.

$$\begin{aligned} P(\text{drawing one white and one black counter}) &= P(W_1 \cap B_2) \\ &\quad + P(B_1 \cap W_2) \\ &= \frac{24}{121} + \frac{24}{121} \\ &= \frac{48}{121} \end{aligned}$$

The probability of drawing one black and one white counter if the counter is replaced after the first draw is $\frac{48}{121}$.

(b) *Without replacement* The events along one branch are no longer independent, but we may still multiply as we go along the branch, using the fact

$$P(W_1 \cap W_2) = P(W_1) \cdot P(W_2 | W_1) \quad \text{and so on}$$

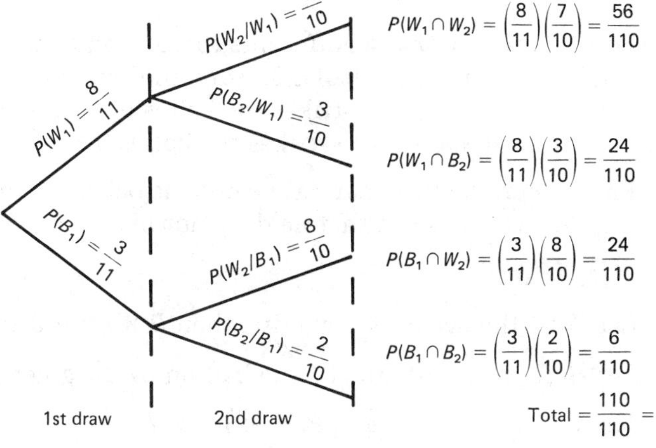

$$P(\text{drawing one white and one black counter}) = P(W_1 \cap B_2) + P(B_1 \cap W_2)$$
$$= \frac{24}{110} + \frac{24}{110}$$
$$= \frac{48}{110}$$
$$= \frac{24}{55}$$

The probability of drawing one black and one white counter if the counter is not replaced after the first draw is $\frac{24}{55}$.

NOTE: the diagram can be made simpler if, instead of writing $P(W_2 | W_1) = \frac{7}{10}$ on the second branch, we write $P(W_2) = \frac{7}{10}$, as the diagram makes it clear that event W_1 *has already occurred.* The diagram then becomes:

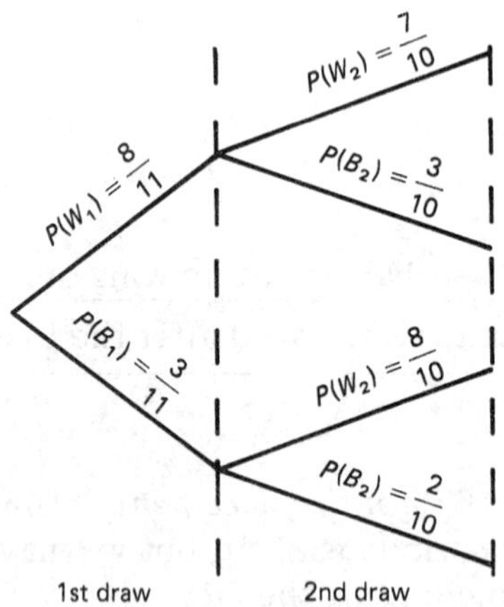

Example 2.28 The probability that a golfer hits the ball on to the green if it is windy as he strikes the ball is 0.4, and the corresponding probability if it is not windy as he strikes the ball is 0.7. The probability that the wind will blow as he strikes the ball is 0.3.

Find the probability that **(a)** he hits the ball on to the green, **(b)** it was not windy, given that he does not hit the ball on to the green.

Solution 2.28 Let W be the event 'it is windy', then $P(W) = 0.3$ and $P(\overline{W}) = 0.7$.

Let H be the event 'he hits the ball on to the green'.

Then $P(H | W) = 0.4$ and $P(H | \overline{W}) = 0.7$.

We can draw a probability tree as follows:

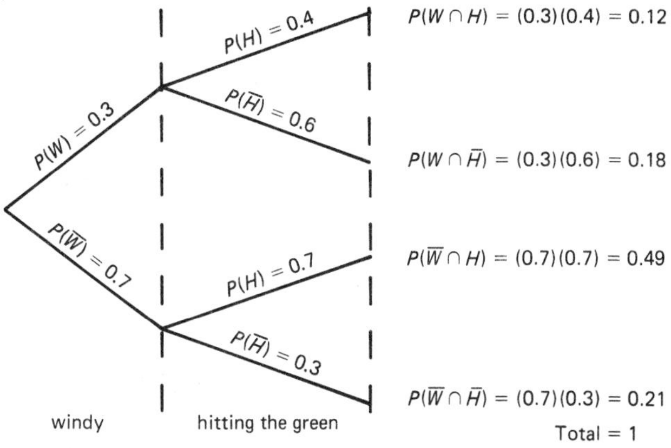

$P(W) = 0.3$, $P(\overline{W}) = 0.7$

$P(H) = 0.4$, $P(\overline{H}) = 0.6$

$P(H) = 0.7$, $P(\overline{H}) = 0.3$

$P(W \cap H) = (0.3)(0.4) = 0.12$

$P(W \cap \overline{H}) = (0.3)(0.6) = 0.18$

$P(\overline{W} \cap H) = (0.7)(0.7) = 0.49$

$P(\overline{W} \cap \overline{H}) = (0.7)(0.3) = 0.21$

Total = 1

windy hitting the green

(a) We require
$$
\begin{aligned}
P(H) &= P(H \cap W) + P(H \cap \overline{W}) \\
&= 0.12 + 0.49 \\
&= 0.61
\end{aligned}
$$

The probability that he hits the ball on to the green is 0.61.

(b) We require
$$
P(\overline{W} \mid \overline{H}) = \frac{P(\overline{W} \cap \overline{H})}{P(\overline{H})}
$$

Now
$$
\begin{aligned}
P(\overline{H}) &= 1 - P(H) \\
&= 1 - 0.61 \\
&= 0.39
\end{aligned}
$$

So
$$
\begin{aligned}
P(\overline{W} \mid \overline{H}) &= \frac{0.21}{0.39} \\
&= 0.54 \quad (2\,\text{d.p.})
\end{aligned}
$$

The probability that it was not windy, given that he does not hit the ball on to the green, is 0.54 (2 d.p.).

Example 2.29 Events A and B are such that $P(A) = \frac{1}{3}$, $P(B \mid A) = \frac{1}{4}$ and $P(\overline{B} \mid \overline{A}) = \frac{4}{5}$.

By drawing a tree diagram, or otherwise, find (a) $P(\overline{B} \mid A)$, (b) $P(A \cap B)$, (c) $P(B)$, (d) $P(A \cup B)$.

Solution 2.29 If we draw a tree diagram and put on it the information given, we have

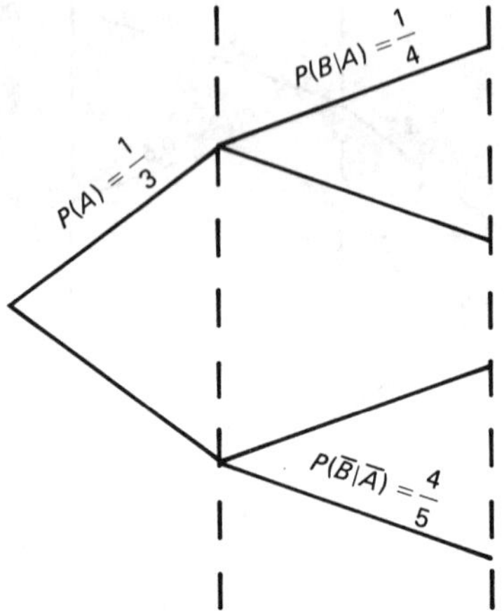

From the diagram, it is obvious that $P(\bar{A}) = \frac{2}{3}$, $P(\bar{B}\,|\,A) = \frac{3}{4}$ and $P(B\,|\,\bar{A}) = \frac{1}{5}$ as the 'total' probability for each set of branches is 1.

The completed tree diagram is

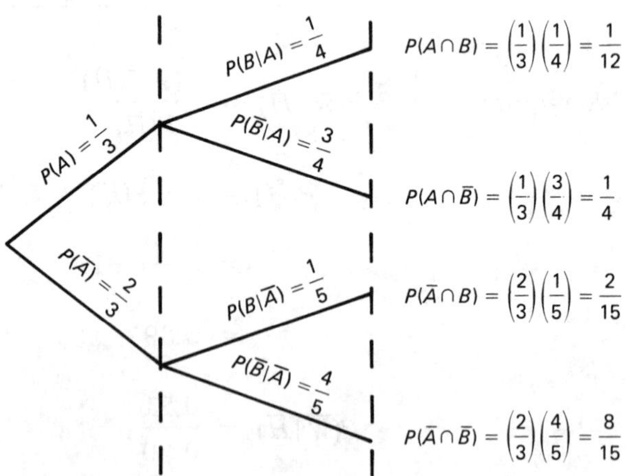

(a) $P(\bar{B}\,|\,A) = \dfrac{3}{4}$

(b) $P(A \cap B) = \dfrac{1}{12}$

(c) $P(B) = P(B \cap A) + P(B \cap \bar{A}) = \dfrac{1}{12} + \dfrac{2}{15} = \dfrac{13}{60}$

(d) $P(A \cup B) = 1 - P(\bar{A} \cap \bar{B}) = 1 - \dfrac{8}{15} = \dfrac{7}{15}$

Example 2.30 A fair coin is tossed three times. What is the probability of obtaining (a) exactly two heads, (b) at least two heads?

Solution 2.30 We now extend the tree to include three events.
Let H_1 be the event 'a head occurs on the first toss', T_2 be the event 'a tail occurs on the second toss', and so on.

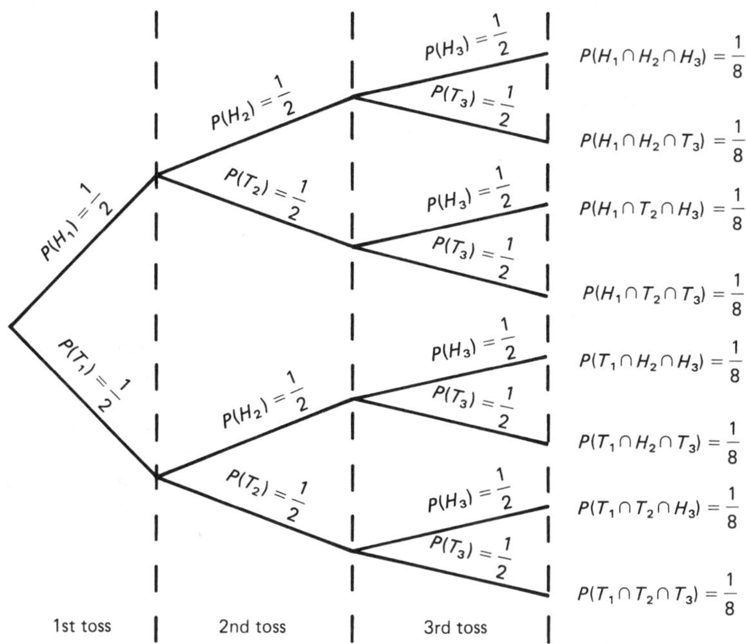

(a)
$$\begin{aligned} P(\text{exactly two heads}) &= P(H_1 \cap H_2 \cap T_3) + P(H_1 \cap T_2 \cap H_3) \\ &\quad + P(T_1 \cap H_2 \cap H_3) \\ &= \frac{1}{8} + \frac{1}{8} + \frac{1}{8} \\ &= \frac{3}{8} \end{aligned}$$

The probability that exactly two heads are obtained is $\frac{3}{8}$.

(b)
$$\begin{aligned} P(\text{at least two heads}) &= P(\text{two heads and a tail}) \\ &\quad + P(\text{three heads}) \\ &= \frac{3}{8} + \frac{1}{8} \\ &= \frac{1}{2} \end{aligned}$$

The probability that at least two heads are obtained is $\frac{1}{2}$.

Exercise 2h

1. The probability that a biased die falls showing a 6 is $\frac{1}{5}$. This biased die is thrown twice.
(*a*) Draw a tree diagram showing the possible outcomes and the corresponding probabilities, considering the event 'a six is thrown'.
(*b*) Find the probability that exactly one six will be obtained.
An unbiased die is now thrown.
(*c*) Extend the tree diagram to show the possible outcomes, again with regard to whether or not a 6 is thrown.
(*d*) Find the probability that, in the three throws, exactly one 6 will be obtained.

2. In a class of 24 girls, 7 have black hair.
(*a*) If 2 girls are chosen at random from the class, find the probability that (i) they both have black hair, (ii) neither has black hair.
(*b*) If 3 girls are chosen at random, find the probability that more than 1 will have black hair.

3. A coin is biased so that the probability that it lands showing heads is $\frac{2}{3}$. The coin is tossed three times. Find the probability that (*a*) no heads are obtained, (*b*) more heads than tails are obtained.

4. A box contains 6 red pens and 3 blue pens.
(*a*) A pen is selected at random, the colour is noted and the pen is returned to the box. This procedure is performed a second, then a third time. Find the probability of obtaining (i) 3 red pens, (ii) 2 red pens and 1 blue pen, in any order, (iii) more than 1 blue pen.
(*b*) Repeat (*a*) but this time find the probabilities if, at each selection, the pen is not returned to the box.

5. In each round of a certain game a player can score 1, 2 or 3 only. Copy and complete the table which shows the scores and two of the respective probabilities of these being scored in a single round.

Score	1	2	3
Probability	$\frac{4}{7}$		$\frac{1}{7}$

Draw a tree diagram to show all the possible total scores and their respective probabilities after a player has completed two rounds.
Find the probability that a player has (*a*) a score of 4 after 2 rounds, (*b*) an odd number score after 2 rounds.
(L Additional)

6. Three bags, *A*, *B* and *C* contain counters as follows:

	Red	Yellow
Bag *A*	4	3
Bag *B*	3	6
Bag *C*	2	4

(*a*) A counter is taken at random from each of the bags in turn and kept. Draw a tree diagram to show the possible outcomes and find the probability that more red counters than yellow counters are kept.
(*b*) The counters are now replaced in their original bags. A counter is taken at random from bag *A* and placed in bag *B*. Then a counter is taken at random from bag *B* and placed in bag *C*. What is the probability that a counter now taken from bag *C* is yellow?

7. Events *X* and *Y* are such that $P(\overline{X}) = \frac{3}{5}$, $P(Y|\overline{X}) = \frac{1}{3}$ and $P(\overline{Y}|X) = \frac{1}{4}$. By drawing a tree diagram, or otherwise, find (*a*) $P(Y)$ (*b*) $P(X \cap Y)$, (*c*) $P(X \cup Y)$.

8. A mother and her daughter both enter the cake competition at a show. The probability that the mother wins a prize is $\frac{1}{6}$ and the probability that her daugher wins a prize is $\frac{2}{7}$. Assuming that the two events are independent, find the probability that (*a*) either the mother, or the daughter, but not both, wins a prize, (*b*) at least one of them wins a prize.

9. A bag contains 7 black and 3 white marbles. Three marbles are chosen at random and in succession, each marble being replaced after it has been taken out of the bag.

Draw a tree diagram to show all possible selections.

From your diagram, or otherwise, calculate, to 2 significant figures, the probability of choosing (*a*) three black marbles, (*b*) a white marble, a black marble and a white marble in that order, (*c*) two white marbles and a black marble in any order, (*d*) at least one black marble.

State an event from this experiment which together with the event described in (*d*) would be both exhaustive and mutually exclusive. (L Additional)

BAYES' THEOREM

Now we come to an important extension of the result

$$P(A \mid B) = \frac{P(B \mid A) \cdot P(A)}{P(B)}$$

Suppose $A_1, A_2, \ldots, A_n$ are n mutually exclusive and exhaustive events so that $A_1 \cup A_2 \cup \ldots \cup A_n = S$, the possibility space, and B is an arbitrary event of S, then

$$P(A_i \mid B) = \frac{P(B \mid A_i) \cdot P(A_i)}{P(B \mid A_1) \cdot P(A_1) + P(B \mid A_2) \cdot P(A_2) + \ldots + P(B \mid A_n) \cdot P(A_n)}$$
$$\text{for } i = 1, 2, \ldots, n$$

This is known as Bayes' theorem and is useful when we have to 'reverse the conditions' in a problem.

Proof

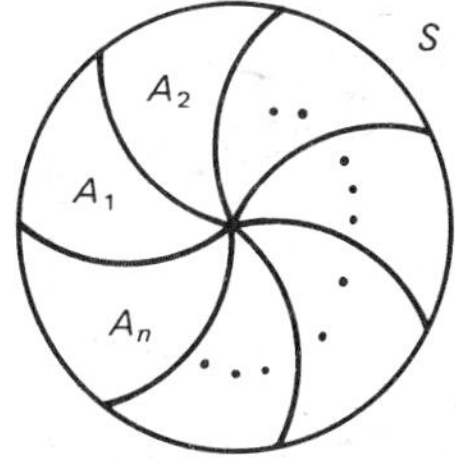

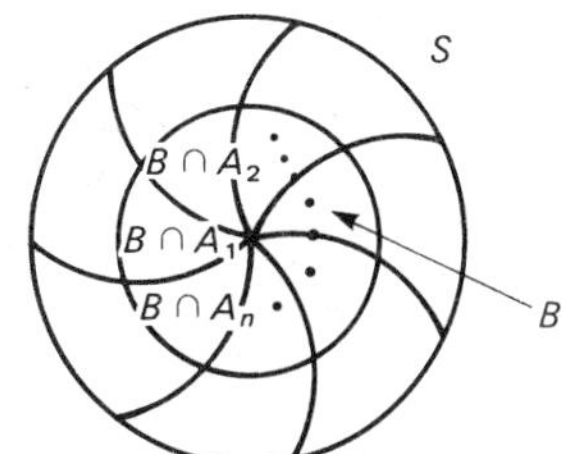

$$P(B) = P(B \cap A_1) + P(B \cap A_2) + \ldots + P(B \cap A_n)$$
$$= P(B \mid A_1) \cdot P(A_1) + P(B \mid A_2) \cdot P(A_2) + \ldots + P(B \mid A_n) \cdot P(A_n)$$

Now

$$P(A_i \mid B) = \frac{P(B \mid A_i) \cdot P(A_i)}{P(B)}$$

so

$$P(A_i \mid B) = \frac{P(B \mid A_i) \cdot P(A_i)}{P(B \mid A_1) \cdot P(A_1) + P(B \mid A_2) \cdot P(A_2) + \ldots + P(B \mid A_n) \cdot P(A_n)}$$

The formula looks very complicated, but in fact it is easy to use if you remember that the denominator is the total probability of B.

Example 2.31 Three girls, Aileen, Barbara and Cathy, pack biscuits in a factory. From the batch allotted to them Aileen packs 55%, Barbara 30% and Cathy 15%. The probability that Aileen breaks some biscuits in a packet is 0.7, and the respective probabilities for Barbara and Cathy are 0.2 and 0.1. What is the probability that a packet with broken biscuits found by the checker was packed by Aileen?

Solution 2.31 Let A be the event 'the packet was packed by Aileen', B be the event 'the packet was packed by Barbara', C be the event 'the packet was packed by Cathy', D be the event 'the packet contains broken biscuits'.

We are given $P(A) = 0.55$, $P(B) = 0.3$, $P(C) = 0.15$ and $P(D \mid A) = 0.7$, $P(D \mid B) = 0.2$, $P(D \mid C) = 0.1$.

We require $P(A \mid D)$. So we use Bayes' theorem to 'reverse the conditions':

$$P(A \mid D) = \frac{P(D \mid A) \cdot P(A)}{P(D)}$$

Now $P(D)$ is the 'total' probability of D, that is the probability that a packet contains broken biscuits. This can be found very easily from the tree diagram. The outcomes resulting in a packet with broken biscuits are shown with an asterisk.

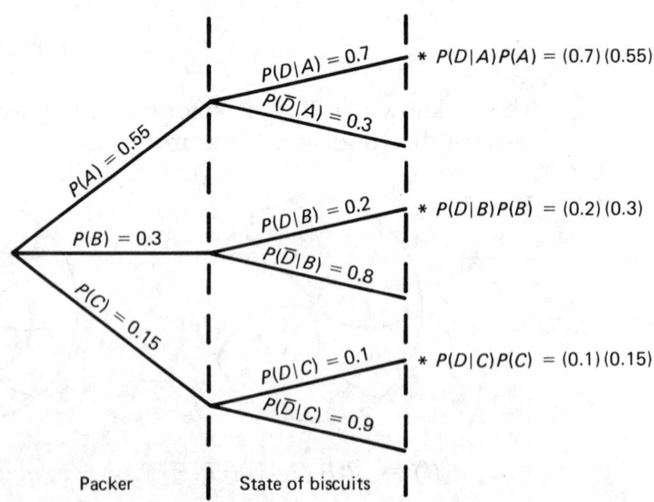

$$\begin{aligned} P(D) &= P(D \mid A) \cdot P(A) + P(D \mid B) \cdot P(B) + P(D \mid C) \cdot P(C) \\ &= (0.7)(0.55) + (0.2)(0.3) + (0.1)(0.15) \\ &= 0.46 \end{aligned}$$

As shown in the tree diagram

$$P(D|A)P(A) = (0.7)(0.55)$$

Therefore

$$\begin{aligned} P(A \mid D) &= \frac{(0.7)(0.55)}{0.46} \\ &= 0.837 \quad (3\text{ d.p.}) \end{aligned}$$

The probability that a packet with broken biscuits was packed by Aileen is 0.837 (3 d.p.).

Example 2.32 Three children, Catherine, Michael and David, have equal plots in a circular patch of garden. The boundaries are marked out by pebbles. Catherine has 80 red and 20 white flowers in her patch, Michael has 30 red and 40 white flowers and David has 10 red and 60 white flowers. Their young sister, Mary, wants to pick a flower for her teacher.

(a) Find the probability that she picks a red flower if she chooses a flower at random from the garden, ignoring the boundaries.

(b) Find the probability that she picks a red flower if she first chooses a plot at random.

(c) If she picks a red flower by the method described in (b), find the probability that it came from Michael's plot.

Solution 2.32 (a) *If the boundaries are ignored*

The possibility space S = (flowers in the garden).

We have $n(S) = 100 + 70 + 70 = 240$

Let R be the event 'a red flower is chosen',

then $n(R) = 80 + 30 + 10 = 120$

Then
$$P(R) = \frac{n(R)}{n(S)} = \frac{120}{240} = \frac{1}{2}$$

The probability that Mary picks a red flower if she ignores the boundaries is $\frac{1}{2}$.

(b) *A plot is chosen first.* Each of the three plots is equally likely to be chosen.

Let C be the event 'Catherine's plot is chosen', then $P(C) = \frac{1}{3}$.

With similar notation $P(M) = \frac{1}{3}$ and $P(D) = \frac{1}{3}$.

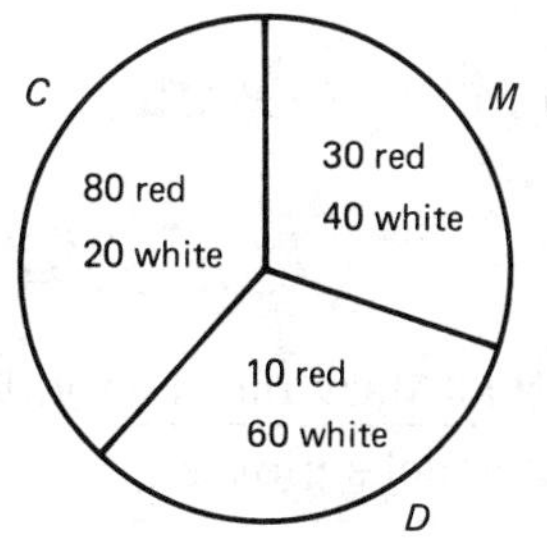

The outcomes resulting in event R are shown with an asterisk on the tree diagram.

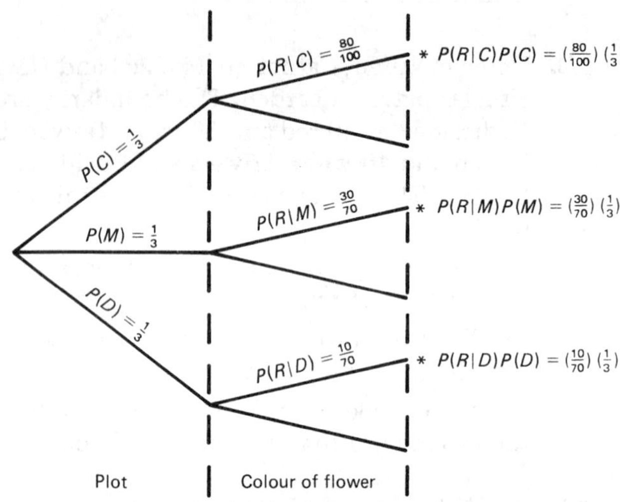

$$\text{Now} \quad P(R) = P(R \mid C) \cdot P(C) + P(R \mid M) \cdot P(M) + P(R \mid D) \cdot P(D)$$

$$= \left(\frac{80}{100}\right)\left(\frac{1}{3}\right) + \left(\frac{30}{70}\right)\left(\frac{1}{3}\right) + \left(\frac{10}{70}\right)\left(\frac{1}{3}\right)$$

$$= \frac{16}{35}$$

The probability that Mary picks a red flower if she chooses a plot at random first is $\frac{16}{35}$.

NOTE: the two different results for part **(a)** and part **(b)** are slightly surprising. In the first case, there is one group of flowers and each flower is equally likely to be chosen. In the second case, even though each plot is equally likely to be chosen, the proportions of red and white flowers within these plots are different.

(c) Using Bayes' theorem:

$$P(M \mid R) = \frac{P(R \mid M) \cdot P(M)}{P(R)}$$

Now $P(R|M) \cdot P(M) = (\frac{30}{70})(\frac{1}{3}) = \frac{1}{7}$ (from tree diagram)

and $P(R) = \frac{16}{35}$ (from part **(b)**)

$$\text{Therefore} \quad P(M|R) = \frac{\frac{1}{7}}{\frac{16}{35}}$$

$$= \frac{5}{16}$$

Given that Mary picks a red flower, the probability that it came from Michael's plot is $\frac{5}{16}$.

Exercise 2i

1. In my bookcase there are four shelves and the number of books on each shelf is as shown in the table:

	Hardback	Paperback
Shelf 1	11	9
Shelf 2	8	12
Shelf 3	16	4
Shelf 4	9	3

(*a*) If I choose a book at random, irrespective of its position in the bookcase, what is the probability that it is a paperback?
(*b*) I am equally likely to choose any shelf. I choose a shelf at random and then choose a book. (i) What is the probability that it is a hardback? (ii) If the book chosen is a hardback, what is the probability that it is from shelf 3?

2. I travel to work by route A or route B. The probability that I choose route A is $\frac{1}{4}$. The probability that I am late for work if I go via route A is $\frac{2}{3}$ and the corresponding probability if I go via route B is $\frac{1}{3}$.
(*a*) What is the probability that I am late for work on Monday?
(*b*) Given that I am late for work, what is the probability that I went via route B?

3. Of the buses leaving the bus station each day, 60% are double deckers and the rest are single deckers; 30% of the double deckers are 'limited stop' buses and 40% of the single deckers are 'limited stop' buses. Draw a tree diagram to represent this information.

Find the probability that a bus leaving the bus station (*a*) is not a 'limited stop' bus, (*b*) is a double decker, given that it is a 'limited stop' bus.

4. Susan takes examinations in mathematics, French and history. The probability that she passes mathematics is 0.7 and the corresponding probabilities for French and history are 0.8 and 0.6. Given that her performances in each subject are independent, draw a tree diagram to show the possible outcomes.

Find the probability that Susan (*a*) fails all three examinations, (*b*) fails just one examination. Given that Susan fails just one examination, (*c*) find the probability that she fails history.

5.

Age group	0–16	17–25	26–64	65 and over
Percentage	25	14	43	18

In a town the percentage of males in particular age groups are as shown in the table above. In a survey it is found that the probability that a male aged 65 or over wearing glasses is 0.8. Similarly if a male is in the age group 0–16, 17–25 and 26–64 the probabilities of his wearing glasses are 0.2, 0.1 and 0.4 respectively. Given that a particular male is wearing glasses, use Bayes' formula

$$P(A_k \mid B) = \frac{P(A_k)\cdot P(B \mid A_k)}{\sum_{r=1}^{n} P(A_r)\cdot P(B \mid A_r)}$$

to calculate the probability that he is 65 or over.

Similarly calculate the probabilities that he is in each of the other age groups and hence state which age group he is most likely to be in. (L Additional)

6. In an experiment two bags A and B, containing red and green marbles are used. Bag A contains 4 red marbles and 1 green marble and bag B contains 2 red marbles and 7 green marbles. An unbiased coin is tossed. If a head turns up, a marble is drawn at random from bag A while if a tail turns up, a marble is drawn at random from bag B. Calculate the probability that a red marble is drawn in a single trial. Given that a red marble is selected, calculate the probability that when the coin was tossed a head was obtained. (L Additional)

7. On a given day a petrol station serves three times as many men as women.

Two types of petrol are available, grade A and grade B. Customers pay by cheque or by cash.

70% of the men and 40% of the women buy grade A petrol.

Of the men buying grade A petrol, 80% pay by cheque, and of the men buying grade B petrol, 60% pay by cheque.

Of the women buying grade A petrol, half pay by cheque and of the women buying grade B petrol, 40% pay by cheque.

Find the probability that (a) a customer buys grade A petrol, (b) a customer pays by cheque, (c) a woman customer pays by cheque, (d) a customer who pays by cheque for grade A petrol is a man.

8. A bag contains 10 counters of which 4 are pink, 3 are green and 3 are yellow. Counters are removed at random, one at a time and without replacement. Find the probability that (a) the first one drawn is green, (b) the first two drawn are both the same colour, (c) the first three drawn are of different colours, (d) the second one drawn is green, given that the first one drawn is pink, (e) the third one drawn is green, given that the first two are the same colour as each other.

9. A and B are two events for which $P(A) = \frac{2}{5}$, $P(\bar{B} \mid A) = \frac{3}{4}$, and $P(B \mid \bar{A}) = \frac{1}{3}$.
(a) Draw a fully labelled tree diagram with A preceding B, that is with A and $\bar{A}$ on the first two branches.
(b) Calculate (i) $P(A \cap B)$, (ii) $P(\bar{A} \cap B)$, (iii) $P(A \cup B)$.
(c) Draw a possibility space diagram to illustrate both the given data and your answers.
(d) Calculate (i) $P(A \mid B)$, (ii) $P(\bar{A} \mid \bar{B})$.
(e) Use your answers to (d) to draw a fully labelled tree diagram with B preceding A. (L Additional)

10.

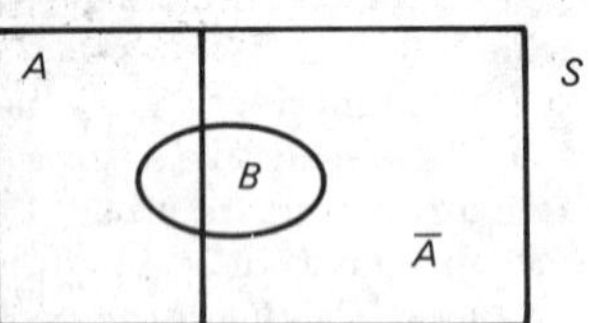

(i) By considering the diagram which represents the sample space S, for $A \cup \bar{A}$, where B is an arbitrary event of S such that $P(B) \neq 0$, show that

$$P(A \mid B) = \frac{P(A) \cdot P(B \mid A)}{P(A) \cdot P(B \mid A) + P(\bar{A}) \cdot P(B \mid \bar{A})}$$

(ii) The probabilities that a boy goes to school by bus, bicycle or on foot on a certain day are 0.2, 0.3 and 0.5 respectively. The probabilities of his being late by these methods are 0.6, 0.3 and 0.1 respectively. If he was late on this particular day, using Bayes' theorem or otherwise, calculate the probability that he travelled by bus. (L Additional)

11. If A_1, A_2 and A_3 are mutually exclusive events whose union is the sample space S of an experiment and B is an arbitrary event of S such that $P(B) \neq 0$, show that

$$P(A_1 \mid B) = \frac{P(A_1) \cdot P(B \mid A_1)}{\sum_{r=1}^{3} P(A_r) \cdot P(B \mid A_r)}$$

and write down the results for $P(A_2 \mid B)$ and $P(A_3 \mid B)$.

A factory has three machines 1, 2 and 3, producing a particular type of item. One item is drawn at random from the factory's production. Let B denote the event that the chosen item is defective and let A_k denote the event that the item was produced on machine k where $k = 1, 2$ or 3. Suppose that machines 1, 2 and 3 produce respectively 35%, 45% and 20% of the total production of items and that $P(B \mid A_1) = 0.02$, $P(B \mid A_2) = 0.01$, $P(B \mid A_3) = 0.03$.

Given that an item chosen at random is defective, find which machine was most likely to have produced it. (L Additional)

SOME USEFUL METHODS

(a) Problems involving an 'at least' situation

Example 2.33 (a) Find the probability of obtaining at least one 6 when 5 dice are thrown.

(b) Find the probability of obtaining at least one 6 when n dice are thrown.

(c) How many dice must be thrown so that the probability of obtaining at least one 6 is at least 0.99?

Solution 2.33 (a) In one throw $P(6) = \frac{1}{6}$ and $P(\bar{6}) = \frac{5}{6}$.

When 5 dice are thrown,

$$\begin{aligned} P(\text{at least one } 6) &= 1 - P(\text{no 6s}) \\ &= 1 - P(\bar{6}\bar{6}\bar{6}\bar{6}\bar{6}) \\ &= 1 - \left(\frac{5}{6}\right)^5 \\ &= 0.598 \quad (3 \text{ d.p.}) \end{aligned}$$

The probability of obtaining at least one 6 when 5 dice are thrown is 0.598 (3 d.p.).

(b) When n dice are thrown, $P(\text{at least one } 6) = 1 - (\frac{5}{6})^n$.

(c) We require n such that

$$1 - \left(\frac{5}{6}\right)^n \geqslant 0.99$$

i.e.

$$\left(\frac{5}{6}\right)^n \leqslant 0.01$$

Taking logs of both sides

$$n \log\left(\frac{5}{6}\right) \leqslant \log(0.01)$$

Dividing both sides by $\log(\frac{5}{6})$ and reversing the inequality sign since $\log(\frac{5}{6})$ is negative, we have

$$n \geqslant \frac{\log(0.01)}{\log(\frac{5}{6})}$$

$$n \geqslant 25.3 \quad (1 \text{ d.p.})$$

so

$$\text{least } n = 26$$

Therefore 26 dice must be thrown so that the probability of obtaining at least one 6 is at least 0.99.

(b) Problems involving the use of an infinite geometric progression (G.P.)

Many probability examples involve the use of G.P.s and the following formula is required.

If $S = a + ar + ar^2 + ar^3 + \ldots$ (to infinity)

Then

$$S = \frac{a}{1-r} \quad \text{for } |r| < 1$$

where a is the first term
r is the common ratio

Example 2.34 A, B and C, in that order, throw a tetrahedral die. The first one to throw a 4 wins. The game is continued indefinitely until someone wins. Find the probability that **(a)** A wins, **(b)** B wins, **(c)** C wins.

Solution 2.34 With a tetrahedral die the number 'thrown' is the number on which the die lands. Therefore $P(4 \text{ is thrown}) = \frac{1}{4}$.

(a) Let A_1 be the event 'A wins on his first throw', A_2 be the event 'A wins on his second throw', and so on.

Now

$$P(A \text{ wins}) = P(A_1) + P(A_2) + P(A_3) + \dots \quad \text{(mutually exclusive events)}$$

$$P(A_1) = \frac{1}{4} \quad \text{and} \quad P(\bar{A}_1) = \frac{3}{4}$$

$$P(A_2) = P(\bar{A}_1\bar{B}_1\bar{C}_1A_2) = \left(\frac{3}{4}\right)^3\left(\frac{1}{4}\right)$$

$$P(A_3) = P(\bar{A}_1\bar{B}_1\bar{C}_1\bar{A}_2\bar{B}_2\bar{C}_2A_3) = \left(\frac{3}{4}\right)^6\left(\frac{1}{4}\right) \quad \text{and so on}$$

Therefore

$$\begin{aligned}
P(A \text{ wins}) &= \left(\frac{1}{4}\right) + \left(\frac{3}{4}\right)^3\left(\frac{1}{4}\right) + \left(\frac{3}{4}\right)^6\left(\frac{1}{4}\right) + \dots \\
&= \left(\frac{1}{4}\right)\left[1 + \left(\frac{3}{4}\right)^3 + \left(\frac{3}{4}\right)^6 + \dots\right] \\
&= \left(\frac{1}{4}\right)S \quad \text{where } S \text{ is the sum of an infinite G.P. with } a = 1 \text{ and } r = \left(\frac{3}{4}\right)^3 \\
&= \left(\frac{1}{4}\right)\left(\frac{1}{1-(\frac{3}{4})^3}\right) \\
&= \left(\frac{1}{4}\right)\left(\frac{1}{1-\frac{27}{64}}\right) \\
&= \frac{16}{37}
\end{aligned}$$

The probability that A wins is $\frac{16}{37}$.

(b)

$$P(B \text{ wins}) = P(B_1) + P(B_2) + P(B_3) + \dots \quad \text{to infinity (mutually exclusive events)}$$

Now $P(B_1) = P(\bar{A}_1B_1) = \left(\frac{3}{4}\right)\left(\frac{1}{4}\right)$

$$P(B_2) = P(\bar{A}_1\bar{B}_1\bar{C}_1\bar{A}_2B_2) = \left(\frac{3}{4}\right)^4\left(\frac{1}{4}\right)$$

$$P(B_3) = P(\bar{A}_1\bar{B}_1\bar{C}_1\bar{A}_2\bar{B}_2\bar{C}_2\bar{A}_3B_3) = \left(\frac{3}{4}\right)^7\left(\frac{1}{4}\right) \quad \text{and so on}$$

So

$$\begin{aligned} P(B \text{ wins}) &= \left(\frac{1}{4}\right)\left(\frac{3}{4}\right)+\left(\frac{1}{4}\right)\left(\frac{3}{4}\right)^4+\left(\frac{1}{4}\right)\left(\frac{3}{4}\right)^7+\dots \\ &= \left(\frac{1}{4}\right)\left(\frac{3}{4}\right)\left[1+\left(\frac{3}{4}\right)^3+\left(\frac{3}{4}\right)^6+\dots\right] \\ &= \left(\frac{3}{16}\right)\left(\frac{1}{1-(\frac{3}{4})^3}\right) \\ &= \frac{12}{37} \end{aligned}$$

Therefore the probability that B wins is $\frac{12}{37}$.

(c)

$$\begin{aligned} P(C \text{ wins}) &= 1-P(A \text{ wins})-P(B \text{ wins}) \\ &= 1-\frac{28}{37} \\ &= \frac{9}{37} \end{aligned}$$

Therefore the probability that C wins is $\frac{9}{37}$.

Exercise 2j

1. A coin is biased so that the probability that it falls showing tails is $\frac{3}{4}$.
(*a*) Find the probability of obtaining at least one head when the coin is tossed five times.
(*b*) How many times must the coin be tossed so that the probability of obtaining at least one head is greater than 0.98?

2. A missile is fired at a target and the probability that the target is hit is 0.7.
(*a*) Find how many missiles should be fired so that the probability that the target is hit at least once is greater than 0.995.
(*b*) Find how many missiles should be fired so that the probability that the target is not hit is less than 0.001.

3. A die is biased so that the probability of obtaining a 3 is p. When the die is thrown four times the probability that there is at least one 3 is 0.9375. Find the value of p.
How many times should the die be thrown so that the probability that there are no threes is less than 0.03?

4. On a safe there are four alarms which are arranged so that any one will sound when someone tries to break into the safe. If the probability that each alarm will function properly is 0.85, find the probability that at least one alarm will sound when someone tries to break into the safe.

5. Two people, A and B, play a game. An

ordinary die is thrown and the first person to throw a 4 wins. A and B take it in turns to throw the die, starting with A. Find the probability that B wins.

6. A, B, C and D throw a coin, in turn, starting with A. The first to throw a head wins. The game can continue indefinitely until a head is thrown. However, D objects because the others have their first turn before him. Compare the probability that D wins with the probability that A wins.

7. A box contains five black balls and one white ball. Alan and Bill take turns to draw a ball from the box, starting with Alan. The first boy to draw the white ball wins the game. Assuming that they do not replace the balls as they draw them out, find the probability that Bill wins the game.
If the game is changed, so that, in the new game, they replace each ball after it has been drawn out, find the probabilities that:
(*a*) Alan wins at his first attempt;
(*b*) Alan wins at his second attempt;
(*c*) Alan wins at his third attempt.
Show that these answers are terms in a Geometric Progression. Hence find the probability that Alan wins the new game.
(SUJB Additional)

ARRANGEMENTS

In order to calculate the number of possible outcomes in a possibility space and the number of sample points for an event, the following results are often used.

Result 1 The number of ways of arranging n unlike objects in a line is $n!$

NOTE: $n! = n(n-1)(n-2)\dots(3)(2)(1)$.

For example, consider the letters A, B, C, D.

The first letter can be chosen in 4 ways (either A or B or C or D),

Then

the second letter can be chosen in 3 ways,
the third letter can be chosen in 2 ways,
the fourth letter can be chosen in only 1 way.

Therefore the number of ways of arranging the 4 letters is $(4)(3)(2)(1) = 4! = 24$.

These are

ABCD	ABDC	ACBD	ACDB	ADCB	ADBC
BCDA	BCAD	BDAC	BDCA	BACD	BADC
CDBA	CDAB	CABD	CADB	CBAD	CBDA
DABC	DACB	DBCA	DBAC	DCAB	DCBA

Example 2.35 How many different number plates can be formed if each is to contain the three letters A, C and E followed by the three digits 4, 7, 8?

Solution 2.35 There are 3! ways of arranging the letters A, C and E, and 3! ways of arranging the digits 4, 7 and 8.

Therefore the total number of different plates $= (3!)(3!)$

$= 36$

36 different number plates can be formed.

Result 2 The number of ways of arranging in a line n objects, of which p are alike, is $\dfrac{n!}{p!}$.

For example, if, instead of the letters A, B, C, D we have the letters A, A, A, D then the 24 arrangements listed previously reduce to the following:

AAAD AADA ADAA DAAA

So, the number of ways of arranging the 4 objects, of which 3 are alike $= \dfrac{4!}{3!} = \dfrac{(4)(3)(2)(1)}{(3)(2)(1)} = 4$.

The result can be extended as follows:

The number of ways of arranging in a line n objects of which p of one type are alike, q of a second type are alike, r of a third type are alike, and so on, is $\dfrac{n!}{p!q!r!\ldots}$.

Example 2.36 (a) In how many ways can the letters of the word STATISTICS be arranged?

(b) If the letters of the word MINIMUM are arranged in a line at random, what is the probability that the three M's are together at the beginning of the arrangement?

Solution 2.36 (a) Consider the word STATISTICS.

There are 10 letters and S occurs 3 times,
T occurs 3 times,
I occurs twice.

Therefore

$$\text{number of ways} = \frac{10!}{3!3!2!}$$

$$= \frac{(10)(9)(8)(7)(6)(5)(4)(3)(2)(1)}{(3)(2)(1)(3)(2)(1)(2)(1)}$$

$$= 50\,400$$

There are 50 400 ways of arranging the letters in the word STATISTICS.

(b) Consider the word MINIMUM.

The possibility space S = (arrangements of MINIMUM).

Now $$n(S) = \frac{7!}{3!2!}$$

$$= \frac{(7)(6)(5)(4)(3)(2)(1)}{(3)(2)(1)(2)(1)}$$

$$= 420$$

Let E be the event 'the three M's are placed together at the beginning of the arrangement'.

So we must have MMM.... There is only one way of arranging MMM, then the remaining 4 letters can be arranged in $\frac{4!}{2!}$ ways = 12 ways.

Therefore $$n(E) = 12$$

So $$P(E) = \frac{n(E)}{n(S)}$$

$$= \frac{12}{420}$$

$$= \frac{1}{35}$$

The probability that the three M's are together at the beginning of an arrangement is $\frac{1}{35}$.

Example 2.37 Ten pupils are placed at random in a line. What is the probability that the two youngest pupils are separated?

Solution 2.37 Let the possibility space be S, then $n(S) = 10!$

Let E be the event 'the two youngest pupils are together'.

Now treat these two together as one 'item' and so we have 9 'items' to arrange.

The 9 items can be arranged in 9! ways.

But the two youngest can be arranged in 2! ways ($Y_1 Y_2$ or $Y_2 Y_1$).

Therefore $n(E) = 2!9!$

So
$$P(E) = \frac{n(E)}{n(S)} = \frac{2!9!}{10!} = \frac{2}{10} = \frac{1}{5}$$

Now $\overline{E}$ is the event 'the two youngest are not together'.

So
$$P(\overline{E}) = 1-P(E) = 1-\tfrac{1}{5} = \tfrac{4}{5}$$

The probability that the two youngest are separated is $\frac{4}{5}$.

Example 2.38 If a four-digit number is formed from the digits 1, 2, 3 and 5 and repetitions are not allowed, find the probability that the number is divisible by 5.

Solution 2.38 Let S be the possibility space, then $n(S) = 4! = 24$.

Let E be the event 'the number is divisible by 5'.

If the number is divisible by 5 then it must end with the digit 5.

Therefore

$$n(E) = \text{number of ways of arranging the digits } 1, 2, 3 = 3!$$

So
$$P(E) = \frac{n(E)}{n(S)} = \frac{3!}{24} = \frac{1}{4}$$

The probability that the number is divisible by 5 is $\frac{1}{4}$.

Result 3 The number of ways of arranging n unlike objects in a ring when clockwise and anticlockwise arrangements are different is $(n-1)!$

For example, consider 4 people A, B, C and D, who are to be seated at a round table. The following four arrangements are the same, as A always has D on his immediate right and B on his immediate left.

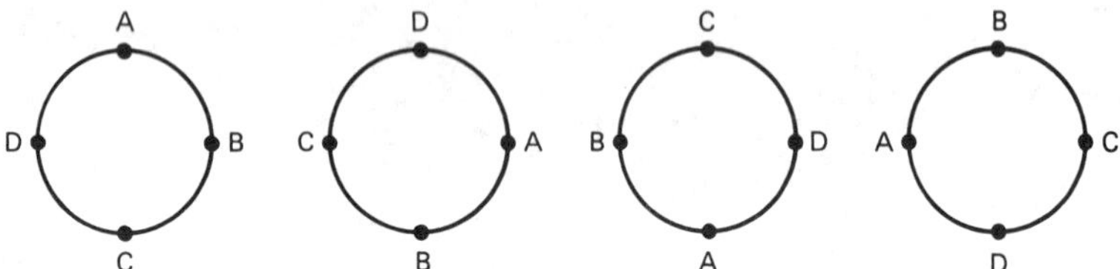

To find the number of different arrangements, we fix A and then consider the number of ways of arranging B, C and D.

Therefore the number of different arrangements of 4 people around the table is 3!

Result 4 The number of ways of arranging n unlike objects in a ring, when clockwise and anticlockwise arrangements are the same, is $\frac{(n-1)!}{2}$.

For example, if A, B, C and D are 4 different coloured beads which are threaded on a ring, then the following two arrangements are the same—the one is the other viewed from the other side.

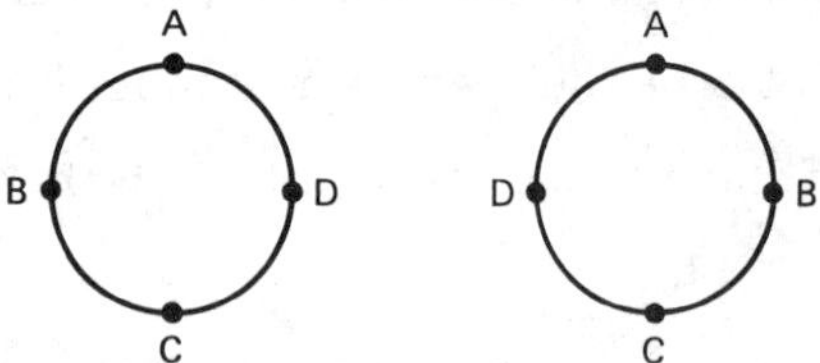

Therefore the number of arrangements of 4 beads on a ring is $\frac{3!}{2} = 3$.

Example 2.39 Six bulbs are planted in a ring and two do not grow. What is the probability that the two that do not grow are next to each other?

Solution 2.39 Let S be the possibility space, then $n(S) = 5!$

Let E be the event 'the bulbs that do not grow are next to each other'. Consider the two bulbs that do not grow as one 'item'. They can be arranged in 2! ways. There are now five 'items' to be arranged in a ring and this can be done in 4! ways.

Therefore $n(E) = 2!4!$

So
$$P(E) = \frac{n(E)}{n(S)} = \frac{2!4!}{5!} = \frac{2}{5}$$

The probability that the bulbs that do not grow are next to each other is $\frac{2}{5}$.

Example 2.40 One white, one blue, one red and two yellow beads are threaded on a ring to make a bracelet. Find the probability that the red and white beads are next to each other.

Solution 2.40 Let S be the possibility space.

If all the objects are unlike, the number of ways of arranging five beads on a ring is $\frac{4!}{2}$, but as there are two yellows

$$n(S) = \frac{4!}{(2)(2!)} = 6$$

Let E be the event 'the red and the white beads are next to each other'.

Then
$$n(E) = \frac{2!3!}{2!2}$$

(2! in numerator: red and white can be arranged in 2! ways; 3!: number of ways of arranging four objects in a ring; 2 in denominator: anticlockwise and clockwise arrangements are the same; 2! in denominator: there are two yellows)

So $n(E) = 3$

$$P(E) = \frac{n(E)}{n(S)} = \frac{3}{6} = \frac{1}{2}$$

The probability that the red and white beads are next to each other is $\frac{1}{2}$.

This result can be shown diagrammatically:

Ways of arranging the beads

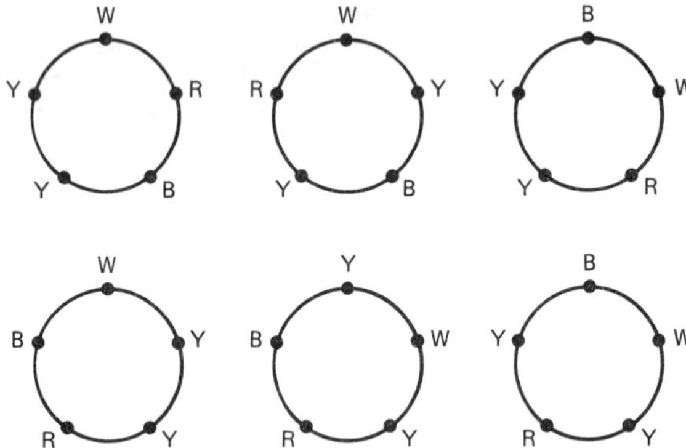

NOTE: in three of the six arrangements the red and white beads are next to each other.

Exercise 2k

1. In how many ways can the letters of the word FACETIOUS be arranged in a line? What is the probability that an arrangement begins with F and ends with S?

2. (*a*) In how many ways can 7 people sit at a round table?
(*b*) What is the probability that a husband and wife sit together?

3. On a shelf there are 4 mathematics books and 8 English books.
(*a*) If the books are to be arranged so that the mathematics books are together, in how many ways can this be done?
(*b*) What is the probability that all the mathematics books will not be together?

4. If the letters of the word PROBABILITY are arranged at random, find the probability that the two I's are separated.

5. Nine children play a party game and hold hands in a circle.
(*a*) In how many different ways can this be done?
(*b*) What is the probability that Mary will be holding hands with her friends Natalie and Sarah?

6. If the letters in the word ABSTEMIOUS are arranged at random, find the probability that the vowels and consonants appear alternately.

7. (*a*) In how many different ways can the letters in the word ARRANGEMENTS be arranged?
(*b*) Find the probability that an arrangement chosen at random begins with the letters EE.

PERMUTATIONS OF *r* OBJECTS FROM *n* OBJECTS

Consider the number of ways of placing 3 of the letters A, B, C, D, E, F, G in 3 empty spaces.

The first space can be filled in 7 ways. The second space can be filled in 6 ways. The third space can be filled in 5 ways. Therefore there are (7)(6)(5) ways of arranging 3 letters taken from 7 letters.

This is the number of permutations of 3 objects taken from 7 and it is written ${}^7P_3$.

So $${}^7P_3 = (7)(6)(5) = 210$$

Now (7)(6)(5) could be written $\dfrac{(7)(6)(5)(4)(3)(2)(1)}{(4)(3)(2)(1)}$,

i.e. $${}^7P_3 = \frac{7!}{4!} = \frac{7!}{(7-3)!}$$

NOTE: the order in which the letters are arranged is important — ABC is a different permutation from ACB.

In general, the number of permutations, or ordered arrangements, of r objects taken from n unlike objects is written ${}^nP_r$ where

$${}^nP_r = \frac{n!}{(n-r)!}$$

NOTE: ${}^nP_n = \dfrac{n!}{(n-n)!} = \dfrac{n!}{0!}$

But we know that the number of ways of arranging n unlike objects is $n!$

So we must define 0! to be 1.

So $$0! = 1$$

COMBINATIONS OF r OBJECTS FROM n OBJECTS

When considering the number of combinations of r objects from n objects, the order in which they are placed is not important.

For example, the one combination ABC gives rise to 3! permutations

ABC, ACB, BCA, BAC, CAB, CBA

So, if the number of combinations of 3 letters from the 7 letters A, B, C, D, E, F, G, is denoted by ${}^7C_3$ then

$${}^7C_3(3!) = {}^7P_3$$

$${}^7C_3 = \frac{{}^7P_3}{3!}$$

$$= \frac{7!}{3!4!}$$

$$= \frac{(7)(6)(5)\cancel{(4)}\cancel{(3)}\cancel{(2)}\cancel{(1)}}{(3)(2)(1)\cancel{(4)}\cancel{(3)}\cancel{(2)}\cancel{(1)}}$$

$$= 35$$

In general, the number of combinations of r objects from n unlike objects is ^nC_r where $^nC_r = \dfrac{n!}{r!(n-r)!}$.

NOTE: ^nC_r is sometimes written $_nC_r$ or $\dbinom{n}{r}$.

Example 2.41 In how many ways can a hand of 4 cards be dealt from an ordinary pack of 52 playing cards?

Solution 2.41 We need to consider combinations, as the order in which the cards are dealt is not important.

Now
$$\begin{aligned} ^{52}C_4 &= \frac{52!}{4!48!} \\ &= \frac{(52)(51)(50)(49)}{(4)(3)(2)(1)} \\ &= 270\,725 \text{ ways} \end{aligned}$$

The number of ways of dealing the hand of 4 cards is 270 725.

Example 2.42 Four letters are chosen at random from the word RANDOMLY. Find the probability that all four letters chosen are consonants.

Solution 2.42 Let S be the possibility space, then

$$\begin{aligned} n(S) &= {}^8C_4 \\ &= \frac{8!}{4!4!} \\ &= \frac{(8)(7)(6)(5)}{(4)(3)(2)(1)} \\ &= 70 \end{aligned}$$

Let E be the event 'four consonants are chosen'. As there are six consonants

$$\begin{aligned} n(E) &= {}^6C_4 \\ &= \frac{6!}{4!2!} \\ &= \frac{(6)(5)}{(2)(1)} \\ &= 15 \end{aligned}$$

Now $$P(E) = \frac{n(E)}{n(S)} = \frac{15}{70} = \frac{3}{14}$$

The probability that the four letters chosen are consonants is $\frac{3}{14}$.

Example 2.43 A team of 4 is chosen at random from 5 girls and 6 boys.

(a) In how many ways can the team be chosen if (i) there are no restrictions; (ii) there must be more boys than girls?

(b) Find the probability that the team contains only one boy.

Solution 2.43 (a) (i) There are 11 people, from whom 4 are chosen. The order in which they are chosen is not important.

$$\text{Number of ways of choosing the team} = {}^{11}C_4 = \frac{11!}{4!7!} = \frac{(11)(10)(9)(8)}{(4)(3)(2)(1)} = 330$$

If there are no restrictions, the team can be chosen in 330 ways.

(ii) If there are to be more boys than girls, then there must be 3 boys and 1 girl or 4 boys.

$$\text{Number of ways of choosing 3 boys and 1 girl} = ({}^6C_3)({}^5C_1) = \left(\frac{6!}{3!3!}\right)\left(\frac{5!}{1!4!}\right) = \frac{(6)(5)(4)(5)}{(3)(2)(1)(1)} = 100$$

$$\text{Number of ways of choosing 4 boys} = {}^6C_4 = \frac{6!}{4!2!} = \frac{(6)(5)}{(2)(1)} = 15$$

Therefore the number of ways of choosing the team if there are more boys than girls $= 100 + 15 = 115$ ways.

(**b**) The possibility space $S =$ (all possible teams of 4) and $n(S) = 330$.

Let E be the event 'only one boy is chosen'.

Now

$$n(E) = ({}^6C_1)({}^5C_3) \quad \text{(if 1 boy is chosen, then 3 girls must be chosen)}$$

$$= \left(\frac{6!}{1!5!}\right)\left(\frac{5!}{3!2!}\right)$$

$$= \frac{(6)(5)(4)}{(2)(1)}$$

$$= 60$$

So $$P(E) = \frac{n(E)}{n(S)}$$

$$= \frac{60}{330}$$

$$= \frac{2}{11}$$

The probability that the team contains only one boy is $\frac{2}{11}$.

Example 2.44 Four items are taken at random from a box of 12 items and inspected. The box is rejected if more than 1 item is found to be faulty. If there are 3 faulty items in the box, find the probability that the box is accepted.

Solution 2.44 The box is accepted if (*a*) there are no faulty items in the sample of 4, or (*b*) there is one faulty item in the sample of 4.

Let S be the possibility space, then

$$n(S) = {}^{12}C_4$$

$$= \frac{12!}{4!8!}$$

$$= \frac{(12)(11)(10)(9)}{(4)(3)(2)(1)}$$

$$= 495$$

There are 9 items that are not faulty, so the number of ways of choosing 4 items that are not faulty

$$= {}^9C_4$$
$$= \frac{9!}{4!5!}$$
$$= \frac{(9)(8)(7)(6)}{(4)(3)(2)(1)}$$
$$= 126$$

The number of ways of choosing 1 faulty item and 3 good items

$$= ({}^3C_1)({}^9C_3)$$
$$= \left(\frac{3!}{1!2!}\right)\left(\frac{9!}{3!6!}\right)$$
$$= \frac{(3)(9)(8)(7)}{(3)(2)(1)}$$
$$= 252$$

Let E be the event 'the number of faulty items chosen is 0 or 1'.

Then $\quad n(E) = 126 + 252 = 378$

So $\quad P(E) = \frac{n(E)}{n(S)}$

$$= \frac{378}{495}$$
$$= 0.76 \quad (2\,\text{d.p.})$$

If the number of faulty items is 0 or 1 then the box is accepted, so the probability that the box is accepted is 0.76 (2 d.p.).

Example 2.45 If a diagonal of a polygon is defined to be a line joining any two non-adjacent vertices, how many diagonals are there in a polygon of (*i*) 5 sides, (*ii*) 6 sides, (*iii*) n sides? SUJB (P)

Solution 2.45 (*i*) Number of ways to choose 2 points from 5

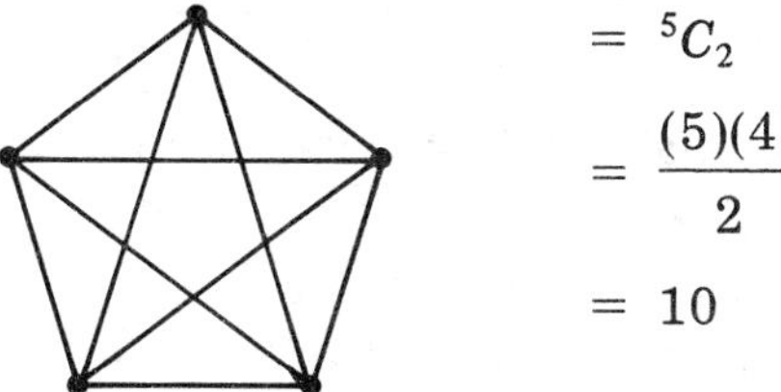

$$= {}^5C_2$$
$$= \frac{(5)(4)}{2}$$
$$= 10$$

So there are 10 possible lines to draw, but as there are 5 sides, 5 of these are joining adjacent vertices.

Therefore the number of diagonals $= 10 - 5 = 5$.

(*ii*) Similarly, for the polygon of 6 sides,

$$\text{the number of diagonals} = {}^6C_2 - 6 = \frac{(6)(5)}{2} - 6 = 9$$

The number of diagonals for a polygon with 6 sides is 9.

(*iii*) For a polygon with n sides,

$$\text{the number of diagonals} = {}^nC_2 - n = \frac{(n)(n-1)}{2} - n = \frac{n^2 - n - 2n}{2} = \frac{n(n-3)}{2}$$

The number of diagonals for a polygon with n sides is $\frac{n(n-3)}{2}$.

Example 2.46 A certain family consists of Mother, Father and their ten sons.

(a) They are invited to send a group of four representatives to a wedding. Evaluate the number of ways in which the group can be formed, if it must contain (*i*) both parents; (*ii*) one and only one parent; (*iii*) neither parent.

(b) On another occasion, the ten sons decide to play five-a-side football. Evaluate the number of ways in which the teams can be made up. Determine the probability that the two eldest brothers are in the same team. (SUJB Additional)

Solution 2.46 (a) (*i*) If the group contains both parents,

$$\text{number of ways to choose remaining 2 from 10} = {}^{10}C_2 = \frac{(10)(9)}{(2)(1)} = 45$$

If the group is to contain both parents, then it can be chosen in 45 ways.

(*ii*) Number of ways to choose one parent = 2.

$$\text{Number of ways to choose remaining 3 from 10} = {}^{10}C_3 = \frac{(10)(9)(8)}{(3)(2)(1)} = 120$$

Therefore number of ways to choose the group of 4 $= (2)(120)$

$= 240$

If the group is to contain one, and only one parent, then it can be chosen in 240 ways.

(*iii*) Number of ways to choose 4 from 10 $= {}^{10}C_4$

$$= \frac{(10)(9)(8)(7)}{(4)(3)(2)(1)}$$

$$= 210$$

If the group is to contain neither parent, then the number of ways in which it can be chosen is 210.

(**b**) Number of ways to choose 5 from 10 $= {}^{10}C_5$

$$= \frac{(10)(9)(8)(7)(6)}{(5)(4)(3)(2)(1)}$$

$$= 252$$

When one team has been chosen, the other team is formed automatically. But, since the pairs of teams are interchangeable, e.g. ABCDE versus FGHIJ is the same as FGHIJ versus ABCDE, the total number of ways in which the two teams can be formed is $\frac{1}{2}(252) = 126$.

The two teams can be formed in 126 ways.

If the two eldest are the in the same team,

the number of ways in which the remaining 3 can be chosen $= {}^{8}C_3$

$$= \frac{(8)(7)(6)}{(3)(2)(1)}$$

$$= 56$$

Let E be the event 'the two eldest are in the same team', then $n(E) = 56$.

If S is the possibility space, then $n(S) = 126$.

$$P(\text{two eldest are in the same team}) = P(E) = \frac{n(E)}{n(S)} = \frac{56}{126} = \frac{4}{9}$$

The probability that the two eldest are in the same team is $\frac{4}{9}$.

SUMMARY — ARRANGEMENTS, PERMUTATIONS AND COMBINATIONS

The number of ways of arranging n unlike objects in a line	$n!$
The number of ways of arranging in a line n objects of which p of one type are alike, q of another type are alike, r of a third type are alike, and so on	$\dfrac{n!}{p!q!r!\ldots}$
The number of ways of arranging n unlike objects in a ring when clockwise and anticlockwise arrangements are different	$(n-1)!$
The number of ways of arranging n unlike objects in a ring when clockwise and anticlockwise arrangements are the same	$\dfrac{(n-1)!}{2}$
The number of permutations of r objects taken from n unlike objects	${}^nP_r = \dfrac{n!}{(n-r)!}$
The number of combinations of r objects taken from n unlike objects	${}^nC_r = \dfrac{n!}{r!(n-r)!}$

Exercise 2I

1. From a group of 10 boys and 8 girls, 2 pupils are chosen at random. Find the probability that they are both girls.

2. From a group of 6 men and 8 women, 5 people are chosen at random. Find the probability that there are more men chosen than women.

3. From a bag containing 6 white counters and 8 blue counters, 4 counters are chosen at random. Find the probability that 2 white counters and 2 blue counters are chosen.

4. From a group of 10 people, 4 are to be chosen to serve on a committee.
(*a*) In how many different ways can the committee be chosen?
(*b*) Among the 10 people there is one married couple. Find the probability that both the husband and the wife will be chosen.
(*c*) Find the probability that the 3 youngest people will be chosen.

5. Four persons are chosen at random from a group of ten persons consisting of four men and six women. Three of the women are sisters. Calculate the probabilities that the four persons chosen will: (i) consist of four women, (ii) consist of two women and two men, (iii) include the three sisters. (JMB)

6. A touring party of 20 cricketers consists of 9 batsmen, 8 bowlers and 3 wicket keepers. A team of 11 players must have at least 5 batsmen, 4 bowlers and 1 wicket keeper. How many different teams can be selected, (*a*) if all the players are available for selection, (*b*) if 2 batsmen and 1 bowler are injured and cannot play?

7. Find the number of ways in which 10 different books can be shared between a boy and a girl if each is to receive an even number of books.

8. Four letters are picked from the word BREAKDOWN. What is the probability that there is at least one vowel among the letters?

9. Eight people sit in a minibus: 4 on the sunny side and 4 on the shady side. If 2 people want to sit on opposite sides to each other, another 2 people want to sit on the shady side, in how many ways can this be done?

10. Disco lights are arranged in a vertical line. How many different arrangements can be made from 2 green, 3 blue and 4 red lights (*a*) if all 9 lights are used, (*b*) if at least 8 lights are used?

11. A group consisting of 10 boys and 11 girls attends a course for special games coaching.
(*a*) When they are introduced, each person hands a card containing his or her photograph, name and address to every other member of the group. State the total number of cards which are exchanged.
(*b*) Five boys are selected for basketball and six girls for netball. Find the number of different possible selections for each of these.
(*c*) Five particular boys and five particular girls are selected and placed in mixed pairs for tennis. Find the total number of different mixed pairs which can be made using these ten children.
(*d*) If 4 children are chosen at random from the whole group find the probability that there is a majority of girls in the 4 selected. (L Additional)

12. To enter a cereal competition, competitors have to choose the 8 most important features of a new car, from a possible 12 features, then list the 8 in order of preference. Each cereal packet entry form contains space for 5 entries. A correct entry wins a new car.
(*a*) What is the probability that a housewife wins a new car if she completes the entry form from one packet?
(*b*) How many entry forms would she need to complete, each entry showing different arrangements, if the probability that she wins a car is to be at least 0.8?

13. Three letters are selected at random from the word BIOLOGY. Find the probability that the selection (*a*) does not contain the letter O, (*b*) contains both the letter O's.

14. How many even numbers can be formed with the digits 3, 4, 5, 6, 7 by using some or all of the numbers (repetitions are not allowed)?

15.

Different coloured pegs, each of which is painted in one and only one of the six colours red, white, black, green, blue and yellow, are to be placed in four holes, as shown in the figure, with one peg in each hole. Pegs of the same colour are indistinguishable. Calculate how many different arrangements of pegs placed in the four holes so that they are all occupied can be made from
(*a*) six pegs, all of different colours,
(*b*) two red and two white pegs,
(*c*) two red, one white and one black peg,
(*d*) twelve pegs, two of each colour.
(L Additional)

16. (*a*) Calculate how many different numbers altogether can be formed by taking one, two, three and four digits from the digits 9, 8, 3 and 2, repetitions not being allowed.
(*b*) Calculate how many of the numbers in part (*a*) are odd and greater than 800.
(*c*) If one of the numbers in part (*a*) is chosen at random, calculate the probability that it will be greater than 300.
(L Additional)

17. The positions of nine trees which are to be planted along the sides of a road, five on the north side and four on the south side, are shown in the figure.

N

S

(*a*) Find the number of ways in which this can be done if the trees are all of different species.
(*b*) If the trees in (*a*) are planted at random, find the probability that two particular trees are next to each other on the same side of the road.
(*c*) If there are 3 cupressus, 4 prunus and 2 magnolias, find the number of different ways in which these could be planted assuming that trees of the same species are identical.
(*d*) If the trees in (*c*) are planted at random, find the probability that the 2 magnolias are on the opposite sides of the road. (L Additional)

18. A committee consisting of 6 persons is to be selected from 5 women and 6 men.

(*a*) Calculate the number of ways in which the chosen committee will contain exactly two men.

(*b*) Given that the committee is to contain at least 2 men, show that it can be selected in 456 ways.

(*c*) Given that these 456 ways are equally likely, calculate the probability that there will be more men than women on the committee.

(*d*) At a meeting the members of the chosen committee sit at a rectangular table in the fixed seats illustrated in the diagram:

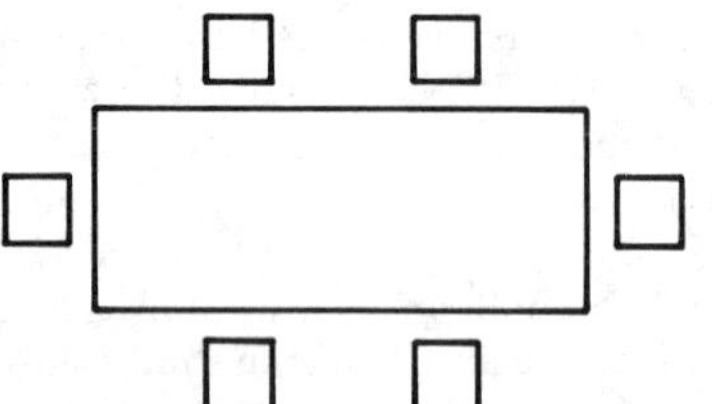

(i) Given that each may sit in any of the six places, calculate the number of different ways they may be seated at the table.

(ii) Given that the committee consists of 3 men and 3 women and that the men and women must sit alternately round the table, calculate in how many different ways they may be seated. (L Additional)

MISCELLANEOUS WORKED EXAMPLES

Example 2.47 The events A and B are such that $P(A) = \frac{1}{3}$, $P(B) = \frac{2}{5}$ and $P(B \mid \bar{A}) = \frac{11}{20}$. Find

(a) $P(A \cap B)$,

(b) $P(A \cup B)$,

(c) $P(\bar{A} \mid B)$,

(d) $P(A \mid B)$.

State whether A and B are (i) independent, (ii) mutually exclusive.

Solution 2.47 Consider a possibility space S and let $n(S) = n$.

Consider events A and B such that $n(A) = r$ and $n(B) = s$, $n(A \cap B) = t$.

First, draw a Venn diagram, showing this information:

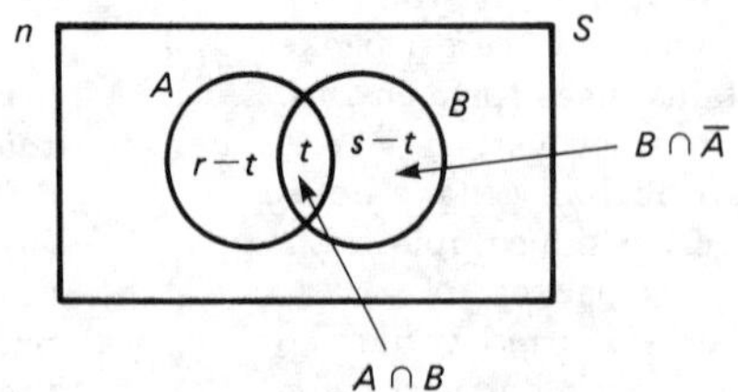

(a) Now

$$P(A) = \frac{n(A)}{n(S)} = \frac{r}{n}, \quad \text{but} \quad P(A) = \frac{1}{3} \quad \text{so} \quad \frac{r}{n} = \frac{1}{3}$$

$$P(B) = \frac{n(B)}{n(S)} = \frac{s}{n}, \quad \text{but} \quad P(B) = \frac{2}{5} \quad \text{so} \quad \frac{s}{n} = \frac{2}{5}$$

Now

$$P(B \mid \bar{A}) = \frac{P(B \cap \bar{A})}{P(\bar{A})} = \frac{n(B \cap \bar{A})}{n(\bar{A})} = \frac{s-t}{n-r} = \frac{s/n - t/n}{1 - r/n}$$

But we are given that $P(B \mid \bar{A}) = \frac{11}{20}$, therefore

$$\frac{11}{20} = \frac{s/n - t/n}{1 - r/n}$$

$$= \frac{2/5 - t/n}{1 - 1/3}$$

$$\left(\frac{2}{3}\right)\left(\frac{11}{20}\right) = \frac{2}{5} - \frac{t}{n}$$

So

$$\frac{t}{n} = \frac{2}{5} - \frac{11}{30}$$

$$= \frac{1}{30}$$

Now

$$P(A \cap B) = \frac{n(A \cap B)}{n(S)}$$

$$= \frac{t}{n}$$

Therefore $P(A \cap B) = \frac{1}{30}$.

(b)

$$P(A \cup B) = P(A) + P(B) - P(A \cap B)$$

So

$$P(A \cup B) = \frac{1}{3} + \frac{2}{5} - \frac{1}{30}$$

$$= \frac{7}{10}$$

Therefore $P(A \cup B) = \frac{7}{10}$.

(c)

$$P(\bar{A} \mid B) = \frac{P(\bar{A} \cap B)}{P(B)}$$

$$= \frac{n(\bar{A} \cap B)}{n(B)}$$

$$= \frac{s-t}{s}$$

$$= \frac{s/n - t/n}{s/n}$$

$$= \frac{2/5 - 1/30}{2/5}$$

$$= \frac{11}{12}$$

Therefore $P(\bar{A} \mid B) = \frac{11}{12}$.

(d) $$P(A \mid B) = \frac{P(A \cap B)}{P(B)}$$

$$= \frac{1/30}{2/5}$$

$$= \frac{1}{12}$$

Therefore $P(A \mid B) = \frac{1}{12}$.

NOTE: $P(A \mid B) + P(\bar{A} \mid B) = 1$.

(i) If two events are independent then $P(A \cap B) = P(A) \cdot P(B)$.

Now $P(A \cap B) = \dfrac{1}{30}$ and $P(A) \cdot P(B) = \left(\dfrac{1}{3}\right)\left(\dfrac{2}{5}\right) = \dfrac{2}{15}$

So $P(A \cap B) \neq P(A) \cdot P(B)$ and the events A and B are not independent.

(ii) If two events are mutually exclusive, then $P(A \cap B) = 0$.

So, as $P(A \cap B) \neq 0$, A and B are not mutually exclusive.

Example 2.48 Tung-Pong and Ping-Ho play a game of table tennis. The score reaches 20–20. The game continues until one player has scored two more points than the other.

The probability that Tung-Pong wins each point is 0.6. What are the probabilities that:

(a) Tung-Pong wins the game after 2 further points?

(b) Ping-Ho wins the game after 2 further points?

(c) The score is 21–21 after 2 further points?

(d) Tung-Pong wins the game after 3 further points?

(e) Tung-Pong wins the game after 4 further points?

(f) Tung-Pong wins the game after 6 further points?

If the game can continue indefinitely, for each player what is the probability that he will ultimately win? (SUJB Additional)

Solution 2.48 Let W be the event 'Tung wins a point'.

Then $P(W) = 0.6$ and $P(\overline{W}) = 0.4$.

(a) $P(\text{Tung wins after 2 further points}) = P(WW)$

$$= (0.6)(0.6)$$
$$= 0.36$$

The probability that Tung wins after 2 further points is 0.36.

(b) $P(\text{Ping wins after 2 further points}) = P(\overline{W}\overline{W})$

$$= (0.4)(0.4)$$
$$= 0.16$$

The probability that Ping wins after 2 further points is 0.16.

(c) $P(\text{score is 21-21 after 2 further points}) = P(W\overline{W}) + P(\overline{W}W)$

$$= 2(0.6)(0.4)$$
$$= 0.48$$

The probability that the score is 21-21 after 2 further points is 0.48.

(d) To consider the situation after 3 further points we look first at the situation after 2 further points. Now after 2 further points either Tung has won, or Ping has won, or the score is 21-21. If the score is 21-21 then there is no way that Tung can win after just one more point.

So the probability that Tung wins after 3 further points is 0.

(e) After 4 further points, if Tung wins, the sequence of points must be

$$\left.\begin{array}{ll} & (W\overline{W})(WW) \\ \text{or} & (\overline{W}W)(WW) \end{array}\right\} \text{2 ways}$$

So

$$P(\text{Tung wins after 4 further points}) = P(W\overline{W}\,WW) + P(\overline{W}W\,WW)$$
$$= 2(0.6)^3(0.4)$$
$$= 0.1728$$

The probability that Tung wins after 4 further points is 0.1728.

(f) If Tung wins after 6 further points, the sequence must be

$$\begin{array}{ccc} (W\overline{W}) & (W\overline{W}) & (WW) \\ \uparrow & \uparrow & \uparrow \\ \text{(2 ways)} & \text{(2 ways)} & \text{(1 way)} \end{array}$$

So $P(\text{Tung wins after 6 further points}) = 4P(W\overline{W}\ W\overline{W}\ WW)$
$= 4(0.6)^4(0.4)^2$
$= 0.0829 \quad (3\text{ S.F.})$

So the probability that Tung wins after 6 further points is 0.0829 (3 S.F.).

If the game can continue indefinitely,

$$\begin{aligned}P(\text{Tung wins}) &= (0.6)^2 + 2(0.6)^3(0.4) + 4(0.6)^4(0.4)^2 + \ldots \\ &= (0.6)^2(1 + 2(0.6)(0.4) + 4(0.6)^2(0.4)^2 + \ldots) \\ &= (0.6)^2(1 + 0.48 + 0.48^2 + \ldots) \\ &= (0.6)^2\left(\frac{1}{1-0.48}\right) \quad \text{(sum of an infinite G.P., common ratio 0.48)} \\ &= \frac{0.36}{0.52} \\ &= \frac{9}{13}\end{aligned}$$

Therefore

$$\begin{aligned}P(\text{Ping wins}) &= 1 - \frac{9}{13} \\ &= \frac{4}{13}\end{aligned}$$

The probability that Tung wins is $\frac{9}{13}$ and the probability that Ping wins is $\frac{4}{13}$.

Example 2.49 (a) A bag contains 5 red and 4 blue balls. 3 balls are picked out, one at a time, and are not replaced. Find the probability that at least 1 of the 3 balls is blue.

(b) One letter is selected from each of the names: SIMMS, SMITH, THOMPSON. What is the probability that 2, and only 2 are the same?

(c) A candidate attempts a question to which 5 possible answers have been given, one of them correct. For any question, there is a probability of $\frac{1}{3}$ that he knows the correct answer. If he does not know the correct answer he will mark one of the answers at random. He does, in fact, mark the correct answer. What is the probability that he knew the correct answer? (SUJB)

Solution 2.49 (a) Using an obvious notation,

$$P(3\text{ red balls}) = P(R_1 \cap R_2 \cap R_3) = \left(\frac{5}{9}\right)\left(\frac{4}{8}\right)\left(\frac{3}{7}\right) \quad \text{(non-independent events)} = \frac{5}{42}$$

So

$$P(\text{at least one ball is blue}) = 1 - P(3\text{ balls are red}) = 1 - \frac{5}{42} = \frac{37}{42}$$

The probability that at least one of the 3 balls is blue is $\frac{37}{42}$.

(b) SIMMS SMITH THOMPSON

Let the event S_1 be 'choosing an S from the first name', and so on.

If one letter is selected from each name, then if 2 and only 2 letters are the same the possible outcomes are listed below, with their respective probabilities:

$$P(S_1S_2\bar{S}_3) = \left(\frac{2}{5}\right)\left(\frac{1}{5}\right)\left(\frac{7}{8}\right) = 0.07$$

$$P(S_1\bar{S}_2S_3) = \left(\frac{2}{5}\right)\left(\frac{4}{5}\right)\left(\frac{1}{8}\right) = 0.04$$

$$P(\bar{S}_1S_2S_3) = \left(\frac{3}{5}\right)\left(\frac{1}{5}\right)\left(\frac{1}{8}\right) = 0.015$$

$$P(I_1I_2\bar{I}_3) = \left(\frac{1}{5}\right)\left(\frac{1}{5}\right)(1) = 0.04$$

$$P(M_1M_2\bar{M}_3) = \left(\frac{2}{5}\right)\left(\frac{1}{5}\right)\left(\frac{7}{8}\right) = 0.07$$

$$P(M_1\bar{M}_2M_3) = \left(\frac{2}{5}\right)\left(\frac{4}{5}\right)\left(\frac{1}{8}\right) = 0.04$$

$$P(\bar{M}_1M_2M_3) = \left(\frac{3}{5}\right)\left(\frac{1}{5}\right)\left(\frac{1}{8}\right) = 0.015$$

$$P(\bar{H}_1H_2H_3) = (1)\left(\frac{1}{5}\right)\left(\frac{1}{8}\right) = 0.025$$

$$P(\bar{T}_1T_2T_3) = (1)\left(\frac{1}{5}\right)\left(\frac{1}{8}\right) = 0.025$$

Total 0.340

Therefore $P(2\text{ and only 2 letters are the same}) = 0.34$.

(c) Let K be the event 'he knows the correct answer', then $P(K) = \frac{1}{3}$.

Let M be the event 'he marks the correct answer'.

Now $P(M \mid \bar{K}) = \frac{1}{5}$, as he marks an answer at random if he does not know the correct answer.

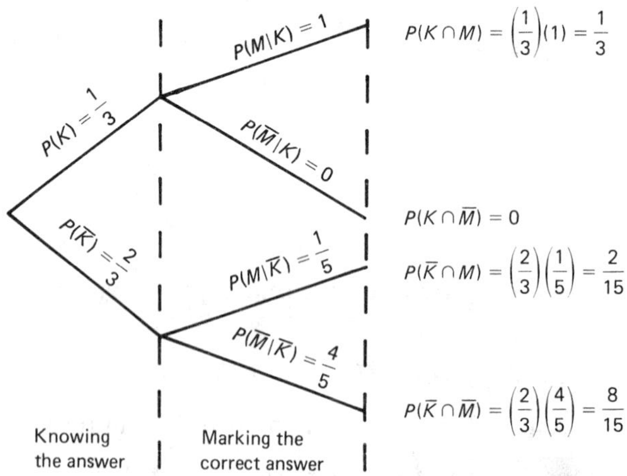

We require $P(K \mid M) = \dfrac{P(K \cap M)}{P(M)}$.

Now

$$P(M) = P(M \cap K) + P(M \cap \bar{K}) = \frac{1}{3} + \frac{2}{15} = \frac{7}{15}$$

So

$$P(K \mid M) = \frac{1/3}{7/15} = \frac{5}{7}$$

<u>The probability that he knew the correct answer, given that he marked the correct answer, is $\frac{5}{7}$.</u>

Example 2.50 (a) A bag contains a number of counters, alike in shape and size, but x are red and y are green. Counters are to be chosen at random from the bag. Prove that the probability that the second counter chosen will be red is the same, whether the first counter is replaced or not before the second is drawn.

(b) A three-figure number, not less than 100, is to be made up using three digits selected at random from the digits 0, 1, 2, 3, 4,

5, 6, 7, 8, 9 WITHOUT using the same digit twice in any number. Show that the total possible number of numbers is 648. Calculate the probabilities: (*i*) that the number is even, (*ii*) that the number is divisible by 5, (*iii*) that the number is greater than 600, (*iv*) that the number is even and greater than 600.

What are the corresponding results for (*i*) and (*ii*) if the same digit may be used two or three times in the same number?

(SUJB)

Solution 2.50 (a) The bag contains x red counters and y green counters.

When the first counter is replaced

With obvious notation $\quad P(R_2) = \dfrac{x}{x+y}$

When the first counter is not replaced

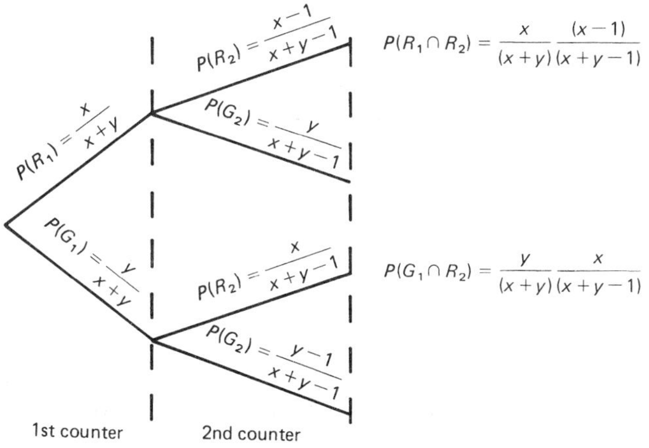

$P(R_1) = \dfrac{x}{x+y}$

$P(R_2) = \dfrac{x-1}{x+y-1}$ $\quad P(R_1 \cap R_2) = \dfrac{x}{(x+y)}\dfrac{(x-1)}{(x+y-1)}$

$P(G_2) = \dfrac{y}{x+y-1}$

$P(G_1) = \dfrac{y}{x+y}$

$P(R_2) = \dfrac{x}{x+y-1}$ $\quad P(G_1 \cap R_2) = \dfrac{y}{(x+y)}\dfrac{x}{(x+y-1)}$

$P(G_2) = \dfrac{y-1}{x+y-1}$

1st counter 2nd counter

$$\begin{aligned} P(R_2) &= P(R_2 \cap R_1) + P(R_2 \cap G_1) \quad \text{(mutually exclusive events)} \\ &= \frac{x}{(x+y)}\frac{(x-1)}{(x+y-1)} + \frac{y}{(x+y)}\frac{x}{(x+y-1)} \\ &= \frac{x}{(x+y)(x+y-1)}(x-1+y) \\ &= \frac{x}{x+y} \end{aligned}$$

Therefore the probability that the second counter is red is $\dfrac{x}{(x+y)}$ whether or not the first counter is replaced. This is because there is no condition placed on the first counter so that any one of the $x+y$ counters is equally likely to be the second counter.

(b) 1st digit can be chosen in 9 ways (0 not included)
2nd digit can be chosen in 9 ways (0 included here)
3rd digit can be chosen in 8 ways

So, total number of ways $= (9)(9)(8)$
$= 648$

Therefore the possibility space is 648 equally likely outcomes.

(i) *If the number is even*

Either the *3rd digit is 0*
3rd digit chosen in 1 way
2nd digit chosen in 9 ways
1st digit chosen in 8 ways

Number of ways $= (1)(9)(8)$
$= 72$

or the *3rd digit is 2, 4, 6 or 8*
3rd digit chosen in 4 ways
1st digit chosen in 8 ways (0 excluded)
2nd digit chosen in 8 ways

Number of ways $= (4)(8)(8)$
$= 256$

Therefore the number of ways in which the number is even = 328.

$$P(\text{number is even}) = \frac{328}{648} = \frac{41}{81}$$

(ii) *If the number is divisible by 5*

Either *the 3rd digit is 0*
3rd digit chosen in 1 way
1st digit chosen in 9 ways
2nd digit chosen in 8 ways

Number of ways $= (1)(9)(8)$
$= 72$

or *the 3rd digit is 5*
3rd digit chosen in 1 way
1st digit chosen in 8 ways (0 excluded)
2nd digit chosen in 8 ways

Number of ways $= (1)(8)(8)$
$= 64$

Therefore the number of ways in which the number is divisible by 5 = 136.

$$P(\text{number is divisible by 5}) = \frac{136}{648} = \frac{17}{81}$$

(iii) *If the number is greater than 600*

1st digit can be chosen in 4 ways (from 6, 7, 8, 9)
2nd digit can be chosen in 9 ways
3rd digit can be chosen in 8 ways

Number of ways $= (4)(9)(8)$
$= 288$

$$P(\text{number is greater than 600}) = \frac{288}{648} = \frac{4}{9}$$

(iv) If the number is even and greater than 600

Either *the 1st digit is 6 or 8*	or *the 1st digit is 7 or 9*
1st digit chosen in 2 ways	1st digit chosen in 2 ways
3rd digit chosen in 4 ways	3rd digit chosen in 5 ways
2nd digit chosen in 8 ways	2nd digit chosen in 8 ways
Number of ways $= (2)(4)(8)$	Number of ways $= (2)(5)(8)$
$= 64$	$= 80$

The number of ways in which the number is even and greater than 600 = 144.

$$P(\text{number is even and greater than } 600) = \frac{144}{648} = \frac{2}{9}$$

Now consider the case when the same digit may be used two or three times:

If the number is even

We are concerned with the 3rd digit, which can be chosen in 5 ways. If there was no restriction, this could be chosen in 10 ways.

$$P(\text{number is even}) = \frac{5}{10} = \frac{1}{2}$$

If the number is divisible by 5

We are concerned with the 3rd digit which can be chosen in 2 ways.

$$P(\text{number is divisible by } 5) = \frac{2}{10} = \frac{1}{5}$$

Example 2.51 (a) A and B play a game as follows: an ordinary die is rolled and if a six is obtained then A wins and if a one is obtained then B wins. If neither a six nor a one is obtained then the die is rolled again until a decision can be made. What is the probability that A wins on (*i*) the first roll, (*ii*) the second roll, (*iii*) the rth roll? What is the probability that A wins?

(b) A bag contains 4 red and 3 yellow balls and another bag contains 3 red and 4 yellow. A ball is taken from the first bag and placed in the second, the second bag is shaken and a ball taken from it and placed in the first bag. If a ball is now taken from the first bag what is the probability that it is red?

(You are advised to draw a tree diagram.) (SUJB)

Solution 2.51 (a) $P(6 \text{ is obtained}) = \frac{1}{6}$ and $P(1 \text{ is obtained}) = \frac{1}{6}$.

(i) $P(A \text{ wins on the first roll}) = P(6 \text{ is obtained}) = \dfrac{1}{6}$

(ii) $P(A \text{ wins on the second roll}) = P(\text{neither a 6 nor a 1 on the first roll}) \cdot P(6 \text{ on second roll})$

$$= \left(\frac{4}{6}\right)\left(\frac{1}{6}\right)$$

$$= \frac{1}{9}$$

(iii) $P(A \text{ wins on the } r\text{th roll}) = P(\text{neither a 6 nor a 1 on 1st } r-1 \text{ rolls}) \cdot P(6 \text{ on the } r\text{th roll})$

$$= \left(\frac{4}{6}\right)^{r-1}\left(\frac{1}{6}\right)$$

$$= \left(\frac{2}{3}\right)^{r-1}\left(\frac{1}{6}\right)$$

These are mutually exclusive events.

So $P(A \text{ wins}) = \dfrac{1}{6} + \left(\dfrac{2}{3}\right)\left(\dfrac{1}{6}\right) + \left(\dfrac{2}{3}\right)^2\left(\dfrac{1}{6}\right) + \ldots + \left(\dfrac{2}{3}\right)^{r-1}\left(\dfrac{1}{6}\right)$

$+ \ldots$ to infinity

$$= \frac{1}{6}\left[1 + \left(\frac{2}{3}\right) + \left(\frac{2}{3}\right)^2 + \ldots\right]$$

$$= \frac{1}{6}S$$

where S is the sum of an infinite G.P. with $a = 1, r = \frac{2}{3}$.

$$P(A \text{ wins}) = \frac{1}{6}\left(\frac{1}{1-\frac{2}{3}}\right)$$

$$= \frac{1}{6}(3)$$

$$= \frac{1}{2}$$

The probability that A wins is $\frac{1}{2}$.

(b) First we show the possibilities diagrammatically:

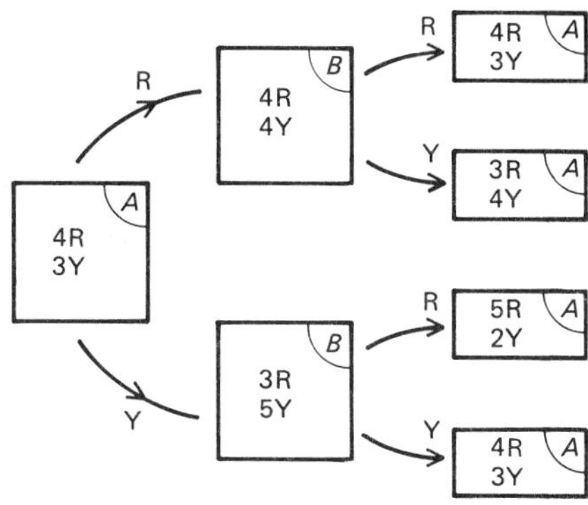

The tree diagram to show the possible outcomes and the probabilities is as follows:

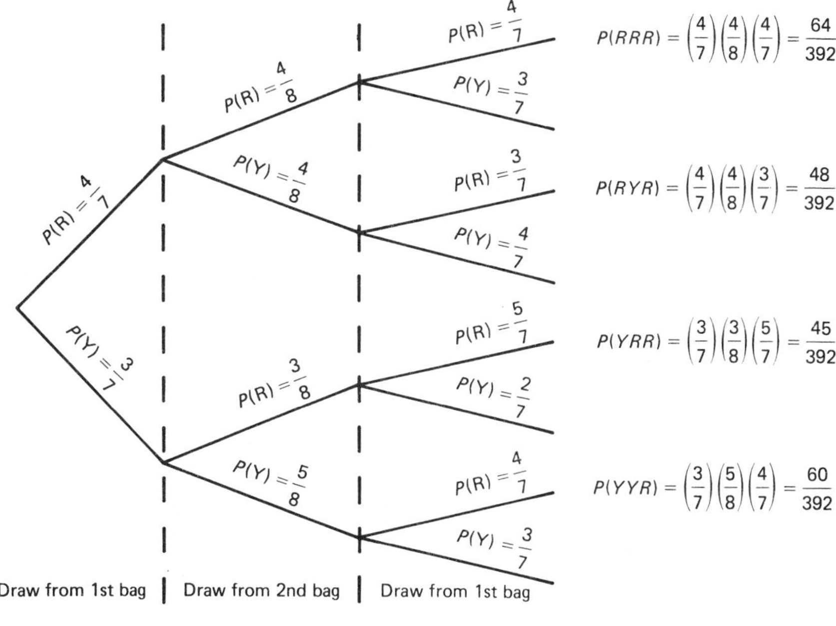

$$
\begin{aligned}
P(\text{red from 1st bag}) &= \frac{1}{392}(64+48+45+60) \\
&= \frac{217}{392} \\
&= 0.554 \quad (3\,\text{d.p.})
\end{aligned}
$$

The probability that the ball is red is 0.554 (3 d.p.).

Example 2.52 (a) A pack of 52 playing cards is cut at random into three piles. Find the probability that the top cards are all (*i*) black, (*ii*) hearts, (*iii*) aces.

After the top cards have been examined and found not to be picture cards, calculate the probability that the three bottom cards are all queens.

(b) A bag contains eight black counters and two white ones. Each of two players, A and B, draws one counter in turn, without replacement, until one of them wins by drawing a white counter. A draws first. Calculate his chance of winning. (AEB 1975)

Solution 2.52 (a) The possibility space S = (the pack of 52 cards).

There are 26 black cards in the pack, so with obvious notation,

$$(i) \quad P(B_1B_2B_3) = \left(\frac{26}{52}\right)\left(\frac{25}{51}\right)\left(\frac{24}{50}\right) \quad \text{(non-independent events)}$$

$$= \frac{2}{17}$$

The probability that the top cards are all black is $\frac{2}{17}$.

$$(ii) \quad P(H_1H_2H_3) = \left(\frac{13}{52}\right)\left(\frac{12}{51}\right)\left(\frac{11}{50}\right) \quad \text{(non-independent events)}$$

$$= \frac{11}{850}$$

The probability that the top cards are all hearts is $\frac{11}{850}$.

$$(iii) \quad P(A_1A_2A_3) = \left(\frac{4}{52}\right)\left(\frac{3}{51}\right)\left(\frac{2}{50}\right) \quad \text{(non-independent events)}$$

$$= \frac{1}{5525}$$

The probability that the top cards are all aces is $\frac{1}{5525}$.

If the three top cards have been examined and found not to be picture cards, the possibility space has been reduced by 3 members so that S consists of the 49 remaining cards.

The number of remaining queens = 4.

$$\text{Therefore} \quad P(Q_1Q_2Q_3) = \left(\frac{4}{49}\right)\left(\frac{3}{48}\right)\left(\frac{2}{47}\right) = \frac{1}{4606}$$

The probability that the three bottom cards are queens is $\frac{1}{4606}$.

(b) Let A_1 be the event 'A wins on the first draw',
A_2 be the event 'A wins on the second draw', and so on.

Then, with obvious notation for the black and white counters,

$$P(A_1) = P(\text{W}) = \frac{2}{10} = \frac{1}{5}$$

$$P(A_2) = P(\text{BBW}) = \left(\frac{8}{10}\right)\left(\frac{7}{9}\right)\left(\frac{2}{8}\right) = \frac{7}{45}$$

$$P(A_3) = P(\text{BBBBW}) = \left(\frac{8}{10}\right)\left(\frac{7}{9}\right)\left(\frac{6}{8}\right)\left(\frac{5}{7}\right)\left(\frac{2}{6}\right) = \frac{1}{9}$$

$$P(A_4) = P(\text{BBBBBBW})$$
$$= \left(\frac{8}{10}\right)\left(\frac{7}{9}\right)\left(\frac{6}{8}\right)\left(\frac{5}{7}\right)\left(\frac{4}{6}\right)\left(\frac{3}{5}\right)\left(\frac{2}{4}\right) = \frac{1}{15}$$

$$P(A_5) = P(\text{BBBBBBBBW})$$
$$= \left(\frac{8}{10}\right)\left(\frac{7}{9}\right)\left(\frac{6}{8}\right)\left(\frac{5}{7}\right)\left(\frac{4}{6}\right)\left(\frac{3}{5}\right)\left(\frac{2}{4}\right)\left(\frac{1}{3}\right)\left(\frac{2}{2}\right) = \frac{1}{45}$$

$$P(A \text{ wins}) = P(A_1) + P(A_2) + P(A_3) + P(A_4) + P(A_5)$$
(mutually exclusive events)
$$= \frac{1}{5} + \frac{7}{45} + \frac{1}{9} + \frac{1}{15} + \frac{1}{45}$$
$$= \frac{5}{9}$$

Therefore the probability that A wins is $\frac{5}{9}$.

Example 2.53 (a) Ruby Welloff, the daugher of a wealthy jeweller, is about to get married. Her father decides that as a wedding present she can select one of two similar boxes. Each box contains three stones. In one box two of the stones are real diamonds, and the other is a worthless imitation; and in the other box one is a real diamond, and the other two are worthless imitations. She has no idea which box is which. If the daughter were to choose randomly between the two boxes, her chance of getting two real diamonds would be $\frac{1}{2}$. Mr Welloff, being a sporting type, allows his daughter to draw one stone from one of the boxes and to examine it to see if it is a real diamond. The daughter decides to take the box that the stone she tested came from if the tested stone is real, and to take the other box otherwise. Now what is the probability that the daughter will get two real diamonds as her wedding present?

(b) A fair die is cast; then n fair coins are tossed, where n is the number shown on the die. What is the probability of exactly two heads?

(c) A fair die is thrown for as long as necessary for a 6 to turn up. Given that 6 does not turn up at the first throw, what is the probability that more than four throws will be necessary?

(AEB 1979)

Solution 2.53 (a) Let A be the event 'she chooses the box with 2 diamonds',
B be the event 'she chooses the box with 1 diamond',
D be the event 'she chooses a diamond from the box'.

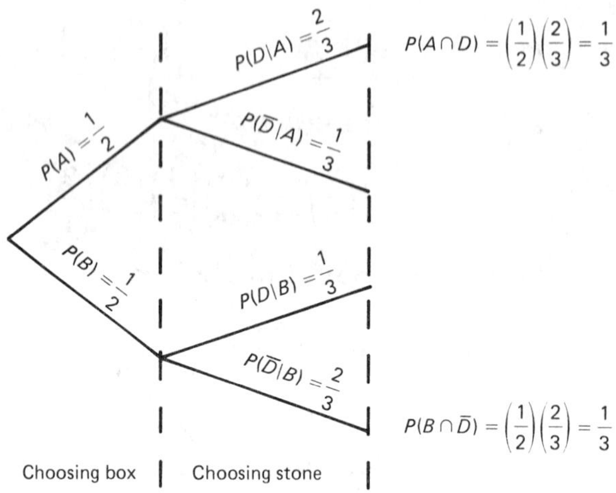

Now, she takes the box if the tested stone is real,

she takes the other box if the tested stone is worthless.

So the probability of getting 2 real diamonds $= P(A \cap D) + P(B \cap \bar{D})$

$$= \frac{1}{3} + \frac{1}{3}$$

$$= \frac{2}{3}$$

Therefore the probability that she has 2 diamonds for her wedding present is $\frac{2}{3}$.

(b)

	Number on the die	Number of heads and tails required
(*i*)	2	2H
(*ii*)	3	2H, 1T
(*iii*)	4	2H, 2T
(*iv*)	5	2H, 3T
(*v*)	6	2H, 4T

We will consider first the probability of obtaining the required number of heads and tails for each of the situations:

(*i*) *Two coins are tossed*

The possibility space consists of 2^2 equally likely outcomes.

So $$P(2\text{H}) = \frac{1}{4}$$

(*ii*) *Three coins are tossed*

The possibility space consists of 2^3 equally likely outcomes.

Number of ways of arranging H, H, T $= \dfrac{3!}{2!} = 3$

So $$P(2\text{H}, 1\text{T}) = \frac{3}{8}$$

(iii) Four coins are tossed

The possibility space consists of 2^4 equally likely outcomes.

Number of ways of arranging H, H, T, T $= \dfrac{4!}{2!2!} = 6.$

So $$P(2\text{H}, 2\text{T}) = \frac{6}{16} = \frac{3}{8}$$

(iv) Five coins are tossed

The possibility space consists of 2^5 equally likely outcomes.

Number of ways of arranging H, H, T, T, T $= \dfrac{5!}{2!3!} = 10.$

So $$P(2\text{H}, 3\text{T}) = \frac{10}{32} = \frac{5}{16}$$

(v) Six coins are tossed

The possibility space consists of 2^6 equally likely outcomes.

Number of ways of arranging H, H, T, T, T, T $= \dfrac{6!}{2!4!} = 15.$

So $$P(2\text{H}, 4\text{T}) = \frac{15}{64}$$

Let $P(n)$ be the probability that n is shown on the die

$$P(\text{exactly 2 heads}) = \frac{1}{4}P(2) + \frac{3}{8}P(3) + \frac{3}{8}P(4) + \frac{5}{16}P(5) + \frac{15}{64}P(6)$$

But $$P(2) = P(3) = P(4) = P(5) = P(6) = \frac{1}{6}$$

Therefore

$$P(\text{exactly 2 heads}) = \frac{1}{6}\left(\frac{1}{4} + \frac{3}{8} + \frac{3}{8} + \frac{5}{16} + \frac{15}{64}\right)$$
$$= \frac{33}{128}$$

The probability that exactly two heads are obtained is $\frac{33}{128}$.

(c) Let 6_1 be the event 'a 6 is obtained on the first throw' and so on.

We require
$$P(\bar{6}_1\bar{6}_2\bar{6}_3\bar{6}_4 \mid \bar{6}_1) = \frac{P(\bar{6}_1\bar{6}_2\bar{6}_3\bar{6}_4 \cap \bar{6}_1)}{P(\bar{6}_1)}$$
$$= \frac{(\frac{5}{6})(\frac{5}{6})(\frac{5}{6})(\frac{5}{6})}{\frac{5}{6}}$$
$$= \left(\frac{5}{6}\right)^3$$
$$= \frac{125}{216}$$

The probability that more than four throws will be necessary is $\frac{125}{216}$.

Example 2.54 **(a)** When a person needs a minicab, it is hired from one of three firms, X, Y and Z. Of the hirings 40% are from X, 50% are from Y and 10% are from Z. For cabs hired from X, 9% arrive late, the corresponding percentages for cabs hired from firms Y and Z being 6% and 20% respectively. Calculate the probability that the next cab hired

(*i*) will be from X and will not arrive late,

(*ii*) will arrive late.

Given that a call is made for a minicab and that it arrives late, find, to 3 decimal places, the probability that it came from Y.

(b) For a certain strain of wallflower, the probability that, when sown, a seed produces a plant with yellow flowers is $\frac{1}{6}$. Find the minimum number of seeds that should be sown in order that the probability of obtaining at least one plant with yellow flowers is greater than 0.98. (L)

Solution 2.54 **(a)**

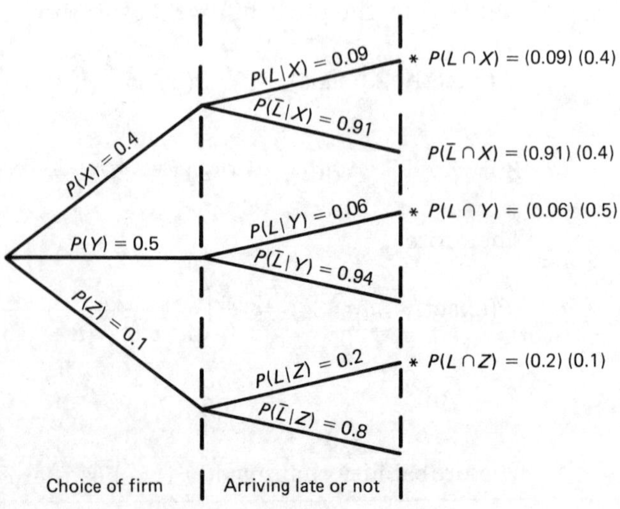

(*i*) From the diagram

$$P(\bar{L} \cap X) = P(\bar{L} \mid X) \cdot P(X) = (0.91)(0.4) = 0.364$$

Therefore the probability that the next cab hired will be from X and will not arrive late is 0.364.

(*ii*)
$$P(L) = P(L \cap X) + P(L \cap Y) + P(L \cap Z)$$
$$= P(L \mid X) \cdot P(X) + P(L \mid Y) \cdot P(Y) + P(L \mid Z) \cdot P(Z) \quad \text{(marked with * on diagram)}$$
$$= (0.09)(0.4) + (0.06)(0.5) + (0.2)(0.1)$$
$$= 0.086$$

Therefore the probability that the next cab hired will arrive late is 0.086.

We require
$$P(Y \mid L) = \frac{P(L \mid Y) \cdot P(Y)}{P(L)} = \frac{(0.06)(0.5)}{0.086} = 0.349 \quad \text{(3 d.p.)}$$

Therefore, given that the cab arrives late, the probability that it came from Y is 0.349 (3 d.p.).

(b) When a seed is sown, $P(\text{yellow flower}) = \frac{1}{6}$

When n seeds are sown,

$$P(\text{at least one yellow flower}) = 1 - P(\text{no yellow flowers}) = 1 - (\tfrac{5}{6})^n$$

Now we need

$$P(\text{at least one yellow flower}) > 0.98$$

so
$$1 - (\tfrac{5}{6})^n > 0.98$$
$$(\tfrac{5}{6})^n > 0.02$$

Taking logs of both sides $\quad n \log(\tfrac{5}{6}) < \log 0.02$

Dividing both sides by $\log(\frac{5}{6})$ and reversing the inequality since $\log(\frac{5}{6})$ is negative, we have

$$n > \frac{\log 0.02}{\log(\frac{5}{6})}$$
$$n > 21.45\ldots$$

Therefore the minimum number of seeds that should be sown is 22.

Miscellaneous Exercise 2m

1. (*a*) Two dice are thrown together, and the scores added. What is the probability that (i) the total score exceeds 8? (ii) the total score is 9, or the individual scores differ by 1, or both?
(*b*) A bag contains 3 red balls and 4 black ones. 3 balls are picked out, one at a time and not replaced. What is the probability that there will be 2 red and 1 black in the sample?
(*c*) A committee of 4 is to be chosen from 6 men and 5 women. One particular man and one particular woman refuse to serve if the other person is on the committee. How many different committees may be formed? (SUJB)

2. In Camelot it never rains on Friday, Saturday, Sunday or Monday. The probability that it rains on a given Tuesday is $\frac{1}{5}$. For each of the remaining two days, Wednesday and Thursday, the conditional probability that it rains, given that it rained the previous day, is α, and the conditional probability that it rains, given that it did not rain the previous day, is β.
(*a*) Show that the (unconditional) probability of rain on a given Wednesday is $\frac{1}{5}(\alpha + 4\beta)$, and find the probability of rain on a given Thursday.
(*b*) If X is the event that, in a randomly chosen week, it rains on Thursday, Y is the event that it rains on Tuesday, and $\overline{Y}$ is the event that it does not rain on Tuesday, show that
$$P(X\mid Y)-P(X\mid \overline{Y}) = (\alpha-\beta)^2$$
(*c*) Explain the implications of the case $\alpha = \beta$. (Cambridge)

3. Mass-produced glass bricks are inspected for defects. The probability that a brick has air bubbles is 0.002. If a brick has air bubbles the probability that it is also cracked is 0.5 while the probability that a brick free of air bubbles is cracked is 0.005. What is the probability that a brick chosen at random is cracked?
The probability that a brick is discoloured is 0.006. Given that discolouration occurs independently of the other two defects, find the probability that a brick chosen at random has no defects. (O & C)

4. A bag contains red, blue and green counters of equal size and shape. A counter is taken at random from the bag. The probability that it is red is 1.5 times the probability that it is blue, and the probability that it is blue is twice the probability that it is green. Find the probabilities that the counter is (*a*) red, (*b*) blue, (*c*) green.
A counter is taken at random from the bag, its colour is noted and it is then replaced in the bag. The process continues until at least one of each colour has been seen. Considering the order in which the colours are first seen, find the probabilities that (*d*) red is seen before green, (*e*) the order is green, blue and finally red. (O & C)

5. There are eight girls and ten boys in the upper sixth form of a small school. Six prefects are to be selected. In how many ways can this be done if (*a*) there must be three girl and three boy prefects; (*b*) there must be at least four boy prefects?
Amongst the eighteen pupils, there is a pair of twins, one girl and one boy. The Headmaster has decided that there must be three girl and three boy prefects. Find the probability that both of the twins are selected. (SUJB Additional)

6. The events A and B are such that
$$P(A) = \frac{1}{2}$$
$$P(A'\mid B) = \frac{1}{3}$$
$$P(A\cup B) = \frac{3}{5}$$
where A' is the event 'A does not occur'. Using a Venn diagram, or otherwise, determine $P(B\mid A')$, $P(B\cap A)$ and $P(A\mid B')$.
The event C is independent of A and $P(A\cap C) = \frac{1}{8}$. Determine $P(C\mid A')$.
State, with a reason in each case, whether (*a*) A and B are independent, (*b*) A and C are mutually exclusive. (Cambridge)

7. If p_1 and p_2 are the probabilities of two independent events, show that the probability of the simultaneous occurrence of these two events is p_1p_2.
In 18 games of chess between A and B, A wins 8, B wins 6 and 4 are drawn. A and B play a tournament of 3 games.
On the basis of the above data, estimate the probability that: (*a*) A wins all three

games, (*b*) A and B win alternately, (*c*) two games are drawn, (*d*) A wins at least one game. (AEB 1972)

8. (*a*) From an ordinary pack of 52 cards two are dealt face downwards on a table. What is the probability that (i) the first card dealt is a heart, (ii) the second card dealt is a heart, (iii) both cards are hearts, (iv) at least one card is a heart?
(*b*) Bag A contains 3 white counters and 2 black counters whilst bag B contains 2 white and 3 black. One counter is removed from bag A and placed in bag B without its colour being seen. What is the probability that a counter removed from bag B will be white?
(*c*) A box of 24 eggs is known to contain 4 old and 20 new eggs. If 3 eggs are picked at random determine the probability that (i) 2 are new and the other old, (ii) they are all new. (SUJB)

9. The probabilities of A, B or C winning a game in which all three take part are 0.5, 0.3 and 0.2 respectively. A match is won by a player who first wins two games. Find the probability that A will win a game involving all three players.
When the players are joined by a fourth player, D, the probabilities of A, B or C winning a game, in which all four take part, are reduced to 0.3, 0.2 and 0.1 respectively. A match is played in which all four players take part; again, the first player to win two games wins the match. Find the probabilities that D wins in fewer than (i) four games, (ii) five games, (iii) six games. (JMB)

10. A bag contains 5 red, 4 orange and 3 yellow sweets. One after another 3 children select and eat one sweet each. When the bag contains n sweets, the probability of any one child choosing any particular sweet is $1/n$. What are the probabilities that (*a*) they all choose red sweets, (*b*) at least one orange sweet is chosen, (*c*) each chooses a different colour, (*d*) all choose the same colour? Answers may be left as fractions in their lowest terms. (O & C)

11. Three men, A, B and C agree to meet at the theatre. The man A cannot remember whether they agreed to meet at the Palace or the Queen's and tosses a coin to decide which theatre to go to. The man B also tosses a coin to decide between the Queen's and the Royalty. The man C tosses a coin to decide whether to go to the Palace or not and in this latter case he tosses again to decide between the Queen's and the Royalty. Find the probability that (*a*) A and B meet, (*b*) B and C meet, (*c*) A, B and C all meet, (*d*) A, B and C all go to different places, (*e*) at least two meet. (C)

12. In a game, three cubical dice are thrown by a player who attempts to throw the same number on all three. What is the chance of the player
(*a*) throwing the same number on all three?
(*b*) throwing the same number on just two?
If the first throw results in just two dice showing the same number, then the third is thrown again. If no two dice show the same number, then all are thrown again. The player then comes to the end of his turn. What is the chance of the player succeeding in throwing three identical numbers in a complete turn?
What is the chance that all the numbers are different at the end of a turn? (O & C)

13. Alec and Bill frequently play each other in a series of games of table tennis. Records of the outcomes of these games indicate that whenever they play a series of games, Alec has the probability 0.6 of winning the first game and that in every subsequent game in the series, Alec's probability of winning the game is 0.7 if he won the preceding game but only 0.5 if he lost the preceding game. A game cannot be drawn. Find the probability that Alec will win the third game in the next series he plays with Bill. (JMB)

14. The events A and B are such that

$$P(A) = \frac{1}{2}$$

$$P(A \text{ or } B \text{ but not both } A \text{ and } B) = \frac{1}{3}$$

$$P(B) = \frac{1}{4}$$

Calculate $P(A \cap B)$, $P(A' \cap B)$, $P(A|B)$ and $P(B|A')$, where A' is the event 'A does not occur'. State, with reasons, whether A and B are (*a*) independent, (*b*) mutually exclusive. (C)

15. (*a*) Two men each have a set of 7 cards, numbered 1 to 7. Each shows a card drawn at random. Find the probability that the total of the two numbers is (i) even, (ii) odd, (iii) greater than 5.
(*b*) A signal consisting of 7 dots and/or dashes is to be given. The probability of a dot in any position is 2/5 and of a dash is

3/5. Find the probability that, in a signal, no two consecutive characters are the same.
(*c*) A die is loaded so that the chance of throwing a one is $x/4$, the chance of a two is 1/4 and the chance of a six is $(1-x)/4$. The chance of a three, four or five is 1/6. The die is thrown twice.

Prove that the chance of throwing a total of 7 is $\dfrac{9x-9x^2+10}{72}$.

Find the value of x which will make this chance a maximum, and find this maximum probability. (SUJB)

16. 4 girls and 3 boys plan to meet together on the following Saturday. The probability that each boy will be present is $\frac{2}{3}$ independently of the other boys. Find the probability that (*a*) 0, (*b*) 1, (*c*) 2, (*d*) 3 boys will be present.

The probability that each girl will be present is $\frac{1}{2}$ independently of the other girls and of the boys.
(*e*) Find the probability that the number of girls present will equal the number of boys.
(*f*) Find the probability that both sexes will be present.
(*g*) Afterwards it was reported that the gathering had included at least one boy and at least one girl. What is the probability that there were equal numbers of boys and girls in the light of this additional information?

(Answers may be left as fractions in their lowest terms.) (O & C)

17. A sailing competition between two boats, *A* and *B*, consists of a series of independent races, the competition being won by the first boat to win three races. Every race is won by either *A* or *B*, and their respective probabilities of winning are influenced by the weather. In rough weather the probability that *A* will win is 0.9; in fine weather the probability that *A* will win is 0.4. For each race the weather is either rough or fine, the probability of rough weather being 0.2. Show that the probability that *A* will win the first race is 0.5.

Given that the first race was won by *A*, determine the conditional probability that (*a*) the weather for the first race was rough, (*b*) *A* will win the competition. (C)

18. Six fuses, of which two are defective and four are good, are to be tested one after another in random order until both defective fuses are identified. Find the probability that the number of fuses that will be tested is
(*a*) three,
(*b*) four or fewer. (L)P

19. In this question you may leave the answers as fractions. Your arguments must be carefully explained in both parts.

(*a*) A pack of ten cards consists of two marked with the letter A, three with E, four with S and one with T. The pack is well shuffled and six cards are dealt. Find the probability that (i) they form the word ASSETS, the letters appearing in that order; (ii) the letters either form or can be made to form the word ASSETS.
(*b*) A manufacturer of tea inserts one of five types of picture card into each packet. Equal numbers of each type are distributed randomly. Estimate the probability that a person buying three packets will have (i) three cards of the same type, (ii) just two the same. If a person buys five packets, estimate the probability of obtaining five different types of card. (SUJB)

20. A census of married couples showed that 50% of the couples had no car, 40% had one car and the remaining 10% had two cars. Three of the married couples are chosen at random.
(*a*) Find the probability that one couple had no car, one has one car and one has two cars.
(*b*) Find the probability that the three couples have a combined total of three cars.
The census also showed that both the husband and the wife were in full-time employment in 16% of those couples having no car, in 45% of those having one car and in 60% of those having two cars.
(*c*) For a randomly chosen married couple find the probability that both the husband and wife are in full-time employment.
(*d*) Given that a randomly chosen married couple is one where both the husband and wife are in full-time employment, find the conditional probability that the couple has no car. (JMB)

21. Three machines *A*, *B* and *C* produce 25%, 25% and 50% respectively of the output of a factory manufacturing a certain article. A sample of 3 articles is selected at random from the total output. Find

the probabilities that (a) they are all from C, (b) at least 2 are from B.

If a second independent sample of 3 articles is selected, find the probability that both samples have the same number of articles produced by A.

Of the articles produced by A, B and C, 1%, 2% and 5% respectively are defective.

A single article is selected at random. If D denotes the event 'defective' and C the event 'produced by machine C', find $p(D)$ and $p(C \text{ and } D)$.

An article is examined and found to be defective. What is the probability that it was produced by C? (SMP)

22. (a) The events A and B are such that $P(A)=0.6, P(B)=0.25, P(A\cup B)=0.725$. Show that the events A and B are neither mutually exclusive nor independent. Calculate the values of $P(\bar{A}\cup\bar{B})$ and $P(A\,|\,\bar{B})$.
(b) One red card and two black cards are removed from a pack of cards. From the remainder, three cards are taken at random without replacement. Show that the probability that they are all of the same colour is $\frac{23}{98}$. Assuming that this event occurs, find the probability that a fourth card drawn from the remaining 46 cards will be of the same colour as the previous three. (L Additional)

23. Events A and B are such that $P(A)=\frac{5}{12}$, $P(A\,|\,\bar{B})=\frac{7}{12}$, $P(A\cap B)=\frac{1}{8}$. Find (a) $P(B)$, (b) $P(A\,|\,B)$, (c) $P(B\,|\,A)$, (d) $P(A\cup B)$.

State whether events A and B are (a) mutually exclusive, (b) independent.

24. The following are three of the classical problems in probability.
(a) Compare the probability of a total of 9 with the probability of a total of 10 when three fair dice are tossed once (Galileo and Duke of Tuscany).
(b) Compare the probability of at least one six in four tosses of a fair die with the probability of at least one double-six in twenty-four tosses of two fair dice (Chevalier de Méré).
(c) Compare the probability of at least one 6 when six dice are tossed with the probability of at least two sixes when 12 dice are tossed (Pepys to Newton).

Solve each of these problems. (AEB 1978)

25. A set consists of 12 observations no two of which are equal. Five of the observations are selected at random. What are the probabilities that
(a) the five observations include the largest and the least among the 12 observations,
(b) the second largest and the second smallest will be included,
(c) the five smallest observations are included,
(d) at least three of the smallest five observations are included? (MEI)

26. In a class of 30 pupils, 12 walk to school, 10 travel by bus, 6 cycle and 2 travel by car. If 4 pupils are picked at random, obtain the probabilities that (a) they all travel by bus, (b) they all travel by the same means.

If 2 are picked at random from the class, find the probability that they travel by different means.

In picking out pupils from the class, find the probability that more than three trials are needed before a pupil who walks to school is selected. (JMB)

27. Four ball-point pen refills are to be drawn at random without replacement from a bag containing ten refills, of which 5 are red, 3 are green and 2 are blue. Find
(a) the probability that both blue refills will be drawn,
(b) the probability that at least one refill of each colour will be drawn. (JMB)

28. At the ninth hole on a certain golf course there is a pond. A golfer hits a grade B ball into the pond. Including the golfer's ball there are then 6 grade C, 10 grade B and 4 grade A balls in the pond. The golfer uses a fishing net and 'catches' four balls. The events X, Y and Z are defined as follows:

X: the catch consists of two grade A balls and two grade C balls

Y: the catch consists of two grade B balls and two other balls

Z: the catch includes the golfer's own ball

Assuming that the catch is a random selection from the balls in the pond, determine

(a) $P(X)$, (b) $P(Y)$, (c) $P(Z)$, (d) $P(Z\,|\,Y)$.

For each of the pairs X and Y, Y and Z, state, with a brief reason, whether the two events are (i) mutually exclusive, (ii) independent. (C)

29. A committee of 8 members consists of one married couple together with 4 other men and 2 other women. From the committee a working party of 4 persons is to be formed. Find the number of

different working parties which can be formed.

Find also the number if the working party
(*a*) may not contain *both* the husband and his wife,
(*b*) must contain 2 men and 2 women,
(*c*) must contain at least one man and at least one women.

The 8 committee members sit round an octagonal table, their positions being decided by drawing lots. Find the probability of
(*d*) the man sitting next to his wife,
(*e*) the man sitting opposite to his wife,
(*f*) the 3 women sitting together. (AEB)

30. In a game of chançe, a player's turn starts by drawing a card at random from a pack of playing cards. If he draws a black card which is not an ace, his turn ends. If he draws a black ace he throws a black die, and if he draws a red card he throws a red die. After a die has been thrown, the card that was drawn is replaced in the pack which is then shuffled and the player draws again with the same conditions leading to the throwing of a die. This continues until the player draws a black card, which is not an ace, when his turn ends. A player's score in any turn is the sum of the scores thrown with the red die plus three times the sum of the scores thrown with the black die. Calculate the probability that in a turn a player will score (*a*) zero, (*b*) exactly three. (L)

31. (*a*) Two cards are drawn at random without replacement, from an ordinary pack of 52 cards. Find the probability that they are: (i) of the same suit, (ii) of the same value (both aces, both kings, etc.), (iii) either of the same suit or of the same value.
(*b*) Two cards are drawn at random, one from each of two ordinary packs. Find the probability that they are (i) of the same suit, (ii) of the same value, (iii) either of the same suit, the same value, or both.
(*c*) Three fair cubical dice are thrown. Find the probability that the sum of the number of spots on the upper faces is a perfect square. (AEB 1976)

32. A card is drawn from a full pack of 52 playing cards. If the card drawn is an Ace, King, Queen or Jack, two dice are thrown and the sum of the scores on the dice noted. If any other card is drawn, one die only is thrown and the sum of the score on the card and die noted. X denotes the event 'Both dice are thrown', and Y denotes the event 'The score noted is less than five.' Calculate the probabilities
(*a*) $P(X)$, (*b*) $P(X \cap Y)$, (*c*) $P(Y)$, (*d*) $P(Y|X)$, (*e*) $P(X|Y)$. (C)

33. In a constituency containing many elderly inhabitants there are twice as many women as men. At an election seven-eighths of the women and half the men cast a vote. Show that the probability that an adult inhabitant (selected at random) casts a vote is 3/4. For a random group of four inhabitants, find
(*a*) the probability that just one of them votes:
(*b*) the probability that two or more vote.

It is further found that for married couples the probability that a man votes is $\frac{4}{5}$, the probability that a woman votes is $\frac{6}{7}$, the probability that a woman votes given that her husband votes is $\frac{9}{10}$, and the probability that a man votes given that his wife votes is $\frac{21}{25}$ (you may assume that this information is consistent). Find
(*c*) the probability that a husband and wife both vote;
(*d*) the probability that a husband votes and his wife does not vote;
(*e*) the expected number of votes per married couple.* (SMP)

34. In a single round of a general knowledge contest, each competitor is first asked a question. If the competitor answers correctly, then that competitor is asked another question. This continues until either the competitor has answered five questions correctly, in which case the competitor scores six points (including a bonus point), or until the competitor answers a question incorrectly, in which case the competitor's score in that round is equal to the number of correct answers given.

One of the competitors is named Smith. The probability that Smith answers a question correctly is p, independent of all previous answers. Determine the probability distribution of Smith's score in a single round, and show that Smith's expected score is $p(1+p+p^2+p^3+2p^4)$*.

At the start of the final round of the contest Smith is 3 points ahead of Jones, and Smith and Jones are then the only

*Expectation required — see p. 171.

competitors who can win the contest. The probability that Jones answers a question correctly is also p, independent of all previous answers. Show that the probability that Jones wins the contest is $p^4(1-p)(1+p^2+p^3)$.

Given that Jones wins the contest, determine the probability that he scored 6 in the final round. (C)

35. Two men are walking directly towards each other on a wide pavement, along the same line. When they are six paces apart, they realise that they are in danger of colliding. With each of his next three steps forward therefore, each pedestrian adopts the following strategy: if the two are still in line with each other each independently steps half-left with probability p, or steps half-right with probability p, or keeps straight on with probability $1-2p$; if they are not still in line, each keeps walking straight forward (the diagram illustrates one possible version of the encounter). Calculate the probability that they are still in line after each has taken his first step, and deduce that the probability of a collision is $(1-4p+6p^2)^3$.

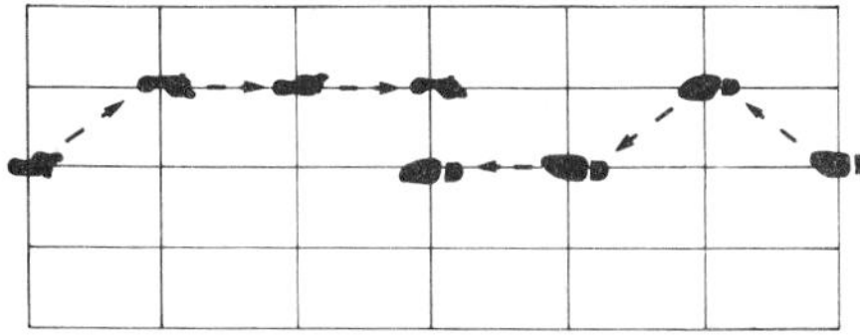

(SMP)

36. A college has 750 women students and 2250 male students. There is a higher proportion of male students in engineering, physics and similar subjects so that 60% of male students study mathematics and only 30% of women students study it. If one student studying mathematics is chosen at random from all the students studying it, what is the probability that the student will be a woman?

If three students studying mathematics are chosen, what is the probability that there will be at least two men?

25% of all the students study French. The proportion of male students of mathematics who also study French is 20% and the proportion of women students of French who also study mathematics is 20%. There are 500 male students of French. If four students are selected what is the probability that at least one is male and at least one is studying mathematics and at least one is studying French? (MEI)

37. A company makes a certain type of fan heater (called an X-heater) at each of its two factories F_1 and F_2. The factory F_1 produces one quarter and F_2 three quarters of the total output. X-heaters are coloured either red or blue. One third of the X-heaters produced at F_1 are red and seven-ninths of the X-heaters produced at F_2 are red.

A customer goes into a shop and selects an X-heater at random. Show the probability is $\frac{2}{3}$ that when he unpacks it he will find that it is red.

Two shops A and B stock X-heaters. Shop A has four and shop B has three. Find

(*a*) the probability that neither shop has a red X-heater;

(*b*) the probability that there are at least 3 X-heaters in shop A;

(*c*) the probability that there are the same number of red X-heaters in each shop;

(*d*) the probability that there are two red X-heaters in each shop, given that all the X-heaters in shop A come from F_1 and that all the X-heaters in shop B come from F_2.

(You may leave all your answers as fractions with powers of 3 as denominators.) (SMP)

38. (*a*) Find the number of ways in which 10 people can be divided into

(i) two groups consisting of 7 and 3 people,

(ii) three groups consisting of 4, 3 and 2 people with 1 person rejected.

(*b*) Seven coins of which 3 are silver and 4 are copper are in a box. A random selection of 3 coins is made and the coins selected are placed in a purse (purse A). The remaining coins are placed in a second purse (purse B). Find the probabilities associated with each of the possible numbers of silver coins (ranging from 0 to 3) in purse A.

On a particular occasion it is known that purse A has in it 2 silver coins and 1 copper coin, and that the remaining coins are in purse B. If one coin is then drawn at random from a purse selected at random, find the probability that the coin is silver. (C)

39. X_1, X_2 and X_3 are three independent events with probabilities of occurrence $Pr(X_i) = p_i$, $i = 1, 2, 3$. Give the probabilities of the occurrence of 0, 1, 2 and 3 events respectively and verify that these

probabilities satisfy the conditions for a distribution.

If

$$Pr(\text{at least one event}) = 0.664$$
$$Pr(\text{at least two events}) = 0.212,$$

and

$$Pr(\text{at most two events}) = 0.976,$$

find the probabilities of exactly 0, 1, 2 and 3 events respectively.

By considering a linear combination of Pr(one event) and Pr(two events) and Pr(three events) find the value of $p_1 + p_2 + p_3$. (MEI)

40. (*a*) Two cards are drawn from a well shuffled pack of 52 playing cards. If Jacks, Queens and Kings count 10 points, aces count 1 point, and the rest count points equal to their face values, what is the chance that the total points of the two cards will be 12?
(*b*) If three cards are drawn in succession from a complete pack, what is the probability that the first two cards score 12 points and the total points will be less than 21? (O & C)

41. A hand of 13 cards is dealt from a standard pack of 52 cards (which consists of 4 suits, clubs, diamonds, hearts, and spades, each of 13 cards).
(*a*) Write down, but do not calculate, an expression for the probability that the hand consists of 3 spades, 4 hearts, and 6 cards from the other suits.
(*b*) Calculate, to 3 d.p., the conditional probability that the hand contains exactly 3 diamonds, given that it contains exactly 3 spades and 4 hearts.
(*c*) Calculate, again to 3 d.p., the probability that the hand contains at least 2 diamonds given the same conditions as in part (*b*). (O)

42. Show that the total number of random samples of size r that can be drawn from a population of size n is

$$\frac{n!}{r!(n-r)!}$$

In a class of 10 boys and 10 girls there are 5 children with blue eyes. A random sample of 4 children is taken. Find the probabilities that in this sample there are exactly (*a*) 2 boys, (*b*) 2 children with blue eyes.

Half the children living in a big city are boys, and one quarter of the children have blue eyes. A random sample of 4 children is taken. Estimate the probabilities that in this sample there are exactly (*c*) 2 boys, (*d*) 2 children with blue eyes. (MEI)

43. A committee has 22 members, of whom 7 have black hair, do not smoke and do not wear glasses; 5 have white hair, do not smoke and do not wear glasses; 4 have white hair, smoke and wear glasses; 3 have black hair, smoke and do not wear glasses, 2 have white hair, do not smoke and wear glasses; 1 has black hair, smokes and wears glasses.
(*a*) One committee member is chosen at random. Let W be the event that this member has white hair, G be the event that this member wears glasses and S the event that the member smokes. Find (i) $P(W)$, (ii) $P(W|S)$, (iii) $P(W|G)$, (iv) the probability that this member has either white hair or glasses (but not both), given that this member smokes. Are the events W and S independent? Are the events W and G independent? Give a reason for each answer.
(*b*) Two committee members are chosen at random. Let W_2 be the event that both have white hair. Let S_2 be the event that both smoke. Find (i) $P(W_2)$, (ii) $P(W_2|S_2)$. (C)

44. (*a*) How many odd numbers can be formed from the figures 1, 2, 3 and 5 if repetitions are not allowed?
(*b*) *See worked example*, p. 135.
(*c*) Six different books lie on a table, and a boy is told that he can take away as many as he likes but he must not leave empty handed. How many different selections can he make? One of the books is a Bible. How many of these selections will include this Bible? (SUJB)P

45. (*a*) Of the households in Edinburgh, 35% have a freezer and 60% have a colour TV set. Given that 25% of the households have both a freezer and a colour TV set, calculate the probability that a household has either a freezer or a colour TV set but not both.

State, with your reasons, whether the events of having a freezer and of having a colour TV set are or are not independent.

(*b*) State in words the meaning of the symbol $P(B|A)$, where A and B are two events.

A shop stocks tinned cat food of two makes, A and B, and two sizes, large and small. Of the stock, 70% is of brand A, 30% is of brand B. Of the tins of brand A, 30% are small size whilst of the tins of brand B, 40% are small size. Using a tree diagram, or otherwise, find the probability that

(i) a tin chosen at random from the stock will be of small size,

(ii) a small tin chosen at random from the stock will be of brand A. (L)

46. During an epidemic of a certain disease a doctor is consulted by 110 people suffering from symptoms commonly associated with the disease. Of the 110 people, 45 are female of whom 20 actually have the disease and 25 do not. Fifteen males have the disease and the rest do not.
(*a*) A person is selected at random. The event that this person is female is denoted by A and the event that this person is suffering from the disease is denoted by B. Evaluate (i) $P(A)$, (ii) $P(A \cup B)$, (iii) $P(A \cap B)$, (iv) $P(A|B)$.
(*b*) If three different people are selected at random without replacement, what is the probability of (i) all three having the disease, (ii) exactly one of the three having the disease, (iii) one of the three being a female with the disease, one a male with the disease and one a female without the disease?
(*c*) Of people with the disease 96% react positively to a test for diagnosing the disease as do 8% of people without the disease. What is the probability of a person selected at random (i) reacting positively, (ii) having the disease given that he or she reacted positively? (AEB 1987)

47. In a simple model of the weather in October, each day is classified as either fine or rainy. The probability that a fine day is followed by a fine day is 0.8. The probability that a rainy day is followed by a fine day is 0.4. The probability that 1 October is fine is 0.75.
(*a*) Find the probability that 2 October is fine and the probability that 3 October is fine.
(*b*) Find the conditional probability that 3 October is rainy, given that 1 October is fine.
(*c*) Find the conditional probability that 1 October is fine, given that 3 October is rainy. (C)

48. Two archers A and B shoot alternately at a target until one of them hits the centre of the target and is declared the winner. Independently, A and B have probabilities of $\frac{1}{3}$ and $\frac{1}{4}$, respectively, of hitting the centre of the target on each occasion they shoot.
(*a*) Given that A shoots first, find (i) the probability that A wins on his second shot, (ii) the probability that A wins on his third shot, (iii) the probability that A wins.
(*b*) Given that the archers toss a fair coin to determine who shoots first, find the probability that A wins. (JMB)

49. (*a*) Explain in words the meaning of the symbol $P(A|B)$ where A and B are two events. State the relationship between A and B when (i) $P(A|B) = 0$, (ii) $P(A|B) = P(A)$.
When a car owner needs her car serviced she phones one of three garages, A, B, or C. Of her phone calls to them, 30% are to garage A, 10% to B and 60% to C. The percentages of occasions when the garage phoned can take the car in on the day of phoning are 20% for A, 6% for B and 9% for C. Find the probability that the garage phoned will *not* be able to take the car in on the day of phoning.
Given that the car owner phones a garage and the garage can take her car in on that day, find the probability that she phoned garage B.
(*b*) A shelf contains ten box files of which four are empty and six contain papers. Five files are chosen at random one after another from the shelf. Find, to 3 decimal places, the probability that exactly two of the chosen files will be empty when the files are chosen (i) with replacement, (ii) without replacement. (L)

50. Show that, for any two events E and F

$$P(E \cup F) = P(E) + P(F) - P(E \cap F)$$

Express in words the meaning of $P(E|F)$.
Given that E and F are independent events, express $P(E \cap F)$ in terms of $P(E)$ and $P(F)$, and show that E' and F are also independent.
In a college, 60 students are studying one or more of the three subjects Geography, French and English. Of these, 25 are studying Geography, 26 are studying French, 44 are studying English, 10 are studying Geography and French, 15 are studying French and English, and 16 are studying Geography and English. Write down the probability that a student chosen at random from those studying English is also studying French. Determine whether or not the events 'studying Geography' and 'studying French' are independent.
A student is chosen at random from all 60 students. Find the probability that the chosen student is studying all three subjects. (L)

51. Explain, by suitably defining events A and B, what is meant by 'the probability of A occurring given that B has occurred'.
A local greengrocer sells conventionally grown and organically grown vegetables. Conventionally grown vegetables constitute 80% of his sales; carrots constitute 12% of the conventional sales and 30% of the organic sales.
Display this information in an appropriately and accurately labelled tree diagram.
One day a customer emerges from the shop and is questioned about her purchases. What is the probability that she bought
(*a*) conventionally grown carrots,
(*b*) carrots?
Given that she did buy carrots, what is the probability that they were organically grown? What assumptions have you made in answering this question? (O)

52. (*a*) In a group of 200 people, each individual is classified as either male or female and according to whether or not he or she wears glasses. The numbers falling into each category are as tabulated.

	Not wearing glasses	Wearing glasses
Male	90	24
Female	66	20

Suppose one of this group is chosen at random. Let A be the event that the person chosen is male and B the event that the person chosen is not wearing glasses.
(i) Define the events A' and $A \cup B'$.
(ii) Calculate the probability of occurrence of each of the events in (i).
(iii) Given that the person chosen is not wearing glasses, calculate the probability that this person is male.
(iv) Use the available data to determine whether not wearing glasses is independent of sex *within the group*. Give a practical interpretation to your finding.
(*b*) After advertising for an assistant, a manager decides to interview suitable applicants. The interview of an applicant will take place during the morning or the afternoon with probabilities 0.45 and 0.55 respectively. Each applicant is informed by telephone and in each case a message has to be left. A morning interview is wrongly transmitted to the applicant as an afternoon interview with probability 0.2, and an afternoon interview is wrongly transmitted to the applicant as a morning interview with probability 0.1. Find the probability that an applicant arriving
(i) for a morning interview is expected for a morning interview,
(ii) for an afternoon interview is expected for an afternoon interview. (AEB 1988)

53. (*a*) A bag contains 4 red, 6 white and 5 blue balls. If a random sample of 6 balls is selected (without replacement) what is the probability that there are two balls of each colour?
(*b*) A number N consists of n digits each of which can be 0 or 1. It is copied onto a sheet of paper by A and the probability that A transcribes any digit wrongly is p. The sheet of paper is then passed to B who copies the number onto another sheet of paper. The probability that B transcribes any digit wrongly is p'. What is the probability that the number written by B contains no error?
(*c*) An assembly plant receives 60% of its resistors from supplier X and 40% from supplier Y. 5% of X's resistors and 6% of Y's are defective. If a resistor is tested at the plant and found to be defective, what is the probability that it was supplied by X? (SUJB)

54. A and B are mutually exclusive and exhaustive events in a sample space S and C is any event in S for which $P(C) \neq 0$. Show, by considering a Venn diagram or otherwise, that

$$P(A|C) = \frac{P(C|A)\cdot P(A)}{P(C|A)\cdot P(A) + P(C|B)\cdot P(B)}$$

When an insurance company sells a policy, it classifies the client as either class A (good risk) or as class B. 25% of the clients are class A. Records show that the probability of a client making a claim during any year is 0.08 for class A and 0.35 for class B.
(*a*) Mr Smith buys a policy and makes a claim during his first year. Calculate, each to 3 decimal places, the probability that Mr Smith was originally classified A or was originally classified B.
(*b*) Mrs Jones bought a policy two years ago and has not made a claim during that time. Show that she is more likely to be class B.

Show, also, that if she does not make a claim for a further two years she is more likely to be class A than class B. What do you need to assume for your calculations to be valid? (SUJB)

3

PROBABILITY DISTRIBUTIONS I — DISCRETE RANDOM VARIABLES

DISCRETE RANDOM VARIABLE

Let X have the following properties:

(a) it is a discrete variable,

(b) it can only assume values $x_1, x_2, \ldots, x_n$,

(c) the probabilities associated with these values are $p_1, p_2, \ldots, p_n$,

where
$$P(X = x_1) = p_1$$
$$P(X = x_2) = p_2$$
$$\vdots$$
$$P(X = x_n) = p_n$$

then X is a discrete **random** variable if $p_1 + p_2 + \ldots + p_n = 1$.

This can be written

$$\sum p_i = 1, \qquad i = 1, 2, \ldots, n$$

or
$$\sum_{\text{all } x} P(X = x) = 1$$

We usually denote a random variable (r.v.) by a capital letter (X, Y, R, etc.) and the particular value it takes by a small letter (x, y, r, etc.).

Example 3.1 Let X be the discrete variable 'the number of fours obtained when two dice are thrown'. Show that X is a random variable.

Solution 3.1 With regard to the number of fours thrown, the outcome could be one of the following: 0 fours, or 1 four, or 2 fours.

Therefore X can assume the values 0, 1 and 2 only.

Then, with obvious notation,

$$P(X=0) = P(\bar{4}\bar{4}) = \left(\frac{5}{6}\right)\left(\frac{5}{6}\right) = \frac{25}{36}$$

$$P(X=1) = P(4\bar{4})+P(\bar{4}4) = \left(\frac{1}{6}\right)\left(\frac{5}{6}\right)+\left(\frac{5}{6}\right)\left(\frac{1}{6}\right) = \frac{10}{36}$$

$$P(X=2) = P(4\,4) = \left(\frac{1}{6}\right)\left(\frac{1}{6}\right) = \frac{1}{36}$$

Now
$$\sum_{\text{all } x} P(X=x) = \frac{25}{36}+\frac{10}{36}+\frac{1}{36} = \frac{36}{36} = 1$$

Therefore X is a discrete random variable.

PROBABILITY DENSITY FUNCTION

We can write the results obtained in Example 3.1 in table form:

x	0	1	2
$P(X=x)$	$\frac{25}{36}$	$\frac{10}{36}$	$\frac{1}{36}$

This is known as the **probability distribution** of X.

The function which is responsible for allocating probabilities is known as the **probability density function** (p.d.f.) of X.

In Example 3.1, the p.d.f. of X is given by $P(X=x)$ for $x=0,1,2$.

Sometimes the p.d.f. can be expressed as a formula, as in the following example.

Example 3.2 Two tetrahedral dice, each with faces labelled 1, 2, 3 and 4 are thrown and the score noted, where the score is the sum of the two numbers on which the dice land. If X is the r.v. 'the score when two tetrahedral dice are thrown', find the p.d.f. of X.

Solution 3.2 The score for each possible outcome is shown in the table:

'Score'

Second die				
4	5	6	7	8
3	4	5	6	7
2	3	4	5	6
1	2	3	4	5
	1	2	3	4
	First die			

From the table we can see that X can assume the values 2, 3, 4, 5, 6, 7, 8 only.

The probabilities can be found from the table, as each outcome shown is equally likely.

For example $P(X=5)=\frac{4}{16}$ as 4 out of the total of 16 outcomes result in a score of 5.

The probability distribution is formed:

x	2	3	4	5	6	7	8
$P(X=x)$	$\frac{1}{16}$	$\frac{2}{16}$	$\frac{3}{16}$	$\frac{4}{16}$	$\frac{3}{16}$	$\frac{2}{16}$	$\frac{1}{16}$

This can be written as a formula, giving the p.d.f. of X as

$$P(X=x) = \frac{x-1}{16} \quad \text{for} \quad x = 2,3,4,5$$

$$P(X=x) = \frac{9-x}{16} \quad \text{for} \quad x = 6,7,8$$

NOTE: $\sum_{\text{all } x} P(X=x) = \frac{1}{16}(1+2+3+4+3+2+1) = 1,$
confirming that X is a random variable.

Example 3.3 The p.d.f. of a discrete random variable Y is given by $P(Y=y)=cy^2$, for $y=0,1,2,3,4$. Given that c is a constant, find the value of c.

Solution 3.3 The probability distribution of Y is

y	0	1	2	3	4
$P(Y=y)$	0	c	$4c$	$9c$	$16c$

As Y is a random variable, $\sum_{\text{all } y} P(Y=y)=1$.

So

$$1 = c+4c+9c+16c$$

$$30c = 1$$

$$c = \frac{1}{30}$$

Therefore if Y is a random variable, then $c = \frac{1}{30}$.

Example 3.4 The p.d.f. of the discrete r.v. X is given by $P(X = x) = a(\frac{3}{4})^x$ for $x = 0, 1, 2, 3, \ldots$. Find the value of the constant, a.

Solution 3.4 As X is a random variable $\sum_{\text{all } x} P(X = x) = 1$.

Now
$$P(X = 0) = a(\tfrac{3}{4})^0$$
$$P(X = 1) = a(\tfrac{3}{4})^1$$
$$P(X = 2) = a(\tfrac{3}{4})^2$$
$$P(X = 3) = a(\tfrac{3}{4})^3$$

and so on

So
$$\sum_{\text{all } x} P(X = x) = a + a(\tfrac{3}{4}) + a(\tfrac{3}{4})^2 + a(\tfrac{3}{4})^3 + \ldots$$
$$= a(1 + \tfrac{3}{4} + (\tfrac{3}{4})^2 + (\tfrac{3}{4})^3 + \ldots)$$
$$= a\left(\frac{1}{1 - \frac{3}{4}}\right)$$ (sum of an infinite G.P. with first term 1 and common ratio $\frac{3}{4}$)
$$= (a)(4)$$

We have
$$4a = 1$$

Therefore
$$a = \frac{1}{4}$$

If X is a random variable, then $a = \frac{1}{4}$.

Exercise 3a

1. For each of the following random variables write out the probability distributions. Check that the variables are random and for parts (*b*), (*d*) and (*f*) write the formula for the p.d.f.
   (*a*) The number of heads obtained when two fair coins are tossed.
   (*b*) The sum of the scores when two ordinary dice are thrown.
   (*c*) The number of threes obtained when two tetrahedral dice are thrown.
   (*d*) The numerical value of a digit chosen from a set of random number tables.
   (*e*) The number of tails obtained when three fair coins are tossed.
   (*f*) The difference between the numbers when two ordinary dice are thrown.

2. The probability density function of a discrete random variable X is given by $P(X = x) = kx$ for $x = 12, 13, 14$. Find the value of the constant k.

3. The discrete random variable R has p.d.f. given by $P(R = r) = c(3 - r)$ for $r = 0, 1, 2, 3$. Find the value of the constant c.

4. A game consists of throwing tennis balls into a bucket from a given distance. The probability that William will get the tennis ball in the bucket is 0.4. A 'go' consists of three attempts. (*a*) Construct the probability distribution for X, the number of tennis balls that land in the bucket in a go. William wins a prize if, at the end of his go, there are two or more tennis balls in the bucket. (*b*) What is the probability that William does not win a prize?

5. A drawer contains 8 brown socks and 4 blue socks. A sock is taken from the drawer at random, its colour is noted and it is then replaced. This procedure is performed twice more. If X is the r.v. 'the number of brown socks taken', find the probability distribution for X.

6. The r.v. X has p.d.f. $P(X=x)=c(\frac{4}{5})^x$ for $x=0,1,2,3,\ldots$. Find the value of the constant, c.

EXPECTATION, $E(X)$

Experimental approach

Suppose we throw an unbiased die 120 times and record the results:

Score, x	1	2	3	4	5	6	
Frequency, f	15	22	23	19	23	18	Total 120

Then we can calculate the mean score obtained where

$$\bar{x} = \frac{\Sigma fx}{\Sigma f} = \frac{15+44+69+76+115+108}{120} = 3.558 \quad \text{(3 d.p.)}$$

Theoretical approach

The probability distribution for the r.v. X where X is 'the number on the die' is as shown:

Score, x	1	2	3	4	5	6
$P(X=x)$	$\frac{1}{6}$	$\frac{1}{6}$	$\frac{1}{6}$	$\frac{1}{6}$	$\frac{1}{6}$	$\frac{1}{6}$

We can obtain a value for the 'expected mean' by multiplying each score by its corresponding probability and summing, so that

$$\begin{aligned}\text{expected mean} &= 1\left(\frac{1}{6}\right)+2\left(\frac{1}{6}\right)+3\left(\frac{1}{6}\right)+4\left(\frac{1}{6}\right)+5\left(\frac{1}{6}\right)+6\left(\frac{1}{6}\right)\\ &= \frac{21}{6}\\ &= 3.5\end{aligned}$$

So, the expected mean = 3.5.

If we have a statistical experiment:

a practical approach results in a frequency distribution and a mean value,

a theoretical approach results in a probability distribution and an expected value.

The **expectation** of X (or expected value), written $E(X)$ is given by

$$E(X) = \sum_{\text{all } x} xP(X = x)$$

This can also be written

$$E(X) = \sum x_i p_i \qquad i = 1, 2, \ldots, n$$

Example 3.5 A r.v. X has a p.d.f. defined as shown. Find $E(X)$.

x	-2	-1	0	1	2
$P(X = x)$	0.3	0.1	0.15	0.4	0.05

Solution 3.5 Now

$$\begin{aligned} E(X) &= \sum_{\text{all } x} xP(X = x) \\ &= (-2)(0.3) + (-1)(0.1) + 0(0.15) + 1(0.4) + 2(0.05) \\ &= -0.2 \end{aligned}$$

Therefore $E(X) = -0.2$.

NOTE: an important property which some probability distributions possess is that of **symmetry**.

For example, **(a)** Consider the r.v. with probability distribution:

x	1	2	3	4	5
$P(X = x)$	0.1	0.2	0.4	0.2	0.1

It can be seen from the table that the distribution is symmetrical about the central value $X = 3$, so $E(X) = 3$.

Check: $E(X) = \sum_{\text{all } x} xP(X = x) = 1(0.1) + 2(0.2) + 3(0.4) + 4(0.2) + 5(0.1) = 3$

(b) Consider the r.v. with p.d.f. $P(X = x) = \frac{1}{8}$ for $x = 1, 2, \ldots, 8$.

The probability distribution for X is:

x	1	2	3	4	5	6	7	8
$P(X = x)$	$\frac{1}{8}$	$\frac{1}{8}$	$\frac{1}{8}$	$\frac{1}{8}$	$\frac{1}{8}$	$\frac{1}{8}$	$\frac{1}{8}$	$\frac{1}{8}$

The distribution is symmetrical about the central value, mid-way between 4 and 5, so $E(X) = 4.5$.

Example 3.6 A fruit machine consists of three windows, each of which shows pictures of fruits—lemons or oranges or cherries or plums. The probability that a window shows a particular fruit is as follows:

$P(\text{lemons}) = 0.4, \quad P(\text{oranges}) = 0.1, \quad P(\text{cherries}) = 0.2,$
$P(\text{plums}) = 0.3$

The windows operate independently.

Anyone wanting to play the fruit machine pays 10p for a turn.

The winning combinations are as follows:

Oranges in 3 windows	£1.00
Cherries in 3 windows	£0.50
Oranges in 2 windows and cherries in 1 window	£0.80
Lemons in 3 windows	£0.40

Find the expected gain/loss per turn.

Solution 3.6

$P(\text{oranges in 3 windows}) = (0.1)^3 = 0.001$ (independent events)

$P(\text{cherries in 3 windows}) = (0.2)^3 = 0.008$

$P(\text{oranges in 2 and cherries in 1}) = 3(0.1)^2(0.2) = 0.006$

$P(\text{lemons in 3 windows}) = (0.4)^3 = 0.064$

Therefore

$$P(\text{combination will not win a prize}) = 1-(0.001+0.008+0.006+0.064) = 0.921$$

Let X be the r.v. 'the amount gained per turn in pence'.

Now the amount paid out by the fruit machine could be 100p, 80p, 50p, 40p or 0p.

So considering the initial payment of 10p for a turn, X can assume the values $90, 70, 40, 30, -10$.

The probability distribution for X is

x	90	70	40	30	-10
$P(X=x)$	0.001	0.006	0.008	0.064	0.921

$$\text{Now } E(X) = \sum_{\text{all } x} xP(X=x)$$

$$= 90(0.001)+70(0.006)+40(0.008)+30(0.064) + (-10)(0.921)$$

$$= -6.46$$

So, the expected loss per turn is 6.46p.

Example 3.7 (a) Three dice are thrown. If a 1 or a 6 turns up, you will be paid 1p, but, if neither a 1 nor a 6 turns up, you will pay 5p. How much would you expect to lose in 9 games?

You are now given the opportunity to change the rule for payment when a 1 or a 6 appears. To make the game worthwhile to yourself, what is the minimum amount in everyday currency that you would suggest?

(b) Three coins are thrown. If one head turns up, 1p is paid. If two heads turn up, 3p is paid, and if three heads turn up 5p is paid.

If the game is to be regarded as fair (i.e. neither the player nor the bank should lose in the long run), what should be the penalty if no heads turn up?

(c) A bag contains 3 red balls and 1 blue ball. A second bag contains 1 red ball and 1 blue ball. A ball is picked out of each bag and and is then placed in the other bag. What is the expected number of red balls in the first bag? (SUJB)

Solution 3.7 (a) $P(1 \text{ or } 6 \text{ on die}) = \frac{2}{6} = \frac{1}{3}$.

If three dice are thrown, $P(\text{neither a 1 nor a 6 on all three}) = (\frac{2}{3})^3$.
So $P(\text{a 1 or a 6 turns up}) = 1 - (\frac{2}{3})^3 = \frac{19}{27}$.

Let X be the r.v. 'the number of pence won in a game'. Then X can assume the values -5 and 1 only. Now

$$P(X = -5) = P(\text{neither a 1 nor a 6}) = \frac{8}{27}$$

$$P(X = 1) = P(\text{a 1 or a 6}) = \frac{19}{27}$$

The probability distribution for X is

x	-5	1
$P(X = x)$	$\frac{8}{27}$	$\frac{19}{27}$

So
$$E(X) = \sum_{\text{all } x} xP(X = x)$$

$$= (-5)\left(\frac{8}{27}\right) + (1)\left(\frac{19}{27}\right)$$

$$= -\frac{7}{9}$$

Therefore the expected *loss* after one game is $\frac{7}{9}$p,

so, after 9 games, the expected loss is 7p.

If we change the rule for payment to y pence when a 1 or a 6 turns up then the probability distribution becomes

x	-5	y
$P(X = x)$	$\frac{8}{27}$	$\frac{19}{27}$

We now have
$$E(X) = (-5)\left(\frac{8}{27}\right) + (y)\left(\frac{19}{27}\right)$$
$$= \frac{-40 + 19y}{27}$$

To make the game worthwhile, we require $E(X) > 0$, i.e. $-40 + 19y > 0$.

So
$$y > 2.1\text{p}$$

Therefore the minimum amount we require to be paid, in everyday currency, is 3p.

(**b**) Three coins are thrown. Let X be the r.v. 'the number of pence paid in 1 game'.

Then

$$P(X = 1) = P(1\text{ head}) = P(\text{HTT}) + P(\text{TTH}) + P(\text{THT})$$
$$= 3\left(\frac{1}{2}\right)^3 = \frac{3}{8}$$

$$P(X = 3) = P(2\text{ heads}) = P(\text{HHT}) + P(\text{HTH}) + P(\text{THH})$$
$$= 3\left(\frac{1}{2}\right)^3 = \frac{3}{8}$$

$$P(X = 5) = P(3\text{ heads}) = P(\text{HHH}) = \left(\frac{1}{2}\right)^3 = \frac{1}{8}$$

If the penalty if 0 heads turn up is y p,

$$P(X = -y) = P(0\text{ heads}) = P(\text{TTT}) = \left(\frac{1}{2}\right)^3 = \frac{1}{8}$$

The probability distribution of X is

x	1	3	5	$-y$
$P(X = x)$	$\frac{3}{8}$	$\frac{3}{8}$	$\frac{1}{8}$	$\frac{1}{8}$

So
$$E(X) = \sum_{\text{all } x} xP(X = x)$$
$$= 1\left(\frac{3}{8}\right) + 3\left(\frac{3}{8}\right) + 5\left(\frac{1}{8}\right) + (-y)\left(\frac{1}{8}\right)$$
$$= \frac{17-y}{8}$$

Now, if the game is to be fair, the expected winnings must be zero.

So
$$\frac{17-y}{8} = 0$$

i.e.
$$y = 17$$

Therefore the penalty if 0 heads turn up should be 17p.

(c) Assume that the balls are taken from each bag simultaneously.

If a red ball is picked from each bag and placed in the other then the number of red balls in the first bag is now 3, etc.

Let X be the r.v. 'the final number of red balls in the first bag'.

Then X can assume the values 2, 3 or 4 only.

$$P(X = 2) = P(\text{red from first bag and blue from second bag})$$
$$= P(R_1B_2) \quad \text{with obvious notation}$$
$$= \left(\frac{3}{4}\right)\left(\frac{1}{2}\right)$$
$$= \frac{3}{8}$$

$$P(X = 3) = P(R_1R_2) + P(B_1B_2)$$
$$= \left(\frac{3}{4}\right)\left(\frac{1}{2}\right) + \left(\frac{1}{4}\right)\left(\frac{1}{2}\right)$$
$$= \frac{1}{2}$$

$$P(X = 4) = P(B_1R_2)$$
$$= \left(\frac{1}{4}\right)\left(\frac{1}{2}\right)$$
$$= \frac{1}{8}$$

The probability distribution for X is

x	2	3	4
$P(X = x)$	$\frac{3}{8}$	$\frac{1}{2}$	$\frac{1}{8}$

So
$$E(X) = \sum_{\text{all } x} xP(X = x)$$
$$= 2\left(\frac{3}{8}\right) + 3\left(\frac{1}{2}\right) + 4\left(\frac{1}{8}\right)$$
$$= 2\frac{3}{4}$$

The expected number of red balls in the first bag after the exchange is $2\frac{3}{4}$ balls.

Exercise 3b

1. The probability distribution for the r.v. X is shown in the table:

x	0	1	2	3	4
$P(X=x)$	$\frac{1}{6}$	$\frac{1}{12}$	$\frac{1}{4}$	$\frac{1}{3}$	$\frac{1}{6}$

Find $E(X)$.

2. The r.v. X has p.d.f. $P(X = x)$ for $x = 5, 6, 7, 8, 9$ as defined in the table:

x	5	6	7	8	9
$P(X=x)$	$\frac{3}{11}$	$\frac{2}{11}$	$\frac{1}{11}$	$\frac{2}{11}$	$\frac{3}{11}$

Find $E(X)$.

3. The probability distribution of a r.v. X is as shown in the table:

x	1	2	3	4	5
$P(X=x)$	0.1	0.3	y	0.2	0.1

Find (*a*) the value of y, (*b*) $E(X)$.

4. Find the expected number of heads when two fair coins are tossed.

5. Find the expected number of ones when three ordinary fair dice are thrown.

6. A bag contains 5 black counters and 6 red counters. Two counters are drawn, one at a time, and not replaced. Let X be the r.v. 'the number of red counters drawn'. Find $E(X)$.

7. An unbiased tetrahedral die has faces marked 1, 2, 3, 4. If the die lands on the face marked 1, the player has to pay 10p. If it lands on a face marked with a 2 or a 4, the player wins 5 p and if it lands on a 3, the player wins 3 p. Find the expected gain in one throw.

8. A discrete r.v. X can assume values 10 and 20 only. If $E(X) = 16$, write the p.d.f. of X in table form.

9. The discrete r.v. X can assume values 0, 1, 2 and 3 only. Given $P(X \leqslant 2) = 0.9$, $P(X \leqslant 1) = 0.5$ and $E(X) = 1.4$, find (*a*) $P(X = 1)$, (*b*) $P(X = 0)$.

10. In a game, a player rolls two balls down an inclined plane so that each ball finally settles in one of five slots and scores the number of points allotted to that slot as shown in the diagram below:

2	4	7	4	2

It is possible for both balls to settle in one slot and it may be assumed that each slot is equally likely to accept either ball.

The player's score is the sum of the points scored by each ball.

Draw up a table showing all the possible scores and the probability of each.
If the player pays 10p for each game and receives back a number of pence equal to his score, calculate the player's expected gain or loss per 50 games.
(C Additional)

11. In a game a player tosses three fair coins. He wins £10 if 3 heads occur, £x if 2 heads occur, £3 if 1 head occurs and £2 if no heads occur. Express in terms of x his expected gain from each game.

Given that he pays £4.50 to play each game, calculate
(*a*) the value of x for which the game is fair,
(*b*) his expected gain or loss over 100 games if $x = 4.90$. (C Additional)

12. A committee of 3 is to be chosen from 4 girls and 7 boys. Find the expected number of girls on the committee, if the members of the committee are chosen at random.

13. The discrete r.v. X has p.d.f. given by $P(X = x) = kx$ for $x = 1, 2, 3, 4, 5$ where k is constant. Find $E(X)$.

14. In an examination a candidate is given the four answers to four questions but is not told which answer applies to which question. He is asked to write down each of the four answers next to its appropriate question.
(*a*) Calculate in how many different ways he could write down the four answers.
(*b*) Explain why it is impossible for him to have just three answers in the correct places and show that there are six ways of getting just two answers in the correct places.
(*c*) If a candidate guesses at random where the four answers are to go and X is the number of correct guesses he makes, draw up the probability distribution for X in tabular form.
(*d*) Calculate $E(X)$. (L Additional)

15.

x	0	1	2	3
$P(X = x)$	c	c^2	$c^2 + c$	$3c^2 + 2c$

The above table shows the probability distribution for a random variable X. Calculate (*a*) c, (*b*) $E(X)$. (L Additional)

16. A box contains 9 discs of which 4 are red, 3 are white and 2 are blue. Three discs are to be drawn at random without replacement from the box. Calculate
(*a*) the probability that the discs, in the order drawn, will be coloured red, white and blue respectively,
(*b*) the probability that one disc of each colour will be drawn,
(*c*) the probability that the third disc drawn will be red,
(*d*) the probability that no red disc will be drawn,
(*e*) the most probable number of red discs that will be drawn,
(*f*) the expected number of red discs that will be drawn, and state the probability that this expected number of red discs will be drawn. (JMB)

17. A woman has 3 keys on a ring, just one of which opens the front door. As she approaches the front door she selects one key after another at random without replacement. Draw a tree diagram to illustrate the various selections before she finds the correct key. Use this diagram to calculate the expected number of keys that she will use before opening the door. (L Additional)

18. An urn containing 4 black balls and 8 white balls is used for two experiments. In experiment 1, two balls are to be drawn at random from the urn, one after the other, without replacement. In experiment 2, one ball is to be drawn at random from the 12 balls in the urn and replaced before a second ball is drawn at random. Copy and complete the following two tables, which give the probabilities for the different compound events in the two experiments.

		Second ball	
		Black	White
First ball	Black		$\frac{8}{33}$
	White		

Experiment 1

		Second ball	
		Black	White
First ball	Black		
	White	$\frac{2}{9}$	

Experiment 2

For each of the two experiments, calculate the expected number of black balls which will be drawn.

If in experiment 2, the urn contains b black balls and w white balls, where $b + w = 12$, calculate the expected number of black balls which will be drawn. (L Additional)

THE EXPECTATION OF ANY FUNCTION OF *X*, $E[g(X)]$

The definition of expectation can be extended to any function of the random variable such as $10X$, X^2, $(X-4)^4$, etc.

In general, if $g(X)$ is any function of the discrete random variable X then

$$E[g(X)] = \sum_{\text{all } x} g(x)P(X=x)$$

Example 3.8 In a game a turn consists of a tetrahedral die being thrown three times. The faces on the die are marked 1, 2, 3, 4 and the number on which the die falls is noted. A man wins £x^2 whenever x fours occur in a turn. Find his average win per turn.

Solution 3.8 Let X be the r.v. 'the number of fours obtained when the die is thrown three times'. Then X can assume the values 0, 1, 2, 3 only.

We have $$P(X=0) = P(\bar{4}\,\bar{4}\,\bar{4}) = \left(\frac{3}{4}\right)^3 = \frac{27}{64}$$

$$P(X=1) = 3P(4\,\bar{4}\,\bar{4}) = 3\left(\frac{1}{4}\right)\left(\frac{3}{4}\right)^2 = \frac{27}{64}$$

$$P(X=2) = 3P(4\,4\,\bar{4}) = 3\left(\frac{1}{4}\right)^2\left(\frac{3}{4}\right) = \frac{9}{64}$$

$$P(X=3) = P(4\,4\,4) = \left(\frac{1}{4}\right)^3 = \frac{1}{64}$$

The average win is given by $E(X^2)$ so we write out the probability distribution for X, but add a row showing the values of X^2 to help make the calculations easier:

x	0	1	2	3
x^2	0	1	4	9
$P(X=x)$	$\frac{27}{64}$	$\frac{27}{64}$	$\frac{9}{64}$	$\frac{1}{64}$

Now $$E(X^2) = \sum_{\text{all } x} x^2P(X=x)$$

$$= 0\left(\frac{27}{64}\right)+1\left(\frac{27}{64}\right)+4\left(\frac{9}{64}\right)+9\left(\frac{1}{64}\right)$$

$$= 1.125$$

Therefore his average win per turn is £1.13 (nearest p).

Example 3.9 The random variable X has p.d.f. $P(X=x)$ for $x = 1, 2, 3$.

x	1	2	3
$P(X=x)$	0.1	0.6	0.3

Calculate (a) $E(3)$, (b) $E(X)$, (c) $E(5X)$, (d) $E(5X+3)$, (e) $5E(X)+3$, (f) $E(X^2)$, (g) $E(4X^2-3)$, (h) $4E(X^2)-3$.

Comment on your answers to parts (d) and (e) and parts (g) and (h).

Solution 3.9 We have

x	1	2	3
$5x$	5	10	15
$5x+3$	8	13	18
x^2	1	4	9
$4x^2-3$	1	13	33
$P(X=x)$	0.1	0.6	0.3

Now $$E[g(X)] = \sum_{\text{all } x} g(x)P(X=x)$$

(a)
$$\begin{aligned} E(3) &= \sum_{\text{all } x} 3P(X=x) \\ &= 3(0.1)+3(0.6)+3(0.3) \\ &= 3 \end{aligned}$$
$$\underline{E(3) = 3}$$

(b)
$$\begin{aligned} E(X) &= \sum_{\text{all } x} xP(X=x) \\ &= 1(0.1)+2(0.6)+3(0.3) \\ &= 2.2 \end{aligned}$$
$$\underline{E(X) = 2.2}$$

(c)
$$\begin{aligned} E(5X) &= \sum_{\text{all } x} 5xP(X=x) \\ &= 5(0.1)+10(0.6)+15(0.3) \\ &= 11 \end{aligned}$$
$$\underline{E(5X) = 11}$$

(d)
$$\begin{aligned} E(5X+3) &= \sum_{\text{all } x} (5x+3)P(X=x) \\ &= 8(0.1)+13(0.6)+18(0.3) \\ &= 14 \\ \underline{E(5X+3)} &\underline{= 14} \end{aligned}$$

(e)
$$\begin{aligned} 5E(X)+3 &= 5(2.2)+3 \\ &= 14 \\ \underline{5E(X)+3} &\underline{= 14} \end{aligned}$$

(f)
$$\begin{aligned} E(X^2) &= \sum_{\text{all } x} x^2P(X=x) \\ &= 1(0.1)+4(0.6)+9(0.3) \\ &= 5.2 \\ \underline{E(X^2)} &\underline{= 5.2} \end{aligned}$$

(g)
$$\begin{aligned} E(4X^2-3) &= \sum_{\text{all } x} (4x^2-3)P(X=x) \\ &= 1(0.1)+13(0.6)+33(0.3) \\ &= 17.8 \\ \underline{E(4X^2-3)} &\underline{= 17.8} \end{aligned}$$

(h)
$$\begin{aligned} 4E(X^2)-3 &= 4(5.2)-3 \\ &= 17.8 \\ \underline{4E(X^2)-3} &\underline{= 17.8} \end{aligned}$$

We note that
$$\begin{aligned} E(5X+3) &= 5E(X)+3 \\ E(4X^2-3) &= 4E(X^2)-3 \end{aligned}$$

In general, the following results hold when X is a discrete random variable:

Result 1 $E(a) = a$, where a is any constant.

Proof:
$$\begin{aligned} E(a) &= \sum_{\text{all } x} aP(X=x) \\ &= a\sum_{\text{all } x} P(X=x) \\ &= a \quad \text{since} \quad \sum_{\text{all } x} P(X=x) = 1 \end{aligned}$$

Result 2 $E(aX) = aE(X)$, where a is any constant.

Proof:

$$E(aX) = \sum_{\text{all } x} axP(X = x)$$
$$= a\sum_{\text{all } x} xP(X = x)$$
$$= aE(X)$$

Result 3 $E(aX + b) = aE(X) + b$, where a and b are any constants.

Proof:

$$E(aX + b) = \sum_{\text{all } x} (ax + b)P(x = x)$$
$$= \sum_{\text{all } x} axP(X = x) + \sum_{\text{all } x} bP(X = x)$$
$$= aE(X) + b$$

Result 4 $E[f_1(X) + f_2(X)] = E[f_1(X)] + E[f_2(X)]$ where f_1 and f_2 are functions of X.

Proof:

$$E[f_1(X) + f_2(X)] = \sum_{\text{all } x} [f_1(x) + f_2(x)]P(X = x)$$
$$= \sum_{\text{all } x} f_1(x)P(X = x) + \sum_{\text{all } x} f_2(x)P(X = x)$$
$$= E[f_1(X)] + E[f_2(X)]$$

Exercise 3c

1. The discrete r.v. X has p.d.f. $P(X = x)$ for $x = 1, 2, 3$.

x	1	2	3
$P(X = x)$	0.2	0.3	0.5

Find (*a*) $E(X)$, (*b*) $E(X^2)$.
(*c*) Verify that $E(3X - 1) = 3E(X) - 1$.
(*d*) Verify that $E(2X^2 + 4) = 2E(X^2) + 4$.

2. The discrete r.v. X has p.d.f. $P(X = 0) = 0.05$, $P(X = 1) = 0.45$, $P(X = 2) = 0.5$. Verify that $E(5X^2 + 2X - 3) = 5E(X^2) + 2E(X) - 3$.

3. The discrete r.v. X has p.d.f. given by $P(X = x) = \frac{1}{6}$ for $x = 1, 2, 3, 4, 5, 6$.
Find (*a*) $E(X)$, (*b*) $E(X^2)$, (*c*) $E(3X + 4)$.
Verify that $E(2X^2 + X - 4) = 2E(X^2) + E(X) - 4$

4. The discrete r.v. X has p.d.f. given by $P(X = x) = \dfrac{3x + 1}{22}$ for $x = 0, 1, 2, 3$.
Find (*a*) $E(X)$, (*b*) $E(X^2)$, (*c*) $E(3X - 2)$, (*d*) $E(2X^2 + 4X - 3)$.

5. A roulette wheel is divided into 6 sectors of unequal area, marked with the numbers 1, 2, 3, 4, 5 and 6. The wheel is spun and X is the r.v. 'the number on which the wheel stops'. The probability distribution of X is as follows:

x	1	2	3	4	5	6
$P(X=x)$	$\frac{1}{16}$	$\frac{3}{16}$	$\frac{1}{4}$	$\frac{1}{4}$	$\frac{3}{16}$	$\frac{1}{16}$

Calculate (*a*) $E(X)$, (*b*) $E(X^2)$, (*c*) $E(3X-5)$, (*d*) $E(6X^2)$, (*e*) $E(6X^2+6X-10)$.

6. The r.v. X has p.d.f. $P(X=x)$ as shown in the table:

x	-2	-1	0	1	c
$P(X=x)$	0.1	0.1	0.3	0.4	0.1

Find the value of c (*a*) if $E(X)=0.3$, (*b*) if $E(X^2)=1.8$.

VARIANCE, Var(X)

Consider the discrete r.v. X and let $E(X)=\mu$ (pronounced 'mew').

The **variance** of X, written Var(X), is given by

$$\text{Var}(X) = E(X-\mu)^2$$

For a frequency distribution we found that the formula for the variance

$$s^2 = \frac{\Sigma f(x-\bar{x})^2}{\Sigma f} \quad \text{simplified to} \quad s^2 = \frac{\Sigma fx^2}{\Sigma f} - \bar{x}^2$$

In the same way, we can find an alternative form for the formula for Var(X). Now

$$\begin{aligned}\text{Var}(X) &= E(X-\mu)^2 \\ &= E(X^2-2\mu X+\mu^2) \\ &= E(X^2)-2\mu E(X)+E(\mu^2) \\ &= E(X^2)-2\mu^2+\mu^2 \\ &= E(X^2)-\mu^2\end{aligned}$$

So we have $\quad \text{Var}(X) = E(X^2)-\mu^2$

NOTE: $\mu^2=[E(X)]^2$.

We write $[E(X)]^2$ as $E^2(X)$ in a similar way to the notation used in trigonometry where $(\sin A)^2$ is written $\sin^2 A$.

So we have $\quad \text{Var}(X) = E(X^2)-E^2(X)$

Example 3.10 The r.v. X has probability distribution as shown in the table:

x	1	2	3	4	5
$P(X=x)$	0.1	0.3	0.2	0.3	0.1

Find
(a) $\mu = E(X)$,
(b) Var(X), using the formula $\text{Var}(X) = E(X-\mu)^2$,
(c) $E(X^2)$,
(d) Var(X), using the formula $\text{Var}(X) = E(X^2)-\mu^2$.

Solution 3.10 (a) By symmetry, $\mu = E(X) = 3$.

(b)
$$E(X-\mu)^2 = E(X-3)^2 = \sum_{\text{all } x} (x-3)^2 P(X=x)$$

x	1	2	3	4	5
$(x-3)$	-2	-1	0	1	2
$(x-3)^2$	4	1	0	1	4
$P(X=x)$	0.1	0.3	0.2	0.3	0.1

So
$$E(X-3)^2 = 4(0.1)+1(0.3)+0(0.2)+1(0.3)+4(0.1) = 1.4$$

Therefore $\text{Var}(X) = E(X-\mu)^2 = 1.4$.

(c)
$$E(X^2) = \sum_{\text{all } x} x^2 P(X=x) = 1(0.1)+4(0.3)+9(0.2)+16(0.3)+25(0.1) = 10.4$$

So $E(X^2) = 10.4$.

(d) Now
$$\text{Var}(X) = E(X^2)-\mu^2 = 10.4-9 = 1.4$$

Therefore $\text{Var}(X) = 1.4$, as before.

Usually we use the most convenient form of the formula for the variance.

Example 3.11 Two discs are drawn, without replacement, from a box containing 3 red discs and 4 white discs. The discs are drawn at random. If X is the r.v. 'the number of red discs drawn', find **(a)** $E(X)$, **(b)** the standard deviation of X.

Solution 3.11 X is the r.v. 'the number of red discs drawn'.

Now X can assume the values 0, 1, 2 only. We have

$$P(X=0) = P(W_1W_2) = \left(\frac{4}{7}\right)\left(\frac{3}{6}\right) = \frac{12}{42} = \frac{2}{7}$$

$$P(X=1) = P(W_1R_2)+P(R_1W_2) = \left(\frac{4}{7}\right)\left(\frac{3}{6}\right) + \left(\frac{3}{7}\right)\left(\frac{4}{6}\right) = \frac{24}{42} = \frac{4}{7}$$

$$P(X=2) = P(R_1R_2) = \left(\frac{3}{7}\right)\left(\frac{2}{6}\right) = \frac{6}{42} = \frac{1}{7}$$

The probability distribution for X is as follows:

x	0	1	2
$P(X=x)$	$\frac{2}{7}$	$\frac{4}{7}$	$\frac{1}{7}$

(a) Now

$$\begin{aligned} E(X) &= \sum_{\text{all } x} xP(X=x) \\ &= 0\left(\frac{2}{7}\right) + 1\left(\frac{4}{7}\right) + 2\left(\frac{1}{7}\right) \\ &= \frac{6}{7} \end{aligned}$$

So $E(X) = \frac{6}{7}$, or the expected number of red discs is $\frac{6}{7}$.

(b) Standard deviation of $X = \sqrt{\text{Var}(X)}$.

Now $\text{Var}(X) = E(X^2) - E^2(X)$

We have

$$\begin{aligned} E(X^2) &= \sum_{\text{all } x} x^2P(X=x) \\ &= 0\left(\frac{2}{7}\right) + 1\left(\frac{4}{7}\right) + 4\left(\frac{1}{7}\right) \\ &= \frac{8}{7} \end{aligned}$$

So

$$\begin{aligned} \text{Var}(X) &= \frac{8}{7} - \left(\frac{6}{7}\right)^2 \\ &= \frac{20}{49} \\ &= 0.408 \quad \text{(3 d.p.)} \end{aligned}$$

Therefore the standard deviation of $X = \sqrt{0.408} = 0.639$ (3 d.p.).

The following results are useful.

Result 1 Var(a) = 0 where a is any constant.

Proof:

$$\begin{aligned}\text{Var}(a) &= E(a^2)-E^2(a)\\ &= a^2-a^2\\ &= 0\end{aligned}$$

NOTE: this is as expected, as a constant does not vary.

Result 2 Var(aX) = a^2 Var(X) where a is any constant.

Proof:

$$\begin{aligned}\text{Var}(aX) &= E(aX)^2-E^2(aX)\\ &= a^2E(X^2)-a^2E^2(X)\\ &= a^2[E(X^2)-E^2(X)]\\ &= a^2\text{Var}(X)\end{aligned}$$

Result 3 Var($aX+b$) = a^2 Var(X) where a and b are any constants.

Proof:

$$\begin{aligned}\text{Var}(aX+b) &= E(aX+b)^2-E^2(aX+b)\\ &= E(a^2X^2+2abX+b^2)-[aE(X)+b]^2\\ &= a^2E(X^2)+2abE(X)+b^2-a^2E^2(X)-2abE(X)-b^2\\ &= a^2E(X^2)-a^2E^2(X)\\ &= a^2[E(X^2)-E^2(X)]\\ &= a^2\text{Var}(X)\end{aligned}$$

Example 3.12 The discrete r.v. X has the probability distribution shown in the table.

Verify that Var($2X+3$) = 4 Var(X).

x	1	2	3	4
$P(X=x)$	$\frac{3}{8}$	$\frac{1}{8}$	$\frac{1}{4}$	$\frac{1}{4}$

Solution 3.12 First, we need to find $E(X)$. Now

$$E(X) = \sum_{\text{all } x} xP(X=x)$$

$$= 1\left(\frac{3}{8}\right) + 2\left(\frac{1}{8}\right) + 3\left(\frac{1}{4}\right) + 4\left(\frac{1}{4}\right)$$

$$= \frac{19}{8}$$

Now $\quad \mathrm{Var}(X) = E(X^2) - E^2(X)$

We have $\quad E(X^2) = \sum_{\text{all } x} x^2P(X=x)$

$$= 1\left(\frac{3}{8}\right) + 4\left(\frac{1}{8}\right) + 9\left(\frac{1}{4}\right) + 16\left(\frac{1}{4}\right)$$

$$= \frac{57}{8}$$

So $\quad \mathrm{Var}(X) = \frac{57}{8} - \left(\frac{19}{8}\right)^2$

$$= \frac{95}{64}$$

Now consider $2X+3$.

x	1	2	3	4
$2x+3$	5	7	9	11
$(2x+3)^2$	25	49	81	121
$P(X=x)$	$\frac{3}{8}$	$\frac{1}{8}$	$\frac{1}{4}$	$\frac{1}{4}$

We require

$$E(2X+3) = \sum_{\text{all } x} (2x+3)\,P(X=x)$$

$$= 5\left(\frac{3}{8}\right) + 7\left(\frac{1}{8}\right) + 9\left(\frac{1}{4}\right) + 11\left(\frac{1}{4}\right)$$

$$= \frac{31}{4}$$

$$E[(2X+3)^2] = \sum_{\text{all } x} (2x+3)^2P(X=x)$$

$$= 25\left(\frac{3}{8}\right) + 49\left(\frac{1}{8}\right) + 81\left(\frac{1}{4}\right) + 121\left(\frac{1}{4}\right)$$

$$= 66$$

Therefore

$$\begin{aligned}\text{Var}(2X+3) &= E[(2X+3)^2]-E^2(2X+3)\\ &= 66-\left(\frac{31}{4}\right)^2\\ &= \frac{95}{16}\\ &= 4\left(\frac{95}{64}\right)\\ &= 4\,\text{Var}(X)\end{aligned}$$

Therefore $\text{Var}(2X+3) = 4\text{Var}(X)$.

Exercise 3d

1. The probability distribution for the r.v. X is as shown:

x	2	4	6	8	10
$P(X=x)$	$\frac{1}{24}$	$\frac{5}{24}$	$\frac{1}{2}$	$\frac{5}{24}$	$\frac{1}{24}$

Find (*a*) $\mu=E(X)$, (*b*) $E(3X+4)$.
(*c*) Verify that $E(X-\mu)^2=E(X^2)-\mu^2$.
(*d*) Verify that $\text{Var}(3X)=9\text{Var}(X)$.
(*e*) Verify that $\text{Var}(3X+4)=9\text{Var}(X)$.

2. If X is the r.v. 'the sum of the scores on two tetrahedral dice', where the 'score' is the number on which the die lands, find (*a*) $E(X)$, (*b*) $\text{Var}(X)$, (*c*) $\text{Var}(2X)$, (*d*) $\text{Var}(2X+3)$.

3. Find $\text{Var}(X)$ for each of the following probability distributions:

(*a*)

x	-3	-2	0	2	3
$P(X=x)$	0.3	0.3	0.2	0.1	0.1

(*b*)

x	1	3	5	7	9
$P(X=x)$	$\frac{1}{6}$	$\frac{1}{4}$	$\frac{1}{6}$	$\frac{1}{4}$	$\frac{1}{6}$

(*c*)

x	0	2	5	6
$P(X=x)$	0.11	0.35	0.46	0.08

4. If X is the r.v. 'the number on a biased die', and the p.d.f. of X is shown,

x	1	2	3	4	5	6
$P(X=x)$	$\frac{1}{6}$	$\frac{1}{6}$	$\frac{1}{5}$	y	$\frac{1}{5}$	$\frac{1}{6}$

find (*a*) the value of y, (*b*) $E(X)$, (*c*) $E(X^2)$, (*d*) $\text{Var}(X)$, (*e*) $\text{Var}(4X)$.

5. A team of 3 is to be chosen from 4 boys and 5 girls. If X is the r.v. 'the number of girls in the team', find (*a*) $E(X)$, (*b*) $E(X^2)$, (*c*) $\text{Var}(X)$.

6. The r.v. X has p.d.f. as shown:

x	1	2	3	4	5
$P(X=x)$	0.11	0.28	0.33	0.18	0.10

(*a*) Find $E(3X^2-5X+7)$.
(*b*) Verify that $\text{Var}(2X-1)=4\text{Var}(X)$.

7. Two discs are drawn without replacement from a box containing 3 red and 4 white discs. If X is the r.v. 'the number of white discs drawn', construct a probability distribution table.

Find (*a*) $E(X)$, (*b*) $E(X^2)$, (*c*) $\text{Var}(X)$, (*d*) $\text{Var}(3X-4)$.

8. For the following probability distribution find (*a*) $\mu=E(X)$, (*b*) $E(X^2)$, (*c*) $E(X-\mu)^2$. Verify that

$$E(X-\mu)^2 = E(X^2)-\mu^2$$

x	-3	-2	1
$P(X=x)$	$\frac{1}{5}$	$\frac{3}{10}$	$\frac{1}{2}$

9. Ten identically shaped discs are in a bag; two of them are black, the rest white. Discs are drawn at random from the bag in turn and not replaced.

Let X be the number of discs drawn up to and including the first black one.

List the values of X and the associated theoretical probabilities.

Calculate the mean value of X and its standard deviation. What is the most likely value of X?

If instead each disc is replaced before the next is drawn, construct a similar list of values and point out the chief differences between the two lists. (SUJB)

10. The discrete r.v. X has p.d.f.

$$P(X=x) = k|x|$$

where x takes the values $-3, -2, -1, 0, 1, 2, 3$. Find (*a*) the value of the constant k, (*b*) $E(X)$, (*c*) $E(X^2)$, (*d*) the standard deviation of X.

11. The random variable X takes integer values only and has p.d.f.

$$P(X=x) = kx \qquad x = 1, 2, 3, 4, 5$$
$$P(X=x) = k(10-x) \qquad x = 6, 7, 8, 9$$

Find (*a*) the value of the constant k, (*b*) $E(X)$, (*c*) $\text{Var}(X)$, (*d*) $E(2X-3)$, (*e*) $\text{Var}(2X-3)$.

THE CUMULATIVE DISTRIBUTION FUNCTION

When we had a frequency distribution, the corresponding cumulative frequencies were obtained by summing all the frequencies up to a particular value. In the same way, if X is a discrete random variable, the corresponding cumulative probabilities are obtained by summing all the probabilities up to a particular value.

If X is a discrete random variable with p.d.f. $P(X=x)$ for $x = x_1, x_2, \dots, x_n$, then the cumulative distribution function is given by $F(t)$ where

$$F(t) = P(X \leqslant t)$$
$$= \sum_{x=x_1}^{t} P(X=x) \qquad t = x_1, x_2, \dots, x_n$$

The cumulative distribution function is sometimes known just as the distribution function.

Example 3.13 Find the cumulative distribution function for the r.v. X where X is 'the score on an unbiased die'.

Solution 3.13 The probability distribution for X is shown in the table:

x	1	2	3	4	5	6
$P(X=x)$	$\frac{1}{6}$	$\frac{1}{6}$	$\frac{1}{6}$	$\frac{1}{6}$	$\frac{1}{6}$	$\frac{1}{6}$

$$F(1) = P(X \leqslant 1) = \frac{1}{6}$$

$$F(2) = P(X \leqslant 2) = P(X=1)+P(X=2) = \frac{1}{6}+\frac{1}{6} = \frac{2}{6}$$

$$F(3) = P(X \leqslant 3) = \frac{3}{6}$$

$$F(4) = P(X \leqslant 4) = \frac{4}{6}$$

$$F(5) = P(X \leqslant 5) = \frac{5}{6}$$

$$F(6) = P(X \leqslant 6) = \frac{6}{6}$$

Therefore $F(t) = \dfrac{t}{6}$ for $t = 1, 2, 3, \ldots, 6$.

NOTE: **(a)** $F(6) = 1$, as expected.
(b) Although we work with the variable t we often write the final answer in terms of x; i.e. $F(x) = \dfrac{x}{6}$, $x = 1, 2, \ldots, 6$.

Example 3.14 The probability distribution for the r.v. X is shown in the table. Construct the cumulative distribution table.

x	0	1	2	3	4	5	6
$P(X = x)$	0.03	0.04	0.06	0.12	0.4	0.15	0.2

Solution 3.14 Now $\quad F(t) = \sum_{x=0}^{t} P(X = x) \qquad t = 0, 1, 2, \ldots, 6$

So
$$F(0) = P(X \leqslant 0) = 0.03$$
$$F(1) = P(X \leqslant 1) = 0.03 + 0.04 = 0.07$$
$$F(2) = P(X \leqslant 2) = 0.03 + 0.04 + 0.06 = 0.13$$

and so on

So we have the cumulative distribution table:

x	0	1	2	3	4	5	6
$F(x)$	0.03	0.07	0.13	0.25	0.65	0.8	1

NOTE: it is not possible to write a formula for the cumulative distribution function in Example 3.13.

Example 3.15 For a discrete r.v. X the cumulative distribution function $F(x)$ is as shown:

x	1	2	3	4	5
$F(x)$	0.2	0.32	0.67	0.9	1

Find **(a)** $P(X = 3)$, **(b)** $P(X > 2)$.

Solution 3.15 (a) From the table,

$$F(3) = P(X \leqslant 3) = P(X=1)+P(X=2)+P(X=3) = 0.67$$

$$F(2) = P(X \leqslant 2) = P(X=1)+P(X=2) = 0.32$$

Now $P(X=3) = F(3)-F(2)$

Therefore $P(X=3) = 0.67-0.32$

$$= 0.35$$

(b) $P(X>2) = 1-P(X \leqslant 2)$

$$= 1-F(2)$$

$$= 1-0.32$$

$$= 0.68$$

So, $P(X=3) = 0.35$ and $P(X>2) = 0.68$.

Exercise 3e

1. Construct the cumulative distribution tables for the following discrete random variables:
(*a*) the number of sixes obtained when two ordinary dice are thrown,
(*b*) the smaller number when two ordinary dice are thrown,
(*c*) the number of heads when three fair coins are tossed.

2. The probability distribution for the r.v. Y is shown in the table:

y	0.1	0.2	0.3	0.4	0.5
$P(Y=y)$	0.05	0.25	0.3	0.15	0.25

Construct the cumulative distribution table.

3. For a discrete r.v. R the cumulative distribution function $F(r)$ is as shown in the table:

r	1	2	3	4
$F(r)$	0.13	0.54	0.75	1

Find (*a*) $P(R=2)$, (*b*) $P(R>1)$, (*c*) $P(R \geqslant 3)$, (*d*) $P(R<2)$, (*e*) $E(R)$.

4. For the discrete r.v. X the cumulative distribution function $F(x)$ is as shown:

x	3	4	5	6	7
$F(x)$	0.01	0.23	0.64	0.86	1

Construct the probability distribution of X, and find $\mathrm{Var}(X)$.

5. For a discrete r.v. X the cumulative distribution function is given by $F(x) = \frac{x^2}{9}$ for $x = 1, 2, 3$. Find (*a*) $F(2)$, (*b*) $P(X=2)$, (*c*) Write out the probability distribution of X, (*d*) Find $E(2X-3)$.

6. For a discrete r.v. X the cumulative distribution function is given by $F(x) = kx$, $x = 1, 2, 3$. Find (*a*) the value of the constant k, (*b*) $P(X<3)$, (*c*) the probability distribution of X, (*d*) the standard deviation of X.

7. The discrete r.v. X has distribution function $F(x)$ where
$F(x) = 1-(1-\frac{1}{4}x)^x$ for $x = 1, 2, 3, 4$
(*a*) Show that $F(3) = \frac{63}{64}$ and $F(2) = \frac{3}{4}$.
(*b*) Obtain the probability distribution of X.
(*c*) Find $E(X)$ and $\mathrm{Var}(X)$.
(*d*) Find $P(X>E(X))$.

TWO INDEPENDENT RANDOM VARIABLES

If X and Y are any two random variables, then

$$E(X+Y) = E(X)+E(Y)$$

If X and Y are independent random variables, then

$$\text{Var}(X+Y) = \text{Var}(X)+\text{Var}(Y)$$

Example 3.16 X is the r.v. 'the score on a tetrahedral die', Y is the r.v. 'the number of heads obtained when two coins are tossed'.

(a) Obtain the probability distributions of X and of Y.

(b) Find $E(X)$ and $E(Y)$.

(c) Find Var(X) and Var(Y).

(d) Obtain the probability distribution for the r.v. $X+Y$.

(e) Find $E(X+Y)$ and Var$(X+Y)$ using the probability distribution for $X+Y$; comment on your results.

Solution 3.16

(a) The probability distributions are as follows:

x	1	2	3	4
$P(X=x)$	$\frac{1}{4}$	$\frac{1}{4}$	$\frac{1}{4}$	$\frac{1}{4}$

y	0	1	2
$P(Y=y)$	$\frac{1}{4}$	$\frac{1}{2}$	$\frac{1}{4}$

(b) By symmetry, $E(X) = 2\frac{1}{2}$

$E(Y) = 1$

(c)

$$E(X^2) = \sum_{\text{all } x} x^2P(X=x)$$

$$= 1\left(\frac{1}{4}\right)+4\left(\frac{1}{4}\right)+9\left(\frac{1}{4}\right)+16\left(\frac{1}{4}\right)$$

$$= 7\frac{1}{2}$$

$$\text{Var}(X) = E(X^2)-E^2(X)$$

$$= 7\frac{1}{2}-6\frac{1}{4}$$

$$= 1\frac{1}{4}$$

So Var(X) $= 1\frac{1}{4}$.

$$E(Y^2) = \sum_{\text{all } y} y^2P(Y=y)$$

$$= 0\left(\frac{1}{4}\right)+1\left(\frac{1}{2}\right)+4\left(\frac{1}{4}\right)$$

$$= 1\frac{1}{2}$$

$$\text{Var}(Y) = E(Y^2)-E^2(Y)$$

$$= 1\frac{1}{2}-1$$

$$= \frac{1}{2}$$

Var(Y) $= \frac{1}{2}$.

(d) Consider the r.v. $X+Y$.

$X+Y$ can assume values 1, 2, 3, 4, 5 and 6.

$$P(X+Y=1) = P(\text{1 on die, 0 heads}) = \left(\frac{1}{4}\right)\left(\frac{1}{4}\right) = \frac{1}{16}$$

$$\begin{aligned} P(X+Y=2) &= P(\text{2 on die, 0 heads}) + P(\text{1 on die, 1 head}) \\ &= \left(\frac{1}{4}\right)\left(\frac{1}{4}\right) + \left(\frac{1}{4}\right)\left(\frac{1}{2}\right) = \frac{3}{16} \end{aligned}$$

$$\begin{aligned} P(X+Y=3) &= P(\text{3 on die, 0 heads}) + P(\text{2 on die, 1 head}) \\ &\qquad + P(\text{1 on die, 2 heads}) \\ &= \left(\frac{1}{4}\right)\left(\frac{1}{4}\right) + \left(\frac{1}{4}\right)\left(\frac{1}{2}\right) + \left(\frac{1}{4}\right)\left(\frac{1}{4}\right) = \frac{4}{16} \end{aligned}$$

$$\begin{aligned} P(X+Y=4) &= P(\text{4 on die, 0 heads}) + P(\text{3 on die, 1 head}) \\ &\qquad + P(\text{2 on die, 2 heads}) \\ &= \left(\frac{1}{4}\right)\left(\frac{1}{4}\right) + \left(\frac{1}{4}\right)\left(\frac{1}{2}\right) + \left(\frac{1}{4}\right)\left(\frac{1}{4}\right) \\ &= \frac{4}{16} \end{aligned}$$

$$\begin{aligned} P(X+Y=5) &= P(\text{4 on die, 1 head}) + P(\text{3 on die, 2 heads}) \\ &= \left(\frac{1}{4}\right)\left(\frac{1}{2}\right) + \left(\frac{1}{4}\right)\left(\frac{1}{4}\right) \\ &= \frac{3}{16} \end{aligned}$$

$$\begin{aligned} P(X+Y=6) &= P(\text{4 on die, 2 heads}) \\ &= \left(\frac{1}{4}\right)\left(\frac{1}{4}\right) \\ &= \frac{1}{16} \end{aligned}$$

The probability distribution is as follows:

$x+y$	1	2	3	4	5	6
$P(X+Y=x+y)$	$\frac{1}{16}$	$\frac{3}{16}$	$\frac{4}{16}$	$\frac{4}{16}$	$\frac{3}{16}$	$\frac{1}{16}$

(e) By symmetry $\quad E(X+Y) = 3\frac{1}{2}$

But from **(b)** $\quad E(X)+E(Y) = 2\frac{1}{2}+1 = 3\frac{1}{2}$

Therefore $\quad \underline{E(X+Y) = E(X)+E(Y)}$

Now

$$\begin{aligned}
\text{Var}(X+Y) &= E[(X+Y)^2]-E^2(X+Y) \\
E(X+Y)^2 &= 1\left(\frac{1}{16}\right)+4\left(\frac{3}{16}\right)+9\left(\frac{4}{16}\right)+16\left(\frac{4}{16}\right)+25\left(\frac{3}{16}\right)+36\left(\frac{1}{16}\right) \\
&= \frac{224}{16} \\
&= 14 \\
\text{Var}(X+Y) &= 14-12\frac{1}{4} \\
&= 1\frac{3}{4}
\end{aligned}$$

Therefore $\text{Var}(X+Y) = 1\frac{3}{4}$.

NOTE: $\text{Var}(X)+\text{Var}(Y) = 1\frac{1}{4}+\frac{1}{2} = 1\frac{3}{4}$.

So $\text{Var}(X+Y) = \text{Var}(X)+\text{Var}(Y)$.

In this example, the variables X and Y are **independent.**

In general, for random variables X and Y and constants a and b,

$$E(aX+bY) = aE(X)+bE(Y)$$

If X and Y are independent, then

$$\text{Var}(aX+bY) = a^2\text{Var}(X)+b^2\text{Var}(Y)$$

An important application of this occurs when $a = 1$ and $b = -1$. In this case, the r.v. is $X+(-1)Y$, i.e. $X-Y$, and

$$\begin{aligned}
E(X-Y) &= E(X)-E(Y) \\
\text{Var}(X-Y) &= \text{Var}(X)+(-1)^2\text{Var}(Y) \\
&= \text{Var}(X)+\text{Var}(Y)
\end{aligned}$$

Example 3.17 X and Y are independent random variables with p.d.f. as shown:

x	0	1	2
$P(X=x)$	0.2	0.6	0.2

y	1	2	3
$P(Y=y)$	0.3	0.4	0.3

Construct the probability distribution for $X-Y$ and find (**a**) $E(X-Y)$, (**b**) $\text{Var}(X-Y)$.

Given that $E(X) = 1$, $\text{Var}(X) = 0.4$, $E(Y) = 2$, $\text{Var}(Y) = 0.6$, comment on your answers.

Solution 3.17 $X-Y$ can take the values $-3, -2, -1, 0, 1$

$$P(X-Y=-3) = P(X=0)\cdot P(Y=3) = 0.06$$

$$P(X-Y=-2) = P(X=0)\cdot P(Y=2)+P(X=1)\cdot P(Y=3) = 0.08+0.18 = 0.26$$

$$P(X-Y=-1) = P(X=0)\cdot P(Y=1)+P(X=1)\cdot P(Y=2) + P(X=2)\cdot P(Y=3) = 0.06+0.24+0.06 = 0.36$$

$$P(X-Y=0) = P(X=1)\cdot P(Y=1)+P(X=2)\cdot P(Y=2) = 0.18+0.08 = 0.26$$

$$P(X-Y=1) = P(X=2)\cdot P(Y=1) = 0.06$$

$x-y$	-3	-2	-1	0	1
$P(X-Y=x-y)$	0.06	0.26	0.36	0.26	0.06

(a) $E(X-Y) = -1$ (by symmetry)

(b)
$$\text{Var}(X-Y) = E[(X-Y)^2]-(-1)^2 = 9(0.06)+4(0.26)+1(0.36)+0+1(0.06)-1 = 1$$

$\text{Var}(X-Y) = 1.$

Now $E(X)-E(Y) = 1-2 = -1$

so $E(X-Y) = E(X)-E(Y)$

$\text{Var}(X)+\text{Var}(Y) = 0.4+0.6 = 1$

so $\text{Var}(X-Y) = \text{Var}(X)+\text{Var}(Y)$

Example 3.18 The r.v. X is such that $E(X)=2$, $\text{Var}(X)=0.5$; the r.v. Y is such that $E(Y)=5$, $\text{Var}(Y)=2$; X and Y are independent.

Find (a) $E(3X+4Y)$, (b) $\text{Var}(3X+4Y)$, (c) $\text{Var}(5X-2Y)$.

Solution 3.18 (a)
$$E(3X+4Y) = 3E(X)+4E(Y) = 3(2)+4(5) = 26$$

So $E(3X+4Y) = 26.$

(b)
$$\text{Var}(3X+4Y) = 9\text{Var}(X)+16\text{Var}(Y) = 9(0.5)+16(2) = 36.5$$

So $\text{Var}(3X+4Y) = 36.5.$

(c)

$$\begin{aligned}\text{Var}(5X-2Y) &= 25\text{Var}(X)+4\text{Var}(Y)\\ &= 25(0.5)+4(2)\\ &= 20.5\end{aligned}$$

So $\text{Var}(5X-2Y)=20.5$.

Example 3.19 The table gives the joint probability distribution of two random variables X and Y:

	$X=0$	$X=1$
$Y=1$	0.2	0.4
$Y=2$	0.3	0.1

Calculate (a) $E(X)$, (b) $E(Y)$, (c) $E(X+Y)$.

Solution 3.19 Consider the r.v. X

$$\begin{aligned}P(X=0) &= P(X=0)\cdot P(Y=1)+P(X=0)\cdot P(Y=2)\\ &= 0.2+0.3\\ &= 0.5\end{aligned}$$

$$\begin{aligned}P(X=1) &= P(X=1)\cdot P(Y=1)+P(X=1)\cdot P(Y=2)\\ &= 0.4+0.1\\ &= 0.5\end{aligned}$$

The probability distribution for X is

x	0	1
$P(X=x)$	0.5	0.5

By symmetry $E(X)=\frac{1}{2}$.

Consider the r.v. Y

$$\begin{aligned}P(Y=1) &= P(Y=1)\cdot P(X=0)+P(Y=1)\cdot P(X=1)\\ &= 0.2+0.4\\ &= 0.6\end{aligned}$$

$$\begin{aligned}P(Y=2) &= P(Y=2)\cdot P(X=0)+P(Y=2)\cdot P(X=1)\\ &= 0.3+0.1\\ &= 0.4\end{aligned}$$

The probability distribution for Y is

y	1	2
$P(Y=y)$	0.6	0.4

$$E(Y) = \sum_{\text{all } y} yP(Y = y)$$

$$= 1(0.6) + 2(0.4)$$

$$= 1.4$$

Therefore $E(Y) = 1.4$.

Now $E(X+Y) = E(X) + E(Y)$

$$= 0.5 + 1.4$$

$$= 1.9$$

Therefore $E(X+Y) = 1.9$.

THE DISTRIBUTION OF $X_1 + X_2$

We now consider the distribution of $X_1 + X_2$, where X_1, X_2 are two independent observations *from the same distribution X*.

Now $E(X_1 + X_2) = E(X_1) + E(X_2)$

$$= E(X) + E(X)$$

$$= 2E(X)$$

and $\text{Var}(X_1 + X_2) = \text{Var}(X_1) + \text{Var}(X_2)$

$$= \text{Var}(X) + \text{Var}(X)$$

$$= 2\text{Var}(X)$$

For the distribution $X_1 + X_2$, where X_1 and X_2 are *independent* observations from the distribution X

$$E(X_1 + X_2) = 2E(X)$$

$$\text{Var}(X_1 + X_2) = 2\text{Var}(X)$$

For n independent observations

$$E(X_1 + X_2 + \ldots + X_n) = nE(X)$$

$$\text{Var}(X_1 + X_2 + \ldots + X_n) = n\,\text{Var}(X)$$

Example 3.20 X has p.d.f. as shown

x	2	3	4
$P(X = x)$	0.3	0.5	0.2

(a) Find $E(X)$ and $\text{Var}(X)$.

(b) Two independent observations are made from X. Construct the probability distribution for $X_1 + X_2$ and find the expectation and variance. Comment on your results.

Solution 3.20 (a)

$$E(X) = \Sigma x P(X = x)$$
$$= 2(0.3) + 3(0.5) + 4(0.2)$$
$$= 2.9$$

$$E(X^2) = \Sigma x^2 P(X = x)$$
$$= 4(0.3) + 9(0.5) + 16(0.2)$$
$$= 8.9$$

$$\text{Var}(X) = E(X^2) - E^2(X)$$
$$= 8.9 - 2.9^2$$
$$= 0.49$$

Therefore $E(X) = 2.9$ and $\text{Var}(X) = 0.49$.

(b) Consider the distribution of $X_1 + X_2$. To show the possible outcomes it is useful to draw a probability tree.

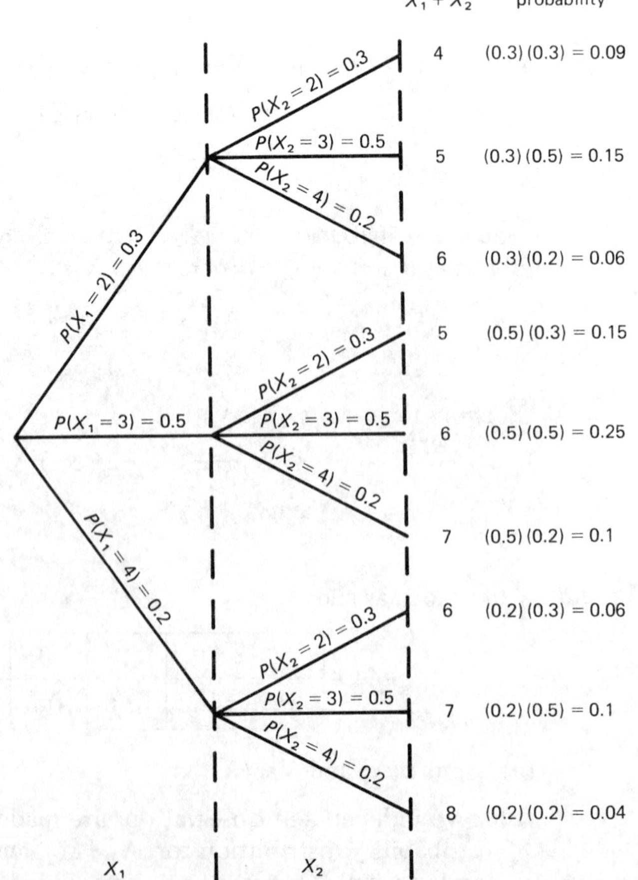

Now $P(X_1+X_2=4)=0.09$, $P(X_1+X_2=5)=0.15+0.15=0.3$, and so on. We see that X_1+X_2 can take values 4, 5, 6, 7, 8 where

x_1+x_2	4	5	6	7	8
$P(X_1+X_2=x_1+x_2)$	0.09	0.3	0.37	0.2	0.04

$$\begin{aligned} E(X_1+X_2) &= 4(0.09)+5(0.3)+6(0.37)+7(0.2)+8(0.04) \\ &= 5.8 \\ \operatorname{Var}(X_1+X_2) &= 16(0.09)+25(0.3)+36(0.37)+49(0.2) \\ &\quad +64(0.04)-5.8^2 \\ &= 0.98 \end{aligned}$$

Therefore $E(X_1+X_2)=5.8$ and $\operatorname{Var}(X_1+X_2)=0.98$.

We note that, as expected,

$$\begin{aligned} E(X_1+X_2) &= 5.8 = 2(2.9) = 2E(X) \\ \operatorname{Var}(X_1+X_2) &= 0.98 = 2(0.49) = 2\operatorname{Var}(X) \end{aligned}$$

In practice, if we need the expectation and variance, but not the actual probability distribution we can just quote these results.

Example 3.21 Find the expectation and variance of the number of heads obtained when 6 coins are tossed.

Solution 3.21 Let X be the r.v. 'the number of heads when a coin is tossed'. Then X can take the values 0, 1

x	0	1
$P(X=x)$	0.5	0.5

Now
$$\begin{aligned} E(X) &= 0.5 \quad \text{(by symmetry)} \\ E(X^2) &= 1(0.5) = 0.5 \end{aligned}$$
so
$$\begin{aligned} \operatorname{Var}(X) &= E(X^2)-E^2(X) \\ &= 0.5-0.5^2 \\ &= 0.25 \end{aligned}$$

Now consider $Y=X_1+X_2+\ldots+X_6$ where Y is the r.v. 'the number of heads when 6 coins are tossed'.

$$\begin{aligned} E(Y) &= 6E(X) & \operatorname{Var}(Y) &= 6\operatorname{Var}(X) \\ &= 6(0.5) & &= 6(0.25) \\ &= 3 & &= 1.5 \end{aligned}$$

So the expected number of heads is 3, and the variance is 1.5.

COMPARING THE DISTRIBUTIONS OF $2X$ AND $X_1 + X_2$

Confusion often arises over the different random variables $2X$ and X_1+X_2, where X_1, X_2 are two independent observations of X. For example, if X is the r.v. 'the number on which a tetrahedral die lands' then $2X$ is the r.v. '*double* the number on which a tetrahedral die lands', whereas X_1+X_2 is the r.v. 'the *sum* of the two numbers when a tetrahedral die is thrown twice'. We will see from the following example that the distributions of the two random variables are very different.

Example 3.22 (a) A tetrahedral die is thrown and the number of the face on which it lands is noted. Find the expectation and the variance.

(b) The 'score' is double the number on which it lands. Find the expectation and the variance of the 'score'.

(c) A new experiment is set up, where the 'score' is the sum of the numbers obtained when the die is thrown twice. Find the expectation and the variance of this new 'score'.

Solution 3.22 (a) Let X be the r.v. 'the number on which the die lands'.

The probability distribution for X is:

x	1	2	3	4
$P(X=x)$	$\frac{1}{4}$	$\frac{1}{4}$	$\frac{1}{4}$	$\frac{1}{4}$

Now, by symmetry, $E(X) = 2.5$.

$$\begin{aligned}\text{Var}(X) &= E(X^2) - E^2(X) \\ &= \sum_{\text{all } x} x^2 P(X=x) - E^2(X) \\ &= \frac{1}{4}(1+4+9+16) - (2.5)^2 \\ &= 1.25\end{aligned}$$

So $E(X) = 2.5$, $\text{Var}(X) = 1.25$.

(b) Now consider the r.v. R, where R is 'double the number on which the die lands', i.e. $R = 2X$. The probability distribution for R is:

r	2	4	6	8
$P(R = r)$	$\frac{1}{4}$	$\frac{1}{4}$	$\frac{1}{4}$	$\frac{1}{4}$

By symmetry $E(R) = 5$.

$$
\begin{aligned}
\text{Var}(R) &= E(R^2) - E^2(R) \\
&= \sum_{\text{all } r} r^2 P(R = r) - 25 \\
&= \frac{1}{4}(4 + 16 + 36 + 64) - 25 \\
&= 5
\end{aligned}
$$

Therefore $E(R) = 5$ and $\text{Var}(R) = 5$, where $R = 2X$.

We note that

$E(2X) = 2(2.5) = 2E(X)$ and $\text{Var}(2X) = 4(1.25) = 4\,\text{Var}(X)$.

(c) Consider the r.v. S where S is the sum of the two numbers on which the die lands when it is thrown twice. Therefore $S = X_1 + X_2$.

Now S can assume the values $2, 3, 4, 5, 6, 7, 8$ and the outcomes (all equally likely) are shown in the diagram:

Second throw				
4	5	6	7	8
3	4	5	6	7
2	3	4	5	6
1	2	3	4	5
	1	2	3	4
	First throw			

The probability distribution for S is:

s	2	3	4	5	6	7	8
$P(S = s)$	$\frac{1}{16}$	$\frac{2}{16}$	$\frac{3}{16}$	$\frac{4}{16}$	$\frac{3}{16}$	$\frac{2}{16}$	$\frac{1}{16}$

By symmetry, $E(S) = 5$.

$$\begin{aligned}\text{Var}(S) &= E(S^2) - E^2(S)\\ &= \sum_{\text{all } s} s^2 P(S = s) - 25\\ &= \frac{1}{16}[4(1) + 9(2) + 16(3) + 25(4) + 36(3) + 49(2)\\ &\qquad + 64(1)] - 25\\ &= 2.5\end{aligned}$$

Therefore $E(S) = 5$, $\text{Var}(S) = 2.5$, where $S = X_1 + X_2$.

We note that

$E(X_1 + X_2) = 2(2.5) = 2E(X)$,

$\text{Var}(X_1 + X_2) = 2(1.25) = 2\,\text{Var}(X)$.

We can see that the distribution for R, double the number on which the die lands, is very different from the distribution for S, the sum of the numbers on which the die lands when it is thrown twice.

r	2	4	6	8
$P(R = r)$	$\frac{1}{4}$	$\frac{1}{4}$	$\frac{1}{4}$	$\frac{1}{4}$

s	2	3	4	5	6	7	8
$P(S = s)$	$\frac{1}{16}$	$\frac{2}{16}$	$\frac{3}{16}$	$\frac{4}{16}$	$\frac{3}{16}$	$\frac{2}{16}$	$\frac{1}{16}$

Although the means of the two distributions are the same, the variances are not, with the r.v. 'double the number' having the greater variance.

Summarising, we have

MULTIPLES	SUMS
$E(2X) = 2E(X)$ $\text{Var}(2X) = 4\,\text{Var}(X)$	$E(X_1 + X_2) = 2E(X)$ $\text{Var}(X_1 + X_2) = 2\,\text{Var}(X)$
In general $E(nX) = nE(X)$ $\text{Var}(nX) = n^2\,\text{Var}(X)$	$E(X_1 + X_2 + \ldots + X_n) = nE(X)$ $\text{Var}(X_1 + X_2 + \ldots + X_n) = n\,\text{Var}(X)$

Exercise 3f

1. Independent random variables X and Y have probability distributions as shown in the tables:

x	0	1	2	3
$P(X=x)$	0.3	0.2	0.4	0.1

y	0	1	2
$P(Y=y)$	0.4	0.2	0.4

(*a*) Find $E(X)$, $E(Y)$, $\text{Var}(X)$, $\text{Var}(Y)$.
(*b*) Construct the probability distribution for the r.v. $X+Y$.
(*c*) Verify that $E(X+Y)=E(X)+E(Y)$.
(*d*) Verify that $\text{Var}(X+Y)=\text{Var}(X)+\text{Var}(Y)$
(*e*) Construct the probability distribution for the r.v. $X-Y$.
(*f*) Verify that $E(X-Y)=E(X)-E(Y)$.
(*g*) Verify that $\text{Var}(X-Y)=\text{Var}(X)+\text{Var}(Y)$.

2. Independent random variables X and Y are such that $E(X)=4$, $E(Y)=5$, $\text{Var}(X)=1$, $\text{Var}(Y)=2$. Find
(*a*) $E(4X+2Y)$, (*b*) $E(5X-Y)$,
(*c*) $\text{Var}(3X+2Y)$, (*d*) $\text{Var}(5Y-3X)$,
(*e*) $\text{Var}(3X-5Y)$.

3.

	$Y=0$	$Y=1$	$Y=2$
$X=1$	$\frac{1}{12}$	$\frac{1}{6}$	$\frac{1}{12}$
$X=2$	$\frac{1}{6}$	$\frac{1}{3}$	$\frac{1}{6}$

The above table gives the joint probability distribution of two random variables X and Y. Calculate (*a*) $P(Y=1)$, (*b*) $P(XY=2)$, (*c*) $E(X+Y)$.
(L Additional)

4. Independent random variables X and Y are such that $E(X^2)=14$, $E(Y^2)=20$, $\text{Var}(X)=10$, $\text{Var}(Y)=11$. Find (*a*) $E(3X-2Y)$, (*b*) $\text{Var}(5X+2Y)$.

5. Independent random variables X and Y are such that $E(X)=3$, $E(X^2)=12$, $E(Y)=4$, $E(Y^2)=18$. Find the value of
(*a*) $E(3X-2Y)$, (*b*) $E(2Y-3X)$,
(*c*) $E(6X+4Y)$, (*d*) $\text{Var}(2X-Y)$,
(*e*) $\text{Var}(2X+Y)$, (*f*) $\text{Var}(3Y+2X)$.

6. Two ordinary dice are thrown, a red and a green die. Let R be the r.v. 'the score on the red die' and let G be the r.v. 'the score on the green die'.
(*a*) Construct the probability distribution for $R+G$, the r.v. 'the sum of the two scores', and find (i) $E(R+G)$, (ii) $\text{Var}(R+G)$.
(*b*) Construct the probability distribution for $R-G$ and find (i) $E(R-G)$, (ii) $\text{Var}(R-G)$.
(*c*) Given that $E(R)=3.5$ and $\text{Var}(R)=\frac{35}{12}$, comment on your answers.

7. X has probability distribution as shown:

x	0	1	2
$P(X=x)$	0.1	0.6	0.3

(*a*) Find $E(X)$ and $\text{Var}(X)$.
(*b*) Find $P(X_1+X_2=4)$ where X_1, X_2 are two independent observations of X.
(*c*) Find $E(X_1+X_2)$ and $\text{Var}(X_1+X_2)$.
(*d*) Find $P(2X=4)$.
(*e*) Find $E(2X)$ and $\text{Var}(2X)$.

8. Rods of length 2 m or 3 m are selected at random with probabilities 0.4 and 0.6 respectively.
(*a*) Find the expectation and variance of the length of a rod.
(*b*) Two lengths are now selected at random. Find the expectation and variance of the sum of the two lengths.
(*c*) Three lengths are now selected at random. Show that the probability distribution of Y, the sum of the three lengths, is

y	6	7	8	9
$P(Y=y)$	0.064	0.288	0.432	0.216

and find $E(Y)$ and $\text{Var}(Y)$. Comment on your results.

9. Find the variance of the sum of the scores when an ordinary die is thrown 10 times.

10. X has a p.d.f. given by $P(X=x)=kx$, $x=1, 2, 3, 4$. Find (*a*) k, (*b*) $E(X)$, (*c*) $\text{Var}(X)$, (*d*) $P(X_1+X_2=5)$, (*e*) $E(4X)$ (*f*) $\text{Var}(X_1+X_2+X_3)$.

SUMMARY — DISCRETE RANDOM VARIABLES

For the discrete random variable X with probability density function $P(X = x)$ for $x = x_1, x_2, \ldots, x_n$,

(1)	$\sum_{\text{all } x} P(X = x) = 1$
(2)	$F(t) = \sum_{x = x_1}^{t} P(X = x)$ where $F(t)$ is the cumulative distribution function
(3)	$E(X) = \sum_{\text{all } x} xP(X = x)$
(4)	$\text{Var}(X) = E(X^2) - E^2(X)$ $= \sum_{\text{all } x} x^2P(X = x) - E^2(X)$

For the random variable X and constants a and b,

$E(a) = a$	$\text{Var}(a) = 0$
$E(aX) = aE(X)$	$\text{Var}(aX) = a^2\text{Var}(X)$
$E(aX + b) = aE(X) + b$	$\text{Var}(aX + b) = a^2\text{Var}(X)$

For any two random variables X and Y and constants a and b,

$E(X + Y) = E(X) + E(Y)$

$E(X - Y) = E(X) - E(Y)$

$E(aX + bY) = aE(X) + bE(Y)$

For independent random variables X and Y and constants a and b,

$\text{Var}(X + Y) = \text{Var}(X) + \text{Var}(Y)$

$\text{Var}(X - Y) = \text{Var}(X) + \text{Var}(Y)$

$\text{Var}(aX + bY) = a^2\text{Var}(X) + b^2\text{Var}(Y)$

$\text{Var}(aX - bY) = a^2\text{Var}(X) + b^2\text{Var}(Y)$

If $X_1, X_2, \ldots, X_n$ are n independent observations of the r.v. X then,

$E(X_1 + X_2 + \ldots + X_n) = nE(X)$

$\text{Var}(X_1 + X_2 + \ldots + X_n) = n\,\text{Var}(X)$

Miscellaneous Exercise 3g

1. Two tetrahedral dice are thrown and the score is the product of the numbers on which the dice fall. What is the expected score for a throw?

2. A housewife removes the labels from three tins of peaches and a tin of baked beans in order to enter a competition and then puts the tins in a cupboard. She discovers that the tins are outwardly identical. Let X be the number of tins she now needs to open in order to have baked beans. List the values that X can take and determine the probabilities for each of these values of X. Calculate the expected value of X.

   Her neighbour has five tins of peaches and two tins of baked beans, again outwardly identical once the labels are removed. This woman removes the labels and puts the tins away. Find the probability that this woman later requires to open at least three tins to have baked beans. (SUJB Additional)

3. On a long train journey, a statistician is invited by a gambler to play a dice game. The game uses two ordinary dice which the statistician is to throw. If the total score is 12, the statistician is paid £6 by the gambler. If the total score is 8, the statistician is paid £3 by the gambler. However if both or either dice show a 1, the statistician pays the gambler £2. Let £X be the amount paid to the statistician by the gambler after the dice are thrown once.

   Determine the probability that (*a*) $X = 6$, (*b*) $X = 3$, (*c*) $X = -2$.

   Find the expected value of X and show that, if the statistician played the game 100 times, his expected loss would be £2.78, to the nearest penny.

   Find the amount, £a, that the £6 would have to be changed to in order to make the game unbiased. (SUJB)

4. A box contains nine numbered balls. Three balls are numbered 3, four balls are numbered 4 and two balls are numbered 5.

   Each trial of an experiment consists of drawing two balls without replacement and recording the sum of the numbers on them, which is denoted by X. Show that the probability that $X = 10$ is $\frac{1}{36}$, and find the probabilities of all other possible values of X.

   Use your results to show that the mean of X is $\frac{70}{9}$, and find the standard deviation of X.

   Two trials are made. (The two balls in the first trial are replaced in the box before the second trial.) Find the probability that the second value of X is greater than or equal to the first value of X. (MEI)

5. A man stakes £2 to play a game in which he rolls an ordinary (fair) die. If he scores 1 or 2 he wins £3 (plus his stake) and loses his stake if he scores 3, 4 or 5. If he scores a six he may roll the die once again, winning if he scores 1, 2 or 6, losing if he scores 3, 4 or 5. Find
   (*a*) the probability that the man wins the game by rolling (i) once, (ii) twice.
   (*b*) his expectation,
   (*c*) the expected number of times he will roll the die.

   If the rules are changed so that the winning scores are 1 and 2 but that every time he scores 6 he may roll the die again, find
   (*d*) the probability that he wins on his rth roll of the die,
   (*e*) the probability that he wins the game. (SUJB)

6. A and B each roll a fair die simultaneously. Construct a table for the difference in their scores showing the associated probabilities. Calculate the mean of the distribution. If the difference in scores is 1 or 2, A wins; if it is 3, 4 or 5, B wins and if it is zero, they roll their dice again. The game ends when one of the players has won. Calculate the probabilitity that A wins on (*a*) the first, (*b*) the second, (*c*) the rth roll. What is the probability that A wins?

   If B stakes £1 what should A stake for the game to be fair? (SUJB)

7. A gambler has 4 packs of cards each of which is well shuffled and has equal numbers of red, green and blue cards. For each turn he pays £2 and draws a card from each pack. He wins £3 if he gets 2 red cards, £5 if he gets 3 red cards and £10 if he gets 4 red cards.
   (*a*) What are the probabilities of his drawing 0, 1, 2, 3, 4 red cards?

(*b*) What is the expectation of his winnings (to the nearest 10p)? (SUJB)

8. During winter a family requests 4 bottles of milk every day, and these are left on the door-step. Three of the bottles have silver tops and the fourth has a gold top. A thirsty blue-tit attempts to remove the tops from these bottles. The probability distribution of X, the number of silver tops removed by the blue-tit, is the same each day and is given by

$$P(X=0) = \tfrac{5}{15}, \quad P(X=1) = \tfrac{6}{15},$$
$$P(X=2) = \tfrac{3}{15}, \quad P(X=3) = \tfrac{1}{15}$$

The blue-tit finds the gold top particularly attractive, and the probability that this top is removed is $\frac{3}{5}$, independent of the number of silver tops removed. Determine the expectation and variance of
(*a*) the number of silver tops removed in a day,
(*b*) the number of gold tops removed in a day,
(*c*) the total number of tops (silver and gold) removed in 7 days.

Find also the probability distribution of the total number of tops (silver and gold) removed in a day. (C)

9. The probability of there being X unusable matches in a full box of Surelite matches is given by $P(X=0)=8k$, $P(X=1)=5k$, $P(X=2)=P(X=3)=k$, $P(X \geqslant 4)=0$. Determine the constant k and the expectation and variance of X.

Two full boxes of Surelite matches are chosen at random and the total number Y of unusable matches is determined. Calculate $P(Y>4)$, and state the values of the expectation and variance of Y. (C)

10. A player throws a die whose faces are numbered 1 to 6 inclusive. If the player obtains a six he throws the die a second time, and in this case his score is the sum of 6 and the second number; otherwise his score is the number obtained. The player has no more than two throws.

Let X be the random variable denoting the player's score. Write down the probability distribution of X, and determine the mean of X.

Show that the probability that the sum of two successive scores is 8 or more is $\frac{17}{36}$. Determine the probability that the first of two successive scores is 7 or more, given that their sum is 8 or more. (C)

11. The faces of an ordinary die are renumbered so that the faces are 1, 2, 2, 3, 3 and 3. This die and an ordinary, unaltered die are thrown at the same time. The score, X, is the sum of the numbers on the uppermost faces of the two dice. Show that the probability of X being 3 is $\frac{1}{12}$ and of being 4 is $\frac{1}{6}$.

List the values that X can take and determine their respective probabilities. Hence obtain the expected value of X, correct to 3 decimal places.

If the dice are thrown 3 times, determine the probability, correct to 3 significant figures, that none of the three values of X exceeds 3. (SUJB)

12. Alan and his younger brother Bill play a game each day. Alan throws three darts at a dartboard and for each dart that scores a bull (which happens with probability p) Bill gives him a penny, while for each dart which misses the bull (which happens with probability $1-p$) Alan gives Bill twopence. By considering all possible outcomes for the three throws, or otherwise, find the distribution of the number of pence (positive or negative) that Bill receives each day. Show that, when $p=\frac{1}{3}$, the mean is 3 and the variance 6.

The game takes place on 150 days. What is the mean and standard deviation of Bill's total winnings when $p=\frac{1}{3}$? (O)

13. In a certain field, each puffball which is growing in one year gives rise to a number, X, of new puffballs in the following year. None of the original puffballs is present in the following year. The probability distribution of the random variable X is as follows:

$$P(X=0) = P(X=2) = 0.3,$$
$$P(X=1) = 0.4.$$

Find the probability distribution of Y, the number of puffballs resulting from there being two puffballs in the previous year, and show that the variance of Y is 1.2.

Hence, or otherwise, determine the probability distribution of the number, Z, of puffballs present in year 3, given that there was a single puffball present in year 1. Find also the mean and variance of Z. (C)

14. A discrete random variable X can take only the values 0, 1, 2 or 3, and its probability distribution is given by $P(X = 0) = k$, $P(X = 1) = 3k$, $P(X = 2) = 4k$, $P(X = 3) = 5k$, where k is a constant. Find
(*a*) the value of k,
(*b*) the mean and variance of X. (JMB)

15. A random variable R takes the integer value r with probability $P(r)$ where

$$P(r) = kr^3, \quad r = 1, 2, 3, 4,$$
$$P(r) = 0, \quad \text{otherwise.}$$

Find
(*a*) the value of k, and display the distribution on graph paper,
(*b*) the mean and the variance of the distribution,
(*c*) the mean and the variance of $5R - 3$. (L)P

16. A gambling machine works in the following way. The player inserts a penny into one of five slots, which are coloured Blue, Red, Orange, Yellow and Green corresponding to five coloured light bulbs. The player can choose which ever coloured slot he likes. After the penny has been inserted one of the five bulbs lights up. If the bulb lit up is the same colour as the slot selected by the player, then the player wins and receives from the machine R pennies, where

$$P(R = 2) = \tfrac{1}{2}, \quad P(R = 4) = \tfrac{1}{4}$$
$$P(R = 6) = \tfrac{3}{20}, \quad \text{and}$$
$$P(R = 8) = P(R = 10) = \tfrac{1}{20}$$

If the colour of the bulb lit up and the slot selected are not the same, the player receives nothing from the machine. In either case the player does not get back the penny that he inserted. Assuming that each of the colours is equally likely to light up, and that the machine selects the bulbs at random, determine
(*a*) the probability that the player receives nothing from the machine,
(*b*) the expected value of the amount gained by the player from a single try,
(*c*) the variance of the amount gained by the player from a single try. (C)

17. Four rods of lengths 1, 2, 3 and 4 units are placed in a bag from which one rod is selected at random. The probability of selecting a rod of length l is kl. Find the value of k.

Show that the expected value of X, the length of the selected rod, is 3 units and find the variance of X.

After a rod has been selected it is not replaced. The probabilities of selection for each of the three rods that remain are in the same ratio as they were before the first selection. A second rod is now selected from the bag. Defining Y to be the length of this rod and writing $P_1 = P(Y = 1 | X = 2), P_2 = P(Y = 2 | X = 1)$ show that $16P_1 = 9P_2$.

Show also that $P(X + Y = 3) = \frac{17}{360}$. (C)

18. A game is played in which a complete throw consists of three fair coins being tossed once each and any which have landed tails being tossed a second time; no coin is tossed more than twice. The score for the complete throw is the total number of heads showing at the end of the throw.
(*a*) Find the respective probabilities that the score after a complete throw is (i) 0, (ii) 1, (iii) 2, (iv) 3.
(*b*) Show that the average score over a large number of complete throws is 9/4.

(You may leave your answers as fractions in their lowest terms.) (O & C)

19. The random variable X takes values -2, 0, 2 with probabilities $\frac{1}{4}, \frac{1}{2}, \frac{1}{4}$ respectively. Find $\text{Var}(X)$ and $E(|X|)$.

The random variable Y is defined by $Y = X_1 + X_2$, where X_1 and X_2 are two independent observations of X. Find the probability distribution of Y. Find $\text{Var}(Y)$ and $E(Y + 3)$. (C)

20. The discrete random variable X can take only the values 0, 1, 2, 3, 4, 5. The probability distribution of X is given by the following:

$$P(X = 0) = P(X = 1) = P(X = 2) = a$$
$$P(X = 3) = P(X = 4) = P(X = 5) = b$$
$$P(X \geqslant 2) = 3P(X < 2)$$

where a and b are constants.
(i) Determine the values of a and b.
(ii) Show that the expectation of X is $\frac{23}{8}$ and determine the variance of X.
(iii) Determine the probability that the sum of two independent observations from this distribution exceeds 7. (C)

21. A random variable R takes the integer values $1, 2, \ldots, n$ each with probability $1/n$. Find the mean and variance of R.

A pack of 15 cards bearing the numbers 1 to 15 is shuffled. Find the probability that the number on the top card is larger than that on the bottom card, giving reasons for your answer.
If the sum of these two numbers is S, find
(*a*) the probability that $S \leqslant 4$,
(*b*) the expected value of S.

(Answers may be left as fractions in their lowest terms.) (O & C)

22. A discrete random variable X has the distribution function

x	1	2	4	5
$F(x)$	$\frac{1}{12}$	$\frac{1}{2}$	$\frac{5}{6}$	1

(*a*) Write down the probability distribution of X.
(*b*) Find the probability distribution of the sum of two independent observations from X and find the mean and variance of the distribution of this sum.

23. A random variable R takes the integer value r with probability $P(r)$ defined by

$$P(r) = kr^2, \qquad r = 1, 2, 3,$$
$$P(r) = k(7-r)^2, \qquad r = 4, 5, 6,$$
$$P(r) = 0, \qquad \text{otherwise}.$$

Find the value of k and the mean and variance of the probability distribution. Exhibit this distribution by a suitable diagram.

Determine the mean and the variance of the variable Y where $Y \equiv 4R - 2$. (L)P

4

SPECIAL DISCRETE PROBABILITY DISTRIBUTIONS

THE BINOMIAL DISTRIBUTION

Consider an experiment which has two possible outcomes, one which may be termed 'success' and the other 'failure'. A binomial situation arises when n independent trials of the experiment are performed, for example

toss a coin 6 times; consider obtaining a head on a single toss as a success, and obtaining a tail as a failure;

throw a die 10 times; consider obtaining a 6 on a single throw as a success, and not obtaining a 6 as a failure.

Example 4.1 A coin is biased so that the probability of obtaining a head is $\frac{2}{3}$. The coin is tossed four times. Find the probability of obtaining exactly two heads.

Solution 4.1 We will consider 'obtaining a head' as success.

Now $P(\text{H}) = \frac{2}{3}$ and $P(\overline{\text{H}}) = \frac{1}{3}$.

The probability of obtaining two tails and two heads, in that order, is given by

$$P(\overline{\text{H}}\overline{\text{H}}\text{HH}) = \left(\frac{1}{3}\right)^2\left(\frac{2}{3}\right)^2 \quad \text{(independent events)}$$

But the result 'two heads and two tails' can be obtained in $\dfrac{4!}{2!2!}$ ways.

This is the number of ways of choosing the 2 places for the heads from the 4 places, i.e. ${}^4C_2$ ways. The arrangements are:

$$\overline{\text{H}}\overline{\text{H}}\text{HH} \quad \text{H}\overline{\text{H}}\overline{\text{H}}\text{H} \quad \text{HH}\overline{\text{H}}\overline{\text{H}} \quad \overline{\text{H}}\text{H}\overline{\text{H}}\text{H} \quad \text{H}\overline{\text{H}}\text{H}\overline{\text{H}} \quad \overline{\text{H}}\text{HH}\overline{\text{H}}$$

Therefore $$P(\text{2 heads exactly}) = {}^4C_2\left(\frac{1}{3}\right)^2\left(\frac{2}{3}\right)^2$$
$$= 6\left(\frac{1}{3}\right)^2\left(\frac{2}{3}\right)^2$$
$$= \frac{8}{27}$$

The probability of obtaining exactly two heads when the biased coin is tossed four times is $\frac{8}{27}$.

Example 4.2 An ordinary die is thrown seven times. Find the probability of obtaining exactly three sixes.

Solution 4.2 We will consider 'obtaining a 6' as success.

Now $P(6) = \frac{1}{6}$ and $P(\bar{6}) = \frac{5}{6}$.

$$P(\bar{6}\bar{6}\bar{6}\bar{6}666) = \left(\frac{5}{6}\right)^4\left(\frac{1}{6}\right)^3$$

But the result 'four numbers which are not 6 and three sixes' can be obtained in $\dfrac{7!}{4!3!}$ ways, i.e. ${}^7C_3$ ways (the number of ways of choosing the 3 places for the sixes from the 7 places).

So $$P(\text{exactly three sixes}) = {}^7C_3\left(\frac{5}{6}\right)^4\left(\frac{1}{6}\right)^3$$
$$= 0.078 \quad (3\text{ d.p.})$$

The probability of obtaining exactly three sixes when a die is thrown seven times is 0.078 (3 d.p.).

Example 4.3 The probability that a marksman hits a target is p and the probability that he misses is q, where $q = 1-p$. Write an expression for the probability that, in 10 shots, he hits the target 6 times.

Solution 4.3 We will consider 'obtaining a hit' as success.

$$P(\text{success}) = p \quad \text{and} \quad P(\text{failure}) = q = 1-p$$

We require 4 failures and 6 successes, in any order, so

$$P(\text{6 successes}) = {}^{10}C_6 q^4 p^6$$

Therefore the probability that he hits the target exactly 6 times in 10 shots is ${}^{10}C_6 q^4 p^6$.

In general:

If the probability that an experiment results in a successful outcome is p and the probability that the outcome is a failure is q, where $q = 1-p$, and if X is the r.v. 'the number of successful outcomes in n independent trials', then the p.d.f. of X is given by

$$P(X = x) = {}^nC_x q^{n-x} p^x \qquad x = 0, 1, 2, \ldots, n$$

Example 4.4 If p is the probability of success and $q = 1-p$ is the probability of failure, find the probability of $0, 1, 2, \ldots, 5$ successes in 5 independent trials of the experiment. Comment on your answer.

Solution 4.4 Let X be the r.v. 'the number of successful outcomes'. Then $P(X = x) = {}^nC_x q^{n-x} p^x$, $x = 0, 1, \ldots, 5$ and $n = 5$. So

$$\begin{aligned}
P(X=0) &= {}^5C_0 q^5 p^0 = q^5 \\
P(X=1) &= {}^5C_1 q^4 p^1 = 5q^4p \\
P(X=2) &= {}^5C_2 q^3 p^2 = 10q^3p^2 \\
P(X=3) &= {}^5C_3 q^2 p^3 = 10q^2p^3 \\
P(X=4) &= {}^5C_4 q^1 p^4 = 5qp^4 \\
P(X=5) &= {}^5C_5 q^0 p^5 = p^5
\end{aligned}$$

We note that $q^5, 5q^4p, \ldots, p^5$ are the terms in the binomial expansion of $(q+p)^5$ and we have

$$\begin{array}{ccccccccccccc}
(q+p)^5 & = & q^5 & + & 5q^4p & + & 10q^3p^2 & + & 10q^2p^3 & + & 5qp^4 & + & p^5 \\
\uparrow & & \uparrow & & \uparrow & & \uparrow & & \uparrow & & \uparrow & & \uparrow \\
1 & = & P(X=0) & + & P(X=1) & + & P(X=2) & + & P(X=3) & + & P(X=4) & + & P(X=5)
\end{array}$$

In general:

The values $P(X = x)$ for $x = 0, 1, \ldots, n$ can be obtained by considering the terms in the binomial expansion of $(q+p)^n$, noting that $q + p = 1$

$$\begin{array}{ccccccccccccc}
(q+p)^n & = & {}^nC_0q^np^0 & + & {}^nC_1q^{n-1}p^1 & + & {}^nC_2q^{n-2}p^2 & + \ldots + & {}^nC_rq^{n-r}p^r & + \ldots + & {}^nC_nq^0p^n \\
\uparrow & & \uparrow & & \uparrow & & \uparrow & & \uparrow & & \uparrow \\
1 & = & P(X=0) & + & P(X=1) & + & P(X=2) & + \ldots + & P(X=r) & + \ldots + & P(x=n)
\end{array}$$

If X is distributed in this way, we write

$X \sim \text{Bin}(n, p)$ where n is the number of independent trials and p is the probability of a successful outcome in one trial

n and p are called the parameters of the distribution.

So we read the statement $X \sim \text{Bin}(n, p)$ thus: X follows a binomial distribution with parameters n and p.

Example 4.5 The probability that a person supports Party A is 0.6. Find the probability that in a randomly selected sample of 8 voters there are **(a)** exactly 3 who support Party A, **(b)** more than 5 who support Party A.

Solution 4.5 We will consider 'supporting Party A' as success. Then $p = 0.6$ and $q = 1-p = 0.4$. Let X be the r.v. 'the number of Party A supporters'. Then $X \sim \text{Bin}(n, p)$ with $n = 8$ and $p = 0.6$.

So $$X \sim \text{Bin}(8, 0.6)$$

and

$$P(X = x) = {}^nC_x q^{n-x}p^x = {}^8C_x(0.4)^{8-x}(0.6)^x, x = 0, 1, \ldots, 8$$

(a) We require

$$P(X = 3) = {}^8C_3(0.4)^5(0.6)^3 = 0.124 \quad (3\text{ d.p.})$$

The probability that there are exactly 3 Party A supporters is 0.124 (3 d.p.).

(b) We require

$$\begin{aligned} P(X > 5) &= P(X = 6) + P(X = 7) + P(X = 8) \\ &= {}^8C_6(0.4)^2(0.6)^6 + {}^8C_7(0.4)(0.6)^7 + {}^8C_8(0.6)^8 \\ &= 28(0.4)^2(0.6)^6 + 8(0.4)(0.6)^7 + (0.6)^8 \\ &= (0.6)^6(4.48 + 1.92 + 0.36) \\ &= 0.315 \quad (3\text{ d.p.}) \end{aligned}$$

The probability that there are more than 5 Party A supporters is 0.315 (3 d.p.).

Example 4.6 A box contains a large number of red and yellow tulip bulbs in the ratio 1 : 3. Bulbs are picked at random from the box. How many bulbs must be picked so that the probability that there is at least one red tulip bulb among them is greater than 0.95?

Solution 4.6 Consider 'obtaining a red tulip bulb' as 'success'.

Then $p = P(\text{success}) = \frac{1}{4}$ and $q = \frac{3}{4}$.

Let X be the r.v. 'the number of red tulip bulbs'.

Then $X \sim \text{Bin}(n, p)$ where $p = \frac{1}{4}$ and n is unknown.

Now $$P(X = x) = {}^nC_x q^{n-x}p^x \qquad x = 0, 1, 2, \ldots, n$$

We require $P(X \geqslant 1) > 0.95$.

Now

$$\begin{aligned} P(X \geqslant 1) &= 1 - P(X = 0) \\ &= 1 - \left(\frac{3}{4}\right)^n \end{aligned}$$

$$\text{So } 1-\left(\frac{3}{4}\right)^n > 0.95$$

$$0.05 > \left(\frac{3}{4}\right)^n$$

$$\log 0.05 > n \log 0.75 \qquad \text{(taking logs to base 10)}$$

i.e.

$$n \log 0.75 < \log 0.05$$

$$(-0.125)n < -1.301$$

$$n > \frac{-1.301}{-0.125} \qquad \text{(change inequality when dividing by a negative quantity)}$$

So $n > 10.4$, and the least value of n is 11.

Therefore at least 11 bulbs must be picked out of the box to ensure that the probability that there is at least one red tulip bulb among them is greater than 0.95.

Exercise 4a

Give answers to 3 S.F. where appropriate.

1. If $X \sim \text{Bin}(6, \frac{1}{3})$, find (*a*) $P(X = 4)$, (*b*) $P(X \leqslant 2)$.

2. If $X \sim \text{Bin}(8, 0.4)$, find (*a*) $P(X = 2)$, (*b*) $P(X = 0)$, (*c*) $P(X > 6)$.

3. The probability that a pen drawn at random from a box of pens is defective is 0.1. If a sample of 6 pens is taken, find the probability that it will contain (*a*) no defective pens, (*b*) 5 or 6 defective pens, (*c*) less than 3 defective pens.

4. Find the probability of throwing at least five sixes in seven throws of an unbiased die.

5. Find the probability of throwing not more than four heads in six tosses of a fair coin.

6. Assuming that a couple are equally likely to produce a girl or a boy, find the probability that in a family of 5 children there will be more boys than girls.

7. The probability that a housewife will buy Soapysuds Powder is 0.65. Find the probability that in a sample of 8 housewives who have each bought a packet of soap powder (*a*) exactly 3 have bought Soapysuds, (*b*) more than 5 have bought Soapysuds.

8. Describe an experiment in which the probabilities involved are the terms of the binomial expansion of $(\frac{4}{5} + \frac{1}{5})^6$. In terms of this experiment describe the event whose probability is given by the fourth term of the expansion, and calculate this probability.

9. A coin is biased so that it is twice as likely to show heads as tails. Find the probability that in five tosses of the coin (*a*) exactly three heads are obtained, (*b*) more than three heads are obtained.

10. The probability that a marksman scores a bull when he shoots at a target is 0.6. Find the probability that in 7 attempts he scores less than 3 bulls. Assume that the outcome of each shot is independent of any other.

11. (*a*) A coin is biased so that the probability of obtaining a head is p. The coin is tossed three times. Show the possible outcomes on a tree diagram and compare

the probabilities of obtaining 0, 1, 2, 3 heads with the terms in the binomial expansion of $(q+p)^3$ where $q = 1-p$.
(*b*) The coin is now tossed four times. Compare the probabilities $P(X = x)$ for $x = 0, 1, 2, 3, 4$ given in the tree diagram with the binomial expansion of $(q+p)^4$. X is the r.v. 'the number of heads obtained in four tosses'.

12. If $X \sim \text{Bin}(n, 0.6)$ and $P(X < 1) = 0.0256$, find n.

13. 1% of a box of light bulbs are faulty. What is the largest sample size which can be taken if it is required that the probability that there are no faulty bulbs in the sample is greater than 0.5?

14. If $X \sim \text{Bin}(n, 0.3)$ and $P(X \geqslant 1) > 0.8$, find the least possible value of n.

15. The probability that a target is hit is 0.3. Find the least number of shots which should be fired if the probability that the target is hit at least once is greater than 0.95.

16. In a multiple choice test there are 10 questions and for each question there is a choice of 4 answers, only one of which is correct. If a student guesses at each of the answers, find the probability that he gets (*a*) none correct, (*b*) more than 7 correct. If he needs to obtain over half marks to pass, and the questions carry equal weight, find the probability that he passes.

17. Of the pupils in a school, 30% travel to school by bus. From a sample of 10 pupils chosen at random, find the probability that (*a*) only 3 travel by bus, (*b*) more than 8 travel by bus.

EXPECTATION AND VARIANCE

If the random variable X is such that $X \sim \text{Bin}(n, p)$

then $E(X) = np$

and $\text{Var}(X) = npq$ where $q = 1-p$

Proof Now

$$P(X = x) = {}^nC_x q^{n-x} p^x \qquad x = 0, 1, 2, \ldots, n$$

So X has the probability distribution shown in the table:

x	0	1	2	3	...	n
$P(X = x)$	q^n	$nq^{n-1}p$	$\frac{n(n-1)}{2!}q^{n-2}p^2$	$\frac{n(n-1)(n-2)}{3!}q^{n-3}p^3$	...	p^n

$$E(X) = \sum_{\text{all } x} xP(X = x)$$

$$= (0)q^n + (1)nq^{n-1}p + \frac{(2)n(n-1)}{2!}q^{n-2}p^2 + \frac{(3)n(n-1)(n-2)}{3!}q^{n-3}p^3 + \ldots + np^n$$

$$= np[q^{n-1} + (n-1)q^{n-2}p + \frac{(n-1)(n-2)}{2!}q^{n-3}p^2 + \ldots + p^{n-1}]$$

$$= np[(q+p)^{n-1}]$$

$$= np \quad \text{since} \quad q+p = 1$$

Therefore

$$E(X) = np$$

Now $\mathrm{Var}(X) = E(X^2) - E^2(X)$.

$$E(X^2) = \sum_{\text{all } x} x^2 P(X = x)$$

$$= (0)q^n + (1)nq^{n-1}p + \frac{(4)n(n-1)}{2!}q^{n-2}p^2 + \frac{9n(n-1)(n-2)}{3!}q^{n-3}p^3 + \ldots + n^2p^n$$

$$= np[q^{n-1} + 2(n-1)q^{n-2}p + \frac{3(n-1)(n-2)}{2!}q^{n-3}p^2 + \ldots + np^{n-1}]$$

$$= np[q^{n-1} + (n-1)q^{n-2}p + \frac{(n-1)(n-2)}{2!}q^{n-3}p^2 + \ldots + p^{n-1} + (n-1)q^{n-2}p + \frac{2(n-1)(n-2)}{2!}q^{n-3}p^2 + \ldots + (n-1)p^{n-1}]$$

Now the first row of terms is, as before, the expansion of $(q+p)^{n-1}$.

$$\text{So } E(X^2) = np\{(q+p)^{n-1} + (n-1)p[q^{n-2} + (n-2)q^{n-3}p + \ldots + p^{n-2}]\}$$

$$= np[1 + (n-1)p(q+p)^{n-2}]$$

$$= np[1 + (n-1)p]$$

$$= np(1-p) + n^2p^2$$

Therefore $\quad \mathrm{Var}(X) = np(1-p) + n^2p^2 - (np)^2$

$$= npq \quad \text{where} \quad q = 1-p$$

Therefore $\quad \mathrm{Var}(X) = npq$

Example 4.7 If the probability that it is a fine day is 0.4, find the expected number of fine days in a week, and the standard deviation.

Solution 4.7 Let 'fine day' be 'success'. Then $p = 0.4$ and $q = 0.6$. Let X be the r.v. 'the number of fine days in a week'.

Then $X \sim \mathrm{Bin}(n, p)$ where $n = 7$ and $p = 0.4$.

Now $\quad E(X) = np = (7)(0.4) = 2.8$

$$\mathrm{Var}(X) = npq = (7)(0.4)(0.6) = 1.68$$

Therefore the standard deviation of $X = \sqrt{1.68} = 1.30$ days (2 d.p.).

The expected number of fine days in a week is 2.8 and the standard deviation is 1.30 days (2 d.p.).

Example 4.8 The r.v. X is such that $X \sim \text{Bin}(n, p)$ and $E(X) = 2$, $\text{Var}(X) = \frac{24}{13}$. Find the values of n and p, and $P(X = 2)$.

Solution 4.8 If $X \sim \text{Bin}(n, p)$ then $E(X) = np$ and $\text{Var}(X) = npq$.

Now $E(X) = 2$, so $np = 2$ (i)

$\text{Var}(X) = \frac{24}{13}$, so $npq = \frac{24}{13}$ (ii)

Substituting for np in (ii) we have

$$2q = \frac{24}{13}$$

$$q = \frac{12}{13}$$

Therefore

$$p = 1-q = 1-\frac{12}{13} = \frac{1}{13}$$

Now substituting for p in (i) we have

$$n\left(\frac{1}{13}\right) = 2$$

$$n = 26$$

Therefore $n = 26$ and $p = \frac{1}{13}$, so that $X \sim \text{Bin}(26, \frac{1}{13})$.

Now $P(X = x) = {}^nC_x q^{n-x} p^x$

$$= {}^{26}C_x \left(\frac{12}{13}\right)^{26-x} \left(\frac{1}{13}\right)^x \qquad x = 0, 1, \ldots, 26$$

so

$$P(X = 2) = {}^{26}C_2 \left(\frac{12}{13}\right)^{24} \left(\frac{1}{13}\right)^2 = \frac{(26)(25)(12)^{24}}{(1)(2)(13)^{26}} = 0.282 \quad (3\text{ d.p.})$$

Therefore $P(X = 2) = 0.282$ (3 d.p.).

Exercise 4b

1. Of the articles from a certain production line, 10% are defective. If a sample of 25 articles is taken, find the expected number of defective articles and the standard deviation.

2. The probability that an apple, picked at random from a sack, is bad is 0.05. Find the standard deviation of the number of bad apples in a sample of 15 apples.

3. X is a r.v. such that $X \sim \text{Bin}(n, p)$. Given that $E(X) = 2.4$ and $p = 0.3$, find n and the standard deviation of X.

4. In a group of people the expected number who wear glasses is 2 and the variance is 1.6. Find the probability that (*a*) a person chosen at random from the group wears glasses, (*b*) 6 people in the group wear glasses.

5. If the r.v. X is such that $X \sim \text{Bin}(10, p)$ where $p < \frac{1}{2}$ and $\text{Var}(X) = 1\frac{7}{8}$, find (*a*) p, (*b*) $E(X)$, (*c*) $P(X = 2)$.

6. A die is biased and the probability, p, of throwing a six is known to be less than $\frac{1}{6}$. An experiment consists of recording the number of sixes in 25 throws of the die. In a large number of experiments the standard deviation of the number of sixes is 1.5. Calculate the value of p and hence determine, to two places of decimals, the probability that exactly three sixes are recorded during a particular experiment. (C Additional)

7. (i) For each of the experiments described below, state, giving a reason, whether a binomial distribution is appropriate.

   Experiment 1. A bag contains black, white and red marbles which are selected at random, one at a time with replacement. The colour of each marble is noted.

   Experiment 2. This experiment is a repeat of Experiment 1 except that the bag contains black and white marbles only.

   Experiment 3. This experiment is a repeat of Experiment 2 except that marbles are not replaced after selection.

   (ii) On average 20% of the bolts produced by a machine in a factory are faulty. Samples of 10 bolts are to be selected at random each day. Each bolt will be selected and replaced in the set of bolts which have been produced on that day.
   (*a*) Calculate, to 2 significant figures, the probability that, in any one sample, two bolts or less will be faulty.
   (*b*) Find the expected value and the variance of the number of bolts in a sample which will not be faulty. (L Additional)

8. In two binomial distributions the ratio of the number of independent trials is 5 : 6, the ratio of the arithmetic means is 2 : 9 and the ratio of the variances is 32 : 45. For each distribution, find the probability of success.

DIAGRAMMATIC REPRESENTATION OF THE BINOMIAL DISTRIBUTION

Consider $X \sim \text{Bin}(5, p)$ for various values of p. The probability distributions are illustrated on the following page. It is useful to compare, for example, the distributions of $X \sim \text{Bin}(5, 0.1)$ and $X \sim \text{Bin}(5, 0.9)$ and these have been printed side by side to facilitate this.

Consider $X \sim \text{Bin}(5, 0.1)$ and $X \sim \text{Bin}(5, 0.9)$

Notice that

$$P(X = 0 \,|\, X \sim \text{Bin}(5, 0.1)) = P(X = 5 \,|\, X \sim \text{Bin}(5, 0.9))$$

$$P(X = 1 \,|\, X \sim \text{Bin}(5, 0.1)) = P(X = 4 \,|\, X \sim \text{Bin}(5, 0.9))$$

and so on.

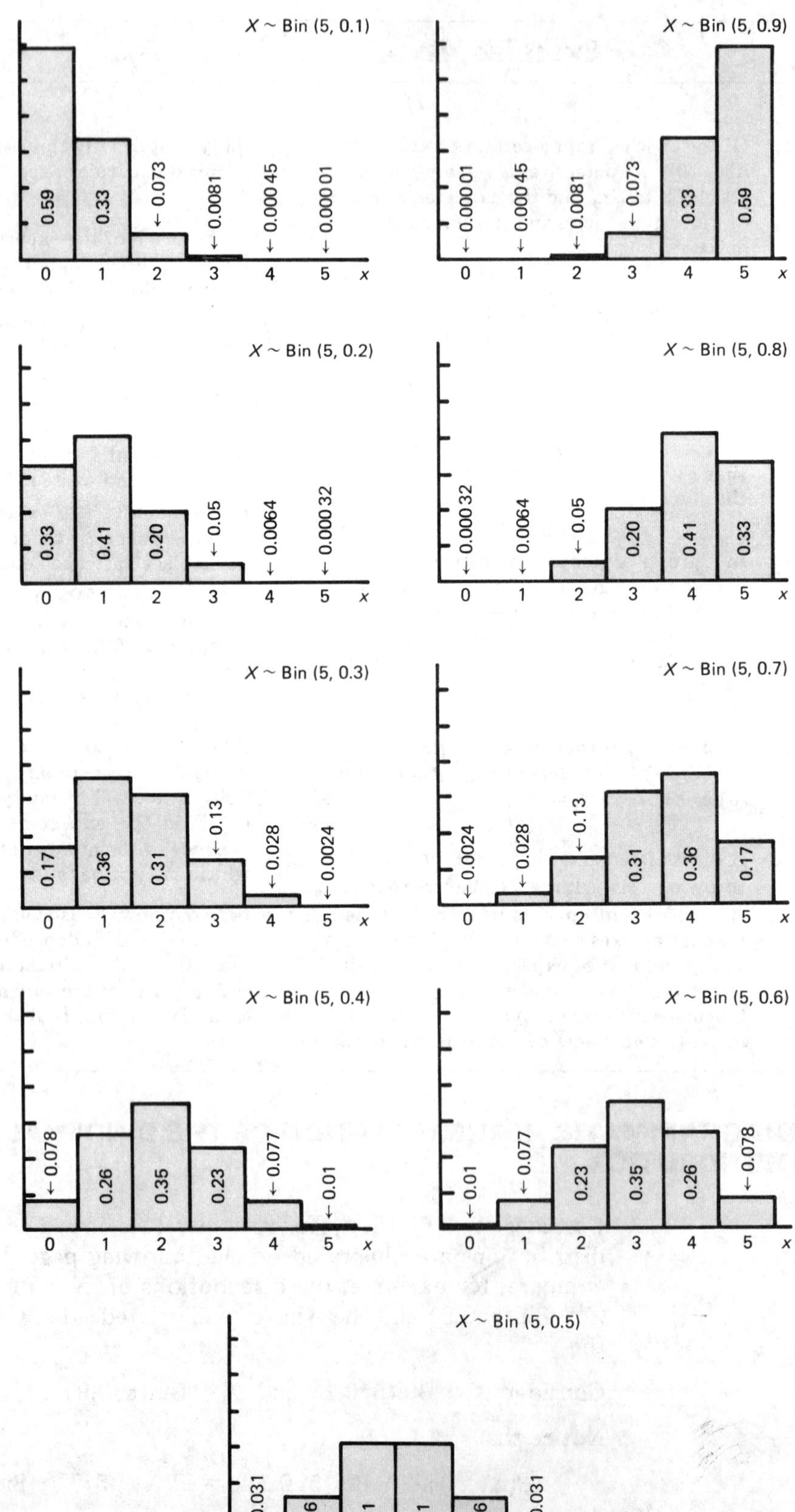
X ~ Bin (5, 0.1)
0.59
0.33
0.073
0.0081
0.000 45
0.000 01
0 1 2 3 4 5 x
X ~ Bin (5, 0.9)
0.000 01
0.000 45
0.0081
0.073
0.33
0.59
0 1 2 3 4 5 x
X ~ Bin (5, 0.2)
0.33
0.41
0.20
0.05
0.0064
0.000 32
0 1 2 3 4 5 x
X ~ Bin (5, 0.8)
0.000 32
0.0064
0.05
0.20
0.41
0.33
0 1 2 3 4 5 x
X ~ Bin (5, 0.3)
0.17
0.36
0.31
0.13
0.028
0.0024
0 1 2 3 4 5 x
X ~ Bin (5, 0.7)
0.0024
0.028
0.13
0.31
0.36
0.17
0 1 2 3 4 5 x
X ~ Bin (5, 0.4)
0.078
0.26
0.35
0.23
0.077
0.01
0 1 2 3 4 5 x
X ~ Bin (5, 0.6)
0.01
0.077
0.23
0.35
0.26
0.078
0 1 2 3 4 5 x
X ~ Bin (5, 0.5)
0.031
0.16
0.31
0.31
0.16
0.031
0 1 2 3 4 5 x

Also, considering $X \sim \text{Bin}(5, 0.2)$ and $X \sim \text{Bin}(5, 0.8)$

$$P(X = 2 \mid X \sim \text{Bin}(5, 0.2)) = P(X = 3 \mid X \sim \text{Bin}(5, 0.8))$$

In general

$$P(X = r \mid X \sim \text{Bin}(n, p)) = P(X = n - r \mid X \sim \text{Bin}(n, 1 - p))$$

NOTE: If $p = 0.5$ then the distribution of X, where $X \sim \text{Bin}(n, 0.5)$, is symmetrical.

CUMULATIVE BINOMIAL PROBABILITY TABLES

The task of finding binomial probabilities is made much easier if tables are available. These give the cumulative probabilities $F(r) = P(X \leqslant r)$ for the possible values of r. The tables are printed on p. 630 and an extract is shown below. In this we are considering $X \sim \text{Bin}(5, 0.3)$.

	$p = 0.3$
$n = 5,\ r = 0$	0.1681
$r = 1$	0.5282
$r = 2$	0.8369
$r = 3$	0.9692
$r = 4$	0.9976
$r = 5$	1.0000

Example 4.9 If $X \sim \text{Bin}(5, 0.3)$ find (a) $P(X \leqslant 4)$, (b) $P(X = 2)$, (c) $P(X < 3)$, (d) $P(X > 1)$, (e) $P(X \geqslant 3)$.

Solution 4.9 (a) $P(X \leqslant 4) = 0.9976$ (directly from the tables)

(b)
$$\begin{aligned} P(X = 2) &= P(X \leqslant 2) - P(X \leqslant 1) \\ &= 0.8369 - 0.5282 \\ &= 0.3087 \end{aligned}$$

(c)
$$\begin{aligned} P(X < 3) &= P(X \leqslant 2) \\ &= 0.8369 \end{aligned}$$

(d)
$$\begin{aligned} P(X > 1) &= 1 - P(X \leqslant 1) \\ &= 1 - 0.5282 \\ &= 0.4718 \end{aligned}$$

(e)
$$\begin{aligned} P(X \geqslant 3) &= 1 - P(X \leqslant 2) \\ &= 1 - 0.8369 \\ &= 0.1631 \end{aligned}$$

In the tables values of p are given from 0.1 to 0.5. However we are still able to use them for $p = 0.6, 0.7, 0.8$ and 0.9 by using the fact that

$$P(X = r \mid X \sim \text{Bin}(n, p)) = P(X = n - r \mid X \sim \text{Bin}(n, 1-p))$$

Consider again the probability distributions for $X \sim \text{Bin}(5, 0.3)$ and $X \sim \text{Bin}(5, 0.7)$.

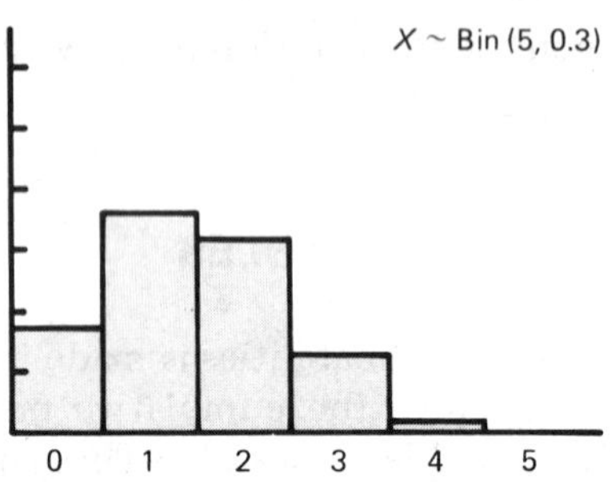

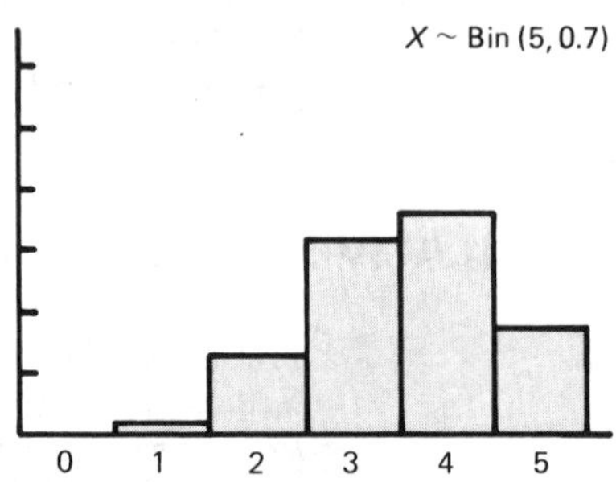

We see that $P(X \leqslant 3 \mid p = 0.3) = P(X \geqslant 2 \mid p = 0.7)$

and $P(X \geqslant 4 \mid p = 0.3) = P(X \leqslant 1 \mid p = 0.7)$

In general

$$P(X \leqslant r \mid X \sim \text{Bin}(n, p)) = P(X \geqslant n - r \mid X \sim \text{Bin}(n, 1-p))$$
$$P(X \geqslant r \mid X \sim \text{Bin}(n, p)) = P(X \leqslant n - r \mid X \sim \text{Bin}(n, 1-p))$$

Example 4.10 If $X \sim \text{Bin}(5, 0.7)$ find (a) $P(X \geqslant 3)$, (b) $P(X \leqslant 4)$, (c) $P(X = 4)$.

Solution 4.10 Using the column headed $p = 0.3$, with $n = 5$:

(a) $P(X \geqslant 3 \mid p = 0.7) = P(X \leqslant 2 \mid p = 0.3)$
$= 0.8369$

(b) $P(X \leqslant 4 \mid p = 0.7) = P(X \geqslant 1 \mid p = 0.3)$
$= 1 - P(X \leqslant 0 \mid p = 0.3)$
$= 1 - 0.1681$
$= 0.8319$

(c) $P(X = 4 \mid p = 0.7) = P(X = 1 \mid p = 0.3)$
$= P(X \leqslant 1 \mid p = 0.3) - P(X \leqslant 0 \mid p = 0.3)$
$= 0.5282 - 0.1681$
$= 0.3601$

NOTE: Obviously there are times when it is not advantageous to use the tables and it is quicker to calculate the probabilities directly. However they are particularly useful when finding the probability distribution $P(X = x)$ for all values of x.

Example 4.11 Given $X \sim \text{Bin}(8, 0.2)$ write out the probability distribution of X.

Solution 4.11 Using the cumulative probability tables on p. 630 with $n = 8$ and $p = 0.2$, and writing $P(X \leqslant r)$ as $F(r)$,

$$\begin{aligned}
P(X=0) &= F(0) = 0.1678 \\
P(X=1) &= F(1)-F(0) = 0.5033-0.1678 = 0.3355 \\
P(X=2) &= F(2)-F(1) = 0.7969-0.5033 = 0.2936 \\
P(X=3) &= F(3)-F(2) = 0.9437-0.7969 = 0.1468 \\
P(X=4) &= F(4)-F(3) = 0.9896-0.9437 = 0.0459 \\
P(X=5) &= F(5)-F(4) = 0.9988-0.9896 = 0.0092 \\
P(X=6) &= F(6)-F(5) = 0.9999-0.9988 = 0.0011 \\
P(X=7) &= F(7)-F(6) = 1.0000-0.9999 = 0.0001 \\
P(X=8) &= F(8)-F(7) = 1-1.0000 = 0.0000
\end{aligned}$$

NOTE: $P(X = 8) = (0.2)^8 = 0.000\,002\,56$, but since the tables give values to 4 d.p. they will give $P(X = 8) = 0.0000$.

Exercise 4c

1. Use cumulative binomial probability tables to find the following:
   (*a*) $X \sim \text{Bin}(6, 0.2)$, find
   (i) $P(X \leqslant 3)$, (ii) $P(X \geqslant 4)$, (iii) $P(X = 5)$.
   (*b*) $X \sim \text{Bin}(10, 0.45)$, find
   (i) $P(X = 6)$, (ii) $P(X \leqslant 3)$, (iii) $P(X < 5)$, (iv) $P(X \geqslant 3)$.
   (*c*) $X \sim \text{Bin}(4, 0.9)$, find
   (i) $P(X \leqslant 1)$, (ii) $P(X < 2)$, (iii) $P(X = 3)$.
   (*d*) $X \sim \text{Bin}(7, 0.75)$, find
   (i) $F(5)$, (ii) $F(3)$, (iii) $P(X \geqslant 4)$, (iv) $P(X = 6)$.
2. Given that $X \sim \text{Bin}(6, 0.4)$ write out the probability distribution of X.
3. Given that $X \sim \text{Bin}(5, 0.65)$ write out the probability distribution of X.

THE RECURRENCE FORMULA FOR THE BINOMIAL DISTRIBUTION

If cumulative probability tables are not available, then calculations can be performed more easily with the help of the recurrence formula, especially when a calculator with a memory is being used. Now, if $X \sim \text{Bin}(n, p)$ then

$$\begin{aligned}
P(X = x) &= {}^nC_x q^{n-x} p^x \\
&= \frac{n!}{(n-x)!x!} q^{n-x} p^x
\end{aligned}$$

and
$$\begin{aligned}
P(X = x+1) &= {}^nC_{x+1} q^{n-x-1} p^{x+1} \\
&= \frac{n!}{(n-x-1)!(x+1)!} q^{n-x-1} p^{x+1}
\end{aligned}$$

Dividing these, we have

$$\frac{P(X=x+1)}{P(X=x)} = \frac{n!}{(n-x-1)!(x+1)!}\,\frac{(n-x)!x!}{n!}\,\frac{q^{n-x-1}}{q^{n-x}}\,\frac{p^{x+1}}{p^x}$$

$$= \frac{(n-x)p}{(x+1)q}$$

So

$$P(X=x+1) = \frac{(n-x)p}{(x+1)q}P(X=x)$$

This is often written

$$p_{x+1} = \frac{(n-x)p}{(x+1)q}p_x \quad \text{where} \quad p_{x+1} = P(X=x+1), \quad p_x = P(X=x)$$

Example 4.12 If $X \sim \text{Bin}(8, 0.3)$ use the recurrence formula to calculate $P(X \leqslant 4)$.

Solution 4.12 We require $P(X \leqslant 4) = p_0 + p_1 + p_2 + p_3 + p_4$.

Now $p = 0.3$, $q = 0.7$ and $n = 8$.

$$p_0 = P(X=0) = (0.7)^8$$

If you are using a calculator with a memory system, there is no need to write out the numerical answer for p_0; it can be stored in the memory straight away. However, the values are written out so that you can check them on your calculator.

Now $p_{x+1} = \dfrac{(8-x)(0.3)}{(x+1)(0.7)}p_x$	(check each value as it is stored in the memory)
and $p_0 = (0.7)^8$	0.057 648
When $x = 0$	
$p_1 = \dfrac{8(0.3)}{1(0.7)}p_0$	0.197 650 3
When $x = 1$	
$p_2 = \dfrac{7(0.3)}{2(0.7)}p_1$	0.296 475 4
When $x = 2$	
$p_3 = \dfrac{6(0.3)}{3(0.7)}p_2$	0.254 121 8
When $x = 3$	
$p_4 = \dfrac{5(0.3)}{4(0.7)}p_3$	0.136 136 7

So $P(X \leqslant 4) = p_0 + p_1 + p_2 + p_3 + p_4$

$= 0.942\,032\,3$ (from 'memory' on calculator)

Therefore $P(X \leqslant 4) = 0.942$ (3 S.F.).

NOTE: if you are using a calculator, it is well worth practising this method. However, you need to set out all your working first, then use the calculator when you are ready.

Example 4.13 A pottery produces royal souvenir mugs. It is known that 6% are defective. If 20 mugs are selected at random, find the probability that the sample contains less than 5 defective mugs.

Solution 4.13 Let 'obtaining a defective mug' be 'success'.
Then $P(\text{success}) = p = 0.06$ and $q = 0.94$.

Let X be the r.v. 'the number of defective mugs'.
Then $X \sim \text{Bin}(n, p)$ where $n = 20$ and $p = 0.06$.

We have

$$P(X = x) = {}^nC_x q^{n-x} p^x \qquad x = 0, 1, \ldots, 20$$

$$= {}^{20}C_x (0.94)^{20-x} (0.06)^x$$

We require

$$P(X < 5) = P(X = 0) + P(X = 1) + P(X = 2) + P(X = 3) + P(X = 4)$$

We show the two methods of performing the calculations:

Method 1

$$P(X = 0) = (0.94)^{20} = 0.290\,106 \text{ (6 d.p.)}$$

$$P(X = 1) = 20(0.94)^{19}(0.06) = 0.370\,348$$

$$P(X = 2) = \frac{(20)(19)}{(2)(1)}(0.94)^{18}(0.06)^2 = 0.224\,573$$

$$P(X = 3) = \frac{(20)(19)(18)}{(3)(2)(1)}(0.94)^{17}(0.06)^3 = 0.086\,007$$

$$P(X = 4) = \frac{(20)(19)(18)(17)}{(4)(3)(2)(1)}(0.94)^{16}(0.06)^4 = 0.023\,332$$

$$0.994\,366$$

So $P(X < 5) = 0.994$ (3 S.F.).

Method 2 using the recurrence formula

Now

$$p_{x+1} = \frac{(n-x)p}{(x+1)q} p_x$$

In the example $n = 20, p = 0.06, q = 0.94$. Therefore

$$p_{x+1} = \frac{(20-x)}{(x+1)}\frac{(0.06)}{(0.94)}p_x \qquad \text{(store each value in the memory)}$$

Now

$$p_0 = (0.94)^{20} \qquad 0.290\,106 \text{ (as above)}$$

When $x = 0$

$$p_1 = \frac{20}{1}\left(\frac{0.06}{0.94}\right)p_0 \qquad 0.370\,348$$

When $x = 1$

$$p_2 = \frac{19}{2}\left(\frac{0.06}{0.94}\right)p_1 \qquad \text{and so on}$$

When $x = 2$

$$p_3 = \frac{18}{3}\left(\frac{0.06}{0.94}\right)p_2$$

When $x = 3$

$$p_4 = \frac{17}{4}\left(\frac{0.06}{0.94}\right)p_3$$

$$P(X < 5) = p_0 + p_1 + p_2 + p_3 + p_4$$
$$= 0.994 \quad (3 \text{ S.F.}).$$

The probability that there are less than 5 defective mugs in a sample of 20 is 0.994 (3 S.F.).

To find the value of X that is most likely to occur

The value of X that is most likely to occur is the one with the highest probability. It is very tedious to work through finding $P(X = x)$ for all x. Instead the recurrence formula can be used.

Example 4.14 Of the inhabitants of a certain African village, 80% are known to have a particular eye disorder. If 12 people are waiting to see the nurse, what is the most likely number of them to have the eye disorder?

Solution 4.14 Let X be the r.v. 'the number of people with the eye disorder'. Then $X \sim \text{Bin}(n, p)$ with $n = 12$ and $p = 0.8$.

Therefore $X \sim \text{Bin}(12, 0, 8)$
and $P(X = x) = {}^{12}C_x(0.2)^{12-x}(0.8)^x$, $x = 0, 1, \ldots, 12$.

Using the recurrence formula

$$p_{x+1} = \frac{(n-x)p}{(x+1)q}p_x$$

So $p_{x+1} > p_x$ when $(n-x)p > (x+1)q$,

i.e. $$(12-x)(0.8) > (x+1)(0.2)$$

$$4(12-x) > x+1$$

i.e. $p_{x+1} > p_x$ when $$x < 9.4$$

So that $$p_{10} > p_9 > p_8 \ldots > p_1 > p_0$$

But $$p_{x+1} < p_x \quad \text{when} \quad x > 9.4$$

So that $$p_{10} > p_{11} > p_{12}.$$

p_{10} is the highest probability, so the most likely number of people with the eye disorder is 10.

NOTE: the reader should verify that $P(X = 10)$ does give the highest probability by calculating $P(X = x)$ for $x = 0, 1, \ldots, 12$.

Exercise 4d

1. (*a*) If the r.v. X is such that $X \sim \text{Bin}(6, \frac{1}{4})$, use the recurrence formula to find the most likely value of X.
(*b*) Now use the formula to find $P(X = x)$ for $x = 0, 1, \ldots, 6$. Check whether your answer is consistent with your answer to part (*a*).

2. The r.v. X is such that $X \sim \text{Bin}(9, 0.35)$. Find $P(X < 6)$ (*a*) without using the recurrence formula, (*b*) using the recurrence formula. Compare your answers.

3. In a bag there are 6 red counters, 8 yellow counters and 6 green counters. A counter is drawn at random from the bag, its colour is noted and it is then replaced. This procedure is carried out ten times in all. Find (*a*) the expected number of red counters drawn,
(*b*) the most likely number of green counters drawn,
(*c*) the probability that no more than 4 yellow counters are drawn.

4. The random variable X is distributed binomially with mean 2 and variance 1.6. Find (*a*) the most likely value of X, (*b*) $P(X < 6)$.

5. The probability that a student is awarded a pass in the mathematics examination is 0.75. Find the probability that in a group of 10 students more than half pass the mathematics examination.

6. The random variable X is such that $X \sim \text{Bin}(8, 0.4)$. Find (*a*) the most likely value of X, (*b*) $P(X \leqslant 4)$, (*c*) $P(X \geqslant 4)$.

FITTING A THEORETICAL DISTRIBUTION

It is sometimes useful to compare experimental results with a theoretical distribution.

Example 4.15 A biased coin is tossed 4 times and the number of heads noted. The experiment is performed 500 times in all. The results obtained are shown in the table:

Number of heads	0	1	2	3	4
Frequency	12	50	151	200	87

(a) Find the probability of obtaining a head when the coin is tossed.

(b) Calculate the theoretical frequencies of 0, 1, 2, 3, 4 heads, using the associated theoretical binomial distribution.

Solution 4.15 (a) For the frequency distribution

$$\text{mean, } \bar{x} = \frac{\Sigma fx}{\Sigma f}$$

$$= \frac{(0)(12)+(1)(50)+(2)(151)+(3)(200)+(4)(87)}{500}$$

$$= \frac{1300}{500}$$

$$= 2.6$$

Let X be the r.v. 'the number of heads obtained in 4 tosses'. Then $X \sim \text{Bin}(n, p)$ with $n = 4$. So the mean, $E(X) = np$.

Therefore $np = 2.6$

$4p = 2.6$

So $p = 0.65$

Therefore the probability that the coin will show heads is 0.65.

(b) $X \sim \text{Bin}(4, 0.65)$. To find the values of $P(X = x)$ for $x = 0, 1, 2, 3, 4$:

Method 1 — using $P(X = x) = {}^4C_x(0.35)^{4-x}(0.65)^x$

$$P(X = 0) = (0.35)^4 = 0.015\,006\,25$$

$$P(X = 1) = 4(0.35)^3(0.65) = 0.111\,475$$

$$P(X = 2) = 6(0.35)^2(0.65)^2 = 0.310\,5375$$

$$P(X = 3) = 4(0.35)(0.65)^3 = 0.384\,475$$

$$P(X = 4) = (0.65)^4 = 0.178\,506\,2$$

Method 2 — using $p_{x+1} = \dfrac{(4-x)(0.65)}{(x+1)(0.35)} p_x$

$$p_0 = (0.35)^4 = 0.015\,006\,25$$

$$p_1 = \frac{4(0.65)}{1(0.35)} p_0 = (7.428\,571\,4)(0.015\,006\,2)$$

$$= 0.111\,475$$

$$p_2 = \frac{3(0.65)}{2(0.35)}p_1 = (2.785\,714\,3)(0.111\,475)$$
$$= 0.310\,537\,5$$

$$p_3 = \frac{2(0.65)}{3(0.35)}p_2 = (1.238\,095\,2)(0.310\,537\,5)$$
$$= 0.384\,475$$

$$p_4 = \frac{1(0.65)}{4(0.35)}p_3 = (0.464\,285\,7)(0.384\,475)$$
$$= 0.178\,506\,25$$

Method 3 — Using cumulative probability tables.

$X \sim \text{Bin}(4, 0.65)$ so we will need to consider $X \sim \text{Bin}(4, 0.35)$ in order to use the tables.

$$P(X = 0 \mid X \sim \text{Bin}(4, 0.65)) = P(X = 4 \mid X \sim \text{Bin}(4, 0.35))$$
$$= 1 - 0.9850 = 0.015$$

$$P(X = 1 \mid X \sim \text{Bin}(4, 0.65)) = P(X = 3 \mid X \sim \text{Bin}(4, 0.35))$$
$$= 0.9850 - 0.8735 = 0.1115$$

$$P(X = 2 \mid X \sim \text{Bin}(4, 0.65)) = P(X = 2 \mid X \sim \text{Bin}(4, 0.35))$$
$$= 0.8735 - 0.5630 = 0.3105$$

$$P(X = 3 \mid X \sim \text{Bin}(4, 0.65)) = P(X = 1 \mid X \sim \text{Bin}(4, 0.35))$$
$$= 0.5630 - 0.1785 = 0.3845$$

$$P(X = 4 \mid X \sim \text{Bin}(4, 0.65)) = P(X = 0 \mid X \sim \text{Bin}(4, 0.35))$$
$$= 0.1785$$

To obtain the theoretical distribution, multiply each of the probabilities by the total frequency, 500.

Therefore the theoretical binomial frequencies (rounded to the nearest integer) are as follows:

Number of heads	0	1	2	3	4
Frequency	8	56	155	192	89

NOTE: this compares reasonably well with the original frequency distribution.

A statistical test to compare the two sets of data is illustrated on p. 540 (chi-squared test).

Exercise 4e

1. A biased die is thrown 3 times and the number of fours is noted. The procedure is performed 180 times in all and the results are shown in the table.

Number of 4's	0	1	2	3
Frequency	50	69	36	25

(*a*) What is the mean of this distribution?
(*b*) What is the probability of obtaining a 4 when the die is thrown?
(*c*) Calculate the theoretical probabilities of obtaining 0, 1, 2, 3 fours, using the binomial distribution.
(*d*) Calculate the corresponding theoretical frequencies.

2. In a large batch of items from a production line the probability that an item is faulty is p. 400 samples, each of size 5, are taken and the number of faulty items in each batch is noted. From the frequency distribution below estimate p and work out the expected frequencies of 0, 1, 2, 3, 4, 5 faulty items per batch for a theoretical binomial distribution having the same mean.

Number of faulty items	0	1	2	3	4	5
Frequency	297	90	10	2	1	0

3. In an experiment a certain number of dice are thrown and the number of sizes obtained is recorded. The dice are all biased and the probability of obtaining a six with each individual die is p. In all there were 60 experiments and the results are shown in the table.

Number of sixes obtained in an experiment	0	1	2	3	4	More than 4
Frequency	19	26	12	2	1	0

Calculate the mean and the standard deviation of these data.

By comparing these answers with those expected for a binomial distribution, estimate (*a*) the number of dice thrown in each experiment, (*b*) the value of p.
(C Additional)

4. Fit a theoretical binomial distribution to the following frequency distribution, given $n = 4$:

x	0	1	2	3	4
f	7	20	35	30	8

5. Seeds are planted in rows of six and after 14 days the number of seeds which have germinated in each of the 100 rows is noted. The results are shown in the table:

Number of seeds germinating	0	1	2	3	4	5	6
Number of rows	2	1	2	10	30	35	20

Find the theoretical frequencies of 0, 1, . . . , 6 seeds germinating in a row, using the associated theoretical binomial distribution.

6. Derive the mean and variance of the binomial distribution.

Mass production of miniature hearing aids is a particularly difficult process and so the quality of these products is monitored carefully. Samples of size six are selected regularly and tested for correct operation. The number of defectives in each sample is recorded. During one particular week 140 samples are taken and the distribution of the number of defectives per sample is given in the following table.

Number of defectives per sample (x)	0	1	2	3	4	5	6
Number of samples with x defectives (f)	27	36	39	22	10	4	2

Find the frequencies of the number of defectives per sample given by a binomial distribution having the same mean and total as the observed distribution.
(AEB 1978)

WORKED EXAMPLE

Example 4.16 70% of the passengers who travel on the 8.17 to London buy the 'Daily Doom' at the bookstall before boarding the train. The train is full and each compartment holds eight passengers.

(a) What is the probability that all the passengers in a compartment have bought the 'Daily Doom'?

(b) What is the probability that none of the passengers in a compartment has bought the 'Daily Doom'?

(c) What is the probability that exactly three of the passengers in a compartment bought the 'Daily Doom'?

(d) What is the most likely number of passengers in a compartment to have bought the 'Daily Doom'?

(e) If there are 40 compartments on the train in how many of them would you expect there to be exactly three copies of the 'Daily Doom'?

(f) The train is so full that in each carriage ten people are standing in the corridor. What is the probability that the third passenger I pass in the corridor of a carriage is the first I meet who has bought the 'Daily Doom'?

(g) What is the mean number of buyers of the 'Daily Doom' standing in a corridor? (SUJB)

Solution 4.16 Let 'buying the Daily Doom' be termed 'success'. Therefore $p = 0.7$ and $q = 1 - p = 0.3$.

Let X be the r.v. 'the number of passengers who have bought the Daily Doom'. Then $X \sim \text{Bin}(n, p)$ where $n = 8$ and $p = 0.7$, i.e. $X \sim \text{Bin}(8, 0.7)$.

$$\begin{aligned} P(X = x) &= {}^nC_x q^{n-x} p^x \qquad x = 0, 1, \ldots, 8 \\ &= {}^8C_x (0.3)^{8-x} (0.7)^x \end{aligned}$$

(a)

$$\begin{aligned} P(X = 8) &= (0.7)^8 \\ &= 0.0576 \quad (3\text{ S.F.}) \end{aligned}$$

The probability that all the passengers in a compartment have bought the Daily Doom is 0.0576 (3 S.F.).

(b)

$$\begin{aligned} P(X = 0) &= (0.3)^8 \\ &= 6.561 \times 10^{-5} \end{aligned}$$

The probability that none of the passengers has bought the Daily Doom is 6.561×10^{-5}.

(c)
$$\begin{aligned} P(X=3) &= {}^8C_3(0.3)^5(0.7)^3 \\ &= \frac{(8)(7)(6)}{(1)(2)(3)}(0.3)^5(0.7)^3 \\ &= 0.0467 \quad (3\text{ S.F.}) \end{aligned}$$

The probability that exactly three passengers have bought the Daily Doom is 0.0467 (3 S.F.).

(d) To find the most likely number of people who have bought the Daily Doom, use the recurrence formula to find the term with the highest probability.

Recurrence formula:

$$\frac{p_{x+1}}{p_x} = \frac{(n-x)p}{(x+1)q} \quad \text{for} \quad x = 0, 1, \ldots, 7$$

So $p_{x+1} > p_x$ when $(8-x)(0.7) > (x+1)(0.3)$,

i.e.
$$\begin{aligned} 5.6 - 0.7x &> 0.3x + 0.3 \\ x &< 5.3 \end{aligned}$$

Therefore $p_6 > p_5 > p_4 > p_3 > p_2 > p_1 > p_0$ but $p_6 > p_7 > p_8$.

So the most likely number of passengers to have bought the Daily Doom is 6.

(e) In one compartment $P(X = 3) = 0.0467$. Let Y be the r.v. 'the number of compartments where there are exactly three copies of the Daily Doom'.

Then $Y \sim \text{Bin}(n, p)$ where $n = 40$ and $p = 0.0467$.

So
$$\begin{aligned} E(Y) &= np \\ &= (40)(0.0467) \\ &= 1.87 \quad (3\text{ S.F.}) \end{aligned}$$

Therefore the expected number of compartments where there are exactly three copies of the Daily Doom on the train of 40 compartments is 1.87 (3 S.F.).

(f) P(third passenger is the first to have a copy) $= P(\bar{D}\bar{D}D)$ where D is the event 'the person has a copy of the Daily Doom' and $P(D) = 0.7$.

Now
$$\begin{aligned} P(\bar{D}\bar{D}D) &= (0.3)(0.3)(0.7) \qquad \text{(independent events)} \\ &= 0.063 \end{aligned}$$

The probability that the third person is the first to have a copy is 0.063.

(g) Let C be the r.v. 'the number of people in the corridor to have bought the Daily Doom'. Then $C \sim \text{Bin}(10, 0.7)$.

$$\begin{aligned} E(C) &= (10)(0.7) \\ &= 7 \end{aligned}$$

Therefore the expected number of buyers standing in a corridor is 7.

Example 4.17 (a) State in words the meaning of $P(E')$ and of $P(E \mid F)$ for two events E and F.

(b) All the letters in a particular office are typed either by Pat, a trainee typist, or by Lyn, who is a fully trained typist. The probability that a letter typed by Pat will contain one or more errors is 0.3. Find the probability that a random sample of 4 letters typed by Pat will include exactly one letter free from error.

(c) The probability that a letter typed by Lyn will contain one or more errors is 0.05. Using the tables provided, or otherwise, find, to 3 decimal places, the probability that in a random sample of 20 letters typed by Lyn, not more than 2 letters will contain one or more errors.

(d) On any one day, 6% of the letters typed in the office are typed by Pat. One letter is chosen at random from those typed on that day. Show that the probability that it will contain one or more errors is 0.065.

(e) Given that each of 2 letters chosen at random from the day's typing contains one or more errors, find, to 4 decimal places, the probability that one was typed by Pat and the other by Lyn. (L)

Solution 4.17 (a) $P(E')$ is the probability that event E does not occur. $P(E \mid F)$ is the probability that E occurs, given that F has occurred.

(b) $P(\text{Pat's letter contains errors}) = 0.3$.
$P(\text{Pat's letter is free from errors}) = 1 - 0.3 = 0.7$.
Let X be the r.v. 'the number of letters typed by Pat which are free from errors'. Then $X \sim \text{Bin}(4, 0.7)$.

$$\begin{aligned} P(X = 1) &= {}^4C_3(0.3)^3(0.7) \\ &= 0.0756 \end{aligned}$$

Therefore the probability that a random sample of 4 letters typed by Pat will include exactly one letter free from error is 0.0756.

(c) $P(\text{Lyn's letter contains errors}) = 0.05$.
Let Y be the r.v. 'the number of letters typed by Lyn containing errors'. Then $Y \sim \text{Bin}(20, 0.05)$.

$$\begin{aligned} P(Y \leqslant 2) &= P(Y=0)+P(Y=1)+P(Y=2) \\ &= (0.95)^{20}+20(0.95)^{19}(0.05) \\ &\qquad +190(0.95)^{18}(0.05)^2 \\ &= 0.925 \quad (3\text{ d.p.}) \end{aligned}$$

Therefore the probability that a random sample of 20 letters typed by Lyn will contain not more than two with errors is 0.925 (3 d.p.).

NOTE: $P(Y \leqslant 2)$ can be found directly from cumulative binomial probability tables, p. 629, with $n = 20$, $p = 0.05$.

$$P(Y \leqslant 2) = 0.925 \text{ (3 d.p.)}$$

(d) Let E be the event 'a letter contains one or more errors'.

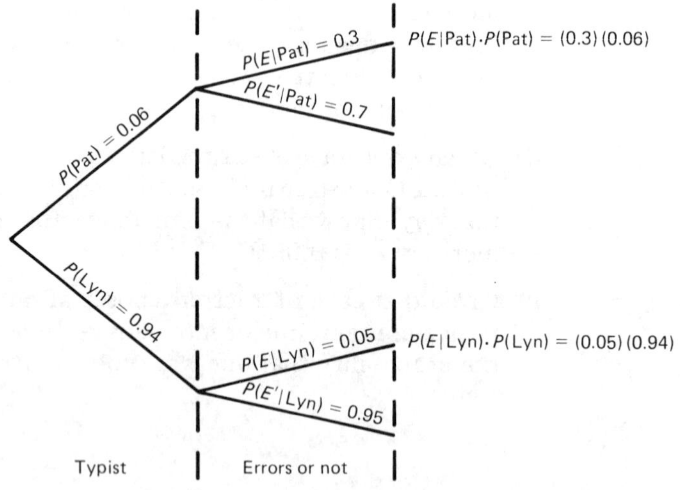

$$\begin{aligned} P(E) &= P(E\,|\,\text{Pat})\cdot P(\text{Pat})+P(E\,|\,\text{Lyn})\cdot P(\text{Lyn}) \\ &= (0.3)(0.6)+(0.05)(0.94) \\ &= 0.065 \end{aligned}$$

Therefore the probability that a letter will contain errors is 0.065, as required.

(e) Now

$$\begin{aligned} P(\text{Pat}\,|\,E) &= \frac{P(E\,|\,\text{Pat})\cdot P(\text{Pat})}{P(E)} \\ &= \frac{(0.3)(0.06)}{0.065} \\ &= \frac{18}{65} \end{aligned}$$

$$P(\text{Lyn} \mid E) = \frac{P(E \mid \text{Lyn}) \cdot P(\text{Lyn})}{P(E)}$$
$$= \frac{(0.05)(0.94)}{0.065}$$
$$= \tfrac{47}{65}$$

$P(1 \text{ typed by Lyn, } 1 \text{ typed by Pat} \mid 2 \text{ letters contained errors})$

$$= 2\left(\tfrac{18}{65}\right)\left(\tfrac{47}{65}\right)$$
$$= 0.4005 \quad (4 \text{ d.p.})$$

Therefore the probability that one was typed by Pat and one by Lyn, given that the two letters contained errors, is 0.4005 (4 d.p.).

SUMMARY — BINOMIAL DISTRIBUTION

If $X \sim \text{Bin}(n, p)$ then $P(X = x) = {}^nC_x q^{n-x} p^x$, $x = 0, 1, 2, \ldots, n$, where $q = 1 - p$.

$E(X) = np$

$\text{Var}(X) = npq$

Recurrence formula:

$$p_{x+1} = \frac{(n-x)p}{(x+1)q} p_x \quad \text{where} \quad p_{x+1} = P(X = x+1), \quad p_x = P(X = x)$$

$$P(X = r \mid X \sim \text{Bin}(n, p)) = P(X = n - r \mid X \sim \text{Bin}(n, 1-p))$$
$$P(X \geqslant r \mid X \sim \text{Bin}(n, p)) = P(X \leqslant n - r \mid X \sim \text{Bin}(n, 1-p))$$
$$P(X \leqslant r \mid X \sim \text{Bin}(n, p)) = P(X \geqslant n - r \mid X \sim \text{Bin}(n, 1-p))$$

Miscellaneous Exercise 4f

1. Find the probability of throwing three sixes twice in five throws of six dice.

2. In a large city 1 person in 5 is left handed.
(*a*) Find the probability that in a random sample of 10 people
   (i) exactly 3 will be left handed,
   (ii) more than half will be left-handed.
(*b*) Find the most likely number of left-handed people in a random sample of 12 people.
(*c*) Find the mean and the standard deviation of the number of left-handed people in a random sample of 25 people.
(*d*) How large must a random sample be if the probability that it contains at least one left-handed person is to be greater than 0.95?

3. A crossword puzzle is published in *The Times* each day of the week, except Sunday. A man is able to complete, on average, 8 out of 10 of the crossword puzzles.
(*a*) Find the expected value and the standard deviation of the number of completed crosswords in a given week.
(*b*) Show that the probability that he will complete at least 5 in a given week is 0.655 (to 3 significant figures).
(*c*) Given that he completes the puzzle on Monday, find, to three significant figures, the probability that he will complete at least 4 in the rest of the week.
(*d*) Find, to three significant figures, the probability that, in a period of four weeks, he completes 4 or less in only one of the four weeks. (C)

4. Samples, each of 8 articles, are taken at random from a large consignment in which 20% of articles are defective. Find the number of defective articles which is most likely to occur in a single sample, and find the probability of obtaining this number.

If 100 samples of 8 articles are to be examined, calculate the number of samples in which you would expect to find 3 or more defective articles. (C)

5. A small boy plays a game in which he has to guess in which hand his uncle is hiding a toffee. The first time he chooses 'left'. For the next three times he chooses 'same hand as previous time' with probability s, and 'different hand from previous time' with probability d, where $s+d=1$. Find the probability that he will choose 'left' on the last time.
By adding together the binomial expansions for $(s+d)^3$ and $(s-d)^3$, deduce that the probability that he chooses 'left' on the last time can be written as $\frac{1}{2}\{1+(s-d)^3\}$. (SMP)

6. In an inspection scheme a sample of 20 items is selected at random from a large batch and the number of defective items is noted. If this number is more than 2 the batch is rejected; if it is less than 2 the batch is accepted. If the number of defective items is exactly 2, a further sample of 10 items is taken and the batch is rejected if this second sample has any defective items, but otherwise the batch is accepted.

If the proportion of defective items in a particular batch is 2%, evaluate, to two decimal places, the probabilities that
(*a*) the batch is accepted as a result of inspection of the first sample,
(*b*) a second sample is taken and the batch accepted as a result of inspection of the second sample,
(*c*) the batch is rejected. (C)

7. Thatcher's Pottery produces large batches of coffee mugs decorated with the faces of famous politicians. They are considering adopting one of the following sampling plans for batch inspection.

Method A (single sample plan) Select 10 mugs from the batch at random and accept the batch if there are 2 or less defectives, otherwise reject the batch.

Method B (double sample plan) Select 5 mugs from the batch at random and accept the batch if there are no defectives, reject the batch if there are 2 or more defectives, otherwise select another 5 mugs at random. When the second sample is drawn count the number of defectives in the combined sample of 10 and accept the batch if the number of defectives is 2 or less, otherwise reject the batch.
(*a*) If the proportion of defectives in a batch is p, find, in terms of p, for each method in turn, the probability that the batch will be accepted.
(*b*) Evaluate *both* the above probabilities for $p=0.2$ and $p=0.5$.
(*c*) Hence, or otherwise, decide which of these two plans is more appropriate, and why. (AEB 1981)

8. A trial may have two outcomes, success or failure. If in n such independent trials, the probability p of a success remains constant from trial to trial, write down the probability of r successes in the n trials.

When two friends A and B play chess, the probability that A wins any game is $\frac{2}{5}$, and if A does not win the game, the probabilities then of B winning and of a draw are equal. In the course of an evening they play four games. Calculate the probabilities (*a*) that A does not win a game, (*b*) that he wins more than two games.
If it is known that A has won exactly two of these four games, write down the probability distribution of the number of games that B has won.

Calculate the probability that A wins more games than B when four games are played. (JMB)

9. At a certain university in Cambford students attending a first course in statistics are asked by the lecturer, Professor Thomas Bayes, to complete 10 example sheets during the course. At the end of the course each student sits an examination as a result of which he either passes or fails. Assuming that

(I) the number, N, of example sheets completed by any student has a binomial distribution given by

$$P(N=n) = {}^{10}C_n\left(\frac{2}{3}\right)^n\left(\frac{1}{3}\right)^{(10-n)}$$

$$n = 0, 1, \ldots, 10$$

and (II) the probability of a student passing the examination *given* that he completed n sheets during the course, is $n/10$,

(*a*) what is the (unconditional) probability that a student passes the examination?

(*b*) What is the probability that a student selected at random from the examination pass list had in fact completed four example sheets or less? (AEB 1977)

10. Two random variables X_1 and X_2 have independent binomial probability distributions, where both X_1 and X_2 can only take the values 0, 1 and 2. If $P(X_1=2)=p_1^2$ and $P(X_2=2)=p_2^2$, show that $E(X_1X_2)=4p_1p_2$.
(L Additional)

In an experiment two bags, A and B, contain a very large number of white and black balls. In Bag A, 20% of the balls are white and in Bag B, 70% of the balls are white. Two balls are selected at random from each bag.
(*a*) Find the expected number of white balls selected from Bag A.
(*b*) If Z denotes the total number of white balls selected, calculate (i) $P(Z=2)$, (ii) $E(Z)$. (L Additional)

THE GEOMETRIC DISTRIBUTION

A geometric distribution arises when we have a sequence of independent trials, each with a definite probability p of success and probability q of failure, where $q = 1-p$. Let X be the r.v. 'the number of trials up to and including the first success'.

Now

$$P(X=1) = P(\text{success on the first trial}) = p$$

$$P(X=2) = P(\text{failure on first trial, success on second}) = qp$$

$$P(X=3) = q^2p$$

$$P(X=4) = q^3p$$

$$\vdots$$

$$P(X=x) = q^{x-1}p$$

A discrete r.v. X having p.d.f. of the form $P(X=x)=q^{x-1}p$, where $0 \leqslant p \leqslant 1$ and $q=1-p$, is said to follow a geometric distribution, with $x = 1, 2, 3, \ldots$

p is the parameter of the distribution.

If X is defined in this way, we write

$$X \sim \text{Geo}(p)$$

EXPECTATION AND VARIANCE

If $X \sim \text{Geo}(p)$ then $E(X) = \dfrac{1}{p}$ and $\text{Var}(X) = \dfrac{q}{p^2}$ where $q = 1 - p$.

x	1	2	3	4	...
$P(X = x)$	p	qp	q^2p	q^3p	...

$$\begin{aligned}
E(X) &= \sum_{\text{all } x} xP(X = x) \\
&= p + 2qp + 3q^2p + 4q^3p + \ldots \\
&= p(1 + 2q + 3q^2 + 4q^3 + \ldots) \\
&= p(1-q)^{-2} \quad \text{since} \quad (1-q)^{-2} = 1 + 2q + 3q^2 + 4q^3 + \ldots \\
&= \frac{p}{p^2} \\
&= \frac{1}{p}
\end{aligned}$$

Now

$$\begin{aligned}
E(X^2) &= \sum_{\text{all } x} x^2P(X = x) \\
&= p + 4qp + 9q^2p + 16q^3p + \ldots \\
&= p(1 + 4q + 9q^2 + 16q^3 + \ldots) \\
&= p(1 + 2q + 3q^2 + 4q^3 + \ldots \\
&\qquad + 2q + 6q^2 + 12q^3 + \ldots) \\
&= p((1-q)^{-2} + 2q(1 + 3q + 6q^2 + \ldots)) \\
&= p\left(\frac{1}{p^2} + 2q(1-q)^{-3}\right) \quad \text{since} \quad (1-q)^{-3} = 1 + 3q + 6q^2 + \ldots \\
&= p\left(\frac{1}{p^2} + \frac{2q}{p^3}\right) \\
&= \frac{1}{p} + \frac{2q}{p^2}
\end{aligned}$$

$$\begin{aligned}\text{Var}(X) &= E(X^2)-E^2(X)\\ &= \frac{1}{p}+\frac{2q}{p^2}-\frac{1}{p^2}\\ &= \frac{p+2q-1}{p^2}\\ &= \frac{q}{p^2} \quad \text{since} \quad p=1-q\end{aligned}$$

Therefore $E(X)=\dfrac{1}{p}$ and $\text{Var}(X)=\dfrac{q}{p^2}$.

Example 4.18 The probability that a marksman hits the bull's eye is 0.4 for each shot, and each shot is independent of all others. Find

(a) the probability that he hits the bull's eye for the first time on his fourth attempt,

(b) the mean number of throws needed to hit the bull's eye, and the standard deviation,

(c) the most common number of throws until he hits the bull's eye.

Solution 4.18 (a) $P(\text{hits bull's eye on fourth attempt}) = (0.6)^3(0.4) = 0.0864.$

(b) Let X be the r.v. 'the number of attempts up to and including the first bull's eye'. Now X follows a geometric distribution with $p = 0.4$.

So $P(X=x)=q^{x-1}p$ with $q=0.6$, $p=0.4$.

Now $E(X)=\dfrac{1}{p}=\dfrac{1}{0.4}=2.5$ and $\text{Var}(X)=\dfrac{q}{p^2}=\dfrac{0.6}{(0.4)^2}=3.75$

and s.d. of $X = \sqrt{3.75} = 1.94$ (3 S.F.)

So the mean number of attempts is 2.5 and the standard deviation is 1.94 (3 S.F.).

(c)
$$P(X=1) = 0.4$$
$$P(X=2) = 0.6 \times 0.4 = 0.24$$
$$P(X=3) = 0.6^2 \times 0.4 = 0.144$$

The probabilities are decreasing, and therefore the most common number of throws is 1.

Result 1 **If $X \sim \text{Geo}(p)$ then $P(X > r) = q^r$ with $q = 1-p$,**

$$P(X=r) = q^{r-1}p$$
$$P(X \leqslant r) = p + qp + \ldots + q^{r-1}p$$
$$= p(1 + q + \ldots + q^{r-1})$$
$$= p\frac{(1-q^r)}{(1-q)}$$
$$= 1-q^r$$

Therefore
$$P(X > r) = 1-(1-q^r)$$
$$= q^r$$

Example 4.19 A coin is biased so that the probability of obtaining a head is 0.6. If X is the r.v. the number of tosses up to and including the first head, find

(a) $P(X \leqslant 4)$,

(b) $P(X > 5)$,

(c) the probability that more than 8 tosses will be required to obtain a head, given that more than 5 tosses are required.

Solution 4.19 $P(X = x) = q^{x-1}p$, $x = 1, 2, 3, \ldots$ with $p = 0.6$ and $q = 0.4$.

(a)
$$P(X > 4) = q^4$$

Therefore
$$P(X \leqslant 4) = 1-q^4$$
$$= 1-(0.4)^4$$
$$= 0.9744$$

Therefore $P(X \leqslant 4) = 0.9744$.

(b)
$$P(X > 5) = q^5 = (0.4)^5 = 0.010\,24$$

Therefore $P(X > 5) = 0.010\,24.$

(c)
$$P(X > 8 \mid X > 5) = \frac{P(X > 8 \cap X > 5)}{P(X > 5)} = \frac{q^8}{q^5} = q^3 = (0.4)^3 = 0.064$$

The probability that more than 8 tosses will be required given that more than 5 tosses are required is 0.064.

In general,
$$P(X > a + b \mid X > a) = \frac{q^{a+b}}{q^a} = q^b = P(X > b)$$

Result 2 $\quad P(X > a + b \mid X > a) = P(X > b)$

Example 4.20 In a particular board game a player can get out of jail only by obtaining two heads when she tosses two coins.

(a) Find the probability that more than 6 attempts are needed to get out of jail.

(b) What is the smallest value of n if there is to be at least a 90% chance of getting out of jail on or before the nth attempt.

Solution 4.20 $P(\text{2 heads when 2 coins are tossed}) = \frac{1}{4}$.

So $p = P(\text{success}) = \frac{1}{4}$ and $q = \frac{3}{4}$.

Let X be the r.v. 'the number of attempts required to get out of jail'. Then X follows a geometric distribution, $X \sim \text{Geo}(\frac{1}{4})$.

(a) $P(X > 6) = q^6$

$= (\frac{3}{4})^6$

$= 0.178 \quad (3\text{ S.F.})$

The probability that a player needs more than 6 attempts before getting out of jail is 0.178 (3 S.F.).

(b) $P(X > n) = (\frac{3}{4})^n$.

So $P(X \leqslant n) = 1 - (\frac{3}{4})^n$.

We require $P(X \leqslant n) \geqslant 0.9$.

So

$$1 - (\tfrac{3}{4})^n \geqslant 0.9$$

$$0.1 \geqslant (\tfrac{3}{4})^n$$

$$(\tfrac{3}{4})^n \leqslant 0.1$$

(Take logs to base 10.)

$$n \log(0.75) \leqslant \log 0.1$$

(Divide by negative quantity, so reverse the inequality.)

$$n \geqslant \frac{\log 0.1}{\log 0.75}$$

$$n \geqslant 8.0039$$

Therefore the smallest value of n is 9.

Check: $P(X \leqslant 6) = 1 - (\frac{3}{4})^6 = 0.8220 \ldots < 90\%$

$P(X \leqslant 7) = 1 - (\frac{3}{4})^7 = 0.8665 \ldots < 90\%$

$P(X \leqslant 8) = 1 - (\frac{3}{4})^8 = 0.8998 \ldots < 90\%$

$P(X \leqslant 9) = 1 - (\frac{3}{4})^9 = 0.9249 \ldots > 90\%$

So least value of n is 9.

Exercise 4g

1. (*a*) The probability distribution of a random variable X is geometric, that is

$P(X = R) = (1-p)p^R, \quad R = 0, 1, 2, \ldots$

where $0 < p < 1$.

Given that $\sum_{n=1}^{\infty} np^{n-1} = 1/(1-p)^2$, show that $E(X) = p/(1-p)$.

(*b*) When two dice are thrown together a 'double' is obtained when the scores on the two dice are the same. Assuming the two dice to be unbiased, calculate the probabilities that

(i) in two throws of the two dice, a double will be obtained on the first throw but not on the second throw,

(ii) in three throws of the two dice, doubles will be obtained on the first two throws but not on the third throw.

(*c*) Suppose the two dice are thrown until a double is not obtained. Using the result of part (*a*) find the expected number of doubles. (L Additional)

2. An unbiased coin is tossed repeatedly until a tail appears. Find the expected number of tosses.

NOTE: $(1-x)^{-2} = 1 + 2x + 3x^2 + \ldots.$

3. A darts player practises throwing a dart at the bull's-eye on a dart board. Independently for each throw, her probability of hitting the bull's-eye is 0.2. Let X be the number of throws she makes, up to and including her first success.

(*a*) Find the probability that she is successful for the first time on her third throw.

(*b*) Write down the distribution of X, and give the name of this distribution.

(*c*) Find the probability that she will have at least 3 failures before her first success.

(*d*) Show that the mean value of X is 5.

(You may assume the result

$$\sum_{r=1}^{\infty} rq^{r-1} = \frac{1}{(1-q)^2} \text{ when } |q| < 1.)$$

On another occasion the player throws the dart at the bull's-eye until she has 2 successes. Let Y be the number of throws she makes up to and including her second success. Given that $\text{Var}(X) = 20$, determine the mean and the variance of Y, and find the probability that $Y = 4$. (L)

4. A random variable R has probability function $P(R = r)$ defined by

$$P(R = r) = \frac{4}{5}\left(\frac{1}{5}\right)^{r-1},$$

for $r = 1, 2, 3, \ldots$

$$P(R = r) = 0, \text{ otherwise.}$$

Given that $\sum_{n=1}^{\infty} np^n = \frac{p}{(1-p)^2}$, find $E(R)$.

Given that the variance of R is $\frac{5}{16}$, determine the mean and variance of S where $S = 3R - 2$.

The probability that a telephone box is occupied is $\frac{1}{5}$. Find, to 2 significant figures, the probability that a person wishing to make a phone call will find a telephone box which is not occupied only at the sixth box tried.

Write down the mean number of occupied boxes which will have to be tried before the person finds a box which is not occupied. (L)P

5. (*a*) Describe an experiment you may have carried out which can be modelled by a geometric distribution. State any assumptions you may have made.

(*b*) In many board games it is necessary to 'throw a six with an ordinary die' before a player can start the game. Write down, as a fraction, the probability of a player

(i) starting on his first attempt,

(ii) not starting until his third attempt,

(iii) requiring more than three attempts before starting.

What is

(iv) the most common number of throws required to obtain a six,

(v) the mean number of throws required to obtain a six,

Prove that the probability of a player requiring more than n attempts before starting is $(\frac{5}{6})^n$.

(*c*) What is the smallest value of n if there is to be at least a 95% chance of starting on or before the nth attempt? (O)

6. State conditions which give rise to a geometric distribution whose probability function is $P(X=r)=(1-p)^{r-1}p$, $r=1, 2, 3, \ldots$, where $0<p<1$.

Prove that $P(X\leqslant r)=1-(1-p)^r$.

Hence prove that, for any two positive integers s and t,

$$P(X>s+t\,|\,X>s) = P(X>t)$$

and explain in words the meaning of this result.
During the winter in Glen Shee, the probability that snow will fall on any given day is 0.1. Taking November 1st as the first day of winter and assuming independence from day to day, find, to 2 significant figures, the probability that the first snow of winter will fall in Glen Shee on the last day of November (30th).

Given that no snow has fallen at Glen Shee during the whole of November, a teacher decides not to wait any longer to book a skiing holiday. The teacher decides to book for the earliest date for which the probability that snow will have fallen on or before that date is at least 0.9. Find the date of the booking. (L)

7. In a sales campaign, a petrol company gives each motorist who buys their petrol a card with a picture of a film star on it. There are 10 different picture cards, one of each of ten different film stars, and any motorist who collects a complete set of all ten pictures gets a free gift. On any occasion when a motorist buys petrol, the card received is equally likely to carry any of the ten pictures in the set.

(*a*) Find the probability that the first four cards the motorist receives all carry different pictures.

(*b*) Find the probability that the first four cards received result in the motorist having exactly three different pictures.

(*c*) Two of the ten film stars in the set are X and Y. Find the probability that the first four cards received result in the motorist having a picture of X or of Y (or both).

(*d*) At a certain stage the motorist has collected nine of the ten pictures. Find the least value of n such that

P(at most n more cards are needed to complete the set) > 0.99.

(C)

8. A marksman fires at a target. The probability of his hitting the bull's-eye is p for each shot, and each shot is independent of all others. The random variable X denotes the number of shots previous to that on which the bull's-eye is first hit. Show that

$$\Pr(X=x) = q^x p$$

where $q=1-p$. Find the mean of X and show that the variance is q/p^2.

(O & C)

THE POISSON DISTRIBUTION

A discrete r.v. X having p.d.f. of the form

$$P(X=x) = e^{-\lambda}\frac{\lambda^x}{x!} \quad \text{for} \quad x = 0, 1, 2, 3, \ldots \text{ to infinity}$$

where λ can take any positive value, is said to follow the Poisson distribution.

NOTE: λ is the parameter of the distribution.

If X is distributed in this way, then $X \sim \text{Po}(\lambda)$.

Example 4.21 Verify that if $X \sim \text{Po}(\lambda)$, then X is a random variable.

Solution 4.21 If $\sum_{\text{all } x} P(X = x) = 1$ then X is a random variable.

Now
$$\sum_{\text{all } x} P(X = x) = \sum_{x=0}^{\infty} e^{-\lambda} \frac{\lambda^x}{x!}$$
$$= e^{-\lambda} \sum_{x=0}^{\infty} \frac{\lambda^x}{x!}$$
$$= e^{-\lambda}\left(1 + \lambda + \frac{\lambda^2}{2!} + \frac{\lambda^3}{3!} + \ldots\right)$$

But $e^{\lambda} = 1 + \lambda + \frac{\lambda^2}{2!} + \frac{\lambda^3}{3!} + \ldots$

So
$$\sum_{\text{all } x} P(X = x) = (e^{-\lambda})(e^{\lambda})$$
$$= 1$$

Therefore X is a random variable.

EXPECTATION AND VARIANCE

Example 4.22 If $X \sim \text{Po}(\lambda)$ find **(a)** $E(X)$, **(b)** $E(X^2)$, **(c)** $\text{Var}(X)$.

Solution 4.22
$$P(X = x) = e^{-\lambda} \frac{\lambda^x}{x!}, \quad x = 0, 1, 2, 3, \ldots$$

The probability distribution can be written as follows:

x	0	1	2	3	4	...
$P(X = x)$	$e^{-\lambda}$	$e^{-\lambda}\lambda$	$e^{-\lambda}\frac{\lambda^2}{2!}$	$e^{-\lambda}\frac{\lambda^3}{3!}$	$e^{-\lambda}\frac{\lambda^4}{4!}$	...

(a) Now
$$E(X) = \sum_{\text{all } x} xP(X = x)$$
$$E(X) = \lambda e^{-\lambda}\left(1 + \lambda + \frac{\lambda^2}{2!} + \frac{\lambda^3}{3!} + \ldots\right)$$
$$= \lambda e^{-\lambda}(e^{\lambda})$$
$$= \lambda$$

Therefore $E(X) = \lambda$.

(b) Now

$$\begin{aligned}
E(X^2) &= \sum_{\text{all } x} x^2 P(X=x) \\
&= e^{-\lambda}\left(0+\lambda+\frac{4\lambda^2}{2!}+\frac{9\lambda^3}{3!}+\frac{16\lambda^4}{4!}+\ldots\right) \\
&= \lambda e^{-\lambda}\left(1+2\lambda+\frac{3\lambda^2}{2!}+\frac{4\lambda^3}{3!}+\ldots\right) \\
&= \lambda e^{-\lambda}\left(1+\lambda+\frac{\lambda^2}{2!}+\frac{\lambda^3}{3!}+\ldots+\lambda+\frac{2\lambda^2}{2!}+\frac{3\lambda^3}{3!}+\ldots\right) \\
&= \lambda e^{-\lambda}\left[e^{\lambda}+\lambda\left(1+\lambda+\frac{\lambda^2}{2!}+\ldots\right)\right] \\
&= \lambda e^{-\lambda}(e^{\lambda}+\lambda e^{\lambda}) \\
&= \lambda+\lambda^2
\end{aligned}$$

Therefore $E(X^2) = \lambda + \lambda^2$.

(c) Now

$$\begin{aligned}
\text{Var}(X) &= E(X^2) - E^2(X) \\
&= \lambda+\lambda^2-\lambda^2 \\
&= \lambda
\end{aligned}$$

Therefore $\text{Var}(X) = \lambda$.

Therefore

if $X \sim \text{Po}(\lambda)$, then	$E(X) = \lambda$
and	$\text{Var}(X) = \lambda$

Example 4.23 If $X \sim \text{Po}(2)$, find (a) $P(X=4)$, (b) $P(X \geqslant 3)$.

Solution 4.23 $X \sim \text{Po}(2)$, so $P(X=x) = e^{-2}\dfrac{2^x}{x!}$ $\quad x = 0, 1, \ldots.$

(a) $$P(X=4) = e^{-2}\frac{2^4}{4!} = 0.0902 \quad (3\text{ S.F.})$$

(b)
$$\begin{aligned}
P(X \geqslant 3) &= 1-P(X<3) \\
&= 1-[P(X=0)+P(X=1)+P(X=2)] \\
&= 1-\left(e^{-2}+e^{-2}2+e^{-2}\frac{2^2}{2!}\right) \\
&= 1-5e^{-2} \\
&= 0.323 \quad (3\text{ S.F.})
\end{aligned}$$

Exercise 4h

1. If $X \sim \text{Po}(3.5)$, find (*a*) $P(X=0)$, (*b*) $P(X=1)$, (*c*) $P(X=2)$, (*d*) $P(X=3)$, (*e*) $P(X<3)$, (*f*) $P(X\geqslant 4)$.

2. If $X \sim \text{Po}(1.8)$, find (*a*) $P(X=6)$, (*b*) $P(X=8)$, (*c*) $P(X\leqslant 2)$, (*d*) $P(X\geqslant 4)$.

3. If $X \sim \text{Po}(2.4)$ and $F(X)$ is the cumulative distribution, find (*a*) $F(0)$, (*b*) $F(1)$, (*c*) $F(2)$, (*d*) $F(3)$.

4. $X \sim \text{Po}(\lambda)$ and $P(X=0)=0.2019$. Find (*a*) λ, (*b*) $P(X\leqslant 4)$.

5. Find the first 5 terms of the Poisson distribution if (*a*) $\lambda=0.5$, (*b*) $\lambda=2.8$, (*c*) $\lambda=3.6$.

6. If $X \sim \text{Po}(\lambda)$ and $E(X^2)=6$, find (*a*) λ, (*b*) $P(X=2)$.

7. The random variable X follows a Poisson distribution with standard deviation 2. Find $P(X\leqslant 3)$.

USES OF THE POISSON DISTRIBUTION

There are two main practical uses of the Poisson distribution:

(1) when considering the distribution of random events,

(2) as an approximation to the binomial distribution.

We shall now look at these in more detail.

(1) The distribution of random events

If an event is randomly scattered in time (or space) and has mean number of occurrences λ in a given interval of time (or space) and if X is the r.v. 'the number of occurrences in the given interval', then $X \sim \text{Po}(\lambda)$.

Examples of events which might follow a Poisson distribution:

The number of

(a) flaws in a given length of material,

(b) car accidents on a particular stretch of road in one day,

(c) accidents in a factory in one week,

(d) telephone calls made to a switchboard in a given minute,

(e) insurance claims made to a company in a given time,

(f) particles emitted by a radioactive source in a given time.

Example 4.24 The mean number of bacteria per millilitre of a liquid is known to be 4. Assuming that the number of bacteria follows a Poisson distribution, find the probability that, in 1 ml of liquid, there will be (a) no bacteria, (b) 4 bacteria, (c) less than 3 bacteria.

Solution 4.24 Let X be the r.v. 'the number of bacteria in 1 ml of liquid'.

Then $X \sim \text{Po}(4)$, so that $P(X = x) = e^{-4}\dfrac{4^x}{x!}, \qquad x = 0, 1, 2, \ldots$.

(a)
$$\begin{aligned} P(X = 0) &= e^{-4} \\ &= 0.0183 \quad (3\text{ S.F.}) \end{aligned}$$

The probability that there will be no bacteria in 1 ml of liquid is 0.0183 (3 S.F.).

(b)
$$\begin{aligned} P(X = 4) &= e^{-4}\frac{4^4}{4!} \\ &= 0.195 \quad (3\text{ S.F.}) \end{aligned}$$

The probability that there will be 4 bacteria in 1 ml of liquid is 0.195 (3 S.F.).

(c)
$$\begin{aligned} P(X < 3) &= P(X = 0) + P(X = 1) + P(X = 2) \\ &= e^{-4} + e^{-4}4 + e^{-4}\frac{4^2}{2!} \\ &= e^{-4}(1 + 4 + 8) \\ &= e^{-4}13 \\ &= 0.238 \quad (3\text{ S.F.}) \end{aligned}$$

The probability that there are less than 3 bacteria in 1 ml of liquid is 0.238 (3 S.F.).

UNIT INTERVAL

In Example 4.24 we have considered 1 ml of liquid as the 'unit' interval.

The number of bacteria in 1 ml of liquid follows a Poisson distribution with parameter 4.

It follows that the number of bacteria in 2 ml follows a Poisson distribution with parameter 8, the number in 3 ml follows a Poisson distribution with parameter 12, and so on.

Example 4.25 Using the data of Example 4.24, find the probability that

(a) in 3 ml of liquid there will be less than 2 bacteria,

(b) in $\frac{1}{2}$ ml of liquid there will be more than 2 bacteria.

Solution 4.25 (a) In 1 ml of liquid we 'expect' to find 4 bacteria, so in 3 ml of liquid we 'expect' to find 12 bacteria.

Let Y be the r.v. 'the number of bacteria in 3 ml of liquid'.

So $\quad Y \sim \text{Po}(12) \quad$ and $\quad P(Y = y) = e^{-12}\dfrac{12^y}{y!} \qquad y = 0, 1, 2, \ldots$

Now, we require
$$\begin{aligned} P(Y < 2) &= P(Y = 0) + P(Y = 1) \\ &= e^{-12} + e^{-12}12 \\ &= e^{-12}13 \\ &= 7.99 \times 10^{-5} \quad (3\text{ S.F.}) \end{aligned}$$

Therefore the probability that there are less than 2 bacteria in 3 ml of liquid is 7.99×10^{-5} (3 S.F.).

(b) In 1 ml of liquid we 'expect' 4 bacteria, so in $\frac{1}{2}$ ml of liquid we 'expect' 2 bacteria.

Let R be the r.v. 'the number of bacteria in $\frac{1}{2}$ ml of liquid'.

Then $\quad R \sim \text{Po}(2) \quad$ and $\quad P(R = r) = e^{-2}\dfrac{2^r}{r!} \qquad r = 0, 1, 2, \ldots$

We require
$$\begin{aligned} P(R > 2) &= 1 - [P(R = 0) + P(R = 1) + P(R = 2)] \\ &= 1 - \left(e^{-2} + e^{-2}2 + e^{-2}\frac{2^2}{2!}\right) \\ &= 1 - e^{-2}(5) \\ &= 0.323 \quad (3\text{ S.F.}) \end{aligned}$$

The probability that there are more than 2 bacteria in $\frac{1}{2}$ ml of liquid is 0.323 (3 S.F.).

Exercise 4i

1. A book containing 750 pages has 500 misprints. Assuming that the misprints occur at random, find the probability that a particular page contains (*a*) no misprints, (*b*) exactly 4 misprints, (*c*) more than 2 misprints.

2. An insurance company receives on average 2 claims per week from a certain factory. Assuming that the number of claims follows a Poisson distribution, find the probability that (*a*) it receives more than 3 claims in a given week, (*b*) it receives more than 2 claims in a given fortnight, (*c*) it receives no claims on a given day, assuming that the factory operates on a 5 day week.

3. Cars arrive at a petrol station at an average rate of 30 per hour. Assuming that the number of cars arriving at the petrol station follows a Poisson distribution, find the probability that
(*a*) no cars arrive during a particular 5 minute interval,
(*b*) more than 3 cars arrive during a 5 minute interval,
(*c*) more than 5 cars arrive in a 15 minute interval,
(*d*) in a period of half an hour, 10 cars arrive,
(*e*) less than 3 cars arrive during a 10 minute interval.

4. If the number of bacterial colonies on a petri dish follows a Poisson distribution with average number 2.5 per cm^2, find the probability that
(*a*) in 1 cm^2 there will be no bacterial colonies,
(*b*) in 1 cm^2 there will be more than 4 bacterial colonies,
(*c*) in 2 cm^2 there will be less than 4 bacterial colonies,
(*d*) in 4 cm^2 there will be 6 bacterial colonies.

5. The mean number of flaws per 100 m of material produced on a certain machine at Blanktown Fabrics is 2. If flaws occur randomly, find the probability that (*a*) in a 200 m length of material there will be more than 3 flaws, (*b*) in 50 m of material there will be exactly 2 flaws.

6. The number of goals scored in a match by Random Rovers follows a Poisson distribution with mean λ. If the probability that the team scores no goals in a match is 0.301 (3 d.p.) find (*a*) the value of λ, (*b*) the probability that the team scores less than 3 goals in a match, (*c*) the probability that the team scores less than 3 goals in 2 matches.

7. The number of telephone calls made to the school office during a 5 minute interval follows a Poisson distribution with mean 0.5. Find the probability that (*a*) no calls will be received between 10 05 and 10 10, (*b*) more than 4 calls will be received during a particular period of 30 minutes.

8. The number of accidents per week in a certain factory follows a Poisson distribution with variance 3.2. Find the probability that (*a*) no accidents occur in a particular week, (*b*) more than 4 accidents occur in a particular week, (*c*) less than 3 accidents occur in a particular fortnight, (*d*) exactly 7 accidents occur in a particular fortnight.

(2) Using the Poisson distribution as an approximation to the binomial distribution

A binomial distribution with parameters n and p can be approximated by a Poisson distribution, with parameter $\lambda = np$, if n is large (>50 say) and p is small (<0.1 say). The approximation gets better as $n \to \infty$ and $p \to 0$.

For a binomial distribution $P(X = x)$, $x = 0, 1, \ldots, n$ are given by the terms of the expansion of $(q + p)^n$.

For a Poisson distribution $P(X = x)$, $x = 0, 1, 2, \ldots$, are given by the terms $e^{-\lambda}\left(1 + \lambda + \frac{\lambda^2}{2!} + \ldots\right)$.

So we wish to show that

$$(q+p)^n \to e^{-\lambda}\left(1 + \lambda + \frac{\lambda^2}{2!} + \ldots\right) \quad \text{as} \quad n \to \infty$$

We will need the following theory, relating to the binomial theorem.

By the binomial theorem,

$$\left(1 + \frac{x}{n}\right)^n = 1 + n\left(\frac{x}{n}\right) + \frac{n(n-1)}{2!}\frac{x^2}{n^2} + \frac{n(n-1)(n-2)}{3!}\frac{x^3}{n^3} + \ldots$$

$$= 1 + x + \frac{x^2}{2!}\left[\frac{n}{n}\frac{(n-1)}{n}\right] + \frac{x^3}{3!}\left[\frac{n}{n}\frac{(n-1)}{n}\frac{(n-2)}{n}\right] + \ldots$$

$$= 1 + x + \frac{x^2}{2!}\left(1 - \frac{1}{n}\right) + \frac{x^3}{3!}\left(1 - \frac{1}{n}\right)\left(1 - \frac{2}{n}\right) + \ldots$$

Now, as $n \to \infty$, $\left(1-\dfrac{1}{n}\right) \to 1$, and

$$\left(1+\frac{x}{n}\right)^n \to 1+x+\frac{x^2}{2!}+\frac{x^3}{3!}+\ldots$$
$$= e^x$$

i.e.
$$\lim_{n\to\infty}\left(1+\frac{x}{n}\right)^n = e^x$$

Similarly
$$\lim_{n\to\infty}\left(1-\frac{x}{n}\right)^n = e^{-x} \qquad \text{(a)}$$

Now

$$(q+p)^n = q^n + nq^{n-1}p + \frac{n(n-1)}{2!}q^{n-2}p^2 + \frac{n(n-1)(n-2)}{3!}q^{n-3}p^3 + \ldots$$

But $q = 1-p$ and $p = \dfrac{\lambda}{n}$ so $q = 1-\dfrac{\lambda}{n}$. So

$$(q+p)^n = \left(1-\frac{\lambda}{n}\right)^n + n\left(1-\frac{\lambda}{n}\right)^{n-1}\left(\frac{\lambda}{n}\right) + \frac{n(n-1)}{2!}\left(1-\frac{\lambda}{n}\right)^{n-2}\left(\frac{\lambda}{n}\right)^2 + \ldots$$

$$= \left(1-\frac{\lambda}{n}\right)^n\left[1+\frac{n}{\left(1-\frac{\lambda}{n}\right)}\left(\frac{\lambda}{n}\right) + \frac{n(n-1)}{2!\left(1-\frac{\lambda}{n}\right)^2}\left(\frac{\lambda}{n}\right)^2 + \ldots\right]$$

$$= \left(1-\frac{\lambda}{n}\right)^n\left[1+\frac{\lambda}{\left(1-\frac{\lambda}{n}\right)} + \frac{\lambda^2\left(1-\frac{1}{n}\right)}{2!\left(1-\frac{\lambda}{n}\right)^2} + \ldots\right]$$

As $n \to \infty$ we have $\left(1-\dfrac{\lambda}{n}\right)^n \to e^{-\lambda}$ from (a) and $\dfrac{\lambda}{n} \to 0$.

$(q+p)^n \to e^{-\lambda}\left(1+\lambda+\dfrac{\lambda^2}{2!}+\ldots\right)$ as required.

Example 4.26 A factory packs bolts in boxes of 500. The probability that a bolt is defective is 0.002. Find the probability that a box contains 2 defective bolts.

Solution 4.26 Let X be the r.v. 'the number of defective bolts in a box'.

This is a binomial situation, with $n = 500$, $p = 0.002$.

So $\qquad X \sim \text{Bin}(500, 0.002)$

Method 1 Using the binomial distribution,

$$P(X=x) = {}^nC_x q^{n-x} p^x, \qquad x = 0, 1, 2, \ldots, n.$$

We have $n = 500$, $p = 0.002$ and $q = 0.998$, so

$$P(X=x) = {}^{500}C_x (0.998)^{500-x}(0.002)^x$$

$$P(X=2) = {}^{500}C_2 (0.998)^{498}(0.002)^2$$

$$= \frac{(500)(499)}{(2)(1)}(0.998)^{498}(0.002)^2$$

$$= 0.184 \quad (3\text{ d.p.})$$

Method 2 Since n is large and p is small, we use the Poisson approximation.

The parameter $\lambda = np = 500(0.002) = 1$.

So $\quad X \sim \text{Po}(1) \quad$ and $\quad P(X=x) = \text{e}^{-1}\dfrac{1^x}{x!} \qquad x = 0, 1, \ldots$

We require $\quad P(X=2) = \text{e}^{-1}\dfrac{1^2}{2!} = 0.184 \quad (3\text{ d.p.})$

Therefore the probability that a box contains 2 defective bolts is 0.184 (3 d.p.).

NOTE: the answers agree to 3 d.p. and the calculations were much easier in method 2.

Example 4.27 Find the probability that at least two double sixes are obtained when two dice are thrown 90 times.

Solution 4.27 Throw two dice, $P(\text{double } 6) = (\frac{1}{6})(\frac{1}{6}) = \frac{1}{36}$.

Let X be the r.v. 'the number of double sixes obtained when two dice are thrown 90 times', then $X \sim \text{Bin}(90, \frac{1}{36})$ and $np = (90)(\frac{1}{36}) = 2.5$.

Using the Poisson approximation,

$X \sim \text{Po}(2.5) \quad$ and $\quad P(X=x) = \text{e}^{-2.5}\dfrac{(2.5)^x}{x!} \qquad x = 0, 1, \ldots$

Now

$$P(X \geqslant 2) = 1 - [P(X=0) + P(X=1)]$$

$$= 1 - (\text{e}^{-2.5} + \text{e}^{-2.5}2.5)$$

$$= 1 - \text{e}^{-2.5}(3.5)$$

$$= 0.713 \quad (3\text{ d.p.})$$

The probability that at least two double sixes are obtained when two dice are thrown 90 times is 0.713 (3 d.p.).

Exercise 4j

1. If $X \sim \text{Bin}(100, 0.03)$, use (*a*) the binomial distribution, (*b*) the Poisson distribution to evaluate (i) $P(X=0)$, (ii) $P(X=2)$, (iii) $P(X=4)$.

2. If $X \sim \text{Bin}(200, 0.006)$, use the Poisson distribution to find (*a*) $P(X<3)$, (*b*) $P(X>5)$.

3. On average one in 200 cars breaks down on a certain stretch of road per day. Find the probability that on a certain day (*a*) none of a sample of 250 cars breaks down, (*b*) more than 2 of a sample of 300 cars break down.

4. The probability that a particular make of light bulb is faulty is 0.01. The light bulbs are packed in boxes of 100.
(*a*) Find the probability that in a certain box there are (i) no faulty light bulbs, (ii) 2 faulty light bulbs, (iii) more than 3 faulty light bulbs.
(*b*) A buyer accepts a consignment of 50 boxes if, when he chooses two boxes at random, he finds that they contain no more than two faulty light bulbs altogether. Find the probability that he accepts the consignment.

5. Eggs are packed in boxes of 500. On average, 0.8% of the eggs are found to be broken when the eggs are unpacked.
(*a*) Find the probability that in a box of 500 eggs (i) exactly 3 will be broken, (ii) less than 2 will be broken.
(*b*) A hypermarket unpacks 100 boxes of eggs. What is the probability that there will be exactly 4 boxes containing no broken eggs?

6. An aircraft has 116 seats. The airline has found, from long experience, that on average 2.5% of people with tickets for a particular flight do not arrive for that flight. If the airline sells 120 seats for a particular flight determine, using a suitable approximation, the probability that more than 116 people arrive for that flight. Determine also the probability that there are empty seats on the flight. (C)

7. A firm selling electrical components packs them in boxes of 60. On average 2% of the components are faulty. What is the chance of getting more than 2 defective components in a box? (SUJB)P

CUMULATIVE POISSON PROBABILITY TABLES

The task of finding Poisson probabilities can be made much easier if tables are available. These give $P(X \leqslant r)$ for given values of λ. The tables are printed on p. 632 and an extract is shown below. In the extract, $X \sim \text{Po}(2.4)$.

	$\lambda = 2.4$
$r=0$	0.0907
1	0.3084
2	0.5697
3	0.7787
4	0.9041
5	0.9643
6	0.9884
7	0.9967
8	0.9991
9	0.9998
10	1.0000

Example 4.28 If $X \sim \text{Po}(2.4)$ find (a) $P(X \leqslant 6)$, (b) $P(X \geqslant 3)$, (c) $P(X < 8)$, (d) $P(X > 7)$, (e) $P(X = 4)$.

Solution 4.28 (a) $P(X \leqslant 6) = 0.9884$ (directly from the tables).

(b) $P(X \geqslant 3) = 1 - P(X \leqslant 2) = 1 - 0.5697 = 0.4303$.

(c) $P(X < 8) = P(X \leqslant 7) = 0.9967$.

(d) $P(X > 1) = 1 - P(X \leqslant 1) = 1 - 0.3084 = 0.6916$.

(e) $P(X = 4) = P(X \leqslant 4) - P(X \leqslant 3) = 0.9041 - 0.7787 = 0.1254$.

DIAGRAMMATIC REPRESENTATION OF THE POISSON DISTRIBUTION

The following diagrams show the probability distribution of $X \sim \text{Po}(\lambda)$ for various values of λ. The horizontal axis gives values of x and the vertical axis gives values of $P(X = x)$.

Notice that for small values of λ the distribution is very skew, but it becomes more symmetrical as λ increases.

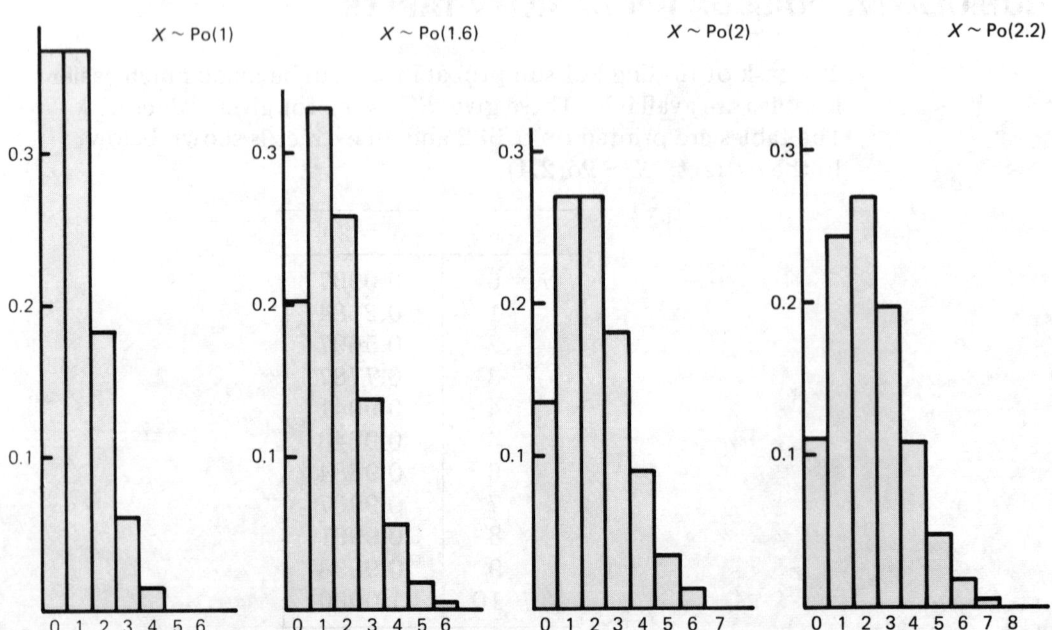

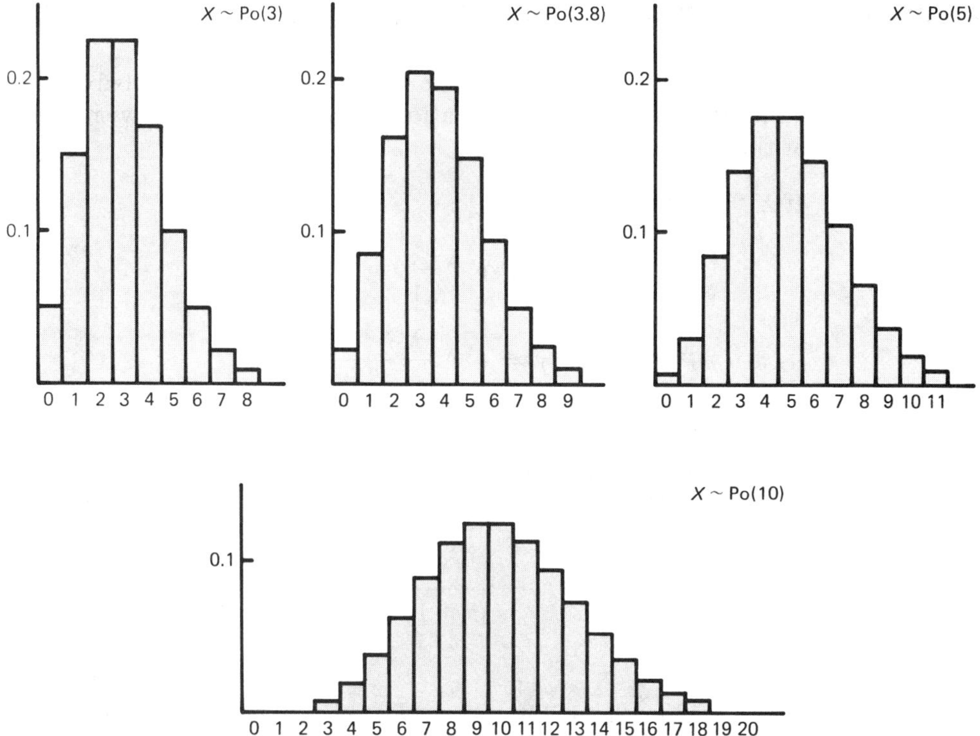

THE MODE OF THE POISSON DISTRIBUTION

The mode is the value which is most likely to occur, that is the one with the highest probability. Consider $X \sim \mathrm{Po}(\lambda)$. Considering the diagrams, we see that

when $\lambda = 1$, there are two modes, 0 and 1

when $\lambda = 2$, there are two modes, 1 and 2

when $\lambda = 3$, there are two modes, 2 and 3

In general, if λ is an integer then there are two modes and these occur when $x = \lambda - 1$, $x = \lambda$.

Notice that

when $\lambda = 1.6$, the mode is 1

when $\lambda = 2.2$, the mode is 2

when $\lambda = 3.8$, the mode is 3.

In general, if λ is not an integer, then the mode m is the integer such that

$$\lambda - 1 < m < \lambda.$$

THE RECURRENCE FORMULA FOR THE POISSON DISTRIBUTION

As with the binomial distribution, when cumulative probability tables are not available calculations can be made easier with the use of the recurrence formula.

If $X \sim \text{Po}(\lambda)$ then

$$P(X=x) = e^{-\lambda}\frac{\lambda^x}{x!}$$

so $$P(X=x+1) = e^{-\lambda}\frac{\lambda^{x+1}}{(x+1)!}$$

Therefore

$$\frac{P(X=x+1)}{P(X=x)} = \frac{e^{-\lambda}\lambda^{x+1}x!}{(x+1)!e^{-\lambda}\lambda^x} = \frac{\lambda}{(x+1)}$$

So $$P(X=x+1)) = \frac{\lambda}{(x+1)}P(X=x) \quad \text{for} \quad x = 0, 1, \ldots$$

Sometimes this is written $p_{x+1} = \dfrac{\lambda}{(x+1)}p_x$ where $p_x = P(X=x)$.

This is known as the recurrence formula for the Poisson distribution with parameter λ.

Example 4.29 If $X \sim \text{Po}(2.3)$, use the recurrence formula to find $P(X \geqslant 5)$.

Solution 4.29

$$p_{x+1} = \frac{2.3}{(x+1)}p_x$$

Now

$$p_0 = e^{-2.3} \qquad (0.100\,258\,8)$$

$$p_1 = \frac{2.3}{1}p_0 \qquad (0.230\,595\,3)$$

$$p_2 = \frac{2.3}{2}p_1 \qquad (0.265\,184\,6)$$

$$p_3 = \frac{2.3}{3}p_2 \qquad (0.203\,308\,2)$$

$$p_4 = \frac{2.3}{4}p_3 \qquad (0.116\,902\,2)$$

We require $P(X \geqslant 5) = 1-(p_0+p_1+p_2+p_3+p_4)$

$= 1-0.916\,249\,2$ (from memory store)

$= 0.0838$ (3 S.F.)

Therefore $P(X \geqslant 5) = 0.0838$ (3 S.F.).

The recurrence formula can be used to find the value of X which is most likely to occur, i.e. the value of X with the highest probability.

Example 4.30 If $X \sim \text{Po}(3.4)$, find the most likely value of X.

Solution 4.30 The recurrence formula for this Poisson distribution is

$$p_{x+1} = \frac{3.4}{(x+1)}p_x \quad \text{for} \quad x = 0, 1, 2, \ldots$$

Now $p_{x+1} > p_x$ when $\dfrac{3.4}{(x+1)}p_x > p_x$,

i.e. $3.4 > x+1$

or $x < 2.4$

i.e. $p_{x+1} > p_x$ when $x < 2.4$

so that $p_3 > p_2 > p_1 > p_0$

But $p_{x+1} < p_x$ when $x > 2.4$

so that $p_3 > p_4 > p_5$

p_3 is the greatest term in the sequence, so the most likely number to occur is 3.

NOTE: $\lambda = 3.4$ and the mode m is such that $2.4 < m < 3.4$, giving $m = 3$ as above.

Exercise 4k

1. Use (*a*) cumulative Poisson probability tables, (*b*) the recurrence formula, to find the first six terms of the following Poisson distributions. Sketch the distributions and verify that the mode m is the integer such that $\lambda - 1 < m < \lambda$.
(i) $X \sim \text{Po}(1.8)$ (ii) $X \sim \text{Po}(2.6)$
(iii) $X \sim \text{Po}(4.5)$ (iv) $X \sim \text{Po}(3.8)$.

FITTING A THEORETICAL DISTRIBUTION

As with the binomial distribution, it is possible to fit a theoretical Poisson distribution to experimental data.

Example 4.31 I recorded the number of phone calls I received over a period of 150 days:

Number of calls	0	1	2	3	4
Number of days	51	54	36	6	3

(a) Find the average number of calls per day.

(b) Calculate the frequencies of the comparable Poisson distribution. (SUJB)P

Solution 4.31 (a) For the data given

$$\begin{aligned}\bar{x} &= \frac{\Sigma fx}{\Sigma f}\\ &= \frac{0(51)+1(54)+2(36)+3(6)+4(3)}{150}\\ &= 1.04\end{aligned}$$

<u>The average number of calls per day is 1.04.</u>

(b) We use the mean of 1.04 as the parameter λ of the Poisson distribution.

Let X be the r.v. 'the number of calls per day'.

Then, if $X \sim \text{Po}(1.04)$ then

$$P(X=x) = e^{-1.04}\frac{(1.04)^x}{x!} \qquad x = 0, 1, 2, \ldots$$

To make the calculations easier, use the recurrence formula.

Then, to find the expected frequencies, multiply each probability by 150 and give the answer to the nearest day.

The recurrence formula for this Poisson distribution is

$$p_{x+1} = \frac{1.04}{(x+1)}p_x$$

	x	Expected frequency ($150\,p_x$)
$p_0 = e^{-1.04} = 0.353\,454\,6$	0	53
$p_1 = 1.04\,p_0 = 0.367\,592\,8$	1	55
$p_2 = \frac{1.04}{2}p_1 = 0.191\,148\,2$	2	29
$p_3 = \frac{1.04}{3}p_2 = 0.066\,264\,7$	3	10
$p_4 = \frac{1.04}{4}p_3 = 0.017\,228\,8$	4	3
		Total 150

NOTE: the sum of the above probabilities is 0.995 689 4, not 1. This is because we have stopped calculating at $P(X = 4)$. As the sum of the remaining probabilities is only 0.004 310 6, we ignore the error involved.

The theoretical Poisson distribution gives frequencies as follows:

Number of calls, x	0	1	2	3	4
Number of days, f	53	55	29	10	3

This compares reasonably well with the original frequency distribution.

A statistical test to compare the two sets of data is illustrated on p. 543 (chi-squared test).

Exercise 4I

1. For each of the following sets of data, fit a theoretical Poisson distribution:

(*a*)

x	0	1	2	3	4
f	45	44	20	8	3

(*b*)

x	0	1	2	3	4	5
f	110	50	20	12	7	1

2. An inn caters for overnight travellers and during its busy season of 100 days the number of requests each day for rooms has a Poisson distribution with mean 4. The inn has four rooms for hire.

Draw up a table to show the expected frequencies of 0, 1, 2, . . . , 12 requests each day for rooms during the 100 days.

Obtain an estimate of the number of requests which will have to be refused during the period.

The cost of building an extra room is estimated at £1000 which the owner would pay from capital invested to yield a net $8\frac{1}{2}$% per annum. If each room let yields on average £3 net per day, estimate the annual gain or loss of income (excluding the capital outlay) were the owner to have the room built. It may be assumed that during the rest of the year fewer than five rooms are let. (SUJB)

3. A firm investigated the number of employees suffering injuries whilst at work. The results recorded below were obtained for a 52-week period:

Number of employees injured in a week	0	1	2	3	4 or more
Number of weeks	31	17	3	1	0

Give reasons why one might expect this distribution to approximate to a Poisson distribution. Evaluate the mean and variance of the data and explain why this gives further evidence in favour of a Poisson distribution.

Using the calculated value of the mean, find the theoretical frequencies of a Poisson distribution for the number of weeks in which 0, 1, 2, 3, 4 or more, employees were injured. (C)

4. State the conditions under which the binomial distribution approximates to the Poisson distribution. Hence derive the Poisson distribution of mean m and show that its variance is also m.

Tests for defects are carried out in a textile factory on a lot comprising 400 pieces of cloth. The results of the tests are shown in Table A below.

Table A

Number of faults per piece	0	1	2	3	4	5	6	Total
Number of pieces	92	142	96	46	18	6	0	400

Show that this is approximately a Poisson distribution and calculate the frequencies on this assumption.

How many pieces from a sample of 1000 pieces may be expected to have 4 or more faults? (AEB 1972)

THE DISTRIBUTION OF TWO INDEPENDENT POISSON VARIABLES

The sum of two independent Poisson variables with parameters m and n, respectively, is a Poisson variable with parameter $(m+n)$,

i.e. if $X \sim \text{Po}(m)$ and $Y \sim \text{Po}(n)$, then $X+Y \sim \text{Po}(m+n)$

Proof

$X \sim \text{Po}(m)$ so $P(X=x) = e^{-m}\dfrac{m^x}{x!}$	$Y \sim \text{Po}(Y)$ so $P(Y=y) = e^{-n}\dfrac{n^x}{x!}$
$P(X=0) = e^{-m}$ $P(X=1) = e^{-m}m$ $P(X=2) = e^{-m}\dfrac{m^2}{2!}$ and so on	$P(Y=0) = e^{-n}$ $P(Y=1) = e^{-n}n$ $P(Y=2) = e^{-n}\dfrac{n^2}{2!}$ and so on

Now

$$\begin{aligned}
P(X+Y=0) &= P(X=0)\cdot P(Y=0)\\
&= (e^{-m})\cdot(e^{-n})\\
&= e^{-(m+n)}\\
P(X+Y=1) &= P(X=0)\cdot P(Y=1)+P(X=1)\cdot P(Y=0)\\
&= (e^{-m})(e^{-n}n)+(e^{-m}m)(e^{-n})\\
&= e^{-(m+n)}(m+n)\\
P(X+Y=2) &= P(X=0)\cdot P(Y=2)+P(X=1)\cdot P(Y=1)\\
&\qquad +P(X=2)\cdot P(Y=0)\\
&= (e^{-m})\left(e^{-n}\frac{n^2}{2!}\right)+(e^{-m}m)(e^{-n}n)+e^{-m}\frac{m^2}{2!}(e^{-n})\\
&= \frac{e^{-(m+n)}}{2!}(m^2+2mn+n^2)\\
&= e^{-(m+n)}\frac{(m+n)^2}{2!}
\end{aligned}$$

and so on.

The probability distribution for $X+Y$ is

$x+y$	0	1	2	...
$P(X+Y=x+y)$	$e^{-(m+n)}$	$e^{-(m+n)}(m+n)$	$e^{-(m+n)}\dfrac{(m+n)^2}{2!}$	...

From the distribution we see that $X+Y \sim \text{Po}(m+n)$, as required.

Example 4.32 Two identical racing cars are being tested on a circuit. For each car, the number of mechanical breakdowns follows a Poisson distribution with a mean of one breakdown in 100 laps.

The first car does 20 laps and the second does 40 laps. What is the probability that there will be **(a)** no breakdowns, **(b)** one breakdown, **(c)** more than two breakdowns altogether? Assume that breakdowns are attended and the cars continue on the circuit.

Solution 4.32 In 100 laps we 'expect' 1 breakdown. So, in 20 laps we 'expect' 0.2 breakdowns and in 40 laps we 'expect' 0.4 breakdowns.

Let X be the r.v. 'the number of breakdowns for the first car'.

Then $$X \sim \text{Po}(0.2)$$

Let Y be the r.v. 'the number of breakdowns for the second car'

Then $$Y \sim \text{Po}(0.4)$$

Let T be the r.v. 'the total number of breakdowns', so $T = X+Y$

$$T \sim \text{Po}(0.2+0.4)$$

i.e. $$T \sim \text{Po}(0.6)$$

(a) $$P(T=0) = e^{-0.6} = 0.549 \quad (3\text{ d.p.})$$

Therefore the probability that there are no breakdowns is 0.549 (3 d.p.).

(b) $$P(T=1) = e^{-0.6}(0.6) = 0.329 \quad (3\text{ d.p.})$$

The probability that there will be one breakdown is 0.329 (3 d.p.).

$$\begin{aligned}
\text{(c)}\ P(T>2) &= 1-[P(T=0)+P(T=1) \\
&\qquad +P(T=2)] \\
&= 1-\left[e^{-0.6}+e^{-0.6}(0.6)+e^{-0.6}\frac{(0.6)^2}{2!}\right] \\
&= 1-e^{-0.6}(1+0.6+0.18) \\
&= 1-e^{-0.6}(1.78) \\
&= 0.023 \quad (3\text{ d.p.})
\end{aligned}$$

The probability that there will be more than two breakdowns is 0.023 (3 d.p.).

Example 4.33 The centre pages of the 'Weekly Sentinel' consist of 1 page of film and theatre reviews and 1 page of classified advertisements. The number of misprints in the reviews has a Poisson distribution with mean 2.3 and the number of misprints in the classified section has a Poisson distribution with mean 1.7.

(a) Find the probability that, on the centre pages, there will be (*i*) no misprints, (*ii*) more than 5 misprints.

(b) Find the smallest integer n such that the probability that there are more than n misprints on the centre pages is less than 0.1.

Solution 4.33 Let X be the r.v. 'the number of misprints on the review page'.

Then $\quad X \sim \text{Po}(2.3)$

Let Y be the r.v. 'the number of misprints on the classified page'.

Then $\quad Y \sim \text{Po}(1.7)$

Let T be the r.v. 'the number of misprints on the centre pages'.

Therefore $T = X+Y$ and $T \sim \text{Po}(2.3+1.7)$

i.e. $T \sim \text{Po}(4)$

(a) (*i*) $P(T=0) = e^{-4} = 0.018\,315\,6 = 0.018 \quad (3\text{ d.p.})$

The probability that there will be no misprints on the centre pages is 0.018 (3 d.p.).

(*ii*) Using cumulative Poisson probability tables, with $\lambda = 4$, $r = 5$,

$$\begin{aligned}
P(T>5) &= 1-P(T\leqslant 5) \\
&= 1-0.7851 \\
&= 0.215 \quad (3\text{ d.p.})
\end{aligned}$$

If tables are not available, consider

$$P(T>5) = 1-[p_0+p_1+p_2+p_3+p_4+p_5]$$

Using the recurrence formula:

$$p_0 = e^{-4} \qquad (0.018\,315\,6)$$

$$p_1 = 4p_0 \qquad (0.073\,262\,5)$$

$$p_2 = \frac{4}{2}p_1 \qquad (0.146\,525\,1)$$

$$p_3 = \frac{4}{3}p_2 \qquad (0.195\,366\,8)$$

$$p_4 = \frac{4}{4}p_3 \qquad (0.195\,366\,8)$$

$$p_5 = \frac{4}{5}p_4 \qquad (0.156\,293\,4)$$

So that $P(T > 5) = 1 - 0.785$ (from memory store)

$= 0.215$ (3 d.p.)

The probability that there will be more than 5 misprints on the centre pages is 0.215 (3 d.p.).

(b) Now $P(T > 5) = 0.215 > 0.1$

and from tables we find that

$$P(T > 6) = 0.111 > 0.1$$

$$P(T > 7) = 0.051 < 0.1$$

So the smallest integer n, such that the probability that there are more than n misprints on the centre pages is less than 0.1, is 7.

Exercise 4m

1. Telephone calls reach a secretary independently and at random, internal ones at a mean rate of 2 in any 5 minute period, and external ones at a mean rate of 1 in any 5 minute period. Calculate the probability that there will be more than 2 calls in any period of 2 minutes. (O & C)

2. During a weekday, heavy lorries pass a census point P on a village high street independently and at random times. The mean rate for westward travelling lorries is 2 in any 30-minute period, and for eastward travelling lorries is 3 in any 30-minute period.

   Find the probability
   (*a*) that there will be no lorries passing P in a given 10-minute period,
   (*b*) that at least one lorry from each direction will pass P in a given 10-minute period,
   (*c*) that there will be exactly 4 lorries passing P in a given 20-minute period. (O & C)

3. A large number of screwdrivers from a trial production run is inspected. It is found that the cellulose acetate handles are defective on 1% and that the chrome steel blades are defective on $1\frac{1}{2}$% of the screwdrivers, the defects occurring independently.
   (*a*) What is the probability that a sample of 80 contains more than two defective screwdrivers?
   (*b*) What is the probability that a sample of 80 contains at least one screwdriver with both defective handle and a defective blade? (O & C)

4. A restaurant kitchen has 2 food mixers, *A* and *B*. The number of times per week that *A* breaks down has a Poisson distribution with mean 0.4, while independently the number of times that *B* breaks down in a week has a Poisson distribution with mean 0.1. Find, to 3 decimal places, the probability that in the next 3 weeks
(*a*) *A* will not break down at all,
(*b*) each mixer will break down exactly once,
(*c*) there will be a total of 2 breakdowns. (L)P

MISCELLANEOUS WORKED EXAMPLES

Example 4.34 Along a stretch of motorway, breakdowns requiring the summoning of the breakdown services occur with a frequency of 2.4 per day, on average. Assuming that the breakdowns occur randomly and that they follow a Poisson distribution, find

(a) the probability that there will be exactly 2 breakdowns on a given day,

(b) the smallest integer n such that the probability of more than n breakdowns in a day is less than 0.03.

Solution 4.34 (a) Let X be the r.v. 'the number of breakdowns a day requiring the breakdown services'.

Then

$$X \sim \text{Po}(2.4) \quad \text{and} \quad P(X = x) = e^{-2.4}\frac{(2.4)^x}{x!} \qquad x = 0, 1, 2, \ldots$$

So $$P(X = 2) = e^{-2.4}\frac{(2.4)^2}{2!} = 0.261 \quad (3\text{ S.F.})$$

<u>The probability that there will be exactly 2 breakdowns on a given day is 0.261 (3 S.F.).</u>

(b) We require the least integer n such that $P(X > n) < 0.03$.

Now if $$P(X > n) < 0.03$$

then $$P(X \leqslant n) > 0.97$$

From cumulative Poisson probability tables, with $\lambda = 2.4$

$$P(X \leqslant 5) = 0.9643 < 0.97$$

$$P(X \leqslant 6) = 0.9884 > 0.97$$

<u>So the least integer n such that $P(X > n) < 0.03$ is 6.</u>

If tables are not available then consider

$$1 - (p_0 + p_1 + p_2 + \ldots + p_n) < 0.03 \quad \text{where} \quad p_x = P(X = x)$$

i.e. $$p_0 + p_1 + p_2 + \ldots + p_n > 0.97$$

Using the recurrence formula $p_{x+1} = \dfrac{2.4}{(x+1)} p_x$ we will need to find $p_0, p_1, p_2, \ldots$, and keep a record of the cumulative probabilities so that we know when the cumulative probability is greater than 0.97.

	Cumulative probability
Now $p_0 = e^{-2.4} = 0.090\,717\,9$	$F(0) = 0.090\,717\,9$
$p_1 = \dfrac{2.4}{1} p_0 = 0.217\,723$	$F(1) = 0.308\,441$
$p_2 = \dfrac{2.4}{2} p_1 = 0.261\,267\,7$	$F(2) = 0.569\,708\,7$
$p_3 = \dfrac{2.4}{3} p_2 = 0.209\,014\,1$	$F(3) = 0.778\,722\,9$
$p_4 = \dfrac{2.4}{4} p_3 = 0.125\,408\,5$	$F(4) = 0.904\,131\,4$
$p_5 = \dfrac{2.4}{5} p_4 = 0.060\,196\,0$	$F(5) = 0.964\,327\,4$
$p_6 = \dfrac{2.4}{6} p_5 = 0.024\,078\,4$	$F(6) = 0.988\,405\,9$

By trial, $p_0 + p_1 + p_2 + p_3 + p_4 + p_5 + p_6 = 0.988\,405\,9$

So the least integer n such that $P(X > n) < 0.03$ is 6.

Example 4.35 A random variable X has a Poisson distribution given by

$$P(X = r) = p_r = e^{-\lambda} \frac{\lambda^r}{r!}, \qquad r = 0, 1, 2, \ldots$$

Prove that the mean of X is λ. Give two examples (other than that suggested below) of situations where you would expect a Poisson distribution to occur.

The number of white corpuscles on a slide has a Poisson distribution with mean 3.2. By considering the values of r for which $\dfrac{p^{r+1}}{p_r} > 1$ find the most likely number of white corpuscles on a slide. Calculate correct to 3 decimal places the probability of obtaining this number. If two such slides are prepared what is the probability, correct to 3 decimal places, of obtaining at least two white corpuscles in total on the two slides? (SUJB)

Solution 4.35 For the answer to the first part of the question, see page 243.

Let X be the r.v. 'the number of white corpuscles on a slide'. Then $X \sim \text{Po}(3.2)$.

So $$P(X=r) = e^{-3.2}\frac{3.2^r}{r!}$$

and $$P(X=r+1) = e^{-3.2}\frac{3.2^{r+1}}{(r+1)!}$$

Dividing $$\frac{P(X=r+1)}{P(X=r)} = \frac{e^{-3.2}\,3.2^{r+1}}{e^{-3.2}\,3.2^r}\,\frac{r!}{(r+1)!}$$

$$= \frac{3.2}{(r+1)}$$

So $$\frac{p_{r+1}}{p_r} = \frac{3.2}{(r+1)} \quad \text{where} \quad p_r = P(X=r)$$

Hence $p_{r+1} > p_r$ when $3.2 > r+1$

$r < 2.2$

i.e. $p_{r+1} > p_r$ when $r = 0, 1, 2$

i.e. $p_0 < p_1 < p_2 < p_3$,

but $p_{r+1} < p_r$ when $r = 3, 4, 5, \ldots$ so $p_3 > p_4 > p_5$.

Therefore the most likely number of white corpuscles on a slide is 3.

Now $$P(X=3) = e^{-3.2}\frac{3.2^3}{3!}$$

$$= 0.223 \quad (3\text{ d.p.})$$

The probability of obtaining 3 white corpuscles on a slide is 0.223 (3 d.p.).

Let X_1 be the r.v. 'the number of white corpuscles on the first slide'. Then $X_1 \sim \text{Po}(3.2)$.

Let X_2 be the r.v. 'the number of white corpuscles on the second slide'. Then $X_2 \sim \text{Po}(3.2)$.

Let $Y = X_1 + X_2$, then $Y \sim \text{Po}(3.2+3.2)$

i.e. $Y \sim \text{Po}(6.4)$

We require $$P(Y \geqslant 2) = 1 - [P(Y=0) + P(Y=1)]$$

$$= 1 - (e^{-6.4} + e^{-6.4}6.4)$$

$$= 1 - 7.4\,e^{-6.4}$$

$$= 0.988 \ (3\text{ d.p.})$$

The probability of obtaining at least two white corpuscles in total on the two slides is 0.988 (3 d.p.).

We now show an alternative approach to the last part:

If two slides are prepared, we require P(total number of white corpuscles $\geqslant 2$).

Now

$$\begin{aligned} P(\text{total} \geqslant 2) &= 1-[P(\text{total}=0)+P(\text{total}=1)] \\ &= 1-[P(X_1=0)P(X_2=0)+P(X_1=0)P(X_1=1) \\ &\qquad +P(X_1=1)P(X_2=0)] \\ &= 1-[(e^{-3.2})(e^{-3.2})+e^{-3.2}(e^{-3.2}3.2)+(e^{-3.2}3.2)e^{-3.2}] \\ &= 1-e^{-6.4}(1+3.2+3.2) \\ &= 1-7.4e^{-6.4} \\ &= 0.988 \quad \text{as before} \end{aligned}$$

Example 4.36 Derive the mean and the variance of the Poisson distribution.

In a large town, one person in 80, on the average, has blood of type X. If 200 blood donors are taken at random, find an approximation to the probability that they include at least five persons having blood of type X.

How many donors must be taken at random in order that the probability of including at least one donor of type X shall be 0.9 or more? (AEB)

Solution 4.36 We have already shown (p. 243–4) that if R is a r.v. such that $R \sim \text{Po}(\lambda)$, then

$$E(R) = \lambda \quad \text{and} \quad \text{Var}(R) = \lambda$$

Let R be the r.v. 'the number of blood donors of type X'.

Then $R \sim \text{Bin}(n, p)$ where $n = 200$ and $p = P(\text{blood type X}) = \frac{1}{80}$.

Now, as n is large and p is small, we use the Poisson approximation to the binomial distribution.

The parameter $\lambda = np = (200)(\frac{1}{80}) = 2.5$

The probability that there are at least five donors of type X is

$$\begin{aligned} P(R \geqslant 5) &= 1-P(R \leqslant 4) \\ &= 1-0.8912 \quad \text{(from tables)} \\ &= 0.109 \quad (3\,\text{d.p.}) \end{aligned}$$

The probability that the sample will contain at least five people having blood of type X is 0.109 (3 d.p.).

Suppose n donors are taken, then $\lambda = n(\frac{1}{80})$.

So $R \sim \text{Po}\left(\frac{n}{80}\right)$.

We require n such that $P(R \geqslant 1) \geqslant 0.9$,

i.e. $$1 - P(R = 0) \geqslant 0.9$$
$$P(R = 0) \leqslant 0.1$$

Now $P(R = 0) = e^{-n/80}$

So $$e^{-n/80} \leqslant 0.1$$
$$e^{n/80} \geqslant \frac{1}{0.1}$$

i.e. $$e^{n/80} \geqslant 10$$

So, taking logs to the base e,

$$\frac{n}{80} \geqslant \ln(10)$$
$$\frac{n}{80} \geqslant 2.30$$
$$n \geqslant (80)(2.30)$$
$$n \geqslant 184.2$$

So we need to take 185 donors in order that the probability of including at least one donor of type X is 0.9 or more.

CHECK: If $n = 184$, $\lambda = (184)(\frac{1}{80}) = 2.3$

So $$P(R \geqslant 1) = 1 - e^{-2.3}$$
$$= 0.8997 \quad (4\text{ d.p.})$$

We have $P(R \geqslant 1) < 0.9$ when $n = 184$.

Now consider $n = 185$, then $\lambda = (185)(\frac{1}{80}) = 2.3125$
$$P(R \geqslant 1) = 1 - e^{-2.3125}$$
$$= 0.901 \quad (3\text{ d.p.})$$

So $P(R \geqslant 1) > 0.9$ when $n = 185$.

Example 4.37 In the Growmore Market Garden plants are inspected for the presence of the deadly red angus leaf bug. The number of bugs per leaf is known to follow a Poisson distribution with mean one. What is the probability that any one leaf on a given plant will have been attacked (at least one bug is found on it)?

A random sample of twelve plants is taken. For each plant ten leaves are selected at random and inspected for these bugs. If more than eight leaves on any particular plant have been attacked then the plant is destroyed. What is the probability that exactly two of these twelve plants are destroyed? (AEB 1977)

Solution 4.37 Let X be the r.v. 'the number of bugs per leaf'.

Then $X \sim \text{Po}(1)$ and $P(X=x) = e^{-1}\frac{1^x}{x!}$ $\qquad x = 0, 1, 2, \ldots$

We require
$$\begin{aligned} P(X \geqslant 1) &= 1 - P(X=0) \\ &= 1 - e^{-1} \\ &= 1 - 0.368 \\ &= 0.632 \quad (3 \text{ S.F.}) \end{aligned}$$

The probability that any one leaf has been attacked is 0.632 (3 S.F.).

Now let Y be the r.v. 'the number of leaves that have been attacked on a plant'.

Then $Y \sim \text{Bin}(10, 0.632)$ since $n = 10$, and the probability that a leaf has been attacked is 0.632.

We have
$$P(Y=y) = {}^{10}C_y(0.368)^{10-y}(0.632)^y \qquad y = 0, 1, 2, \ldots, 10$$

We require
$$\begin{aligned} P(Y>8) &= P(Y=9) + P(Y=10) \\ &= 10(0.368)^1(0.632)^9 + (0.632)^{10} \\ &= 0.069 \ (3 \text{ d.p.}) \end{aligned}$$

The probability that any one plant is destroyed is 0.069 (3 d.p.).

Now let R be the r.v. 'the number of plants that are destroyed'.

Then $R \sim \text{Bin}(12, 0.069)$ since 12 plants are inspected, and the probability that a plant is destroyed is 0.069.

$$P(R=r) = {}^{12}C_r(0.931)^{12-r}(0.069)^r$$

We require
$$\begin{aligned} P(R=2) &= {}^{12}C_2(0.931)^{10}(0.069)^2 \\ &= \frac{(12)(11)}{(2)(1)}(0.931)^{10}(0.069)^2 \\ &= 0.154 \quad (3 \text{ d.p.}) \end{aligned}$$

The probability that exactly two of the twelve plants will be destroyed is 0.154 (3 d.p.).

SUMMARY — POISSON DISTRIBUTION

If $X \sim \text{Po}(\lambda)$ then $P(X = x) = e^{-\lambda}\dfrac{\lambda^x}{x!}$ $\quad x = 0, 1, 2, \ldots$

$E(X) = \lambda$

$\text{Var}(X) = \lambda$

Recurrence formula: $\quad p_{x+1} = \dfrac{\lambda}{(x+1)} p_x$

If $X \sim \text{Po}(m)$ and $Y \sim \text{Po}(n)$ then $X + Y \sim \text{Po}(m+n)$
(X, Y independent)

Miscellaneous Exercise 4n

1. Lemons are packed in boxes, each box containing 200. It is found that, on average, 0.45% of the lemons are bad when the boxes are opened. Use the Poisson distribution to find the probabilities of 0, 1, 2, and more than 2 bad lemons in a box.

   A buyer who is considering buying a consignment of several hundred boxes checks the quality of the consignment by having a box opened. If the box opened contains no bad lemons he buys the consignment. If it contains more than 2 bad lemons he refuses to buy, and if it contains 1 or 2 bad lemons he has another box opened and buys the consignment if the second box contains fewer than 2 bad lemons. What is the probability that he buys the consignment?

   Another buyer checks consignments on a different basis. He has one box opened; if that box contains more than 1 bad lemon he asks for another to be opened and does not buy if the second also contains more than 1 bad lemon. What is the probability that he refuses to buy the consignment? (SUJB)

2. A manufacturer produces an integrated electronic unit which contains 36 separate pressure sensors. Due to difficulties in manufacture, it happens very often that not all the sensors in a unit are operational. 100 units are tested and the numbers N of pressure sensors which function correctly are distributed according to Table A below.
   Calculate the mean number of sensors which are faulty.

   The manufacturer only markets those units which have at least 32 of their 36 sensors operational. Estimate, using the Poisson distribution, the percentage of units produced which are not marketed. (O & C)

3. Show that, for the Poisson distribution in which the probabilities of 0, 1, 2, ... successes are $e^{-m}, me^{-m}, \dfrac{m^2e^{-m}}{2!}, \ldots,$ the mean number of successes is equal to m. State the variance.
   A sales manager receives 6 telephone calls on average between 9.30 a.m. and 10.30 a.m. on a weekday. Find the probability that
   (*a*) he will receive 2 or more calls between 9.30 and 10.30 on a certain weekday;
   (*b*) he will receive exactly 2 calls between 9.30 and 9.40;
   (*c*) during a normal 5-day working week, there will be exactly 3 days on which he will receive no calls between 9.30 and 9.40. (SUJB)

Table A

N	36	35	34	33	32	31	30	29	28	< 28
Number of units	5	15	22	22	17	11	5	2	1	0

4. X is a random variable having a Poisson distribution given by

$$f(x) = e^{-m} m^x/x!, \quad x = 0, 1, 2, \ldots$$

Prove that the mean of X is m and state the variance of X.

The number of telephone calls received per minute at the switchboard of a certain office was logged during the period 10 a.m. to noon on a working day. The results were as follows:

Calls per min. (x)	0	1	2	3	4	5	6	7	8
f	7	18	27	28	20	11	5	3	1

f is the number of minutes with x calls per minute.

By consideration of the mean and variance of this distribution show that a possible model is a Poisson distribution.

Using the calculated mean and on the assumption of a Poisson distribution calculate (*a*) the probability that two or more calls were received during any one minute, (*b*) the probability that no calls were received during any two consecutive minutes. (SUJB)

5. Customers enter an antique shop independently of one another and at random intervals of time at an average rate of four per hour throughout the five days of a week on which the shop is open. The owner has a coffee-break of fifteen minutes each morning; if one or more customers arrive during this period then his coffee goes cold, otherwise he drinks it while it is hot.

Let X be the random variable denoting the number of customers arriving during a Monday coffee-break, and let Y be the random variable denoting the number of days during a week on which the owner's coffee goes cold. Assuming that X has a Poisson distribution, determine (correct to three significant figures) (*a*) $P(X = 0)$, (*b*) $P(X \geqslant 2)$, (*c*) $E(Y)$, (*d*) $P(Y = 2)$. (C)

6. A hire company has two electric lawnmowers which it hires out by the day. The number of demands per day for a lawnmower has the form of a Poisson distribution with mean 1.50. In a period of 100 working days, how many times do you expect

(*a*) neither of the lawnmowers to be used,

(*b*) some requests for the lawnmowers to have to be refused?

If each lawnmower is to be used an equal amount, on how many days in a period of 100 working days would you expect a particular lawnmower not to be in use? (MEI)

7. An experimenter marked out ten neighbouring plots of land, all of the same area, and examined them for the occurrence of a certain species of plant. The numbers, f_r, of plots in which r plants were found were as follows:

r	0	1	2	3	4	5
f_r	3	1	1	2	1	2

Calculate the mean number of plants per plot.

Assuming that the plants were scattered randomly with this same mean number of plants per plot, find

(*a*) the probability of a given plot containing no plants;

(*b*) the probability of at least three plots being found which contain no plants.

What conclusions, if any, can be drawn from the observed number of plots in which the experimenter found no plants?

(You may assume that $e^{-2.3} \approx 0.1$) (SMP)

8. Derive the Poisson distribution as the limiting form of the Binomial distribution when n becomes very large and p becomes very small in such a way that np remains constant. Write down the mean and the variance of this distribution.

The mean number of bacteria per millilitre of a liquid is known to be 3. Ten samples of the liquid, chosen at random and each of volume 1 ml, are examined. Assuming the Poisson distribution is applicable, obtain expressions for the probabilities

(*a*) that each of the ten samples contains at least one bacterium,

(*b*) that exactly eight of the samples contain at least one bacterium,

If 3 ml of the liquid is examined show that it is rather improbable that it will contain fewer than 3 or more than 15 bacteria. (MEI)

9. A discrete random variable, X, can take values $0, 1, \ldots$ and has a Poisson distribution such that the probability that $X = r$ is $e^{-m} m^r/r!$. Prove that the mean of X is m.

During each working day in a certain factory a number of accidents occur independently according to a Poisson distribution with mean 0.5.

Calculate the probability that
(*a*) during any one day there are 2 or more accidents,
(*b*) during two consecutive days there are exactly three accidents altogether.

Out of 50 consecutive five-day weeks how many would you expect to be accident-free?

Give two further situations where you would expect a Poisson distribution to apply. (SUJB)

10. Prove that for the Poisson distribution in which the probability of r successes is

$$\frac{e^{-m} m^r}{r!} \quad (r \geqslant 0)$$

the expected number of successes is equal to m.

The telephone exchange inside an office building has a number of outside lines of which, on average, 3 are being used at any instant. Assuming that the number of lines in use at any instant follows a Poisson distribution, find
(*a*) the probability that, at any given instant, not more than 3 lines are in use,
(*b*) the minimum number of outside lines required if there is to be a probability of more than 0.9 that, at any given instant, at least one of the lines is not being used. (C)

11. The monthly demand for a certain magazine at a small newsagent's shop has a Poisson distribution with mean 3. The newsagent always orders 4 copies of the magazine for sale each month; any demand for the magazine in excess of 4 is not met.
(*a*) Calculate the probability that the newsagent will not be able to meet the demand in a given month.
(*b*) Find the most probable number of magazines *sold* in one month.
(*c*) Find the expected number of magazines *sold* in one month.
(*d*) Determine the least number of copies of the magazine that the newsagent should order each month so as to meet the demand with a probability of at least 0.95. (JMB)

12. A random sample of 500 people born in 1961 is being studied. It can be assumed that birthdays are uniformly distributed throughout the year.
(*a*) Use the Poisson distribution to find, to 3 decimal places, the probabilities that there are (i) exactly two people, and (ii) no more than two people, with birthdays on 1 January.
(*b*) Also find, to 3 decimal places, the probability that, if two of the sample are chosen at random, they have birthdays in the same month. (In 1961 there were 7 months with 31 days, 4 months with 30 days and 1 month with 28 days.) (MEI)

13. Define the Poisson distribution and derive its mean. State the circumstances under which it is appropriate to use the Poisson distribution as an approximation to the binomial distribution.

A lottery has a very large number of tickets, one in every 500 of which entitles the purchaser to a prize. An agent sells 1000 tickets for the lottery. Using the Poisson distribution, find, to three decimal places, the probabilities that the number of prize-winning tickets sold by the agent is (*a*) less than three, (*b*) more than five.

Calculate the minimum number of tickets the agent must sell to have a 95% chance of selling at least one prize-winning ticket. (JMB)

14. Define the Poisson distribution and derive its mean and variance.

The number of telephone calls received at a switchboard in any time interval of length T minutes has a Poisson distribution with mean $\frac{1}{2}T$. The operator leaves the switchboard unattended for five minutes. Calculate to three decimal places the probabilities that there are (*a*) no calls, (*b*) four or more calls in her absence.

Find to three significant figures the maximum length of time in seconds for which the operator could be absent with a 95% probability of not missing a call. (JMB)

15. Define the Poisson distribution and derive its mean and variance.

In the first year of the life of a certain type of machine, the number of times a maintenance engineer is required has a Poisson distribution with mean four. Find the probability that more than four calls are necessary.

The first call is free of charge and subsequent calls cost £20 each. Find the mean cost of maintenance in the first year. (JMB)

16. The number of oil tankers arriving at a port between successive high tides has a Poisson distribution with mean 2. The depth of the water is such that loaded

vessels can enter the dock area only on the high tide. The port has dock space for only three tankers, which are discharged and leave the dock area before the next tide. Only the first three loaded tankers waiting at any high tide go into the dock area; any others must await another high tide.

Starting from an evening high tide after which no ships remain waiting their turn, find (to three decimal places) the probabilities that after the next morning's high tide (*a*) the three dock berths remain empty, (*b*) the three berths are all filled. Find (to two decimal places) the probability that no tankers are left waiting outside the dock area after the following evening's high tide. (JMB)

17. The random variable X has a Poisson distribution with parameter λ.
(*a*) Prove that $E(X) = \lambda$.
(*b*) If $P(X = k) = P(X = k + 1)$, where k is some integer, show that λ must also be an integer.
(*c*) If λ is not an integer, show that the mode, m, of the distribution is such that $\lambda - 1 < m < \lambda$.

In the manufacture of commercial carpet, small faults occur at random in the carpet at an average rate of 0.95 per 20 m^2. Find the probability that in a randomly selected 20 m^2 area of this carpet
(*d*) there are no faults,
(*e*) there are at most 2 faults.

The ground floor of a new office block has 10 rooms. Each room has an area of 80 m^2 and has been carpeted using the same commercial carpet described above. For any one of these rooms, determine the probability that the carpet in that room
(*f*) contains at least 2 faults,
(*g*) contains exactly 3 faults,
(*h*) contains at most 5 faults.

Find the probability that in exactly half of these 10 rooms the carpets will contain exactly 3 faults. (AEB 1988)

18. A randomly chosen doctor in general practice sees, on average, one case of a broken nose per year and each case is independent of other similar cases.
(*a*) Regarding a month as a twelfth part of a year,
(i) show that the probability that, between them, three such doctors see no cases of a broken nose in a period of one month is 0.779, correct to three significant figures,
(ii) find the variance of the number of cases seen by three such doctors in a period of six months.
(*b*) Find the probability that, between them, three such doctors see at least three cases in one year.
(*c*) Find the probability that, of three such doctors, one sees three cases and the other two see no cases in one year. (C)

19. State, giving your reasons, the distribution which you would expect to be appropriate in describing
(*a*) the number of heads in 10 throws of a penny,
(*b*) the number of blemishes per m^2 of sheet metal.

A building has an automatic telephone exchange. The number X of wrong connections in any one day is a Poisson variable with parameter λ. Find, in terms of λ, the probability that in any one day there will be
(*c*) exactly 3 wrong connections,
(*d*) 3 or more wrong connections.

Evaluate, to 3 decimal places, these probabilities when $\lambda = 0.5$. Find, to 3 decimal places, the largest value of λ for the probability of one or more wrong connections in any day to be at most $\frac{1}{6}$. (L)

5

PROBABILITY DISTRIBUTIONS II — CONTINUOUS RANDOM VARIABLES

A continuous random variable (r.v.) is a theoretical representation of a continuous variable such as height, mass or time.

PROBABILITY DENSITY FUNCTION

A continuous r.v. is specified by its probability density function (p.d.f.) written $f(x)$.

If X is a continuous r.v. with p.d.f. $f(x)$ valid over the range $a \leqslant x \leqslant b$. then

(i) $$\int_{\text{all } x} f(x)\,\mathrm{d}x = 1$$ since X is a r.v.

i.e. $$\int_a^b f(x)\,\mathrm{d}x = 1$$

The area under the curve $y = f(x)$ between $x = a$ and $x = b$ is 1.

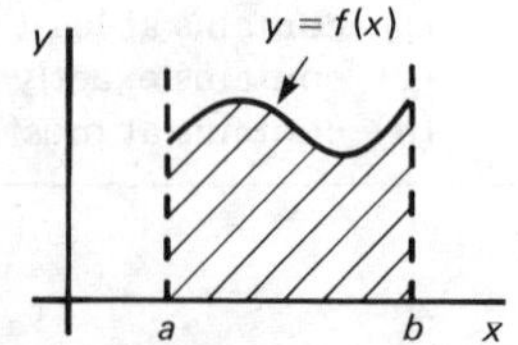

(ii) If $a \leqslant x_1 \leqslant x_2 \leqslant b$ then

$$P(x_1 \leqslant X \leqslant x_2) = \int_{x_1}^{x_2} f(x)\,\mathrm{d}x$$

$P(x_1 \leqslant X \leqslant x_2)$ is given by the area under the curve $y = f(x)$ between $x = x_1$ and $x = x_2$.

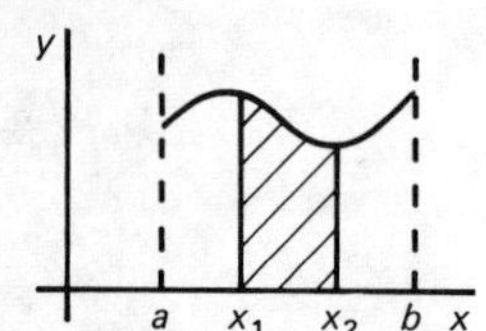

NOTE: in an experimental approach, the area under the histogram represents frequency. In a theoretical approach, the area under the curve $y = f(x)$ represents probability.

Example 5.1 A continuous r.v. has p.d.f. $f(x)$ where $f(x) = kx,\ 0 \leqslant x \leqslant 4$.

(a) Find the value of the constant k,

(b) sketch $y = f(x)$,

(c) find $P(1 \leqslant X \leqslant 2\frac{1}{2})$.

Solution 5.1 (a) Since X is a r.v. $\qquad \displaystyle\int_{\text{all } x} f(x)\,dx = 1$

So
$$\int_0^4 kx\,dx = 1$$
$$\left[\frac{kx^2}{2}\right]_0^4 = 1$$
$$8k = 1$$
$$k = \frac{1}{8}$$

Therefore $f(x) = \frac{1}{8}x,\ 0 \leqslant x \leqslant 4$.

(b) Sketch of $y = f(x)$

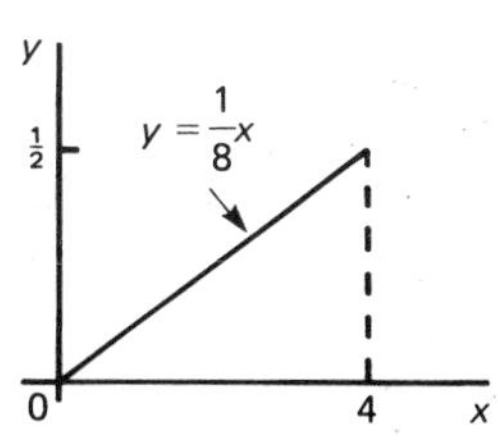

(c)
$$P(1 \leqslant X \leqslant 2\tfrac{1}{2}) = \int_1^{2\frac{1}{2}} \frac{1}{8}x\,dx$$
$$= \left[\frac{x^2}{16}\right]_1^{2\frac{1}{2}}$$
$$= 0.328 \quad (3\text{ S.F.})$$

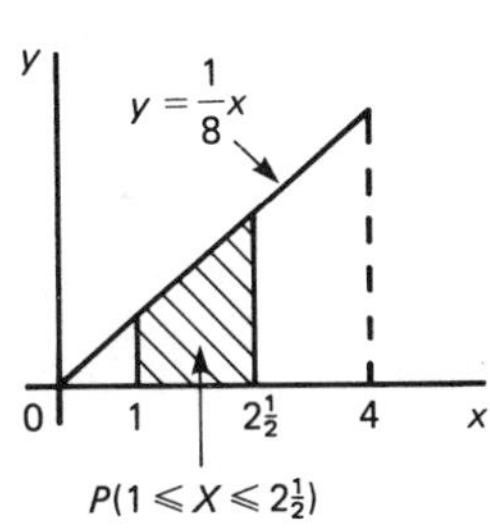

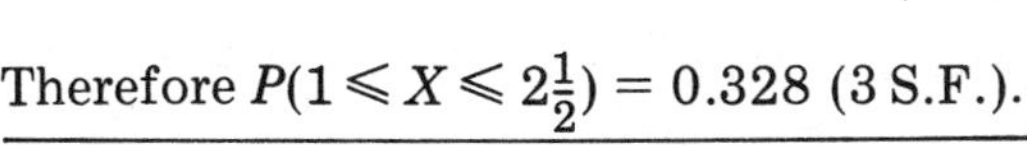
Therefore $P(1 \leqslant X \leqslant 2\frac{1}{2}) = 0.328$ (3 S.F.).

Example 5.2 A continuous r.v. X has p.d.f. $f(x)$ where

$$f(x) = \begin{cases} kx & 0 \leqslant x < 2 \\ k(4-x) & 2 \leqslant x \leqslant 4 \\ 0 & \text{otherwise} \end{cases}$$

(a) Find the value of the constant k.
(b) Sketch $y = f(x)$.
(c) Find $P(\frac{1}{2} \leqslant X \leqslant 2\frac{1}{2})$.

Solution 5.2 (a) Since X is a r.v. $\int_{\text{all } x} f(x)\,dx = 1$

Therefore $\int_0^2 kx\,dx + \int_2^4 k(4-x)\,dx = 1$

$$k\left[\frac{x^2}{2}\right]_0^2 + k\left[4x - \frac{x^2}{2}\right]_2^4 = 1$$

$$2k + k\{16-8-(8-2)\} = 1$$

$$4k = 1$$

$$k = \frac{1}{4}$$

So the p.d.f. for X is given by

$$f(x) = \begin{cases} \frac{1}{4}x & 0 \leqslant x < 2 \\ \frac{1}{4}(4-x) & 2 \leqslant x \leqslant 4 \\ 0 & \text{otherwise} \end{cases}$$

(b) Sketch of $y = f(x)$

NOTE: this is known as a triangular distribution.

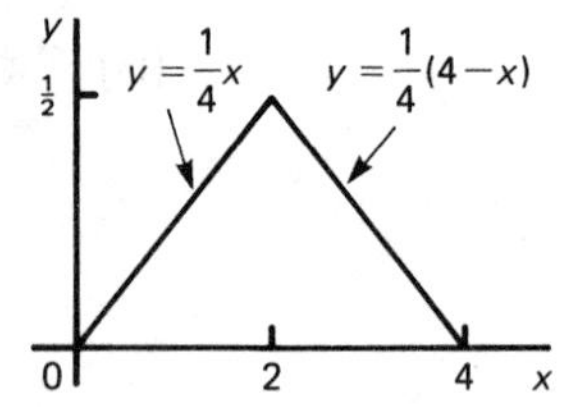

(c) We require $P(\frac{1}{2} \leqslant X \leqslant 2\frac{1}{2})$.

This is given by the shaded areas in the diagram.

NOTE: we find the required area in two stages:

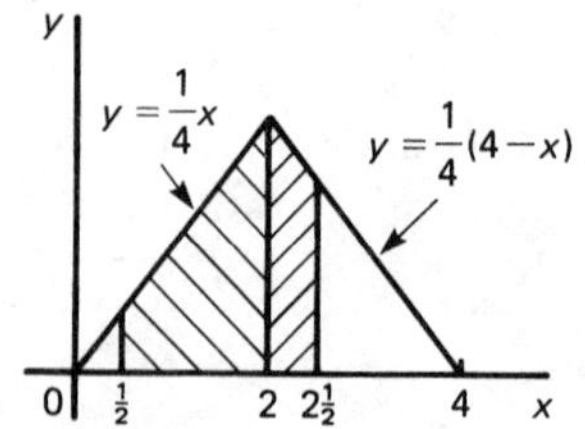

$$P(\tfrac{1}{2} \leqslant X \leqslant 2\tfrac{1}{2}) = P(\tfrac{1}{2} \leqslant X \leqslant 2) + P(2 \leqslant X \leqslant 2\tfrac{1}{2})$$

$$= \int_{\frac{1}{2}}^{2} \tfrac{1}{4}x\,dx + \int_2^{2\frac{1}{2}} \tfrac{1}{4}(4-x)\,dx$$

$$= \frac{1}{4}\left[\frac{x^2}{2}\right]_{\frac{1}{2}}^{2} + \frac{1}{4}\left[4x - \frac{x^2}{2}\right]_2^{2\frac{1}{2}}$$

$$= \frac{1}{4}\left(2 - \frac{1}{8}\right) + \frac{1}{4}\left\{10 - \frac{25}{8} - (8-2)\right\}$$

$$= \frac{15}{32} + \frac{7}{32}$$

$$= \frac{11}{16}$$

Therefore $P(\frac{1}{2} \leqslant X \leqslant 2\frac{1}{2}) = \frac{11}{16}$.

Exercise 5a

The following distributions will be used in Exercises 5c and 5d and it will be useful to refer to these answers. Start each answer on a fresh sheet of paper so that you can add to it later.

1. The continuous r.v. X has p.d.f. $f(x)$ where $f(x) = kx^2$, $0 \leqslant x \leqslant 2$.
   (*a*) Find the value of the constant k.
   (*b*) Sketch $y = f(x)$.
   (*c*) Find $P(X \geqslant 1)$.
   (*d*) Find $P(0.5 \leqslant X \leqslant 1.5)$.

2. The continuous r.v. X has p.d.f. $f(x)$ where $f(x) = k$, $-2 \leqslant x \leqslant 3$. (*a*) Find the value of the constant k. (*b*) Sketch $y = f(x)$. (*c*) Find $P(-1.6 \leqslant X \leqslant 2.1)$.

3. The continuous r.v. X has p.d.f. $f(x)$ where $f(x) = k(4-x)$, $1 \leqslant x \leqslant 3$. (*a*) Find the value of the constant k. (*b*) Sketch $y = f(x)$. (*c*) Find $P(1.2 \leqslant X \leqslant 2.4)$.

4. The continuous r.v. X has p.d.f. $f(x)$ where $f(x) = k(x+2)^2$, $0 \leqslant x \leqslant 2$. (*a*) Find the value of the constant k. (*b*) Sketch $y = f(x)$. (*c*) Find $P(0 \leqslant X \leqslant 1)$ and hence find $P(X > 1)$.

5. The continuous r.v. X has p.d.f. $f(x)$ where $f(x) = kx^3$, $0 \leqslant x \leqslant c$ and $P(X \leqslant \frac{1}{2}) = \frac{1}{16}$. Find the values of the constants c and k and sketch $y = f(x)$.

6. The continuous r.v. X has p.d.f. $f(x)$ where

$$f(x) = \begin{cases} k & 0 \leqslant x \leqslant 2 \\ k(2x-3) & 2 \leqslant x \leqslant 3 \\ 0 & \text{otherwise} \end{cases}$$

   (*a*) Find the value of the constant k.
   (*b*) Sketch $y = f(x)$.
   (*c*) Find $P(X \leqslant 1)$.
   (*d*) Find $P(X > 2.5)$.
   (*e*) Find $P(1 \leqslant X \leqslant 2.3)$.

7. The continuous r.v. X has p.d.f. $f(x)$ where

$$f(x) = \begin{cases} k(x+2)^2 & -2 \leqslant x < 0 \\ 4k & 0 \leqslant x \leqslant 1\frac{1}{3} \\ 0 & \text{otherwise} \end{cases}$$

   (*a*) Find the value of the constant k.
   (*b*) Sketch $y = f(x)$.
   (*c*) Find $P(-1 \leqslant X \leqslant 1)$.
   (*d*) Find $P(X > 1)$.

EXPECTATION

If X is a continuous r.v. with p.d.f. $f(x)$, then the expectation of X is $E(X)$ where

$$E(X) = \int_{\text{all } x} x\, f(x)\, \mathrm{d}x$$

NOTE: $E(X)$ is often denoted by μ and referred to as the mean of X.

Example 5.3 If X is a continuous r.v. with p.d.f. $f(x) = \dfrac{3x^2}{64}$, $0 \leqslant x \leqslant 4$, find $E(X)$.

Solution 5.3 Now

$$E(X) = \int_{\text{all } x} x\,f(x)\,dx$$

$$= \int_0^4 x\left(\frac{3x^2}{64}\right)dx$$

$$= \frac{3}{64}\int_0^4 x^3\,dx$$

$$= \frac{3}{64}\left[\frac{x^4}{4}\right]_0^4$$

$$= 3$$

Therefore $E(X) = 3$.

Example 5.4 If the continuous r.v. X has p.d.f. $f(x) = \frac{3}{4}(3-x)(x-5)$, $3 \leqslant x \leqslant 5$, find $E(X)$.

Solution 5.4

$$E(X) = \int_{\text{all } x} x\,f(x)\,dx$$

$$= \int_3^5 \tfrac{3}{4}x(3-x)(x-5)\,dx$$

$$= \frac{3}{4}\int_3^5 (8x^2 - 15x - x^3)\,dx$$

$$= \frac{3}{4}\left[\frac{8x^3}{3} - \frac{15x^2}{2} - \frac{x^4}{4}\right]_3^5$$

$$= \frac{3}{4}\left(\frac{16}{3}\right)$$

$$= 4$$

Therefore $E(X) = 4$.

NOTE: in this example it would have been advantageous to have drawn a sketch of $y = f(x)$, thus:

From the sketch, we see that there is **symmetry** about the line $x = 4$.

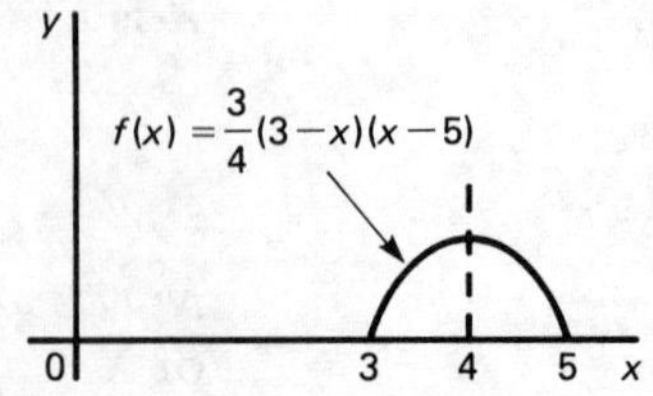

Therefore, by inspection, $E(X) = 4$.

So, whenever possible, draw a sketch of $y = f(x)$ and look for symmetry when finding $E(X)$.

The expectation of any function of X

If $g(x)$ is any function of the continuous r.v. X having p.d.f. $f(x)$, then

$$E[g(X)] = \int_{\text{all } x} g(x)\, f(x)\, dx$$

In particular $$E(X^2) = \int_{\text{all } x} x^2 f(x)\, dx$$

As in the case of the discrete r.v. (see p. 181), the following results hold when X is a continuous r.v.

Result 1 $E(a) = a$

$$\begin{aligned} E(a) &= \int_{\text{all } x} a\, f(x)\, dx \\ &= a \int_{\text{all } x} f(x)\, dx \\ &= a \end{aligned}$$

Result 2 $E(aX) = aE(X)$

$$\begin{aligned} E(aX) &= \int_{\text{all } x} ax\, f(x)\, dx \\ &= a \int_{\text{all } x} x\, f(x)\, dx \\ &= aE(X) \end{aligned}$$

Result 3 $E(aX + b) = aE(X) + b$

Result 4 $E[f_1(X) + f_2(X)] = E[f_1(X)] + E[f_2(X)]$

Example 5.5 The continuous r.v. X has p.d.f. $f(x)$ where $f(x) = \frac{1}{20}(x+3)$, $0 \leqslant x \leqslant 4$.

(a) Find $E(X)$. (b) Verify that $E(2X+5) = 2E(X)+5$.

(c) Find $E(X^2)$. (d) Find $E(X^2 + 2X - 3)$.

Solution 5.5 Sketch of $y = f(x)$

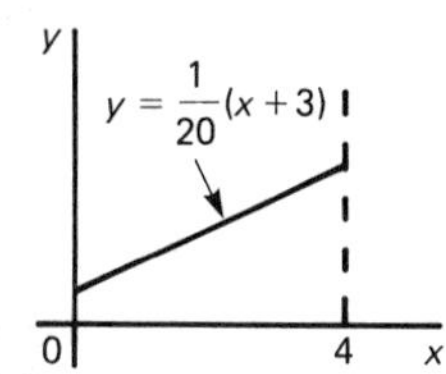

(a) We note from the sketch that there is no symmetry.

Now
$$\begin{aligned} E(X) &= \int_{\text{all } x} x\,f(x)\,dx \\ &= \int_0^4 \frac{1}{20}x(x+3)\,dx \\ &= \frac{1}{20}\int_0^4 (x^2+3x)\,dx \\ &= \frac{1}{20}\left[\frac{x^3}{3}+\frac{3x^2}{2}\right]_0^4 \\ &= \frac{1}{20}\left(\frac{64}{3}+24\right) \\ &= \frac{34}{15} \end{aligned}$$

Therefore $E(X) = \frac{34}{15}$.

(b)
$$\begin{aligned} E(2X+5) &= \int_{\text{all } x} (2x+5)\,f(x)\,dx \\ &= \int_0^4 \frac{1}{20}(2x+5)(x+3)\,dx \\ &= \frac{1}{20}\int_0^4 (2x^2+11x+15)\,dx \\ &= \frac{1}{20}\left[\frac{2x^3}{3}+\frac{11x^2}{2}+15x\right]_0^4 \\ &= \frac{1}{20}\left(\frac{128}{3}+88+60\right) \\ &= \frac{572}{60} \\ &= \frac{143}{15} \end{aligned}$$

Therefore $E(2X+5) = \frac{143}{15}$.

Now
$$\begin{aligned} 2E(X)+5 &= (2)\left(\frac{34}{15}\right)+5 \\ &= \frac{143}{15} \end{aligned}$$

So $E(2X+5) = 2E(X)+5$.

(c)
$$\begin{aligned}
E(X^2) &= \int_{\text{all } x} x^2 f(x)\,dx \\
&= \frac{1}{20}\int_0^4 x^2(x+3)\,dx \\
&= \frac{1}{20}\int_0^4 (x^3+3x^2)\,dx \\
&= \frac{1}{20}\left[\frac{x^4}{4}+x^3\right]_0^4 \\
&= \frac{1}{20}(64+64) \\
&= \frac{32}{5}
\end{aligned}$$

Therefore $E(X^2) = \frac{32}{5}$.

(d)
$$\begin{aligned}
E(X^2+2X-3) &= E(X^2)+E(2X)-E(3) \\
&= E(X^2)+2E(X)-3 \\
&= \frac{32}{5}+(2)\left(\frac{34}{15}\right)-3 \\
&= \frac{119}{15}
\end{aligned}$$

Therefore $E(X^2+2X-3) = \frac{119}{15}$.

Example 5.6 The continuous r.v. X has p.d.f. $f(x)$ where

$$f(x) = \begin{cases} \frac{6}{7}x & 0 \leqslant x \leqslant 1 \\ \frac{6}{7}x(2-x) & 1 \leqslant x \leqslant 2 \\ 0 & \text{otherwise} \end{cases}$$

Find (a) $E(X)$, (b) $E(X^2)$

Solution 5.6 From the sketch, we see that there is no symmetry.

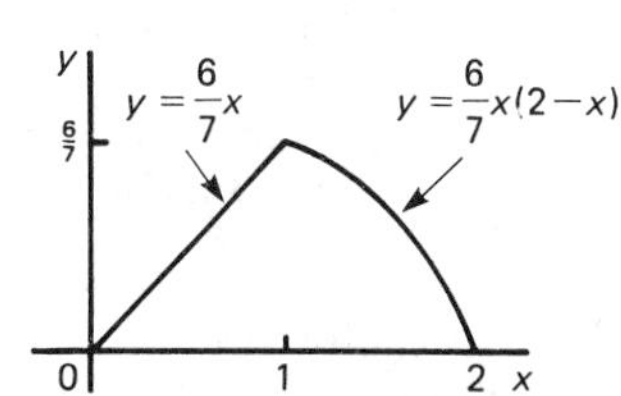

(a)
$$\begin{aligned}
E(X) &= \int_{\text{all } x} x\,f(x)\,dx \\
&= \int_0^1 \frac{6}{7}x^2\,dx + \int_1^2 \frac{6}{7}x^2(2-x)\,dx
\end{aligned}$$

$$= \frac{6}{7}\left[\frac{x^3}{3}\right]_0^1 + \frac{6}{7}\left[\frac{2x^3}{3} - \frac{x^4}{4}\right]_1^2$$

$$= \frac{6}{7}\left(\frac{1}{3}\right) + \frac{6}{7}\left\{\frac{16}{3} - 4 - \left(\frac{2}{3} - \frac{1}{4}\right)\right\}$$

$$= \frac{6}{7}\left(\frac{5}{4}\right)$$

$$= \frac{15}{14}$$

Therefore $E(X) = \frac{15}{14}$.

(b) $$E(X^2) = \int_{\text{all } x} x^2 f(x)\,dx$$

$$= \int_0^1 \tfrac{6}{7}x^3\,dx + \int_1^2 \tfrac{6}{7}x^3(2-x)\,dx$$

$$= \frac{6}{7}\left[\frac{x^4}{4}\right]_0^1 + \frac{6}{7}\left[\frac{x^4}{2} - \frac{x^5}{5}\right]_1^2$$

$$= \frac{6}{7}\left(\frac{1}{4}\right) + \frac{6}{7}\left\{8 - \frac{32}{5} - \left(\frac{1}{2} - \frac{1}{5}\right)\right\}$$

$$= \frac{6}{7}\left(\frac{31}{20}\right)$$

$$= \frac{93}{70}$$

Therefore $E(X^2) = \frac{93}{70}$.

Exercise 5b

1. The continuous r.v. X has p.d.f. $f(x)$ where $f(x) = 3x^2$, $0 \leqslant x \leqslant 1$. (*a*) Find $E(X)$. (*b*) Find $E(X^2)$. (*c*) Verify that $E(3X-1) = 3E(X) - 1$. (*d*) Find $E(2X^2 + 3X + 3)$.

2. The continuous r.v. X has p.d.f. $f(x)$ where $f(x) = \frac{3}{4}x(2-x)$, $0 \leqslant x \leqslant 2$. (*a*) Find $E(X)$. (*b*) Find $E(X^2)$.

3. The continuous r.v. X has p.d.f. $f(x)$ where $f(x) = \frac{1}{18}(6-x)$, $0 \leqslant x \leqslant 6$. Find (*a*) $E(X)$, (*b*) $E(2X-1)$, (*c*) $E(X^2)$, (*d*) $E(X^2 - 4X + 3)$.

4. The continuous r.v. X has p.d.f. $f(x)$ where $f(x) = kx^3$, $0 \leqslant x \leqslant 2$. (*a*) Find the value of the constant k. (*b*) Find $\mu = E(X)$. (*c*) Find $E(3X)$. (*d*) Find $E(X^2 - 4X + 3)$.

5. The continuous r.v. X has p.d.f. $f(x)$ where

$$f(x) = \begin{cases} \frac{3}{8} & \frac{2}{3} \leqslant x < 2 \\ \frac{3}{32}x(4-x) & 2 \leqslant x \leqslant 4 \\ 0 & \text{otherwise} \end{cases}$$

(*a*) Find $E(X)$. (*b*) Find $E(5X-2)$. (*c*) Find $E(X^2)$.

6. The continuous r.v. X has p.d.f. $f(x)$ where

$$f(x) = \begin{cases} kx & 0 \leqslant x < 1 \\ k & 1 \leqslant x < 3 \\ k(4-x) & 3 \leqslant x \leqslant 4 \\ 0 & \text{otherwise} \end{cases}$$

(*a*) Find k. (*b*) Calculate $E(X)$. (*c*) Now sketch $y = f(x)$. (*d*) Check $E(X)$ from the sketch. (*e*) Find $E(X^2)$.

7. In a game a wooden block is propelled with a stick across a flat deck. On each attempt the distance, x metre, reached by the block lies between 0 and 10 m, and the variation is modelled by the probability density function

$$\phi(x) = 0.0012x^2(10-x).$$

Calculate the mean distance reached by the block. (SMP)

8. The continuous random variable X has the probability density function f given by $f(x) = kx$, $5 < x < 10$, $f(x) = 0$ otherwise.

(*a*) Find the value of k. (*b*) Find the expected value of X. (*c*) Find the probability that $X > 8$.

The annual income from money invested in a Unit Trust Fund is X per cent of the amount invested, where X has the above distribution. Suppose that you have a sum of money to invest and that you are prepared to leave the money invested over a period of several years.

State, with your reasons, whether you would invest in the Unit Trust Fund or in a Money Bond offering a guaranteed annual income of 8 per cent on the money invested. (JMB)

9. The lifetime X in *tens of hours* of a torch battery is a random variable with probability density function

$$f(x) = \begin{cases} \frac{3}{4}\{1-(x-2)^2\} & 1 \leqslant x \leqslant 3, \\ 0 & \text{otherwise} \end{cases}$$

Calculate the mean of X.

A torch runs on two batteries, both of which have to be working for the torch to function. If two new batteries are put in the torch, what is the probability that the torch will function for at least 22 hours, on the assumption that the lifetimes of the batteries are independent? (O & C)

10. A random variable X has a probability density function f given by

$$f(x) = \begin{cases} cx(5-x) & 0 \leqslant x \leqslant 5 \\ 0 & \text{otherwise} \end{cases}$$

Show that $c = 6/125$ and find the mean of X.

The lifetime X in years of an electric light bulb has this distribution. Given that a lamp standard is fitted with two such new bulbs and that their failures are independent, find the probability that neither bulb fails in the first year and the probability that exactly one bulb fails within two years. (MEI)

11. The mass X kg of a particular substance produced per hour in a chemical process is a continuous random variable whose probability density function is given by

$$f(x) = 3x^2/32 \qquad 0 \leqslant x < 2$$
$$f(x) = 3(6-x)/32 \qquad 2 \leqslant x \leqslant 6$$
$$f(x) = 0 \qquad \text{otherwise}$$

(*a*) Find the mean mass produced per hour.
(*b*) The substance produced is sold at £2 per kg and the total running cost of the process is £1 per hour. Find the expected profit per hour and the probability that in an hour the profit will exceed £7. (JMB)

12. A continuous random variable X has the probability density function f defined by

$$f(x) = \frac{c}{3}x \qquad 0 \leqslant x < 3$$
$$f(x) = c \qquad 3 \leqslant x \leqslant 4$$
$$f(x) = 0 \qquad \text{otherwise}$$

where c is a positive constant. Find
(i) the value of c,
(ii) the mean of X,
(iii) the value, a, for there to be a probability of 0.85 that a randomly observed value of X will exceed a. (JMB)

VARIANCE

For a random variable X,

$$\text{Var}(X) = E(X-\mu)^2 \quad \text{where} \quad \mu = E(X)$$

As in the discrete case (see p. 183) the formula can be written:

$$\text{Var}(X) = E(X^2) - E^2(X)$$

or $\quad \text{Var}(X) = E(X^2) - \mu^2$

If X is a continuous r.v. with p.d.f. $f(x)$, then

$$\text{Var}(X) = \int_{\text{all } x} x^2 f(x)\,\text{d}x - \mu^2$$

where $\quad \mu = E(X) = \int_{\text{all } x} x\,f(x)\,\text{d}x$

The standard deviation of X is often written as σ, so $\sigma = \sqrt{\text{Var}(X)}$.

As in the discrete case (see p. 186), the following results also hold when X is a continuous r.v.:

(1) $\text{Var}(a) = 0$

(2) $\text{Var}(aX) = a^2\text{Var}(X)$

(3) $\text{Var}(aX + b) = a^2\text{Var}(X)$ where a and b are any constants.

Example 5.7 The continuous r.v. X has p.d.f. $f(x)$ where $f(x) = \frac{1}{8}x,\ 0 \leqslant x \leqslant 4$.

Find **(a)** $E(X)$, **(b)** $E(X^2)$, **(c)** $\text{Var}(X)$, **(d)** the standard deviation σ of X, **(e)** $\text{Var}(3X + 2)$.

Solution 5.7 From the sketch of $y = f(x)$ we note that there is no symmetry.

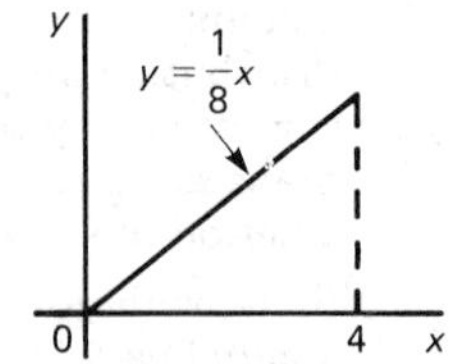

(a)
$$\begin{aligned} E(X) &= \int_{\text{all } x} x\,f(x)\,\text{d}x \\ &= \int_0^4 \tfrac{1}{8}x^2\,\text{d}x \\ &= \frac{1}{8}\left[\frac{x^3}{3}\right]_0^4 \\ &= \frac{8}{3} \end{aligned}$$

Therefore $E(X) = \frac{8}{3}$.

(b)
$$\begin{aligned} E(X^2) &= \int_{\text{all } x} x^2 f(x)\,\text{d}x \\ &= \int_0^4 \tfrac{1}{8}x^3\,\text{d}x \end{aligned}$$

$$= \frac{1}{8}\left[\frac{x^4}{4}\right]_0^4$$

$$= \frac{1}{8}(64)$$

$$= 8$$

Therefore $E(X^2) = 8$.

(c) $$\text{Var}(X) = E(X^2) - E^2(X)$$

$$= 8 - \left(\frac{8}{3}\right)^2$$

$$= \frac{8}{9}$$

Therefore $\text{Var}(X) = \frac{8}{9}$.

(d) Standard deviation, $$\sigma = \sqrt{\text{Var}(X)}$$

$$= \sqrt{\frac{8}{9}}$$

$$= \frac{2\sqrt{2}}{3}$$

Therefore $\sigma = \dfrac{2\sqrt{2}}{3}$.

(e) $$\text{Var}(3X+2) = 9\text{Var}(X)$$

$$= 9\left(\frac{8}{9}\right)$$

$$= 8$$

Therefore $\text{Var}(3X+2) = 8$.

Example 5.8 The continuous random variable X has p.d.f. given by $f(x)$ where

$$f(x) = \begin{cases} \frac{1}{27}x^2 & 0 \leqslant x < 3 \\ \frac{1}{3} & 3 \leqslant x \leqslant 5 \\ 0 & \text{otherwise} \end{cases}$$

(a) Sketch $y = f(x)$. (b) Find $E(X)$. (c) Find $E(X^2)$. (d) Find the standard deviation σ of X.

Solution 5.8 (a) Sketch of $y = f(x)$

NOTE: there is no symmetry.

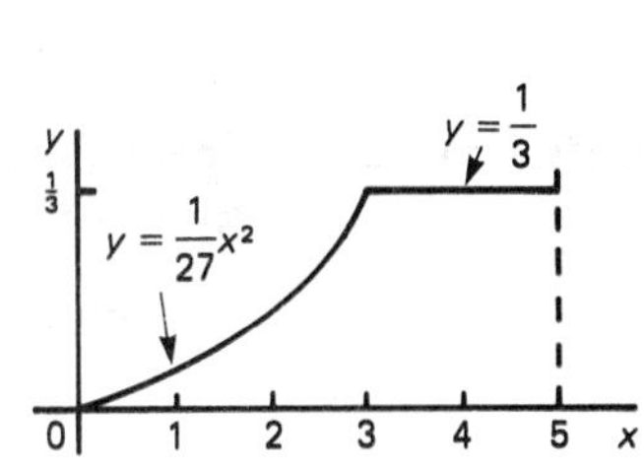

(b)
$$\begin{aligned} E(X) &= \int_{\text{all } x} x\,f(x)\,dx \\ &= \int_0^3 \tfrac{1}{27}x^3\,dx + \int_3^5 \tfrac{1}{3}x\,dx \\ &= \frac{1}{27}\left[\frac{x^4}{4}\right]_0^3 + \left[\frac{x^2}{6}\right]_3^5 \\ &= \frac{3}{4} + \frac{8}{3} \\ &= \frac{41}{12} \end{aligned}$$

Therefore $E(X) = \frac{41}{12}$.

(c)
$$\begin{aligned} E(X^2) &= \int_{\text{all } x} x^2 f(x)\,dx \\ &= \int_0^3 \tfrac{1}{27}x^4\,dx + \int_3^5 \tfrac{1}{3}x^2\,dx \\ &= \frac{1}{27}\left[\frac{x^5}{5}\right]_0^3 + \frac{1}{3}\left[\frac{x^3}{3}\right]_3^5 \\ &= \frac{9}{5} + \frac{98}{9} \\ &= \frac{571}{45} \end{aligned}$$

Therefore $E(X^2) = \frac{571}{45}$.

(d)
$$\begin{aligned} \sigma^2 &= \text{Var}(X) \\ &= E(X^2) - E^2(X) \\ &= \tfrac{571}{45} - (\tfrac{41}{12})^2 \\ &= 1.0153 \end{aligned}$$

Therefore $\sigma = 1.008$ (3 d.p.)

The standard deviation of X is 1.008 (3 d.p.)

Example 5.9 The continuous r.v. X has p.d.f. $f(x)$ where $f(x) = \frac{3}{4}(1+x^2)$, $0 \leqslant x \leqslant 1$. If $E(X) = \mu$ and $\text{Var}(X) = \sigma^2$, find $P(|X - \mu| < \sigma)$.

Solution 5.9 From the sketch of $y = f(x)$ we see that there is no symmetry.

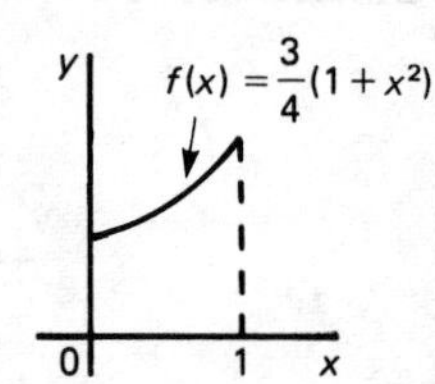

We will need to find $\mu = E(X)$.

Now
$$\begin{aligned} E(X) &= \int_{\text{all } x} x\,f(x)\,\mathrm{d}x \\ &= \int_0^1 \tfrac{3}{4}x(1+x^2)\,\mathrm{d}x \;=\; \frac{3}{4}\int_0^1 (x+x^3)\,\mathrm{d}x \\ &= \frac{3}{4}\left[\frac{x^2}{2}+\frac{x^4}{4}\right]_0^1 \\ &= \frac{3}{4}\left(\frac{3}{4}\right) \\ &= \frac{9}{16} \end{aligned}$$

Therefore $\mu = \frac{9}{16} = 0.5625$.

To find $\sigma^2 = \mathrm{Var}(X)$, first consider $E(X^2)$.

$$\begin{aligned} E(X^2) &= \int_{\text{all } x} x^2 f(x)\,\mathrm{d}x \\ &= \int_0^1 \tfrac{3}{4}x^2(1+x^2)\,\mathrm{d}x \\ &= \frac{3}{4}\int_0^1 (x^2+x^4)\,\mathrm{d}x \\ &= \frac{3}{4}\left[\frac{x^3}{3}+\frac{x^5}{5}\right]_0^1 \\ &= \frac{3}{4}\left(\frac{8}{15}\right) \\ &= \frac{2}{5} \end{aligned}$$

So
$$\begin{aligned} \mathrm{Var}(X) &= E(X^2)-\mu^2 \\ &= \frac{2}{5}-\left(\frac{9}{16}\right)^2 \\ &= 0.0836 \quad (3\text{ S.F.}) \end{aligned}$$

and
$$\begin{aligned} \sigma &= \sqrt{\mathrm{Var}(X)} \\ &= 0.289 \quad (3\text{ S.F.}) \end{aligned}$$

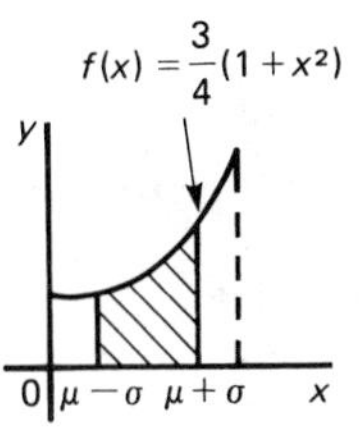

We require
$$\begin{aligned} P(|X-\mu|<\sigma) &= P(-\sigma<X-\mu<\sigma) \\ &= P(\mu-\sigma<X<\mu+\sigma) \end{aligned}$$

$$= P(0.5625 - 0.289 < X < 0.5625 + 0.289)$$
$$= P(0.2735 < X < 0.8515)$$
$$= \int_{0.2735}^{0.8515} \tfrac{3}{4}(1 + x^2)\, dx$$
$$= \frac{3}{4}\left[x + \frac{x^3}{3}\right]_{0.2735}^{0.8515}$$
$$= \frac{3}{4}\left\{0.8515 + \frac{(0.8515)^3}{3} - \left(0.2735 + \frac{(0.2735)^3}{3}\right)\right\}$$
$$= 0.583 \quad (3\text{ S.F.})$$

Therefore $P(|X - \mu| < \sigma) = 0.583$ (3 S.F.).

THE MODE

The mode is the value of X for which $f(x)$ is greatest, in the given range of X. It is usually necessary to draw a sketch of $y = f(x)$ and this will give an idea of the location of the mode.

For some probability density functions it is possible to determine the mode by finding the maximum point on the curve $y = f(x)$ from the relationship $f'(x) = 0$, where $f'(x) = \dfrac{d}{dx}f(x)$.

Example 5.10 The continuous r.v. X has p.d.f. $f(x)$ where $f(x) = \frac{3}{80}(2 + x)(4 - x)$, $0 \leqslant x \leqslant 4$. (a) Sketch $y = f(x)$. (b) Find the mode.

Solution 5.10 (a) Sketch of $y = f(x)$.

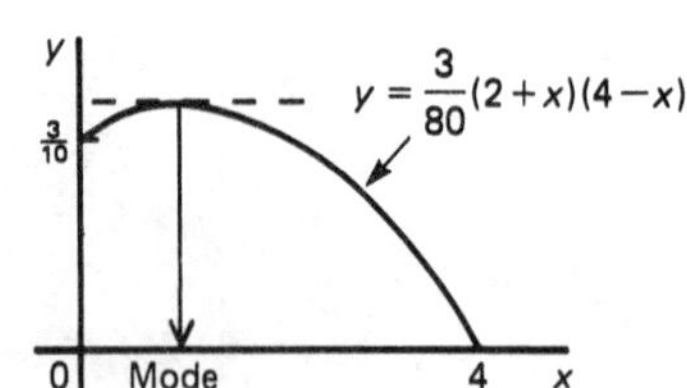

(b)
$$f(x) = \frac{3}{80}(2 + x)(4 - x) = \frac{3}{80}(8 + 2x - x^2)$$
$$f'(x) = \frac{3}{80}(2 - 2x)$$

Now $f'(x) = 0$ when $2 - 2x = 0$

$$x = 1$$

To check that this is a maximum, consider $f''(x) = \frac{3}{80}(-2) < 0$.

i.e. $f''(x)$ is negative for all values of x, therefore there is a maximum at $x = 1$.

So the mode is 1.

Exercise 5c

For each of the following probability density functions of the continuous r.v. X, find (*a*) $E(X)$, (*b*) $E(X^2)$, (*c*) $\text{Var}(X)$, (*d*) the standard deviation of X. It is assumed that the value of the function is zero outside the range(s) stated. Do not forget to look for symmetry when considering $E(X)$.

NOTE: These functions were given in Exercise 5a and you may wish to refer to your previous sketches. You will need them again for Exercise 5d.

1. $f(x) = \frac{3}{8}x^2 \qquad 0 \leqslant x \leqslant 2$
2. $f(x) = \frac{1}{5} \qquad -2 \leqslant x \leqslant 3$
3. $f(x) = \frac{1}{4}(4-x) \qquad 1 \leqslant x \leqslant 3$
4. $f(x) = \frac{3}{56}(x+2)^2 \qquad 0 \leqslant x \leqslant 2$
5. $f(x) = 4x^3 \qquad 0 \leqslant x \leqslant 1$
6. $f(x) = \begin{cases} \frac{1}{4} & 0 \leqslant x \leqslant 2 \\ \frac{1}{4}(2x-3) & 2 \leqslant x \leqslant 3 \end{cases}$
7. $f(x) = \begin{cases} \frac{1}{8}(x+2)^2 & -2 \leqslant x \leqslant 0 \\ \frac{1}{2} & 0 \leqslant x \leqslant 1\frac{1}{3} \end{cases}$
8. A continuous r.v. X has p.d.f. $f(x) = kx^2$, $0 \leqslant x \leqslant 4$.
   (*a*) Find the value of k, and sketch $y = f(x)$.
   (*b*) Find $E(X)$ and $\text{Var}(X)$.
   (*c*) Find $P(1 < x < 2)$.
9. A continuous r.v. X has p.d.f. $f(x)$ where

$$f(x) = \begin{cases} kx & 0 \leqslant x < 1 \\ k(2-x) & 1 \leqslant x \leqslant 2 \\ 0 & \text{otherwise} \end{cases}$$

   Find (*a*) the value of the constant k, (*b*) $E(X)$, (*c*) $\text{Var}(X)$, (*d*) $P(\frac{3}{4} \leqslant X \leqslant 1\frac{1}{2})$, (*e*) the mode.

CUMULATIVE DISTRIBUTION FUNCTION, $F(x)$

When considering a frequency distribution the corresponding cumulative frequencies were obtained by summing all the frequencies up to a particular value.

In the same way, if X is a continuous random variable with p.d.f. $f(x)$ defined for $a \leqslant x \leqslant b$, then the cumulative distribution function is given by $F(t)$ where

$$F(t) = P(X \leqslant t) = \int_a^t f(x)\,dx$$

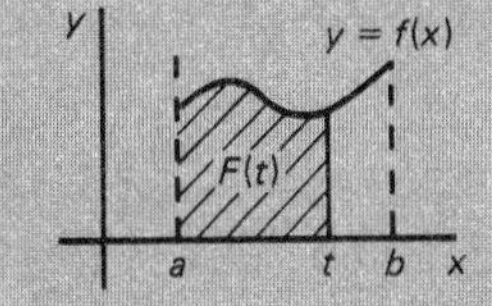

NOTE: (1) $F(b) = \int_a^b f(x)\,dx = 1$.

(2) If $f(x)$ is valid for $-\infty \leqslant x \leqslant \infty$ then $F(t) = \int_{-\infty}^t f(x)\,dx$, where the interval is taken over all values of $x \leqslant t$.

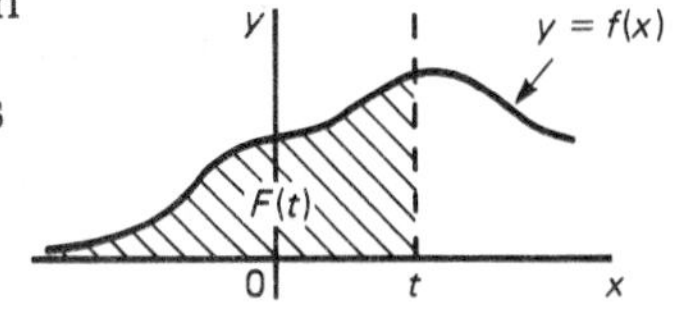

(3) The cumulative distribution function is sometimes known just as the distribution function.

USING $F(x)$ TO FIND $P(x_1 \leqslant X \leqslant x_2)$

The cumulative distribution function can be used to find $P(x_1 \leqslant X \leqslant x_2)$ as follows:

$$P(X \leqslant x_1) = F(x_1)$$
$$P(X \leqslant x_2) = F(x_2)$$

So
$$P(x_1 \leqslant X \leqslant x_2) = F(x_2) - F(x_1)$$

The median

The median splits the area under the curve $y = f(x)$ into two halves. So, if the value of the median is m,

$$\int_a^m f(x)\,dx = 0.5$$

i.e.
$$F(m) = 0.5$$

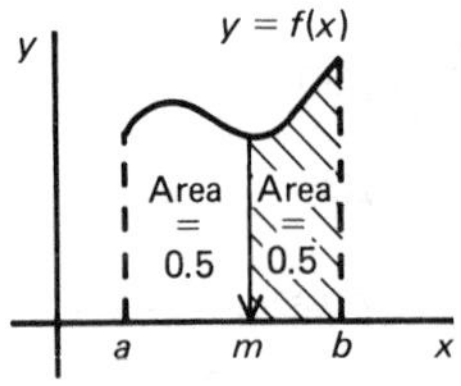

Example 5.11 If X is a continuous r.v. with p.d.f. $f(x) = \frac{1}{8}x$, $0 \leqslant x \leqslant 4$, **(a)** find the cumulative distribution function $F(x)$ and sketch $y = F(x)$, **(b)** find the median m, **(c)** find $P(0.3 \leqslant X \leqslant 1.8)$.

Solution 5.11 (a) Now

$$F(t) = \int_0^t \tfrac{1}{8}x\,dx = \left[\frac{x^2}{16}\right]_0^t = \frac{t^2}{16}$$

So that
$$F(t) = \frac{t^2}{16}, \qquad 0 \leqslant t \leqslant 4$$

NOTE: (1) $F(4) = \dfrac{4^2}{16} = 1$ (as expected).

(2) $F(t) = 1,\ t \geqslant 4$.

(3) we usually calculate the answer in terms of t and then write out the cumulative distribution in terms of x as follows:

$$F(x) = \begin{cases} 0 & x \leqslant 0 \\ \dfrac{x^2}{16} & 0 \leqslant x \leqslant 4 \\ 1 & x \geqslant 4 \end{cases}$$

Sketch of $y = F(x)$

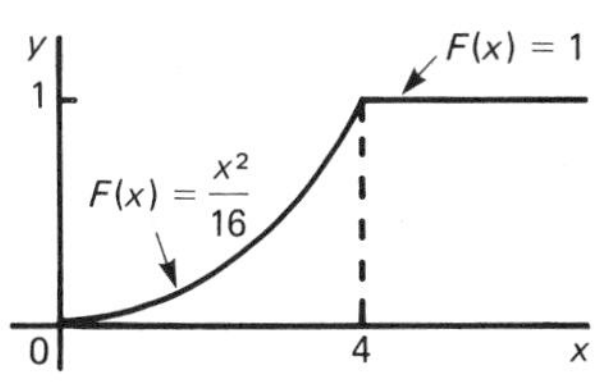

(**b**) For the median m,

$$F(m) = 0.5$$

i.e.

$$\frac{m^2}{16} = 0.5$$

$$m^2 = 8$$

$$m = 2.83 \quad (2\text{ d.p.})$$

The median $m = 2.83$ (2 d.p.).

NOTE: $0 < m < 4$, so m cannot be negative.

(**c**) $P(0.3 \leqslant x \leqslant 1.8) = F(1.8) - F(0.3)$.

Now
$$F(1.8) = \frac{1.8^2}{16} = 0.2025$$

and
$$F(0.3) = \frac{0.3^2}{16} = 0.005\,625$$

Therefore
$$P(0.3 \leqslant x \leqslant 1.8) = 0.2025 - 0.005\,625 = 0.196\,875 = 0.197 \quad (3\text{ d.p.})$$

So $P(0.3 \leqslant x \leqslant 1.8) = 0.197$ (3 d.p.).

Example 5.12 X is a continuous r.v. with p.d.f. $f(x)$ where

$$f(x) = \begin{cases} \dfrac{x}{3} & 0 \leqslant x \leqslant 2 \\ -\dfrac{2x}{3} + 2 & 2 \leqslant x \leqslant 3 \\ 0 & \text{otherwise} \end{cases}$$

(**a**) Sketch $y = f(x)$. (**b**) Find the cumulative distribution function $F(x)$. (**c**) Sketch $y = F(x)$. (**d**) Find $P(1 \leqslant X \leqslant 2.5)$. (**e**) Find the median, m.

Solution 5.12 (a) Sketch of $y = f(x)$.

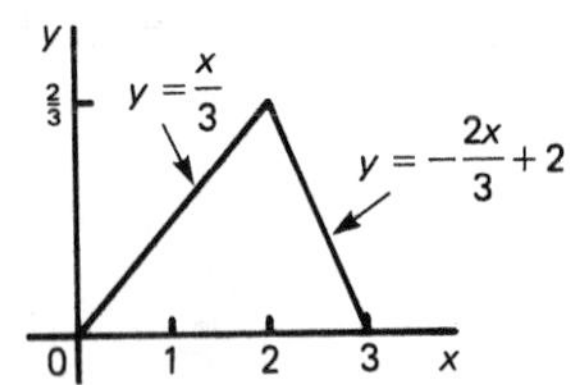

(b) Now $F(t) = \int_0^t f(x)\,dx$.

But, as $f(x)$ is given in two parts, we must find $F(x)$ in two stages:

Consider $0 \leqslant t \leqslant 2$

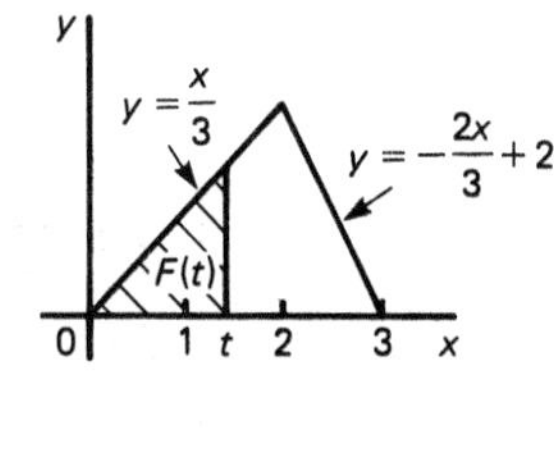

$$F(t) = \int_0^t \frac{x}{3}\,dx = \left[\frac{x^2}{6}\right]_0^t = \frac{t^2}{6}$$

So, for $0 \leqslant x \leqslant 2$, $F(x) = \dfrac{x^2}{6}$.

NOTE: $F(2) = \frac{4}{6} = \frac{2}{3}$.

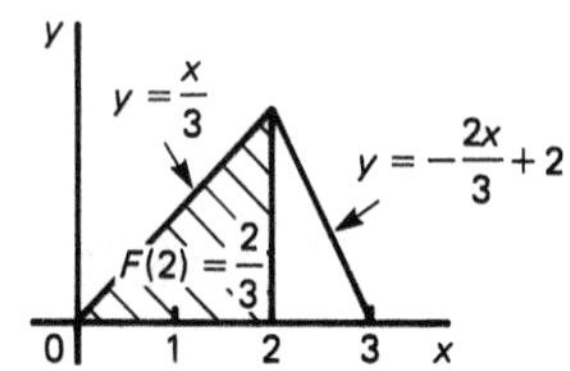

Now, for t in the range $2 \leqslant t \leqslant 3$ we see from the diagram that

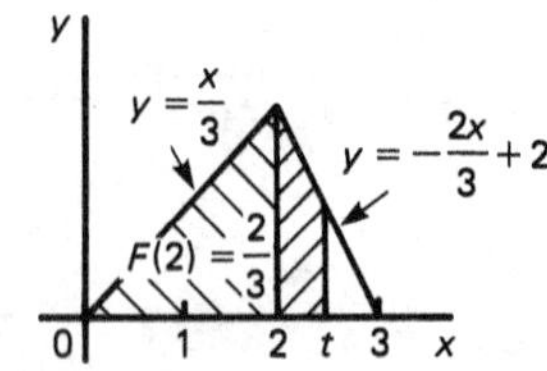

$F(t) = F(2) +$ (area under the curve $y = -\dfrac{2x}{3} + 2$ between 2 and t)

So

$$F(t) = F(2) + \int_2^t \left(-\frac{2x}{3} + 2\right) dx$$

$$= F(2) + \left[-\frac{x^2}{3} + 2x\right]_2^t$$

$$= \frac{2}{3} + \left\{-\frac{t^2}{3} + 2t - \left(-\frac{4}{3} + 4\right)\right\}$$

$$= -\frac{t^2}{3} + 2t - 2 \qquad 2 \leqslant t \leqslant 3$$

Now check the value of $F(3)$

$$F(3) = -\frac{9}{3}+6-2 = 1, \quad \text{as required}$$

For any value of $t \geqslant 3$, $F(t) = 1$.

Writing the answer in terms of x, we have

$$F(x) = \begin{cases} \dfrac{x^2}{6} & 0 \leqslant x \leqslant 2 \\ -\dfrac{x^2}{3}+2x-2 & 2 \leqslant x \leqslant 3 \\ 1 & x \geqslant 3 \end{cases}$$

(c) Sketch of $y = F(x)$.

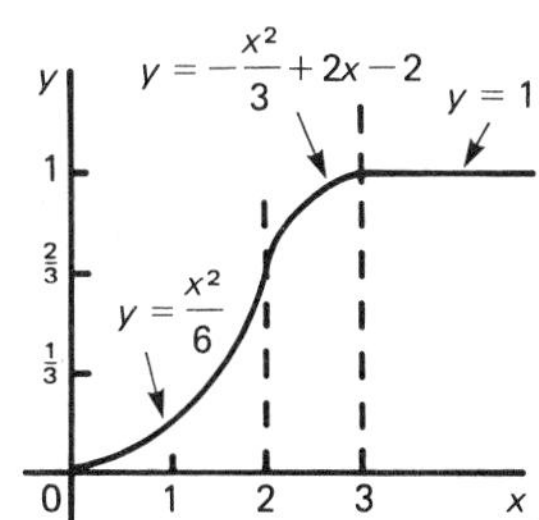

(d) $P(1 \leqslant X \leqslant 2.5) = F(2.5)-F(1)$.

$$\text{Now} \quad F(2.5) = -\frac{x^2}{3}+2x-2 \quad (\text{as } 2.5 \text{ is in the range } 2 \leqslant x \leqslant 3)$$

$$= -\frac{(2.5)^2}{3}+2(2.5)-2$$

$$= \frac{11}{12}$$

$$F(1) = \frac{x^2}{6} \quad (\text{as } 1 \text{ is in the range } 0 \leqslant x \leqslant 2)$$

$$= \frac{1}{6}$$

$$\text{Therefore} \quad P(1 \leqslant X \leqslant 2.5) = F(2.5)-F(1)$$

$$= \frac{11}{12}-\frac{1}{6}$$

$$= 0.75$$

So $P(1 \leqslant X \leqslant 2.5) = 0.75$.

(e) $F(m) = 0.5$, where m is the median.

Now $F(2) = \frac{2}{3}$, so the median must lie in the range $0 \leqslant x \leqslant 2$.

$$\text{Therefore} \quad F(m) = \frac{m^2}{6}$$

So
$$\frac{m^2}{6} = 0.5$$
$$m^2 = 3$$
$$m = 1.73 \quad (2\text{ d.p.})$$

The median m is 1.73 (2 d.p.).

Example 5.13 A continuous r.v. X takes values in the range $0 \leqslant x \leqslant 1$ and has p.d.f.

$$f(x) = \begin{cases} 3.75x + 0.1 & 0 \leqslant x \leqslant 0.4 \\ 1.6 & 0.4 \leqslant x \leqslant 0.6 \\ 3.85 - 3.75x & 0.6 \leqslant x \leqslant 1 \end{cases}$$

(a) Sketch $y = f(x)$. **(b)** Find the mean μ. **(c)** Find the cumulative distribution function, $F(x)$ and sketch $y = F(x)$. **(d)** Find $P(|X - \mu| \leqslant 0.2)$.

Solution 5.13 (a) Sketch of $y = f(x)$.

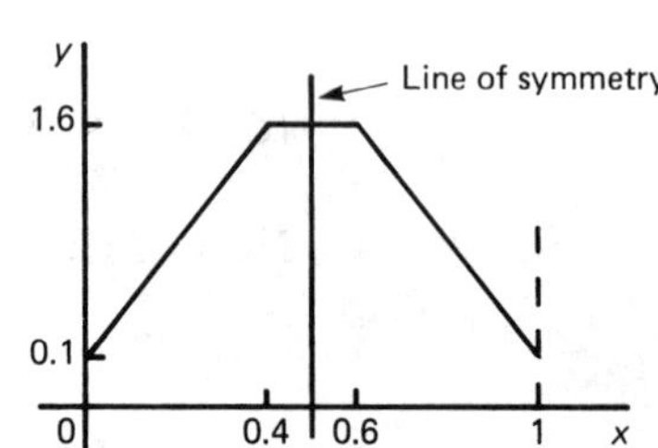

(b) By symmetry, $\mu = E(X) = 0.5$.

(c) We must consider $F(x)$ in three stages:

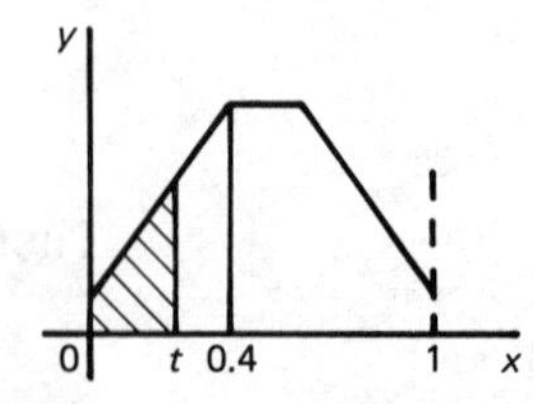

$0 \leqslant t \leqslant 0.4$

$$F(t) = \int_0^t (3.75x + 0.1)\,dx$$
$$= \left[3.75\frac{x^2}{2} + 0.1x\right]_0^t$$
$$= 1.875t^2 + 0.1t$$

Now

$$F(0.4) = (1.875)(0.4)^2 + (0.1)(0.4)$$
$$= 0.34$$

$0.4 \leqslant t \leqslant 0.6$

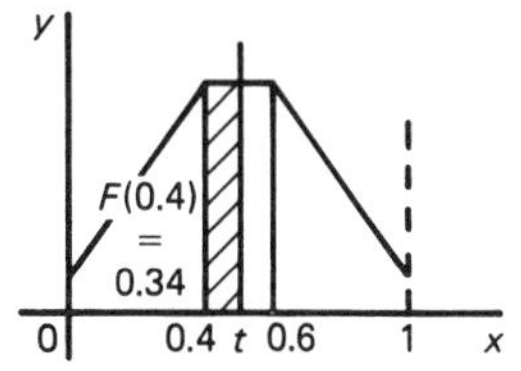

$$\begin{aligned} F(t) &= F(0.4) + \int_{0.4}^{t} 1.6 \, dx \\ &= F(0.4) + [1.6x]_{0.4}^{t} \\ &= 0.34 + 1.6t - 0.64 \\ &= 1.6t - 0.3 \end{aligned}$$

Now $F(0.6) = (1.6)(0.6) - 0.3$

$= 0.66$

$0.6 \leqslant x \leqslant 1$

$$\begin{aligned} F(t) &= F(0.6) + \int_{0.6}^{t} (3.85 - 3.75x) \, dx \\ &= F(0.6) + \left[3.85x - 3.75 \frac{x^2}{2}\right]_{0.6}^{t} \\ &= 0.66 + 3.85t - 1.875t^2 - 2.31 + 0.675 \\ &= 3.85t - 1.875t^2 - 0.975 \end{aligned}$$

Check $F(1)$:

$$\begin{aligned} F(1) &= 3.85 - 1.875 - 0.975 \\ &= 1 \text{ as required} \end{aligned}$$

Writing the answer in terms of x we have:

$$F(x) = \begin{cases} 0 & x \leqslant 0 \\ 1.875x^2 + 0.1x & 0 \leqslant x \leqslant 0.4 \\ 1.6x - 0.3 & 0.4 \leqslant x \leqslant 0.6 \\ 3.85x - 1.875x^2 - 0.975 & 0.6 \leqslant x \leqslant 1 \\ 1 & x \geqslant 1 \end{cases}$$

Sketch of $y = F(x)$.

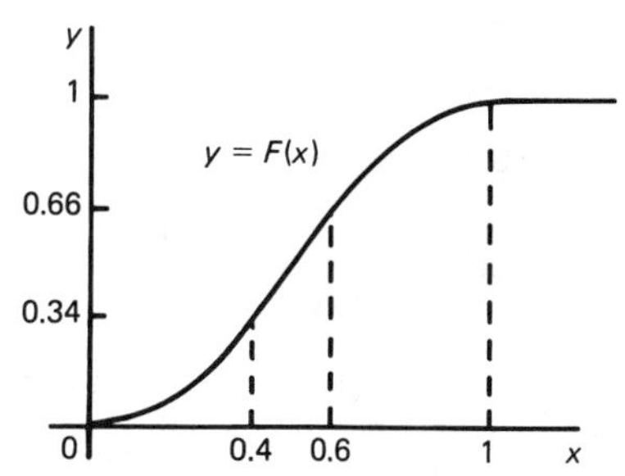

(d) $$\begin{aligned} P(|X - \mu| \leqslant 0.2) &= P(-0.2 \leqslant X - \mu \leqslant 0.2) \\ &= P(-0.2 \leqslant X - 0.5 \leqslant 0.2) \\ &= P(0.3 \leqslant X \leqslant 0.7) \\ &= F(0.7) - F(0.3) \end{aligned}$$

Now 0.7 lies in the range $0.6 \leqslant x \leqslant 1$, so

$$\begin{aligned} F(0.7) &= 3.85x - 1.875x^2 - 0.975 \\ &= (3.85)(0.7) - (1.875)(0.7)^2 - 0.975 \\ &= 0.801\,25 \end{aligned}$$

0.3 lies in the range $0 \leqslant x \leqslant 0.4$, so

$$\begin{aligned} F(0.3) &= 1.875x^2 + 0.1x \\ &= (1.875)(0.3)^2 + (0.1)(0.3) \\ &= 0.198\,75 \end{aligned}$$

Therefore

$$\begin{aligned} P(0.3 \leqslant X \leqslant 0.7) &= F(0.7) - F(0.3) \\ &= 0.801\,25 - 0.198\,75 \\ &= 0.6025 \end{aligned}$$

So $P(|X-\mu| \leqslant 0.2) = 0.6025$.

Exercise 5d

For each of the following probability density functions of the continuous r.v. X, find (*a*) the cumulative distribution function $F(x)$, and for questions 1, 2, 5 and 7 find also (*b*) the median, m.

NOTE: These are the functions used in Exercise 5a and 5c.

1. $f(x) = \frac{3}{8}x^2 \qquad 0 \leqslant x \leqslant 2$
2. $f(x) = \frac{1}{5} \qquad -2 \leqslant x \leqslant 3$
3. $f(x) = \frac{1}{4}(4-x) \qquad 1 \leqslant x \leqslant 3$
4. $f(x) = \frac{3}{56}(x+2)^2 \qquad 0 \leqslant x \leqslant 2$
5. $f(x) = 4x^3 \qquad 0 \leqslant x \leqslant 1$
6. $f(x) = \begin{cases} \frac{1}{4} & 0 \leqslant x \leqslant 2 \\ \frac{1}{4}(2x-3) & 2 \leqslant x \leqslant 3 \end{cases}$
7. $f(x) = \begin{cases} \frac{1}{8}(x+2)^2 & -2 \leqslant x \leqslant 0 \\ \frac{1}{2} & 0 \leqslant x \leqslant 1\frac{1}{3} \end{cases}$
8. The continuous r.v. X has p.d.f. $f(x) = \frac{1}{3}$, $0 \leqslant x \leqslant 3$. Find (*a*) $E(X)$, (*b*) $\text{Var}(X)$, (*c*) $F(x)$ and sketch $y = F(x)$, (*d*) $P(X \geqslant 1.8)$, (*e*) $P(1.1 \leqslant X \leqslant 1.7)$.
9. X is the continuous r.v. with p.d.f. $f(x) = kx^2$, $1 \leqslant x \leqslant 2$. Find (*a*) the constant k and sketch $y = f(x)$, (*b*) the standard deviation σ, (*c*) the cumulative distribution function $F(x)$, (*d*) the median, m.
10. The continuous r.v. X has continuous p.d.f. $f(x)$ where

$$f(x) = \begin{cases} \dfrac{x}{3} - \dfrac{2}{3} & 2 \leqslant x \leqslant 3 \\ \alpha & 3 \leqslant x \leqslant 5 \\ 2 - \beta x & 5 \leqslant x \leqslant 6 \\ 0 & \text{otherwise} \end{cases}$$

Find (*a*) α and β, (*b*) $F(x)$ and sketch $y = F(x)$, (*c*) $P(2 \leqslant X \leqslant 3.5)$, (*d*) $P(X \geqslant 5.5)$, (*e*) $E(X)$, (*f*) $\text{Var}(X)$.

11. The continuous r.v. X has probability density function given by

$$f(x) = \begin{cases} \dfrac{k}{x} & \text{for } 1 \leqslant x \leqslant 9, \\ 0 & \text{otherwise,} \end{cases}$$

where k is a constant. Giving your answers correct to three significant figures where appropriate, find (*a*) the value of k, and find also the median value of X, (*b*) the mean and variance of X, (*c*) the cumulative distribution function, F, of X, and sketch the graph of $y = F(x)$. (C)

12. The continuous r.v. X has probability density function f given by

$$f(x) = \begin{cases} k(4-x^2) & \text{for } 0 \leqslant x \leqslant 2. \\ 0 & \text{otherwise,} \end{cases}$$

where k is a constant. Show that $k = \frac{3}{16}$ and find the values of $E(X)$ and $\text{Var}(X)$.

Find the cumulative distribution function of X, and verify by calculation that the median value of X is between 0.69 and 0.70.

Find also $P(0.69 < X < 0.70)$, giving your answer correct to one significant figure. (C)

13. A continuous random variable X has probability density function, f, defined by

$$f(x) = \tfrac{1}{4}, \qquad 0 \leqslant x < 1,$$

$$f(x) = \frac{x^3}{5}, \qquad 1 \leqslant x \leqslant 2,$$

$$f(x) = 0, \qquad \text{otherwise.}$$

Obtain the distribution function and hence, or otherwise, find, to 3 decimal places, the median and the interquartile range of the distribution (L)P

14. Define the probability density function $f(x)$ and the distribution function $F(x)$ of a continuous random variable X.

A factory is supplied with flour at the beginning of each week. The weekly demand, X thousand tonnes, for flour from this factory is a continuous random variable having the probability density function

$$f(x) = k(1-x)^4, \qquad 0 \leqslant x \leqslant 1,$$

$$f(x) = 0, \qquad \text{elsewhere.}$$

Find
(*a*) the value of k,
(*b*) the mean value of X,
(*c*) the variance of X, to 3 decimal places.

Sketch the probability density function.

Find, to the nearest tonne, the quantity of flour that the factory should have in stock at the beginning of a week in order that there is a probability of 0.98 that the demand in that week will be met. (L)

OBTAINING THE p.d.f. FROM THE CUMULATIVE DISTRIBUTION

The probability density function can be obtained from the cumulative distribution as follows:

Now $$F(t) = \int_a^t f(x)\,dx \qquad a \leqslant t \leqslant b$$

So $$f(x) = \frac{d}{dx}F(x) = F'(x)$$

NOTE: the gradient of the $F(x)$ curve gives the value of $f(x)$.

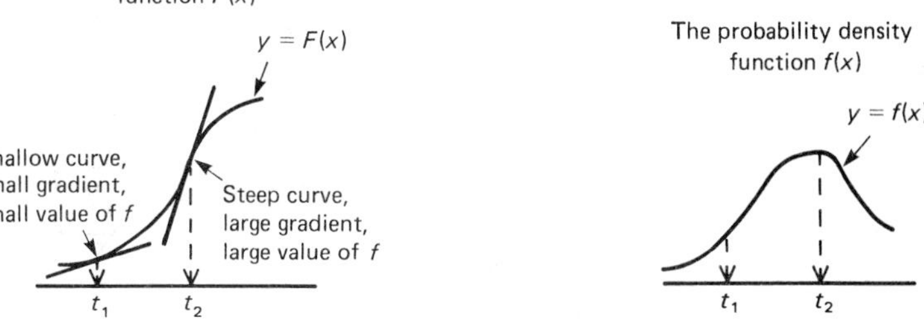

Example 5.14 The continuous r.v. X has cumulative distribution function $F(x)$ where

$$F(x) = \begin{cases} 0 & x \leqslant 0 \\ \dfrac{x^3}{27} & 0 \leqslant x \leqslant 3 \\ 1 & x \geqslant 3 \end{cases}$$

Find the p.d.f. of X, $f(x)$, and sketch $y = f(x)$.

Solution 5.14

$$f(x) = \frac{\mathrm{d}}{\mathrm{d}x} F(x) = \frac{\mathrm{d}}{\mathrm{d}x}\left(\frac{x^3}{27}\right) = \frac{3x^2}{27} = \frac{x^2}{9}$$

The p.d.f. for X is $f(x)$ where

$$f(x) = \begin{cases} \dfrac{x^2}{9} & 0 \leqslant x \leqslant 3 \\ 0 & \text{otherwise} \end{cases}$$

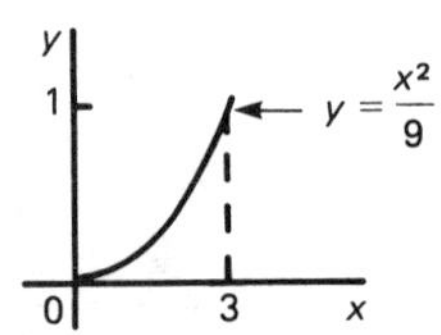

Example 5.15 The continuous r.v. X has cumulative distribution function $F(x)$ where

$$F(x) = \begin{cases} 0 & x < -2 \\ \frac{1}{12}(2+x) & -2 \leqslant x < 0 \\ \frac{1}{6}(1+x) & 0 \leqslant x < 4 \\ \frac{1}{12}(6+x) & 4 \leqslant x < 6 \\ 1 & x \geqslant 6 \end{cases}$$

Find the p.d.f. of X, $f(x)$, and sketch $y = f(x)$.

Solution 5.15 Now $f(x) = \dfrac{\mathrm{d}}{\mathrm{d}x} F(x)$.

So, for $-2 \leqslant x < 0$ $\quad f(x) = \dfrac{\mathrm{d}}{\mathrm{d}x}\dfrac{1}{12}(2+x) = \dfrac{1}{12}$

for $0 \leqslant x < 4$ $\quad f(x) = \dfrac{\mathrm{d}}{\mathrm{d}x}\dfrac{1}{6}(1+x) = \dfrac{1}{6}$

for $4 \leqslant x < 6$ $\quad f(x) = \dfrac{\mathrm{d}}{\mathrm{d}x}\dfrac{1}{12}(6+x) = \dfrac{1}{12}$

The sketch of $y = f(x)$ is shown:

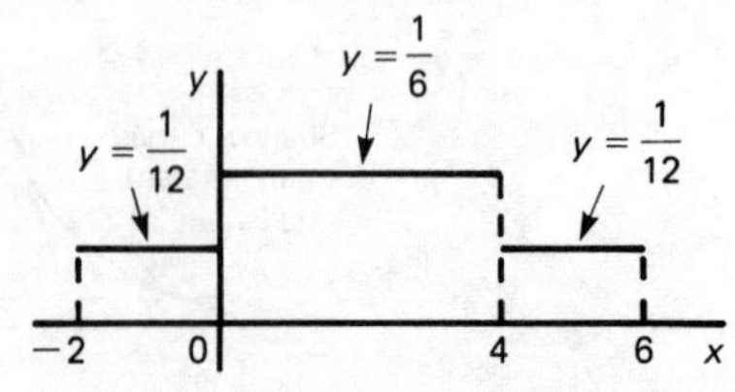

Example 5.16 The cumulative distribution function $F(x)$ for a continuous r.v. X is defined as

$$F(x) = \begin{cases} 0 & x < 0 \\ \frac{1}{2}x - \frac{1}{8}x^2 & 0 \leqslant x \leqslant 1 \\ a + \frac{1}{4}x & 1 \leqslant x \leqslant 2 \\ b + \frac{1}{8}x^2 - \frac{1}{4}x & 2 \leqslant x \leqslant 3 \\ 1 & x > 3 \end{cases}$$

(**a**) Find a and b and sketch $F(x)$. (**b**) Find and sketch the p.d.f. $f(x)$. (**c**) Find the mean μ and the standard deviation σ.

Solution 5.16 (a) Now it is obvious that $F(3) = 1$ as $F(x) = 1$ when $x \geqslant 3$.

But $$F(x) = b + \frac{1}{8}x^2 - \frac{1}{4}x \qquad 2 \leqslant x \leqslant 3$$

Therefore $$F(3) = b + \frac{9}{8} - \frac{3}{4}$$

$$= b + \frac{3}{8}$$

So $$b + \frac{3}{8} = 1$$

$$\underline{b = \frac{5}{8}}$$

So we have for $2 \leqslant x \leqslant 3$

$$F(x) = \frac{5}{8} + \frac{1}{8}x^2 - \frac{1}{4}x$$

Therefore $$F(2) = \frac{5}{8} + \frac{4}{8} - \frac{2}{4}$$

$$= \frac{5}{8}$$

But for the range $1 \leqslant x \leqslant 2$

$$F(x) = a + \frac{1}{4}x$$

$$F(2) = a + \frac{1}{2}$$

So we have $$a+\frac{1}{2} = \frac{5}{8}$$

Therefore $$\underline{a = \frac{1}{8}}$$

So the cumulative distribution function $F(x)$ is as follows:

$$F(x) = \begin{cases} 0 & x \leqslant 0 \\ \frac{1}{2}x-\frac{1}{8}x^2 & 0 \leqslant x \leqslant 1 \\ \frac{1}{8}+\frac{1}{4}x & 1 \leqslant x \leqslant 2 \\ \frac{5}{8}+\frac{1}{8}x^2-\frac{1}{4}x & 2 \leqslant x \leqslant 3 \\ 1 & x \geqslant 3 \end{cases}$$

Sketch of $y = F(x)$.

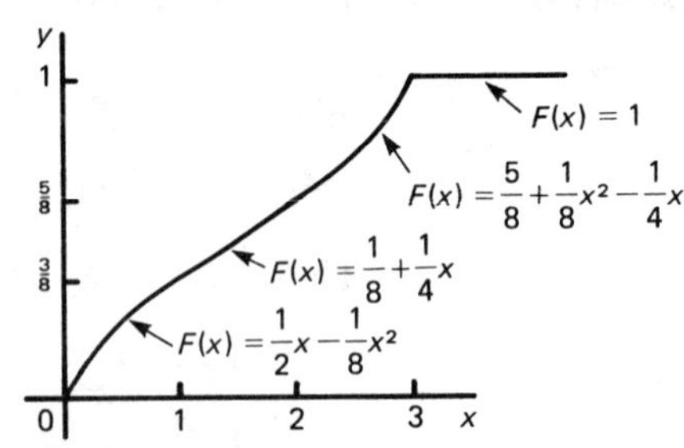

(b) $$f(x) = \frac{\mathrm{d}}{\mathrm{d}x}F(x)$$

Therefore

$$f(x) = \begin{cases} 0 & x < 0 \\ \frac{1}{2}-\frac{1}{4}x & 0 \leqslant x \leqslant 1 \\ \frac{1}{4} & 1 \leqslant x \leqslant 2 \\ \frac{1}{4}x-\frac{1}{4} & 2 \leqslant x \leqslant 3 \\ 0 & x > 3 \end{cases}$$

Sketch of $y = f(x)$.

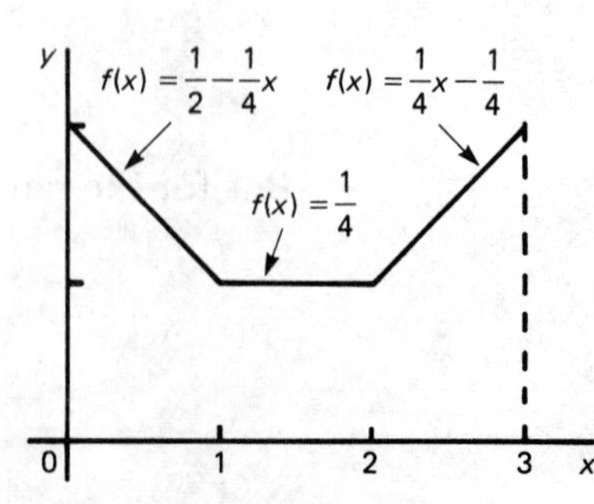

(c) By symmetry $\mu = E(X) = 1.5$.

$$
\begin{aligned}
E(X^2) &= \int_{\text{all } x} x^2 f(x)\,dx \\
&= \int_0^1 \left(\tfrac{1}{2}x^2 - \tfrac{1}{4}x^3\right) dx + \int_1^2 \tfrac{1}{4}x^2\,dx + \int_2^3 \left(\tfrac{1}{4}x^3 - \tfrac{1}{4}x^2\right) dx \\
&= \left[\frac{x^3}{6} - \frac{x^4}{16}\right]_0^1 + \left[\frac{x^3}{12}\right]_1^2 + \left[\frac{x^4}{16} - \frac{x^3}{12}\right]_2^3 \\
&= \frac{19}{6}
\end{aligned}
$$

Now

$$
\begin{aligned}
\text{Var}(X) &= E(X^2) - E^2(X) \\
&= \frac{19}{6} - (1.5)^2 \\
&= \frac{11}{12}
\end{aligned}
$$

The standard deviation

$$
\begin{aligned}
\sigma &= \sqrt{\text{Var}(X)} \\
&= \sqrt{\frac{11}{12}} \\
&= 0.957 \quad (3\text{ S.F.})
\end{aligned}
$$

Therefore the mean $\mu = 1.5$ and the standard deviation $\sigma = 0.957$ (3 S.F.).

Example 5.17 A continuous random variable X takes values in the interval 0 to 3. It is given that $P(X > x) = a + bx^3$, $0 \leqslant x \leqslant 3$.

(a) Find the values of the constants a and b.

(b) Find the cumulative distribution function $F(x)$.

(c) Find the probability density function $f(x)$.

(d) Show that $E(X) = 2.25$.

(e) Find the standard deviation.

Solution 5.17 (a) $P(X>x)=a+bx^3, 0 \leqslant x \leqslant 3.$

$$\begin{aligned} \text{So} \quad P(X>0) &= 1 \quad \text{and} \quad P(X>3) = 0 \\ \text{i.e.} \quad a+b(0) &= 1 \quad \text{and} \quad a+b(27) = 0 \\ \text{Therefore} \quad \underline{a} &\underline{= 1}, \quad \text{and} \quad 1+27b = 0 \\ & \qquad \underline{b = -\frac{1}{27}} \end{aligned}$$

So $P(X>x)=1-\dfrac{x^3}{27}, 0 \leqslant x \leqslant 3.$

(b) Now $$P(X \leqslant x) = \frac{x^3}{27}$$

i.e. $$F(x) = \begin{cases} \dfrac{x^3}{27} & 0 \leqslant x \leqslant 3 \\ 1 & x>3 \end{cases}$$

(c) $$\begin{aligned} f(x) &= \frac{\mathrm{d}}{\mathrm{d}x}F(x) \\ &= \frac{x^2}{9} \end{aligned}$$

Therefore the p.d.f. of X is $f(x)$ where $f(x)=\dfrac{x^2}{9}, 0 \leqslant x \leqslant 3.$

(d) $$\begin{aligned} E(X) &= \int_{\text{all } x} x\, f(x)\, \mathrm{d}x \\ &= \int_0^3 \frac{x^3}{9}\mathrm{d}x \\ &= \left[\frac{x^4}{36}\right]_0^3 \\ &= 2.25 \end{aligned}$$

Therefore $E(X)=2.25$, as required.

(e) $$\begin{aligned} \mathrm{Var}(X) &= \int_{\text{all } x} x^2 f(x)\, \mathrm{d}x - E^2(X) \\ &= \int_0^3 \frac{x^4}{9}\mathrm{d}x - 2.25^2 \\ &= \left[\frac{x^5}{45}\right]_0^3 - 5.0625 \\ &= 0.3375 \end{aligned}$$

So $$\begin{aligned}\sigma &= \sqrt{0.3375}\\ &= 0.581 \quad (3 \text{ S.F.})\end{aligned}$$

The standard deviation of X is 0.581 (3 S.F.).

Exercise 5e

1. The continuous r.v. X has cumulative distribution function $F(x)$ where

$$F(x) = \begin{cases} 0 & x \leqslant 0 \\ \dfrac{2x}{3} & 0 \leqslant x \leqslant 1 \\ \dfrac{x}{3} + k & 1 \leqslant x \leqslant 2 \\ 1 & x \geqslant 2 \end{cases}$$

Find (*a*) the value of k, (*b*) the p.d.f. $f(x)$ and sketch it, (*c*) the mean μ, (*d*) the standard deviation σ, (*e*) $P(|X-\mu| \leqslant \sigma)$.

2. The continuous r.v. X has cumulative distribution function $F(x)$ where

$$F(x) = \begin{cases} 0 & x \leqslant 1 \\ \dfrac{(x-1)^2}{12} & 1 \leqslant x \leqslant 3 \\ \dfrac{(14x - x^2 - 25)}{24} & 3 \leqslant x \leqslant 7 \\ 1 & x \geqslant 7 \end{cases}$$

(*a*) Find and sketch $f(x)$. (*b*) Find $E(X)$ and $\text{Var}(X)$. (*c*) Find the median m. (*d*) Find $P(2.8 \leqslant X \leqslant 5.2)$.

3. A random variable X has cumulative (distribution) function $F(x)$ where

$$F(x) = \begin{cases} 0 & x < -1 \\ \alpha x + \alpha & -1 \leqslant x < 0 \\ 2\alpha x + \alpha & 0 \leqslant x < 1 \\ 3\alpha & 1 \leqslant x \end{cases}$$

Determine
(*a*) the value of α,
(*b*) the frequency function $f(x)$ of X,
(*c*) the expected value μ of X,
(*d*) the standard deviation σ of X,
(*e*) the probability that $|X-\mu|$ exceeds $\frac{1}{3}$. (C)

4. The length X of an offcut of wooden planking is a random variable which can take any value up to 0.5 m. It is known that the probability of the length being not more than x metres $(0 \leqslant x \leqslant 0.5)$ is equal to kx. Determine
(*a*) the value of k,
(*b*) the probability density function of X,
(*c*) the expected value of X,
(*d*) the standard deviation of X (correct to 3 significant figures). (C)

5. A continuous random variable X takes values in the interval 0 to 4. The probability that X takes a value greater than x is equal to $\alpha x^2 + \beta$, $(0 \leqslant x \leqslant 4)$.
(*a*) Determine the values of α and β.
(*b*) Determine the probability density function $f(x)$ of X.
(*c*) Show that the expected value μ of X is $\frac{8}{3}$.
(*d*) Show that the standard deviation σ of X is $\frac{2}{3}\sqrt{2}$.
(*e*) Show that the probability that

$$(\mu-\sigma) \leqslant X \leqslant (\mu+\sigma) \text{ is } \tfrac{4}{9}\sqrt{2}$$ (C)

6. The continuous random variable X has (cumulative) distribution function given by

$$F(x) = \begin{cases} (1+x)/8 & (-1 \leqslant x \leqslant 0) \\ (1+3x)/8 & (0 \leqslant x \leqslant 2) \\ (5+x)/8 & (2 \leqslant x \leqslant 3) \end{cases}$$

with $F(x) = 0$ for $x < -1$, and $F(x) = 1$ for $x > 3$.
(*a*) Sketch the graph of the probability density function $f(x)$.
(*b*) Determine the expectation of X and the variance of X.
(*c*) Determine $P(3 \leqslant 2X \leqslant 5)$. (C)

THE RECTANGULAR DISTRIBUTION

A continuous r.v. X having p.d.f. $f(x)$ where

$$f(x) = \frac{1}{b-a} \quad \text{for} \quad a \leqslant x \leqslant b$$

where a and b are constants, is said to follow a **rectangular** (or **uniform**) **distribution.**

a and b are the parameters of the distribution.

If X is distributed in this way, we write

$$X \sim \mathrm{R}(a, b)$$

The graph of $y = f(x)$ is

X is a random variable, since

$$\begin{aligned}\int_{\text{all } x} f(x)\,\mathrm{d}x &= \int_a^b \frac{1}{b-a}\,\mathrm{d}x \\ &= \frac{1}{b-a}[x]_a^b \\ &= \frac{b-a}{b-a} \\ &= 1\end{aligned}$$

This can be seen more easily from the diagram

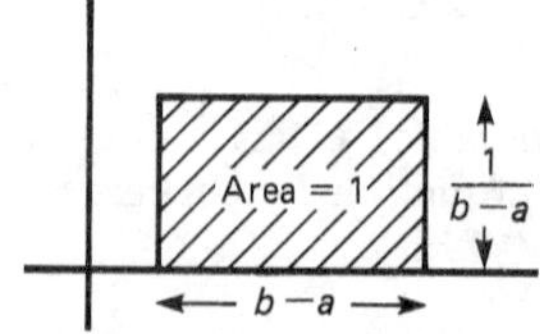

For a rectangular distribution, the probability of the variable lying in one particular range of length l (say) is exactly the same as the probability that it lies in another range of the same length l.

In each case, the probability is $l\left(\frac{1}{b-a}\right)$.

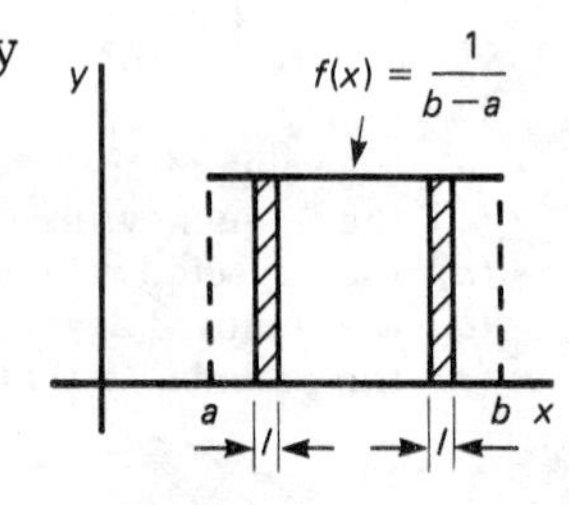

For example, if $X \sim \mathrm{R}(1, 6)$ then $f(x) = \frac{1}{5}$ and

$$P(1.8 \leqslant X \leqslant 1.9) = 0.1\left(\frac{1}{5}\right) = 0.02$$

$$P(5.1 \leqslant X \leqslant 5.2) = 0.1\left(\frac{1}{5}\right) = 0.02$$

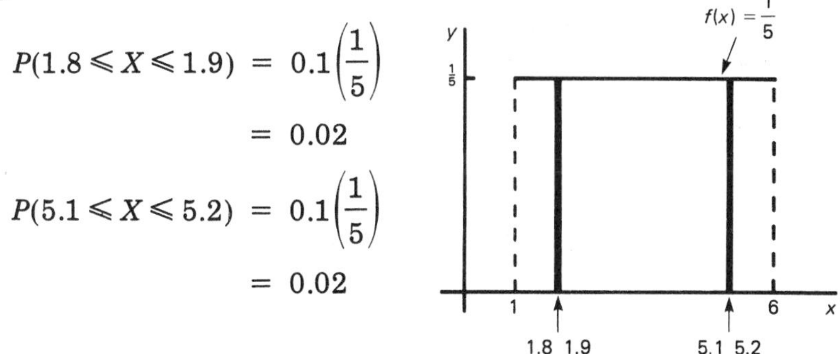

Example 5.18 The rounding error made when measuring the lengths of metal rods to the nearest 5 mm is a random variable E. What is the distribution of E?

Solution 5.18 The error is the difference between the true length and the recorded length after rounding to the nearest 5 mm.

Suppose we have recorded a length to be 75 mm, to the nearest 5 mm. Now the true length could have been any length in the interval

$$72.5\,\text{mm} \leqslant l < 77.5\,\text{mm}$$

So, the error, E, could be anywhere in the interval $-2.5 \leqslant E < 2.5$.

All points in this interval are equally likely 'stopping places' for E, so E is uniformly distributed in the interval.

We write $$E \sim \mathrm{R}(-2.5, 2.5)$$

Example 5.19 A child spins a 'Spinning Jenny' at a fair. When the wheel stops, the shorter distance of an arrow measured along the circumference from the child is denoted by C. What is the distribution of C?

Solution 5.19 All the points on the circumference are equally likely stopping places for the arrow, so C is uniformly distributed between 0 (when the arrow is next to the child) and πr (when the arrow is diametrically opposite the child).

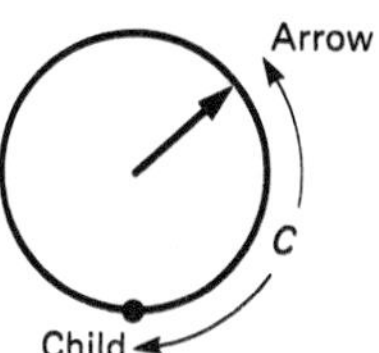

So $$C \sim \mathrm{R}(0, \pi r)$$

EXPECTATION AND VARIANCE

If $X \sim R(a, b)$ then

$$E(X) = \tfrac{1}{2}(a+b)$$

$$\text{Var}(X) = \tfrac{1}{12}(b-a)^2$$

The graph of $y = f(x)$ is as shown.

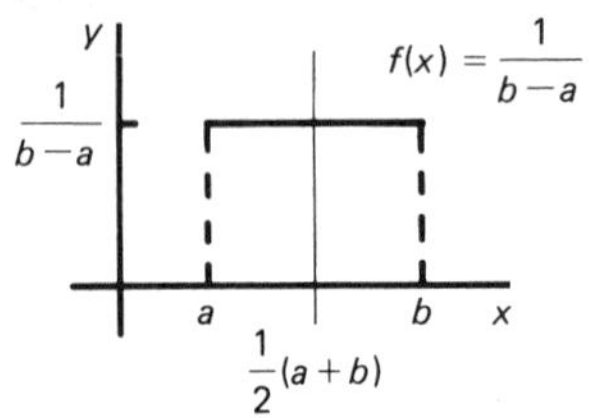

By symmetry, $E(X) = \frac{1}{2}(a+b)$.

Now $\quad \text{Var}(X) = E(X^2) - E^2(X)$

where
$$E(X^2) = \int_{\text{all } x} x^2 f(x)\,dx$$
$$= \int_a^b x^2\left(\frac{1}{b-a}\right)dx$$
$$= \frac{1}{b-a}\left[\frac{x^3}{3}\right]_a^b$$
$$= \frac{1}{3(b-a)}(b^3-a^3)$$
$$= \frac{1}{3(b-a)}(b-a)(b^2+ab+a^2)$$
$$= \frac{(b^2+ab+a^2)}{3}$$

So
$$\text{Var}(X) = E(X^2) - E^2(X)$$
$$= \frac{(b^2+ab+a^2)}{3} - \frac{(a^2+2ab+b^2)}{4}$$
$$= \frac{1}{12}\{4(b^2+ab+a^2) - 3(a^2+2ab+b^2)\}$$
$$= \frac{1}{12}(b^2-2ab+a^2)$$
$$= \frac{1}{12}(b-a)^2$$

Therefore $E(X) = \frac{1}{2}(a+b)$ and $\text{Var}(X) = \frac{1}{12}(b-a)^2$

Example 5.20 The random variable X has a rectangular distribution over the interval $(0, 8)$. Find the probability density function of Y where $Y = X^{1/3}$.

Solution 5.20 The p.d.f. of X is $f(x) = \frac{1}{8}$.

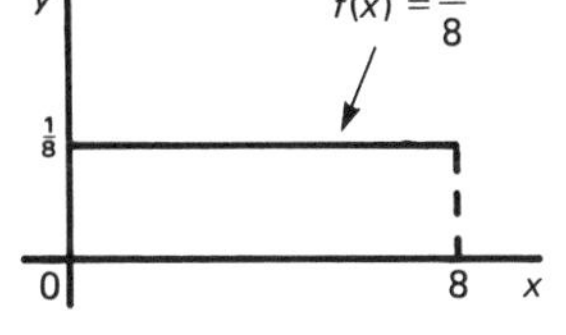

Now, since X is a random variable,

$$1 = \int_0^8 \tfrac{1}{8}\,dx$$

Now, we require some function of y, $g(y)$ say, such that

$$1 = \int_{y_1}^{y_2} g(y)\,dy$$

$g(y)$ then gives the p.d.f. of Y.

Now $x^{1/3} = y$, so that

$$\tfrac{1}{3}x^{-2/3}\,dx = dy$$

i.e.

$$dx = 3x^{2/3}\,dy = 3y^2\,dy$$

Also, when $x = 0$, $y = 0$ and when $x = 8$, $y = 2$. So, the first integral becomes

$$1 = \int_0^2 \frac{1}{8}\,3y^2\,dy$$

i.e.

$$1 = \int_0^2 \frac{3y^2}{8}\,dy$$

Thus, the p.d.f. of Y is $g(y) = \frac{3}{8}y^2$, $0 \leqslant y \leqslant 2$.

Example 5.21 A rectangle, with one side of length x cm and perimeter 12 cm, has area A cm^2. If X is uniformly distributed between 0 and 2, find the probability density function of A.

Solution 5.21 Let the p.d.f. of A be $f(a)$

One side of the rectangle has length x cm.

If the perimeter is 12 cm, then the length of the other side is $(6-x)$.

Now

$$a = x(6-x)$$

So

$$x^2 - 6x = -a$$
$$(x-3)^2 = 9-a$$
$$x-3 = \pm\sqrt{9-a}$$

Now

$$dx = \mp\frac{da}{2\sqrt{9-a}}$$

We are given that X is uniformly distributed between 0 and 2, therefore

$$1 = \int_0^2 \frac{1}{2}\,dx$$

and when $x = 0$, $a = 0$; when $x = 2$, $a = 8$.

Therefore
$$1 = \int_0^8 \frac{1}{2}\,\frac{da}{2\sqrt{9-a}}$$

so A is distributed with p.d.f. $f(a) = \dfrac{1}{4\sqrt{9-a}}$, $0 \leqslant a \leqslant 8$.

Exercise 5f

1. If the continuous r.v. X is such that $X \sim R(3, 6)$ find (*a*) the p.d.f. of X, (*b*) $E(X)$, (*c*) $\text{Var}(X)$, (*d*) $P(X > 5)$.

2. If the continuous r.v. X has p.d.f. $f(x)$ where $f(x) = k$ and $X \sim R(-5, -2)$ find (*a*) the value of the constant k, (*b*) $P(-4.3 < X < -2.8)$, (*c*) $E(X)$, (*d*) $\text{Var}(X)$.

3. The continuous r.v. X has p.d.f. $f(x)$ as shown in the diagram:

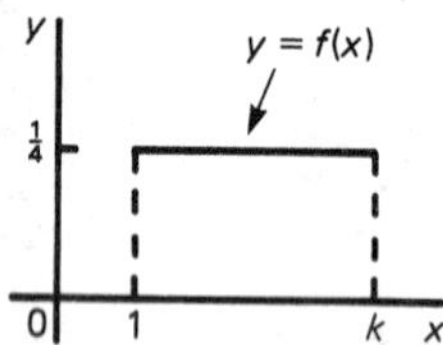

Find (*a*) the value of k,
(*b*) $P(2.1 < X < 3.4)$,
(*c*) $E(X)$,
(*d*) $\text{Var}(X)$.

4. The length X of a side of a square is rectangularly distributed between 1 and 4. Find the probability density function of A, the area of the square, and calculate the mean and variance of the area of the square.

5. The radius of a circle follows a rectangular distribution between 1 and 3. Find the probability density function of A, the area of the circle and calculate the mean and the variance of the area of the circle.

6. The random variable X has probability density function given by

$$f(x) = \begin{cases} \dfrac{1}{(b-a)} & a \leqslant x \leqslant b \text{ where } b > a \\ 0 & \text{otherwise} \end{cases}$$

Show that the mean is $(b+a)/2$, and the variance is $(b-a)^2/12$ for this distribution.

Given that the mean equals 1 and the variance equals 4/3 find
(i) $P(X < 0)$,
(ii) the value of z such that

$$P(X > z + \sigma_X) = \tfrac{1}{4},$$

where σ_X is the standard deviation of X. (AEB 1979)

7. A rectangle of area A square metres has a perimeter of 20 metres and a side of length X metres, where X is uniformly distributed between 0 and 2. Show that the probability density function of A is

$$\frac{1}{4\sqrt{(25-A)}} \quad (0 \leqslant A \leqslant 16)$$

Find the mean and variance of A. (JMB)

8. A child rides on a roundabout and his father waits for him at the point where he started. His journey may be regarded as a circular route of radius six metres and the father's position as a fixed point on the circle. When the roundabout stops, the shorter distance of the child from the father, measured alone the circular path, is S metres. All points on the circle are equally likely stopping points, so that S is uniformly distributed between 0 and 6π. Find the mean and variance of S.

The direct linear distance of the child's stopping point from the father is D metres. Show that the probability density function of D is $\dfrac{2}{\pi\sqrt{(144-D^2)}}$ for D between 0 and 12 and zero outside this range.

The father's voice can be heard at a distance of up to ten metres. Find to two decimal places the probability that the child can hear his father shout to him when the roundabout stops. (JMB)

9. The object distance U and the image distance V for a concave mirror are related to the focal distance f by the formula

$$\frac{1}{u}+\frac{1}{v} = \frac{1}{f}$$

U is a random variable uniformly distributed over the interval $(2f, 3f)$. Show that V is distributed with probability density function

$$\frac{f}{(v-f)^2}$$

and state the range of corresponding values for V. Obtain the mean and median of V. (C)

10. The line $y + 2x = k$ crosses the coordinate axes Ox and Oy at P and Q respectively. Given that the area of $\triangle OPQ$ is A, show that $A = k^2/4$.

A random variable takes values k such that $0 \leqslant k \leqslant 5$ and is rectangularly distributed in this interval.
(*a*) Show that the expected value of A is $25/12$.
(*b*) Calculate the variance of A.
(L Additional)

THE EXPONENTIAL DISTRIBUTION

A continuous r.v. X having p.d.f. $f(x)$ where

$$f(x) = \lambda e^{-\lambda x} \quad \text{for} \quad x \geqslant 0,$$

where λ is a positive constant, is said to follow an exponential distribution.

λ is the parameter of the distribution.

The graph of $y = f(x)$ is

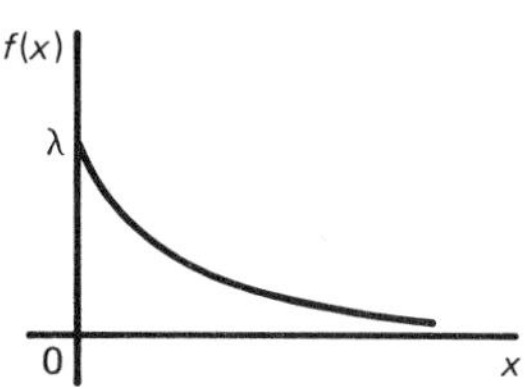

Now X is a random variable, since

$$\begin{aligned}\int_{\text{all } x} f(x)\,\mathrm{d}x &= \int_0^\infty \lambda \mathrm{e}^{-\lambda x}\,\mathrm{d}x \\ &= -[\mathrm{e}^{-\lambda x}]_0^\infty \\ &= 1 \quad (\text{since } \lim_{x\to\infty} \mathrm{e}^{-\lambda x} = 0)\end{aligned}$$

EXPECTATION AND VARIANCE

If X has p.d.f. given by $f(x) = \lambda e^{-\lambda x}$, $x \geqslant 0$, then

$$E(X) = \frac{1}{\lambda}$$

$$\text{Var}(X) = \frac{1}{\lambda^2}$$

Now
$$\begin{aligned}
E(X) &= \int_{\text{all } x} xf(x)\,dx \\
&= \int_0^\infty x(\lambda e^{-\lambda x})\,dx \\
&= [x(-e^{-\lambda x})]_0^\infty - \int_0^\infty (-e^{-\lambda x})\,dx \\
&= 0 + \int_0^\infty e^{-\lambda x}\,dx \quad (\text{since } \lim_{x\to\infty} xe^{-\lambda x} = 0) \\
&= -\frac{1}{\lambda}[e^{-\lambda x}]_0^\infty \\
&= -\frac{1}{\lambda}(0-1) \\
&= \frac{1}{\lambda}
\end{aligned}$$

To find $\text{Var}(X)$ consider first $E(X^2)$.

Now
$$\begin{aligned}
E(X^2) &= \int_{\text{all } x} x^2 f(x)\,dx \\
&= \int_0^\infty x^2(\lambda e^{-\lambda x})\,dx \\
&= [x^2(-e^{-\lambda x})]_0^\infty - \int_0^\infty 2x(-e^{-\lambda x})\,dx \\
&= 0 + 2\int_0^\infty x\,e^{-\lambda x}\,dx \quad (\text{since } \lim_{x\to\infty} x^2 e^{-\lambda x} = 0) \\
&= \frac{2}{\lambda^2} \qquad (\text{since } \int_0^\infty \lambda x\,e^{-\lambda x}\,dx = \frac{1}{\lambda})
\end{aligned}$$

$$\begin{aligned}\text{Var}(X) &= E(X^2) - E^2(X)\\ &= \frac{2}{\lambda^2} - \left(\frac{1}{\lambda}\right)^2\\ &= \frac{1}{\lambda^2}\end{aligned}$$

Therefore $E(X) = \dfrac{1}{\lambda}$ and $\text{Var}(X) = \dfrac{1}{\lambda^2}$.

If the continuous r.v. X has p.d.f. $f(x) = \lambda e^{-\lambda x}$, $x \geqslant 0$, then

$$\begin{aligned}P(X > a) &= e^{-\lambda a}\\ P(X > a + b \mid X > a) &= P(X > b).\end{aligned}$$

To show this, consider

$$\begin{aligned}P(X > a) &= \int_a^{\infty} \lambda e^{-\lambda x}\,dx\\ &= \left[-e^{-\lambda x}\right]_a^{\infty}\\ &= e^{-\lambda a}\end{aligned}$$

Also
$$\begin{aligned}P(X > a + b \mid X > a) &= \frac{P(X > a + b \cap X > a)}{P(X > a)}\\ &= \frac{e^{-\lambda(a+b)}}{e^{-\lambda a}}\\ &= e^{-\lambda b}\\ &= P(X > b).\end{aligned}$$

Example 5.22 The lifetime in years of a television tube of a certain make is a random variable T and its probability density function $f(t)$ is given by

$$\begin{aligned}f(t) &= A e^{-kt} \quad \text{for} \quad 0 \leqslant t \leqslant \infty \quad (k > 0)\\ &= 0 \qquad\quad \text{elsewhere}\end{aligned}$$

Obtain A in terms of k.

(a) If the manufacturer, after some research, finds that out of 1000 such tubes 371 failed within the first two years of use, estimate the value of k.

(b) Using this value of k correct to 3 significant figures, calculate the mean and variance of T, giving answers correct to 2 significant figures ($t^r e^{-kt} = 0$ when $t = \infty$ for finite r).

(c) If two such tubes are bought, what is the probability that one fails within its first year and the other lasts longer than six years? (SUJB)

Solution 5.22 Since T is a random variable $\int_{\text{all } t} f(t)\,dt = 1$.

So
$$\int_0^\infty A\,e^{-kt}\,dt = 1$$

$$A\left[-\frac{1}{k}e^{-kt}\right]_0^\infty = 1$$

$$\frac{A}{k}e^{-0} = 1 \qquad \text{since } e^{-x} \to 0 \text{ as } x \to \infty$$

$$A = k$$

So $f(t) = ke^{-kt},\ 0 \leqslant t \leqslant \infty$. This is an exponential distribution.

(a) $P(T \leqslant 2) = 0.371$.

i.e.
$$\int_0^2 k\,e^{-kt}\,dt = 0.371$$

$$[-e^{-kt}]_0^2 = 0.371$$

$$-e^{-2k} + e^0 = 0.371$$

$$e^{-2k} = 1 - 0.371$$

$$= 0.629$$

So
$$e^{2k} = \frac{1}{0.629}$$

Taking logs to base e
$$2k = \ln\left(\frac{1}{0.629}\right)$$

$$= 0.464 \quad (3 \text{ S.F.})$$

Therefore we estimate that $k = 0.232$ (3 S.F.).

(b)
$$E(T) = \frac{1}{k}$$ see p. 308

$$= \frac{1}{0.232}$$

$$= 4.3 \quad (2 \text{ S.F.})$$

Therefore $E(T) = 4.3$ years (2 S.F.).

Now $$\text{Var}(T) = \frac{1}{k^2} = \frac{1}{(0.232)^2} = 19 \quad (2\text{ S.F.})$$

So, putting $k = 0.232$, we have $\text{Var}(T) = 19$ years^2 (2 S.F.).

(c) $$P(T<1) = \int_0^1 k\,\mathrm{e}^{-kt}\,\mathrm{d}t = [-\mathrm{e}^{-kt}]_0^1 = -\mathrm{e}^{-k}+1 = 1-0.793 = 0.207 \quad (3\text{ S.F.})$$

and $$P(T>6) = 1-P(T\leqslant 6) = 1-\int_0^6 k\,\mathrm{e}^{-kt}\,\mathrm{d}t = 1-[-\mathrm{e}^{-kt}]_0^6 = 1+\mathrm{e}^{-6k}-1 = 0.249 \quad (3\text{ S.F.})$$

Therefore, if two tubes are bought,

$$P[(T_1<1)\cap(T_2>6)]+P[(T_2<1)\cap(T_1>6)] = 2(0.207)(0.249) = 0.103 \quad (3\text{ S.F.})$$

Therefore the probability that one fails within its first year and the other lasts longer than 6 years is 0.103 (3 S.F.).

Example 5.23 The continuous random variable X has the negative exponential distribution whose probability density function is given by

$$f(x) = \lambda\mathrm{e}^{-\lambda x}, \quad x\geqslant 0,$$
$$f(x) = 0, \quad \text{otherwise},$$

where λ is a positive constant. Obtain expressions, in terms of λ, for

(a) the mean, $E(X)$, of the distribution,

(b) $F(x)$, the (cumulative) distribution function.

Television sets are hired out by a rental company. The time in months, X, between major repairs has the above negative exponential distribution with $\lambda = 0.05$. Find, to 3 significant figures, the probability that a television set hired out by the company will not require a major repair for at least a 2-year period. Find also the median value of X.

The company agrees to replace any set for which the time between major repairs is less than M months. Given that the company does not want to have to replace more than one set in 5, find M. (L)

Solution 5.23 $f(x) = \lambda e^{-\lambda x}, \quad x \geqslant 0$

(a) $E(X) = \dfrac{1}{\lambda}$ (see page 308)

(b)

$$\begin{aligned} F(t) &= \int_0^t \lambda e^{-\lambda x}\, dx \qquad t \geqslant 0 \\ &= [-e^{-\lambda x}]_0^t \\ &= -(e^{-\lambda t} - 1) \\ &= 1 - e^{-\lambda t} \end{aligned}$$

Therefore $F(x) = 1 - e^{-\lambda x}, \quad x \geqslant 0.$

Let X be the r.v. 'the time, in months, between major repairs'.

$$\begin{aligned} f(x) &= 0.05e^{-0.05x} \\ P(X > 24) &= 1 - F(24) \\ &= e^{-0.05(24)} \\ &= e^{-1.2} \\ &= 0.301 \quad \text{(3 S.F.)} \end{aligned}$$

The probability that a television set will not need major repair in a 2-year period is 0.301 (3 S.F.).

Let m be the median value, then

$$\begin{aligned} F(m) &= 0.5 \\ \text{So} \qquad 1 - e^{-\lambda m} &= 0.5 \\ e^{-\lambda m} &= 0.5 \\ -\lambda m &= \ln 0.5 \\ m &= -\frac{1}{0.05} \ln 0.5 \\ &= 13.9 \text{ months} \quad \text{(3 S.F.)} \end{aligned}$$

The median is 13.9 months (3 S.F.).

We require $P(X < M) \leqslant 0.2$

Therefore
$$1 - e^{-0.05M} \leqslant 0.2$$
$$e^{-0.05M} \geqslant 0.8$$
$$-0.05M \geqslant \ln 0.8$$
$$M \leqslant -\frac{\ln 0.8}{0.05}$$
$$M \leqslant 4.46$$

Since M is an integer $M = 4$

The company agrees to replace any set for which the time between major repairs is less than 4 months.

Example 5.24 The lifetime of a particular type of lightbulb has a negative exponential distribution with mean lifetime 1000 hours.

(a) Find the probability that a bulb is still working after 1300 hours.

(b) Given that it is still working after 1300 hours, find the probability that it is still working after 1500 hours.

Solution 5.24 Let X be the r.v. 'the lifetime of a lightbulb in hours',

Then $f(x) = \lambda e^{-\lambda x} \quad x \geqslant 0.$

Now $E(X) = \dfrac{1}{\lambda}$

But $E(X) = 1000$, therefore $\dfrac{1}{\lambda} = 1000$

$$\lambda = 0.001$$

So $f(x) = 0.001e^{-0.001x}$

(a) $P(X > x) = e^{-\lambda x}$ (see p. 309)

$$P(X > 1300) = e^{-0.001(1300)}$$
$$= e^{-1.3}$$
$$= 0.273 \quad \text{(3 S.F.)}$$

The probability that a bulb is still working after 1300 hours is 0.273 (3 S.F.).

(b) $P(X > 1500 \mid X > 1300) = P(X > 200)$ (see p. 309)

$$= e^{-0.001(200)}$$

$$= e^{-0.2}$$

$$= 0.819 \quad (3 \text{ S.F.})$$

The probability that the bulb is still working after 1500 hours, given that it is still working after 1300 hours, is 0.819 (3 S.F.).

THE LINK BETWEEN THE EXPONENTIAL DISTRIBUTION AND THE POISSON DISTRIBUTION

The exponential distribution can be regarded as the 'waiting time' between events following a Poisson distribution. This is illustrated as follows:

Cars arrive at a petrol station at an average rate of λ per minute. Let X be the r.v. 'the number of cars arriving in t minutes'. Assuming that the number of cars arriving follows a Poisson distribution and we 'expect' λt to arrive in t minutes, then $X \sim \text{Po}(\lambda t)$.

Now $\quad P(X = 0) = e^{-\lambda t}$

and $\quad P(X \geqslant 1) = 1 - e^{-\lambda t}$

So $\quad P(\text{at least one car arrives}) = 1 - e^{-\lambda t}$

Now let T be the r.v. 'the length of time before a car arrives'.

We have $\quad P(\text{waiting time is less than } t \text{ minutes})$

$$= P(\text{at least one car arrives in } t \text{ minutes})$$

So $\quad P(T \leqslant t) = 1 - e^{-\lambda t}$

Therefore $\quad F(t) = 1 - e^{-\lambda t}$ (cumulative distribution function)

Now $\quad f(t) = F'(t)$

$$= \lambda e^{-\lambda t}$$ (probability density function)

This is the exponential distribution, with parameter λ.

NOTE: the parameter λ is the same value as the respective Poisson parameter, and the units of time are the same in both distributions.

Example 5.25 On a stretch of road, breakdowns occur at an average rate of 2 per day, and the number of breakdowns follows a Poisson distribution. Find

(a) the mean time between breakdowns,

(b) the median time between breakdowns.

Solution 5.25 Let T be the r.v. 'the time between breakdowns'. Then T follows an exponential distribution with parameter 2 where

$$f(t) = 2e^{-2t} \quad t \geqslant 0 \text{ and } t \text{ is in days}$$

(a) $E(T) = \dfrac{1}{\lambda} = \dfrac{1}{2}$

The mean time between breakdowns is half a day.

(b) The median is m, where

$$\begin{aligned} 0.5 &= \int_0^m 2e^{-2t}\,dt \\ &= [-e^{-2t}]_0^m \\ &= -(e^{-2m} - 1) \\ &= 1 - e^{-2m} \end{aligned}$$

Therefore

$$\begin{aligned} e^{-2m} &= 0.5 \\ -2m &= \ln 0.5 \qquad \text{(taking logs to base e)} \\ m &= -\tfrac{1}{2}\ln 0.5 \\ &= 0.3465 \text{ days} \quad \text{(4 d.p.)} \\ &= 8 \text{ hours} \quad \text{(approx.)} \end{aligned}$$

Therefore the median time between breakdowns is approximately 8 hours.

Exercise 5g

1. A continuous r.v. X has p.d.f. $f(x)$ where $f(x) = 5e^{-5x}$, $x \geqslant 0$. Find (*a*) $P(X > 0.5)$, (*b*) $E(X)$, (*c*) $P(X < E(X))$, (*d*) the standard deviation of X, (*e*) the median, (*f*) the mode.

2. The lifetime, in thousands of hours, of Extralight lightbulbs follows an exponential distribution with p.d.f. $f(x) = 0.5e^{-0.5x}$.
(*a*) Find the mean lifetime.
(*b*) A bulb is selected at random. Find the probability that it lasts (i) more than 2500 hours, (ii) less than 1800 hours.
(*c*) Two lightbulbs are selected at random. Find the probability that one lasts more than the mean number of hours and the other lasts for less than the mean number of hours.
(*d*) A random sample of 6 lightbulbs is chosen. Find the probability that exactly 4 will each last more than 2500 hours.

3. A batch of high-power light bulbs is such that the probability that any bulb fails before x hours, when kept on continuously, is $F(x) = 1 - e^{-x/10}$, $(x \geqslant 0)$. Find
(*a*) the median time to failure,
(*b*) the density function of the distribution of the time to failure,
(*c*) the mean and the variance of the distribution
(*d*) the probability that a bulb will fail between five and ten hours. (O)

4. The lifetime T, in years, of articles produced by a manufacturer can be modelled by the probability density function given by

$$f(t) = ae^{-at} \cdot t \geqslant 0,$$
$$f(t) = 0, \ t < 0.$$

Prove that the mean of T is $\frac{1}{a}$ and its median is $\frac{\ln 2}{a}$.

The articles are produced at a unit cost of £10 and sold for £25. Research shows that 50% of those produced fail within the first five years of life. Find the value of a.

After some time in business the manufacturer decides to guarantee free replacement of items which fail during their first year, but at the same time he raises the price so that the increase covers the expected cost of providing the guarantee. What should the new price be?

If two items are purchased what is the probability that just one will be replaced under guarantee? (SUJB)

5. Describe the conditions under which it is appropriate to use the Exponential Distribution, supporting your answer with reference to an experiment you may have carried out.

A major road construction project is underway. In the site supervisor's office, there is an average of two telephone calls every 5 minutes. Stating any assumptions you make, write down the probability that in a period of t minutes there is
(*a*) no telephone call,
(*b*) at least one telephone call.

Presenting a carefully reasoned argument, give the cumulative distribution function, $F(t)$, for the length of time between telephone calls. Hence establish that the probability density function, $f(t)$, is

$$f(t) = 0.4e^{-0.4t} \quad t > 0.$$

Calculate
(*c*) the mean time between calls,
(*d*) the median time between calls.

Given that the supervisor has had no call in the last 3 minutes, what is the probability that he could leave the office for 5 minutes without missing a call? (O)

6. A continuous random variable T has a negative exponential distribution given by

$$f(t) = Ae^{-t/a} \quad t \geqslant 0$$
$$= 0 \quad \text{elsewhere.}$$

Show that $A = 1/a$ and that the mean and variance of T are a and a^2 respectively.

(You may assume that $\int_0^{\infty} t^n e^{-t/a} \, dt = \alpha^{n+1} n!$ for integral values of n.)

The life in hours of a type of electric battery can be modelled by the above distribution and when a sample of 800 is tested the mean life is found to be 92.2 h. What are the values of A and a based on this figure?
(*a*) What is the probability that a battery will last for at least 200 h?
(*b*) If a battery has lasted 200 h what is the probability that it will last for at least a further 100 h?
(*c*) If two batteries are bought what is the probability that one fails before 200 h and the other after 200 h? (SUJB)

7. The random variable X can take all values between 0 and a inclusive, where $a > 0$. Its probability density function $f(x)$ is zero for $x < 0$ and $x > a$, and, for $0 \leqslant x \leqslant a$, satisfies

$$f(x) = (A/a)\exp(-x/a),$$

where A is a positive constant. Show by integration that $A = 1.582$ to 3 decimal places.

Also use integration to find to 2 decimal places
(i) the probability that X is less than $\frac{1}{2}a$;
(ii) the number λ for which there is a probability $\frac{1}{2}$ that X is less than λa. (MEI)

8. Explain briefly, from your projects if possible, a real-life situation that can be modelled by an exponential distribution.

An archer shoots arrows at a target. The distance X cm from the centre of the target at which an arrow strikes the target has probability density function, f, defined by

$$f(x) = \tfrac{1}{10} e^{-x/10} \quad x > 0.$$
$$f(x) = 0 \quad \text{otherwise.}$$

An arrow scores 8 points if $X \leqslant 2$, 5 points if $2 < X \leqslant 5$, one point if $5 < X \leqslant 15$ and no points otherwise. Find, to 3 decimal places, the expected score when one arrow is shot at the target. (L)

9. Find the mean of the random variable X which has an exponential distribution with probability density function

$f(x) = \lambda e^{-\lambda x}$ for $x \geqslant 0$ where $\lambda > 0$

$f(x) = 0$ for $x < 0$

For people suffering from a mental illness, the time in days from the end of a treatment to the occurrence of renewed symptoms is an exponential random variable with parameter $\lambda > 0$. Find, in terms of λ and t, the probability that neither of two randomly chosen sufferers from the illness will show renewed symptoms for a time t days after a treatment.

Given that two patients have no renewed symptoms for a time t days after a treatment, find, in terms of λ and t, the probability that both will remain free of symptoms for a further t days.

During a routine check at time t days after his treatment, another patient is found to be showing renewed symptoms. Find, in terms of λ, k and t, the probability that the renewed symptoms first showed in this patient less than kt days before the day of the routine check, where $0 \leqslant k \leqslant 1$. (L)

10. A random variable X has the probability density function f given by

$$f(x) = ce^{-2x} \quad x > 0$$
$$f(x) = 0 \quad \text{otherwise.}$$

Find the value of c. Find also the mean and the variance of X.

[You may assume that

$\int_0^\infty x^2 e^{-2x}\,dx = \frac{1}{4}$.]

Find the distribution function of X. Hence, or otherwise, show that, for positive t and k,

$$P(X > t + k \mid X > k) = P(X > t)$$

Given that X is the lifetime in years of a particular type of indicator lamp that is alight continuously, explain in words the meaning of the above result.

Given that 2 such lamps, A and B, have already been alight for 3 months and 4 months respectively, find the probability that both will still be alight in 3 months time. (L)

THE NORMAL DISTRIBUTION

This distribution will be used extensively in the following chapters and it is a very important one in statistics. Here we consider some of its mathematical properties.

A continuous r.v. X having p.d.f. $f(x)$ where

$$f(x) = \frac{1}{\sigma\sqrt{2\pi}} e^{-(x-\mu)^2/2\sigma^2} \quad (-\infty < x < \infty)$$

is said to follow a normal distribution.

μ and σ^2 are the parameters of the distribution.

We write $X \sim N(\mu, \sigma^2)$.

EXPECTATION AND VARIANCE

If $X \sim N(\mu, \sigma^2)$ then

$$E(X) = \mu$$
$$\text{Var}(X) = \sigma^2$$

In the following, we assume that $\frac{1}{\sqrt{2\pi}}\int_{-\infty}^{\infty} e^{-\frac{1}{2}t^2}\,dt = 1$.)

$$E(X) = \int_{\text{all } x} x\, f(x)\, dx$$

$$= \frac{1}{\sigma\sqrt{2\pi}}\int_{-\infty}^{\infty} x\, e^{-(x-\mu)^2/2\sigma^2}\, dx$$

Now, let $t = \frac{x-\mu}{\sigma}$ so that $dt = \frac{1}{\sigma}dx$ and when $x = \infty$, $t = \infty$,
$x = -\infty$, $t = -\infty$.

So $$E(X) = \frac{1}{\sigma\sqrt{2\pi}}\int_{-\infty}^{\infty} (\mu + \sigma t)\, e^{-\frac{1}{2}t^2}\, \sigma\, dt$$

$$= \frac{\mu}{\sqrt{2\pi}}\int_{-\infty}^{\infty} e^{-\frac{1}{2}t^2}\, dt + \frac{\sigma}{\sqrt{2\pi}}\int_{-\infty}^{\infty} t\, e^{-\frac{1}{2}t^2}\, dt$$

$$= \mu + \frac{\sigma}{\sqrt{2\pi}}\left[-e^{-\frac{1}{2}t^2}\right]_{-\infty}^{\infty}$$

$$= \mu$$

Therefore $E(X) = \mu$.

$$\text{Var}(X) = \int_{\text{all } x} x^2 f(x)\, dx - \mu^2$$

$$= I - \mu^2 \quad \text{where} \quad I = \frac{1}{\sigma\sqrt{2\pi}}\int_{-\infty}^{\infty} (\mu + \sigma t)^2 e^{-\frac{1}{2}t^2}\, \sigma\, dt$$

$$I = \frac{1}{\sqrt{2\pi}}\left\{\mu^2\int_{-\infty}^{\infty} e^{-\frac{1}{2}t^2}\, dt + 2\mu\sigma\int_{-\infty}^{\infty} t\, e^{-\frac{1}{2}t^2}\, dt + \sigma^2\int_{-\infty}^{\infty} t^2 e^{-\frac{1}{2}t^2}\, dt\right\}$$

Now

$$\int_{-\infty}^{\infty} t^2 e^{-\frac{1}{2}t^2}\, dt = \int_{-\infty}^{\infty} t(t\, e^{-\frac{1}{2}t^2})\, dt$$

$$= \left[t(-e^{-\frac{1}{2}t^2})\right]_{-\infty}^{\infty} - \int_{-\infty}^{\infty} -e^{-\frac{1}{2}t^2}\, dt$$

$$= 0 + \sqrt{2\pi}$$

So $$I = \frac{1}{\sqrt{2\pi}}\left(\mu^2\sqrt{2\pi} + 2\mu\sigma\left[-e^{-\frac{1}{2}t^2}\right]_{-\infty}^{\infty} + \sigma^2\sqrt{2\pi}\right)$$

$$= \mu^2 + \sigma^2 \quad \text{since} \quad e^{-\frac{1}{2}t^2} \to 0 \text{ as } t \to \pm\infty$$

Therefore $\quad \mathrm{Var}(X) = \mu^2 + \sigma^2 - \mu^2$

$$= \sigma^2$$

So, $E(X) = \mu$ and $\mathrm{Var}(X) = \sigma^2$.

The following results are also important

Result 1 If $X \sim \mathrm{N}(\mu, \sigma^2)$, the maximum value of $f(x)$ occurs when $x = \mu$.

We consider
$$f(x) = \frac{1}{\sigma\sqrt{2\pi}} e^{-(x-\mu)^2/2\sigma^2}$$

$$f'(x) = \frac{1}{\sigma\sqrt{2\pi}} \left\{-\frac{(x-\mu)}{\sigma^2}\right\} e^{-(x-\mu)^2/2\sigma^2}$$

$$= -\frac{1}{\sigma^3\sqrt{2\pi}}(x-\mu)\, e^{-(x-\mu)^2/2\sigma^2}$$

Now $\quad f'(x) = 0 \quad$ when $\quad x-\mu = 0$

i.e. $\quad x = \mu$

Now

$$f''(x) = -\frac{1}{\sigma^3\sqrt{2\pi}}\left\{-\frac{(x-\mu)^2}{\sigma^2} e^{-(x-\mu)^2/2\sigma^2} + e^{-(x-\mu)^2/2\sigma^2}\right\}$$

$$= -\frac{1}{\sigma^3\sqrt{2\pi}} e^{-(x-\mu)^2/2\sigma^2}\left\{-\frac{(x-\mu)^2}{\sigma^2} + 1\right\}$$

When $x = \mu$, $f''(x) < 0$.

There is a maximum value of $f(x)$ when $x = \mu$.

Result 2 If $X \sim \mathrm{N}(\mu, \sigma^2)$ then $f(x)$ has points of inflexion at $x = \mu + \sigma$ and $x = \mu - \sigma$.

To show this, consider $f''(x)$.

$$f''(x) = 0 \quad \text{when} \quad (x-\mu)^2 = \sigma^2$$

$$x - \mu = \pm\sigma$$

$$x = \mu + \sigma \quad \text{or} \quad x = \mu - \sigma$$

There are points of inflexion at $x = \mu + \sigma$ and $x = \mu - \sigma$.

NOTE: Sketch of $y = f(x)$.

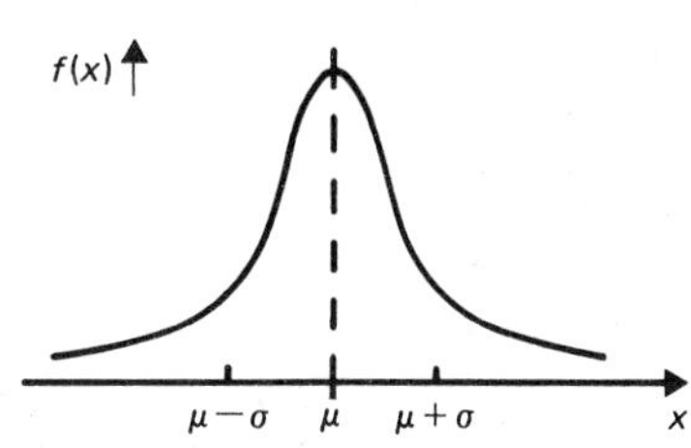

MISCELLANEOUS WORKED EXAMPLES

Example 5.26 A random variable X has a probability density function

$$f(x) = Ax(6-x)^2 \quad 0 \leqslant x \leqslant 6$$
$$= 0 \text{ elsewhere.}$$

Find the value of the constant A.

Calculate the arithmetic mean, mode, variance and standard deviation of X. (AEB)

Solution 5.26 Since X is a r.v. $\int_{\text{all } x} f(x)\,dx = 1$

Therefore $1 = \int_0^6 Ax(6-x)^2\,dx$

$$= A\int_0^6 (36x - 12x^2 + x^3)\,dx$$

$$= A[18x^2 - 4x^3 + \tfrac{1}{4}x^4]_0^6$$

$$= 108A$$

So $A = \dfrac{1}{108}$

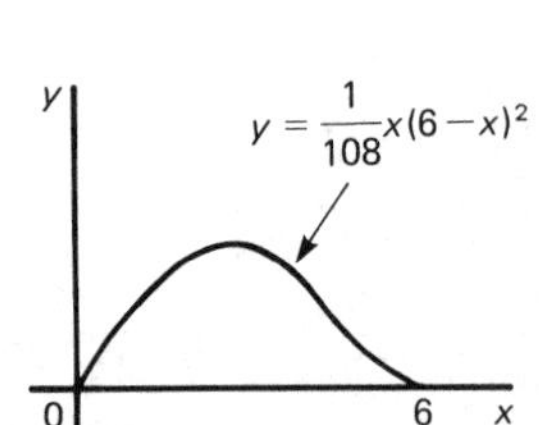

Therefore $f(x) = \dfrac{1}{108}x(6-x)^2 \quad 0 \leqslant x \leqslant 6$

The arithmetic mean is $E(X)$ where $E(X) = \int_{\text{all } x} x\,f(x)\,dx$

$$E(X) = \frac{1}{108}\int_0^6 x^2(6-x)^2\,dx$$

$$= \frac{1}{108}\int_0^6 (36x^2 - 12x^3 + x^4)\,dx$$

$$= \frac{1}{108}\left[12x^3 - 3x^4 + \frac{x^5}{5}\right]_0^6$$

$$= 2.4$$

The arithmetic mean is 2.4.

To find the mode we consider the maximum point on $y = f(x)$.

$$f(x) = \frac{1}{108}(36x - 12x^2 + x^3)$$

$$f'(x) = \frac{1}{108}(36 - 24x + 3x^2)$$

$$= \frac{3}{108}(6-x)(2-x)$$

$$f'(x) = 0 \quad \text{when} \quad x = 2 \quad \text{and when} \quad x = 6$$

Consider $f''(x) = \frac{3}{108}(6x - 24)$.

When $x = 2$, $f''(x) < 0$ and when $x = 6$, $f''(x) > 0$.

Therefore $x = 2$ gives a maximum value of $f(x)$, i.e. the mode is 2.

$$\text{Var}(X) = E(X^2) - E^2(X)$$

Now
$$E(X^2) = \int_{\text{all } x} x^2 f(x)\,dx$$

$$= \frac{1}{108}\int_0^6 (36x^3 - 12x^4 + x^5)\,dx$$

$$= \frac{1}{108}\left[9x^4 - \frac{12x^5}{5} + \frac{x^6}{6}\right]_0^6$$

$$= 7.2$$

$$\text{Var}(X) = 7.2 - (2.4)^2$$

$$= 1.44$$

$$\text{Standard deviation of } X = \sqrt{\text{Var}(X)}$$

$$= 1.2$$

Therefore the variance of X is 1.44 and the standard deviation is 1.2.

Example 5.27 A continuous random variable, X, has probability density function given by

$$f(x) = ax - bx^2 \quad \text{for} \quad 0 \leqslant x \leqslant 2$$
$$= 0 \quad \text{elsewhere}$$

Observations on X indicate that the mean is 1.

(a) Obtain two simultaneous equations for a and b, show that $a = 1.5$ and find the value of b.

(b) Find the variance of X.

(c) If $F(x)$ is the probability that $X \leqslant x$ find $F(x)$ and verify that $F(2) = 1$.

(d) If two independent observations are made on X what is the probability that at least one of them is less than $\frac{1}{2}$? (SUJB)

Solution 5.27 (a) If X is a random variable $\int_{\text{all } x} f(x)\,dx = 1$. Therefore

$$\int_0^2 (ax - bx^2)\,dx = 1$$

$$\left[\frac{ax^2}{2} - \frac{bx^3}{3}\right]_0^2 = 1$$

$$2a - \frac{8b}{3} = 1$$

so

$$\frac{8b}{3} = 2a - 1 \qquad \text{(i)}$$

Now

$$E(X) = \int_{\text{all } x} x\,f(x)\,dx$$

$$= \int_0^2 (ax^2 - bx^3)\,dx$$

$$= \left[\frac{ax^3}{3} - \frac{bx^4}{4}\right]_0^2$$

$$= \frac{8a}{3} - 4b$$

But, we are given that $E(X) = 1$. So

$$\frac{8a}{3} - 4b = 1$$

$$4b = \frac{8a}{3} - 1$$

so

$$\frac{8b}{3} = \frac{16a}{9} - \frac{2}{3} \qquad \text{(ii)}$$

From equations (i) and (ii) we have

$$\frac{16a}{9} - \frac{2}{3} = 2a - 1$$

$$\frac{1}{3} = \frac{2a}{9}$$

$$a = \frac{3}{2}$$

Substituting for a in (i) we have $\dfrac{8b}{3} = 3-1$

$$b = \frac{3}{4}$$

Therefore $a = 1.5$ and $b = \frac{3}{4}$.

(b)

$$\begin{aligned} E(X^2) &= \int_{\text{all } x} x^2 f(x)\,dx \\ &= \int_0^2 \left(\frac{3}{2}x^3 - \frac{3}{4}x^4\right) dx \\ &= \left[\frac{3}{8}x^4 - \frac{3}{20}x^5\right]_0^2 \\ &= 6-4.8 \\ &= 1.2 \end{aligned}$$

Now

$$\begin{aligned} \text{Var}(X) &= E(X^2) - E^2(X) \\ &= 1.2-1 \\ &= 0.2 \end{aligned}$$

Therefore $\text{Var}(X) = 0.2$.

(c)

$$\begin{aligned} F(t) &= P(X \leqslant t) \\ &= \int_0^t \left(\frac{3}{2}x - \frac{3}{4}x^2\right) dx \\ &= \left[\frac{3}{4}x^2 - \frac{1}{4}x^3\right]_0^t \\ &= \frac{3}{4}t^2 - \frac{1}{4}t^3 \end{aligned}$$

So

$$F(x) = \frac{3}{4}x^2 - \frac{1}{4}x^3$$

and

$$\begin{aligned} F(2) &= \frac{3}{4}(4) - \frac{1}{4}(8) \\ &= 1 \quad \text{as required} \end{aligned}$$

(d)

$$\begin{aligned} P(X < \tfrac{1}{2}) &= F(\tfrac{1}{2}) \\ &= \frac{3}{4}\left(\frac{1}{2}\right)^2 - \frac{1}{4}\left(\frac{1}{2}\right)^3 \\ &= \frac{5}{32} \end{aligned}$$

Therefore $\qquad P(X \geqslant \tfrac{1}{2}) = \dfrac{27}{32}$

So if two independent observations are made,

$$\begin{aligned} P(\text{at least one is less than } \tfrac{1}{2}) &= 1 - P(\text{both} \geqslant \tfrac{1}{2}) \\ &= 1 - \left(\frac{27}{32}\right)\left(\frac{27}{32}\right) \\ &= 0.288 \quad (3\text{ d.p.}) \end{aligned}$$

Therefore if two independent observations are made on X, the probability that at least one of them is less than $\frac{1}{2}$ is 0.288 (3 d.p.).

Example 5.28 The time taken to perform a particular task, t hours, has the probability density function

$$f(t) = \begin{cases} 10ct^2 & 0 \leqslant t < 0.6 \\ 9c(1-t) & 0.6 \leqslant t \leqslant 1.0 \\ 0 & \text{otherwise.} \end{cases}$$

where c is a constant.

(a) Find the value of c and sketch the graph of this distribution.

(b) Write down the most likely time.

(c) Find the expected time.

(d) Determine the probability that the time will be
(i) more than 48 minutes,
(ii) between 24 and 48 minutes.

Solution 5.28 (a) Now

$$\begin{aligned} 1 &= \int_{\text{all } t} f(t)\,dt \\ &= 10c\int_0^{0.6} t^2\,dt + 9c\int_{0.6}^{1.0} (1-t)\,dt \\ &= \frac{10c}{3}[t^3]_0^{0.6} + 9c\left[t - \frac{t^2}{2}\right]_{0.6}^{1.0} \\ &= 0.72c + 0.72c \\ &= 1.44c \end{aligned}$$

Therefore
$$\begin{aligned} c &= \frac{1}{1.44} \\ &= \frac{100}{144} \\ &= \frac{25}{36} \end{aligned}$$

We have

$$f(t) = \begin{cases} \frac{125}{18}t^2 & 0 \leqslant t < 0.6 \\ \frac{25}{4}(1-t) & 0.6 \leqslant t \leqslant 1.0 \\ 0 & \text{otherwise.} \end{cases}$$

(b) $t = 0.6$ gives the maximum value of $f(t)$.

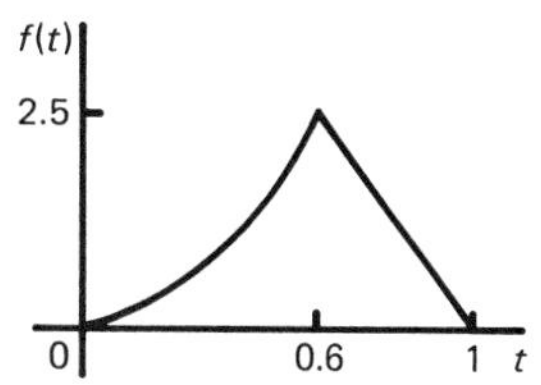

Therefore the mode is 0.6 hours = 36 mins.

The most likely time is 36 minutes.

(c)

$$\begin{aligned} E(t) &= \int_{\text{all } t} tf(t)\,\mathrm{d}t \\ &= 10c \int_0^{0.6} t^3\,\mathrm{d}t + 9c \int_{0.6}^{1.0} (t-t^2)\,\mathrm{d}t \\ &= \frac{10c}{4}[t^4]_0^{0.6} + 9c\left[\frac{t^2}{2} - \frac{t^3}{3}\right]_{0.6}^{1.0} \\ &= 0.225 + 0.366\ldots \\ &= 0.591\ldots \text{ hours} \\ &= 35.5 \text{ minutes} \end{aligned}$$

The expected time is 35.5 minutes.

(d) (i) 48 minutes = 0.8 hours.

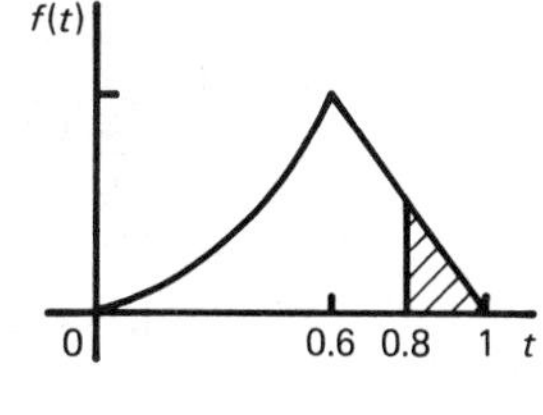

$$\begin{aligned} P(T > 0.8) &= 9c \int_{0.8}^{1.0} (1-t)\,\mathrm{d}t \\ &= 9c\left[t - \frac{t^2}{2}\right]_{0.8}^{1.0} \\ &= 0.125 \end{aligned}$$

The probability that the time will be more than 48 minutes is 0.125.

(ii) 24 minutes = 0.4 hours.

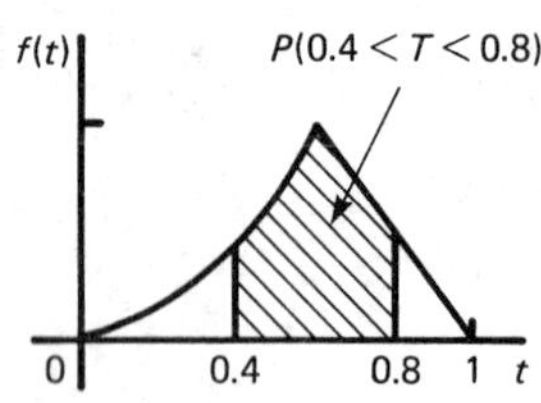

Now

$$P(0.4 < T < 0.8) = 1 - P(T > 0.8) - P(T < 0.4)$$

and

$$P(T < 0.4) = 10c \int_0^{0.4} t^2 \, dt$$

$$= \frac{10c}{3} [t^3]_0^{0.4}$$

$$= 0.1481 \ldots$$

Therefore

$$P(0.4 < T < 0.8) = 1 - 0.125 - 0.1481 \ldots$$

$$= 0.727 \quad (3 \text{ S.F.})$$

The probability that the time will be between 24 and 48 minutes is 0.727 (3 S.F.).

SUMMARY — CONTINUOUS RANDOM VARIABLES

For a continuous random variable X, with p.d.f. $f(x)$ $(a \leqslant x \leqslant b)$

$$\int_{\text{all } x} f(x) \, dx = 1$$

$$P(c \leqslant X \leqslant d) = \int_c^d f(x) \, dx \qquad a \leqslant c < d \leqslant b$$

$$E(X) = \int_{\text{all } x} x f(x) \, dx$$

$$\text{Var}(X) = \int_{\text{all } x} x^2 f(x) \, dx - E^2(X)$$

$$F(t) = \int_a^t f(x) \, dx \quad a \leqslant t \leqslant b$$ where $F(t)$ is the cumulative distribution function

$$f(x) = \frac{d}{dx} F(x)$$

The rectangular distribution

If $f(x) = \dfrac{1}{b-a} \quad a \leqslant x \leqslant b,$ then $X \sim \mathrm{R}(a, b)$

$E(X) = \dfrac{1}{2}(a+b)$

$\mathrm{Var}(X) = \dfrac{1}{12}(b-a)^2$

The exponential distribution

If $f(x) = \lambda e^{-\lambda x} \quad x \geqslant 0$

$E(X) = \dfrac{1}{\lambda}$

$\mathrm{Var}(X) = \dfrac{1}{\lambda^2}$

The normal distribution

If $f(x) = \dfrac{1}{\sigma\sqrt{2\pi}} e^{-(x-\mu)^2/2\sigma^2} \quad -\infty < x < \infty$

then $X \sim \mathrm{N}(\mu, \sigma^2)$

$E(X) = \mu$

$\mathrm{Var}(X) = \sigma^2$

Miscellaneous Exercise 5h

1. (*a*) A continuous variable X is distributed at random between the values 2 and 3 and has a probability density function of $\dfrac{6}{x^2}$. Find the median value of X.
(*b*) A continuous random variable X takes values between 0 and 1, with a probability density function of $Ax(1-x)^3$. Find the value of A, and the mean and standard deviation of X. (SUJB)

2. A continuous variable X is distributed at random between 2 values, $x = 0$ and $x = 2$, and has a probability density function of $ax^2 + bx$. The mean is 1.25.
(i) Show that $b = \frac{3}{4}$, and find the value of a.
(ii) Find the variance of X.
(iii) Verify that the median value of X is approximately 1.3.
(iv) Find the mode. (SUJB)

3. The random variable X is the distance, in metres, that an inexperienced tight-rope walker has moved along a given tight-rope before falling off. It is given that

$$P(X > x) = 1 - \frac{x^3}{64} \quad 0 \leqslant x \leqslant 4$$

(*a*) Show that $E(X) = 3$.
(*b*) Find the standard deviation, σ, of X.
(*c*) Show that $P(|X-3| < \sigma) = \frac{69}{80}\sqrt{\frac{3}{5}}$. (C)

4. The random variable X has a probability density function given by

$$P(x) = \begin{cases} kx(1-x^2) & (0 \leqslant x \leqslant 1) \\ 0 & \text{elsewhere} \end{cases}$$

k being a constant. Find the value of k and find also the mean and variance of this distribution.

Find the median of the distribution. (O & C)

5. An ironmonger is supplied with paraffin once a week. The weekly demand, X hundred litres, has the probability density function f, where

$$f(x) = c(1-x)^7 \quad 0 \leqslant x \leqslant 1$$
$$f(x) = 0 \quad \text{otherwise}$$

where c is a constant. Find the value of c.

Find the mean value of X, and, to the nearest litre, the minimum capacity of his paraffin tank if the probability that it will be exhausted in a given week is not to exceed 0.02. (L)P

6. A continuous random variable X has probability density function $f(x)$ given by $f(x) = 0$ for $x < 0$ and $x > 3$ and between $x = 0$ and $x = 3$ its form is as shown in the graph.

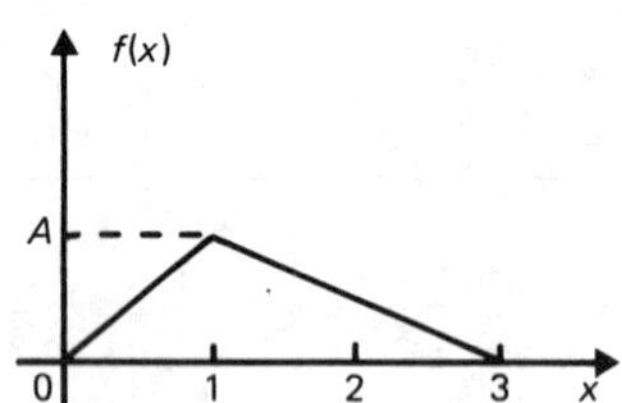

(a) Find the value of A.
(b) Express $f(x)$ algebraically and obtain the mean and variance of X.
(c) Find the median value of X.
A sample X_1, X_2 and X_3 is obtained. What is the probability that at least one is greater than the median value? (SUJB)

7. Determine λ such that

$$f(x) = \begin{cases} 0 & x < 0 \\ \lambda/2 & 0 \leqslant x \leqslant 1 \\ 0 & 1 < x < 2 \\ 3\lambda/2 - 3\lambda(x-3)^2/4 & 2 \leqslant x \leqslant 4 \\ 0 & x > 4 \end{cases}$$

is a probability density function of the distribution of a random variable X. Sketch the density function and find $E(X)$ and $\Pr(X \leqslant 3.5)$. (MEI)

8. The probability density function of X is given by

$$f(x) = \begin{cases} k(ax-x^2) & 0 \leqslant x \leqslant 2 \\ 0 & x < 0, \quad x > 2 \end{cases}$$

where k and a are positive constants.

Show that $a \geqslant 2$ and that $k = \dfrac{3}{6a-8}$.

Given that the mean value of X is 1, calculate the values of a and k.

For these values of a and k sketch the graph of the probability density function and find the variance of X. (JMB)

9. A continuous random variable X has probability density function $f(x)$ defined by

$$f(x) = \begin{cases} 12(x^2-x^3) & 0 \leqslant x \leqslant 1 \\ 0 & \text{otherwise} \end{cases}$$

Find the mean and standard deviation of X; find also its mean deviation about the mean. (O & C)

10. The continuous random variable X has probability density function $f(x)$ defined by

$$f(x) = \begin{cases} \dfrac{c}{x^4} & (x < -1) \\ c(2-x^2) & (-1 \leqslant x \leqslant 1) \\ \dfrac{c}{x^4} & (x > 1) \end{cases}$$

(a) Show that $c = \frac{1}{4}$.
(b) Sketch the graph of $f(x)$.
(c) Determine the cumulative distribution function $F(x)$.
(d) Determine the expected value of X and the variance of X. (C)

11. The random variable X takes all values x in the range $0 \leqslant x \leqslant 1$, and has a continuous probability density function $f(x)$ defined by

$$f(x) = kx^{\theta-1}(1-x)^2 \quad (\theta \geqslant 1)$$

(a) Show that $k = \frac{1}{2}\theta(\theta+1)(\theta+2)$.
(b) Find $E(X)$ and $E(X^2)$.
(c) Deduce the variance of X.
(d) For $\theta = \frac{3}{2}$, find the location of the mode and sketch $f(x)$. (O & C)

12. State the conditions under which the binomial distribution is a suitable model to use in statistical work. Describe briefly how you used, or could have used, a

binomial distribution in a project, giving the parameters of your distribution.

A large store sells a certain size of nail either in a small packet at 50p per packet, or loose at £3 per kg. On any shopping day the number, X, of packets sold is a random variable where $X \sim B(8, 0.6)$, and the weight, Y kg, of nails sold loose is a continuous random variable with probability density function f given by

$$f(y) = \frac{2(y-1)}{25} \quad 1 \leqslant y \leqslant 6,$$
$$f(y) = 0, \quad \text{otherwise.}$$

Find, to 3 decimal places, the probability that, on any shopping day, the number of packets sold will be

(*a*) more than one,

(*b*) seven or fewer.

Find the probability that

(*c*) the weight of nails sold loose on any shopping day will be between 4 kg and 5 kg.

(*d*) on any one shopping day the shop will sell exactly 2 packets of nails and less than 2 kg of nails sold loose, giving your answer to 2 significant figures.

(*e*) Calculate the expected money received on any shopping day from the sale of this size of nail in this store. (L)

13. A beam of electrons is directed at a solid object. The depth, X, to which any given electron will penetrate the object before colliding with an atom is a random variable whose probability density function is f.
(i) If $f(x) = a - bx (a > 0, b > 0, 0 \leqslant x \leqslant a/b)$, find b in terms of a. Find $E(X)$ and $\text{Var}(X)$ in terms of a.
If $f(x) = a\,e^{-cx}$ $(a > 0, c > 0, x \geqslant 0)$ find c in terms of a. Show that, if a fraction $1/N$ of all electrons penetrates to a depth greater than d, then $d = \frac{1}{a}\ln N$. (C)

14. The number of kilograms of metal extracted from 10 kg of ore from a certain mine is a continuous random variable X with probability density function $f(x)$, where $f(x) = cx(2-x)^2$ if $0 \leqslant x \leqslant 2$ and $f(x) = 0$ otherwise, where c is a constant.
Show that $c = \frac{3}{4}$, and find the mean and variance of X. The cost of extracting the metal from 10 kg of ore is £$10x$. Find the expected cost of extracting the metal from 10 kg of ore. (MEI)

15. (*a*) A discrete random variable R takes integer values between 0 and 4 inclusive with probabilities given by

$$P(R = r) = \begin{cases} \dfrac{r+1}{10} & (r = 0, 1, 2) \\ \dfrac{9-2r}{10} & (r = 3, 4) \end{cases}$$

Find the expectation and variance of R.
(*b*) A continuous random variable X takes values in the interval $x \geqslant 0$. The probability density function of X is defined by

$$f(x) = \begin{cases} kx & \text{if} \quad 0 \leqslant x \leqslant 1 \\ \dfrac{k}{x^4} & \text{if} \quad x > 1 \end{cases}$$

Prove that $k = \frac{6}{5}$ and find the expectation and variance of X. (C)

16. A continuous random variable X has probability density function defined by

$$f(x) = \begin{cases} 0 & x < 0, \\ kx & 0 \leqslant x \leqslant 2, \\ \dfrac{16k}{x^3} & x > 2. \end{cases}$$

Calculate the value of k. Find the median value and the expectation of X.
Prove that the standard deviation of X is infinite.
Find the value of a such that

$$P(X > a) = 0.005 \quad \text{(C)}$$

17. The distances x, in miles, travelled by customers to the 'Cheep Supermarket' are distributed with density function

$$f(x) = \begin{cases} \frac{1}{5}e^{-x/5} & 0 \leqslant x < \infty \\ 0 & \text{otherwise} \end{cases}$$

Find the proportion of customers travelling less than 1 mile and the proportion travelling more than 15 miles to the supermarket.
The chance that a customer goes twice to the supermarket on one day is $p(x)$ when the customer has to travel x miles each way and the chance of one visit only is $1-p(x)$, where

$$p(x) = \begin{cases} \frac{1}{4} & 0 \leqslant x < 5 \\ \frac{1}{8} & 5 \leqslant x < \infty \end{cases}$$

Find the expected distance travelled by a customer on one day. (MEI)

18. A person frequently makes telephone calls to destinations for which each call is charged at the rate of 15p per minute or part of a minute. The cost of such a call is X pence and its duration, T minutes, has the exponential probability density function

$$f(t) = \alpha e^{-\alpha t} \quad t \geqslant 0$$
$$f(t) = 0 \quad t < 0$$

Show that

$$P(X = 15r) = e^{-r\alpha}(e^{\alpha} - 1) \quad r = 1, 2, 3, \ldots$$

and that the mean cost per call in pence is

$$15/(1 - e^{-\alpha}).$$

[You may assume that $\sum_{r=1}^{\infty} r x^r = \frac{x}{(1-x)^2}$, $|x| < 1$.]

When a caller telephones a particular company, there is a probability of $\frac{1}{2}$ that he will be asked to hold the line. When he is asked to hold the line and decides to do so, the total time taken for the call has the above exponential distribution with $\alpha = \frac{1}{6}$; when he is not asked to hold the the line the total time for the call has the exponential distribution with $\alpha = \frac{1}{2}$. Calculate the expected cost if he rings the company and is asked to hold the line and he

(*a*) holds and completes the call,
(*b*) rings off and then rings later, completing the second call whether he is asked to hold or not.

Assume that a wasted call costs 15p and take $e^{-1/2} = 0.6065$, $e^{-1/6} = 0.8465$.

(JMB)

6

THE NORMAL DISTRIBUTION

The normal distribution is the most important continuous distribution in statistics. Many measured quantities in the natural sciences follow a normal distribution, for example heights, masses, ages, random errors, I.Q. scores, examination results.

PROBABILITY DENSITY FUNCTION OF NORMAL VARIABLE

A continuous random variable X having p.d.f. $f(x)$ where

$$f(x) = \frac{1}{\sigma\sqrt{2\pi}} e^{-(x-\mu)^2/2\sigma^2} \qquad -\infty < x < \infty$$

is said to have a normal distribution with mean μ and variance σ^2.

μ and σ^2 are the parameters of the distribution.

If X is distributed in this way we write

$$X \sim \mathrm{N}(\mu, \sigma^2)$$

Sketch of $y = f(x)$:

The distribution is bell shaped and symmetrical about $x = \mu$.

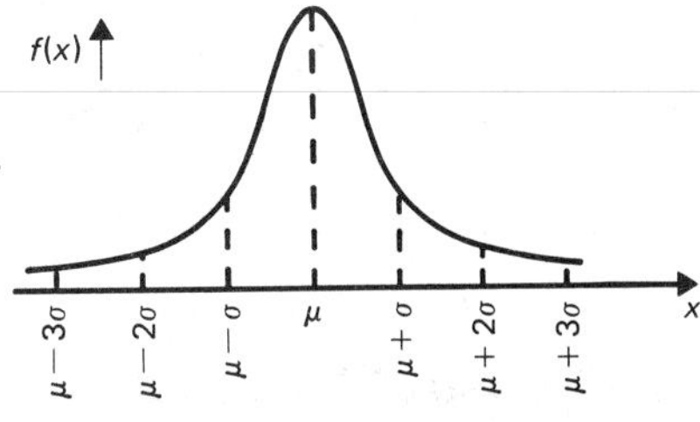

Approximately 95% of the distribution lies within ± 2 standard deviations of the mean.

Approximately 99.8% of the distribution lies within ±3 standard deviations of the mean.

The range of the distribution is therefore approximately 6 standard deviations.

The maximum value of $f(x)$ occurs when $x = \mu$ and is given by $f(x) = \dfrac{1}{\sigma\sqrt{2\pi}}$.

There is a point of inflexion at $x = \mu - \sigma$ and at $x = \mu + \sigma$.

The actual size of the bell-shaped curve depends on the values of μ and σ.

Here are some examples, each drawn to the same scale:

(1) $X \sim N(0, 1)$

(2) $X \sim N(100, 6.25)$

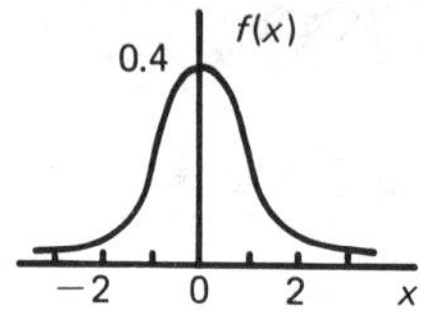

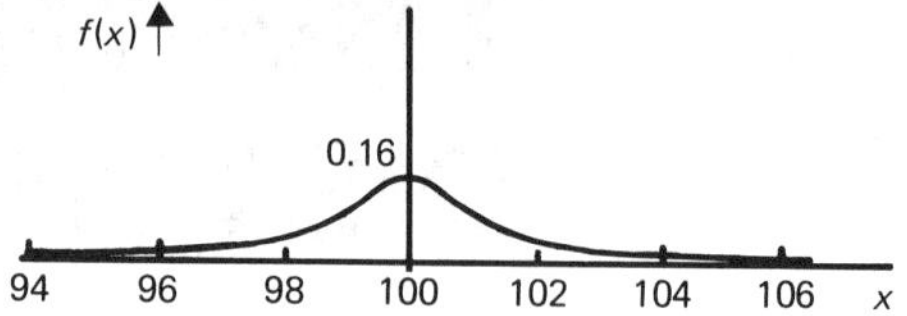

(3) $X \sim N(50, 4)$

(4) $X \sim N(4, \frac{1}{4})$

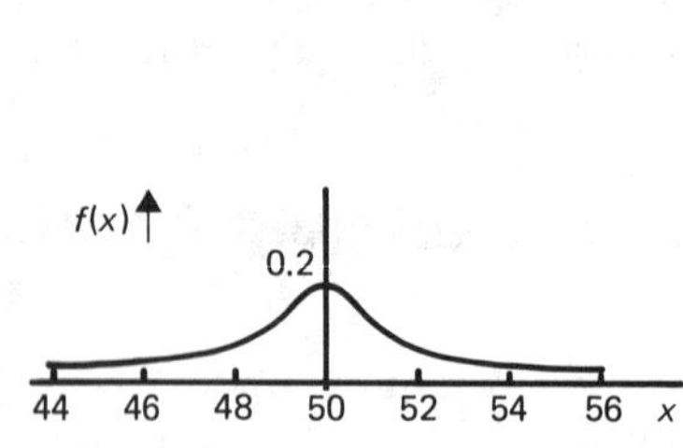

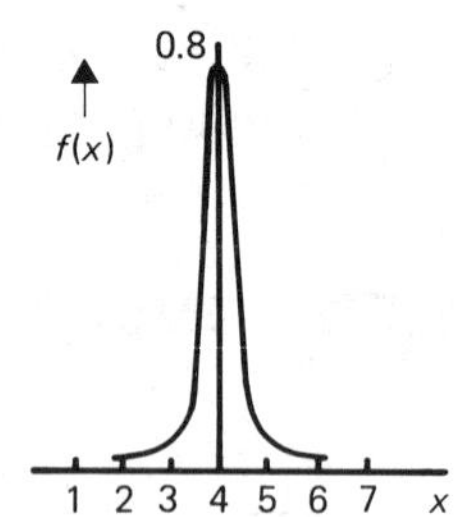

NOTE: the variable X is random. It is possible to show that

$$\int_{-\infty}^{\infty} \frac{1}{\sigma\sqrt{2\pi}} e^{-(x-\mu)^2/2\sigma^2} dx = 1$$

but this is beyond the scope of this book.

The probability that X lies between a and b is given by

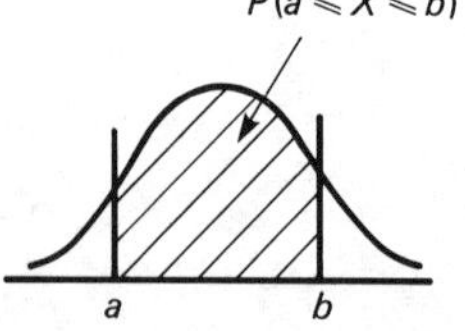

$$P(a \leqslant X \leqslant b) = \int_{a}^{b} \frac{1}{\sigma\sqrt{2\pi}} e^{-(x-\mu)^2/2\sigma^2} dx$$

However, this integral is *very difficult* to evaluate, so we work with the **standard normal** variable Z.

THE STANDARD NORMAL DISTRIBUTION

To standardise X, subtract μ and divide by σ. So

$$Z = \frac{X - \mu}{\sigma}$$

Now, if the r.v. X has a normal distribution with mean μ and variance σ^2, then the r.v. Z has a **standard normal distribution** with mean 0 and variance 1, i.e.

$$\text{if} \quad X \sim \mathrm{N}(\mu, \sigma^2) \quad \text{and} \quad Z = \frac{X-\mu}{\sigma}$$

$$\text{then} \quad Z \sim \mathrm{N}(0,1)$$

Example 6.1 Show that $E(Z) = 0$ and $\mathrm{Var}(Z) = 1$, where Z is the standard normal variable.

Solution 6.1

$$Z = \frac{X-\mu}{\sigma} \quad \text{and} \quad E(X) = \mu, \quad \mathrm{Var}(X) = \sigma^2$$

$$\begin{aligned} E(Z) &= E\left(\frac{X-\mu}{\sigma}\right) \\ &= \frac{1}{\sigma}[E(X)-E(\mu)] \\ &= \frac{1}{\sigma}(\mu-\mu) \\ &= 0 \end{aligned}$$

So $E(Z) = 0$

$$\begin{aligned} \mathrm{Var}(Z) &= \mathrm{Var}\left(\frac{X-\mu}{\sigma}\right) \\ &= \frac{1}{\sigma^2}[\mathrm{Var}(X)+\mathrm{Var}(\mu)] \\ &= \frac{1}{\sigma^2}[\mathrm{Var}(X)+0] \\ &= \frac{\sigma^2}{\sigma^2} \\ &= 1 \end{aligned}$$

So $\mathrm{Var}(Z) = 1$

THE PROBABILITY DENSITY FUNCTION FOR Z, $\phi(z)$

The p.d.f. of the standard normal variable Z is denoted by $\phi(z)$ where

$$\phi(z) = \frac{1}{\sqrt{2\pi}} e^{-\frac{1}{2}z^2} \quad -\infty < z < \infty$$

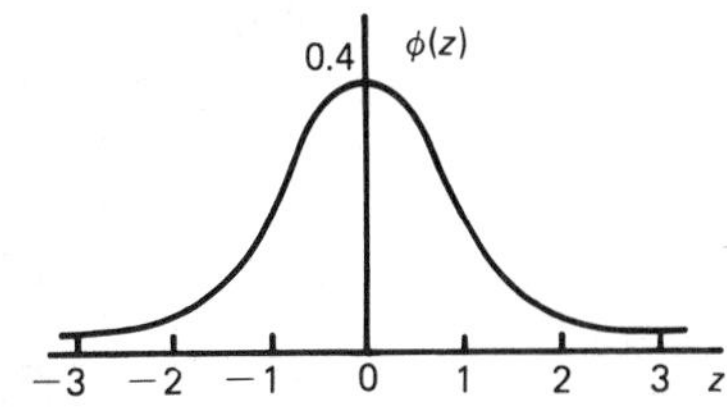

THE CUMULATIVE DISTRIBUTION FUNCTION FOR Z, $\Phi(z)$

The cumulative distribution function of the standard normal variable Z is denoted by $\Phi(z)$ where

$$\Phi(z) = P(Z < z) = \int_{-\infty}^{z} \frac{1}{\sqrt{2\pi}} e^{-\frac{1}{2}z^2} \, dz$$

This integral is still very difficult to evaluate, so we refer to tables.

It should be noted that the tables may be printed in one of two different formats.

They may give the values of $\Phi(z)$, where $\Phi(z) = P(Z < z)$

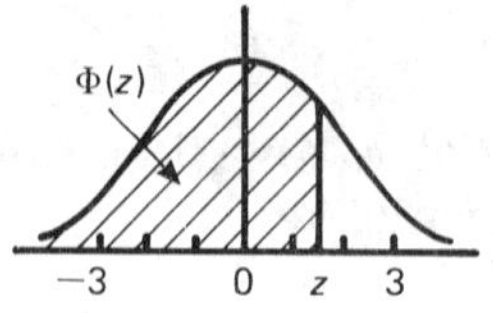

or the values of $Q(z)$, where $Q(z) = P(Z > z)$.

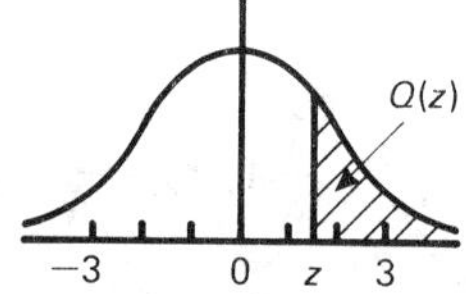

In the main text we will refer to the tables giving $\Phi(z)$, the cumulative probabilities of the standard normal distribution. These are printed on p. 634.

However, you will find instructions for the use of the $Q(z)$ tables in Appendix 2 on pp. 641–53. $Q(z)$ tables are printed on p. 640.

The values of $Q(z)$ are known as the 'Uppertail Probabilities'.

USE OF THE STANDARD NORMAL TABLES USING $\Phi(z)$

Only positive values of z are printed in the tables, so for negative values of z the symmetrical properties of the curve are used:

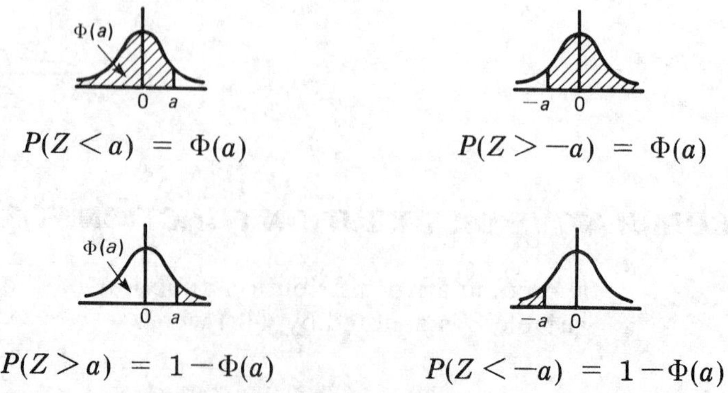

$P(Z < a) = \Phi(a)$ $\quad$ $P(Z > -a) = \Phi(a)$

$P(Z > a) = 1 - \Phi(a)$ $\quad$ $P(Z < -a) = 1 - \Phi(a)$

NOTE: We have $\Phi(-a) = 1 - \Phi(a)$.

Example 6.2 If $Z \sim N(0, 1)$, find from tables (a) $P(Z < 1.377)$, (b) $P(Z > -1.377)$, (c) $P(Z > 1.377)$, (d) $P(Z < -1.377)$.

Solution 6.2 (a)

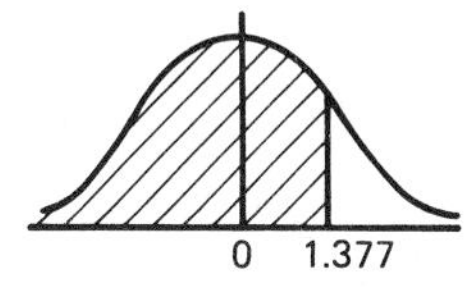

$$P(Z < 1.377) = \Phi(1.377) = 0.9158$$

(b)

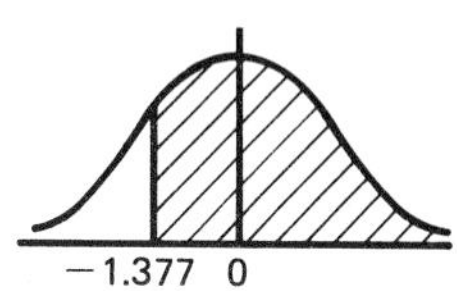

$$P(Z > -1.377) = P(Z < 1.377) = \Phi(1.377) = 0.9158$$

(c)

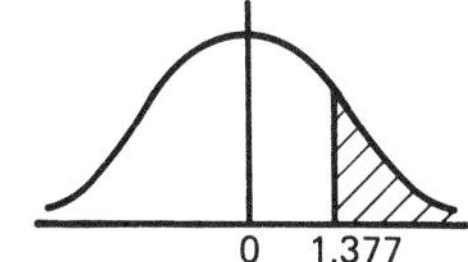

$$P(Z > 1.377) = 1 - \Phi(1.377) = 1 - 0.9158 = 0.0842$$

(b)

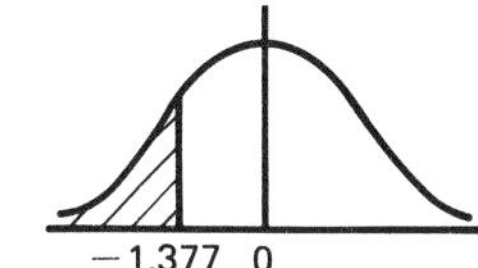

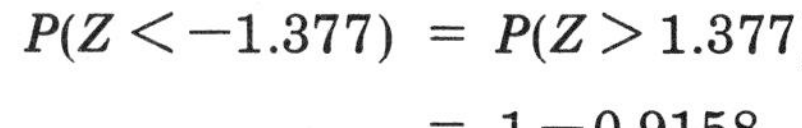

$$P(Z < -1.377) = P(Z > 1.377) = 1 - 0.9158 = 0.0842$$

Example 6.3 If $Z \sim N(0, 1)$, find (a) $P(0.345 < Z < 1.751)$, (b) $P(-2.696 < Z < 1.865)$, (c) $P(-1.4 < Z < -0.6)$, (d) $P(|Z| < 1.433)$, (e) $P(Z > 0.863 \text{ or } Z < -1.527)$.

Solution 6.3 (a)

$$P(0.345 < Z < 1.751) = \Phi(1.751) - \Phi(0.345) = 0.9600 - 0.6350 = 0.3250$$

0 0.345 1.751

So $P(0.345 < Z < 1.751) = 0.3250.$

(b)

$$P(-2.696 < Z < 1.865) = \Phi(1.865) - \Phi(-2.696) = \Phi(1.865) - (1 - \Phi(2.696)) = \Phi(1.865) + \Phi(2.696) - 1 = 0.9690 + 0.99650 - 1 = 0.9655$$

−2.696 0 1.865

So $P(-2.696 < Z < 1.865) = 0.9655.$

(c) $P(-1.4 < Z < -0.6) = \Phi(-0.6) - \Phi(-1.4)$

$= \Phi(1.4) - \Phi(0.6)$

$= 0.9192 - 0.7257$

$= 0.1935$

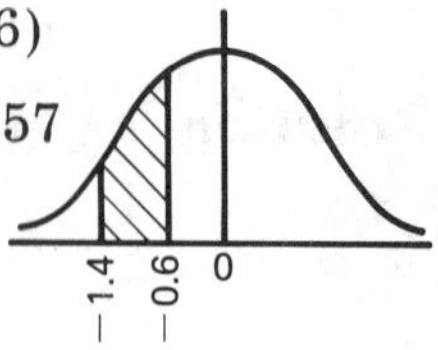

So $P(-1.4 < Z < -0.6) = 0.1935$.

−1.4 −0.6 0

(d) $P(|Z| < 1.433) = P(-1.433 < Z < 1.433)$

$= 2\Phi(1.433) - 1$

$= 2(0.9240) - 1$

$= 0.848$

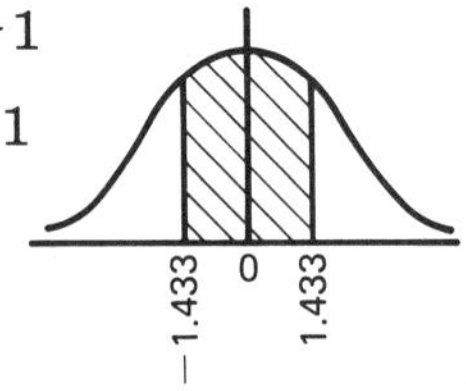

So $P(|Z| < 1.433) = 0.848$.

−1.433 0 1.433

(e)

$P(Z > 0.863 \text{ or } Z < -1.527) = 1 - (\Phi(0.863) + \Phi(1.527) - 1)$

$= 2 - \Phi(0.863) - \Phi(1.527)$

$= 2 - 0.8059 - 0.9365$

$= 0.2576$

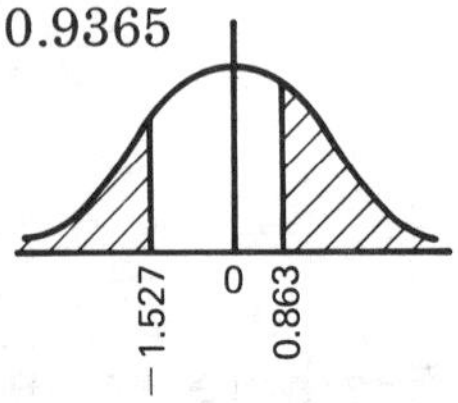

So $P(Z > 0.863 \text{ or } Z < -1.527) = 0.2576$.

−1.527 0 0.863

Example 6.4 If $Z \sim N(0, 1)$, show that **(a)** $P(-1.96 < Z < 1.96) = 0.95$, **(b)** $P(-2.575 < Z < 2.575) = 0.99$.

Solution 6.4 (a) $P(-1.96 < Z < 1.96) = 2\Phi(1.96) - 1$

$= 2(0.975) - 1$

$= 0.95$

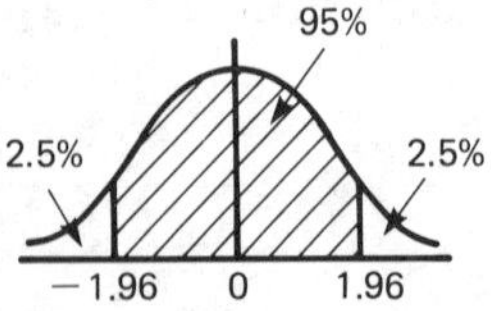

Therefore $P(-1.96 < Z < 1.96) = 0.95$.

95%
2.5% 2.5%
−1.96 0 1.96

NOTE: This is an important result:

The central 95% of the distribution lies between ±1.96.

(b) $P(-2.575 < Z < 2.575) = 2\Phi(2.575) - 1$

$= 2(0.995) - 1$

$= 0.99$

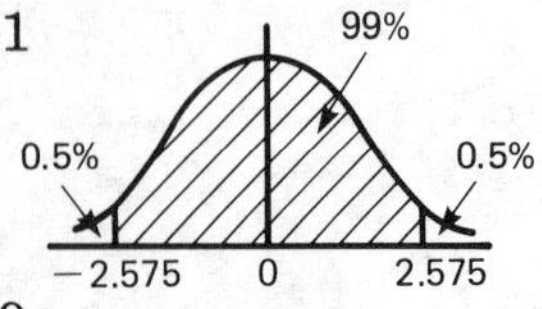

Therefore $P(-2.575 < Z < 2.575) = 0.99$.

99%
0.5% 0.5%
−2.575 0 2.575

The central 99% of the distribution lies between ±2.575.

Exercise 6a

1. If $Z \sim N(0,1)$, find (*a*) $P(Z > 0.874)$, (*b*) $P(Z < 0.874)$, (*c*) $P(Z < -0.874)$, (*d*) $P(Z > -0.874)$.

2. If $Z \sim N(0,1)$, find (*a*) $P(Z > 1.8)$, (*b*) $P(Z < -0.65)$, (*c*) $P(Z > -3.46)$, (*d*) $P(Z < 1.36)$, (*e*) $P(Z > 2.58)$, (*f*) $P(Z > -2.37)$, (*g*) $P(Z < 1.86)$, (*h*) $P(Z < -0.725)$, (*i*) $P(Z > 1.863)$, (*j*) $P(Z < 1.63)$, (*k*) $P(Z > -2.061)$, (*l*) $P(Z < -2.875)$.

3. If $Z \sim N(0,1)$, find (*a*) $P(Z > 1.645)$, (*b*) $P(Z < -1.645)$, (*c*) $P(Z > 1.282)$, (*d*) $P(Z > 1.96)$, (*e*) $P(Z > 2.575)$, (*f*) $P(Z > 2.326)$, (*g*) $P(Z > 2.808)$, (*h*) $P(Z < 1.96)$.

4. If $Z \sim N(0,1)$, find
(*a*) $P(0.829 < Z < 1.843)$,
(*b*) $P(-2.56 < Z < 0.134)$,
(*c*) $P(-1.762 < Z < -0.246)$,
(*d*) $P(0 < Z < 1.73)$,
(*e*) $P(-2.05 < Z < 0)$,
(*f*) $P(-3.08 < Z < 3.08)$,
(*g*) $P(1.764 < Z < 2.567)$,
(*h*) $P(-1.65 < Z < 1.725)$,
(*i*) $P(-0.98 < Z < -0.16)$,
(*j*) $P(Z < -1.97$ or $Z > 2.5)$,
(*k*) $P(|Z| < 1.78)$, (*l*) $P(|Z| > 0.754)$,
(*m*) $P(-1.645 < Z < 1.645)$,
(*n*) $P(|Z| > 2.326)$.

Example 6.5 If $Z \sim N(0, 1)$, find the value of a if (a) $P(Z > a) = 0.3802$, (b) $P(Z > a) = 0.7818$, (c) $P(Z < a) = 0.0793$, (d) $P(Z < a) = 0.9693$, (e) $P(|Z| < a) = 0.9$.

Solution 6.5 (a) $P(Z > a) = 0.3802$.

$$\Phi(a) = 1 - 0.3802 = 0.6198$$

so from tables

$$\underline{a = 0.305.}$$

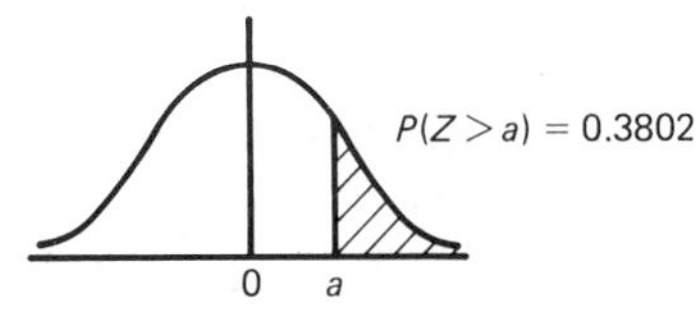

(b) $P(Z > a) = 0.7818$.

Now, since the probability is greater than 0.5, a must be negative.

Now
$$\Phi(-a) = 0.7818$$
$$-a = 0.778$$
$$\underline{a = -0.778}$$

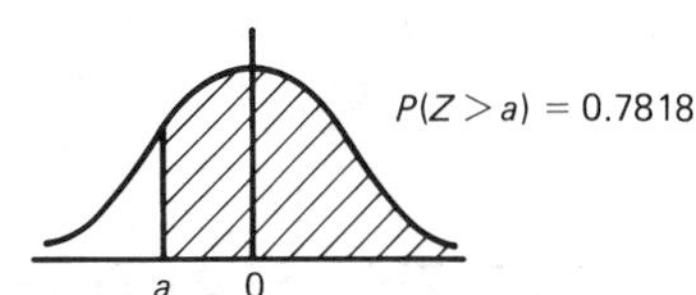

(c) $P(Z < a) = 0.0793$.

From the diagram it is obvious that a must be negative

$$\Phi(-a) = 1 - 0.0793 = 0.9207$$
$$-a = 1.41$$
$$\underline{a = -1.41}$$

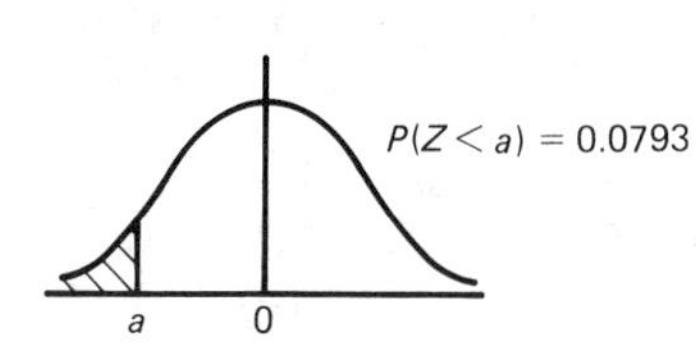

(d) $P(Z < a) = 0.9693$.

$$\Phi(a) = 0.9693$$
$$\underline{a = 1.87}$$

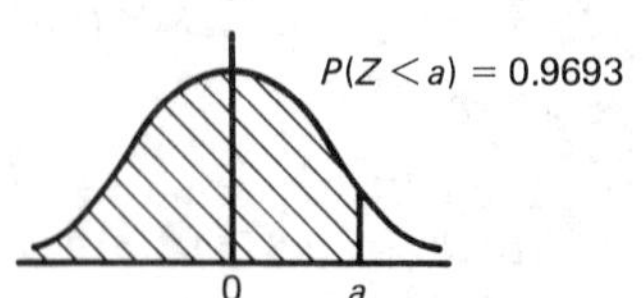

(e) $P(|Z| < a) = 0.9$,

i.e. $P(-a < Z < a) = 0.9$.

From symmetry

$$2\Phi(a) - 1 = 0.9$$
$$2\Phi(a) = 1.9$$
$$\Phi(a) = 0.95$$
$$\underline{a = 1.645}$$

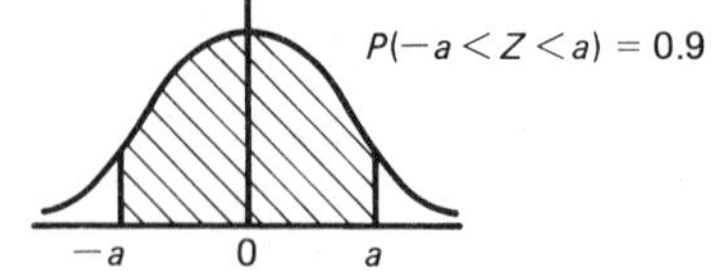

Exercise 6b

1. If $Z \sim N(0, 1)$, find a if
(*a*) $P(Z > a) = 0.001\,22$,
(*b*) $P(Z > a) = 0.0100$,
(*c*) $P(Z > a) = 0.025$, (*d*) $P(Z > a) = 0.198$,
(*e*) $P(Z > a) = 0.481$, (*f*) $P(Z > a) = 0.692$,
(*g*) $P(Z > a) = 0.812$, (*h*) $P(Z > a) = 0.9885$.

2. If $Z \sim N(0, 1)$, find a if
(*a*) $P(Z < a) = 0.0003$,
(*b*) $P(Z < a) = 0.0296$,
(*c*) $P(Z < a) = 0.325$, (*d*) $P(Z < a) = 0.506$,
(*e*) $P(Z < a) = 0.787$, (*f*) $P(Z < a) = 0.891$,
(*g*) $P(Z < a) = 0.8297$,
(*h*) $P(Z < a) = 0.9738$.

3. If $Z \sim N(0, 1)$, find a if
(*a*) $P(|Z| < a) = 0.6372$,
(*b*) $P(|Z| > a) = 0.097$,
(*c*) $P(|Z| < a) = 0.5$,
(*d*) $P(|Z| > a) = 0.0404$.

4. If $Z \sim N(0, 1)$, find the upper quartile and the lower quartile of the distribution. Find also the 70th percentile.

5. If $Z \sim N(0, 1)$, find a if $P(|Z| > a)$ takes the value (*a*) 10%, (*b*) 5%, (*c*) 4%, (*d*) 2%, (*e*) 1%, (*f*) 0.5%.

6. If $Z \sim N(0, 1)$, find a if $P(|Z| < a)$ takes the value (*a*) 80%, (*b*) 96%, (*c*) 97%, (*d*) 99%.

USE OF THE STANDARD NORMAL TABLES FOR ANY NORMAL DISTRIBUTION

We now show how the tables for the standard normal distribution can be adapted for use with any random variable X where $X \sim N(\mu, \sigma^2)$.

Example 6.6 The r.v. $X \sim N(300, 25)$. Find **(a)** $P(X > 305)$, **(b)** $P(X < 291)$, **(c)** $P(X < 312)$, **(d)** $P(X > 286)$.

Solution 6.6 (a) $P(X > 305)$.

First we have to standardise the random variable X by subtracting the mean, 300, and dividing by the standard deviation, (s.d.), 5, so that $Z = \dfrac{X-300}{5}$.

We also use the following properties of inequalities:

$$X > 305 \Rightarrow X - 300 > 305 - 300 \Rightarrow \frac{X-300}{5} > \frac{305-300}{5}$$

$$\begin{aligned} \text{So} \quad P(X > 305) &= P\left(\frac{X-300}{5} > \frac{305-300}{5}\right) \\ &= P(Z > 1) \\ &= 1 - \Phi(1) \\ &= 1 - 0.8413 \\ &= 0.1587 \end{aligned}$$

Standard normal curve

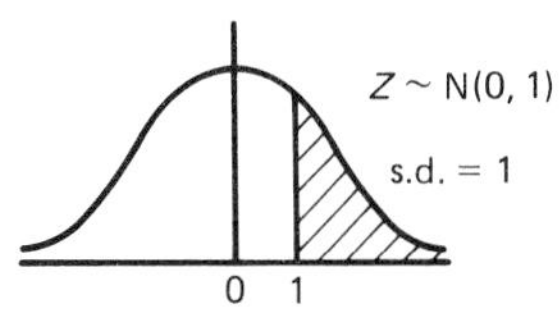

Therefore $P(X > 305) = 0.1587.$

NOTE: if the two curves had been drawn to scale, the curve for X would have been much more spread out and not as steep as the curve for Z. However, for convenience of drawing, we use the same sketch.

Often, again for convenience, we draw one sketch and write the values of the standardised variable underneath the x values. We use the abbreviation S.V. for 'standardised variable'.

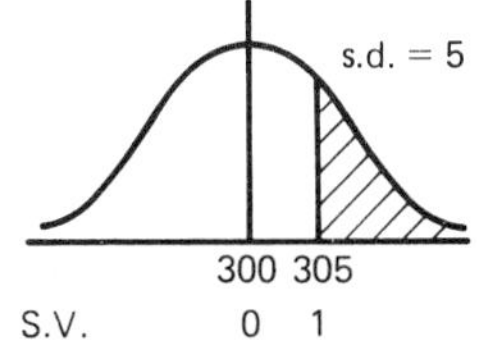

$$\begin{aligned} \text{(b)} \quad P(X < 291) &= P\left(\frac{X-300}{5} < \frac{291-300}{5}\right) \\ &= P(Z < -1.8) \\ &= 1 - \Phi(1.8) \\ &= 1 - 0.9641 \\ &= 0.0359 \end{aligned}$$

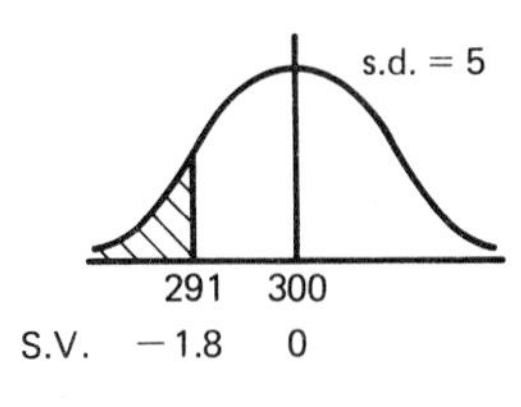

Therefore $P(X < 291) = 0.0359.$

$$\begin{aligned} \text{(c)} \quad P(X < 312) &= P\left(\frac{X-300}{5} < \frac{312-300}{5}\right) \\ &= P(Z < 2.4) \\ &= \Phi(2.4) \\ &= 0.9918 \end{aligned}$$

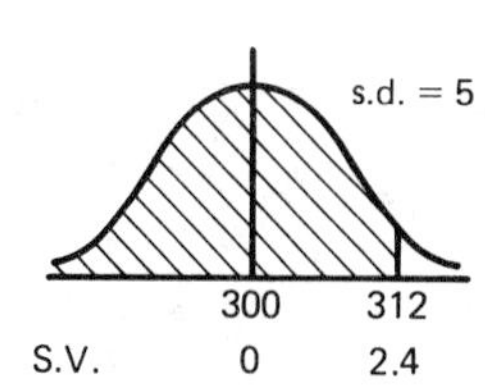

Therefore $P(X < 312) = 0.9918.$

$$\text{(a)}\quad P(X > 286) = P\left(\frac{X-300}{5} > \frac{286-300}{5}\right)$$
$$= P(Z > -2.8)$$
$$= \Phi(2.8)$$
$$= 0.997\,44$$

s.d. = 5
286 300
S.V. −2.8 0

Therefore $P(X > 286) = 0.997\,44.$

Example 6.7 The r.v. X is such that $X \sim \mathrm{N}(50, 8)$. Find **(a)** $P(48 < X < 54)$, **(b)** $P(52 < X < 55)$, **(c)** $P(46 < X < 49)$, **(d)** $P(|X-50| < \sqrt{8})$.

Solution 6.7 Standardise X so that $Z = \dfrac{X-50}{\sqrt{8}}$.

$$\text{(a)}\quad P(48 < X < 54) = P\left(\frac{48-50}{\sqrt{8}} < \frac{X-50}{\sqrt{8}} < \frac{54-50}{\sqrt{8}}\right)$$
$$= P(-0.707 < Z < 1.414)$$
$$= \Phi(1.414) + \Phi(0.707) - 1$$
$$= 0.9213 + 0.7601 - 1$$
$$= 0.6814$$

s.d. = $\sqrt{8}$
48 50 54
S.V. −0.707 0 1.414

Therefore $P(48 < X < 54) = 0.6814.$

$$\text{(b)}\quad P(52 < X < 55) = P\left(\frac{52-50}{\sqrt{8}} < \frac{X-50}{\sqrt{8}} < \frac{55-50}{\sqrt{8}}\right)$$
$$= P(0.707 < Z < 1.768)$$
$$= \Phi(1.768) - \Phi(0.707)$$
$$= 0.9615 - 0.7601$$
$$= 0.2014$$

s.d. = $\sqrt{8}$
50 52 55
S.V. 0 0.707 1.768

Therefore $P(52 < X < 55) = 0.2014.$

$$\text{(c)}\quad P(46 < X < 49) = P\left(\frac{46-50}{\sqrt{8}} < \frac{X-50}{\sqrt{8}} < \frac{49-50}{\sqrt{8}}\right)$$
$$= P(-1.414 < Z < -0.354)$$
$$= \Phi(1.414) - \Phi(0.354)$$
$$= 0.9213 - 0.6383$$
$$= 0.283$$

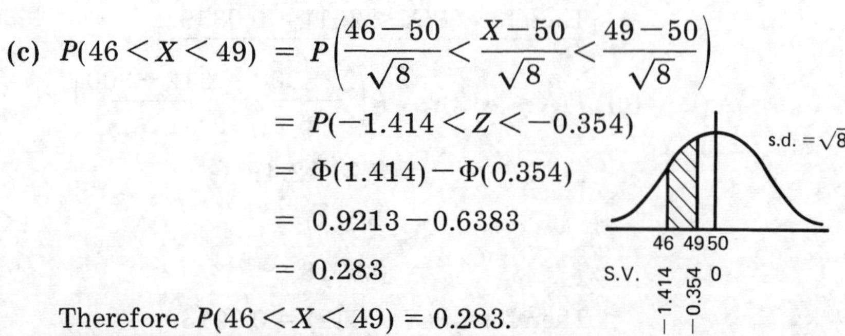

Therefore $P(46 < X < 49) = 0.283.$

(d) $P(|X-50|<\sqrt{8}) = P(-\sqrt{8}<X-50<\sqrt{8})$

$$= P\left(-1<\frac{X-50}{\sqrt{8}}<1\right)$$

$$= P(-1<Z<1)$$

$$= 2\Phi(1)-1$$

$$= 2(0.8413)-1$$

$$= 0.6826$$

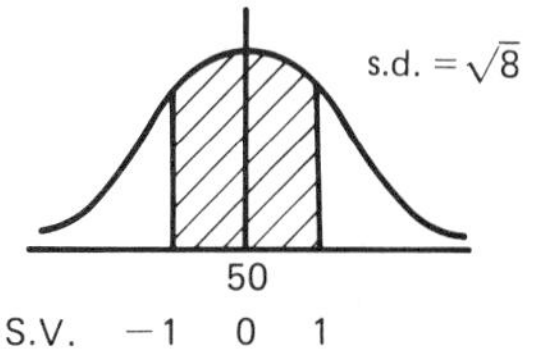

Therefore $P(|X-50|<\sqrt{8}) = 0.6826$.

Example 6.8 The time taken by a milkman to deliver milk to the High Street is normally distributed with mean 12 minutes and standard deviation 2 minutes. He delivers milk every day. Estimate the number of days during the year when he takes **(a)** longer than 17 minutes, **(b)** less than 10 minutes, **(c)** between 9 and 13 minutes.

Solution 6.8 Let X be the r.v. 'the time taken to deliver the milk to the High Street'. Then $X \sim N(12, 2^2)$.

We standardise X so that $Z = \dfrac{X-12}{2}$.

(a) $P(X>17) = P\left(\dfrac{X-12}{2}>\dfrac{17-12}{2}\right)$

$$= P(Z>2.5)$$

$$= 1-\Phi(2.5)$$

$$= 1-0.993\,79$$

$$= 0.006\,21$$

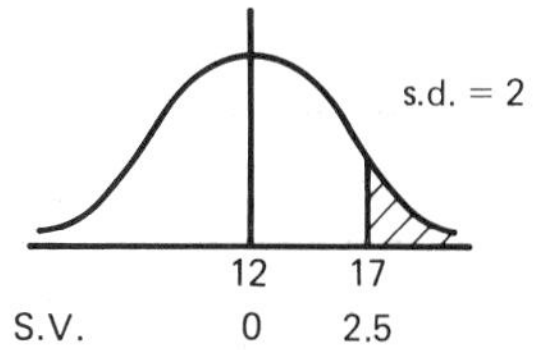

The number of days when he takes longer than 17 minutes

$$= 365(0.006\,21)$$

$$= 2.27$$

$$\approx 2$$

Therefore on approximately 2 days in the year he takes longer than 17 minutes.

(b) $P(X<10) = P\left(\dfrac{X-12}{2}<\dfrac{10-12}{2}\right)$

$$= P(Z<-1)$$

$$= 1-\Phi(1)$$

$$= 1-0.8413$$

$$= 0.1587$$

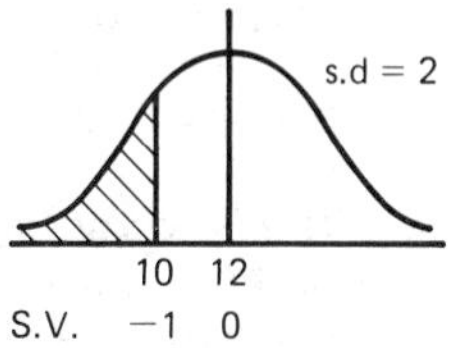

The number of days when he takes less than 10 minutes

$$= 365(0.1587)$$
$$= 57.9$$
$$\approx 58$$

Therefore on approximately 58 days in the year he takes less than 10 minutes.

(c) $P(9 < X < 13) = P\left(\frac{9-12}{2} < \frac{X-12}{2} < \frac{13-12}{2}\right)$

$$= P(-1.5 < Z < 0.5)$$
$$= \Phi(0.5) + \Phi(1.5) - 1$$
$$= 0.6915 + 0.9332 - 1$$
$$= 0.6247$$

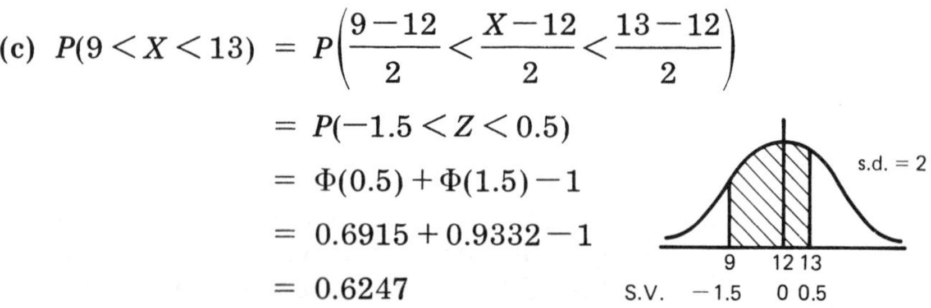

The number of days when he takes between 9 and 13 minutes

$$= 365(0.6247)$$
$$= 228 \text{ days}$$

Therefore on 288 days he takes between 9 and 13 minutes.

Exercise 6c

1. If $X \sim N(300, 25)$, find (*a*) $P(X > 308)$, (*b*) $P(X > 311.5)$, (*c*) $P(X > 294)$, (*d*) $P(X > 290.5)$, (*e*) $P(X < 302)$, (*f*) $P(X < 312)$, (*g*) $P(X < 299.5)$, (*h*) $P(X < 293)$.

2. If $X \sim N(50, 20)$, find (*a*) $P(X > 60.3)$, (*b*) $P(X < 47.3)$, (*c*) $P(X > 48.9)$, (*d*) $P(X > 53.5)$, (*e*) $P(X < 59.8)$, (*f*) $P(X < 62.3)$.

3. If $X \sim N(-8, 12)$, find (*a*) $P(X < -9.8)$, (*b*) $P(X > 0)$, (*c*) $P(X < -3.4)$, (*d*) $P(X > -5.7)$, (*e*) $P(X < -10.8)$, (*f*) $P(X > -1.6)$, (*g*) $P(X > -8.2)$.

4. If $X \sim N(a, a^2)$, find (*a*) $P(X < 0)$, (*b*) $P(X > 0)$, (*c*) $P(X > \frac{1}{2}a)$, (*d*) $P\left(X > \frac{3a}{2}\right)$, (*e*) $P\left(X < \frac{5a}{2}\right)$.

5. If $X \sim N(100, 80)$, find (*a*) $P(85 < X < 112)$, (*b*) $P(105 < X < 115)$, (*c*) $P(85 < X < 92)$, (*d*) $P(|X - 100| < \sqrt{80})$, (*e*) $P(99 < X < 105)$.

6. If $X \sim N(84, 12)$, find (*a*) $P(80 < X < 89)$, (*b*) $P(X < 79 \text{ or } X > 92)$, (*c*) $P(76 < X < 82)$, (*d*) $P(|X - 84| > 2.9)$, (*e*) $P(87 < X < 93)$.

7. If $X \sim N(2, 0.3)$, find (*a*) $P(1.8 < X < 2.9)$, (*b*) $P(2.01 < X < 2.8)$, (*c*) $P(|X - 2| < 2\sqrt{0.3})$.

8. Packages from a packing machine have a mass which is normally distributed with mean 200 g and standard deviation 2 g. Find the probability that a package from the machine weighs (*a*) less than 197 g, (*b*) more than 200.5 g, (*c*) between 198.5 g and 199.5 g.

9. The heights of boys at a particular age follow a normal distribution with mean 150.3 cm and standard deviation 5 cm. Find the probability that a boy picked at random from this age group has height (*a*) less than 153 cm, (*b*) less than 148 cm, (*c*) more than 158 cm, (*d*) more than 144 cm, (*e*) between 147 cm and 149.5 cm, (*f*) between 150 cm and 158 cm.

10. A random variable X is such that $X \sim N(-5, 9)$. Find the probability that (*a*) an item chosen at random will have a positive value, (*b*) out of 10 items chosen at random, just 4 will have a positive value.

11. A certain type of cabbage has a mass which is normally distributed with mean 1 kg and standard deviation 0.15 kg. In a lorry load of 800 of these cabbages, estimate how many will have mass

(*a*) greater than 0.79 kg,
(*b*) less than 1.13 kg,
(*c*) between 0.85 kg and 1.15 kg,
(*d*) between 0.75 kg and 1.29 kg.

De-standardising

Sometimes it is necessary to find a value X which corresponds to the standardised value Z. We use $Z = \dfrac{X-\mu}{\sigma}$, so that $X = \mu + \sigma Z$.

Example 6.9 If $X \sim N(50, 6.8)$, find the value of X which corresponds to a standardised value of **(a)** -1.2, **(b)** 0.6.

Solution 6.9 Now $X = \mu + \sigma Z$, where $\mu = 50$ and $\sigma = \sqrt{6.8}$, so that $X = 50 + \sqrt{6.8}Z$.

(a) when $z = -1.2$,

$$x = 50 + \sqrt{6.8}(-1.2) = 46.87 \quad (2\text{ d.p.})$$

(b) when $z = 0.6$,

$$x = 50 + \sqrt{6.8}(0.6) = 51.56 \quad (2\text{ d.p.})$$

Exercise 6d

Find the value of X which corresponds to a standardised value of (*a*) -2.05, (*b*) 0.86 for each of the following distributions:

(i) $X \sim N(60, 17)$, (ii) $X \sim N(124, 3.2^2)$, (iii) $X \sim N(84.5, 50)$, (iv) $X \sim N(62.3, 38)$, (v) $X \sim N(\mu, \sigma^2)$, (vi) $X \sim N(a, b)$, (vii) $X \sim N(a, a^2)$, (viii) $X \sim N(49, 49)$.

Example 6.10 If $X \sim N(100, 36)$ and $P(X > a) = 0.1093$, find the value of a.

Solution 6.10 As $P(X > a)$ is less than 0.5, a must be greater than the mean, 100.

Now $P(X > a) = 0.1093$

so $P\left(\dfrac{X-100}{6} > \dfrac{a-100}{6}\right) = 0.1093$

i.e. $P\left(Z > \dfrac{a-100}{6}\right) = 0.1093$

We have $\Phi\left(\dfrac{a-100}{6}\right) = 1 - 0.1093 = 0.8907$

But from tables,

$$\Phi(1.23) = 0.8907$$

Therefore $$\frac{a-100}{6} = 1.23$$

$$a = 100+6(1.23) = 107.38$$

Therefore, if $P(X > a) = 0.1093$, then $a = 107.38$.

Example 6.11 If $X \sim N(24, 9)$ and $P(X > a) = 0.974$, find the value of a.

Solution 6.11 As $P(X > a)$ is greater than 0.5, a must be less than the mean 24.

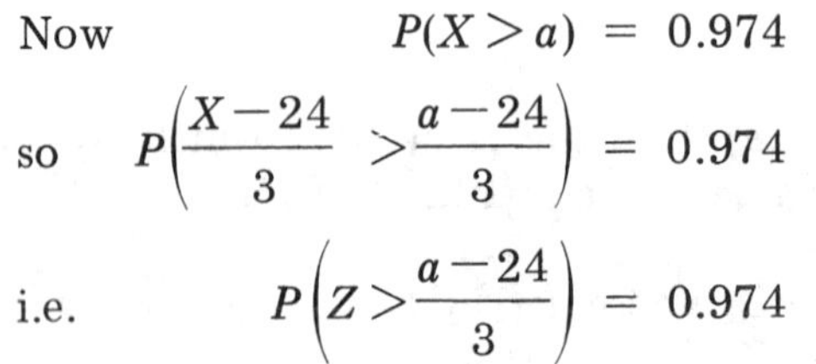

Now $$P(X > a) = 0.974$$

so $$P\left(\frac{X-24}{3} > \frac{a-24}{3}\right) = 0.974$$

i.e. $$P\left(Z > \frac{a-24}{3}\right) = 0.974$$

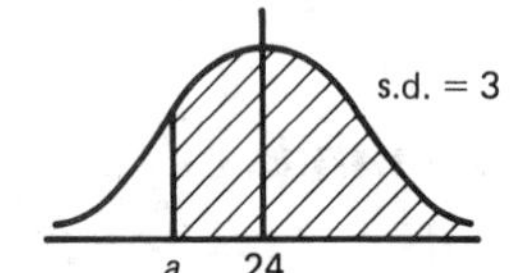

Now $\frac{a-24}{3}$ must be negative and

$$\Phi\left(-\left(\frac{a-24}{3}\right)\right) = 0.974$$

$$-\left(\frac{a-24}{3}\right) = 1.943$$

$$\frac{a-24}{3} = -1.943$$

$$a = 24-(3)(1.943)$$

$$= 18.171$$

Therefore, if $P(X > a) = 0.974$, then $a = 18.171$.

Example 6.12 If $X \sim N(70, 25)$, find the value of a such that $P(|X-70| < a) = 0.8$. Hence find the limits within which the central 80% of the distribution lies.

Solution 6.12 $$P(|X-70| < a) = 0.8$$

Therefore

$$P(-a < X-70 < a) = 0.8$$

$$P\left(-\frac{a}{5} < \frac{X-70}{5} < \frac{a}{5}\right) = 0.8$$

$$P\left(-\frac{a}{5} < Z < \frac{a}{5}\right) = 0.8$$

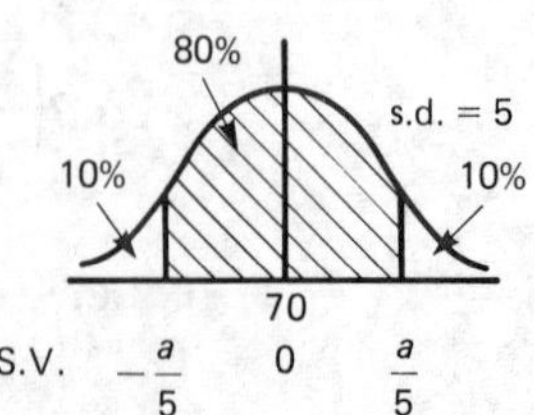

Now, by symmetry

$$2\Phi\left(\frac{a}{5}\right)-1 = 0.8$$

$$2\Phi\left(\frac{a}{5}\right) = 1.8$$

$$\Phi\left(\frac{a}{5}\right) = 0.9$$

Therefore $\frac{a}{5} = 1.282$

$$a = 6.41$$

So $P(-6.41 < X-70 < 6.41) = 0.8$

or $P(63.59 < X < 76.41) = 0.8$

The central 80% of the distribution lies between 63.59 and 76.41.

Exercise 6e

1. If $X \sim N(60, 25)$ and if
   (i) $P(X > a) = 0.2324$, find a,
   (ii) $P(X > b) = 0.0702$, find b,
   (iii) $P(X > c) = 0.837$, find c,
   (iv) $P(X > d) = 0.7461$, find d.

2. If $X \sim N(45, 16)$ and if
   (i) $P(X < a) = 0.0317$, find a,
   (ii) $P(X < b) = 0.895$, find b,
   (iii) $P(X < c) = 0.0456$, find c,
   (iv) $P(X < d) = 0.996$, find d.

3. If $X \sim N(80, 36)$, find c such that $P(|X-80| < c) = 0.9$ and hence find the limits within which the central 90% of the distribution lies.

4. If $X \sim N(400, 64)$, find
   (i) a such that $P(|X-400| < a) = 0.75$,
   (ii) b such that $P(|X-400| < b) = 0.98$,
   (iii) c such that $P(|X-400| < c) = 0.95$,
   (iv) d such that $P(|X-400| < d) = 0.975$,
   (v) the limits within which the central 95% of the distribution lies.

5. The masses of cos lettuces sold at a hypermarket are normally distributed with mean mass 600 g and standard deviation 20 g.
   (*a*) If a lettuce is chosen at random, find the probability that its mass lies between 570 g and 610 g.
   (*b*) Find the mass exceeded by 7% of the lettuces.
   (*c*) In one day, 1000 lettuces are sold. Estimate how many weigh less than 545 g.

6. The marks of 500 candidates in an examination are normally distributed with a mean of 45 marks and a standard deviation of 20 marks.
   (*a*) Given that the pass mark is 41, estimate the number of candidates who passed the examination.
   (*b*) If 5% of the candidates obtain a distinction by scoring x marks or more, estimate the value of x.
   (*c*) Estimate the interquartile range of the distribution. (L Additional)

7. If $X \sim N(k, k^2)$, find
   (i) a such that $P(|X-k| < ak) = 0.9$,
   (ii) b such that $P(|X-k| > bk) = 0.01$,
   (iii) c such that $P(|X-k| > ck) = 0.05$,
   (iv) d such that $P(|X-k| < dk) = 0.995$.

8. Bags of flour packed by a particular machine have masses which are normally distributed with mean 500 g and standard deviation 20 g. 2% of the bags are rejected for being underweight and 1% of the bags are rejected for being overweight. Between what range of values should the mass of a bag of flour lie if it is to be accepted?

9. A sample of 100 apples is taken from a load. The apples have the following distribution of sizes

Diameter to nearest cm	6	7	8	9	10
Frequency	11	21	38	17	13

Determine the mean and standard deviation of these diameters.

Assuming that the distribution is approximately normal with this mean and this standard deviation find the range of size of apples for packing, if 5% are to be rejected as too small and 5% are to be rejected as too large. (O & C)

PROBLEMS THAT INVOLVE FINDING THE VALUE OF μ OR σ OR BOTH

Example 6.13 The lengths of certain items follow a normal distribution with mean μ cm and standard deviation 6 cm. It is known that 4.78% of the items have a length greater than 82 cm. Find the value of the mean μ.

Solution 6.13 Let X be the r.v. 'the length of an item in cm'.

$X \sim N(\mu, 36)$ and $P(X > 82) = 0.0478$.

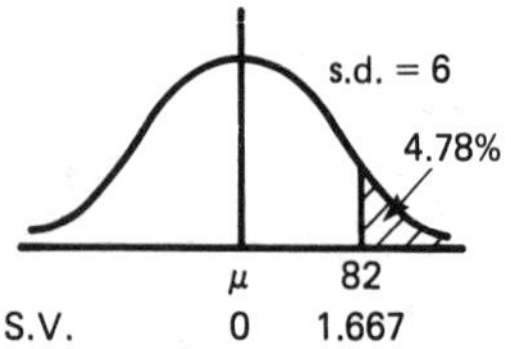

Now
$$P(X > 82) = P\left(\frac{X-\mu}{6} > \frac{82-\mu}{6}\right) = P\left(Z > \frac{82-\mu}{6}\right) = 1-\Phi\left(\frac{82-\mu}{6}\right)$$

so
$$1-\Phi\left(\frac{82-\mu}{6}\right) = 0.0478$$
$$\Phi\left(\frac{82-\mu}{6}\right) = 0.9522$$

But from tables

$$\Phi(1.667) = 0.9522$$

so
$$\frac{82-\mu}{6} = 1.667$$
$$82-\mu = 10.002$$
$$\mu = 72 \quad (2\text{ S.F.})$$

The mean of the distribution is 72 cm.

Example 6.14 $X \sim N(100, \sigma^2)$ and $P(X < 106) = 0.8849$. Find the standard deviation, σ.

Solution 6.14

$$P(X < 106) = 0.8849$$

$$P\left(\frac{X-100}{\sigma} < \frac{106-100}{\sigma}\right) = 0.8849$$

$$P\left(Z < \frac{6}{\sigma}\right) = 0.8849$$

$$\Phi\left(\frac{6}{\sigma}\right) = 0.8849$$

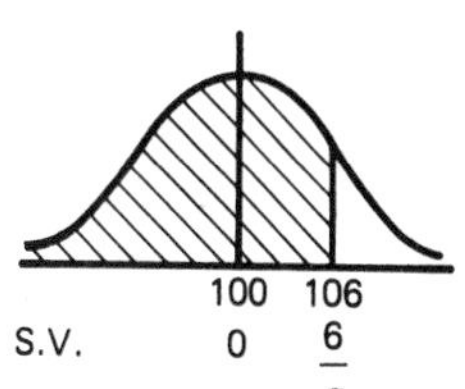

But from tables

$$\Phi(1.2) = 0.8849$$

Therefore $$\frac{6}{\sigma} = 1.2$$

$$\sigma = \frac{6}{1.2} = 5$$

The standard deviation of the distribution is 5.

Example 6.15 The masses of articles produced in a particular workshop are normally distributed with mean μ and standard deviation σ. 5% of the articles have a mass greater than 85 g and 10% have a mass less than 25 g. Find the values of μ and σ, and find the range symmetrical about the mean, within which 75% of the masses lie.

Solution 6.15 Let X be the r.v. 'the mass, in g, of an article'. Then $X \sim N(\mu, \sigma^2)$ where μ and σ are unknown.

Now $$P(X > 85) = 0.05$$

i.e. $$P\left(\frac{X-\mu}{\sigma} > \frac{85-\mu}{\sigma}\right) = 0.05$$

$$P\left(Z > \frac{85-\mu}{\sigma}\right) = 0.05$$

$$\Phi\left(\frac{85-\mu}{\sigma}\right) = 0.95$$

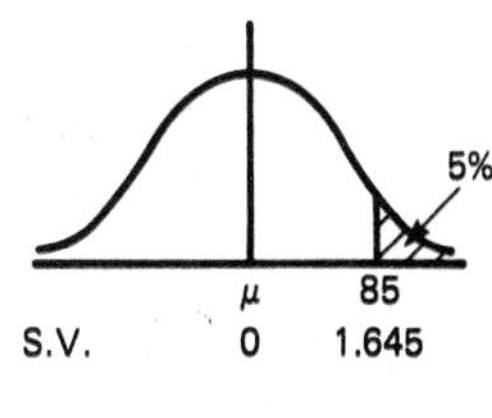

But from tables

$$\Phi(1.645) = 0.95$$

Therefore $$\frac{85-\mu}{\sigma} = 1.645$$

$$85-\mu = 1.645\sigma \qquad \text{(i)}$$

Also $$P(X < 25) = 0.10$$

$$P\left(\frac{X-\mu}{\sigma} < \frac{25-\mu}{\sigma}\right) = 0.10$$

$$P\left(Z < \frac{25-\mu}{\sigma}\right) = 0.10$$

10%
25 μ
S.V. −1.282 0

But $\dfrac{25-\mu}{\sigma}$ is negative, and by symmetry,

$$\Phi\left[-\left(\frac{25-\mu}{\sigma}\right)\right] = 0.9$$

From tables $$\Phi(1.282) = 0.9$$

Therefore $$-\left(\frac{25-\mu}{\sigma}\right) = 1.282$$

i.e. $$\mu-25 = 1.282\sigma \qquad \text{(ii)}$$

Adding (i) and (ii) we have

$$60 = 2.927\sigma$$

$$\sigma = 20.5 \quad \text{(3 S.F.)}$$

Substituting for σ in (ii)

$$\mu = 25+(1.282)(20.5)$$

$$= 51.3 \quad \text{(3 S.F.)}$$

Therefore the distribution has mean mass 51.3 g and standard deviation 20.5 g.

Now consider values a and b such that

$$P(a \leqslant X \leqslant b) = 0.75$$

and a and b are symmetrical about the mean.

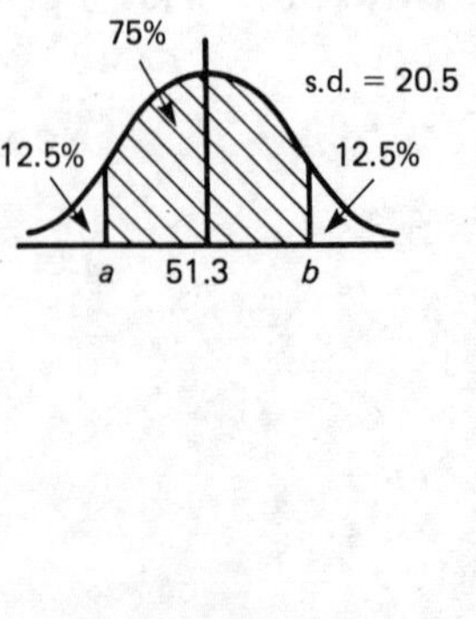

Now $$P(X > b) = 0.125$$

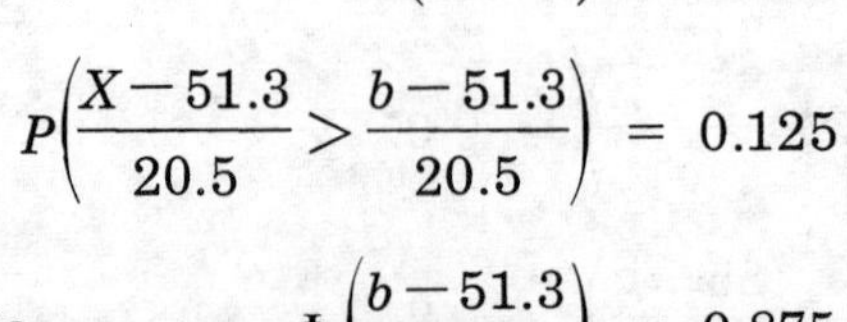

$$P\left(\frac{X-51.3}{20.5} > \frac{b-51.3}{20.5}\right) = 0.125$$

So $$\Phi\left(\frac{b-51.3}{20.5}\right) = 0.875$$

But from tables $\Phi(1.15) = 0.875$

Therefore $\frac{b-51.3}{20.5} = 1.15$

$$b = 51.3+(20.5)(1.15) = 74.9 \quad (3\text{ S.F.})$$

From symmetry $a = 51.3-(20.5)(1.15) = 27.7 \quad (3\text{ S.F.})$

Therefore, the central 75% of the distribution lies between the limits 27.7 g and 74.9 g.

Exercise 6f

1. $X \sim N(45, \sigma^2)$ and $P(X > 51) = 0.288$. Find σ.

2. $X \sim N(21, \sigma^2)$ and $P(X < 27) = 0.9332$. Find σ.

3. $X \sim N(\mu, 25)$ and $P(X < 27.5) = 0.3085$. Find μ.

4. $X \sim N(\mu, 12)$ and $P(X > 32) = 0.8438$. Find μ.

5. $X \sim N(\mu, \sigma^2)$ and $P(X > 80) = 0.0113$, $P(X > 30) = 0.9713$. Find μ and σ.

6. $X \sim N(\mu, \sigma^2)$ and $P(X > 102) = 0.42$, $P(X < 97) = 0.25$. Find μ and σ.

7. $X \sim N(\mu, \sigma^2)$ and $P(X < 57.84) = 0.90$, $P(X > 50) = 0.5$. Find μ and σ.

8. $X \sim N(\mu, \sigma^2)$ and $P(X < 35) = 0.2$, $P(35 < X < 45) = 0.65$. Find μ and σ.

9. The marks in an examination were normally distributed with mean μ and standard deviation σ. 10% of the candidates had more than 75 marks and 20% had less than 40 marks. Find the values of μ and σ.

10. The lengths of rods produced in a workshop follow a normal distribution with mean μ and variance 4. 10% of the rods are less than 17.4 cm long. Find the probability that a rod chosen at random will be between 18 and 23 cm long.

11. A man cuts hazel twigs to make bean poles. He says that a stick is 240 cm long. In fact, the length of the stick follows a normal distribution and 10% are of length 250 cm or more while 55% have a length over 240 cm. Find the probability that a stick, picked at random, is less than 235 cm long.

12. The diameters of bolts produced by a particular machine follow a normal distribution with mean 1.34 cm and standard deviation 0.04 cm. A bolt is rejected if its diameter is less than 1.24 cm or more than 1.40 cm. (*a*) Find the percentage of bolts which are accepted. The setting of the machine is altered so that the mean diameter changes but the standard deviation remains the same. With the new setting, 3% of the bolts are rejected because they are too large in diameter. (*b*) Find the new mean diameter of the bolts produced by the machine. (*c*) Find the percentage of bolts which are rejected because they are too small in diameter.

13. A certain make of car tyre can be safely used for 25 000 km on average before it is replaced. The makers guarantee to pay compensation to anyone whose tyre does not last for 22 000 km. They expect 7.5% of all tyres sold to qualify for compensation. Assuming that the distance, X, travelled before a tyre is replaced has a normal probability distribution, draw a diagram illustrating the facts given above.

    Calculate, to 3 significant figures, the standard deviation of X.

    Estimate the number of tyres per 1000 which will not have been replaced when they have covered 26 500 km.

    (L Additional)

14. A cutting machine produces steel rods which must not be more than 100 cm in length. The mean length of a large batch of rods taken from the machine is found to be 99.80 cm and the standard deviation of these lengths is 0.15 cm.
    (*a*) Assuming that the lengths of the rods

are normally distributed, calculate, to one decimal place, the percentage of rods which are too long.
(*b*) The position of the cut can be adjusted without altering the standard deviation of the lengths. Calculate in cm, to 2 decimal places, how small the mean length should be if no more than 2% of the rods are to be rejected for being longer than 100 cm.
(*c*) If the mean length is maintained at 99.80 cm, calculate, to the nearest tenth of a mm, by how much the standard deviation must be reduced if no more than 4% of the rods are to be rejected for being longer than 100 cm. (L Additional)

15. The continuous random variable X is normally distributed with mean μ and standard deviation σ. Given that $P(X < 53) = 0.04$ and $P(X < 65) = 0.97$, find the interquartile range of the distribution.

16. Tea is sold in packages marked 750 g. The masses of the packages are normally distributed with mean 760 g, standard deviation σ. What is the maximum value of σ if less than 1% of the packages are underweight?

NOTE: In solutions involving the normal distribution we will now omit the line of working involving Φ and the solutions should appear the same whether the reader is using the standard normal tables giving $\Phi(Z)$ or $Q(Z)$.

MISCELLANEOUS WORKED EXAMPLES

Example 6.16 Tests on 2 types of electric light bulb show the following:

Type A, lifetime distributed normally with an average life of 1150 hours and a standard deviation of 30 hours.

Type B, long-life bulb, average lifetime of 1900 hours, with standard deviation of 50 hours.

(a) What percentage of bulbs of type A could be expected to have a life of more than 1200 hours?

(b) What percentage of type B would you expect to last longer than 1800 hours?

(c) What lifetime limits would you estimate would contain the central 80% of the production of type A? (SUJB)

Solution 6.16 (a) Let X be the r.v. 'the length of life in hours of type A bulb'. Then $X \sim N(1150, 30^2)$.

$$P(X > 1200) = P\left(\frac{X-1150}{30} > \frac{1200-1150}{30}\right)$$
$$= P(Z > 1.667)$$
$$= 0.0478$$

Therefore 4.78% of type A bulbs could be expected to have a life of more than 1200 hours.

(b) Let Y be the r.v. 'the length of life in hours of type B bulb'. Then $Y \sim \text{N}(1900, 50^2)$.

$$P(Y > 1800) = P\left(\frac{Y-1900}{50} > \frac{1800-1900}{50}\right)$$
$$= P(Z > -2)$$
$$= 0.9772$$

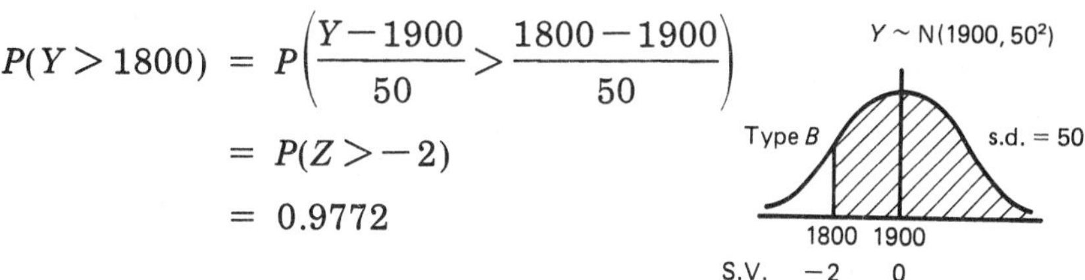

Therefore 97.72% of type B bulbs could be expected to have a life of more than 1800 hours.

(c) Now, from tables $\Phi(1.282) = 0.90$ or $Q(1.282) = 0.10$.

So that the central 80% of the standard normal distribution lies between the limits ± 1.282.

We wish to find the values of X which correspond to the standardised values of ± 1.282, so we use $X = \mu + \sigma Z$.

So the central 80% of the distribution lies between the limits $\mu \pm 1.282\sigma$.

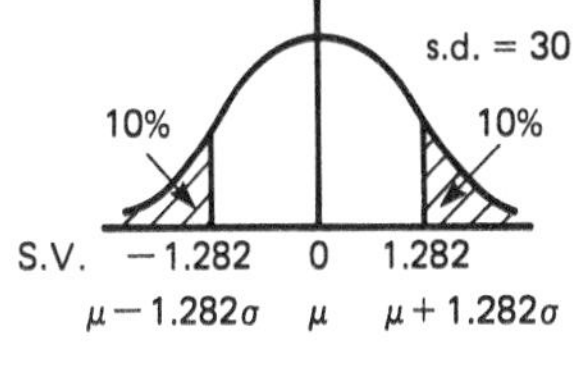

For type A, the limits are

$$1150 \pm (1.282)(30) = 1150 \pm 38.46$$
$$= (1111.54, 1188.46)$$

We estimate that the limits (1111.54 hours, 1188.46 hours) would contain the central 80% of the production of type A.

Example 6.17 A machine is producing components whose lengths are normally distributed about a mean of 6.50 cm. An upper tolerance limit of 6.54 cm has been adopted and, when the machine is correctly set, 1 in 20 components is rejected as exceeding this limit. On a certain day, it is found that 1 in 15 components is rejected for exceeding this limit.

(a) Assuming that the mean has not changed but that the production has become more variable, estimate the new standard deviation.

(b) Assuming that the standard deviation has not changed but that the mean has moved, estimate the new mean.

(c) If 1000 components are produced in a shift, how many of them may be expected to have lengths in the range 6.48 to 6.53 cm if the machine is set as in **(a)**? (AEB 1972)

Solution 6.17 (a) Let X be the r.v. 'the length in cm of a component'. Then $X \sim \mathrm{N}(6.50, \sigma^2)$ where σ is the new standard deviation.

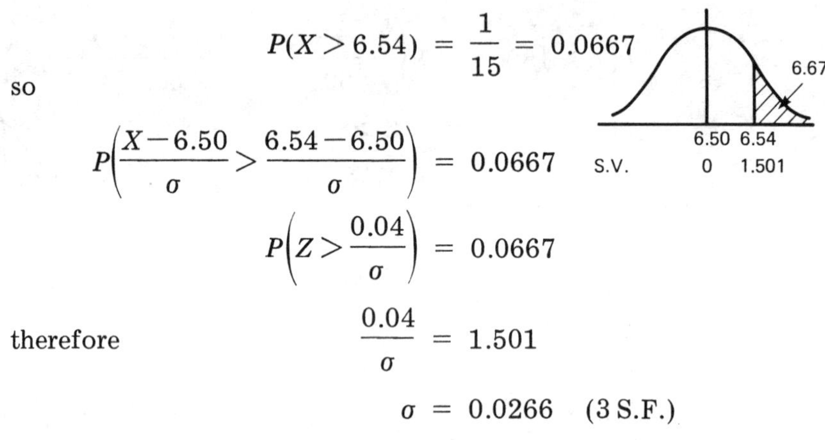

$$P(X > 6.54) = \frac{1}{15} = 0.0667$$

so

$$P\left(\frac{X-6.50}{\sigma} > \frac{6.54-6.50}{\sigma}\right) = 0.0667$$

$$P\left(Z > \frac{0.04}{\sigma}\right) = 0.0667$$

therefore

$$\frac{0.04}{\sigma} = 1.501$$

$$\sigma = 0.0266 \quad \text{(3 S.F.)}$$

The new standard deviation is 0.0266 cm (3 S.F.).

(b) Let the original standard deviation be σ_1. Then $X \sim \mathrm{N}(6.50, {\sigma_1}^2)$ originally.

$$P(X > 6.54) = \frac{1}{20} = 0.05$$

$$P\left(\frac{X-6.50}{\sigma_1} > \frac{6.54-6.50}{\sigma_1}\right) = 0.05$$

$$P\left(Z > \frac{0.04}{\sigma_1}\right) = 0.05$$

therefore

$$\frac{0.04}{\sigma_1} = 1.645$$

$$\sigma_1 = 0.0243 \quad \text{(3 S.F.)}$$

Now suppose that the new mean is μ.

So $X \sim \mathrm{N}(\mu, 0.0243^2)$.

$$P(X > 6.54) = 0.0667$$

$$P\left(\frac{X-\mu}{0.0243} > \frac{6.54-\mu}{0.0243}\right) = 0.0667$$

$$P\left(Z > \frac{6.54-\mu}{0.0243}\right) = 0.0667$$

therefore

$$\frac{6.54-\mu}{0.0243} = 1.501$$

$$\mu = 6.54 - (1.501)(0.0243)$$

$$= 6.504 \quad \text{(3 d.p.)}$$

Therefore the new mean is 6.504 cm (3 d.p.).

(c) If the machine is set as in part (a) then $X \sim N(6.50, 0.0266^2)$.

$$P(6.48 < X < 6.53) = P\left(\frac{6.48-6.50}{0.0266} < \frac{X-6.50}{0.0266} < \frac{6.53-6.50}{0.0266}\right)$$
$$= P(-0.752 < Z < 1.128)$$
$$= 0.6442$$

s.d. = 0.0266
6.48 6.50 6.53
S.V. −0.752 0 1.128

So for 1000 components we expect $(0.644)(1000) = 644$ to have lengths in the range 6.48 to 6.53 cm.

Miscellaneous Exercise 6g

1. Batteries for a transistor radio have a mean life under normal usage of 160 hours, with a standard deviation of 30 hours. Assuming that battery life follows a normal distribution,
   (*a*) calculate the percentage of batteries which have a life between 150 hours and 180 hours;
   (*b*) calculate the range, symmetrical about the mean, within which 75% of the battery lives lie;
   (*c*) if a radio takes four of these batteries and requires all of them to be working, calculate the probability that the radio will run for at least 135 hours. (O & C)

2. (*a*) **Without using a calculator**, calculate the mean and standard deviation of the numbers: 2, 3, 5, 5, 8, 11, 11, 11.

   (*b*) A machine produces components in batches of 20 000, the lengths of which may be considered to be normally distributed.

   At the beginning of production, the machine is set to produce the required mean length of components at 15 mm, and it can then be set to give any one of three standard deviations: 0.06 mm, 0.075 mm, 0.09 mm.

   It costs £850, £550 and £100 respectively to set these deviations.

   Any length produced must lie in the range 14.82 mm to 15.18 mm, otherwise it is classed as defective and costs the company £1.

   Which standard deviation should be used, if the decision is to be made purely on the cost of setting the machine and of the defectives? (SUJB)

3. Six hundred rounds are fired from a gun at a horizontal target 50 m long which extends from 950 m to 1000 m in range from the gun. The trajectories of the rounds all lie in the vertical plane through the gun and the target. It is found that 27 rounds fall short of the target and 69 rounds fall beyond it. Assuming that the range of rounds is normally distributed, find the mean and standard deviation of the range.

   Estimate the number of rounds falling within 5 m of the centre of the target. (C)

4. Machine components are mass-produced at a factory. A customer requires that the components should be 5.2 cm long but they will be acceptable if they are within limits 5.195 cm to 5.205 cm. The customer tests the components and finds that 10.75% of those supplied are over-size and 4.95% are under-size. Find the mean and standard deviation of the lengths of the components supplied assuming that they are normally distributed.

   If three of the components are selected at random what is the probability that one is under-size, one over-size and one satisfactory?

   If the standard deviation of the machine producing the components is altered without altering the mean so that 4.95% are over-size, what will be the new standard deviation and what percentage of components will now be under-size? (SUJB)

5. A marketing organisation grades onions into 3 sizes: small (diameter less than 60 mm), medium (diameter between 60 mm

and 80 mm) and large (diameter greater than 80 mm). A certain grower finds that 61% of his crop falls into the small category and 14% into the large category. Assuming that the distribution of diameters of the onions in his crop is described by a Normal probability function, sketch a graph showing the information given above.

On this basis, calculate the standard deviation and the mean of the diameters of the onions in his crop. (SMP)

6. Packets of semolina are nominally 226 g in weight. The actual weights have a Normal distribution with $\mu = 230.00$ g and $\sigma = 1.50$ g. What is the probability that a packet is underweight?

A decision is taken that the probability of an underweight packet should not exceed 0.001. To change the distribution of weights of the semolina packets to conform to this decision, two methods are considered:
(*a*) to increase μ, leaving σ unaltered;
(*b*) to improve the packing machine, thus reducing σ, while leaving μ unaltered.

Find the new values (of μ and of σ respectively) required for each method to succeed, given that, for the standardised Normal distribution,

$$P(Z > 3.0902) = 0.0010$$

(SMP)

7. A factory is illuminated by 2000 bulbs. The lives of these bulbs are normally distributed with a mean of 550 hours and a standard deviation of 50 hours. It is decided to replace all the bulbs at such intervals of time that only about 20 bulbs are likely to fail during each interval. How frequently should the bulbs be changed?

When the manufacturing process is improved so that the mean life of bulbs is increased to 600 hours and the standard deviation is reduced to 40 hours, the replacement interval is changed to 500 hours. Show that it will now be necessary to tolerate the failure of only about 12 bulbs per interval. (AEB 1973)

8. Before joining the Egghead Society, every candidate is given an intelligence test which, applied to the general public, would give a normal distribution of I.Q.'s with mean 100 and standard deviation 20. The candidate is not admitted unless his I.Q. as given by the test, is at least 130.

Estimate the median I.Q. of the members of the Egghead Society, assuming that their I.Q. distribution is representative of that of the part of the population having I.Q.s greater than, or equal to, 130.

What I.Q. would be expected to be exceeded by one member in ten of the society? (AEB)

9. The acidity of each of 100 random samples of soil from an area of land was measured and the results given in Table A below.

Assuming that the pH values are determined correct to the nearest tenth of a unit construct a cumulative frequency curve to illustrate the distribution.

A possible measure of kurtosis (i.e. flatness) is given by

$$k = \frac{Q}{P_{90} - P_{10}}$$

where Q is the semi-interquartile range, P_{90} the 90th percentile and P_{10} the 10th percentile. Estimate the value of k for the above distribution.

Use the standard normal table (p. 633) to estimate the value of k for a Normal distribution. Is the above distribution flatter than a Normal distribution with the same total frequency? (SUJB)

10. Describe the principal features of a normal distribution. Draw a sketch of the probability density function of the distribution N(0, 1).

A machine is producing a type of circular gasket. The specifications for the use of these gaskets in the manufacture of a certain make of engine are that the thickness should lie between 5.45 mm and 5.55 mm, and the diameter should lie between 8.45 mm and 8.54 mm. The machine is producing the gaskets so that their thicknesses are N(5.5, 0.0004), that is, normally distributed with mean 5.5 mm and variance 0.0004 mm^2, and their diameters are independently distributed N(8.54, 0.0025).

Calculate, to one decimal place, the percentage of gaskets produced which will not meet
(*a*) the specified thickness limits,

Table A

Acidity (pH)	4.6-	4.8-	5.0-	5.2-	5.4-	5.6-	5.8-	6.0-6.2
No. of samples	4	6	16	18	22	21	11	2

(*b*) the specified diameter limits,
(*c*) the specifications.

Find, to 3 decimal places, the probability that, if 6 gaskets made by the machine are chosen at random, exactly 5 of them will meet the specifications. (L)

11. State the conditions under which the binomial distribution is a suitable model to use in statistical work. Describe briefly how a binomial distribution was used, or could have been used, in one of your projects giving the parameters of your distribution.

It is known that bearings produced at a factory have diameters that are normally distributed with mean 14.2 mm and standard deviation 1.2 mm. Find, to 4 decimal places, the probability that a bearing chosen at random from the production will have a diameter less than 13.9 mm.

Six bearings are to be chosen at random from the production. Find, to 2 significant figures, the probability that at least 5 of these bearings will have diameters between 13.9 mm and 14.6 mm. (L)

12. The random variables X_1 and X_2 are both normally distributed such that $X_1 \sim N(\mu_1, \sigma_1^2)$ and $X_2 \sim N(\mu_2, \sigma_2^2)$. Given that $\mu_1 < \mu_2$ and $\sigma_1^2 < \sigma_2^2$, sketch both distributions on the same diagram.

State the '2σ rule' for a normal random variable. Explain how you used, or could have used, a normal distribution in a project.

The weights of vegetable marrows supplied to retailers by a wholesaler have a normal distribution with mean 1.5 kg and standard deviation 0.6 kg. The wholesaler supplies 3 sizes of marrow:

Size 1, under 0.9 kg,
Size 2, from 0.9 kg to 2.4 kg,
Size 3, over 2.4 kg.

Find, to 3 decimal places, the proportions of marrows in the three sizes. Find, in kg to one decimal place, the weight exceeded on average by 5 marrows in every 200 supplied.

The prices of the marrows are 16p for Size 1, 40p for Size 2 and 60p for Size 3. Calculate the expected total cost of 100 marrows chosen at random from those supplied. (L)

THE NORMAL APPROXIMATION TO THE BINOMIAL DISTRIBUTION

Under certain circumstances the normal distribution can be used as an approximation to the binomial distribution. One practical advantage is that calculations are much less tedious to perform.

If $X \sim \text{Bin}(n, p)$ then

$$E(X) = np$$

$$\text{Var}(X) = npq \quad \text{where} \quad q = 1-p$$

Now, for large n and p not too small or too large,

$$X \sim N(np, npq) \quad \text{approximately}$$

Example 6.18 Find the probability of obtaining between 4 and 7 heads inclusive with 12 tosses of a fair coin,

(a) using the binomial distribution,

(b) using the normal approximation to the binomial distribution.

Solution 6.18 Let X be the r.v. 'the number of heads obtained'. Let 'success' be 'obtaining a head'.

Then $X \sim \text{Bin}(n, p)$ where $n = 12$ and $p = P(\text{head}) = \frac{1}{2}$.

(a) So $X \sim \text{Bin}(12, \frac{1}{2})$ and $P(X = x) = {}^{12}C_x \left(\frac{1}{2}\right)^{12-x} \left(\frac{1}{2}\right)^x$

$x = 0, 1, 2, \ldots, 12$

$$\text{Now} \quad P(X=4) = {}^{12}C_4 \left(\frac{1}{2}\right)^8 \left(\frac{1}{2}\right)^4 = {}^{12}C_4 \left(\frac{1}{2}\right)^{12} = 0.121$$

$$P(X=5) = {}^{12}C_5 \left(\frac{1}{2}\right)^{12} = 0.193$$

$$P(X=6) = {}^{12}C_6 \left(\frac{1}{2}\right)^{12} = 0.226$$

$$P(X=7) = {}^{12}C_7 \left(\frac{1}{2}\right)^{12} = 0.193$$

$$\text{So} \quad P(4 \leqslant X \leqslant 7) = 0.121 + 0.193 + 0.226 + 0.193$$

$$= 0.733 \quad (3\text{ d.p.})$$

The probability of obtaining between 4 and 7 heads inclusive is 0.733 (3 d.p.).

The probability distribution for the number of heads in 12 tosses has been calculated and is shown below. The required probability is the sum of the areas of the shaded rectangles. Now this can be approximated by the area under the corresponding normal curve.

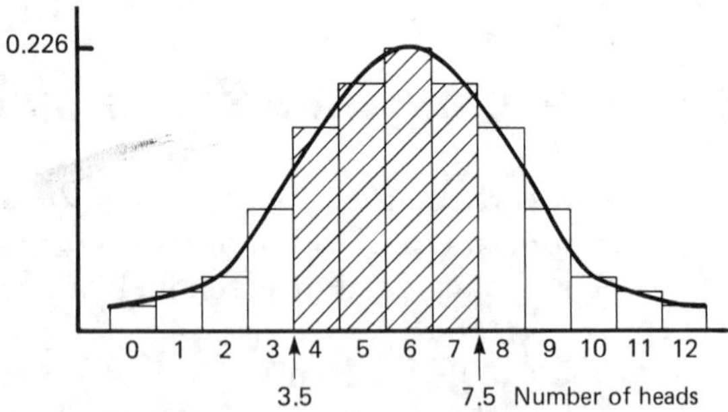

(b) To find the probability of obtaining between 4 and 7 heads using the normal approximation we consider X such that

$$X \sim \text{N}(np, npq) \quad \text{where} \quad n = 12 \quad \text{and} \quad p = \tfrac{1}{2}$$

$$\text{So} \quad X \sim \text{N}(6, 3)$$

However, before using the approximation we must take into account the fact that we are using a *continuous* distribution to approximate a *discrete* variable. So we make a **continuity correction.**

In this example $P(4 \leqslant X \leqslant 7)$ transforms to $P(3.5 < X < 7.5)$.

So,

$$P(3.5 < X < 7.5) = P\left(\frac{3.5-6}{\sqrt{3}} < \frac{X-6}{\sqrt{3}} < \frac{7.5-6}{\sqrt{3}}\right)$$

$$= P(-1.443 < Z < 0.866)$$

$$= 0.732 \quad (3\text{ d.p.})$$

s.d. $= \sqrt{3}$
3.5 6 7.5
S.V. −1.443 0 0.866

The probability of obtaining between 4 and 7 heads inclusive is 0.732 (3 d.p.).

NOTE: this answer compares very well with the answer in part **(a)**, and the working is much quicker to perform.

The approximation is even better for large n and it is preferable that p is close to $\frac{1}{2}$.

Continuity corrections often cause difficulty so we will look at these in more detail. It will be helpful to refer to the diagram showing the distribution for 12 tosses of the coin.

If we require the probability that there are 3 heads or less, i.e. $P(X \leqslant 3)$, then we consider $P(X < 3.5)$.

So $P(X \leqslant 3)$ transforms to $P(X < 3.5)$.

We will use the notation

$$P(X \leqslant 3) \longrightarrow P(X < 3.5)$$

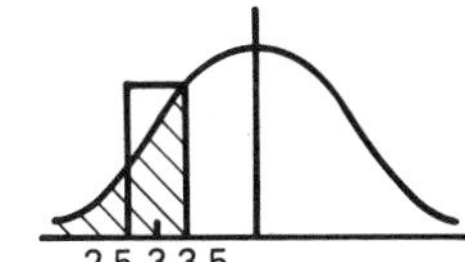

$P(X \leqslant 3)$ — rectangle included

If we require the probability that there are less than 3 heads, i.e. $P(X < 3)$, then we consider $P(X < 2.5)$.

So $\qquad P(X < 3) \longrightarrow P(X < 2.5)$

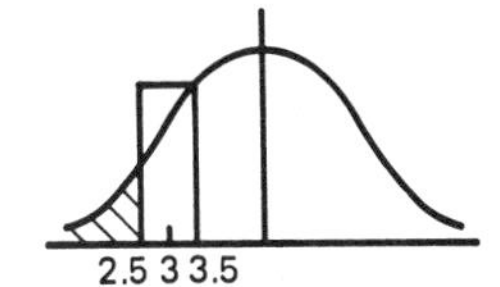

$P(X < 3)$ — rectangle not included

If we require the probability that there are exactly 3 heads, then

$$P(X = 3) \longrightarrow P(2.5 < X < 3.5)$$

Further examples:

$$P(5 \leqslant X \leqslant 8) \longrightarrow P(4.5 < X < 8.5)$$
$$P(5 < X \leqslant 8) \longrightarrow P(5.5 < X < 8.5)$$
$$P(5 \leqslant X < 8) \longrightarrow P(4.5 < X < 7.5)$$
$$P(5 < X < 8) \longrightarrow P(5.5 < X < 7.5)$$

$$P(X < 4) \longrightarrow P(X < 3.5)$$
$$P(X \leqslant 4) \longrightarrow P(X < 4.5)$$
$$P(X \geqslant 4) \longrightarrow P(X > 3.5)$$
$$P(X > 4) \longrightarrow P(X > 4.5)$$
$$P(X = 9) \longrightarrow P(8.5 < X < 9.5)$$
$$P(X = 7) \longrightarrow P(6.5 < X < 7.5)$$

$$P(X \geqslant 0) \longrightarrow P(X > -0.5)$$
$$P(X > 0) \longrightarrow P(X > 0.5)$$
$$P(X = 0) \longrightarrow P(-0.5 < X < 0.5)$$

Example 6.19 It is known that in a sack of mixed grass seeds 35% are ryegrass. Use the normal approximation to the binomial distribution to find the probability that in a sample of 400 seeds there are

(a) less than 120 ryegrass seeds,

(b) between 120 and 150 ryegrass seeds (inclusive),

(c) more than 160 ryegrass seeds.

Solution 6.19 Let X be the r.v. 'the number of ryegrass seeds'. Let 'success' be 'obtaining a ryegrass seed'.

Then $X \sim \text{Bin}(n, p)$ where $n = 400$ and $p = 0.35$.

Now, as n is large, we use the normal approximation to give

$$X \sim N(np, npq) \quad \text{where} \quad np = (400)(0.35) = 140$$
$$npq = (140)(0.65) = 91$$

so $\quad X \sim \text{N}(140, 91)$

(a) We require

$$P(X < 120) \longrightarrow P(X < 119.5) \qquad \text{(continuity correction)}$$

Now

$$P(X < 119.5) = P\left(\frac{X-140}{\sqrt{91}} < \frac{119.5-140}{\sqrt{91}}\right)$$
$$= P(Z < -2.149)$$
$$= 0.0158$$

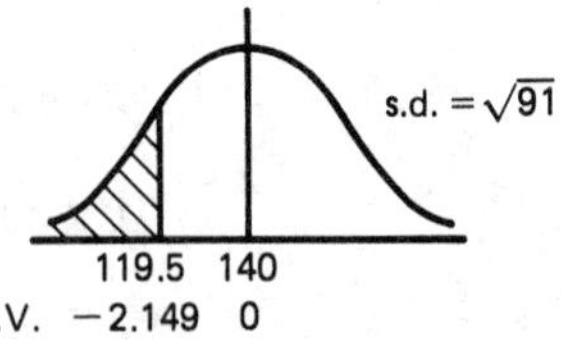

The probability that there are less than 120 ryegrass seeds is 0.0158.

(b) $P(120 \leqslant X \leqslant 150) \longrightarrow P(119.5 < X < 150.5)$ (continuity correction)

$$P(119.5 < X < 150.5) = P\left(\frac{119.5-140}{\sqrt{91}} < \frac{X-140}{\sqrt{91}} < \frac{150.5-140}{\sqrt{91}}\right)$$
$$= P(-2.149 < Z < 1.101)$$
$$= 0.8487$$

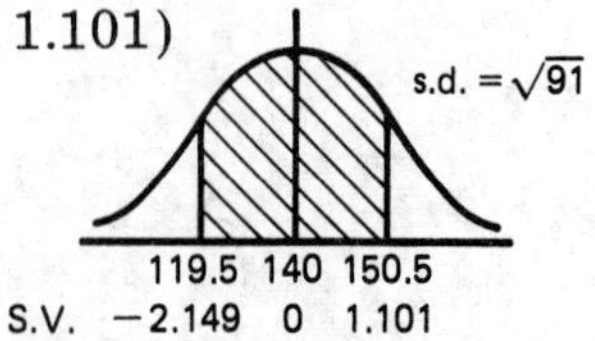

The probability that there are between 120 and 150 ryegrass seeds is 0.8487.

(c) $P(X > 160) \longrightarrow P(X > 160.5)$ (continuity correction)

$$P(X > 160.5) = P\left(\frac{X-140}{\sqrt{91}} > \frac{160.5-140}{\sqrt{91}}\right)$$
$$= P(Z > 2.149)$$
$$= 0.0158$$

s.d. $= \sqrt{91}$

140 160.5

S.V. 0 2.149

The probability that there are more than 160 ryegrass seeds is 0.0158.

Example 6.20 The random variable X has a binomial distribution with parameters n and p. Derive the mean and variance of X.

Show that the probability of obtaining a total of seven when two fair dice are tossed is 1/6. A pair of fair dice is tossed 100 times and the total observed on each occasion. What is the probability of getting more than 25 sevens? How many tosses would be required in order that the probability of getting at least one seven is 0.9 or more. (AEB)

Solution 6.20 If $X \sim \text{Bin}(n, p)$ then $E(X) = np$ and $\text{Var}(X) = npq$ (see p. 214).

P(total of 7 when two dice are tossed)

$$= \frac{6}{36}$$
$$= \frac{1}{6}$$

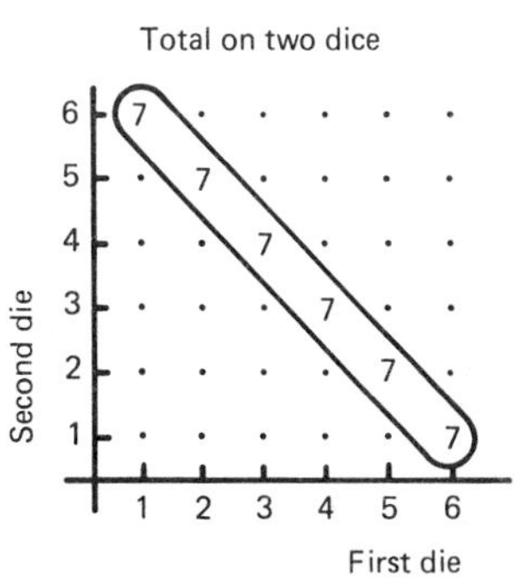

Let X be the r.v. 'the number of sevens when two dice are tossed'. Let 'success' be 'obtaining a total of 7'.

Then $X \sim \text{Bin}(n, p)$ where $n = 100$ and $p = \frac{1}{6}$, so $X \sim \text{Bin}(100, \frac{1}{6})$.

Now n is large and p is not too small, so we use the normal approximation:

$$X \sim \text{N}(np, npq) \quad \text{where} \quad np = (100)\left(\frac{1}{6}\right) = 50/3$$
$$npq = (100)\left(\frac{1}{6}\right)\left(\frac{5}{6}\right) = \frac{125}{9}$$
$$\text{standard deviation} = 5\sqrt{5/3}$$

so $X \sim \text{N}(50/3, 125/9)$.

We require

$$P(X > 25) \longrightarrow P(X > 25.5) \qquad \text{(continuity correction)}$$

$$P(X > 25.5) = P\left(\frac{X - 50/3}{5\sqrt{5/3}} > \frac{25.5 - 50/3}{5\sqrt{5/3}}\right)$$

$$= P(Z > 2.370)$$

$$= 0.00889$$

The probability of obtaining more than 25 sevens is 0.008 89.

Let the number of tosses required be n.

$$X \sim \text{Bin}\left(n, \frac{1}{6}\right)$$

Now
$$P(X = 0) = \left(\frac{5}{6}\right)^n$$

$$P(\text{at least one } 7) = P(X \geqslant 1)$$

$$= 1 - P(X = 0)$$

$$= 1 - \left(\frac{5}{6}\right)^n$$

We require n such that

$$1 - \left(\frac{5}{6}\right)^n \geqslant 0.9$$

i.e.
$$\left(\frac{5}{6}\right)^n \leqslant 0.1$$

Taking logs to base 10

$$n \log\left(\frac{5}{6}\right) \leqslant \log(0.1)$$

$$n \geqslant \frac{\log(0.1)}{\log(5/6)}$$

(when dividing by a negative quantity the inequality is reversed.)

$$n \geqslant 12.63$$

The number of tosses required is 13.

Exercise 6h

1. Continuity corrections — write down the transformations for each of the following: (*a*) $P(3 \leqslant X \leqslant 9)$, (*b*) $P(3 < X < 9)$, (*c*) $P(10 < X \leqslant 24)$, (*d*) $P(2 \leqslant X < 8)$, (*e*) $P(X > 54)$, (*f*) $P(X \geqslant 76)$, (*g*) $P(45 < X < 67)$, (*h*) $P(X < 109)$, (*i*) $P(X \leqslant 45)$, (*j*) $P(X = 56)$, (*k*) $P(400 < X \leqslant 560)$, (*l*) $P(X = 67)$, (*m*) $P(X > 59)$, (*n*) $P(X = 100)$, (*o*) $P(34 \leqslant X < 43)$, (*p*) $P(X = 7)$, (*q*) $P(X \geqslant 509)$, (*r*) $P(X < 7)$, (*s*) $P(27 \leqslant X < 29)$, (*t*) $P(X = 53)$.

2. If $X \sim \text{Bin}(200, 0.7)$, use the normal approximation to find (*a*) $P(X \geqslant 130)$, (*b*) $P(136 \leqslant X < 148)$, (*c*) $P(X < 142)$, (*d*) $P(X > 152)$, (*e*) $P(141 < X < 146)$.

3. 10% of the chocolates produced in a factory are mis-shapes. In a sample of 1000 chocolates find the probability that the number of mis-shapes is (*a*) less than 80, (*b*) between 90 and 115 inclusive, (*c*) 120 or more.

4. Find the probability of obtaining more than 110 ones in 400 tosses of an unbiased tetrahedral die with faces marked 1, 2, 3 and 4.

5. A coin is biased so that the probability that it will come down heads is double the probability that it will come down tails. The coin is tossed 120 times. Find the probability that there will be (*a*) between 42 and 51 tails inclusive, (*b*) 48 tails or less, (*c*) less than 34 tails, (*d*) between 72 and 90 heads inclusive.

6. An experiment consists of tossing two unbiased coins. The outcome is called a success if and only if two heads appear, all other outcomes being called a failure. If the experiment were repeated 27 times, write down the binomial distribution governing this series of experiments in the form $(p+q)^n$, stating the values of p, q and n.

   Find the expected number of successes and the standard deviation of this distribution.

   With the normal curve approximation estimate, using tables and giving your answer to 2 decimal places, the probability of obtaining at least 5 successes. (L Additional)

7. It is estimated that 1/5 of the population of England watched last year's Cup Final on television. If random samples of 100 people are interviewed, calculate the mean and variance of the number of people from these samples who watched the Cup Final on television.

   Use normal distribution tables to estimate, to 2 significant figures, the approximate probability of finding, in a random sample of 100 people, more than 30 people who watched the Cup Final on television. (L Additional)

8. In a series of n independent trials the probability of a 'success' at each trial is p. If R is the random variable denoting the total number of successes, state the probability that $R = r$. State, also, the mean and variance of R.

   A certain variety of flower seed is sold in packets containing about 1000 seeds. The packet claims that 40% will bloom white and 60% red. This may be assumed to be accurate.

   If five seeds are planted estimate the probability that
   (*a*) exactly three will bloom white;
   (*b*) at least one will bloom white.

   If 100 seeds are planted use the normal approximation to estimate the probability of obtaining between 30 and 45 white flowers. (SUJB)

9. A die is biased so that the probability of obtaining a six is $\frac{1}{4}$. The die is thrown 200 times. (*a*) Find the probability of obtaining a six on the die (i) more than 60 times, (ii) less than 45 times, (iii) between 40 and 55 times (inclusive). (*b*) How many throws would be required if the probability of obtaining at least one six is greater than 0.9?

10. Two hundred fair dice are thrown 1000 times. Use the normal approximation to the binomial distribution to find the number of times you would expect to have the following number of sixes (*a*) 30, (*b*) 53, (*c*) more than 38, (*d*) less than 28, (*e*) between 28 and 38 inclusive.

11. A certain tribe is distinguished by the fact that 45% of the males have 6 toes on their right foot. Two explorers discover a group of 200 males from the tribe. Find the probability that the number who have

six toes on their right foot is (*a*) 90, (*b*) less than 85, (*c*) between 82 and 91 inclusive, (*d*) more than 97.

12. Four hundred pupils sit a test which consists of 80 true–false questions. None of the candidates knows any of the answers and so guesses. (*a*) If the pass mark is 38, how many of the candidates would be expected to pass? (*b*) What should the new pass mark be if it is decided that only 115 candidates pass?

13. A lorry load of potatoes has, on average, one rotten potato in 6. A greengrocer tests a random sample of 100 potatoes and decides to turn away the lorry if he finds more than 18 rotten potatoes in the sample. Find the probability that he accepts the consignment.

THE NORMAL APPROXIMATION TO THE POISSON DISTRIBUTION

If $X \sim \text{Po}(\lambda)$ then

$$E(X) = \lambda$$
$$\text{Var}(X) = \lambda$$

Now, for *large* λ

$$X \sim \text{N}(\lambda, \lambda) \quad \text{approximately}$$

Generally, we require $\lambda > 20$ for a good approximation.

Example 6.21 A radioactive disintegration gives counts that follow a Poisson distribution with mean count per second of 25. Find the probability that in 1 second the count is between 23 and 27 inclusive,

(a) using the Poisson distribution,

(b) using the normal approximation to the Poisson distribution.

Solution 6.21 Let X be the r.v. 'the radioactive count in a 1 second interval'. Then $X \sim \text{Po}(25)$.

Now, we require $P(23 \leqslant X \leqslant 27)$,

(a) Using the Poisson distribution:

$$P(X = x) = e^{-25}\frac{25^x}{x!}$$

so

$$P(X = 23) = e^{-25}\frac{(25)^{23}}{23!} = 0.076\,342$$

Using the recurrence formula for ease of calculation:

$$p_{x+1} = \frac{25}{x+1}p_x \quad \text{where} \quad P(X = x) = p_x$$

$$p_{24} = \frac{25}{24}p_{23} \qquad (0.079\,5229)$$

$$p_{25} = \frac{25}{25}p_{24} \qquad (0.079\,5229)$$

$$p_{26} = \frac{25}{26} p_{25} \qquad (0.076\,464\,3)$$

$$p_{27} = \frac{25}{27} p_{26} \qquad (0.070\,800\,3)$$

So $\quad P(23 \leqslant X \leqslant 27) = p_{23} + p_{24} + p_{25} + p_{26} + p_{27}$

$\qquad = 0.383 \quad (3 \text{ d.p.})$

The probability that the count is between 23 and 27 inclusive is 0.383 (3 d.p.).

(**b**) Using the normal approximation, $X \sim \mathrm{N}(25, 25)$.

So $P(23 \leqslant X \leqslant 27) \longrightarrow P(22.5 < X < 27.5)$ (continuity correction)

$$P(22.5 < X < 27.5) = P\left(\frac{22.5-25}{5} < \frac{X-25}{5} < \frac{27.5-25}{5}\right)$$

$$= P(-0.5 < Z < 0.5)$$

$$= 0.383$$

s.d. = 5

22.5 25 27.5

S.V. −0.5 0 0.5

So the probability that the count is between 23 and 27 inclusive is 0.383.

NOTE: this answer compares very well with the answer in part (**a**) and the working is easier.

Exercise 6i

1. If $X \sim \mathrm{Po}(24)$, use the normal approximation to find (*a*) $P(X \leqslant 25)$, (*b*) $P(22 \leqslant X \leqslant 26)$, (*c*) $P(X > 23)$.

2. If $X \sim \mathrm{Po}(35)$, use the normal approximation to find (*a*) $P(X \leqslant 33)$, (*b*) $P(33 < X < 37)$, (*c*) $P(X > 37)$, (*d*) $P(X = 37)$.

3. If $X \sim \mathrm{Po}(60)$, use the normal approximation to find (*a*) $P(50 < X \leqslant 58)$, (*b*) $P(57 \leqslant X < 68)$, (*c*) $P(X > 52)$, (*d*) $P(X \geqslant 70)$.

4. The number of calls received by an office switchboard per hour follows a Poisson distribution with parameter 30. Using the normal approximation to the Poisson distribution, find the probability that, in one hour, (*a*) there are more than 33 calls, (*b*) there are between 25 and 28 calls (inclusive), (*c*) there are 34 calls.

5. In a certain factory the number of accidents occurring in a month follows a Poisson distribution with mean 4. Find the probability that there will be at least 40 accidents during one year.

6. The number of bacteria on a plate viewed under a microscope follows a Poisson distribution with parameter 60. Find the probability that there are between 55 and 75 bacteria on a plate.

   A plate is rejected if less than 38 bacteria are found. If 2000 such plates are viewed, how many will be rejected?

7. In an experiment with a radioactive substance the number of particles reaching a counter over a given period of time follows a Poisson distribution with mean 22. Find the probability that the number of particles reaching the counter over the given period of time is (*a*) less than 22, (*b*) between 25 and 30, (*c*) 18 or more.

8. The number of accidents on a certain railway line occur at an average rate of one every 2 months. Find the probability that (*a*) there are 25 or more accidents in 4 years, (*b*) there are 30 or less accidents in 5 years.

9. The number of eggs laid by an insect follows a Poisson distribution with parameter 200. (*a*) Find the probability that (i) more than 150 eggs are laid, (ii) more than 250 eggs are laid, (iii) between 180 and 240 eggs (inclusive) are laid.

(*b*) If the probability that an egg develops is 0.1, show that the number of survivors follows a Poisson distribution with parameter 20, and find the probability that there are more than 30 survivors.

10. Two towns, Allport and Bunchester, are linked by telephone. There are 2000 subscribers in Allport, but it is too expensive to install 2000 trunk lines between the two towns. In a busy hour, each subscriber in Allport requires a trunk line to Bunchester for an average time of 2 minutes. Show that the number of trunk lines in use follows a Poisson distribution with mean 66.67 per hour. What is the minimum number of trunk lines that should be installed if only 1% of all the calls will fail to find an empty trunk line?

WHEN TO USE THE DIFFERENT APPROXIMATIONS

	Distribution of X	Restrictions on parameters	Approximation
(1)	$X \sim \text{Bin}(n, p)$	n large (say $n > 50$) p small (say $p < 0.1$)	$X \sim \text{Po}(np)$
(2)	$X \sim \text{Bin}(n, p)$	$n > 10$, p close to $\frac{1}{2}$ or $n > 30$(say), p moving away from $\frac{1}{2}$	$X \sim \text{N}(np, npq)$
(3)	$X \sim \text{Po}(\lambda)$	$\lambda > 20$(say)	$X \sim \text{N}(\lambda, \lambda)$

Example 6.22 If $X \sim \text{Bin}(10, \frac{1}{2})$, find the probability that $X = 5$. Then find the approximation to this probability using **(a)** the normal distribution, **(b)** the Poisson distribution.

Solution 6.22

$$X \sim \text{Bin}(n, p) \quad \text{where} \quad n = 10, \quad p = \tfrac{1}{2}$$

$$X \sim \text{Bin}(10, \tfrac{1}{2})$$

Now $$P(X = x) = {}^{10}C_x \left(\frac{1}{2}\right)^{10-x} \left(\frac{1}{2}\right)^x = {}^{10}C_x \left(\frac{1}{2}\right)^{10}$$

so $$P(X = 5) = {}^{10}C_5 \left(\frac{1}{2}\right)^{10} = 0.2461$$

$$\underline{P(X = 5) = 0.2461.}$$

(**a**) Using the normal approximation:

$$X \sim N(np, npq) \quad \text{where} \quad np = (10)\left(\frac{1}{2}\right) = 5$$

$$npq = (5)\left(\frac{1}{2}\right) = 2.5$$

So $X \sim N(5, 2.5)$

Now

$$P(X=5) \longrightarrow P(4.5 < X < 5.5) \qquad \text{(continuity correction)}$$

$$P(4.5 < X < 5.5) = P\left(\frac{4.5-5}{\sqrt{2.5}} < \frac{X-5}{\sqrt{2.5}} < \frac{5.5-5}{\sqrt{2.5}}\right)$$

$$= P(-0.316 < Z < 0.316)$$

$$= 0.2478$$

So the probability that $X = 5$ is 0.2478 (using the normal approximation).

NOTE: this is a fairly good approximation even though $n = 10$. This is because p is exactly $\frac{1}{2}$.

(**b**) Using the Poisson approximation:

$$\lambda = np = 5 \quad \text{so} \quad X \sim Po(5)$$

$$P(X=5) = e^{-5}\frac{5^5}{5!} = 0.175$$

So the probability that $X = 5$ is 0.175 (using the Poisson approximation).

NOTE: this is a poor approximation since we should have $n > 50$ and $p < \frac{1}{10}$ for a good Poisson approximation.

Example 6.23 If $X \sim \text{Bin}(20, 0.4)$, find the probability that $6 \leqslant X \leqslant 10$. Then find the approximations to this probability using (**a**) the normal distribution, (**b**) the Poisson distribution.

Solution 6.23

$$X \sim \text{Bin}(n, p) \quad \text{where} \quad n = 20, \quad p = 0.4$$

i.e. $X \sim \text{Bin}(20, 0.4)$

$$P(X=x) = {}^{20}C_x(0.6)^{20-x}(0.4)^x$$

$$P(X=6) = p_6 = {}^{20}C_6(0.6)^{14}(0.4)^6 = 0.124\,411\,7$$

Now, using the recurrence formula:

$$p_{x+1} = \frac{(n-x)p}{(x+1)q} p_x$$

$$p_7 = \frac{14}{7}\left(\frac{0.4}{0.6}\right) p_6 \qquad (0.165\,882\,2)$$

$$p_8 = \frac{13}{8}\left(\frac{0.4}{0.6}\right) p_7 \qquad (0.179\,705\,7)$$

$$p_9 = \frac{12}{9}\left(\frac{0.4}{0.6}\right) p_8 \qquad (0.159\,738\,4)$$

$$p_{10} = \frac{11}{10}\left(\frac{0.4}{0.6}\right) p_9 \qquad (0.117\,141\,5)$$

$$P(6 \leqslant X \leqslant 10) = p_6 + p_7 + \ldots + p_{10}$$

$$= 0.7469 \quad (4\text{ d.p.})$$

Using the binomial distribution, $P(6 \leqslant X \leqslant 10) = 0.7469$ (4 d.p.).

(a) Using the normal distribution:

$$X \sim \text{N}(np, npq) \quad \text{where} \quad np = (20)(0.4) = 8$$

$$npq = (8)(0.6) = 4.8$$

so $\quad X \sim \text{N}(8, 4.8)$

Now

$$P(6 \leqslant X \leqslant 10) \longrightarrow P(5.5 < X < 10.5) \qquad \text{(continuity correction)}$$

$$P(5.5 < X < 10.5) = P\left(\frac{5.5-8}{\sqrt{4.8}} < \frac{X-8}{\sqrt{4.8}} < \frac{10.5-8}{\sqrt{4.8}}\right)$$

$$= P(-1.141 < Z < 1.141)$$

$$= 0.7462$$

s.d. $= \sqrt{4.8}$

5.5 8 10.5

S.V. −1.141 0 1.141

Therefore $P(6 \leqslant X \leqslant 10) = 0.7462$, using the normal approximation.

NOTE: this is a good approximation. p has moved away from $\frac{1}{2}$ but $n = 20$, which is quite large.

(b) Using the Poisson distribution:

$$\lambda = np = 8$$

So $X \sim \text{Po}(8)$ and $P(X = x) = e^{-8}\dfrac{8^x}{x!}$

$$p_6 = P(X = 6) = e^{-8}\frac{8^6}{6!} \qquad (0.122\,138\,2)$$

Now, using the recurrence formula:

$$p_{x+1} = \frac{\lambda}{(x+1)}p_x$$

we have

$$p_7 = \frac{8}{7}p_6 \qquad (0.139\,586\,5)$$

$$p_8 = \frac{8}{8}p_7 \qquad (0.139\,586\,5)$$

$$p_9 = \frac{8}{9}p_8 \qquad (0.124\,076\,9)$$

$$p_{10} = \frac{8}{10}p_9 \qquad (0.099\,261\,5)$$

$$P(6 \leqslant X \leqslant 10) = p_6 + p_7 + \ldots + p_{10}$$
$$= 0.6246 \quad (4\text{ d.p.})$$

So $P(6 \leqslant X \leqslant 10) = 0.6246$ (4 d.p.) using the Poisson approximation.

NOTE: this is a poor approximation since we should have $n > 50$ and $p < \frac{1}{10}$.

Example 6.24 If $X \sim \text{Bin}(100, 0.05)$ find the probability that $X = 4$. Then find the approximations to the probability using **(a)** the normal distribution, **(b)** the Poisson distribution.

Solution 6.24

$X \sim \text{Bin}(n, p)$ where $n = 100$, $p = 0.05$

i.e. $X \sim \text{Bin}(100, 0.05)$

Now $P(X = x) = {}^{100}C_x(0.95)^{100-x}(0.05)^x$

so $P(X = 4) = {}^{100}C_4(0.95)^{96}(0.05)^4$

$= 0.1781 \quad (4\text{ d.p.})$

Therefore $P(X = 4) = 0.1781$ (4 d.p.) using the binomial distribution.

(a) Using the normal approximation:

$X \sim \text{N}(np, npq)$ where $np = (100)(0.05) = 5$

$npq = (5)(0.95) = 4.75$

so $X \sim \text{N}(5, 4.75)$

Now

$$P(X=4) \longrightarrow P(3.5 < X < 4.5) \qquad \text{(continuity correction)}$$

$$\begin{aligned} P(3.5 < X < 4.5) &= P\left(\frac{3.5-5}{\sqrt{4.75}} < \frac{X-5}{\sqrt{4.75}} < \frac{4.5-5}{\sqrt{4.75}}\right) \\ &= P(-0.6882 < Z < -0.2294) \\ &= 0.1637 \end{aligned}$$

s.d. = $\sqrt{4.75}$

3.5 4.5 5

S.V. −0.6882 −0.2294 0

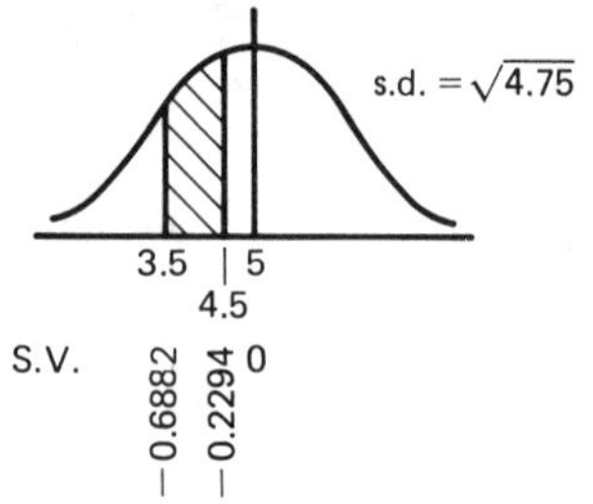

So $P(X=4) = 0.1637$ using the normal approximation.

NOTE: this is a fairly good approximation, even though p is small. This is because n is large.

(**b**) Using the Poisson approximation.

$$\lambda = np = 5 \qquad X \sim \text{Po}(5)$$

$$P(X=x) = \text{e}^{-5}\frac{5^x}{x!}$$

so, $$P(X=4) = \text{e}^{-5}\frac{5^4}{4!} = 0.1755 \quad \text{(4 d.p.)}$$

So $P(X=4) = 0.1755$ (4 d.p.) using the Poisson approximation.

NOTE: this is a good approximation, since n is large and p is small; also note that mean $\approx$ variance.

Example 6.25 If $X \sim \text{Po}(30)$, find $P(28 \leqslant X \leqslant 32)$. Then find the approximation to this probability using the normal distribution.

Solution 6.25

$$X \sim \text{Po}(30)$$

So $$P(X=x) = \text{e}^{-30}\frac{30^x}{x!}$$

and $$P(X=28) = p_{28} = \text{e}^{-30}\frac{30^{28}}{28!} \qquad (0.070\,213\,3)$$

Using the recurrence formula

$$p_{x+1} = \frac{30}{(x+1)}p_x$$

we have
$$p_{29} = \frac{30}{29}p_{28} \qquad (0.072\,634\,5)$$
$$p_{30} = \frac{30}{30}p_{29} \qquad (0.072\,634\,5)$$
$$p_{31} = \frac{30}{31}p_{30} \qquad (0.070\,291\,4)$$
$$p_{32} = \frac{30}{32}p_{31} \qquad (0.065\,898\,2)$$
$$P(28 \leqslant X \leqslant 32) = p_{28} + \ldots + p_{32}$$
$$= 0.3517 \quad (4\text{ d.p.})$$

So $P(28 \leqslant X \leqslant 32) = 0.3517$ (4 d.p.).

Using the normal approximation:
$$X \sim \text{N}(30, 30)$$

We require
$$P(28 \leqslant X \leqslant 32) \longrightarrow P(27.5 < X < 32.5) \qquad \text{(continuity correction)}$$
$$P(27.5 < X < 32.5) = P\left(\frac{27.5-30}{\sqrt{30}} < \frac{X-30}{\sqrt{30}} < \frac{32.5-30}{\sqrt{30}}\right)$$
$$= P(-0.456 < Z < 0.456)$$
$$= 0.3516$$

So $P(28 \leqslant X \leqslant 32) = 0.3516$ using the normal approximation.

NOTE: this is a good approximation since $\lambda > 20$.

Exercise 6j

In Questions 1 to 4 calculate the probabilities using the binomial distribution. Then find the approximations to them using (*a*) the normal distribution, (*b*) the Poisson distribution. Comment on your answers.

1. $X \sim \text{Bin}(15, \frac{1}{2})$, find $P(7 \leqslant X \leqslant 9)$.
2. $X \sim \text{Bin}(60, 0.03)$, find $P(4 \leqslant X \leqslant 6)$.
3. $X \sim \text{Bin}(30, 0.6)$, find $P(X = 17)$.
4. $X \sim \text{Bin}(120, 0.1)$, find $P(X = 8)$.
5. If $X \sim \text{Po}(27)$, (*a*) find $P(X = 30)$, (*b*) find an approximation using the normal distribution. Comment on your answer.
6. If $X \sim \text{Po}(12)$, find $P(10 \leqslant X < 12)$. Then find an approximation using the normal distribution. Comment on your answer.

Miscellaneous Exercise 6k

1. A number of different types of fungi are distributed at random in a field. Eighty per cent of these fungi are mushrooms, and the remainder are toadstools. Five per cent of the toadstools are poisonous. A man, who cannot distinguish between mushrooms and toadstools, wanders across the field and picks a total of 100 fungi. Determine, correct to 2 significant figures, using appropriate approximations, the probability that the man has picked
(*a*) at least 20 toadstools,
(*b*) exactly two poisonous toadstools. (C)

2. An old car is never garaged at night. On the morning following a wet night, the probability that the car does not start is $\frac{1}{3}$. On the morning following a dry night, this probability is $\frac{1}{25}$. The starting performance of the car each morning is independent of its performance on previous mornings.
(*a*) There are 6 consecutive wet nights. Determine the probability that the car does not start on at least 2 of the 6 mornings.
(*b*) During a wet autumn there are 32 wet nights. Using a suitable approximation, determine the probability that the car does not start on less than 16 of the 32 mornings.
(*c*) During a long summer drought there are 100 dry nights. Using a Poisson approximation, determine the probability that the car does not start on 5 or more of the 100 mornings.
(Give 3 decimal places in your answers.) (C)

3. An urn contains 100 balls of which 4 are coloured red and the remainder are coloured white. A ball is drawn at random from the urn, its colour is noted and it is then replaced in the urn.
Write down (but do not evaluate) an expression for the probability that, in a total of 10 such draws, a red ball is drawn exactly once.
Determine, correct to two decimal places, making use of a suitable approximation in each case, the probability that
(*a*) in a total of 100 such draws, a red ball is drawn on exactly four occasions,
(*b*) in a total of 9600 such draws, a red ball is drawn on between 350 and 400 occasions *inclusive*. (C)

4. Henri de Lade regularly travels from his home in the suburbs to his office in Paris. He always tries to catch the same train, the 08.05 from his local station. He walks to the station from his home in such a way that his arrival times form a normal distribution with mean 08.00 hours and standard deviation 6 minutes.
(*a*) Assuming that his train always leaves on time, what is the probability that on any given day Henri misses his train?
(*b*) If Henri visits his office in this way 5 days each week and if his arrival times at the station each day are independent, what is the probability that he misses his train once and only once in a given week?
(*c*) Henri visits his office 46 weeks every year. Assuming that there are no absences during this time, what is the probability that he misses his train less than 35 times in the year? (AEB 1980)

5. The probability of a man aged exactly 85 dying before he is 86 is about 0.211. Write down an expression for p_r, the probability that r of a group of n men aged exactly 85 die before they are 86.
(*a*) Calculate p_0 when $n = 5$.
(*b*) By considering (p_r/p_{r+1}), or otherwise, calculate the most likely value of r for the case $n = 100$.
(*c*) Use the normal approximation to the binomial to estimate the probability that at least 25 of a group of 100 men aged exactly 85 die before they are 86. (MEI)

6. In Urbania, selection for the Royal Flying Corps (RFC) is by means of an aptitude test based on a week's intensive military training. It is known that the scores of potential recruits on this test follow a normal distribution with mean 45 and standard deviation 10.
(*a*) What is the probability that a randomly chosen recruit will score between 40 and 60?
(*b*) What percentage of the recruits is expected to score more than 30?
(*c*) In a particular year 100 recruits take the test. Assuming that the pass mark is 50, calculate the probability that less than 35 recruits qualify for the RFC. (AEB 1978)

7. During an advertising campaign, the manufacturers of Wolfitt, (a dog food) claimed that 60% of dog owners preferred

to buy Wolfitt. Assuming that the manufacturer's claim is correct for the population of dog owners, calculate
(a) using the binomial distribution, and
(b) using a normal approximation to the binomial;
the probability that at least 6 of a random sample of 8 dog owners prefer to buy Wolfitt. Comment on the agreement, or disagreement, between your two values. Would the agreement be better or worse if the proportion had been 80% instead of 60%?

Continuing to assume that the manufacturer's figure of 60% is correct, use the normal approximation to the binomial to estimate the probability that, of a random sample of 100 dog owners, the number preferring Wolfitt is between 60 and 70 inclusive. (MEI)

8. If the probability of a male birth is 0.514, what is the probability that there will be fewer boys than girls in 1000 births? (You may assume that $0.514 \times 0.486 \approx 0.25$.)

How large a sample, to the nearest hundred, should be taken to reduce the probability of fewer boys than girls to less than 5%? (You may assume that the sample size in this part of the question is sufficiently large for a continuity correction to be unnecessary.) (SMP)

9. On the surface of halfpenny postage stamps there are either one or two phosphor bands. Ninety per cent of halfpenny stamps have two bands and the rest have one band. Of those having one band, 95% have the band in the centre of the stamp and the remainder have the band on the left-hand edge of the stamp.
(a) Determine the probability that in a random sample of ten halfpenny stamps there are exactly eight having two phosphor bands.
(b) Determine, using a normal approximation, the probability that in a random sample of 100 halfpenny stamps there are between five and fifteen stamps (inclusive) having one phosphor band.
(c) Determine, using a Poisson approximation, the probability that in a random sample of 100 halfpenny stamps there are less than three stamps which have only a single band, this band being on the left-hand edge of the stamp.

(Any expressions evaluated should be clearly exhibited, and answers should be given correct to three significant figures.) (C)

10. (a) Every year very small numbers of American wading birds lose their way on migration between North and South America and arrive in Great Britain instead, so that in September the proportion of American waders amongst the waders in Great Britain is about one in ten thousand.
At Dunsmere (a bird reserve in Great Britain), one September, there are twenty thousand waders, which may be regarded as a random sample of the waders present in Great Britain. Determine the probability that there are
(i) no American waders present at Dunsmere,
(ii) more than two American waders present at Dunsmere.

(b) Three-quarters of all the sightings in Great Britain of American waders are made in the autumn. Suppose that in 1980 there will be ten sightings of American waders at Dunsmere. Assuming that all sightings are independent of one another, determine the probability that exactly seven of these ten sightings will be made in the autumn. (C)

11. An inter-city telephone exchange has 100 lines and on average 80 are in use at any moment (on a typical business-day morning). Calculate
(a) the probability that all lines are engaged;
(b) the probability that more than 30 lines are free.
We say that a number x of lines is the 'effective minimum level' if the number of lines in use exceeds x for 95% of the time. Find x.

(You may assume that for large n the binomial probability may be approximated by a normal probability with mean na and variance nab.) (SMP)

12. A telephone exchange serves 2000 subscribers, and at any moment during the busiest period there is a probability of 1/30 for each subscriber that he will require a line. Assuming that the needs of subscribers are independent, write down an expression for the probability that exactly N lines will be occupied at any moment during the busiest period.

Use the normal distribution to estimate the minimum number of lines that would ensure that the probability that a call cannot be made because all the lines are occupied is less than 0.01.

Investigate whether the total number of lines needed would be reduced if the subscribers were split into two groups of 1000, each with its own set of lines. (MEI)

13. A population consists of individuals of three types A, B and C occurring in proportions 1:5:14.

(a) A sample of three individuals is drawn at random from the population.

(i) Determine the probability that all three are of different types.

(ii) Determine the probability that all three are of the same type.

(b) A sample of 40 individuals is drawn at random from the population.

(i) Determine the approximate value of the probability that 4 or more are of type A.

(ii) Determine the approximate value of the probability that exactly 10 are of type B. (C)

14. In each of n independent trials of an experiment the probability of an event A occurring is 0.05.

(a) When $n = 10$, determine the probability that A occurs exactly once, giving your answer to three decimal places.

(b) When $n = 200$, use a suitable approximate method to determine the probability that A occurs not more than 10 times. (JMB)

15. A discrete random variable X has the Poisson distribution given by

$$P(X = r) = e^{-a}\frac{a^r}{r!}, \quad r = 0, 1, 2, \ldots$$

Prove that the mean and the variance of X are each equal to a.

When a trainee typist types a document the number of mistakes made on any one page is a Poisson variable with mean 3, independently of the number of mistakes made on any other page. Use tables, or otherwise, to find, to three significant figures,

(a) the probability that the number of mistakes on the first page is less than two,

(b) the probability that the number of mistakes on the first page is more than four.

Find expressions in terms of e for

(c) the probability that the first mistake appears on the second page,

(d) the probability that the first mistake appears on the second page and the second mistake appears on the third page.

Evaluate these expressions, giving your answers to four significant figures.

When the typist types a 48-page document the total number of mistakes made by the typist is a Poisson variable with mean 144. Use a suitable approximate method to find, to three decimal places, the probability that this total number of mistakes is greater than 130. (JMB)

16. Describe, briefly, the conditions under which the binomial distribution Bin (n, p) may be approximated by

(a) a normal distribution,

(b) a Poisson distribution,

giving the parameters of each of the approximate distributions.

Among the blood cells of a certain animal species, the proportion of cells which are of type A is 0.37 and the proportion of cells which are of type B is 0.004. Find, to 3 decimal places, the probability that in a random sample of 8 blood cells at least 2 will be of type A.

Find, to 3 decimal places, an approximate value for the probability that

(c) in a random sample of 200 blood cells the combined number of type A and type B cells is 81 or more,

(d) there will be 4 or more cells of type B in a random sample of 300 blood cells. (L)

17. Manufactured articles are packed in boxes each containing 200 articles, and on average $1\frac{1}{2}\%$ of all articles manufactured are defective. A box which contains 4 or more defective articles is substandard. Using a suitable approximation, show that the probability that a randomly chosen box will be substandard is 0.353, correct to three decimal places.

A lorry-load consists of 16 boxes, randomly chosen. Find the probability that a lorry-load will include at most 2 boxes which are substandard, giving three decimal places in your answer.

A warehouse holds 100 lorry-loads. Show that, correct to two decimal places, the probability that exactly one of the lorry-loads in the warehouse will include at most 2 substandard boxes is 0.06. (C)

18. Answer the following questions using, in each case, tables of the binomial, Poisson or normal distribution according to which you think is most appropriate. In each example draw attention to any feature which either supports or casts doubt on your choice of distribution.

(a) Cars pass a point on a busy city centre road at an average rate of 7 per five second interval. What is the probability that in a particular five second interval the number of cars passing will be

(i) 7 or less,

(ii) exactly 7?

(*b*) Weather records show that for a certain airport during the winter months an average of one day in 25 is foggy enough to prevent landings. What is the probability that in a period of seven winter days landings are prevented on
(i) 2 or more days,
(ii) no days?
(*c*) The working lives of a particular brand of electric light bulb are distributed with mean 1200 hours and standard deviation 200 hours. What is the probability of
(i) a bulb lasting more than 1150 hours,
(ii) the mean life of a sample of 64 bulbs exceeding 1150 hours?*
(AEB 1988)

*See 'Distribution of the sample mean', p. 403.

19. Explain briefly the circumstances under which a normal distribution may be used as an approximation to a binomial distribution. Write down the mean and the variance of the normal approximation to the binomial distribution Bin (n, p). Give an example, from your projects if possible, of the use of this approximation stating the parameters of your binomial distribution and of your normal approximation.

In a multiple-choice examination, candidate Jones picks his answer to each question at random from the list of 3 answers provided, of which only one is correct. A candidate answering 18 or more questions correctly passes the examination.
(*a*) For a paper containing 45 questions, use a normal approximation to find, to 3 decimal places, the probability that Jones passes.
(*b*) It is required that the probability that Jones passes should be less than 0.005. Use a normal approximation to show that the paper should contain at most 31 questions. (L)

20. Explain briefly how you used, or could have used, a binomial distribution in a project.

State the conditions under which a normal distribution may be used as an approximation to the distribution Bin (n, p) and write down, in terms of n and p, the mean and the variance of this normal approximation.

A large bag of seeds contains three varieties in the ratios 4 : 2 : 1 and their germination rates are 50%, 60% and 80% respectively. Show that the probability that a seed chosen at random from the bag will germinate is $\frac{4}{7}$.

Find, to 3 decimal places, the probability that of 4 seeds chosen at random from the bag, exactly two of them will germinate. Given that 150 seeds are chosen at random from the bag, estimate, to 3 decimal places, the probability that less than 90 of them will germinate. (L)

21. Describe a project, or an experiment in your course work, which you conducted to demonstrate a Poisson distribution. State the condition under which a normal distribution may be used as an approximation to the Poisson distribution. Write down the mean and the variance of the normal approximation to the Poisson distribution with mean λ.

Tomatoes from a particular nursery are packed in boxes and sent to a market. Assuming that the number of bad tomatoes in a box has a Poisson distribution with mean 0.44, find, to 3 significant figures, the probability of there being
(*a*) fewer than 2,
(*b*) more than 2 bad tomatoes in a box when it is opened.
Use a normal approximation to find, to 3 decimal places, the probability that in 50 randomly chosen boxes there will be fewer than 20 bad tomatoes in total. (L)

7

RANDOM VARIABLES AND RANDOM SAMPLING

If X and Y are any two random variables, continuous or discrete, then

$$E(X+Y) = E(X)+E(Y)$$
$$E(X-Y) = E(X)-E(Y)$$

Also, if X and Y are independent, then

$$\text{Var}(X+Y) = \text{Var}(X)+\text{Var}(Y)$$
$$\text{Var}(X-Y) = \text{Var}(X)+\text{Var}(Y)$$

SUM AND DIFFERENCE OF TWO INDEPENDENT NORMAL VARIABLES

If X and Y are two independent normal variables such that

$$X \sim \text{N}(\mu_1, \sigma_1^2) \quad \text{and} \quad Y \sim \text{N}(\mu_2, \sigma_2^2)$$

then $$X+Y \sim \text{N}(\mu_1+\mu_2, \sigma_1^2+\sigma_2^2)$$

and $$X-Y \sim \text{N}(\mu_1-\mu_2, \sigma_1^2+\sigma_2^2)$$

Example 7.1 If $X \sim \text{N}(60, 16)$ and $Y \sim \text{N}(70, 9)$, find (a) $P(X+Y<140)$, (b) $P(120<X+Y<135)$, (c) $P(Y-X>7)$, (d) $P(2<Y-X<12)$.

Solution 7.1 (a) $X+Y \sim \text{N}(60+70, 16+9)$,

i.e. $$X+Y \sim \text{N}(130, 25)$$

For convenience, let $R = X+Y$,

so $$R \sim \text{N}(130, 25)$$

We require

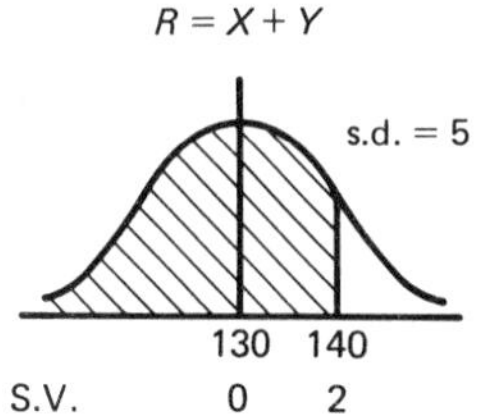

$$P(R < 140) = P\left(\frac{R-130}{5} < \frac{140-130}{5}\right)$$
$$= P(Z < 2)$$
$$= 0.9772$$

Therefore $P(X+Y<140) = 0.9772$.

(b) We require

$$P(120 < R < 135) = P\left(\frac{120-130}{5} < \frac{R-130}{5} < \frac{135-130}{5}\right)$$
$$= P(-2 < Z < 1)$$
$$= 0.8185$$

Therefore $P(120 < X+Y < 135) = 0.8185$.

(c) We need to consider the r.v. $Y-X$.

Now $\quad Y-X \sim \mathrm{N}(70-60, 9+16)$

For convenience, let $T = Y-X$.

So $\quad T \sim \mathrm{N}(10, 25)$

We require

$$P(T > 7) = P\left(\frac{T-10}{5} > \frac{7-10}{5}\right)$$
$$= P(Z > -0.6)$$
$$= 0.7257$$

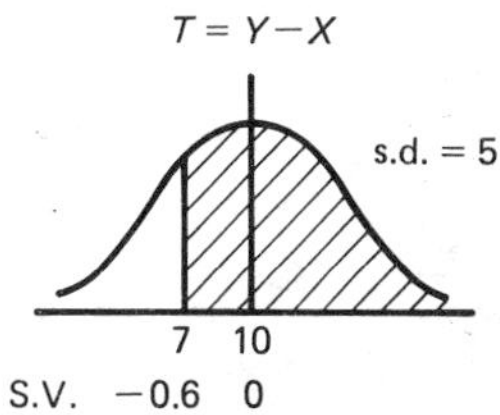

Therefore $P(Y-X>7) = 0.7257$.

(d) We require

$$P(2 < T < 12) = P\left(\frac{2-10}{5} < \frac{T-10}{5} < \frac{12-10}{5}\right)$$
$$= P(-1.6 < Z < 0.4)$$
$$= 0.6006$$

Therefore $P(2 < Y-X < 12) = 0.6006$.

Example 7.2 Each weekday Mr Jones walks to the local library to read the newspapers. The time he takes to walk to and from the library is a normal variable with mean 15 minutes and standard deviation 2 minutes. The time he spends in the library is a normal variable with mean 25 minutes and standard deviation $\sqrt{12}$ minutes. Find the probability that, on a particular day, **(a)** Mr Jones is away from the house for more than 45 minutes, **(b)** Mr Jones spends more time travelling than in the library.

Solution 7.2 Let L be the r.v. 'the time in minutes spent in the library'. Then $L \sim \mathrm{N}(25, 12)$.

Let W be the r.v. 'the time in minutes spent walking to and from the library'. Then $W \sim \mathrm{N}(15, 4)$.

(a) We require the distribution of the total time spent away from the house.

Let $\quad T = L + W$

So $\quad T \sim \mathrm{N}(40, 16)$

We require

$$P(T > 45) = P\left(\frac{T-40}{4} > \frac{45-40}{4}\right)$$

$$= P(Z > 1.25)$$

$$= 0.1056$$

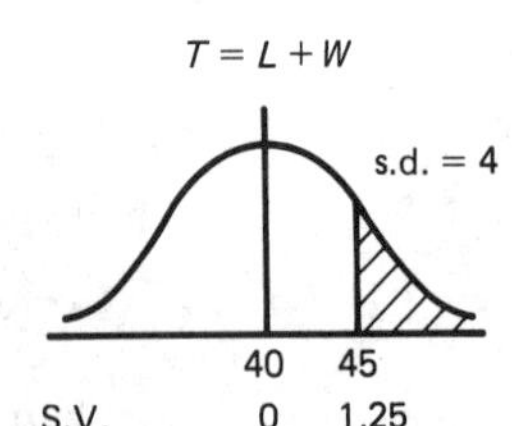

Therefore the probability that Mr Jones is away from the house for more than 45 minutes is 0.1056.

(b) We require $P(W > L)$, i.e. $P(W - L > 0)$.

Let

$U = W - L \quad$ then $\quad U \sim \mathrm{N}(15-25, 12+4)$

i.e. $\quad U \sim \mathrm{N}(-10, 16)$

We require

$$P(U > 0) = P\left(\frac{U-(-10)}{4} > \frac{0-(-10)}{4}\right)$$

$$= P(Z > 2.5)$$

$$= 0.006\,21$$

U = W − L
s.d. = 4
−10
0
S.V.
0
2.5

Therefore the probability that Mr. Jones spends more time travelling than in the library is 0.006 21.

Exercise 7a

1. If $X \sim N(100, 49)$ and $Y \sim N(110, 576)$, find (a) $P(X+Y>200)$, (b) $P(180<X+Y<240)$, (c) $P(Y-X<0)$, (d) $P(-20<Y-X<50)$.

2. If $X \sim N(75, 5)$ and $Y \sim N(78, 20)$, find (a) $P(X+Y>162)$, (b) $P(140<X+Y<150)$, (c) $P(X+Y<155)$, (d) $P(X-Y>0)$, (e) $P(Y-X<15)$.

3. If $A \sim N(3, 0.05)$ and $B \sim N(2, 0.04)$, find (a) $P(A-B>1.9)$, (b) $P(A+B<4.4)$, (c) $P(B>A-0.6)$.

4. If $X \sim N(25, 5)$ and $Y \sim N(30, 4)$, find (a) $P(|X+Y-55|<5)$, (b) $P(Y>X)$, (c) $P(|Y-X-5|<3)$.

5. At a self-service cafeteria a coffee machine is installed which dispenses (a) black coffee in amounts normally distributed with mean 6.1 oz and standard deviation 0.4 oz, (b) white coffee by first releasing a quantity of black coffee normally distributed with mean 4.9 oz and s.d. 0.3 oz and then adding milk normally distributed with mean 1.2 oz and s.d. 0.2 oz. Each cup is marked on the inside to a level of 5.5 oz and if this level is not attained the customer receives the drink without charge.
(i) What percentage of cups of black coffee will fall short of the 5.5 oz?
(ii) What is the mean and s.d. of the amount of white coffee dispensed into each cup?
(iii) What percentage of cups of white coffee will fall short of 5.5 oz?
(iv) If 10% of cups dispensed are black and the cost per cup for the ingredients is 2.1 p per cup for both black and white coffee, whilst the customer is charged 10 p per cup, what will be the gross profit on 1000 dispensed cups?
(v) What price per cup (to the nearest $\frac{1}{2}$p) should the cafeteria charge if the average profit is to be 5 p per cup? (SUJB)

6. Bolts are manufactured which are to fit in holes in steel plates. The diameter of the bolts is normally distributed with mean 2.60 cm and standard deviation 0.03 cm; the diameter of the holes is normally distributed with mean 2.71 cm and standard deviation 0.04 cm.
(a) Find the probability that a bolt selected at random has a diameter greater than 2.65 cm.
(b) Find the probability that a hole selected at random has a diameter less than 2.65 cm.
(c) Prove that, if a bolt and a hole are selected at random, the probability that the bolt will be too large to enter the hole is about 0.0139.
(d) The random selection of a bolt and a hole described in (c) above is carried out five times. Find the probability that in every case the bolt will be able to enter the hole. (C)

7. The diameters of axles supplied by a factory have a mean value of 19.92 mm and a standard deviation of 0.05 mm. The inside diameters of bearings supplied by another factory have a mean of 20.04 mm and a standard deviation of 0.03 mm. What is the mean and standard deviation of the random variable defined to be the diameter of a bearing less the diameter of an axle?
Assuming that both dimensions are normally distributed, what percentage of axles and bearings taken at random will not fit? (O & C)

EXTENSION TO MORE THAN TWO INDEPENDENT NORMAL VARIABLES

We can extend the results on p. 374 as follows:

If $X_1, X_2, \ldots, X_n$ is any set of random variables, then

$$E(X_1+X_2+\ldots+X_n) = E(X_1)+E(X_2)+\ldots+E(X_n)$$

If the random variables are independent, then

$$\text{Var}(X_1+X_2+\ldots+X_n) = \text{Var}(X_1)+\text{Var}(X_2)+\ldots+\text{Var}(X_n)$$

If $X_1, X_2, \ldots, X_n$ are n independent <u>normal</u> variables such that

$$X_1 \sim \mathrm{N}(\mu_1, \sigma_1^2), X_2 \sim \mathrm{N}(\mu_2, \sigma_2^2), \ldots, X_n \sim \mathrm{N}(\mu_n, \sigma_n^2)$$

then

$$X_1 + X_2 + \ldots + X_n \sim \mathrm{N}(\mu_1 + \mu_2 + \ldots + \mu_n, \sigma_1^2 + \sigma_2^2 + \ldots + \sigma_n^2)$$

NOTE: In the special case when $X_1, X_2, \ldots, X_n$ are independent observations from the same normal distribution so that $X_i \sim \mathrm{N}(\mu, \sigma^2)$ for $i = 1, 2, \ldots, n$, then

$$X_1 + X_2 + \ldots + X_n \sim \mathrm{N}(n\mu, n\sigma^2).$$

Example 7.3 If $W \sim \mathrm{N}(100, 8)$, $X \sim \mathrm{N}(120, 10)$ and $Y \sim \mathrm{N}(110, 12)$, find $P(W + X + Y < 320)$.

Solution 7.3 Let $A = W + X + Y$.

Then

$$E(A) = E(W) + E(X) + E(Y) = 100 + 120 + 110 = 330$$
$$\mathrm{Var}(A) = \mathrm{Var}(W) + \mathrm{Var}(X) + \mathrm{Var}(Y) = 8 + 10 + 12 = 30$$

So $A \sim \mathrm{N}(330, 30)$.

We require

$$P(A < 320) = P\left(\frac{A - 330}{\sqrt{30}} < \frac{320 - 330}{\sqrt{30}}\right)$$
$$= P(Z < -1.826)$$
$$= 0.0340$$

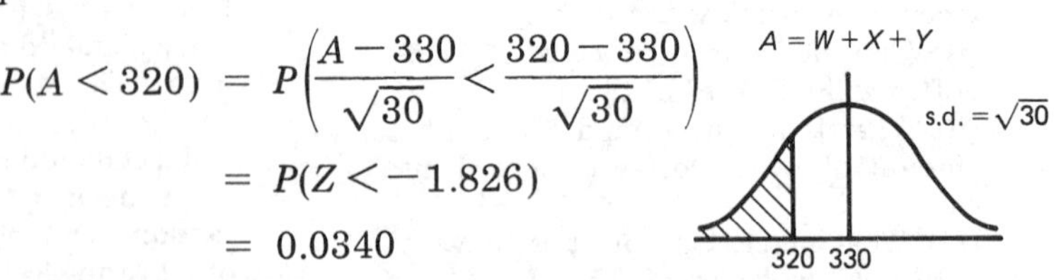

<u>Therefore $P(W + X + Y < 320) = 0.0340$.</u>

Example 7.4 Masses of a particular article are normally distributed with mean 20 g and standard deviation 2 g. If a random sample of 12 such articles is chosen, find the probability that the total mass is less than 230 g.

Solution 7.4 Let X be the r.v. 'the mass, in g, of an article'.

Then

$$X_1 \sim \mathrm{N}(20, 4)$$
$$X_2 \sim \mathrm{N}(20, 4)$$
$$\vdots \qquad \vdots$$
$$X_{12} \sim \mathrm{N}(20, 4)$$

Now, let $B = X_1 + X_2 + \ldots + X_{12}$,

so
$$\begin{aligned} E(B) &= E(X_1) + E(X_2) + \ldots + E(X_{12}) \\ &= 12E(X) \\ &= 240 \end{aligned}$$

and
$$\begin{aligned} \text{Var}(B) &= \text{Var}(X_1) + \text{Var}(X_2) + \ldots + \text{Var}(X_{12}) \\ &= 12\text{Var}(X) \\ &= 48 \end{aligned}$$

We have $\quad B \sim \text{N}(240, 48)$

We require

$$\begin{aligned} P(B < 230) &= P\left(\frac{B-240}{\sqrt{48}} < \frac{230-240}{\sqrt{48}}\right) \\ &= P(Z < -1.443) \\ &= 0.0745 \end{aligned}$$

$B = X_1 + X_2 + \ldots + X_{12}$

s.d. $= \sqrt{48}$

230 240

S.V. −1.443 0

Therefore the probability that the total mass of the articles is less than 230 g is 0.0745.

Example 7.5 If $A \sim \text{N}(10, 4)$, $B \sim \text{N}(12, 9)$ and $C \sim \text{N}(8, 12)$, find
(a) $P(A + B - C < 10)$, (b) $P(B - C - A > 0)$,
(c) $P[A_1 + A_2 - (B_1 + B_2) + C_1 + C_2 > 20]$ where A_1, A_2 are two independent observations from the population of A, etc., (d) the probability that three independent observations from the population of A have a sum which is greater than four independent observations from the population of C.

Solution 7.5 (a) Let $Y = A + B - C$.

Then

$$E(Y) = E(A) + E(B) - E(C) = 10 + 12 - 8 = 14$$
$$\text{Var}(Y) = \text{Var}(A) + \text{Var}(B) + \text{Var}(C) = 4 + 9 + 12 = 25$$

So $Y \sim \text{N}(14, 25)$

We require

$$\begin{aligned} P(Y < 10) &= P\left(\frac{Y-14}{5} < \frac{10-14}{5}\right) \\ &= P(Z < -0.8) \\ &= 0.2119 \end{aligned}$$

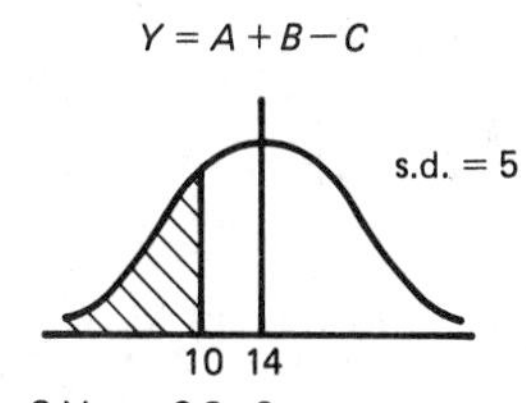

Therefore $P(A + B - C < 10) = 0.2119$.

(b) Let $W = B - C - A$.

Then

$$E(W) = E(B) - E(C) - E(A) = 12 - 8 - 10 = -6$$

$$\text{Var}(W) = \text{Var}(B) + \text{Var}(C) + \text{Var}(A) = 9 + 12 + 4 = 25$$

So $W \sim \text{N}(-6, 25)$

We require

$$P(W > 0) = P\left(\frac{W-(-6)}{5} > \frac{0-(-6)}{5}\right)$$

$$= P(Z > 1.2)$$

$$= 0.1151$$

Therefore $P(B - C - A > 0) = 0.1151$.

(c) Let $V = A_1 + A_2 - (B_1 + B_2) + C_1 + C_2$.

Then
$$E(V) = E(A_1) + E(A_2) - E(B_1) - E(B_2) + E(C_1) + E(C_2)$$
$$= 2E(A) - 2E(B) + 2E(C)$$
$$= 20 - 24 + 16$$
$$= 12$$

$$\text{Var}(V) = 2\text{Var}(A) + 2\text{Var}(B) + 2\text{Var}(C)$$
$$= 8 + 18 + 24$$
$$= 50$$

So $V \sim \text{N}(12, 50)$

We require

$$P(V > 20) = P\left(\frac{V-12}{\sqrt{50}} > \frac{20-12}{\sqrt{50}}\right)$$

$$= P(Z > 1.131)$$

$$= 0.1290$$

Therefore $P[A_1 + A_2 - (B_1 + B_2) + C_1 + C_2 > 20] = 0.1290$.

(d) Let $U = A_1 + A_2 + A_3 - (C_1 + C_2 + C_3 + C_4)$.

Then
$$E(U) = 3E(A) - 4E(C)$$
$$= 30 - 32$$
$$= -2$$

$$\text{Var}(U) = 3\text{Var}(A) + 4\text{Var}(C)$$
$$= 12 + 48$$
$$= 60$$

So $U \sim \text{N}(-2, 60)$

We require

$$P(A_1+A_2+A_3 > C_1+C_2+C_3+C_4)$$

i.e. $$P(U>0) = P\left(\frac{U-(-2)}{\sqrt{60}} > \frac{0-(-2)}{\sqrt{60}}\right)$$
$$= P(Z>0.258)$$
$$= 0.3982$$

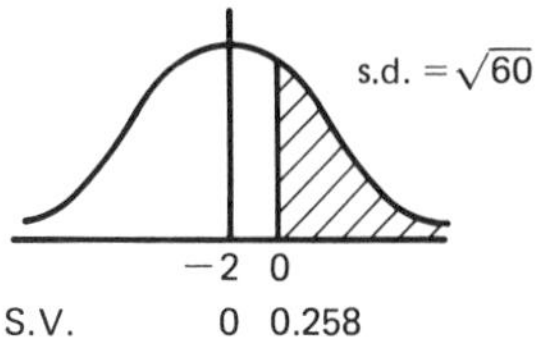

s.d. $=\sqrt{60}$

-2 0

S.V. 0 0.258

Therefore the probability that three observations from the population of A have a sum which is greater than four observations from the population of C is 0.3982.

Example 7.6 In a cafeteria, baked beans are served either in ordinary portions or in children's portions. The quantity given for an ordinary portion is a normal variable with mean 90 g and standard deviation 3 g and the quantity given for a children's portion is a normal variable with mean 43 g and standard deviation 2 g. What is the probability that John, who has two children's portions, is given more than his father, who has an ordinary portion?

Solution 7.6 Let C be the r.v. 'the quantity given, in g, in a children's portion'. Then $C \sim N(43, 4)$.

Let A be the r.v. 'the quantity given, in g, in an ordinary portion'. Then $A \sim N(90, 9)$.

We require $P(C_1+C_2>A)$,

i.e. $P(C_1+C_2-A>0)$.

Now let
$$\begin{aligned} W &= C_1+C_2-A \\ E(W) &= E(C_1)+E(C_2)-E(A) \\ &= 2E(C)-E(A) \\ &= 86-90 \\ &= -4 \end{aligned}$$

and
$$\begin{aligned} \text{Var}(W) &= \text{Var}(C_1)+\text{Var}(C_2)+\text{Var}(A) \\ &= 2\,\text{Var}(C)+\text{Var}(A) \\ &= 8+9 \\ &= 17 \end{aligned}$$

So $$W \sim N(-4, 17).$$

Now

$$\begin{aligned} P(C_1+C_2-A>0) &= P(W>0) \\ &= P\left(\frac{W-(-4)}{\sqrt{17}} > \frac{(0-(-4))}{\sqrt{17}}\right) \quad W = C_1+C_2-A \\ &= P(Z>0.970) \\ &= 0.166 \end{aligned}$$

s.d. $=\sqrt{17}$

-4 0

S.V. 0 0.970

Therefore the probability that John has more than his father is 0.166.

Exercise 7b

1. If $A \sim N(50, 6)$, $B \sim N(30, 8)$ and $C \sim N(80, 11)$ find
   (*a*) $P(A+B+C>170)$,
   (*b*) $P(-6<A+B-C<10)$,
   (*c*) $P(A_1+A_2-(B+C)<0)$.

2. A random sample of 20 items is taken from a normal population with mean 15 and variance 5. Find the probability that the sum of the values in the sample is less than 305.

3. Lengths of rod of type A are normally distributed with mean 5 cm and standard deviation 0.5 cm and lengths of rod of type B are normally distributed with mean 10 cm and standard deviation 1 cm. Find the probability that (*a*) a length consisting of 2 rods of type A and 4 rods of type B is more than 52 cm long, (*b*) a length consisting of 3 rods of type A and 2 rods of type B is between 33 cm and 36 cm long, (*c*) a length consisting of 6 rods of type A is longer than a length consisting of 3 rods of type B.

4. If $X \sim N(5, 4)$ and $Y \sim N(6, 9)$, find the probability that a sample consisting of 3 items from the population with r.v. X and 4 items from the population with r.v. Y will have a sum exceeding 50.

5. Chocolate Delight cakes are sold in packets of 6. The mass of each cake is a normal variable with mean 20 g and standard deviation 2 g. The mass of the packing material is a normal variable with mean 30 g and standard deviation 4 g. Find the probability that the total mass of the packet (*a*) exceeds 162 g, (*b*) is less than 137 g, (*c*) lies between 140 g and 153 g.

6. If $X_i \sim N(2, 2)$ and $Y_i \sim N(1.5, 2.2)$ and if $L = \sum_{i=1}^{20} X_i$ and $M = \sum_{i=1}^{26} Y_i$, find $P(L>M)$.

7. In a certain village the heights of women follow a normal distribution with mean 164 cm and standard deviation 5 cm and the heights of men are normally distributed with mean 173 cm and standard deviation 6 cm. If a man and woman are picked at random, find the probability that (*a*) the woman is taller than the man, (*b*) the man is more than 5 cm taller than the woman.

8. The time taken to carry out a standard service on a car of type A is known, to a good approximation, to be a normal variable with mean 1 hour and standard deviation 10 minutes. Assuming that only one car is serviced at a time, find the probability that it will take more than $6\frac{1}{2}$ hours to service 6 cars.

   The time taken to carry out a standard service on a car of type B is a normal variable with mean $1\frac{1}{2}$ hours and standard deviation 15 minutes. Find the probability that 5 cars of type B can be serviced more quickly than 8 cars of type A. (C)

9. If $X_i \sim N(70, 10)$, find
   $P\left(335 < \sum_{i=1}^{5} X_i < 360\right)$.

10. Four runners, A, B, C and D train to run the distances 100 m, 200 m, 500 m and 800 m respectively, in order to take part in a 1600 m relay race. During training

their individual times (recorded in seconds) are normally distributed as follows: $A \sim N(10.8, 0.2^2)$, $B \sim N(23.7, 0.3^2)$, $C \sim N(62.8, 0.9^2)$, $D \sim N(121.2, 2.1^2)$. Find the probability that the runners take less than 3 minutes 35 seconds to run the relay race.

11. Mr Smith has five dogs, two of which are male and three are female. The masses of food they eat in any given week are normally distributed as follows:

	Mean (kg)	Standard deviation (kg)
Male	3.5	0.4
Female	2.5	0.3

Find the probability that the two males eat more than the three females in a particular week.

12. The process of painting the body-work of a mass-produced lorry consists of giving it 1 coat of paint A, 3 coats of paint B and 2 coats of paint C. A record of the quantity of each type of paint used for each coat is kept for each lorry produced over a long period. The following table gives the means and standard deviations of these quantities measured in litres:

	Mean	Standard deviation
The coat of paint A	3.7	0.42
Each coat of paint B	1.3	0.15
Each coat of paint C	1.0	0.12

Assuming independence of the distribution for each coat, calculate the mean and standard deviation for the total quantity of paint used on each lorry.

Assuming that the quantities of paint used for each coat are normally distributed, calculate

(*a*) the percentage of lorries receiving less than 8.5 litres of paint,

(*b*) the percentage of lorries receiving more than 10.0 litres of paint. (C)

13. The means and variances of independent normal variables X and Y are known. State the means and variances of $X \pm Y$ in terms of those of X and Y.

The values of two types of resistors are normally distributed as follows:

Type A: mean: 100 ohms; standard deviation: 2 ohms

Type B: mean: 50 ohms, standard deviation: 1.3 ohms

(*a*) What tolerances would be permitted for type A if only 0.5% were rejected?

(*b*) 300-ohm resistors are made by connecting together three of the type A resistors, drawn from the total production. What percentage of the 300-ohm resistors may be expected to have resistances greater than 295 ohms?

(*c*) Pairs of resistors, one of 100 ohms and one of 50 ohms, drawn from the total production for types A and B respectively, are connected together to make 150-ohm resistors. What percentage of the resulting resistors may be expected to have resistances in the range 150 to 151.4 ohms? (AEB)

14. The time of departure of my train from Temple Meads Station is distributed normally about the scheduled time of 08 25 with a standard deviation 1 minute. I arrive at Temple Meads Station on another train whose time of arrival is normally distributed about the scheduled time of 08 20 with standard deviation of 1 minute. It takes me 3 minutes to change platforms.

(*a*) Find the probability that I miss the 08 25 and am late for work.

(*b*) Find the probability that this happens every day from Monday to Friday in a given week.

15. The mass of a certain grade of apple is normally distributed with mean mass 120 g and standard deviation 10 g.

(*a*) If an apple of this grade is chosen at random, find the probability that its mass lies between 100.5 g and 124 g.

(*b*) If four apples of this grade are chosen at random, find the probability that their total mass will exceed 505 g.

MULTIPLES OF NORMAL VARIABLES

We have shown previously that, for any constant a,

$$E(aX) = aE(X)$$

$$\text{Var}(aX) = a^2\text{Var}(X)$$

Now, if X is a normal variable such that $X \sim N(\mu, \sigma^2)$ then

$$aX \sim N(a\mu, a^2\sigma^2)$$

If X and Y are two independent normal variables such that $X \sim N(\mu_1, \sigma_1^2)$ and $Y \sim N(\mu_2, \sigma_2^2)$, and a and b are any constants, then

$$aX + bY \sim N(a\mu_1 + b\mu_2, a^2\sigma_1^2 + b^2\sigma_2^2)$$
$$aX - bY \sim N(a\mu_1 - b\mu_2, a^2\sigma_1^2 + b^2\sigma_2^2)$$

Example 7.7 If $X \sim N(50, 25)$, find $P(3X > 160)$.

Solution 7.7 Now

$$E(3X) = 3E(X) = 150$$
$$\text{Var}(3X) = 9\text{Var}(X) = 225$$

So

$$3X \sim N(150, 225)$$

Therefore

$$P(3X > 160) = P\left(\frac{3X - 150}{15} > \frac{160 - 150}{15}\right)$$
$$= P(Z > 0.667)$$
$$= 0.2523$$

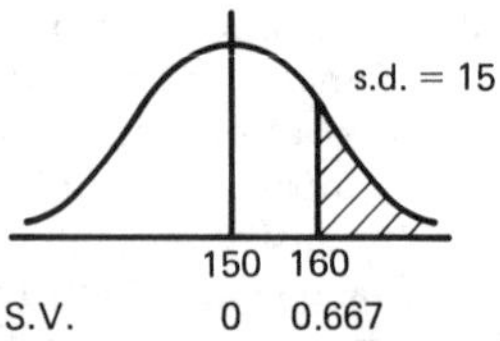

Therefore $P(3X > 160) = 0.2523$.

Example 7.8 If $X \sim N(70, 10)$ and $Y \sim N(50, 8)$, find $P(2X > 3Y)$.

Solution 7.8 We require $P(2X > 3Y)$, i.e. $P(2X - 3Y > 0)$.

Let

$$A = 2X - 3Y$$

then

$$E(A) = 2E(X) - 3E(Y)$$
$$= 140 - 150$$
$$= -10$$
$$\text{Var}(A) = 4\text{Var}(X) + 9\text{Var}(Y)$$
$$= 40 + 72$$
$$= 112$$

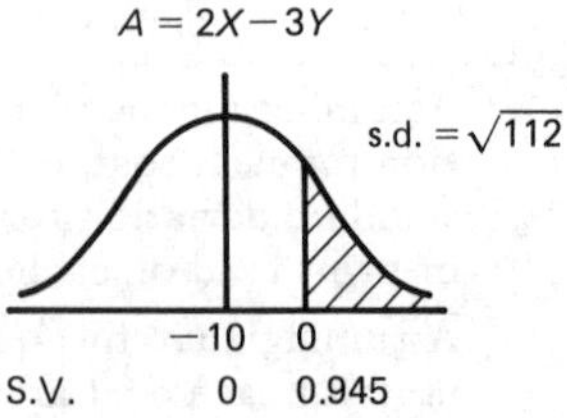

So

$$A \sim N(-10, 112)$$

$$P(A > 0) = P\left(\frac{A - (-10)}{\sqrt{112}} > \frac{0 - (-10)}{\sqrt{112}}\right)$$
$$= P(Z > 0.945)$$
$$= 0.1723$$

Therefore $P(2X > 3Y) = 0.1723$.

DISTINGUISHING BETWEEN MULTIPLES AND SUMS OF RANDOM VARIABLES

Care must be taken to distinguish between the r.v. $2X$ and the r.v. X_1+X_2, where X_1 and X_2 are two independent observations of the r.v. X.

If	$X \sim N(\mu, \sigma^2)$ then	$2X \sim N(2\mu, 4\sigma^2)$
but		$X_1+X_2 \sim N(2\mu, 2\sigma^2)$

NOTE: the means of the two distributions are the same, but the variances are different.

Example 7.9 If $X \sim N(10, 9)$, find (*a*) $P(2X > 23)$, (*b*) $P(X_1+X_2 > 23)$ where X_1 and X_2 are two independent observations from the population of X.

Solution 7.9 Now $X \sim N(10, 9)$.

(a) Let $V = 2X$, then

$$\begin{aligned} E(V) &= E(2X) \quad \text{and} \quad & \mathrm{Var}(V) &= \mathrm{Var}(2X) \\ &= 2E(X) & &= 4\mathrm{Var}(X) \\ &= 20 & &= 36 \end{aligned}$$

So $\qquad V \sim N(20, 36)$

and
$$\begin{aligned} P(V > 23) &= P\left(\frac{V-20}{6} > \frac{23-20}{6}\right) \\ &= P(Z > 0.5) \\ &= 0.3085 \end{aligned}$$

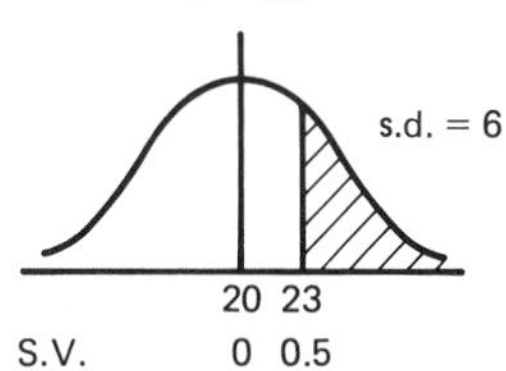

Therefore $P(2X > 23) = 0.3085$.

(b) Let $W = X_1 + X_2$.

Then

$$\begin{aligned} E(W) &= E(X_1)+E(X_2) \quad \text{and} \quad & \mathrm{Var}(W) &= \mathrm{Var}(X_1)+\mathrm{Var}(X_2) \\ &= 2E(X) & &= 2\mathrm{Var}(X) \\ &= 20 & &= 18 \end{aligned}$$

So $\qquad W \sim N(20, 18)$

and
$$\begin{aligned} P(W > 23) &= P\left(\frac{W-20}{\sqrt{18}} > \frac{23-20}{\sqrt{18}}\right) \\ &= P(Z > 0.707) \\ &= 0.2399 \end{aligned}$$

$W = X_1 + X_2$

s.d. $= \sqrt{18}$

20 23

S.V. 0 0.707

Therefore $P(X_1 + X_2 > 23) = 0.2399$.

In general, if $X \sim N(\mu, \sigma^2)$

then $nX \sim N(n\mu, n^2\sigma^2)$

but $X_1 + X_2 + \ldots + X_n \sim N(n\mu, n\sigma^2)$

The following example illustrates the difference between multiples and sums of random variables.

Example 7.10 A soft drinks manufacturer sells bottles of drinks in two sizes. The amount in each bottle, in ml, is normally distributed as shown in the table:

	Mean (ml)	Variance (ml^2)
Small	252	4
Large	1012	25

(a) A bottle of each size is selected at random. Find the probability that the large bottle contains less than four times the amount in the small bottle.

(b) One large and four small bottles are selected at random. Find the probability that the amount in the large bottle is less than the total amount in the four small bottles.

Solution 7.10 Let S be the r.v. 'the amount, in ml, in a small bottle'. Then $S \sim N(252, 4)$.

Let L be the r.v. 'the amount, in ml, in a large bottle'. Then $L \sim N(1012, 25)$.

(a) We need $P(L < 4S) = P(L - 4S < 0)$.

$$\begin{aligned}\text{Now } E(L-4S) &= E(L) - E(4S) \quad \text{(multiple of } S\text{)}\\ &= E(L) - 4E(S)\\ &= 1012 - 1008\\ &= 4\end{aligned}$$

$$\begin{aligned}\text{Var}(L-4S) &= \text{Var}(L) + \text{Var}(4S)\\ &= \text{Var}(L) + 16\text{Var}(S)\\ &= 25 + 64\\ &= 89\end{aligned}$$

So $L - 4S \sim N(4, 89)$

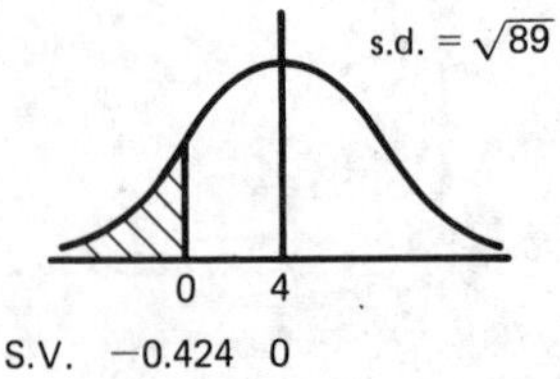

$$\begin{aligned}P(L-4S<0) &= P\left(Z < \frac{0-4}{\sqrt{89}}\right)\\ &= P(Z < -0.424)\\ &= 0.3358\end{aligned}$$

s.d. = $\sqrt{89}$

0 4

S.V. −0.424 0

Therefore the probability that the large bottle contains less than four times the amount of a small bottle is 0.3358.

(b) We need $P(L < S_1 + S_2 + S_3 + S_4) = P(L - (S_1 + \ldots + S_4) < 0)$.

Now

$$\begin{aligned} E(L-(S_1+\ldots+S_4)) &= E(L) - E(S_1+\ldots+S_4) \quad \text{(sum of r.v. } S) \\ &= E(L) - 4E(S) \\ &= 1012 - 1008 \\ &= 4 \end{aligned}$$

$$\begin{aligned} \mathrm{Var}(L-(S_1+\ldots+S_4)) &= \mathrm{Var}(L) + \mathrm{Var}(S_1+\ldots+S_4) \\ &= \mathrm{Var}(L) + 4\mathrm{Var}(S) \\ &= 25 + 16 \\ &= 41 \end{aligned}$$

Therefore $L - (S_1 + \ldots + S_4) \sim \mathrm{N}(4, 41)$

and

$$\begin{aligned} P(L-(S_1+\ldots+S_4) < 0) &= P\left(Z < \frac{0-4}{\sqrt{41}}\right) \\ &= P(Z < -0.625) \\ &= 0.266 \end{aligned}$$

s.d. $= \sqrt{41}$

0 4

S.V. −0.625 0

Therefore the probability that the large bottle contains less than the four small bottles is 0.266.

Exercise 7c

1. If $X \sim \mathrm{N}(40, 12)$ and $Y \sim \mathrm{N}(60, 15)$, find (*a*) $P(2X > 90)$, (*b*) $P(4Y < 270)$, (*c*) $P(3X - 2Y < 20)$, (*d*) $P[\frac{1}{2}(X+Y) > 55]$.

2. If $A \sim \mathrm{N}(82, 1.5^2)$, $B \sim \mathrm{N}(42, 0.3^2)$ and $C \sim \mathrm{N}(85, 0.7^2)$, find (*a*) $P(3A < 250)$, (*b*) $P(6B > 255)$, (*c*) $P(3A > 6B)$, (*d*) $P(2B + A > 2C)$, (*e*) $P[\frac{1}{2}(A+B) < 64]$, (*f*) $P(\frac{1}{3}(A+B+C) > 70)$.

3. The r.v. X is normally distributed with mean μ and variance 6, and the r.v. Y is normally distributed with mean 8 and variance σ^2. If the r.v. $2X - 3Y$ is normally distributed with mean -12 and variance 42, find (*a*) the values of μ and σ^2, (*b*) $P(X > 8)$, (*c*) $P(Y < 9)$, (*d*) $P(-4 < 3X - 2Y < 7)$.

4. The r.v. X is distributed normally with mean 25 and standard deviation 4, the r.v. Y is distributed normally with mean 30 and standard deviation 3, and X and Y are independent. Find the probability that a single observation from the population of X is greater than two-thirds of the value of a single observation from the population of Y.

5. If $X \sim \mathrm{N}(50, 16)$ and $Y \sim \mathrm{N}(40, 9)$, find (*a*) $P(2X + Y > 120)$, (*b*) $P[\frac{1}{2}(X - Y) > 0]$, (*c*) $P(100 < 3X - Y < 130)$.

6. If $X \sim \mathrm{N}(30, 4)$ find (*a*) $P(5X > 160)$, (*b*) $P(Y > 160)$ where $Y = X_1 + \ldots + X_5$.

7. The thickness, P cm, of a randomly chosen paperback book may be regarded as an observation from a normal distribution with mean 2.0 and variance 0.730. The thickness, H cm, of a randomly chosen hardback book may be regarded as an observation from a normal distribution with mean 4.9 and variance 1.920.
(*a*) Determine the probability that the combined thickness of four randomly chosen paperbacks is greater than the

combined thickness of two randomly chosen hardbacks.
(*b*) By considering $X = 2P - H$, or otherwise, determine the probability that a randomly chosen paperback is less than half as thick as a randomly chosen hardback.

(*c*) Determine the probability that a randomly chosen collection of sixteen paperbacks and eight hardbacks will have a combined thickness of less than 70 cm.

(Give 3 decimal places in your answers.)
(C)

MISCELLANEOUS WORKED EXAMPLES

Example 7.11 (a) A certain liquid drug is marketed in bottles containing a nominal 20 ml of drug. Tests on a large number of bottles indicate that the volume of liquid in each bottle is distributed normally with mean 20.42 ml and s.d. 0.429 ml.
(i) Estimate the percentage of bottles which would be expected to contain less than 20 ml of drug.
(ii) Find the level to which the mean should be adjusted (without altering the s.d.) so that only 1% of bottles should contain less than 20 ml.

(b) If the independent random variables X and Y are normally distributed with means μ_1, μ_2 and variances σ_1^2, σ_2^2 respectively, state what you can about the distribution of $Z = X - Y$.

If the capacity of the bottles in (a) is normally distributed with mean 21.77 ml and s.d. 0.210 ml and the liquid with (unadjusted) mean 20.42 ml and s.d. 0.429 ml, estimate what percentage of bottles will overflow during filling. (SUJB)

Solution 7.11 (a) Let X be the r.v. 'the volume in ml of liquid in a bottle'

Then $X \sim N(20.42, 0.429^2)$

(*i*) $$P(X < 20) = P\left(\frac{X - 20.42}{0.429} < \frac{20 - 20.42}{0.429}\right)$$
$$= P(Z < -0.979)$$
$$= 0.1637$$

Therefore 16.37% of bottles would be expected to contain less than 20 ml of drug.

(*ii*) We need to find μ such that

$$P(X < 20) = 0.01$$

i.e. $$P\left(\frac{X - \mu}{0.429} < \frac{20 - \mu}{0.429}\right) = 0.01$$

Now $\dfrac{20-\mu}{0.429}$ must be negative, and by symmetry we find that

$$\frac{\mu-20}{0.429} = 2.326$$

$$\mu = 20+(2.326)(0.429)$$
$$= 21.00$$

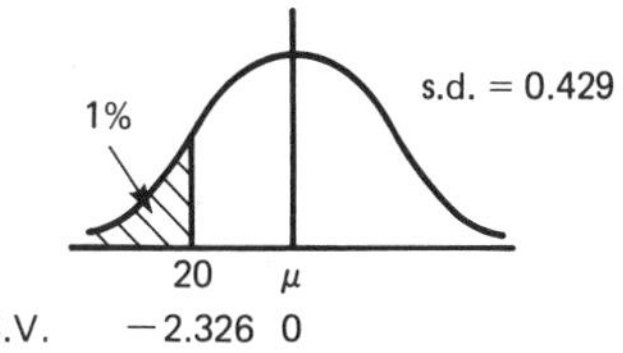

1%
s.d. = 0.429
20 μ
S.V. −2.326 0

The adjusted value of the mean should be 21.00 ml of drug.

(b) $X \sim N(\mu_1, \sigma_1^2)$ and $Y \sim N(\mu_2, \sigma_2^2)$.

If $Z = X - Y$, then

$$E(Z) = E(X) - E(Y) = \mu_1 - \mu_2$$
$$\text{Var}(Z) = \text{Var}(X) + \text{Var}(Y) = \sigma_1^2 + \sigma_2^2$$

and $Z \sim N(\mu_1 - \mu_2, \sigma_1^2 + \sigma_2^2)$.

If X is the r.v. 'the volume in ml of liquid'

then $X \sim N(20.42, 0.429^2)$ as before

If Y is the r.v. 'the capacity in ml of a bottle',

then $Y \sim N(21.77, 0.210^2)$

and $X - Y \sim N(20.42 - 21.77, 0.429^2 + 0.210^2)$

i.e. $X - Y \sim N(-1.35, 0.2281)$

Now, the bottle will overflow if $X > Y$, i.e. if $X - Y > 0$.

$$P(X-Y>0) = P\left(\frac{X-Y-(-1.35)}{\sqrt{0.2281}} > \frac{0-(-1.35)}{\sqrt{0.2281}}\right)$$
$$= P(Z > 2.827)$$
$$= 0.002\,35$$

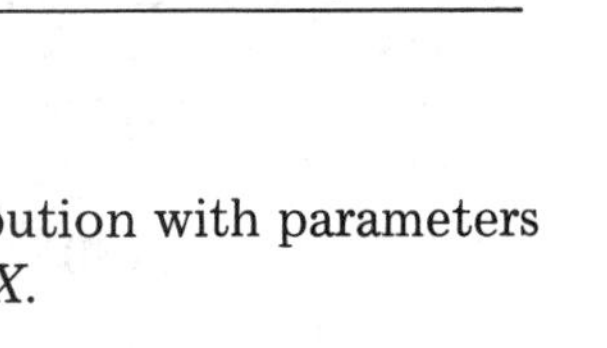

s.d. $=\sqrt{0.2281}$
−1.35 0
S.V. 0 2.827

We estimate that 0.2% of the bottles will overflow during filling.

Example 7.12 The random variable X has a normal distribution with parameters μ and σ^2. Derive the mean and variance of X.

$$\left(\text{You may assume that } \frac{1}{\sqrt{2\pi}}\int_{-\infty}^{\infty} e^{-t^2/2}\,dt = 1\right).$$

Ben Wedgewood and Sons in co-operation with the National Enterprise Commission have just developed a sophisticated new microwave oven. The 'in use' lifetimes of two vital components may be considered to be random variables, such that the lifetime of the quality sensitiser, x, is normal with mean 60 hours and standard deviation 5 hours and the lifetime of the overheat warning mechanism, Y, is normal with mean 70 hours and standard deviation 4 hours.

(a) What value of x should be quoted such that $P(X > x) = 0.99$?

(b) The intensive inspection period for the overheat warning mechanism begins at 60 hours and ends at 75 hours. What is the probability of the mechanism failing in this period?

(c) Assuming that X and Y are independent and that $W = Y - X$, what are $E(W)$ and $V(W)$? Further, what is the probability that the overheat warning mechanism lasts longer than the quality sensitiser? (AEB 1981)

Solution 7.12 For the derivation of the mean and variance of the normal distribution see p. 317.

Let X be the r.v. 'the lifetime of the sensitiser in hours'. Then $X \sim N(60, 5^2)$.

Let Y be the r.v. 'the lifetime of the overheat warning mechanism in hours'. Then $Y \sim N(70, 4^2)$.

(a) If $P(X > x) = 0.99$

then $P\left(\frac{X-60}{5} > \frac{x-60}{5}\right) = 0.99$

i.e. $P\left(Z > \frac{x-60}{5}\right) = 0.99$

therefore $\frac{x-60}{5} = -2.326$

$x = 48.37$

Therefore the value of x which should be quoted is 48 hours (to the nearest hour).

(b)

$$P(60 < Y < 75) = P\left(\frac{60-70}{4} < \frac{Y-70}{4} < \frac{75-70}{4}\right)$$

$$= P(-2.5 < Z < 1.25)$$

$$= 0.8323$$

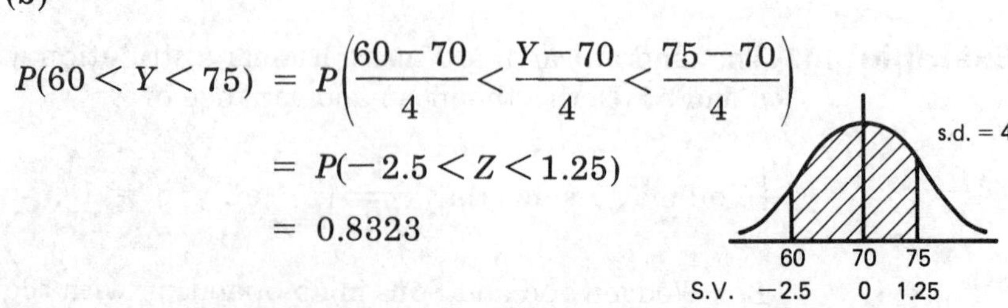

Therefore the probability that the mechanism will fail in the intensive inspection period is 0.8323.

(c) If $W = Y - X$ then

$$E(W) = E(Y) - E(X) = 70 - 60 = 10 \quad \text{and} \quad \text{Var}(W) = \text{Var}(Y) + \text{Var}(X) = 16 + 25 = 41$$

Therefore $E(W) = 10$ and $\text{Var}(W) = 41$.

Now $W \sim N(10, 41)$.

We require $P(Y > X)$, i.e. $P(Y - X > 0)$.

This is $P(W > 0) = P\left(\frac{W-10}{\sqrt{41}} > \frac{0-10}{\sqrt{41}}\right)$

$= P(Z > -1.562)$

$= 0.9408$

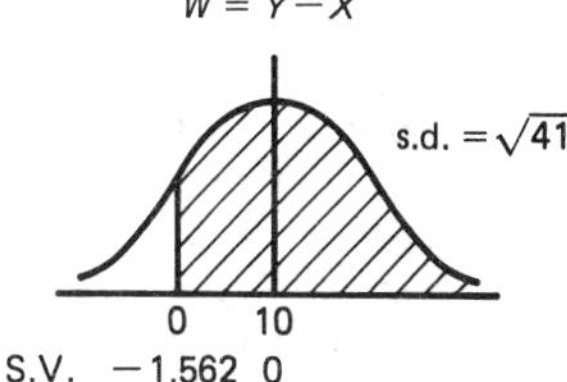

Therefore the probability that the overheat warning mechanism lasts longer than the quality sensitiser is 0.9408.

SUMMARY — SUMS, DIFFERENCES AND MULTIPLES OF INDEPENDENT NORMAL VARIABLES

For two independent normal variables such that $X \sim N(\mu_1, \sigma_1^2)$ and $Y \sim N(\mu_2, \sigma_2^2)$ $$X + Y \sim N(\mu_1 + \mu_2, \sigma_1^2 + \sigma_2^2)$$ $$X - Y \sim N(\mu_1 - \mu_2, \sigma_1^2 + \sigma_2^2)$$
For n independent normal variables such that $X_i \sim N(\mu_i, \sigma_i^2)$ $$X_1 + X_2 + \ldots + X_n \sim N(\mu_1 + \mu_2 + \ldots + \mu_n, \sigma_1^2 + \sigma_2^2 + \ldots + \sigma_n^2)$$
For n independent observations of the r.v. X where $X \sim N(\mu, \sigma^2)$, $$X_1 + X_2 + \ldots + X_n \sim N(n\mu, n\sigma^2)$$
For the normal variable such that $X \sim N(\mu, \sigma^2)$ and for any constant a $$aX \sim N(a\mu, a^2\sigma^2)$$
For two independent normal variables such that $X \sim N(\mu_1, \sigma_1^2)$ and $Y \sim N(\mu_2, \sigma_2^2)$ and for any constants a and b $$aX + bY \sim N(a\mu_1 + b\mu_2, a^2\sigma_1^2 + b^2\sigma_2^2)$$ $$aX - bY \sim N(a\mu_1 - b\mu_2, a^2\sigma_1^2 + b^2\sigma_2^2)$$

Miscellaneous Exercise 7d

1. The weights of grade A oranges are normally distributed with mean 200 g and standard deviation 12 g. Determine, correct to 2 significant figures, the probability that
(*a*) a grade A orange weighs more than 190 g but less than 210 g,
(*b*) a sample of 4 grade A oranges weighs more than 820 g.

The weights of grade B oranges are normally distributed with mean 175 g and standard deviation 9 g. Determine, correct to 2 significant figures, the probability that
(*c*) a grade B orange weighs less than a grade A orange,
(*d*) a sample of 8 grade B oranges weighs more than a sample of 7 grade A oranges. (C)

2. Prints from two types of film C and D have developing times which can be modelled by normal variables, C with mean 16.18 s and standard deviation 0.11 s and D with mean 15.88 s and standard deviation 0.10 s.
(*a*) What is the probability that a type C print will take less than 16 s to develop?
(*b*) A type C print is developed and immediately afterwards a type D print is developed. What is the probability that the total time is greater than 32.5 s?
(*c*) What is the probability of a type C print taking longer to develop than a type D print? (SUJB)

3. In testing the length of life of electric light bulbs of a particular type, it is found that 12.3% of the bulbs tested fail within 800 hours and that 28.1% are still operating 1100 hours after the start of the test. Assuming that the distribution of the length of life is normal, calculate, to the nearest hour in each case, the mean, μ, and the standard deviation, σ, of the distribution.

A light fitting takes a single bulb of this type. A packet of three bulbs is bought, to be used one after the other in this fitting. State the mean and variance of the total life of the three bulbs in the packet in terms of μ and σ and calculate, to two decimal places, the probability that the total life is more than 3300 hours.

Calculate the probability that all three bulbs have lives in excess of 1100 hours, so that again the total life is more than 3300 hours. Explain why this answer should be different from the previous one. (JMB)

4. The weight of a large loaf of bread is a normal variable with mean 420 g and standard deviation 30 g. The weight of a small loaf of bread is a normal variable with mean 220 g and standard deviation 10 g.
(*a*) Find the probability that 5 large loaves weigh more than 10 small loaves.
(*b*) Find the probability that the total weight of 5 large loaves and 10 small loaves lies between 4.25 kg and 4.4 kg. (C)

5. The tensile strengths, measured in newtons (N), of a large number of ropes of equal length are independently and normally distributed such that five per cent are under 706 N and five per cent over 1294 N. Four such ropes are randomly selected and joined end-to-end to form a single rope; the strength of the combined rope is equal to the strength of the weakest of the four selected ropes. Derive the probabilities that this combined rope will not break under tensions of 1000 N and 900 N, respectively.

A further four ropes are randomly selected and attached between two rings, the strength of the arrangement being the sum of the strengths of the four separate ropes. Derive the probabilities that this arrangement will break under tensions of 4000 N and 4200 N, respectively.

Find the smallest number of ropes that should be selected if the probability that at least one of them has a strength greater than 1000 N is to exceed 0.99. (JMB)

6. The independent random variables X_1 and X_2 are normally distributed with means μ_1, μ_2 and variances σ_1^2, σ_2^2 respectively. What is the distribution of the random variable $Y = a_1X_1 + a_2X_2$?

Certain components for a revolutionary new sewing machine are assembled by inserting a part of one type (sprotsil) into a part of another type (weavil). Sprotsils have external dimensions which are normally distributed with mean 2.50 cm and standard deviation 0.018 cm. Weavils have internal dimensions which are normally distributed with mean 2.54 cm and standard deviation 0.024 cm. Under suitable pressure, the two types fit together satisfactorily if the dimensions differ by not more than ± 0.035 cm. Show that, if pairs of parts are chosen at random, the difference

$D =$ internal dimension of a weavil
$\quad -$ external dimension of a sprotsil

is distributed with mean 0.04 cm and standard deviation 0.030 cm. Hence show that approximately 42.8% of randomly selected pairs will fit together satisfactorily.

Now, if it is known that the internal dimension of a given weavil is 2.517 cm, what is the probability that a randomly chosen sprotsil will fit this weavil satisfactorily? (AEB 1980)

7. The mass of a cheese biscuit has a normal distribution with mean 6 g and standard deviation 0.2 g. Determine the probability that
(*a*) a collection of twenty-five cheese biscuits has a mass of more than 149 g,
(*b*) a collection of thirty cheese biscuits has a mass of less than 180 g,
(*c*) twenty-five times the mass of a cheese biscuit is less than 149 g.

The mass of a ginger biscuit has a normal distribution with mean 10 g and standard deviation 0.3 g. Determine the probability that a collection of seven cheese biscuits has a mass greater than a collection of four ginger biscuits.

(It may be assumed that all the biscuits were sampled at random from their respective populations.) (C)

8. In a packaging factory, the empty containers for a certain product have a mean weight of 400 g with a standard deviation of 10 g. The mean weight of the contents of a full container is 800 g with a standard deviation of 15 g. Find the expected total weight of 10 full containers and the standard deviation of this weight, assuming that the weights of containers and contents are independent.

Assuming further that these weights are normally distributed random variables, find the proportion of batches of 10 full containers which weigh more than 12.1 kg.

If 1% of the containers are found to be holding weights of product which are less than the guaranteed minimum amount, deduce this minimum weight. (O & C)

9. Next May, an ornithologist intends to trap one male cuckoo and one female cuckoo. The mass M of the male cuckoo may be regarded as being a normal random variable with mean 116 g and standard deviation 16 g. The mass F of the female cuckoo may be regarded as being independent of M and as being a normal random variable with mean 106 g and standard deviation 12 g. Determine
(*a*) the probability that the mass of the two birds together will be more than 230 g,
(*b*) the probability that the mass of the male will be more than the mass of the female.

By considering $X = 9M - 16F$, or otherwise, determine the probability that the mass of the female will be less than nine-sixteenths of that of the male.

Suppose that one of the two trapped birds escapes. Assuming that the remaining bird will be equally likely to be the male or the female, determine the probability that its mass will be more than 118 g. (C)

10. A train leaves a station punctually at its scheduled time, which is currently 0808 hours (i.e. 8 minutes past 8 a.m.). A bus is due to arrive at that station at 08 00 hours, but in fact its arrival time is normally distributed about the scheduled time with standard deviation 5 minutes. Transfer from bus to train requires 1 minute. What is the probability that the bus–train connection is made?

It is proposed to change the scheduled departure time of the train (it must still be an exact minute, e.g. 0809 hours, 0810 hours). What would be the earliest scheduled departure time in order that the probability of making the bus–train connection should be at least 99%?

The train travels to a junction station, its journey time being normally distributed with mean 15 minutes and standard deviation 1.6 minutes. A connecting train leaves the junction punctually at 08 29 hours. Transfer between the two trains can be regarded as instantaneous. What is the probability that the two trains will connect with the original train schedule?

Find what departure times (exact minutes) of the train from the first station will result in *both* connections being made with probability at least 95%. Find also whether it is possible to arrange for this probability to be at least $97\frac{1}{2}$%. (MEI)

11. The random variables X_1, X_2, X_3 and X_4 are normal, independent and identically distributed with mean μ and variance σ^2. The random variables Y and Z are defined by

$$Y = 4X_1 \quad \text{and} \quad Z = \sum_{i=1}^{4} X_i.$$

Show that $\text{Var}(Y) = 4\text{Var}(Z)$.

The number of hours per week spent in study by both male and female college students is known to be normally distributed. For male students the mean is 28 hours and the standard deviation 6 hours, with corresponding figures for the female students being 30 hours and 4 hours respectively. If a random sample of 6 male and 2 female students is taken, find the probability that in a given week the mean number of hours spent in study by this sample of students will lie between 25 and 31 hours.

Calculate the probability that the number of hours studied that week by the 2 female students will differ by more than 8 hours.

Two of the students in the sample are twins, one male and the other female. Calculate the probability that in a particular week the female twin works less hours than her brother. Comment briefly on the assumption of independence.
(AEB 1987)

12. A dispenser discharges an amount of soft drink which is normally distributed with standard deviation 20 ml. The mean amount may be set to any required value. If the cups into which it is dispensed have a capacity of 500 ml,
(*a*) what proportion of cups will overflow if the mean amount is set to 475 ml,
(*b*) to what value should the mean be set so that only 0.1% of cups will overflow?

A customer requires a double size drink. If the mean is set to 475 ml what is the probability of no overflow occurring if he
(*c*) uses two 500 ml cups,
(*d*) make two discharges into a 1000 ml cup?
If now the capacity of the cups is normally distributed with mean 500 ml and standard deviation 30 ml,

(*e*) what proportion of cups will overflow if the mean amount discharged is 475 ml,
(*f*) to what value should the mean be set so that only 0.1% of cups will overflow?
(AEB 1987)

13. The random variables X_1, X_2 and X_3 are independent and normally distributed with means μ_1, μ_2 and μ_3 respectively and common variance σ^2. State precisely the distribution of $X_1+X_2-X_3$.

Two types of metal bars, A and B, are produced. The lengths of A bars are distributed normally with mean 20 cm and standard deviation 0.05 cm and the lengths of B bars distributed normally with mean 30 cm and standard deviation 0.05 cm. An A bar is welded to a B bar with an overlap whose length is normally distributed with mean 5 cm and standard deviation 0.05 cm. The lengths of the welded bars must lie between 44.9 cm and 45.15 cm in order to be acceptable.

(*a*) Calculate the proportion of welded bars that are unsatisfactory.
(*b*) If the welded bars cost 40p each to produce find the price that the manufacturer should charge in order that the expected profit per article should be 50p.
(*c*) Before testing the lengths of the welded bars, two are selected at random. What is the probability that their lengths differ by more than 0.1 cm? (SUJB)

14. X and Y are independent normally distributed random variables such that X has mean 32 and variance 25, and Y has mean 43 and variance 96. Find
(*a*) $P(X>43)$,
(*b*) $P(X-Y>0)$,
(*c*) $P(2X-Y>0)$. (JMB)

15. The times taken by two runners A and B to run 400 m races are independent and normally distributed with means 45.0 s and 45.2 s, and standard deviations 0.5 s and 0.8 s respectively. The two runners are to compete in a 400 m race for which there is a track record of 44.5 s.
(*a*) Calculate, to three decimal places, the probability of runner A breaking the track record.
(*b*) Show that the probability of runner B breaking the track record is greater than that of runner A.
(*c*) Calculate, to three decimal places, the probability of runner A beating runner B. (JMB)

THE SAMPLE MEAN

If $X_1, X_2, \ldots, X_n$ is a random sample of size n taken from any infinite population (or finite population if sampling is with replacement) with mean μ and variance σ^2, then the sample mean $\overline{X}$ is such that $E(\overline{X}) = \mu$ and $\mathrm{Var}(\overline{X}) = \dfrac{\sigma^2}{n}$.

Proof:

$$\begin{aligned} E(\overline{X}) &= E\left[\frac{1}{n}(X_1 + X_2 + \ldots + X_n)\right] \\ &= \frac{1}{n}[E(X_1) + E(X_2) + \ldots + E(X_n)] \\ &= \frac{1}{n}(\mu + \mu + \ldots \mu) = \frac{1}{n}(n\mu) = \mu \end{aligned}$$

$$\begin{aligned} \mathrm{Var}(\overline{X}) &= \mathrm{Var}\left[\frac{1}{n}(X_1 + X_2 + \ldots + X_n)\right] \\ &= \frac{1}{n^2}[\mathrm{Var}(X_1) + \mathrm{Var}(X_2) + \ldots + \mathrm{Var}(X_n)] \\ &= \frac{1}{n^2}(\sigma^2 + \sigma^2 + \ldots + \sigma^2) = \frac{1}{n^2}(n\sigma^2) = \frac{\sigma^2}{n} \end{aligned}$$

Example 7.13 The discrete r.v. X has probability distribution $P(X = x)$ where $P(X = 0) = 0.5$, $P(X = 1) = 0.3$, $P(X = 2) = 0.2$. The mean μ is 0.7 and the variance σ^2 is 0.61. Random samples of size 2 are taken from the distribution. By considering all possible samples, find the probability distribution of the mean $\overline{X}$ of such samples. Verify that $E(\overline{X}) = \mu$ and $\mathrm{Var}(\overline{X}) = \dfrac{\sigma^2}{2}$.

Solution 7.13 Consider the samples of size 2 from the distribution. For the sample $(2, 1)$ say, $P(X = 2) \cdot P(X = 1) = (0.2)(0.3) = 0.06$ and the sample mean is 1.5. Summarising the results for all the samples:

Samples	(0,0)	(0,1)	(0,2)	(1,0)	(1,1)	(1,2)	(2,0)	(2,1)	(2,2)
Mean	0	0.5	1	0.5	1	1.5	1	1.5	2
Probability	0.25	0.15	0.1	0.15	0.09	0.06	0.1	0.06	0.04

The probability distribution for $\overline{X}$ is therefore

Mean	0	0.5	1	1.5	2
Probability	0.25	0.3	0.29	0.12	0.04

We have

$$E(\overline{X}) = 0+0.15+0.29+0.18+0.08 = 0.7$$

$$E(\overline{X}^2) = 0+0.075+0.29+0.27+0.16$$

$$= 0.795$$

$$\text{Var}(\overline{X}) = E(\overline{X}^2)-E^2(\overline{X}) = 0.795-0.49 = 0.305$$

Since $\mu = 0.7$ and $\sigma^2 = 0.61$, $E(\overline{X}) = \mu$ and $\text{Var}(\overline{X}) = \frac{\sigma^2}{2}$

Example 7.14 (a) For the set of numbers 1, 4, 7 find the mean μ and the variance σ^2.

(b) Draw up a frequency distribution of the means of all possible samples of size 3, where sampling is carried out with replacement. Find the mean and variance of this distribution and comment.

Solution 7.14 (a) $\mu = 4$ and $\sigma^2 = 6$ (calculator).

(b) There are three ways of obtaining a sample containing the numbers 1, 1, 4 i.e. (1, 1, 4), (1, 4, 1), (4, 1, 1) and there are 6 ways of obtaining a sample containing the numbers 1, 4, 7. The frequency distribution of the means of all possible samples of size 3 is shown in the table.

Numbers in sample	Sample mean	Frequency
1, 1, 1	1	1
4, 4, 4	4	1
7, 7, 7	7	1
1, 1, 4	2	3
1, 1, 7	3	3
1, 4, 4	3	3
1, 7, 7	5	3
7, 4, 4	5	3
4, 7, 7	6	3
1, 4, 7	4	6
		27

Now, using a calculator, we find that

$$\text{Mean of sample means} = 4$$

and
$$\text{Variance of sample means} = 2$$

So, when samples of size 3 are taken,

$$\text{Mean of sample means} = \text{population mean}$$

$$\text{Variance of sample means} = \frac{\text{Population variance}}{3}$$

SAMPLING WITHOUT REPLACEMENT

If $X_1, X_2, \ldots, X_n$ is a random sample of size n taken *without replacement* from a finite population of size N with mean μ and variance σ^2, then the sample mean $\overline{X}$ is such that $E(\overline{X}) = \mu$ and $\text{Var}(\overline{X}) = \dfrac{\sigma^2}{n}\left(\dfrac{N-n}{N-1}\right)$.

Example 7.15 Find the mean μ and the variance σ^2 of the population 1, 4, 7. Draw up a frequency distribution of the means of all possible samples of size 2, taken without replacement. Find the mean and variance of this distribution and verify that

$$\text{Var}(\overline{X}) = \frac{\sigma^2}{n}\left(\frac{N-n}{N-1}\right)$$

where $\overline{X}$ is the r.v. 'the sample mean', N is the number in the population and n is the sample size.

What happens as $N \to \infty$?

Solution 7.15 $\mu = 4, \sigma^2 = 6$ (calculator)

Sample	(1, 4)	(1, 7)	(4, 1)	(4, 7)	(7, 1)	(7, 4)
Mean	2.5	4	2.5	5.5	4	5.5

From calculator, mean = 4, variance = 1.5.

Now $\dfrac{\sigma^2}{n}\left(\dfrac{N-n}{N-1}\right) = \dfrac{6}{2}\left(\dfrac{3-2}{3-1}\right)$ with $N = 3,\ n = 2,\ \sigma^2 = 6$

$$= 1.5$$

$$= \text{Var}(\overline{X}) \qquad \text{as required.}$$

Now, as $N \to \infty$, $\dfrac{N-n}{N-1} \to 1$ and $\text{Var}(\overline{X}) \to \dfrac{\sigma^2}{n}$

(sampling from an infinite population).

Exercise 7e

1. For each of the following distributions, (*a*) find the mean μ and the variance σ^2, (*b*) by taking all possible samples of size 2 verify that $E(\overline{X}) = \mu$ and $\text{Var}(\overline{X}) = \dfrac{\sigma^2}{2}$.
   (i) $P(X=0)=0.6$, $P(X=1)=0.3$, $P(X=2)=0.1$.
   (ii) $P(X=0)=0.2$, $P(X=1)=0.3$, $P(X=2)=0.3$, $P(X=3)=0.2$.
   (iii) $P(X=-3)=0.4$, $P(X=2)=0.3$, $P(X=4)=0.3$.

2. The discrete random variable J has the distribution

j	-2	-1	0	1	2
$P(J=j)$	$\frac{1}{12}$	$\frac{1}{4}$	$\frac{1}{3}$	$\frac{1}{4}$	$\frac{1}{12}$

Find the mean, μ, and variance, σ^2, of the distribution.

Random samples of size 2 are taken from the distribution. By considering all possible samples, or otherwise, obtain the probability distribution of the mean of such samples.

Give the mean and variance of the distribution of the mean of random samples of size 3 from the original distribution. (O)

3. Find the mean μ and the variance σ^2 of the population 1, 4, 5, 9. Draw up a frequency distribution of the means of all possible samples of size 2. Find the mean and the variance of the distribution formed and comment on your answers.

4. Find the mean μ and the variance σ^2 of the population 1, 4, 7, 8. Draw up a frequency distribution of the means of all possible samples of size 2, taken *without replacement.*

Find the mean and the variance of this distribution and verify that

$$\text{Var}(\overline{X}) = \frac{\sigma^2}{n}\left(\frac{N-n}{N-1}\right)$$

where $\overline{X}$ is the r.v. 'the sample mean', N is the number in the population and n is the sample size.

What happens as $N \to \infty$?

5. Find the mean μ and the variance σ^2 of the five numbers 0, 3, 3, 6, 6. A sample of three of these numbers is to be drawn at random *without* replacement. By making a list of all such samples, or otherwise, show that the sampling distribution of the sample mean $\overline{X}$ is given by the following table.

$\bar{x}$	2	3	4	5
$P(\overline{X}=\bar{x})$	0.1	0.4	0.3	0.2

Verify that $\overline{X}$ is an unbiased estimator* of μ and calculate the variance. If, instead, the sample is to be taken *with* replacement, state the value of the variance of the sample mean. (JMB)

*See p. 420, Unbiased estimator.

6. A lecturer sets her students an assignment on sampling. Part of the assignment involves the students sampling from a population which consists of 50 peeled pickling onions kept in a large water-filled bowl. The lecturer knows that the mean and standard deviation of the weight of such onions is 24.5 grammes and 7.7 grammes respectively.

One student randomly selects an onion, finds its weight and then returns it to the bowl. This student repeats the process until he has weighed a total of nine onions. What would you expect to be
(*a*) the mean of the weights of the nine onions,
(*b*) the standard deviation of the mean weight?

Another student adopts a different procedure and she selects nine onions at random without replacing them. What would you expect the standard deviation of the mean weight of the nine onions to be in this case?

For each approach, what is the minimum number of onions that have to be selected if the standard deviation of the sample mean is to be less than 3 grammes? (O)

THE DISTRIBUTION OF THE SAMPLE MEAN

(a) From a normal population

If $X_1, X_2, \ldots, X_n$ is a random sample of size n taken from a normal distribution with mean μ and variance σ^2 such that $X \sim N(\mu, \sigma^2)$, then the distribution of $\bar{X}$ is also normal and $\bar{X} \sim N\left(\mu, \frac{\sigma^2}{n}\right)$ where $\bar{X} = \frac{1}{n}(X_1 + X_2 + \ldots + X_n)$.

The distribution of the sample mean ($\bar{X}$) is known as the **sampling distribution of means** and the standard deviation of this distribution $\left(\frac{\sigma}{\sqrt{n}}\right)$ is known as the **standard error of the mean.**

Example 7.16 A random sample of size 15 is taken from a normal distribution with mean 60 and standard deviation 4. Find the probability that the mean of the sample is less than 58.

Solution 7.16 Now $X \sim N(60, 16)$.

So, for samples of size 15, $\bar{X} \sim N\left(60, \frac{16}{15}\right)$.

We require

$$P(\bar{X} < 58) = P\left(\frac{\bar{X}-60}{\sqrt{16/15}} < \frac{58-60}{\sqrt{16/15}}\right)$$
$$= P(Z < -1.936)$$
$$= 0.0264$$

The probability that the mean of the sample is less than 58 is 0.0264.

Example 7.17 The heights of a particular species of plant follow a normal distribution with mean 21 cm and standard deviation $\sqrt{90}$ cm. A random sample of 10 plants is taken and the mean height calculated. Find the probability that this sample mean lies between 18 cm and 27 cm.

Solution 7.17 Let X be the r.v. 'the height in cm of a plant'. Then $X \sim N(21, 90)$.

Now $n = 10$, so $\bar{X} \sim N\left(21, \frac{90}{10}\right)$, i.e. $\bar{X} \sim N(21, 9)$.

We require

$$P(18 < \overline{X} < 27) = P\left(\frac{18-21}{3} < \frac{\overline{X}-21}{3} < \frac{27-21}{3}\right)$$
$$= P(-1 < Z < 2)$$
$$= 0.8185$$

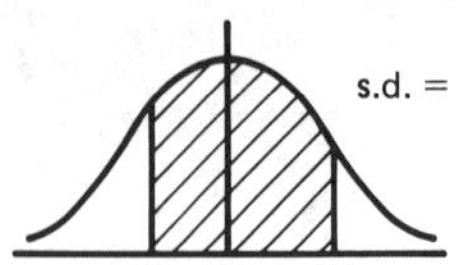

Distribution of $\overline{X}$

s.d. = 3

18 21 27

S.V. −1 0 2

Therefore the probability that the mean height of the sample lies between 18 cm and 27 cm is 0.8185.

Example 7.18 A large number of random samples of size n are taken from the distribution of X where $X \sim N(74, 36)$ and the sample means are calculated. If $P(\overline{X} > 72) = 0.854$, estimate the value of n.

Solution 7.18 $X \sim N(74, 36)$.

Therefore

$$\overline{X} \sim N\left(74, \frac{36}{n}\right)$$

Now $$P(\overline{X} > 72) = P\left(\frac{\overline{X}-74}{6/\sqrt{n}} > \frac{72-74}{6/\sqrt{n}}\right)$$
$$= P\left(Z > \frac{-\sqrt{n}}{3}\right)$$

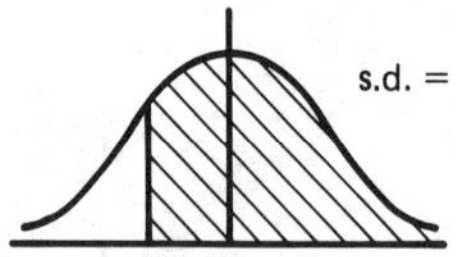

s.d. $= \dfrac{6}{\sqrt{n}}$

72 74

S.V. $\dfrac{-\sqrt{n}}{3}$ 0

So $$P\left(Z > \frac{-\sqrt{n}}{3}\right) = 0.854$$

Therefore $$\frac{\sqrt{n}}{3} = 1.054$$
$$n = 9(1.054)^2$$
$$= 10.0 \quad (3\text{ S.F.})$$

Samples of size 10 are taken.

Example 7.19 (a) If $X_1, X_2, \ldots, X_n$ is a random sample from $N(\mu, 1)$, state the distribution of the sample mean $\overline{X}$.

(b) Find the sample size required to ensure that the probability that $\overline{X}$ is within 0.1 of μ is greater than 0.95.

Solution 7.19 (a) $X \sim N(\mu, 1)$, therefore $\overline{X} \sim N\left(\mu, \dfrac{1}{n}\right)$.

(**b**) We require n such that

$$P(|\overline{X}-\mu|\leqslant 0.1) > 0.95$$

i.e. $$P(-0.1\leqslant \overline{X}-\mu\leqslant 0.1) > 0.95$$

$$P\left(\frac{-0.1}{\sqrt{1/n}}\leqslant\frac{\overline{X}-\mu}{\sqrt{1/n}}\leqslant\frac{0.1}{\sqrt{1/n}}\right) > 0.95$$

$$P(-0.1\sqrt{n}\leqslant Z\leqslant 0.1\sqrt{n} > 0.95$$

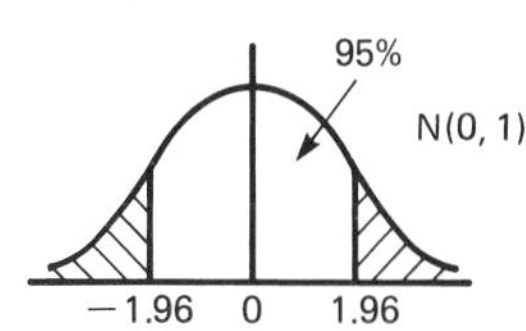

Now $$P(-1.96\leqslant Z\leqslant 1.96) = 0.95$$

So $$0.1\sqrt{n} > 1.96$$

$$\sqrt{n} > \frac{1.96}{0.1}$$

$$n > 384.16$$

Therefore the least sample size required is 385.

Exercise 7f

1. If $X \sim \mathrm{N}(200, 80)$ and a random sample of size 5 is taken from the distribution, find the probability that the sample mean (*a*) is greater than 207, (*b*) lies between 201 and 209.

2. If $X \sim \mathrm{N}(200, 100)$ and a random sample of size 10 is taken from the distribution, find the probability that the sample mean lies outside the range 198 to 205.

3. If $X \sim \mathrm{N}(50, 12)$ and a random sample of size 12 is taken from the distribution, find the probability that the sample mean (*a*) is less than 48.5, (*b*) is less than 52.3, (*c*) lies between 50.7 and 51.7.

4. At a college, the masses of the male students are distributed approximately normally with mean mass 70 kg and standard deviation 5 kg. Four male students are chosen at random. Find the probability that their mean mass is less than 65 kg.

5. A normal distribution has a mean of 40 and a standard deviation of 4. If 25 items are drawn at random, find the probability that their mean is (*a*) 41.4 or more, (*b*) between 38.7 and 40.7, (*c*) less than 39.5.

6. If a large number of samples, size n are taken from a population which follows a normal distribution with mean 74 and standard deviation 6, (*a*) find n if the probability that the sample mean exceeds 75 is 0.282, (*b*) find n if the probability that the sample mean is less than 70.4 is 0.001 35.

7. A normal distribution has a mean of 30 and a variance of 5. Find the probability that (*a*) the average of 10 observations exceeds 30.5, (*b*) the average of 40 observations exceeds 30.5, (*c*) the average of 100 observations exceeds 30.5. Find n such that the probability that the average of n observations exceeds 30.5 is less than 1%.

8. The r.v. X is such that $X \sim \mathrm{N}(\mu, 4)$. A random sample, size n, is taken from the population. Find the least n such that $P(|\overline{X}-\mu| < 0.5) > 0.95$.

9. $\overline{X}$ is the r.v. 'the sample mean of samples, size 15, taken from $\mathrm{N}(30, 18)$' and $\overline{Y}$ is the r.v. 'the sample mean of samples, size 8, taken from $\mathrm{N}(20, 16)$'. Find the distribution of (*a*) $\overline{X}-\overline{Y}$, (*b*) $\overline{X}+\overline{Y}$, (*c*) $\overline{Y}-\overline{X}$, (*d*) $5\overline{X}+3\overline{Y}$, (*e*) $4\overline{X}-2\overline{Y}$.

10. In a certain country the heights of men are normally distributed with mean 175 cm and standard deviation 5 cm and the heights of women are normally distributed with mean 165 cm and standard deviation 6 cm. Find the probability that the mean height of three women chosen

at random is greater than the mean height of four men chosen at random from the population.

11. The continuous random variable X is such that $X \sim N(20, 16)$. If samples of size n are taken and $\bar{X}$ is the random variable 'the mean of the n sample values', find the least value of n such that $P(\bar{X} > 21) \leqslant 0.05$.

12. A random sample X_1, X_2 is drawn from a distribution with mean μ and standard deviation σ. State the mean and standard deviation of the distribution of (a) $X_1 + X_2$, (b) $X_1 - X_2$, (c) $\bar{X}$.
A student's performance is equally good in two subjects. The marks he might be expected to score in each subject may be treated as independent observations drawn from a normal distribution with mean 45 and standard deviation 5. Two procedures might be used to decide whether to give the student an overall pass. One is to demand that he pass separately in each subject, the pass mark being 40; the other is to require that his mean mark in the two subjects exceeds 40. Find the probability that the student will obtain an overall pass by each of these procedures. (O)

13. In a certain nation, men have heights distributed normally with mean 1.70 m and standard deviation 10 cm. Find the probability that a man chosen randomly has height not less than 1.83 m.
What is the probability that the average height of three men chosen randomly is greater than 1.78 m and the probability that all three will have heights greater than 1.83 m?
For the nation, women have heights distributed normally with mean 1.60 m and standard deviation 7.5 cm. Find the probability that a husband and wife have not more than 5 cm difference in heights and state the assumptions that you have made in the calculation. (MEI)

14. X_1 and X_2 are random variables such that X_1 is normally distributed with mean 120 and variance 8 and X_2 is normally distributed with mean 150 and variance 22. A random sample of size 20 is taken from the distribution of $3X_1 + 4X_2$. Find the distribution of the sample mean.

15. Random variables X and Y are such that $X \sim N(100, 10)$ and $Y \sim N(120, 20)$. Random samples of size 50 are taken from each distribution. Find the probability that the sample from the distribution of Y will have a mean which is at least 21 more than the mean of the sample from the distribution of X.

16. Every child in a class does an experiment which consists of measuring V, the volume of water displaced by a solid sphere. The children's values of V are distributed approximately normally with mean 27.4 cm^3 and standard deviation 1 cm^3.
(a) Given that the nominal volume V_0 of the sphere is 27.1 cm^3, estimate to 2 decimal places the probability, p, that the value of V of a child chosen at random exceeds 1.05 V_0.
(b) The nominal radius r_0 of the sphere, calculated from the formula $r_0 = (3V_0/4\pi)^{\frac{1}{3}}$, is 1.86 cm. Each child calculates a value r for the radius of the sphere, using the formula $r = (3V/4\pi)^{\frac{1}{3}}$. Explain why you would expect the probability that a value of r of a child chosen at random exceeds $1.05r_0$ to be different from the value of p you obtained in (a). Show that, in fact, it is extremely unlikely that any child's value of r exceeds $1.05r_0$.
(Hint: express $r > 1.05r_0$ in terms of V and V_0.)
(c) The measured values of V obtained by a second class of children are also distributed approximately normally with the same mean as the first class, but with a standard deviation of 1.5 cm^3. One child from each class is chosen at random. Estimate to 2 decimal places the probability that the mean of their values of V exceeds 1.05 V_0. (MEI)

17. (a) If X and Y are independent random variables with means μ_x, μ_y and variances σ_x^2, σ_y^2 respectively, show from first principles that the mean and variance of $aX + bY$ are $a\mu_x + b\mu_y$ and $a^2\sigma_x^2 + b^2\sigma_y^2$ respectively where a and b are constants.
(b) The diameters x of 110 steel rods were measured in centimetres and the results were summarised as follows:

$$\Sigma x = 36.5, \qquad \Sigma x^2 = 12.49.$$

Find the mean and standard deviation of these measurements.
Assuming these measurements are a sample from a normal distribution with this mean and this variance, find the probability that the mean diameter of a sample of size 110 is greater than 0.345 cm. (O & C)

18. The number of miles travelled per week by a motorist is distributed normally with a mean of 640 and a standard deviation of 50.

(*a*) Calculate the probabilities that in a week he will travel (i) more than 600 miles, (ii) between 600 and 700 miles.

(*b*) Calculate the probability that the average number of miles travelled per week over a complete year of 52 weeks will exceed 650.

(*c*) If the car's petrol consumption is 30 miles per gallon, calculate the probability that the motorist will use less than 80 gallons of petrol over a period of 4 weeks. (JMB)

THE DISTRIBUTION OF THE SAMPLE MEAN (continued)

(b) From any population

The central limit theorem

If $X_1, X_2, \ldots, X_n$ is a random sample of size n from **any** distribution with mean μ and variance σ^2 then, **for large n**, the distribution of the sample mean ($\overline{X}$) is approximately **normal** and $\overline{X} \sim \mathrm{N}\left(\mu, \dfrac{\sigma^2}{n}\right)$ where $\overline{X} = \dfrac{1}{n}(X_1 + X_2 + \ldots + X_n)$.

NOTE: the approximation gets better as n gets larger.

Now if $\overline{X} \sim \mathrm{N}\left(\mu, \dfrac{\sigma^2}{n}\right)$ then $n\overline{X} \sim \mathrm{N}\left(n\mu, n^2\dfrac{\sigma^2}{n}\right)$

But $\quad n\overline{X} = X_1 + X_2 + \ldots + X_n$

therefore $\quad X_1 + X_2 + \ldots + X_n \sim \mathrm{N}(n\mu, n\sigma^2)$

If $X_1, X_2, \ldots, X_n$ is a random sample of size n from **any** distribution with mean μ and variance σ^2 then, **for large n**, the distribution of the sum of the random variables is approximately **normal** with mean $n\mu$ and variance $n\sigma^2$.

The definition in this form is also referred to as the **central limit theorem.**

Example 7.20 If a random sample of size 30 is taken from each of the following distributions, find, for each case, the probability that the sample mean exceeds 5.

(a) $X \sim \mathrm{Po}(4.5)$, (b) $X \sim \mathrm{Bin}(9, 0.5)$, (c) $X \sim \mathrm{R}(3, 6)$.

Solution 7.20 (a) $X \sim \text{Po}(4.5)$, so $E(X) = 4.5$ and $\text{Var}(X) = 4.5$.

The sample size is large, so by the central limit theorem

$$\overline{X} \sim \text{N}\left(4.5, \frac{4.5}{30}\right) \quad \text{approximately}$$

The standard deviation of $\overline{X}$ is $\sqrt{\dfrac{4.5}{30}} = \sqrt{0.15}$.

$$\text{So} \quad P(\overline{X} > 5) = P\left(\frac{\overline{X}-4.5}{\sqrt{0.15}} > \frac{5-4.5}{\sqrt{0.15}}\right)$$
$$= P(Z > 1.291)$$
$$= 0.0983.$$

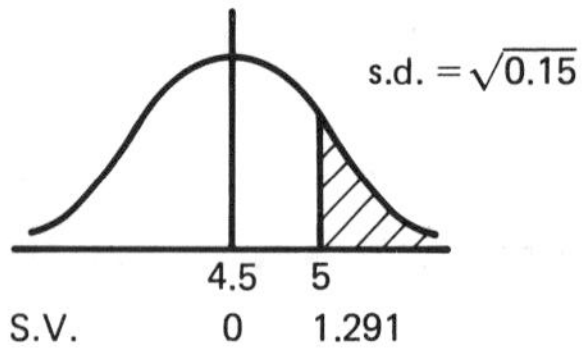

s.d. $= \sqrt{0.15}$

4.5 5

S.V. 0 1.291

So, if $X \sim \text{Po}(4.5)$, then $P(\overline{X} > 5) = 0.0983$.

(b) $X \sim \text{Bin}(9, 0.5)$

so

$$E(X) = (9)(0.5) = 4.5 \quad \text{and} \quad \text{Var}(X) = (9)(0.5)(0.5) = 2.25$$

Now, by the central limit theorem, $\overline{X} \sim \text{N}\left(4.5, \dfrac{2.25}{30}\right)$.

The standard deviation of $\overline{X} = \sqrt{\dfrac{2.25}{30}} = \sqrt{0.075}$.

$$\text{So} \quad P(\overline{X} > 5) = P\left(\frac{\overline{X}-4.5}{\sqrt{0.075}} > \frac{5-4.5}{\sqrt{0.075}}\right)$$
$$= P(Z > 1.826)$$
$$= 0.0340$$

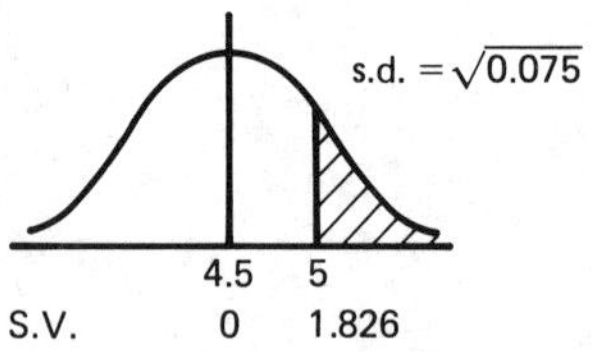

s.d. $= \sqrt{0.075}$

4.5 5

S.V. 0 1.826

Therefore if $X \sim \text{Bin}(9, 0.5)$ then $P(\overline{X} > 5) = 0.034$.

(c) $X \sim \text{R}(3, 6)$.

Then

$$E(X) = \tfrac{1}{2}(3+6) = 4.5 \quad \text{and} \quad \text{Var}(X) = \frac{(6-3)^2}{12} = 0.75$$

$$\text{So} \quad \overline{X} \sim \text{N}\left(4.5, \frac{0.75}{30}\right) \text{ by the central limit theorem}$$

$$\text{i.e.} \quad \overline{X} \sim \text{N}(4.5, 0.025)$$

$$\text{Now} \quad P(\overline{X} > 5) = P\left(\frac{\overline{X}-4.5}{\sqrt{0.025}} > \frac{5-4.5}{\sqrt{0.025}}\right)$$
$$= P(Z > 3.162)$$
$$= 0.000\,783$$

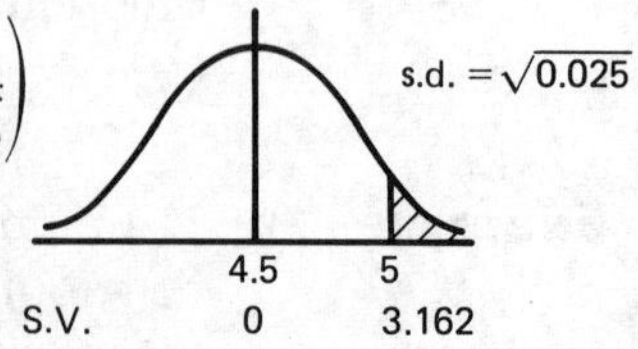

s.d. $= \sqrt{0.025}$

4.5 5

S.V. 0 3.162

Therefore if $X \sim \text{R}(3, 6)$ then $P(\overline{X} > 5) = 0.000\,783$.

Example 7.21 If a large number of samples of size n are taken from Po(2.5) and approximately 5% of the sample means are less than 2.025, estimate n.

Solution 7.21 If $X \sim \text{Po}(2.5)$ then

$$E(X) = \mu = 2.5$$
$$\text{Var}(X) = \sigma^2 = 2.5$$

So, by the central limit theorem

$$\bar{X} \sim \text{N}\left(\mu, \frac{\sigma^2}{n}\right) \quad \text{approximately}$$

i.e. $$\bar{X} \sim \text{N}\left(2.5, \frac{2.5}{n}\right)$$

Now, we require n such that $P(\bar{X} < 2.025) = 0.05$

i.e. $$P\left(\frac{\bar{X}-2.5}{\sqrt{2.5/n}} < \frac{2.025-2.5}{\sqrt{2.5/n}}\right) = 0.05$$

$$P\left(Z < -\frac{0.475}{\sqrt{2.5/n}}\right) = 0.05$$

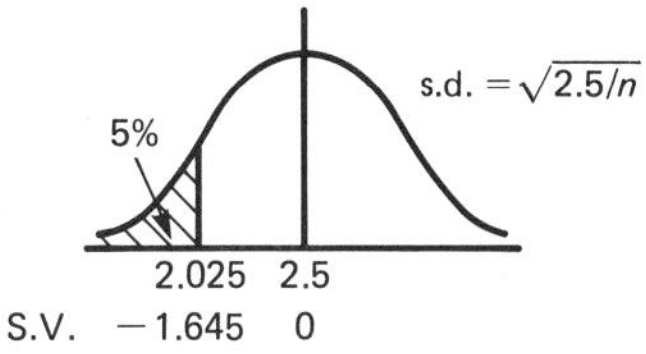

Now, from tables $P(Z < -1.645) = 0.05$

So $$\frac{0.475}{\sqrt{2.5/n}} = 1.645$$

i.e. $$\sqrt{n} = \frac{1.645\sqrt{2.5}}{0.475}$$
$$= 5.476$$

so $$n = 29.98$$

Therefore an approximate value for n is 30.

Exercise 7g

1. To find the mean life and the standard deviation of a particular make of fluorescent light bulbs a large number of samples of 100 bulbs are tested. The mean and the standard deviation of the resulting sampling distribution of means were found to be 1580 hours and 120 hours, respectively. Calculate the mean life and the standard deviation of this make of light bulbs.

2. If a large number of samples, size 30, are taken with replacement from the following distribution, find the mean and standard deviation of the sampling distribution of means. Estimate the probability that a sample mean exceeds 4.

x	0	1	2	3	4	5	6
f	3	10	18	27	21	16	5

3. A random sample of size 100 is taken from Bin(20, 0.6). Find the probability that (*a*) $\bar{X}$ is greater than 12.4, (*b*) $\bar{X}$ is less than 12.2, where $\bar{X}$ is the sample mean.

4. The heights of a new variety of sunflower are normally distributed with mean 2 m and standard deviation 40 cm. 100 samples of 50 flowers each are measured. In how many would you expect the sample mean to be (*a*) greater than 210 cm, (*b*) between 195 cm and 205 cm, (*c*) less than 188 cm?

5. A random sample of size 30 is taken from Po(4). Find (*a*) $P(\bar{X} < 4.5)$, (*b*) $P(\bar{X} > 3.8)$, (*c*) $P(3.8 < \bar{X} < 4.5)$.

6. If a large number of samples, of size n, are taken from Po(4.6) and approximately 2.5% of the sample means are less than 4.005, estimate n.

7. If a large number of samples of size n are taken from Po(2.9) and approximately 1% of the sample means are greater than 3.41, estimate n.

8. If a large number of samples of size n are taken from R(2, 30) and approximately 80% of the sample means are less than 17.15, estimate n.

9. If a large number of samples of size n are taken from Bin(20, 0.2) and approximately 90% of the sample means are less than 4.354, estimate n.

10. If a large number of samples of size n is taken from R(2, 6) and approximately 1% of the sample means are less than 2.8, estimate the value of n.

11. To find the mean life and standard deviation of a certain brand of car tyres a large number of random samples of size 50 were tested. The mean and standard error of the sampling distribution obtained were 20 500 km and 250 km respectively. Estimate the mean life and the standard deviation of this brand of car tyre.

12. The standard deviation of the masses of articles in a large population is 4.55 kg. If random samples of size 100 are drawn from the population, find the probability that a sample mean will differ from the true population mean by less than 0.8 kg.

13. The lifetime, X, in hours of an electrical component is modelled by the following probability density function.

$$f(x) = \begin{cases} \dfrac{a}{(x+10)^2}, & 0 \leqslant x \leqslant 10 \\ 0, & \text{elsewhere} \end{cases}$$

(*a*) Show that $a = 20$ and sketch the graph of $y = f(x)$.
(*b*) Find the mean and variance of X, each correct to three significant figures. (The substitution $\mu = x + 10$ will help with the integrals.)
(*c*) What is the probability that a random sample of 80 components has a mean life of more than 4 hours? (SUJB)

14. A continuous random variable X has probability density function given by

$$f(x) = \begin{cases} 0, & x < -2, \\ -ax, & -2 \leqslant x \leqslant 0, \\ ax, & 0 \leqslant x \leqslant 2, \\ 0, & 2 < x, \end{cases}$$

where a is a constant. Sketch the graph of f and hence, or otherwise, find the value of a.

Show that $\text{Var}(X) = 2$.

A random sample of 200 independent observations of X is taken. Using a suitable approximation, find the probability that the sample mean exceeds 0.2. (C)

THE DISTRIBUTION OF THE SAMPLE PROPORTION

Consider a binomial population in which p is the proportion of 'successes', then the probability of success in any trial is p.

If a random sample of size n is taken from this population and X is the r.v. 'the number of successes',

then $$X \sim \text{Bin}(n, p)$$

and *for large n*, $X \sim N(np, npq)$ approximately, where $q = 1 - p$ (see p. 355).

Now, if P_s is the r.v. 'the proportion of successes in a sample',

then $$P_s = \frac{X}{n}$$

So $$E(P_s) = E\left(\frac{X}{n}\right) = \frac{1}{n}E(X) = \frac{1}{n}(np) = p$$

and

$$\text{Var}(P_s) = \text{Var}\left(\frac{X}{n}\right) = \frac{1}{n^2}\text{Var}(X) = \frac{1}{n^2}(npq) = \frac{pq}{n}$$

Now $$P_s \sim \text{N}\left(p, \frac{pq}{n}\right)$$

NOTE: the larger the sample size, the better the approximation.

The distribution of P_s is known as the **sampling distribution of proportions** and the standard deviation of the sampling distribution $\sqrt{\frac{pq}{n}}$ is known as the **standard error of proportion.**

NOTE: when considering the normal approximation to the binomial distribution, a continuity correction of $\pm\frac{1}{2}$ is used.

Now, since $P_s = \frac{X}{n}$, we use a continuity correction of $\pm\frac{1}{2n}$.

Example 7.22 It is known that 3% of frozen pies arriving at a freezer centre are broken. What is the probability that, on a morning when 500 pies arrive, **(a)** 5% or more will be broken, **(b)** 3% or less will be broken?

Solution 7.22 Let P_s be the r.v. 'the proportion of pies in the sample which are broken'.

Then

$$P_s \sim \text{N}\left(p, \frac{pq}{n}\right) \quad \text{approximately,} \quad \text{where } n = 500, p = 0.03$$

So $$P_s \sim \text{N}\left(0.03, \frac{(0.03)(0.97)}{500}\right)$$

i.e. $$P_s \sim \text{N}(0.03, 0.007\,63^2)$$

(a) We require

$$P(P_s \geqslant 0.05) \to P\left(P_s > 0.05 - \frac{1}{(2)(500)}\right) \quad \text{(continuity correction)}$$

$$= P\left(\frac{P_s - 0.03}{0.007\,63} > \frac{(0.05 - 1/1000) - 0.03}{0.007\,63}\right)$$

$$= P(Z > 2.49)$$

$$= 0.006\,39$$

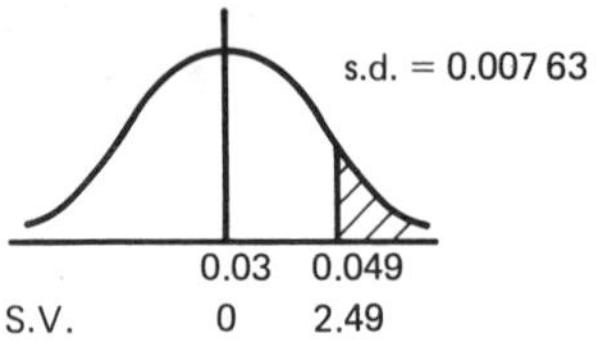

s.d. = 0.007 63

0.03 0.049

S.V. 0 2.49

Therefore the probability that 5% or more will be broken is 0.006 39.

(b) We require

$$P(P_s \leqslant 0.03) \to P\left(P_s < 0.03 + \frac{1}{(2)(500)}\right) \quad \text{(continuity correction)}$$

$$= P\left(\frac{P_s - 0.03}{0.007\,63} < \frac{(0.03 + 1/1000) - 0.03}{0.007\,63}\right)$$

$$= P(Z < 0.131)$$

$$= 0.5521$$

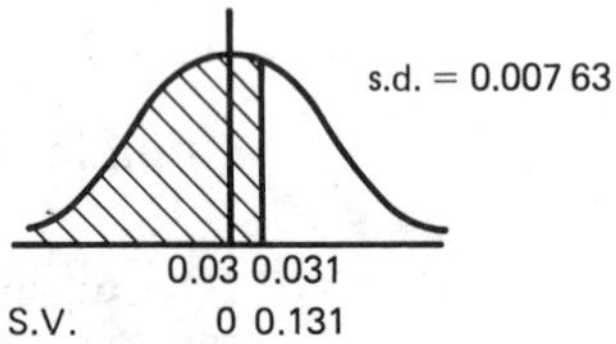

s.d. = 0.007 63

0.03 0.031

S.V. 0 0.131

Therefore the probability that 3% or less are broken is 0.5521.

We now show another method of approaching this problem.
Let X be the r.v. 'the number of broken pies in a sample'.
Then $X \sim N(np, npq)$ where $n = 500, p = 0.03, q = 0.97$.
So $X \sim N(15, 3.814^2)$.

(a) Now 5% of 500 = 25.

So we require

$$P(X \geqslant 25) \to P(X > 24.5) \quad \text{(continuity correction)}$$

$$= P\left(\frac{X - 15}{3.814} > \frac{24.5 - 15}{3.814}\right)$$

$$= P(Z > 2.49)$$

$$= 0.006\,39$$

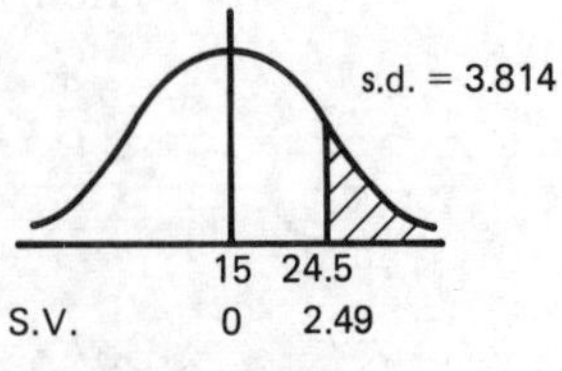

s.d. = 3.814

15 24.5

S.V. 0 2.49

The probability that 5% or more will be broken is 0.006 39, as before.

(**b**) 3% of 500 = 15.

So we require

$$P(X \leqslant 15) \rightarrow P(X < 15.5) \qquad \text{(continuity correction)}$$

$$= P\left(\frac{X-15}{3.814} < \frac{15.5-15}{3.814}\right)$$

$$= P(Z < 0.131)$$

$$= 0.5521$$

s.d. = 3.814

15 15.5

S.V. 0 0.131

The probability that 3% or less will be broken is 0.5521, as before.

NOTE: problems of this type may be solved by considering the distribution of the sample proportion, P_s or by using the normal approximation to the binomial distribution. If the continuity corrections are used in both cases, or omitted in both cases, the standardised values will agree exactly.

Exercise 7h

1. 2% of the trees in a plantation are known to have a certain disease. What is the probability that, in a sample of 300 trees (*a*) less than 1%, (*b*) more than 4% are diseased?

2. A fair coin is tossed 150 times. Find the probability that (*a*) less than 40% of the tosses will result in heads, (*b*) between 40% and 50% (inclusive) are heads, (*c*) more than 55% are heads.

3. A fair coin is tossed 300 times. Work through parts (*a*), (*b*), (*c*) as in Question 2. Why are the results different?

4. Mr Hand gained 48% of the votes in the District Council Elections. What is the probability that a poll of (*a*) 100, (*b*) 1000 randomly selected voters would show over 50% in favour of Mr Hand?

5. Three-quarters of the houseowners in a particular area own a colour television set. Find the probability that at least 73 of a random sample of 100 houseowners in the area own a colour television set.

6. A die is biased so that 1 in 5 throws results in a six. Find the probability that, when the die is thrown 300 times, (*a*) more than 70 throws will result in a six, (*b*) at least 70 throws will result in a six, (*c*) less than 57 throws will result in a six.

7. 70% of the strawberry plants of a particular variety produce more than 10 strawberries per plant. Find the probability that the random sample of 50 plants in my garden consist of more than 37 plants which produce more than 10 strawberries per plant.

SUMMARY — THE SAMPLE MEAN AND THE SAMPLE PROPORTION

Distribution of the sample mean ($\overline{X}$)

If $X_1, X_2, \ldots, X_n$ is a random sample of size n taken from a *normal* distribution such that $X \sim N(\mu, \sigma^2)$ then

$$\overline{X} \sim N\left(\mu, \frac{\sigma^2}{n}\right) \qquad \text{where} \qquad \overline{X} = \frac{1}{n}\sum_{i=1}^{n} X_i$$

The central limit theorem:

For large n, the result holds for a random sample taken from *any* distribution.

Distribution of the sample proportion (P_s)

For large n, $\qquad P_s \sim N\left(p, \frac{pq}{n}\right)$

where p is the proportion of successes in the population,

$q = 1-p$,

n is the number in the sample.

RANDOM SAMPLING

If we are to select an item *at random* from a population then we must ensure that each item in the population has an *equal chance* of being selected.

To obtain a random sample of n items we repeat n times the procedure for selecting one item. However, each selection must be independent of any other.

Example 7.23 Discuss how to select, at random, a sample of two people from a group of six.

Solution 7.23 Write the name of each person on one of six otherwise identical discs and mix them thoroughly in a hat. Without looking, select a disc, note the name and return it to the hat. Draw again. If the first name re-appears, disregard it and repeat the procedure until a different name appears. The sample of two people is then obtained.

An alternative method might be to allocate to each person one of the numbers 1, 2, 3, 4, 5, 6 and then select the people corresponding to the numbers obtained on a die when it is thrown twice, for example (3, 5).

If the population is large then the method of 'drawing out of a hat' is obviously not practical. We can however allocate a number to each item and make the choice by referring to Random Number Tables, shown on p. 629. If you have a random number generator [Ran #] on your calculator you will be able to produce a random 3-digit number every time you press it.

NOTE: Most random number tables are computer-generated. These numbers and the numbers produced on your calculator are known as 'pseudo' random numbers. However, they suit our purposes very well indeed.

RANDOM NUMBER TABLES

Random number tables consist of lists of digits 0, 1, 2, . . . , 9 which are such that each digit has an *equal chance* of appearing at any stage. So each digit has a probability of $\frac{1}{10}$ of occuring.

In random number tables the digits may be listed individually, or grouped in some way. This is solely for convenience of printing. Here are some examples:

List (*a*)	6	8	7	2	5	3	8	1	5	9
	2	5	3	4	7	0	5	4	9	5
	3	2	6	8	7	4	4	7	0	5
List (*b*)	52	74	54	80	68	72	51	96	08	00
	02	52	09	93	60	43	57	42	13	44

List (*c*)	848051	386103	153842
	242330	580007	479971

These tables may be used to represent any number, discrete or continuous.

Example 7.24 Using random number tables, select at random a sample of 8 people from a group of 100.

Solution 7.24 Allocate a two-digit number to each person, for example 01 for the first on the list, 02 for the second, . . . to 98, 99, 00 (calling the hundredth person 00, for convenience).

Using list (*a*) above, we might select people corresponding to the following numbers:

68 72 53 81 59 25 34 70

Example 7.25 Choose 8 people from a group of 60.

Solution 7.25 Allocate each person with a number 01 to 60, then disregard any number outside this range. Using list (*a*)

~~68~~ ~~72~~ 53 ~~81~~ 59 25 34 ~~70~~ 54 ~~95~~
32 ~~68~~ ~~74~~ 47 05

So the people chosen will correspond to the numbers

53, 59, 25, 34, 54, 32, 47, 05.

Example 7.26 Take a random sample of 12 numbers (to 2 d.p.) from the continuous range $0 \leqslant x < 10$.

Solution 7.26 We require the sample values to have 2 d.p. accuracy so we will need to consider groups of 3 digits, inserting the decimal point between the first and second digit. Using list (*b*) on p. 411.

5.27, 4.54, 8.06, 8.72, 5.19, 6.08,
0.00, 2.52, 0.99, 3.60, 4.35, 7.42

Example 7.27 Take a random sample of 4 numbers (to 3 d.p.) from the continuous range $0 \leqslant x < 5$.

Solution 7.27 Using list (*c*) on p. 411 and disregarding any values out of range, we have

~~8.480~~ ~~5.138~~ ~~6.103~~ 1.538 4.224 2.330 ~~5.800~~ 0.747

So the numbers chosen are

1.538, 4.224, 2.330, 0.747.

SAMPLING FROM GIVEN DISTRIBUTIONS

(a) Frequency distributions

Example 7.28 Take a random sample of size 5 from the following distribution, using the random numbers 364294 588330 923918 400300.

x	1	2	3	4	
f	8	12	14	6	Total 40

Solution 7.28 Consider first the cumulative frequencies and then transfer them to proportional frequencies with a total proportion of 1. Random numbers can then be allocated in accordance with the cumulative proportional frequencies as shown:

x	1	2	3	4
f	8	12	14	6
Cumulative frequency	8	20	34	40
Cumulative proportional frequency	$\frac{8}{40} = 0.20$	$\frac{20}{40} = 0.50$	$\frac{34}{40} = 0.85$	$\frac{40}{40} = 1$
Corresponding random numbers	01 to 20	21 to 50	51 to 85	86 to 99 and 00

Since the proportional frequencies are all given to 2 d.p., we consider 2-digit random numbers. Note that 00 was allocated to the x-value of 4 for convenience.

Random numbers: 36, 42, 94, 58, 83
Sample value: 2, 2, 4, 3, 3

So a random sample of size 5 taken from the distribution gives sample values 2, 2, 3, 3, 4.

(b) Probability distributions

Example 7.29 A discrete random variable X has probability distribution

x	0	1	2	3
$P(X = x)$	0.1	0.2	0.4	0.3

Generate a random sample of size 10 from the distribution, using the random numbers 3, 7, 4, 7, 6, 5, 3, 3, 9, 0.

Solution 7.29 Form the cumulative distribution function $F(x)$ and then allocate random numbers in a convenient way:

x	0	1	2	3
$P(X = x)$	0.1	0.2	0.4	0.3
$F(x)$	0.1	0.3	0.7	1
Corresponding random numbers	1	2, 3	4, 5 6, 7	8, 9, 0

Taking 10 sample values, using the random numbers given, we have

Random number: 3, 7, 4, 7, 6, 5, 3, 3, 9, 0
Sample value: 1, 2, 2, 2, 2, 2, 1, 1, 3, 3

NOTE: We could have decided on a different allocation of the random numbers, for example

x	0	1	2	3
$F(x)$	0.1	0.3	0.7	1
Corresponding random numbers	0	1, 2	3, 4 5, 6	7, 8, 9

In this case, the sample generated would have been

Random number:	3, 7, 4, 7, 6, 5, 3, 3, 9, 0
Sample value:	2, 3, 2, 3, 2, 2, 2, 2, 3, 0

NOTE: When sampling from a given p.d.f. remember that every member of the population must have an equal chance of being selected. In each case, work with the cumulative distribution function $F(x)$. When we know $F(x)$ it is easy to allocate the random numbers.

Example 7.30 Take a random sample of four from a binomial distribution with parameters $n = 4$ and $p = 0.2$, using the random numbers 2811, 5747, 6157, 8988.

Solution 7.30 $X \sim \text{Bin}(4, 0.2)$. Since the given random numbers have 4 digits, we will work to 4 d.p.

$P(X = x)$	Cumulative distribution function, $F(x)$
$P(X=0) = (0.8)^4 = 0.4096$	$F(0) = 0.4096$
$P(X=1) = 4(0.8)^3(0.2) = 0.4096$	$F(1) = 0.8192$
$P(X=2) = 6(0.8)^2(0.2)^2 = 0.1536$	$F(2) = 0.9728$
$P(X=3) = 4(0.8)(0.2)^3 = 0.0256$	$F(3) = 0.9984$
$P(X=4) = (0.2)^4 = 0.0016$	$F(4) = 1$ (as expected)

NOTE: We could have used the cumulative binomial probability tables to calculate to values of $F(x)$.

Putting these results in table form, together with the corresponding random number allocation, we have:

x	0	1	2	3	4
$F(x)$	0.4096	0.8192	0.9728	0.9984	1
Corresponding random numbers	0001 to 4096	4097 to 8192	8193 to 9728	9729 to 9984	9985 to 9999, and 0000

The given number 2811 is in the range 0001 to 4096 and corresponds to $x = 0$.

Similarly 5747 corresponds to $x = 1$,
6157 corresponds to $x = 1$,
and 8988 corresponds to $x = 2$.

So the random sample of four is 0, 1, 1, 2.

Example 7.31 Using the random number 8135 take a single random observation from a Poisson distribution with parameter 3.

Solution 7.31 $X \sim \text{Po}(3)$, so that $P(X = x) = e^{-3}\dfrac{3^x}{x!}$, $x = 0, 1, 2, \ldots$

Let $p_x = P(X = x)$ and using the recurrence formula

$$p_{x+1} = \frac{3}{x+1}p_x$$

writing values to 4 d.p. but retaining all figures in the calculator:

$P(X = x)$	Cumulative distribution function, $F(x)$
$p_0 = e^{-3} = 0.0498$	$F(0) = 0.0498$
$p_1 = \frac{3}{1}p_0 = 0.1494$	$F(1) = 0.1991$
$p_2 = \frac{3}{2}p_1 = 0.2240$	$F(2) = 0.4232$
$p_3 = \frac{3}{3}p_2 = 0.2240$	$F(3) = 0.6472$
$p_4 = \frac{3}{4}p_3 = 0.1680$	$F(4) = 0.8153$
$p_5 = \frac{3}{5}p_4 = 0.1008$	$F(5) = 0.9161$
$p_6 = \frac{3}{6}p_5 = 0.0504$	$F(6) = 0.9665$
$p_7 = \frac{3}{7}p_6 = 0.0216$	$F(7) = 0.9881$

The probabilities are now very small, so we end with
$P(X \geqslant 8) = 1 - F(7) = 1 - 0.9881 = 0.0119$
and the corresponding distribution function has the value 1.
Arranging these results in a table, we have

x	$F(x)$	Corresponding random numbers
0	0.0498	0001 to 0498
1	0.1991	0499 to 1991
2	0.4232	1992 to 4232
3	0.6472	4233 to 6472
4	0.8153	6473 to 8153
5	0.9161	8154 to 9161
6	0.9665	9162 to 9665
7	0.9881	9666 to 9881
8 or over	1	9882 to 9999 and 0000

The given random number 8135 is in the range 6473 to 8153, so the random observation corresponds to $x = 4$.

Example 7.32 Using the random numbers 723, 850, take a random sample of size two from the continuous distribution whose p.d.f. is $f(x)$ where $f(x) = \frac{3}{8}x^2 \quad (0 \leqslant x \leqslant 2)$

Solution 7.32 The cumulative distribution function is given by

$$F(x) = \int_0^x \tfrac{3}{8}x^2 \, dx = \frac{x^3}{8}$$

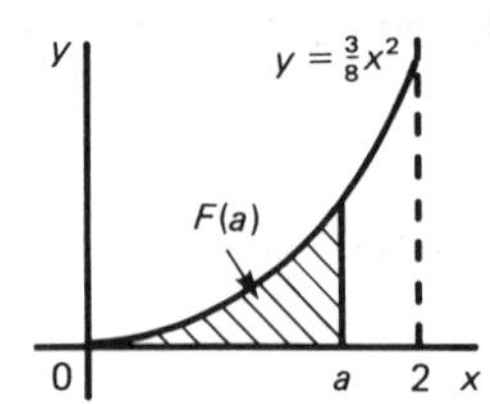

Now, we use the given random numbers in the following way.

If $F(x) = 0.723$, then

$$\frac{x^3}{8} = 0.723$$

and $\quad x = \sqrt[3]{8(0.723)} = 1.80 \qquad (2 \text{ d.p.})$

and if $F(x) = 0.850$, then

$$\frac{x^3}{8} = 0.850$$

and $\quad x = \sqrt[3]{8(0.850)} = 1.89 \qquad (2 \text{ d.p.})$

So the two random observations are $x = 1.80$ and $x = 1.89$.

Example 7.33 Use the random numbers 382 824 to take a random sample of two from the normal distribution N(30, 4).

Solution 7.33 $X \sim \text{N}(30, 4)$.

The cumulative distribution function is given by $\Phi(z)$ where

$$z = \frac{x - 30}{2}$$

NOTE: $\Phi(z) = 1 - Q(z)$.

Now, if $\Phi(a) = 0.382$

then $P(Z \leqslant a) = 0.382$.

and $a = -0.3$

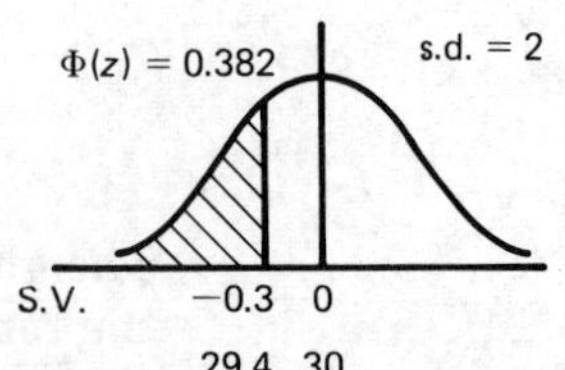

Therefore $\dfrac{x-30}{2} = -0.3$

$x = 30-0.6 = 29.4$

If $\Phi(a) = 0.824$, then

$a = 0.931$

Therefore $\dfrac{x-30}{2} = 0.931$

$x = 30+1.862 = 31.862 = 31.9$ (1 d.p.)

So the two random observations are 29.4 and 31.9.

Exercise 7i

In the following, use the random number tables on p. 629 if random numbers have not been given in the question.

1. Select a random sample of size 10 (to 3 d.p.) from the continuous range $3 \leqslant x < 9$.

2. Draw up a random sample of 100 numbers from the discrete integer range 0 to 9. Find the mean and variance of the sample values and compare them with the theoretical mean and variance.

3. The discrete random variable X has probability distribution

x	5	6	7	8	9
$P(X=x)$	0.15	0.2	0.33	0.21	0.11

Simulate a sample of size 12 from the distribution of X. Compare the mean and variance of this sample with $E(X)$ and $\text{Var}(X)$.

4. The discrete random variable X has distribution function $F(x) = \frac{1}{4}(x-2)$, $x = 3, 4, 5, 6$. Using random number tables, generate 10 observations of X, showing your working clearly.

Describe how you would select a random sample of 30 pupils from a school containing 850 pupils.

5. You wish to select a person at random from a group of 58 people. The following procedure is suggested:

Allocate the numbers 1 to 58 to the people. Choose a line in a table of random numbers and call the first two digits x and y. Let $z = 10x+y$. If $1 \leqslant z \leqslant 58$ then the person who was allocated the number is selected. Otherwise, the person allocated the number $z-58$ is selected. Comment on this method of selection.

6. Take a random sample of size 6 from the distribution:

x	15	16	17	18	19
f	13	15	12	6	4

7. Take a random sample of size 3 from the distribution:

x	2.3	2.4	2.5	2.6	2.7
f	40	60	90	50	60

8. Take a random sample of size 10 from each of the following probability distributions. In each case, find the sample mean and variance and compare with $E(X)$ and $\text{Var}(X)$.

(a)

x	1	2	3	4
$P(X=x)$	0.11	0.2	0.45	0.24

(b)

x	0.1	0.2	0.3
$P(X=x)$	0.175	0.214	0.329

x	0.4	0.5
$P(X=x)$	0.165	0.117

(c) $P(X=x) = kx$, $x = 0, 1, 2, 3$.

9. Take a random sample of size 5 from the distribution of X where $F(x) = \frac{1}{5}x$, $x = 2, 3, 4, 5$.

10. (*a*) The discrete r.v. X is such that $X \sim \text{Bin}(3, 0.4)$. Take a random sample of size 5 from this distribution, using the random numbers

407 315 401 203 972

(*b*) Using the random number 6143 take a single random observation from the Poisson distribution with parameter 4.

11. Using the random numbers 267 394 018 take a random sample of size 3 from the normal distribution with mean 35 and variance 9.

12. Using the random numbers 2654 9342, make two random observations from each of the following distributions:
(*a*) The number of seeds that germinate in a group of 5 selected at random, given that 75% are expected to germinate.
(*b*) The number of goals in a football match, where the number of goals follows a Poisson distribution with variance 2.4.
(*c*) The mass of a bag of sugar, where the mass is normally distributed with mean 1010 g and standard deviation 4.5 g.

13. Using the random number 256 construct a random observation of the continuous r.v. X where

(*a*) $F(x) = \frac{1}{9}x^2,\ 0 \leqslant x \leqslant 3$

(*b*) $f(x) = \frac{4}{15}x^3,\ 1 \leqslant x \leqslant 2$

14. Take 20 samples, each of size 2, from the following distribution:

x	1	2	3	4	5
f	10	15	25	35	15

Calculate the mean of each sample and find the mean and variance of the sample means. Find the mean and variance of the original distribution. Comment.

15. The following table gives the frequency distribution of the number of telephone calls per minute received over a period of 2400 minutes at the switchboard of a solicitor's office.

No. of calls	0	1	2	3	4	5
Frequency	592	844	602	269	91	2

(*a*) Convert the frequencies to probabilities working correct to 4 decimal places. Hence draw up a table of cumulative probabilities.
(*b*) Using the table of random numbers provided, simulate the number of calls arriving at the switchboard in 30 consecutive minutes. Indicate precisely how your values have been obtained.
(*c*) Calculate the sample mean number of calls per minute. Given that the mean and standard deviation of the number of calls per minute obtained from the table are 1.345 and 1.087 respectively, calculate the probability of a random sample of 30 giving a mean value of at least that obtained from your sample. (SUJB)

16. The digits 8453276 are obtained from a table of random digits. Use them to obtain a random observation from each of the following distributions:
(*a*) the number of the winning ticket in a lottery in which there are 500 ticket numbers from 1 to 500 and every ticket has the same chance of being selected.
(*b*) the number of babies born in a cottage hospital in a week, assuming that on average one baby is born every 3 days and that births are independent (and ignoring the possibility of multiple births),
(*c*) the time between successive emissions of a particle from a radioactive substance, assuming that the probability density function of this time is $2e^{-2t} (t > 0)$. (O)

17. You are given the random number 431. Use this number to obtain a sample observation from
(*a*) a Binomial distribution with $n = 12$ and $p = 0.4$.
(*b*) a Normal distribution with mean 6.2 and standard deviation 0.1.
You are expected to explain clearly how you obtain the sample observations. (O)

18. The 25 members of a City Council were asked to record over a twenty day period the number of days on which they made a journey by public transport. The results are given below, c indicating that the councillor was a car owner.

0c	0c	5c	1c	0c
2c	0c	0c	6c	1c
0c	1c	3c	8c	0c
15	12	9	16	17
11	4	3	9	11

(*a*) Calculate the arithmetic mean and the standard deviation of the population.
(*b*) Explaining fully the procedure you have followed use the extract from a table of random sampling numbers at the end of the question to
(i) take an unrestricted random sample (i.e. allow the same person to

be chosen more than once) of size 5 from the population. Calculate the sample mean and state its standard deviation.

(ii) take a simple random sample (i.e. do not allow the same person to be chosen more than once) of size 5 from the population. Calculate the sample mean and state its standard deviation.

(*c*) A councillor suggests that an alternative way of estimating the population mean would be to make up the sample of 5 by taking a simple random sample of size 3 from the car owners and one of size 2 from the rest.

Rank the three methods for estimating the mean in order of preference, explaining your choice.

Extract from table of random sampling numbers
70209 23316 32828 00920 61841 64754
94342 91090 94035 02650 36284 91162

(AEB 1987)

8

ESTIMATION OF POPULATION PARAMETERS

Suppose that a population has an unknown parameter, such as the mean, or the variance, or the proportion of 'successes'. Then an *estimate* of the unknown parameter can be made from the information supplied by a random sample (or samples) taken from the population.

A statistic used to estimate the value of a parameter is called an **estimator** and it is denoted by a capital letter (e.g. U, T, ...). The numerical value taken by the estimator in a particular instance is called an **estimate** and is denoted by a small letter (e.g. u, t, ...).

POINT ESTIMATION — UNBIASED ESTIMATOR

Consider a population with unknown parameter θ.

If U is some statistic derived from a random sample taken from the population, then U is an unbiased estimator for θ if

$$E(U) = \theta$$

There are many estimators which could be formed, but the *best* (*or most efficient*) *estimator* is the one which (i) is unbiased, and (ii) has the smallest variance.

Example 8.1 If X_1, X_2, X_3 is a random sample taken from a population with mean μ and variance σ^2, find which of the following estimators for μ are unbiased, and which is the most efficient of these.

$$T_1 = \frac{X_1+X_2+X_3}{3}, \quad T_2 = \frac{X_1+2X_2}{3}, \quad T_3 = \frac{X_1+2X_2+3X_3}{3}$$

Solution 8.1 Now

$$E(X_i) = \mu \quad \text{for} \quad i = 1, 2, 3$$

So

$$E(T_1) = E\left(\frac{X_1+X_2+X_3}{3}\right)$$
$$= \frac{1}{3}[E(X_1)+E(X_2)+E(X_3)]$$
$$= \frac{1}{3}(3\mu)$$
$$= \mu$$

As $E(T_1) = \mu$, T_1 is an unbiased estimator for μ.

Now

$$E(T_2) = E\left(\frac{X_1+2X_2}{3}\right)$$
$$= \frac{1}{3}[E(X_1)+2E(X_2)]$$
$$= \frac{1}{3}(\mu+2\mu)$$
$$= \mu$$

As $E(T_2) = \mu$, T_2 is an unbiased estimator for μ.

Now

$$E(T_3) = E\left(\frac{X_1+2X_2+3X_3}{3}\right)$$
$$= \frac{1}{3}[E(X_1)+2E(X_2)+3E(X_3)]$$
$$= \frac{1}{3}(\mu+2\mu+3\mu)$$
$$= 2\mu$$

As $E(T_3) \neq \mu$, T_3 is not an unbiased estimator for μ.

The more efficient estimator is the one which has the smaller variance.

Now

$$\text{Var}(T_1) = \text{Var}\left(\frac{X_1+X_2+X_3}{3}\right)$$
$$= \frac{1}{9}[\text{Var}(X_1)+\text{Var}(X_2)+\text{Var}(X_3)]$$
$$= \frac{3\sigma^2}{9}$$

and $$\text{Var}(T_2) = \text{Var}\left(\frac{X_1 + 2X_2}{3}\right) = \frac{1}{9}[\text{Var}(X_1) + 4\text{Var}(X_2)] = \frac{5\sigma^2}{9}$$

As $\text{Var}(T_1) < \text{Var}(T_2)$, T_1 is a more efficient estimator for μ than T_2.

NOTE: the sample X_1, X_2, X_3 is made with replacement and the observations are independent.

Example 8.2 Two random samples of sizes n and $3n$ are taken from normal populations with means μ and 3μ and variances σ^2 and $3\sigma^2$ respectively. If $\overline{X}_1$ and $\overline{X}_2$ are the sample means, show that the estimator $a\overline{X}_1 + b\overline{X}_2$ is an unbiased estimator for μ if $a + 3b = 1$. Also, find the values of a and b if this estimator is to be the most efficient estimator.

Solution 8.2

$$E(\overline{X}_1) = \mu \quad \text{and} \quad \text{Var}(\overline{X}_1) = \frac{\sigma^2}{n}$$

$$E(\overline{X}_2) = 3\mu \quad \text{and} \quad \text{Var}(\overline{X}_2) = \frac{3\sigma^2}{3n} = \frac{\sigma^2}{n}$$

Therefore $$E(a\overline{X}_1 + b\overline{X}_2) = aE(\overline{X}_1) + bE(\overline{X}_2) = a\mu + 3b\mu = \mu(a + 3b)$$

The estimator is unbiased if $E(a\overline{X}_1 + b\overline{X}_2) = \mu$,

i.e. the estimator is unbiased if $a + 3b = 1$.

Also

$$\text{Var}(a\overline{X}_1 + b\overline{X}_2) = a^2\text{Var}(\overline{X}_1) + b^2\text{Var}(\overline{X}_2) = a^2\left(\frac{\sigma^2}{n}\right) + b^2\left(\frac{\sigma^2}{n}\right) = \frac{\sigma^2}{n}(a^2 + b^2)$$

But $a + 3b = 1$.

So

$$\begin{aligned}\text{Var}(a\bar{X}_1 + b\bar{X}_2) &= \frac{\sigma^2}{n}[(1-3b)^2 + b^2] \\ &= \frac{\sigma^2}{n}(1-6b+10b^2) \\ &= \frac{10\sigma^2}{n}\left[\left(b-\frac{3}{10}\right)^2 + \frac{1}{100}\right] \quad \text{completing the square}\end{aligned}$$

The minimum variance occurs when $b = \frac{3}{10}$.

When $b = \frac{3}{10}$, $a = 1 - 3(\frac{3}{10}) = \frac{1}{10}$.

The most efficient estimator is the estimator which is unbiased and which has the minimum variance. To satisfy this, $a = \frac{1}{10}$ and $b = \frac{3}{10}$.

The most efficient estimator for μ is $\frac{1}{10}(\bar{X}_1 + 3\bar{X}_2)$.

Example 8.3 If $X_1, X_2, \ldots, X_n$ is a random sample taken from a population with mean μ and variance σ^2, prove that $T = k_1X_1 + k_2X_2 + \ldots + k_nX_n$ is an unbiased estimator for μ provided that $\sum_{i=1}^{n} k_i = 1$.

Solution 8.3 Now $\quad E(X_i) = \mu \quad$ for $\quad i = 1, 2, \ldots, n$.

So

$$\begin{aligned}E(T) &= E(k_1X_1 + k_2X_2 + \ldots + k_nX_n) \\ &= k_1E(X_1) + k_2E(X_2) + \ldots + k_nE(X_n) \\ &= k_1\mu + k_2\mu + \ldots + k_n\mu \\ &= \mu(k_1 + k_2 + \ldots + k_n)\end{aligned}$$

The estimator is unbiased if $E(T) = \mu$.

Now if $(k_1 + k_2 + \ldots + k_n) = 1$, then $E(T) = \mu$.

So, the estimator is unbiased provided that $\sum_{i=1}^{n} k_i = 1$.

CONSISTENT ESTIMATOR

If U is an estimator for an unknown parameter θ, then U is a **consistent estimator** for θ if $\text{Var}(U) \to 0$ as $n \to \infty$, where n is the size of the sample from which U is obtained.

Example 8.4 If $X_1, X_2, \ldots, X_n$ is a random sample taken from a population with mean μ and variance σ^2, state a condition which will ensure that the estimator $T = k_1X_1 + k_2X_2 + \ldots + k_nX_n$ is a consistent estimator for μ.

Solution 8.4 Now $\text{Var}(X_i) = \sigma^2$ for each $i = 1, 2, \ldots, n$. If T is a consistent estimator for μ, then $\text{Var}(T) \to 0$ as $n \to \infty$.

$$\begin{aligned}\text{Var}(T) &= \text{Var}(k_1X_1 + k_2X_2 + \ldots + k_nX_n) \\ &= k_1^2\text{Var}(X_1) + k_2^2\text{Var}(X_2) + \ldots + k_n^2\text{Var}(X_n) \\ &= \text{Var}(X)[k_1^2 + k_2^2 + \ldots + k_n^2] \\ &= \sigma^2 \sum_{i=1}^{n} k_i^2\end{aligned}$$

So $\text{Var}(T) \to 0$ as $n \to \infty$ if $\sum_{i=1}^{n} k_i^2 \to 0$ as $n \to \infty$.

Exercise 8a

1. If X_1, X_2, X_3 is a random sample taken from a population with mean μ and variance σ^2, find which of the following estimators for μ are unbiased:

   (a) $U_1 = \frac{1}{4}X_1 + \frac{1}{2}X_2 + \frac{1}{4}X_3$

   (b) $U_2 = \frac{1}{3}X_1 + \frac{3}{5}X_2$

   (c) $U_3 = \frac{4}{5}X_1 + \frac{1}{10}X_2 + \frac{1}{10}X_3$

   (d) $U_4 = \frac{1}{6}X_1 + \frac{2}{3}X_2 + \frac{1}{2}X_3$

   (e) $U_5 = \frac{1}{3}(X_1 + X_2 + X_3)$

2. Of the unbiased estimators given in Question 1, which is the most efficient estimator?

3. If $X_1, X_2, \ldots, X_n$ is a random sample taken from a population with mean μ and variance σ^2, show that

   (a) $\frac{1}{n}(X_1 + X_2 + \ldots + X_n)$ is an unbiased and consistent estimator for μ,

   (b) $\dfrac{X_1 + 2X_2 + \ldots + nX_n}{n(n+1)/2}$ is an unbiased and consistent estimator for μ.

4. We wish to estimate K, the breaking strength of elastic bands in a batch. We observe values of a related random variable Z which has density $3z^2/K^3$ for $0 \leqslant z \leqslant K$ and 0 elsewhere. Show that $4Z/3$ is an unbiased estimate of K and that it has variance $K^2/15$.

   We can also observe independent variables X_1, X_2, X_3 each having mean $K/2$ and variance $K^2/12$. Set $Y = X_1 + X_2 + X_3$. What is the value of the constant c such that cY is an unbiased estimate of K? Of the unbiased estimates cY and $4Z/3$, which do you prefer, and why? (O)

5. Two populations have the same mean μ but different variances σ_1^2, σ_2^2. A random observation X_1 is taken from the first population and another, X_2, from the second population. Show that if, in the linear function $T = c_1X_1 + c_2X_2$, c_1 and c_2 are chosen so that T is an unbiased estimator of μ (i.e. $E(T) = \mu$), and T has minimum variance, then $c_1 = \sigma_2^2/(\sigma_1^2 + \sigma_2^2)$ and $c_2 = \sigma_1^2/(\sigma_1^2 + \sigma_2^2)$.

   Two instruments are used in a laboratory to measure a particular physical property of metals. From long experience it has been found that both instruments give unbiased readings and that determinations by Instrument A have a variance that is twice that of determinations by Instrument B. Random samples were taken from an ingot of metal and divided between the two instruments. The mean of 12 determinations by Instrument A was 6.0, in appropriate units, while the mean of 9 determinations by Instrument B was 6.5. Estimate the common mean from these data by using the linear function of the two means that is unbiased and has minimum variance. (O)

6. A random variable X has mean μ and variance 2 and an independent random variable Y has mean 3μ and variance 7. Find the values of a and b if $aX + bY$ has mean μ and minimum variance.

   The values obtained in a single observation of each of X and Y are 10 and 25 respectively. Obtain a best estimate of μ and explain in what sense it is best. (C)

THE MOST EFFICIENT ESTIMATOR OF THE POPULATION MEAN

From a population with *unknown* mean μ take a random sample of size n, and let $\bar{X} = \frac{1}{n}\sum_{i=1}^{n} X_i$

Then the most efficient estimator for μ, which we will write as $\hat{\mu}$, is $\bar{X}$, where $\bar{X}$ is the sample mean

We write $\hat{\mu} = \bar{X}$

NOTE: $\bar{X}$ is an unbiased estimator for μ:

$$E(\bar{X}) = \frac{1}{n}[E(X_1) + E(X_2) + \ldots E(X_n)] = \frac{1}{n}nE(X) = \mu$$

$\bar{X}$ is a consistent estimator for μ:

$$\text{Var}(\bar{X}) = \frac{1}{n^2}[\text{Var}(X_1) + \text{Var}(X_2) + \ldots + \text{Var}(X_n)] = \frac{1}{n^2}n\text{Var}(X)$$

$$= \frac{\sigma^2}{n}, \quad \text{and} \quad \frac{\sigma^2}{n} \to 0 \text{ as } n \to \infty$$

THE MOST EFFICIENT ESTIMATOR OF THE POPULATION VARIANCE

From a population with *unknown* variance σ^2 take a random sample of size n, and let $S^2 = \frac{\Sigma(X_i - \bar{X})^2}{n}$, where S^2 is the sample variance.

The most efficient estimator for σ^2, written $\hat{\sigma}^2$, is $\frac{nS^2}{n-1}$.

We write $\hat{\sigma}^2 = \frac{nS^2}{n-1}$

NOTE: this is a surprising result, as one might expect $\hat{\sigma}^2 = S^2$, but this is not the case.

Example 8.5 Show that $\dfrac{nS^2}{n-1}$ is an unbiased estimator for σ^2.

Solution 8.5 Now

$$S^2 = \frac{\Sigma (X_i - \overline{X})^2}{n} = \frac{\Sigma X_i^2}{n} - \overline{X}^2 \quad i = 1, \ldots, n$$

so $$nS^2 = \sum_1^n X_i^2 - n\overline{X}^2$$

and

$$E(nS^2) = E\left(\sum_1^n X_i^2\right) - E(n\overline{X}^2)$$

$$= E(X_1^2) + E(X_2^2) + \ldots + E(X_n^2) - nE(\overline{X}^2) \qquad \text{(i)}$$

Now $E(X_i) = \mu$ and $\text{Var}(X_i) = \sigma^2$

and $\text{Var}(X_i) = E(X_i^2) - E^2(X_i)$

so $\sigma^2 = E(X_i^2) - \mu^2$

Therefore $\underline{\underline{E(X_i^2) = \mu^2 + \sigma^2}} \quad i = 1, 2, \ldots, n$

Also $E(\overline{X}) = \mu$ and $\text{Var}(\overline{X}) = \dfrac{\sigma^2}{n}$

and $\text{Var}(\overline{X}) = E(\overline{X}^2) - E^2(\overline{X})$

so $\dfrac{\sigma^2}{n} = E(\overline{X}^2) - \mu^2$

Therefore $\underline{\underline{E(\overline{X}^2) = \mu^2 + \dfrac{\sigma^2}{n}}}$

Therefore, from (i)

$$E(nS^2) = n(\mu^2 + \sigma^2) - n\left(\mu^2 + \frac{\sigma^2}{n}\right)$$

$$= (n-1)\sigma^2$$

So that $E\left[\dfrac{nS^2}{(n-1)}\right] = \sigma^2$ as required.

NOTE: $nS^2 = \Sigma (X_i - \overline{X})^2$, so $E[\Sigma(X_i - \overline{X})^2] = (n-1)\sigma^2$.

Also $\dfrac{nS^2}{(n-1)}$ is a consistent estimator for σ^2 (no proof given).

Example 8.6 Obtain the most efficient, or best, estimates of the population mean and variance from which the following sample is drawn:

19.30, 19.61, 18.27, 18.90, 19.14, 19.90, 18.76, 19.10

Solution 8.6 The best estimate of the population mean is $\widehat{\mu}$ where $\widehat{\mu} = \overline{x}$, the sample mean.

Now $$\overline{x} = \frac{\Sigma x}{n} = \frac{152.98}{8} = 19.12 \quad (2\text{ d.p.})$$

So $\widehat{\mu} = 19.12$ (2 d.p.).

The best estimate of the population variance is $\widehat{\sigma}^2$ where $\widehat{\sigma}^2 = \dfrac{ns^2}{(n-1)}$ and s^2 is the sample variance.

Now

$$s^2 = \frac{\Sigma x^2}{n} - \overline{x}^2 = \frac{2927.1}{8} - \left(\frac{152.98}{8}\right)^2 = 0.217 \quad (3\text{ d.p.})$$

So $\widehat{\sigma}^2 = \dfrac{ns^2}{n-1} = \frac{8}{7}s^2 = 0.25$ (2 d.p.).

Example 8.7 Obtain the best unbiased estimates of the population mean and variance from which the following sample is drawn: $n = 12, \overline{x} = 23.5$, $\Sigma(x-\overline{x})^2 = 48.72$.

Solution 8.7 The best estimate of the population mean is $\widehat{\mu}$ where $\widehat{\mu} = \overline{x}$.

Therefore $\widehat{\mu} = 23.5$.

The best estimate of the population variance is $\widehat{\sigma}^2$ where

$$\widehat{\sigma}^2 = \frac{ns^2}{n-1} = \frac{\Sigma(x-\overline{x})^2}{n-1} \quad \text{since} \quad ns^2 = \Sigma(x-\overline{x})^2$$

So $\widehat{\sigma}^2 = \dfrac{48.72}{11} = 4.43$ (2 d.p.).

Exercise 8b

In questions 1 to 11, find the best unbiased estimate of the population mean and of the population variance from which each of the following samples is drawn:

1. 46, 48, 51, 50, 45, 53, 50, 48.
2. 35, 42, 38, 55, 70, 69.
3. 1.684, 1.691, 1.687, 1.688, 1.689, 1.688, 1.690, 1.693, 1.685.
4. $\Sigma x = 120, \Sigma x^2 = 2102, n = 8$.
5. $\Sigma x = 120, \Sigma(x-\overline{x})^2 = 302, n = 8$.
6. $\Sigma x = 100, \Sigma x^2 = 1028, n = 10$.
7. $n = 34, \Sigma x = 330, \Sigma x^2 = 23\,700$.
8. $n = 27, \Sigma x = 1560, \Sigma(x-\overline{x})^2 = 168\,900$.
9.

Interval	1-10	11-20	21-30	31-40	41-50	51-60	61-70	71-80
Frequency	2	4	9	18	21	23	22	1

10.

Interval	0–	4–	8–	12–	16–	20–
Frequency	3	6	24	10	7	0

11.

x	20	21	22	23	24	25
f	4	14	17	26	20	9

12. The concentrations, in mg per litre, of a trace element in 7 randomly chosen samples of water from a spring were:

240.8	237.3	236.7	236.6
234.2	233.9	232.5	

Determine unbiased estimates of the mean and the variance of the concentration of the trace element per litre of water from the spring. (L)P

13. Using the random numbers on p. 636 take a random sample of size 10 from the following distribution:

x	1	2	3	4	5
f	12	18	28	25	17

Use the sample to obtain unbiased estimates of the population mean and variance. Compare these with the true values.

14. The random variable X has probability distribution

x	10	11	12	13
$P(X=x)$	0.3	0.2	0.4	0.1

Use the random numbers given below to generate a random sample of size 20 from the distribution of X and use it to obtain unbiased estimates of the population mean and variance.

Random numbers: 57048 86526
27795 36820

ESTIMATOR OF POPULATION PROPORTION

From a binomial population in which p is the proportion of successes (*unknown*), a random sample of size n is taken.

Let P_s be the r.v. 'the proportion of successes in the sample'.

Then, an unbiased estimator for p is P_s.

NOTE: the estimator is unbiased, since $E(P_s) = p$ (see p. 407).

The estimator is consistent, since $\text{Var}(P_s) = \dfrac{pq}{n}$, where $q = 1-p$ and $\dfrac{pq}{n} \to 0$ as $n \to \infty$.

Example 8.8 A random sample of 50 children from a large school is chosen and the number who are left handed is noted. It is found that 6 are left handed. Obtain an unbiased estimate of the proportion of children in the school who are left handed.

Solution 8.8 From the sample, the proportion of children who are left handed is p_s where $p_s = \frac{6}{50} = 0.12$.

An unbiased estimate of the proportion of children in the school who are left handed is 0.12.

POOLED ESTIMATORS FROM TWO SAMPLES

Estimates of the population mean, variance, proportion, etc., may be made by 'pooling' values from two samples.

Pooled estimators of population mean and of population variance

From a population with *unknown* mean μ and *unknown* variance σ^2 we take two random samples:

	Size	Mean	Variance
Sample I	n_1	$\overline{X}_1$	S_1^2
Sample II	n_2	$\overline{X}_2$	S_2^2

Then
$$\widehat{\mu} = \frac{n_1\overline{X}_1 + n_2\overline{X}_2}{n_1+n_2}$$

where $\widehat{\mu}$ is an unbiased estimator for the population mean μ.

The estimator is unbiased since

$$E\left(\frac{n_1\overline{X}_1 + n_2\overline{X}_2}{n_1+n_2}\right) = \frac{1}{n_1+n_2}E(n_1\overline{X}_1 + n_2\overline{X}_2)$$

$$= \frac{1}{n_1+n_2}[n_1E(\overline{X}_1) + n_2E(\overline{X}_2)]$$

$$= \frac{1}{n_1+n_2}(n_1\mu + n_2\mu)$$

$$= \mu$$

Also
$$\widehat{\sigma}^2 = \frac{n_1S_1^2 + n_2S_2^2}{n_1+n_2-2}$$

where $\widehat{\sigma}^2$ is an unbiased estimator for the population variance σ^2.

The estimator is unbiased since

$$E(n_1S_1^2 + n_2S_2^2) = E(n_1S_1^2) + E(n_2S_2^2)$$

$$= (n_1-1)\sigma^2 + (n_2-1)\sigma^2$$ (see p. 426)

$$= (n_1+n_2-2)\sigma^2$$

So
$$E\left(\frac{n_1S_1^2 + n_2S_2^2}{n_1+n_2-2}\right) = \sigma^2$$

Example 8.9 Two samples, sizes 40 and 50 respectively, are taken from a population with unknown mean μ and unknown variance σ^2. Using the data from the two samples, obtain unbiased estimates of μ and σ^2.

Sample I

x_1	18	19	20	21	22
f	3	7	15	10	5

Sample II

x_2	18	19	20	21	22	23
f	10	21	8	6	3	2

Solution 8.9 Sample I

$$\bar{x}_1 = \frac{\Sigma fx}{\Sigma f} = \frac{807}{40} = 20.175$$

$$s_1^2 = \frac{\Sigma fx^2}{\Sigma f} - \bar{x}_1^2 = \frac{16\,329}{40} - \left(\frac{807}{40}\right)^2 = 1.194$$

Sample II

$$\bar{x}_2 = \frac{\Sigma fx}{\Sigma f} = \frac{977}{50} = 19.54$$

$$s_2^2 = \frac{\Sigma fx^2}{\Sigma f} - \bar{x}_2^2 = \frac{19\,177}{50} - \left(\frac{977}{50}\right)^2 = 1.7284$$

An unbiased estimate for μ is $\hat{\mu}$ where

$$\hat{\mu} = \frac{n_1\bar{x}_1 + n_2\bar{x}_2}{n_1+n_2} = \frac{40(20.175)+50(19.54)}{40+50} = 19.82 \quad (2\text{ d.p.})$$

An unbiased estimate for σ^2 is $\hat{\sigma}^2$ where

$$\hat{\sigma}^2 = \frac{n_1 s_1^2 + n_2 s_2^2}{n_1+n_2-2} = \frac{40(1.194)+50(1.728)}{40+50-2} = 1.52 \quad (2\text{ d.p.})$$

Therefore an unbiased estimate of the population mean is 19.82 (2 d.p.) and an unbiased estimate of the population variance is 1.52 (2 d.p.).

Example 8.10 A count was made of the bacteria in a certain volume of water. Denoting the number of bacteria by $x = 1800 + d$, the results for the first sample were

$$n_1 = 27, \quad \Sigma d = 162, \quad \Sigma(d-\bar{d})^2 = 11\,466$$

The results for the second sample, where $y = 1800 + e$, were

$$n_2 = 25, \quad \Sigma e = 125, \quad \Sigma(e-\bar{e})^2 = 14\,984$$

Obtain unbiased estimates of the population mean and standard deviation (**a**) considering the results of the first sample only, (**b**) considering both samples.

Solution 8.10 (a) Sample I: $x = 1800 + d$,

so
$$\bar{x} = 1800 + \bar{d} = 1800 + \frac{162}{27} = 1806$$

$$\frac{n_1 s_1^2}{n_1 - 1} = \frac{\Sigma(d - \bar{d})^2}{n_1 - 1} = \frac{11\,466}{26} = 441 = 21^2$$

Therefore, an unbiased estimate of the population mean is 1806 and an unbiased estimate of the population standard deviation is 21.

(b) For sample II:

$$\bar{y} = 1800 + \bar{e} = 1800 + \frac{125}{25} = 1805$$

So, for the two samples together,

$$\hat{\mu} = \frac{n_1\bar{x} + n_2\bar{y}}{n_1 + n_2} = \frac{27(1806) + 25(1805)}{52} = 1805.52 \quad \text{(2 d.p.)}$$

and

$$\hat{\sigma}^2 = \frac{\Sigma(d - \bar{d})^2 + \Sigma(e - \bar{e})^2}{n_1 + n_2 - 2} = \frac{11\,466 + 14\,984}{50} = 529 = 23^2$$

So, on the basis of the two samples, $\hat{\mu} = 1805.52$ (2 d.p.) and $\hat{\sigma} = 23$.

Pooled estimator of population proportion

From a binomial population which has *unknown* proportion p of 'successes', we take two samples:

	Size	Proportion
Sample I	n_1	P_{s_1}
Sample II	n_2	P_{s_2}

Then $\hat{p}$, an unbiased estimator for the population proportion p, is given by

$$\hat{p} = \frac{n_1 P_{s_1} + n_2 P_{s_2}}{n_1 + n_2}$$

The estimator is unbiased since

$$\begin{aligned} E\left(\frac{n_1P_{s_1}+n_2P_{s_2}}{n_1+n_2}\right) &= \frac{1}{n_1+n_2}[E(n_1P_{s_1})+E(n_2P_{s_2})] \\ &= \frac{1}{n_1+n_2}[n_1E(P_{s_1})+n_2E(P_{s_2})] \\ &= \frac{1}{n_1+n_2}(n_1p+n_2p) \\ &= p \end{aligned}$$

Example 8.11 An opinion poll in a certain city indicated that 69 people in a random sample of 120 said that they would vote for Mr Jones, while in a second random sample of 160, 93 said that they would vote for Mr. Jones. Find an unbiased estimate of the proportion of people in the city who will vote for Mr Jones.

Solution 8.11

$$n_1 = 120, \quad p_{s_1} = \frac{69}{120}; \qquad n_2 = 160, \quad p_{s_2} = \frac{93}{160}$$

An unbiased estimate $\widehat{p}$ is given by

$$\widehat{p} = \frac{n_1p_{s_1}+n_2p_{s_2}}{n_1+n_2} = \frac{69+93}{120+160} = 0.58 \quad (2\text{ d.p.})$$

So, on the basis of the two samples, it is estimated that approximately 58% of the people in the city will vote for Mr Jones.

Exercise 8c

In each of the following, find unbiased estimates of the population mean and variance, using the data given by the two samples.

1. Sample I 0.68, 0.67, 0.61, 0.78, 0.65
   Sample II 0.64, 0.66, 0.63, 0.69, 0.66, 0.71, 0.64, 0.60

2. Sample I 10.2, 10.1, 10.3, 10.5, 8.9, 9.8
   Sample II 8.7, 10.6, 10.8, 9.6, 9.9, 10.9, 8.4, 8.6, 10.9

3. Sample I

x	1	2	3	4	5
f	2	5	18	12	3

Sample II

x	1	2	3	4	5	6	7
f	3	6	12	26	17	5	1

4. Sample I $n_1 = 13$, $\Sigma x = 109.8$, $\Sigma x^2 = 1110.7$
   Sample II $n_2 = 15$, $\Sigma x = 147.6$, $\Sigma x^2 = 1529.68$

5. Sample I $n_1 = 23$, $\Sigma x = 48$, $\Sigma x^2 = 333.94$
   Sample II $n_2 = 18$, $\Sigma x = 45$, $\Sigma x^2 = 275$

6. Sample I 5.26, 5.89, 5.64, 5.83, 5.81, 5.28, 5.21, 5.26, 5.74, 5.34
   Sample II 5.31, 5.37, 5.41, 5.45, 5.58, 5.29, 5.36, 5.28

7. Sample I $n_1 = 9$, $\Sigma x = 267$, $\Sigma(x-\bar{x})^2 = 100$
   Sample II $n_2 = 11$, $\Sigma x = 336$, $\Sigma(x-\bar{x})^2 = 114.7$

8. Sample I $n_1 = 15$, $\Sigma x = 35.9$, $\Sigma(x-\bar{x})^2 = 0.269$
   Sample II $n_2 = 20$, $\Sigma x = 47.8$, $\Sigma(x-\bar{x})^2 = 0.638$

In the following questions, find an unbiased estimate of the population proportion, based on the data given by the two samples.

9. $n_1 = 200, p_{s_1} = 0.36; n_2 = 300, p_{s_2} = 0.34$

10. $n_1 = 50, p_{s_1} = 0.82; n_2 = 80, p_{s_2} = 0.85.$

11. $n_1 = 10, p_{s_1} = 0.6; n_2 = 20, p_{s_2} = 0.7$

12. A random sample of 600 people from a certain district were questioned and the results indicated that 30% used a particular product. In a second random sample of 300 people, 96 used the product. Find an unbiased estimate of the proportion of people in the district who used the product.

SUMMARY — POINT ESTIMATORS

POPULATION MEAN	
From one sample $\hat{\mu} = \overline{X}$	From two samples $\hat{\mu} = \dfrac{n_1\overline{X}_1 + n_2\overline{X}_2}{n_1 + n_2}$
POPULATION VARIANCE	
From one sample $\hat{\sigma}^2 = \dfrac{nS^2}{(n-1)}$	From two samples $\hat{\sigma}^2 = \dfrac{n_1S_1^2 + n_2S_2^2}{n_1 + n_2 - 2}$
POPULATION PROPORTION	
From one sample $\hat{p} = P_s$	From two samples $\hat{p} = \dfrac{n_1P_{s1} + n_2P_{s2}}{n_1 + n_2}$

INTERVAL ESTIMATION — CONFIDENCE INTERVALS

An **interval estimate** of an unknown population parameter is a random interval constructed so that it has a *given probability of including the parameter.*

Consider a population with unknown parameter θ.

If we can find an interval (a, b) such that $P(a < \theta < b) = 0.95$, we say that (a, b) is a 95% confidence interval for θ.

In this case, 0.95 is the probability that the interval includes θ.

NOTE: it is *not* the probability that θ lies in the interval.

CONFIDENCE INTERVAL FOR THE POPULATION MEAN

Consider a population, with mean μ and variance σ^2. Now take a random sample from the population, $X_1, X_2, \ldots, X_n$ and consider the distribution of $\overline{X}$ where $\overline{X} = \dfrac{1}{n}\Sigma X_i,\ i = 1, 2, \ldots, n.$

(a) Confidence interval for μ, population variance σ^2 known

If X is *normally distributed* such that $X \sim N(\mu, \sigma^2)$ then, for *any n*,

$$\overline{X} \sim N\left(\mu, \frac{\sigma^2}{n}\right)$$

If X does not follow a normal distribution, and *n is large*, then by the central limit theorem

$$\overline{X} \sim N\left(\mu, \frac{\sigma^2}{n}\right)$$

Standardising, we have $Z = \dfrac{\overline{X} - \mu}{\sigma/\sqrt{n}}$ where $Z \sim N(0,1)$.

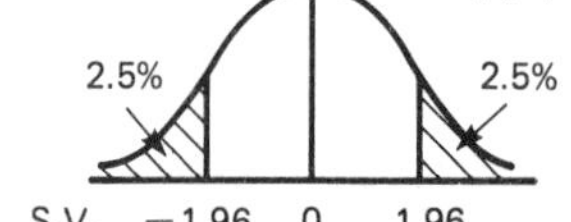

We know that the central 95% of N(0, 1) lies between the values ±1.96 (p. 336).

So $$P\left(-1.96 \leqslant \frac{\overline{X} - \mu}{\sigma/\sqrt{n}} \leqslant 1.96\right) = 0.95$$

$$P\left(-1.96\frac{\sigma}{\sqrt{n}} \leqslant \overline{X} - \mu \leqslant 1.96\frac{\sigma}{\sqrt{n}}\right) = 0.95$$

Now multiply through by -1, so reversing the inequality

$$P\left(1.96\frac{\sigma}{\sqrt{n}} \geqslant \mu - \overline{X} \geqslant -1.96\frac{\sigma}{\sqrt{n}}\right) = 0.95$$

$$P\left(\overline{X} + 1.96\frac{\sigma}{\sqrt{n}} \geqslant \mu \geqslant \overline{X} - 1.96\frac{\sigma}{\sqrt{n}}\right) = 0.95$$

Therefore

$$P\left(\overline{X} - 1.96\frac{\sigma}{\sqrt{n}} \leqslant \mu \leqslant \overline{X} + 1.96\frac{\sigma}{\sqrt{n}}\right) = 0.95$$

So, we have found an interval such that the probability that the interval includes μ is 0.95. This is called the **95% confidence interval for μ.**

> If $\bar{x}$ is the mean of a random sample of size n taken from a *normal* population with *known variance* σ^2, then a central **95% confidence interval for** μ, the population mean, is given by
>
> $$\left(\bar{x} - 1.96\frac{\sigma}{\sqrt{n}},\ \bar{x} + 1.96\frac{\sigma}{\sqrt{n}}\right)$$
>
> This can be written $\quad \bar{x} \pm 1.96\dfrac{\sigma}{\sqrt{n}}$

NOTE: if a large number of intervals are calculated in the same way, then 95% of them will include, or 'trap', μ.

> If the population is *not normal*, then we require n to be large ($n \geqslant 30$, say) for the result to be used.

Similarly, a *central 99% confidence interval for* μ is given by

$$\left(\bar{x} - 2.575\frac{\sigma}{\sqrt{n}},\ \bar{x} + 2.575\frac{\sigma}{\sqrt{n}}\right)$$

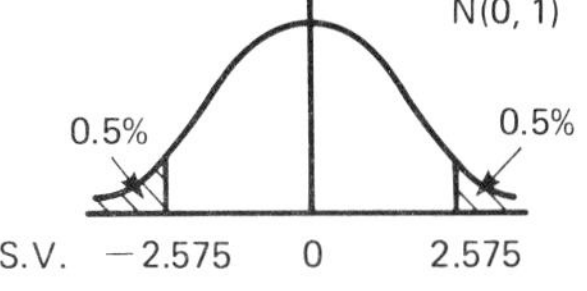

This can be written

$$\bar{x} \pm 2.575\frac{\sigma}{\sqrt{n}}$$

A *central 98% confidence interval for* μ is given by

$$\left(\bar{x} - 2.326\frac{\sigma}{\sqrt{n}},\ \bar{x} + 2.326\frac{\sigma}{\sqrt{n}}\right)$$

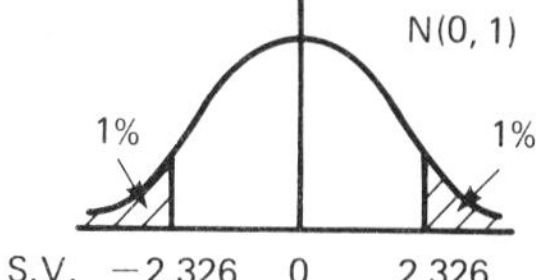

This can be written

$$\bar{x} \pm 2.326\frac{\sigma}{\sqrt{n}}$$

NOTE: often the word 'central' is omitted when considering confidence intervals, but it is assumed, unless otherwise stated, that an interval that is central, or symmetric, about the mean is required. A central 95% confidence interval is sometimes written 95% C.I.

NOTE: One-sided confidence intervals.

A one-sided 95% confidence interval for μ is given by

$$\left(\bar{x} - 1.645\frac{\sigma}{\sqrt{n}},\ \infty\right)$$

and $$P\left(\bar{X} - 1.645\frac{\sigma}{\sqrt{n}} \leqslant \mu \leqslant \infty\right) = 0.95$$

or $$\left(-\infty,\ \bar{x} + 1.645\frac{\sigma}{\sqrt{n}}\right)$$

and $$P\left(-\infty \leqslant \mu \leqslant \bar{x} + 1.645\frac{\sigma}{\sqrt{n}}\right) = 0.95$$

The format of the one-sided confidence interval depends on the information required in a particular situation.

Example 8.12 After a particularly wet night, 12 worms surfaced on the lawn. Their lengths, measured in cm, were:

9.5, 9.5, 11.2, 10.6, 9.9, 11.1, 10.9, 9.8, 10.1, 10.2, 10.9, 11.0

Assuming that this sample came from a normal population with variance 4, calculate a 95% confidence interval for the mean length of all the worms in the garden.

Solution 8.12 $n = 12,\ \Sigma x = 124.7.$

So $$\bar{x} = \frac{\Sigma x}{n} = \frac{124.7}{12} = 10.39 \quad (2\text{ d.p.})$$

We are given the population variance $\sigma^2 = 4$, so $\sigma = 2$.

A symmetric 95% confidence interval for the mean length of the worms is

$$\begin{aligned}\bar{x} \pm 1.96\frac{\sigma}{\sqrt{n}} &= 10.39 \pm (1.96)\frac{2}{\sqrt{12}}\\ &= 10.39 \pm 1.13\\ &= (9.26, 11.52)\end{aligned}$$

The 95% confidence interval for the mean length of the worms in the garden is (9.26 cm, 11.52 cm).

Example 8.13 On the basis of the results obtained from a random sample of 100 men from a particular district, the 95% confidence interval for the mean height of the men in the district is found to be (177.22 cm, 179.18 cm). Find the value of $\bar{x}$, the mean of the sample, and σ, the standard deviation of the normal population from which the sample is drawn. Calculate the 98% confidence interval for the mean height.

Solution 8.13 The 95% confidence interval is given by

$$\bar{x} \pm 1.96\frac{\sigma}{\sqrt{n}} = (177.22, 179.18)$$

Hence $$\bar{x} + 1.96\frac{\sigma}{10} = 179.18 \qquad \text{(i)}$$

$$\bar{x} - 1.96\frac{\sigma}{10} = 177.22 \qquad \text{(ii)}$$

Adding (i) and (ii), $$\begin{aligned}2\bar{x} &= 356.4\\ \bar{x} &= 178.2\end{aligned}$$

Subtracting (i) and (ii),

$$\begin{aligned}2(1.96)\frac{\sigma}{10} &= 1.96\\ \sigma &= \frac{10}{2}\\ \sigma &= 5\end{aligned}$$

Therefore the sample mean $\bar{x}$ is 178.2 cm and the population standard deviation σ is 5 cm.

The 98% confidence interval is given by

$$\bar{x} \pm 2.326\frac{\sigma}{\sqrt{n}} = 178.2 \pm 2.326\left(\frac{5}{10}\right)$$
$$= 178.2 \pm 1.163$$
$$= (177.037, 179.363)$$

N(0, 1)
1% 1%
S.V. −2.326 0 2.326

The 98% confidence interval for the mean height of the men in the district is (177.04 cm, 179.36 cm) (2 d.p.).

(b) Confidence interval for μ, population variance σ^2 unknown

We must consider separately the following cases:

(i) n, the sample size, is large, ($n \geqslant 30$, say),

(ii) n, the sample size, is small.

(b) (i) Sample size large ($n \geqslant 30$, say)

Since σ^2 is *unknown*, it is necessary to use an estimator, $\hat{\sigma}^2$ for it.

Now, $\hat{\sigma}^2 = \dfrac{nS^2}{n-1}$ where S^2 is the sample variance

If $\bar{x}$ and s^2 are the mean and variance of a random sample of size n (where n is large) from a normal population with *unknown* mean μ and *unknown* variance σ^2, then a central 95% confidence interval for μ is given by

$$\left(\bar{x} - 1.96\frac{\hat{\sigma}}{\sqrt{n}},\ \bar{x} + 1.96\frac{\hat{\sigma}}{\sqrt{n}}\right) \text{ where } \hat{\sigma}^2 = \frac{ns^2}{n-1} \approx s^2 \text{ for large } n$$

This can be written $\bar{x} \pm 1.96\dfrac{\hat{\sigma}}{\sqrt{n}}$

NOTE: $P\left(\bar{X} - 1.96\frac{\hat{\sigma}}{\sqrt{n}} \leqslant \mu \leqslant \bar{X} + 1.96\frac{\hat{\sigma}}{\sqrt{n}}\right) = 0.95.$

Similarly,

a 99% confidence interval for μ is given by $\bar{x} \pm 2.575\dfrac{\hat{\sigma}}{\sqrt{n}}$,

a 98% confidence interval for μ is given by $\bar{x} \pm 2.326\dfrac{\hat{\sigma}}{\sqrt{n}}$

Example 8.14 A random sample of 120 measurements taken from a normal population gave the following data:

$$n = 120, \quad \Sigma x = 1008, \quad \Sigma(x-\bar{x})^2 = 172.8$$

Find (a) a 97% confidence interval, (b) a 99% confidence interval for the population mean μ,

Solution 8.14 Now
$$\bar{x} = \frac{\Sigma x}{n} = \frac{1008}{120} = 8.4$$

and
$$s^2 = \frac{\Sigma(x-\bar{x})^2}{n} = \frac{172.8}{120} = 1.44$$

so
$$s = 1.2$$

Using the large sample approximation, with $\hat{\sigma}^2 = \dfrac{ns^2}{n-1} \approx s^2$

(a) a 97% confidence interval for the population mean is given by

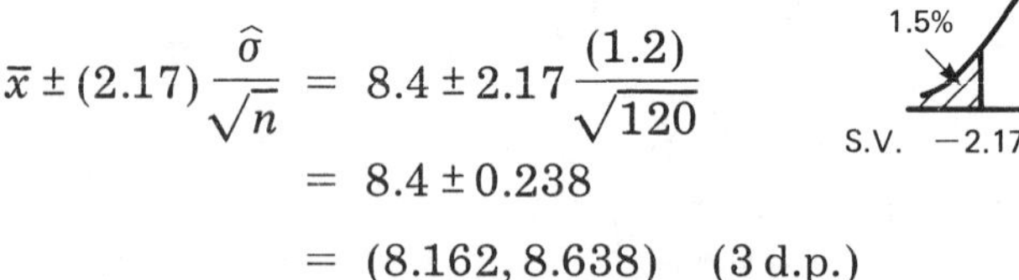

$$\bar{x} \pm (2.17)\frac{\hat{\sigma}}{\sqrt{n}} = 8.4 \pm 2.17\frac{(1.2)}{\sqrt{120}}$$
$$= 8.4 \pm 0.238$$
$$= (8.162, 8.638) \quad (3\text{ d.p.})$$

Therefore a 97% confidence interval for the population mean is (8.162, 8.638).

(b) A 99% confidence interval for the population mean is given by

$$\bar{x} \pm 2.575\frac{\hat{\sigma}}{\sqrt{n}} = 8.4 \pm 2.575\frac{(1.2)}{\sqrt{120}}$$
$$= 8.4 \pm 0.282$$
$$= (8.118, 8.682) \quad (3\text{ d.p.})$$

Therefore a 99% confidence interval for the population is (8.118, 8.682).

Example 8.15 A sample of readings from a normal population with unknown mean μ and unknown variance σ^2 gave the following data:

x	17.4	17.5	17.6	17.7	17.8
f	12	16	19	23	10

A second sample of readings taken from the same population gave

$$n_2 = 72, \quad \Sigma x = 1267.2, \quad \Sigma x^2 = 22\,536$$

Combine the two samples to give estimates of μ and σ^2, and give the appropriate 90% confidence interval for μ.

Solution 8.15 Sample I:

$$\bar{x}_1 = \frac{\Sigma fx}{\Sigma f} = \frac{1408.3}{80} = 17.604$$

$$s_1^2 = \frac{\Sigma fx^2}{\Sigma f} - \bar{x}_1^2 = \frac{24\,792.63}{80} - \left(\frac{1408.3}{80}\right)^2 = 0.0159 \quad (3\text{ S.F.})$$

Sample II:

$$\bar{x}_2 = \frac{\Sigma x}{n_2} = \frac{1267.2}{72} = 17.6$$

$$s_2^2 = \frac{\Sigma x^2}{n_2} - \bar{x}_2^2 = \frac{22\,536}{72} - 17.6^2 = 3.24$$

Now, for the combined sample

$$\hat{\mu} = \frac{n_1\bar{x}_1 + n_2\bar{x}_2}{n_1 + n_2} = \frac{80(17.604) + 72(17.6)}{80 + 72} = 17.602 \quad (3\text{ d.p.})$$

and

$$\hat{\sigma}^2 = \frac{n_1 s_1^2 + n_2 s_2^2}{n_1 + n_2 - 2} = \frac{80(0.0159) + 72(3.24)}{80 + 72 - 2} = 1.564 \quad (3\text{ d.p.})$$

From the two samples, estimates of μ and σ^2 are 17.602 and 1.564 respectively.

A 90% confidence interval for μ based on the combined sample is

$$\bar{x} \pm 1.645\frac{\hat{\sigma}}{\sqrt{n}} \quad \text{where} \quad \bar{x} = 17.602, \quad \hat{\sigma} = \sqrt{1.564}, \quad n = 152$$

90% confidence interval is

$$17.602 \pm 1.645\frac{\sqrt{1.564}}{\sqrt{152}} = 17.602 \pm 0.167$$

$$= (17.435, 17.769)$$

Therefore a 90% confidence interval for μ, based on the two samples is (17.435, 17.769).

Exercise 8d

1. A certain type of tennis ball is known to have a height of bounce which is normally distributed with standard deviation 2 cm. A sample of 60 tennis balls is tested and the mean height of bounce of the sample is 140 cm. Find (*a*) 95%, (*b*) 98% confidence intervals for the mean height of bounce of this type of tennis ball.

2. A random sample of 100 is taken from a population. The sample is found to have a mean of 76.0 and standard deviation 12.0. Find (*a*) 90%, (*b*) 97%, (*c*) 99% confidence intervals for the mean of the population.

3. 150 bags of flour of a particular brand are weighed and the mean mass is found to be 748 g with standard deviation 3.6 g. Find (*a*) 90%, (*b*) 95%, (*c*) 98% confidence intervals for the mean mass of bags of flour of this brand.

4. A random sample of 100 readings taken from a normal population gave the following data: $\bar{x} = 82$, $\Sigma x^2 = 686\,800$. Find (*a*) 98%, (*b*) 99% confidence intervals for the population mean μ.

5. 80 people were asked to measure their pulse rates when they woke up in the morning. The mean was 69 beats and the standard deviation 4 beats. Find (*a*) 95%, (*b*) 99% confidence intervals for the population mean.

6. The 95% confidence interval for the mean length of life of a particular brand of light bulb is (1023.3 h, 1101.7 h). This interval is based on results from a random sample of 36 light bulbs. Find the 99% confidence interval for the mean length of life of this brand of light bulb, assuming that the length of life is normally distributed.

7. A random sample of six items taken from a normal population with variance 4.5 cm^2 gave the following data:

   Sample values: 12.9 cm, 13.2 cm, 14.6 cm, 12.6 cm, 11.3 cm, 10.1 cm

   Find the 94% confidence interval for the population mean μ.

8. The data is from a random sample of 150 readings taken from a population with mean μ and variance σ^2. Estimate μ and σ^2.

   $$n_1 = 150, \quad \Sigma x_1 = 1623, \quad \Sigma x_1^2 = 17\,814.36$$

   A second sample of 100 readings is taken from the same population. For this sample, $n_2 = 100$, $\Sigma x_2 = 1119$, $\Sigma x_2^2 = 12\,585.61$. Calculate estimates of μ and σ^2 from this second sample. Now combine the two samples to give a further estimate of μ, together with its appropriate 96% confidence interval.

9. A sample of 64 readings from a normal population with mean μ and variance σ^2 gave $n_1 = 64$, $\Sigma x = 5452.8$, $\Sigma(x-\bar{x})^2 = 973.44$. Estimate μ and σ^2 from this data.

   A second sample of readings gave:

x	82	83	84	85	86	87
f	6	9	19	27	22	17

   Estimate μ and σ^2 for this second sample. Now combine the two samples to give a further estimate of μ, together with the appropriate 97% confidence interval.

10. The age, X, in years at last birthday, of 250 mothers when their first child was born is given in the following table:

X	18-	20-	22-	24-	26-	28-	30-	32-	34-	36-	38-
No. of mothers	14	36	42	57	48	26	17	7	2	0	1

   (The notation implies that, for example in column 1, there are 14 mothers for whom the continuous variable X satisfies $18 \leqslant X < 20$.)

   Calculate, to the nearest 0.1 of a year, estimates of the mean and the standard deviation of X.

   If the 250 mothers are a random sample from a large population of mothers, find 95% confidence limits for the mean age, μ, of the total population. (C)

11. The distribution of measurements of thicknesses of a random sample of yarns produced in a textile mill is shown in the following table.

Yarn thickness in microns (mid-interval value)	Frequency
72.5	6
77.5	18
82.5	32
87.5	57
92.5	102
97.5	51
102.5	25
107.5	9

   Illustrate these data on a histogram. Estimate to two decimal places the mean and standard deviation of yarn thickness.

   Hence estimate the standard error of the mean to two decimal places, and use it to determine approximate symmetric 95% confidence limits, giving your answer to one decimal place. (MEI)

12. The lifetimes of 200 electrical components were recorded to the nearest hour and classified in the frequency tabulation.

Lifetime	Frequency	Lifetime	Frequency
0-	80	600-	4
100-	48	700-	3
200-	30	800-	2
300-	18	900-	0
400-	10	1000-	0
500-	5		

Draw a histogram of the data and estimate the mean and standard deviation of the distribution.

Calculate a symmetric 90% confidence interval for the population means, using a suitable normal approximation for the distribution of the sample mean. (MEI)

13. A random sample of 250 adult men undergoing a routine medical inspection had their heights (x cm) measured to the nearest centimetre, and the following data were obtained: $\Sigma x = 43\,205$, $\Sigma x^2 = 7\,469\,107$. Calculate an unbiased estimate of the population variance. Calculate also a symmetric 99% confidence interval for the population mean (C)P

14. The time to failure of a sample of 200 batteries is given in Table A below.

(*a*) Draw a histogram of the data.
(*b*) Estimate the sample mean and variance of the time to failure by the usual method of considering all observations in a class as being concentrated at the mid-point of that class. Would you expect the actual sample mean to be greater or less than the estimated value? Give a reason for your answer.
(*c*) Using your calculated values obtain a 95% confidence interval for the population mean.
(*d*) Estimate the median time to failure of the sample. (SUJB)

15. A machine produces plastic balls for use in an industrial process. It incorporates a device which automatically recycles those balls whose mass is outside certain limits. The mass of the balls produced (measured as a deviation in g from the minimum value) may be regarded as a random variable, X, with probability density function

$$f(x) = \begin{cases} k(2-x) & 0 < x < 2 \\ 0 & \text{otherwise} \end{cases}$$

(*a*) Show that $k = 0.5$.
(*b*) Find the mean and the standard deviation of X correct to 3 significant figures.
(*c*) Find the probability that the mass of a ball is less than the mean. Compare this with the result you would have obtained if X had followed a normal distribution.
(*d*) The mean of the distribution may change from day to day but the shape and standard deviation do not. A random sample of size 20 yields a sample mean of 0.9. Calculate a 90% confidence interval for the population mean. Explain the relevance of the Central Limit Theorem to your calculations. (AEB 1988)

16. The lifetime of a shuttlecock is the number of hours of continuous play before it becomes unusable. A random sample of 40 shuttlecocks had a mean lifetime of 4 hours, with standard deviation 1.1 hours. Find the value of c such that $c < \mu < \infty$ is a 95% one-sided confidence interval for μ, the mean lifetime of a shuttlecock.

Table A

Time (hours)	0-20	20-40	40-60	60-80	80-100	100-120	120-140	140-160
Frequency	80	48	29	18	12	7	4	2

Before we consider the case when the sample size is small, we must introduce the t-distribution. It has a very complicated p.d.f. which is included here only for completeness.

THE t-DISTRIBUTION

The r.v. X is said to follow the t-distribution if the p.d.f. of X is

$$f(x) = C_\nu\left(1+\frac{x^2}{\nu}\right)^{-\frac{1}{2}(\nu-1)} \qquad -\infty < x < \infty$$

X has one parameter, ν, known as the number of degrees of freedom. (We use the Greek letter ν, pronounced 'new'.) The constant C_ν depends on ν.

We say that $$X \sim t(\nu)$$

The curve of $y = f(x)$ is shown for $\nu = 2$ and $\nu = 10$.

For large values of ν the t-distribution approximates to the standard normal distribution N(0, 1), shown by the broken line.

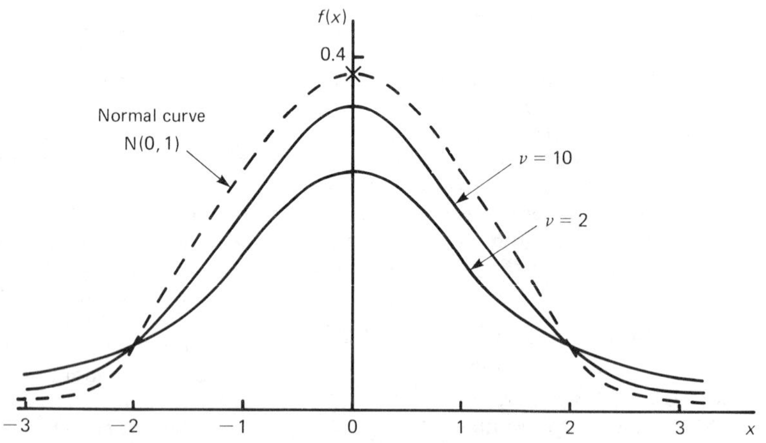

USE OF t-DISTRIBUTION TABLES

These are printed on p. 636. Note that the upper quantiles of the t-distribution are printed.

Referring to the tables, note that each column is headed P, Q, $2Q$.

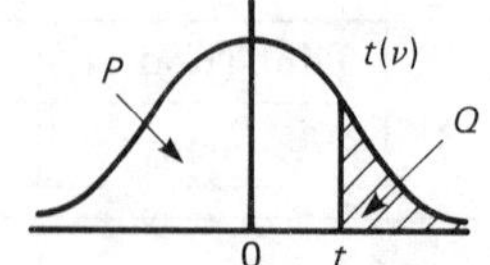

NOTE. We will use this diagram for all values of ν.

This means, for $X \sim t(\nu)$,

$$P(X < t) = P$$

$$P(X > t) = 1 - P = Q$$

$$P(|X| > t) = 2Q$$

Consider column

P	0.975
Q	0.025
$2Q$	0.050

For $X \sim t(6)$, row $\nu = 6$ gives $t = 2.447$.

$P = 0.975 = 97.5\%$, means that 97.5% of the $t(6)$ distribution lies to the left of 2.447.

$Q = 0.025 = 2.5\%$, means that 2.5% of the $t(6)$ distribution lies to the right of 2.447.

$2Q = 0.05 = 5\%$ means that 5% of the $t(6)$ distribution lies outside the range $(-2.447, 2.447)$ (using symmetry properties of the curve).

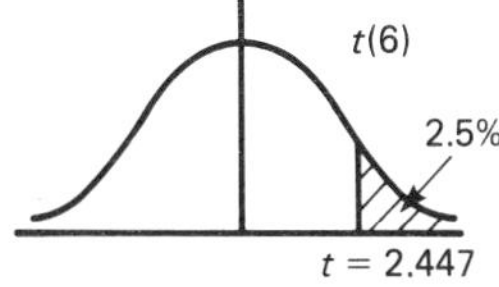

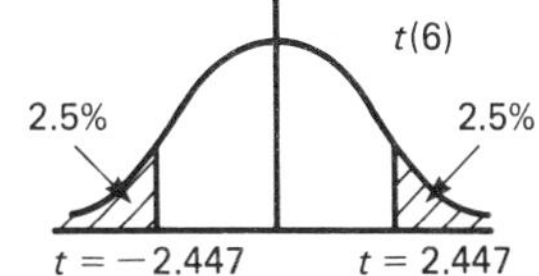

Example 8.16 (a) Find two symmetrically placed values for t outside which 1% of the $t(11)$ distribution lies.

(b) If $X \sim t(4)$, find t such that (*i*) $P(X < t) = 0.99$, (*ii*) $P(X > t) = 0.05$, (*iii*) $P(|X| < t) = 0.95$.

Solution 8.16 (a) Row $\nu = 11$, column $2Q = 0.01$ gives $t = 3.106$, so that 1% of the distribution lies outside $(-3.106, 3.106)$.

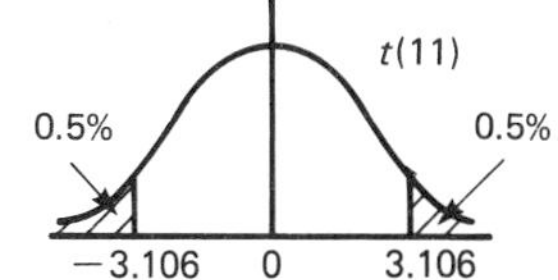

(b) (*i*) Row $\nu = 4$, column $P = 0.99$ gives $t = 3.747$, so $P(X < 3.747) = 0.99.$

(*ii*) Row $\nu = 4$, column $Q = 0.05$ gives $t = 2.132$, so $P(X > 2.132) = 0.05.$

(*iii*) Row $\nu = 4$, column $2Q = 1 - 0.95 = 0.05$ gives $t = 2.776$, so $P(-2.776 < X < 2.776) = 0.95.$

Example 8.17 If the r.v. X is such that $X \sim t(8)$, find (a) $P(X < -2.9)$, (b) $P(X > 3.36)$, (c) $P(-2.9 < X < 3.36)$.

Solution 8.17

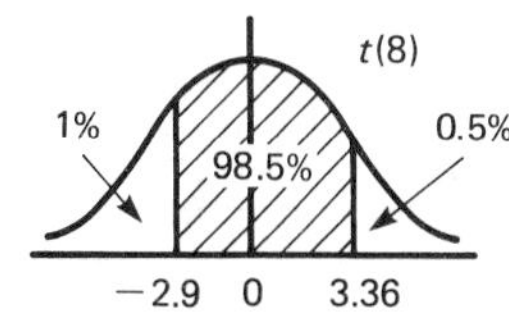

(a) Row $\nu = 8$, $t = 2.896$ gives $Q = 0.01$, so that, by symmetry, approximately 1% of the distribution lies to the left of -2.9, so $P(X < -2.9) = 0.01.$

(b) Row $\nu = 8$, $t = 3.355$ gives $Q = 0.005$, so that 0.5% of the distribution lies to the right of 3.36 and $P(X > 3.36) = 0.005.$

(c)
$$P(-2.9 < X < 3.36) = 1 - (0.01 + 0.005)$$
$$= 0.985$$

So 98.5% of the distribution lies between -2.9 and 3.36.

Exercise 8e

1. Find two symmetrically placed values for t outside which 1% of a $t(6)$ distribution lies.

2. Repeat Question 1 for (*a*) 10%, (*b*) 5%, (*c*) 2%, (*d*) $\frac{1}{2}$% of a $t(6)$ distribution.

3. Repeat Question 1 for 1% of a (*a*) $t(7)$, (*b*) $t(12)$, (*c*) $t(15)$, (*d*) $t(16)$ distribution.

4. If $X \sim t(13)$, find $P(-1.77 < X < 3.012)$.

5. If $X \sim t(10)$, find $P(1.812 < X < 3.169)$.

6. If $X \sim t(6)$, find $P(-2.447 < X < -1.44)$.

7. If $X \sim t(8)$, find the value of a such that $P(X > a) = 0.05$.

8. If $X \sim t(12)$, find the value of a such that $P(|X| < a) = 0.95$.

NOTE: DEGREES OF FREEDOM The number of degrees of freedom associated with a sample statistic is given by

ν = number of variables − number of restrictions involved in calculating the statistic

For example

(a) the sample variance is given by $S^2 = \dfrac{1}{n}\Sigma(X_i - \overline{X})^2$.

When calculating S^2 the variables are $X_1, X_2, \ldots, X_n$ so the number of variables is n. However, these variables are 'restricted' by the fact that $\Sigma X_i = n\overline{X}$, so the number of restrictions is 1.

Therefore $\nu = n - 1$.

(b) Consider

$$\hat{\sigma}^2 = \frac{n_1 S_1^2 + n_2 S_2^2}{n_1 + n_2 - 2}$$

The number of variables $= n_1 + n_2$.

The number of restrictions = 2 (these are $\overline{X}_1$ and $\overline{X}_2$).

Therefore $\nu = n_1 + n_2 - 2$.

(b) (ii) Confidence interval for μ, population variance unknown, sample size small ($n < 30$, say)

Since σ^2 is unknown, we use the estimator $\hat{\sigma}^2$ for it.

Now $\quad \hat{\sigma}^2 = \dfrac{nS^2}{n-1} \quad$ where S^2 is the sample variance

and $\quad \dfrac{\hat{\sigma}}{\sqrt{n}} = \dfrac{\sqrt{n}\,S}{\sqrt{n}\sqrt{n-1}} = \dfrac{S}{\sqrt{n-1}}$

The statistic $\dfrac{\overline{X}-\mu}{\sigma/\sqrt{n}}$ becomes $\dfrac{\overline{X}-\mu}{\hat{\sigma}/\sqrt{n}}$ which is $\dfrac{\overline{X}-\mu}{S/\sqrt{n-1}}$

Now, for small samples this statistic does not follow the normal distribution; the t-distribution must be used instead. The number of degrees of freedom involved in calculating the statistic is $n-1$.

So, if $T=\dfrac{\overline{X}-\mu}{S/\sqrt{n-1}}$ then T follows a t-distribution with $(n-1)$ degree of freedom

i.e. $$T \sim t(n-1)$$

If $\bar{x}$ and s^2 are the mean and variance of a random sample of size n (where n is small) from a normal population with *unknown* mean μ and *unknown* variance σ^2, then a central 95% confidence interval for μ is given by

$$\left(\bar{x}-t\frac{s}{\sqrt{n-1}},\ \bar{x}+t\frac{s}{\sqrt{n-1}}\right)$$

where $$P\left(\overline{X}-t\frac{S}{\sqrt{n-1}}\leqslant\mu\leqslant\overline{X}+t\frac{S}{\sqrt{n-1}}\right) = 0.95$$

i.e. $(-t, t)$ encloses 95% of the $t(n-1)$ distribution.

Example 8.18 Ten packets of a particular brand of biscuits are chosen at random and their masses noted. The results (in grams) are 397.3, 399.6, 401.0, 392.9, 396.8, 400.0, 397.6, 392.1, 400.8, 400.6. Assuming that the sample is taken from a normal population with mean mass μ, calculate (*a*) the 95% confidence interval for μ, (*b*) the 99% confidence interval for μ.

Solution 8.18 From the sample,

$$\bar{x} = \frac{\Sigma x}{n} = \frac{3978.7}{10} = 397.8 \quad (1\text{ d.p.})$$

and

$$s^2 = \frac{\Sigma x^2}{n} - \bar{x}^2 = \frac{1\,583\,098.3}{10} - (397.87)^2 = 9.29 \quad (2\text{ d.p.})$$

As n is small, the small sample approximation is used.

(a) The 95% confidence interval for μ is

$$\bar{x} \pm t\frac{s}{\sqrt{n-1}}$$

where $(-t, t)$ encloses 95% of the $t(n-1)$ distribution, with $n = 10$.

From tables, row $\nu = 9$, column $2Q = 5\%$, we find that $t = 2.262$.

So that the 95% confidence interval is

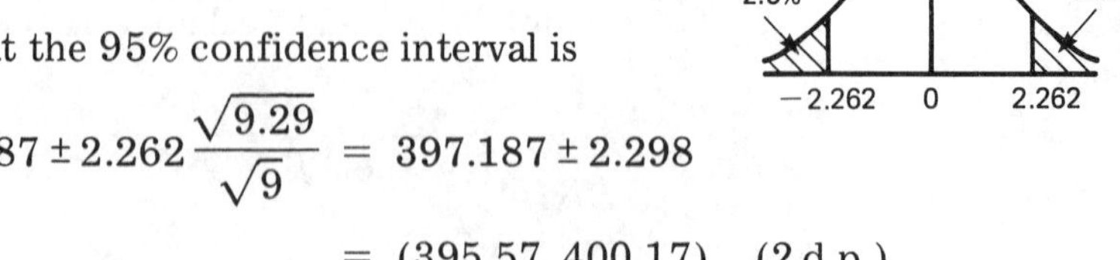

$$397.87 \pm 2.262\frac{\sqrt{9.29}}{\sqrt{9}} = 397.187 \pm 2.298$$

$$= (395.57, 400.17) \quad (2\text{ d.p.})$$

The 95% confidence interval for μ is (395.57 g, 400.17 g).

(b) The 99% confidence interval for μ is

$$\bar{x} \pm t\frac{s}{\sqrt{n-1}}$$

where $(-t, t)$ encloses 99% of the $t(n-1)$ distribution, with $n = 10$.

From tables, row $\nu = 9$, column $2Q = 1\%$, we find that $t = 3.25$.

The 99% confidence interval is

t(9)
0.5% 0.5%
−3.25 0 3.25

$$397.87 \pm 3.25\frac{\sqrt{9.29}}{\sqrt{9}} = 397.87 + 3.302$$

$$= (394.57, 401.17) \quad (2\text{ d.p.})$$

The 99% confidence interval for μ is (394.57 g, 401.17 g).

Example 8.19 Fifteen pupils experimented to find the value of g, the acceleration due to gravity. Their results were as follows:

9.806, 9.807, 9.810, 9.802, 9.805, 9.806, 9.804, 9.811, 9.801, 9.804, 9.805, 9.808, 9.803, 9.809, 9.807

Calculate the mean and the standard deviation of these results.

Give 95% confidence limits for the value of g based upon them.

Estimate the number of experimenters needed to give a confidence interval of less than 0.001. (SUJB)

Solution 8.19

$$\bar{x} = \frac{\Sigma x}{n} = \frac{147.088}{15} = 9.8059 \quad (4\text{ d.p.})$$

$$s^2 = \frac{\Sigma x^2}{n} - \bar{x}^2 = \frac{1442.3254}{15} - \left(\frac{147.088}{15}\right)^2 = 7.72 \times 10^{-6}$$

Therefore $s = 0.0028$ (2 S.F.).

The mean value for g obtained from the results is 9.8059 and the standard deviation is 0.0028.

From the sample: $n = 15, \bar{x} = 9.8059, s = 0.0028$.

As n is small, we use the small sample approximation, so a central 95% confidence interval for g is $\bar{x} \pm t\dfrac{s}{\sqrt{n-1}}$, where $(-t, t)$ encloses 95% of the $t(n-1)$ distribution, with $n = 15$.

From tables, $\nu = 14$, column $2Q = 5\%$, we find that $t = 2.145$.

So the 95% confidence interval is

$$9.8059 \pm 2.145\left(\frac{0.0028}{\sqrt{14}}\right) = 9.8059 \pm 0.0016$$

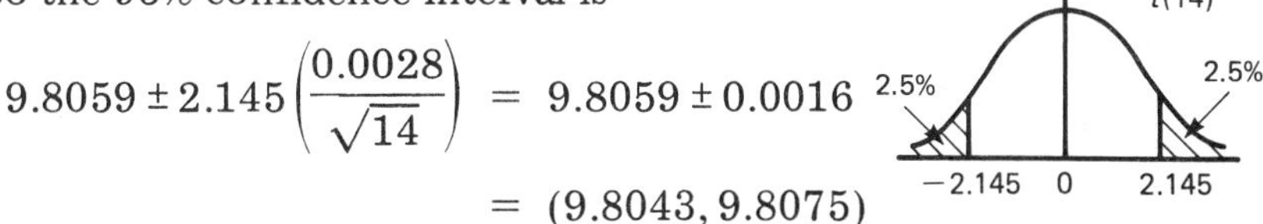

$$= (9.8043, 9.8075)$$

So the 95% confidence limits for the value of g based on the results are (9.8043, 9.8075).

If we require a confidence interval of less than 0.001, then n will be large, so we use the *large sample approximation.*

A central 95% confidence interval for g is given by

$$\bar{x} \pm 1.96\frac{\hat{\sigma}}{\sqrt{n}} \quad \text{where} \quad \hat{\sigma}^2 = \frac{ns^2}{n-1} \approx s^2$$

So we require

$$(2)(1.96)\frac{\hat{\sigma}}{\sqrt{n}} < 0.001 \quad \text{where} \quad \hat{\sigma}^2 \approx 0.0028$$

i.e. $$\sqrt{n} > \frac{(2)(1.96)(0.0028)}{0.001}$$

$$\sqrt{n} > 10.976$$

so $$n > 120.5$$

So the number of experimenters needed would be at least 121.

Exercise 8f

1. The heights (measured in cm) of six policemen were as follows:

   180, 176, 179, 181, 183, 179

   Calculate (*a*) 90%, (*b*) 95%, (*c*) 99% confidence intervals for the mean height of the population of policemen. (Assume that the heights of policemen are normally distributed.)

2. A sample of eight observations of a normally distributed variable gives values

   3.6, 3.9, 4.5, 3.8, 4.4, 4.9, 4.2, 3.8

   Determine a 95% confidence interval for the unknown population mean μ.

3. A normal distribution has variance σ^2, and mean μ. A random sample of ten observations gives values

   0.3, 0.28, 0.27, 0.33, 0.35, 0.33, 0.27
   0.31, 0.37, 0.29

   Find (*a*) 95%, (*b*) 99% confidence intervals for μ.

4. The masses, in grams, of thirteen washers selected at random are

   15.4, 15.2, 14.6, 16.1, 14.8, 15.3, 15.9, 16.0, 15.4, 14.6, 15.0, 15.5, 16.1

   Calculate 98% confidence limits for the mean mass of the population from which the sample is drawn, assuming that the population is normal.

5. Twenty measurements of the life of a candle (measured in hours) gave the following data: $\Sigma x = 172$, $\Sigma x^2 = 1495.4$. By taking a sample of 20 as (*a*) large, (*b*) small, calculate a 99% confidence interval for the mean life of a candle, assuming the length of life to be normally distributed.

6. A random sample of seven independent observations of a normal variable gave $\Sigma x = 35.9$, $\Sigma x^2 = 186.19$. Calculate a 90% confidence interval for the population mean.

7. A random sample of eight observations of a normal variable gave

   $$\Sigma x = 261.2, \quad \Sigma(x-\bar{x})^2 = 3.22$$

   Calculate a 95% confidence interval for the population mean.

CONFIDENCE INTERVAL FOR THE PROPORTION OF SUCCESSES IN A POPULATION

Consider a binomial population where p, the proportion of 'successes' in the population is *unknown*.

Take a random sample of size n from the population and let P_s be the random variable 'the proportion of successes in the sample'.

Then $P_s \sim N\left(p, \dfrac{pq}{n}\right)$ where $q = 1-p$ (see p. 407)

Now, since p is unknown, we use an estimator for it.

An unbiased estimator for p is P_s

It is reasonable to assume that an estimator for $\dfrac{pq}{n}$ is $\dfrac{P_sQ_s}{n}$, where $Q_s = 1-P_s$.

So we have $P_s \sim N\left(p, \dfrac{P_sQ_s}{n}\right)$ approximately

Standardising, we have

$$Z = \frac{P_s-p}{\sqrt{P_sQ_s/n}} \quad \text{where} \quad Z \sim N(0,1)$$

Therefore

$$P\left(-1.96 < \frac{P_s-p}{\sqrt{P_sQ_s/n}} < 1.96\right) = 0.95$$

so, rewriting

$$P\left(P_s - 1.96\sqrt{\frac{P_sQ_s}{n}} < p < P_s + 1.96\sqrt{\frac{P_sQ_s}{n}}\right) = 0.95$$

If in a random sample of size n ($n \geqslant 30$) the proportion with a particular property is p_s, the 95% confidence interval for the population proportion p is given by

$$\left(p_s - 1.96\sqrt{\frac{p_s q_s}{n}},\ p_s + 1.96\sqrt{\frac{p_s q_s}{n}}\right) \quad \text{where} \quad q_s = 1 - p_s$$

This can be written $\quad p_s \pm 1.96\sqrt{\dfrac{p_s q_s}{n}}$

Similarly, a 99% confidence interval for p is $p_s \pm 2.575\sqrt{\dfrac{p_s q_s}{n}}$

and a 98% confidence interval for p is $p_s \pm 2.326\sqrt{\dfrac{p_s q_s}{n}}$.

Example 8.20 A manufacturer wants to assess the proportion of defective items in a large batch produced by a particular machine. He tests a random sample of 300 items and finds that 45 are defective. Calculate (*a*) a 95% confidence interval, (*b*) a 98% confidence interval for the proportion of defective items in the complete batch.

Solution 8.20 The proportion of defective items in the sample, $p_s = \dfrac{45}{300} = 0.15$.

So $q_s = 1 - p_s = 0.85$, $n = 300$.

The 95% confidence interval for the proportion p of defective items in the complete batch is given by

$$p_s \pm 1.96\sqrt{\frac{p_s q_s}{n}} = 0.15 \pm (1.96)\sqrt{\frac{(0.15)(0.85)}{300}}$$
$$= 0.15 \pm 0.0404$$
$$= (0.1096, 0.1904)$$

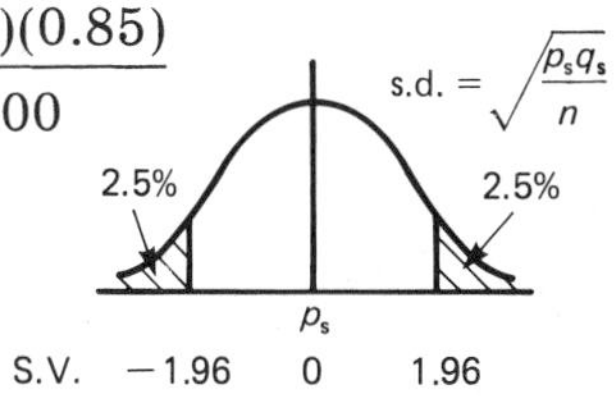

s.d. = $\sqrt{\dfrac{p_s q_s}{n}}$

2.5% 2.5%

p_s

S.V. −1.96 0 1.96

The 95% confidence interval is (0.1096, 0.1904).

The 98% confidence interval for p is given by

$$p_s \pm 2.326\sqrt{\frac{p_s q_s}{n}} = 0.15 \pm (2.326)\sqrt{\frac{(0.15)(0.85)}{300}}$$
$$= 0.15 \pm 0.048$$
$$= (0.101, 0.198)$$

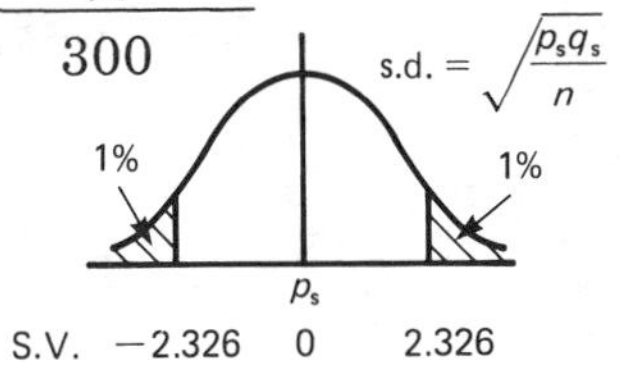

s.d. = $\sqrt{\dfrac{p_s q_s}{n}}$

1% 1%

p_s

S.V. −2.326 0 2.326

The 98% confidence interval is (0.101, 0.198).

Example 8.21 A point whose coordinates are (X, Y) with respect to rectangular axes is chosen at random where $0 < X < 1$ and $0 < Y < 1$. What is the probability that the point lies inside the circle whose equation is $x^2 + y^2 = 1$?

In a computer simulation 1000 such points were generated and 784 of them lay inside the circle. Obtain an estimate for π and give an approximate 90% confidence interval for your estimate. Show that about 290 000 points need to be selected in order to be 90% certain of obtaining a value for π which will be in error by less than 0.005. (SUJB)

Solution 8.21 The point (X, Y) is chosen at random, where $0 < X < 1$ and $0 < Y < 1$.

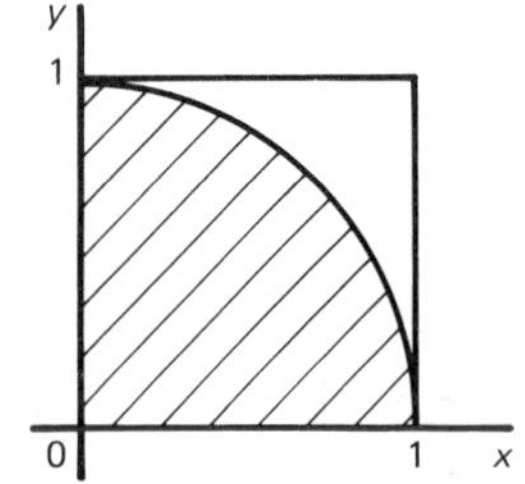

Points thus chosen are spread uniformly over the square with vertices $(0,0)$, $(1,0)$, $(1,1)$, $(0,1)$.

So the probability that the point lies within the circle $x^2 + y^2 = 1$ is equal to the fraction of the area of the square which lies in the region defined by the inequality $x^2 + y^2 < 1$.

So,
$$P(\text{point lies within circle}) = \frac{\text{area quadrant}}{\text{area square}} = \frac{\pi/4}{1} = \pi/4$$

If 1000 points are taken and 784 lie within the circle, then if the true proportion of points lying within the circle is p, an estimate for p is p_s where

$$p_s = \frac{784}{1000} = 0.784$$

So an estimate for $\pi/4 = 0.784$ and hence an estimate for π is $(0.784)(4) = 3.136$.

Now, a 90% confidence interval for p is given by

$$p_s \pm 1.645\sqrt{\frac{p_s q_s}{n}} \quad \text{where } q_s = 1 - p_s$$

$$= 0.784 \pm (1.645)\sqrt{\frac{(0.784)(0.216)}{1000}}$$

$$= 0.784 \pm 0.0214$$

$$= (0.7626, 0.8054)$$

Therefore $P(0.7626 < \pi/4 < 0.8054) = 0.90$

i.e. $P(3.0504 < \pi < 3.2216) = 0.90$

So, a 90% confidence interval for π is (3.0504, 3.2216).

If the value for π is to be in error by less than 0.005 then the value for $\pi/4$ must be in error by less than 0.001 25.

When $n = 1000$, the size of the interval was $p_s \pm 0.0214$.

Now we need to find n such that the size of the interval is $p_s \pm 0.001\,25$,

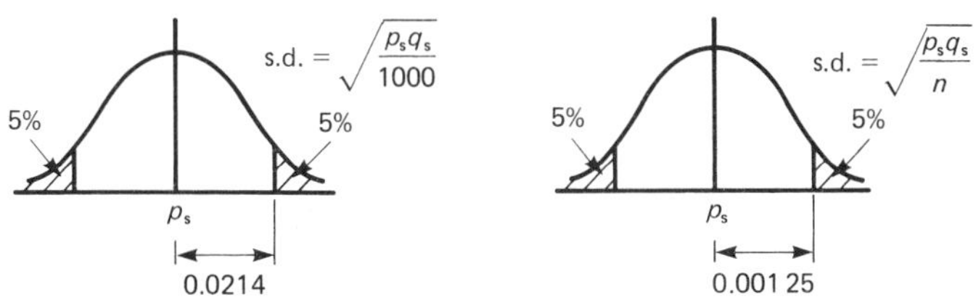

i.e. we require n such that

$$1.645\sqrt{\frac{p_s q_s}{n}} < 0.001\,25$$

so
$$\sqrt{n} > \frac{1.645}{0.001\,25}\sqrt{(0.784)(0.216)}$$
$$\sqrt{n} > 541.55$$
$$n > 293\,279$$

So about 290 000 points need to be selected in order to be 90% certain of obtaining a value for π which will be in error by less than 0.005.

Example 8.22 Derive the mean and variance of the binomial distribution.

In a survey carried out in Funville, 28 children out of a random sample of 80 said that they bought Bopper comic regularly. Find 95% approximate confidence limits for the true proportion of all children in Funville who buy this comic. A similar survey in Funville found that 45 children out of a random sample of 100 said that they bought Shooter comic regularly. Find 95% approximate confidence limits for the true proportion of all children in Funville who buy this comic.

On the basis of these surveys, is there any evidence that the sales of Shooter comic are higher than the sales of Bopper comic in Funville? Justify your reply. (AEB 1980)

Solution 8.22 For the answer to the first part, see p. 214.

Let p_B be the true proportion of all children who buy the Bopper.

In the sample of 80,

$$p_s = \frac{28}{80} = 0.35 \qquad q_s = 1 - p_s = 0.65$$

The 95% confidence interval for p_B is

$$p_s \pm 1.96\sqrt{\frac{p_s q_s}{n}} = 0.35 \pm (1.96)\sqrt{\frac{(0.35)(0.65)}{80}}$$
$$= 0.35 \pm 0.105$$
$$= (0.245, 0.455)$$

The 95% confidence interval for the proportion who buy the Bopper is (0.245, 0.455).

Let p_C be the true proportion who buy the Shooter.

In the sample of 100, $p_s = \dfrac{45}{100} = 0.45$, $q_s = 0.55$.

The 95% confidence interval for p_C is

$$p_s \pm 1.96\sqrt{\frac{p_s q_s}{n}} = 0.45 \pm (1.96)\sqrt{\frac{(0.45)(0.55)}{100}}$$
$$= 0.45 \pm 0.098$$
$$= (0.352, 0.548)$$

The 95% confidence interval for the proportion who buy the Shooter is (0.352, 0.548).

These confidence intervals overlap, so it is possible that $p_B = 0.43$ (say) and $p_C = 0.39$ (say) so that $p_B > p_C$.

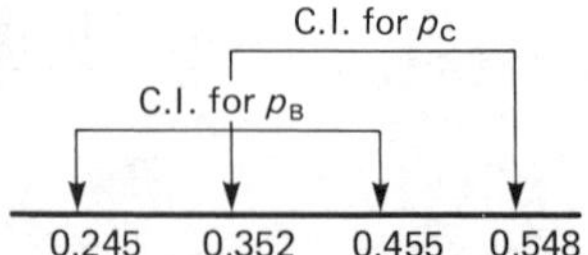

So, on these results there is not sufficient evidence to suggest that the sales of Shooter comic are higher than the sales of Bopper comic.

NOTE: this could have been approached by considering a significance test for the difference between proportions — for method, see p. 497.

Example 8.23 In a sample of 400 shops taken in 1972, it was discovered that 136 of them sold carpets at below the list prices which had been recommended by manufacturers.

(a) Estimate the percentage of all carpet selling shops selling below list price.

(b) Calculate the 95% confidence limits for this estimate, and explain briefly what these mean.
(c) What size sample would have to be taken in order to estimate the percentage to within ± 2%? (SUJB)

Solution 8.23 From the sample, the proportion of shops selling below list price is p_s where $p_s = \dfrac{136}{400} = 0.34$.

(a) An estimate of the percentage of all carpet selling shops selling below list price is $\hat{p}$ where $\hat{p} = p_s$.

So $\hat{p} = 0.34 = 34\%$.

(b) A 95% confidence interval for the true population proportion p is given by

$$p_s \pm 1.96\sqrt{\frac{p_s q_s}{n}} = 0.34 \pm (1.96)\sqrt{\frac{(0.34)(0.66)}{400}}$$
$$= 0.34 \pm 0.046$$
$$= 34\% \pm 4.6\%$$

The 95% confidence interval is (34% ± 4.6%) = (29.4%, 38.6%).

(c) In part (b) the percentage of shops was estimated to within ± 4.6%. So we now require n such that the percentage of shops is estimated to within ± 2% (assuming 95% confidence).

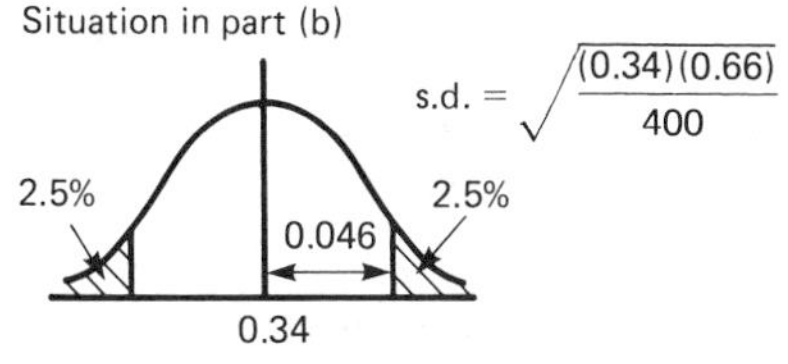

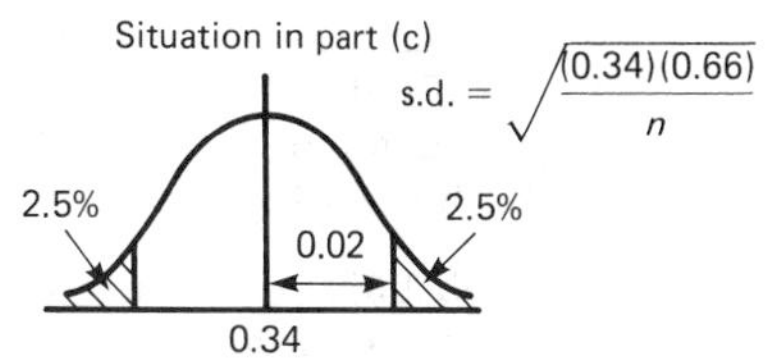

We require n such that

$$p_s \pm 1.96\sqrt{\frac{p_s q_s}{n}} = p_s \pm 0.02$$

i.e. $$1.96\sqrt{\frac{(0.34)(0.66)}{n}} = 0.02$$

so $$\sqrt{n} = \frac{1.96}{0.02}\sqrt{(0.34)(0.66)}$$
$$= 46.42$$
$$n = 2155.14$$

So a sample of size 2156 would have to be taken.

Exercise 8g

1. A sample, of size n, is taken from a population in which the proportion of 'successes' is p. From the value of the sample proportion given, calculate the confidence interval indicated for the proportion p.

	n	p_s	% confidence interval
(a)	100	0.42	95%
(b)	70	0.8	97%
(c)	150	0.32	99%
(d)	50	0.66	95%
(e)	150	0.2	90%
(f)	200	0.77	99%
(g)	1000	0.5	95%

2. In a survey carried out in a large city, 170 households out of a random sample of 250 owned at least one pet. Find 95% confidence limits for the proportion of households in the city who own at least one pet.

3. In order to assess the probability of a successful outcome, an experiment is performed 200 times and the number of successful outcomes is found to be 72. Find (*a*) 95%, (*b*) 99% confidence intervals for p, the probability of a successful outcome.

4. In a market research survey 25 people out of a random sample of 100 from a certain area said that they used a particular brand of soap. Find 97% confidence limits for the proportion of people in the area who use this brand of soap.

5. The probability of success in each of a long series of n independent trials is constant and equal to p. Explain how 95% approximate confidence limits for p may be obtained.

   In an opinion poll carried out before a local election, 501 people out of a random sample of 925 declare that they will vote for a particular one of two candidates contesting the election. Find 95% confidence limits for the true proportion of all voters in favour of this candidate.

   Do you consider there is significant evidence that this candidate will win the election? (AEB 1977)

6. The data in Table A overleaf refer to 144 recoveries of a particular type of sea-bird. The distances given are those between the original colonies where the birds were born and the place of recovery.

   Using an assumed mean of 250.5 obtain the sample mean recovery distance and the sample standard deviation.

   Assuming that the recovery distances accurately reflect the dispersal of birds from their original colonies, estimate the proportion of this type of bird at more than 300 miles from the original colony. Give an approximate symmetric 95% confidence interval for this estimate. (C)

7. A random sample of 600 was chosen from the adults living in a town in order to investigate the number x of days of work lost through illness. Before taking the sample it was decided that certain categories of people would be excluded from the analysis of the number of working days lost although they would not be excluded from the sample. In the sample 180 were found to be from these categories. For the remaining 420 members of the sample $\Sigma x = 1260$ and $\Sigma x^2 = 46\,000$.

   (*a*) Estimate the mean number of days lost through illness, for the restricted population, and give a 95% confidence interval for the mean.

   (*b*) Estimate the percentage of people in the town who fall into the excluded categories, and give a 99% confidence interval for this percentage.

   (*c*) Give two examples, with reasons, of people who might fall into the excluded categories. (O)

8. There are n_0 fish in a lake. A random sample of m of these fish is taken. The fish in this sample are tagged and released unharmed back into the lake. After a suitable interval, a second random sample of size n is taken. The random variable R is the number of fish in this second sample that are found to have been tagged. Assuming that the probability that a fish is captured is independent of whether it has been tagged or not, and that n_0 is sufficiently large for a binomial approximation to be used, obtain the expectation of R in terms of m, n and n_0.

   Suppose that $m = 100$, $n = 4000$ and that the observed value of R is 20. Obtain an approximate symmetric 98% confidence interval for the proportion of fish in the lake which are tagged. Deduce an approximate 98% confidence interval for n_0. (C)

9. A random sample of 500 fish is taken from a lake, marked, and returned to the lake. After a suitable interval a second sample of 500 is taken, and 25 of these are found to be marked. By considering the proportion of marked fish in the second sample, estimate the number of fish in the lake and, by considering a confidence interval for the proportion of marked fish in the lake, obtain a 95% confidence interval for the number of fish. (O)

10. In observations of a particular type of event, the probability of a positive result of any one observation is independent of the results of other observations and has the value θ, the same for all observations. In n observations the proportion giving positive results is p. State the mean and standard deviation of the probability distribution of p. Say also how and in what circumstances this probability distribution can be approximated by a normal distribution. Show that, according to this approximation, the probability that p satisfies the inequality

$$|p-\theta| < 1.96\sqrt{\frac{\theta(1-\theta)}{n}}$$

is 95%.

In a set of 100 observations of this type, 90 gave a positive result. Obtain an inequality of the above form, and by squaring both sides of the inequality calculate from the roots of a quadratic equation an approximate 95% symmetric confidence interval for the value of θ for the type of event observed. (JMB)

Table A

Distance (miles)	1-100	101-200	201-300	301-400	401-500	501-600
Frequency	50	16	30	11	12	9

Distance (miles)	601-700	701-800	801-900	901-1000
Frequency	8	3	4	1

SUMMARY — CONFIDENCE INTERVALS

	95% Confidence interval	99% Confidence interval	
Population mean μ σ^2 *known*	$\bar{x} \pm 1.96\frac{\sigma}{\sqrt{n}}$	$\bar{x} \pm 2.575\frac{\sigma}{\sqrt{n}}$	n is sample size.
σ^2 *unknown,* *n large*	$\bar{x} \pm 1.96\frac{\hat{\sigma}}{\sqrt{n}}$	$\bar{x} \pm 2.575\frac{\hat{\sigma}}{\sqrt{n}}$	n is sample size. $\hat{\sigma}^2 = \frac{n}{n-1}s^2 \approx s^2$, where s^2 is the sample variance.
σ^2 *unknown,* *n small*	$\bar{x} \pm t\frac{s}{\sqrt{n-1}}$ where $(-t, t)$ encloses 95% of the $t(n-1)$ distribution	$\bar{x} \pm t\frac{s}{\sqrt{n-1}}$ where $(-t, t)$ encloses 99% of the $t(n-1)$ distribution	n is sample size. s^2 is the sample variance.
Population proportion p *n large*	$p_s \pm 1.96\sqrt{\frac{p_s q_s}{n}}$	$p_s \pm 2.575\sqrt{\frac{p_s q_s}{n}}$	n is sample size. p_s is the sample proportion and $q_s = 1-p_s$.

SPECIAL NOTE: In some texts the unbiased estimate of the variance is not written as $\hat{\sigma}^2$, but as s^2, the notation that we have used for the sample variance. Therefore care must be taken to ensure that these formulae are fully understood so that they can be used accurately whichever notation is adopted.

Miscellaneous Exercise 8h

1. (*a*) Before its annual overhaul, the mean operating time of an automatic machine was 103 seconds. After the annual overhaul, the following random sample of operating times (in seconds) was obtained.

90	97	101	92	101	95	95	98	96	95

Assuming that the time taken by the machine to perform the operation is a normally distributed random variable with a known standard deviation of 5 seconds, find 98% confidence limits for the mean operating time after the overhaul.

Comment on the magnitude of these limits relative to the mean operating time before the overhaul.

(*b*) The results of a survey showed that 3600 out of 10 000 families regularly purchased a specific weekly magazine. Find the 95% confidence limits for the proportion of the population buying the magazine.

Estimate the additional number of families to be contacted if the probability that the estimated proportion is in error by more than 0.01 is to be at most 1%. (AEB 1987)

2. From a large pile of industrial diamonds, 20 were put through 6 sieves of different mesh sizes and the number of diamonds passing through each sieve was counted. The table shows the mesh size (mm) and corresponding number of diamonds passing through each sieve.

Mesh size	1	2	4	6	8	12
Number of diamonds	1	2	6	8	14	20

Graphically, or otherwise, estimate the mesh size of the sieve if half the diamonds will pass through it, and the mesh size of the sieve if one quarter of the diamonds will pass through it.

Construct a frequency table showing for each mesh size listed in the table the extra number of diamonds which passed through.

Calculate unbiased estimates, in each case to 2 decimal places, of the mean and the variance of the sizes of the diamonds in the original large pile. (L)

3. Describe briefly the empirical evidence that you acquired for the Central Limit theorem.

The amount, to the nearest mg, of a certain chemical in particles in the atmosphere at a meteorological station was measured each day for 300 days. The results are shown in the table.

Amount of chemical (mg)	12	13	14	15	16
Number of days	5	42	210	31	12

Find the mean daily amount of chemical over the 300 days and estimate, to 2 decimal places, its standard error.

Obtain, to 2 decimal places, approximate 98% confidence limits for the mean daily amount of chemical in the atmosphere.

If daily measurements are taken for a further 300 days, estimate, to 2 decimal places, the probability that the mean of these daily measurements will be less than 14. (L)

4. A company manufactures bars of soap. In a random sample of 70 bars, 18 were found to be mis-shaped. Calculate an approximate 99% confidence interval for the proportion of mis-shaped bars of soap.

Explain what you understand by a 99% confidence interval by considering
(*a*) intervals in general based on the above method,
(*b*) the interval you have calculated.

The bars of soap are either pink or white in colour and differently shaped according to colour. The masses of both types of soap are known to be normally distributed, the mean mass of the white bars being 176.2 g. The standard deviation for both bars is 6.46 g. A sample of 12 of the pink bars of soap had masses, measured to the nearest gram, as follows.

174	164	182	169	171	187	176
177	168	171	180	175		

Find a 95% confidence interval for the mean mass of pink bars of soap.

Calculate also an interval within which approximately 90% of the masses of the white bars of soap will lie.

The cost of manufacturing a pink bar of soap of mass x gm is $(15 + 0.065x)$p, and it is sold for 32p. If the company manufactures 9000 bars of pink soap per week, derive a 95% confidence interval for its weekly expected profit from pink bars of soap. (AEB 1988)

9

SIGNIFICANCE TESTING

Often in scientific enquiry a statement concerning a population parameter is put forward as a statistical hypothesis. Its validity is then tested, based on observations made from random samples taken from the population.

NULL AND ALTERNATIVE HYPOTHESES

For example, suppose we wish to test whether a sample value x could have been drawn from a normal population with mean μ and variance σ^2.

We *assume* that the sample *is* drawn from $N(\mu, \sigma^2)$. This hypothesis is called the **null hypothesis** and is denoted by H_0.

If statistical tests show that we should reject the null hypothesis, we do so in favour of the **alternative hypothesis**, denoted by H_1.

For example, if we wish to investigate whether the mean of the population from which the sample value is taken is 25, say, or whether the population mean is not 25, the hypotheses would be written:

H_0: $\mu = 25$ (the population mean μ is 25)

H_1: $\mu \neq 25$ (the population mean μ is not 25)

If H_0 is true, then $X \sim N(\mu, \sigma^2)$.

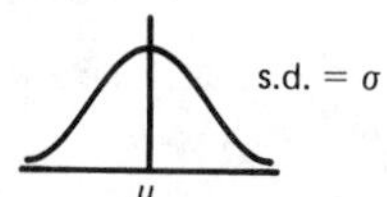

Now, we must decide whether it is likely that the sample value has been drawn from this population. We consider whether it is 'close to μ', or whether it is in the tail end of the distribution.

We consider the standard normal variable Z.

In this case $Z = \dfrac{X-\mu}{\sigma}$ where $Z \sim N(0, 1)$

Z is known as the **test statistic.**

Based on the sample observation, we calculate $z = \dfrac{x-\mu}{\sigma}$.

If z is small (i.e. close to zero), we accept that the sample value could have been taken from a population with mean μ and we do not reject H_0. If z is large (i.e. far from zero), we reject H_0.

CRITICAL REGION AND CRITICAL VALUES

We need to select a set of values for Z which tell us when to reject H_0. This set of values is known as the **critical region** and it depends on the type and the level of the test chosen. The boundaries of the critical region are called the **critical values.**

Often, the critical region is chosen so that the probability that Z falls within it is just 5%.

We have $P(|Z| > 1.96) = 0.05$ so the critical values are ± 1.96.

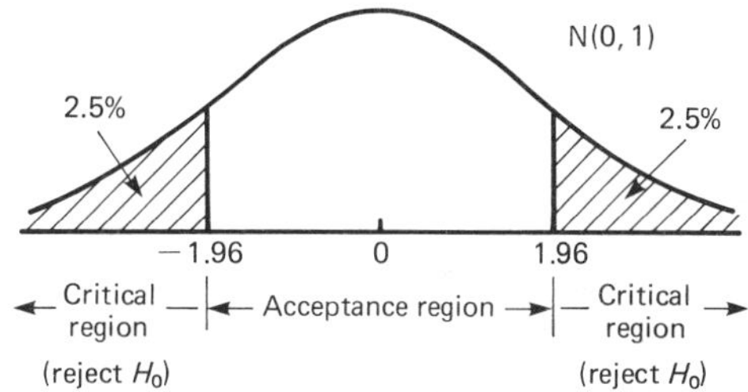

If the test is carried out at the 5% level, we reject H_0 if $z < -1.96$ or if $z > 1.96$, i.e. if $|z| > 1.96$.

If we reject H_0 at this level, we say that 'there is significant evidence, at the 5% level, that the population mean is not μ'.

ONE-TAILED AND TWO-TAILED TESTS

There are two types of test which could be performed, depending on the alternative hypothesis being made. These are **(a)** a two-tailed test, **(b)** a one-tailed test.

(a) Two-tailed test

A two-tailed test looks for *any change* in the parameter, e.g. the hypotheses could be

$$H_0: \quad \mu = 25$$

$$H_1: \quad \mu \neq 25$$

The critical region depends on the level of the test as shown:

Critical region at 5% level:

$$P(|Z| > 1.96) = 0.05$$

N(0, 1)
2.5% 2.5%
−1.96 0 1.96
Reject H_0 Reject H_0

Critical region at 2% level:

$$P(|Z| > 2.326) = 0.02$$

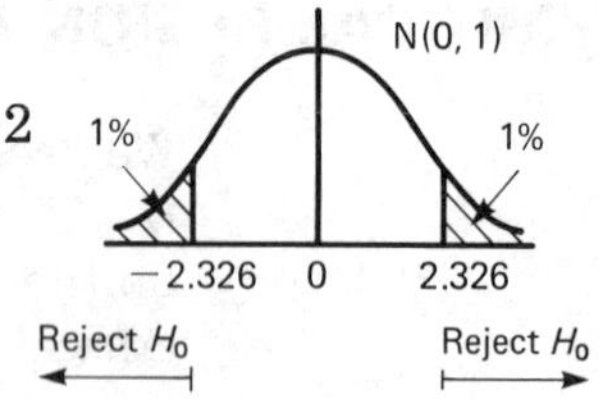

Critical region at 1% level:

$$P(|Z| > 2.575) = 0.01$$

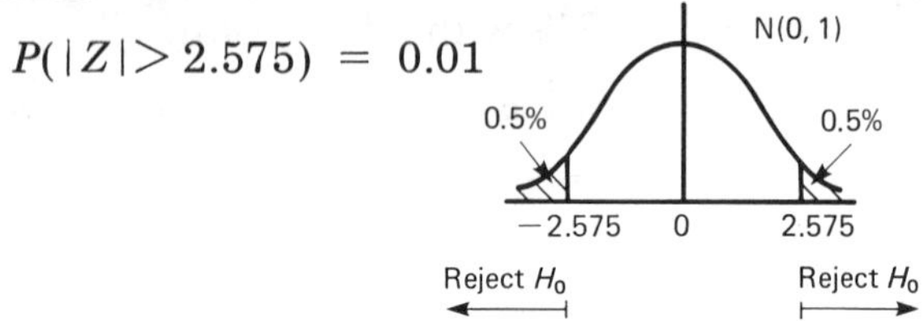

(b) One-tailed test

A one-tailed test looks for a *definite decrease* or a *definite increase* in the parameter. For example, the hypotheses could be:

Definite decrease H_0: $\mu = 25$ H_1: $\mu < 25$	Definite increase H_0: $\mu = 25$ H_1: $\mu > 25$
Critical region at 5% level:	
$P(Z < -1.645) = 0.05$ 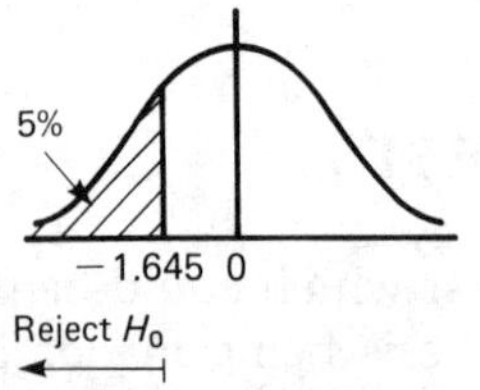	$P(Z > 1.645) = 0.05$ 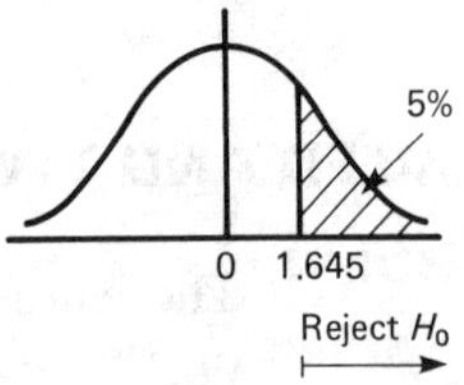
Critical region at 2% level:	
$P(Z < -2.054) = 0.02$ 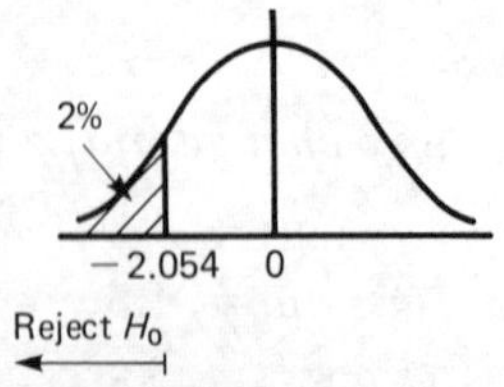	$P(Z > 2.054) = 0.02$ 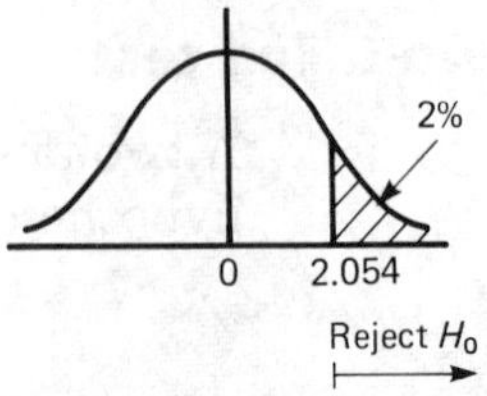
Critical region at 1% level:	
$P(Z < -2.326) = 0.01$ 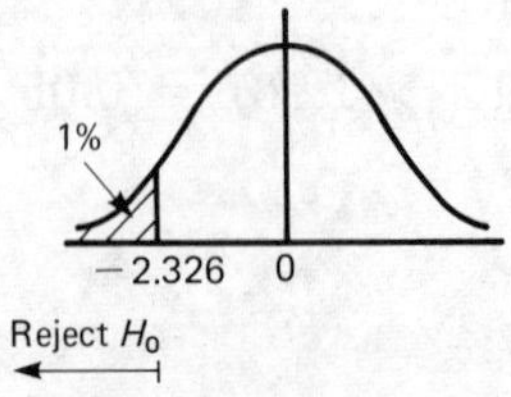	$P(Z > 2.326) = 0.01$ 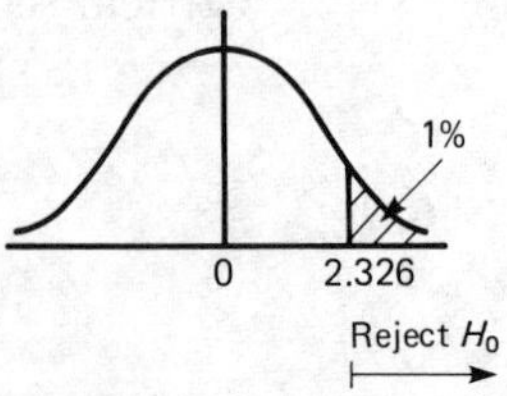

If the distribution given by the null hypothesis H_0 *is* true, then the probability that z lies in the region pronounced as 'critical' is 0.05, 0.02, 0.01, ..., depending on the level of the test (5%, 2%, 1%, ...).

But if z lies in the critical region, we reject H_0. Therefore the probability that we reject H_0, when in fact it is true, is determined by the level of the test chosen. For example, if the test is performed at the 5% level, then the probability of wrongly rejecting H_0 is 0.05.

When performing a significance test it is useful to follow a set procedure:

Before any sample readings are considered:

(1) State the null hypothesis, H_0 and the alternative hypothesis H_1.

If we are looking for a definite increase or a definite decrease in the population parameter, we use a *one-tailed test* and if we are looking for any change we use a *two-tailed test.*

(2) Consider the appropriate distribution given by the null hypothesis.

(3) Decide on the level of the test. This fixes the critical values of the test statistic.

(4) Decide on the rejection criteria.

Now consider the sample values.

(5) Calculate the value of the test statistic.

(6) Make a conclusion: If the value of the test statistic lies in the critical region, reject H_0.
If the value of the test statistic does not lie in the critical region, do not reject H_0.

If H_0 is rejected at the 5% level, we say that the test value is 'significant'.

If H_0 is rejected at the 1% level, then the test value is 'highly significant'.

TEST 1 — TESTING A SINGLE SAMPLE VALUE

Example 9.1 Test, at the 5% level, whether the single sample value of 172 comes from a normal population with mean $\mu = 150$ and variance $\sigma^2 = 100$.

Solution 9.1 *Reminders*:

(1) State H_0 and H_1; decide whether the test is one-tailed or two-tailed	H_0: $\mu = 150$ (the mean of the distribution is 150) H_1: $\mu \neq 150$ (the mean is not 150) (two-tailed test)
(2) Consider the distribution given by H_0	Now, under H_0, $X \sim \mathrm{N}(\mu, \sigma^2)$ with $\mu = 150$, $\sigma = 10$
(3) Decide on the level of the test	Test at the 5% level
(4) Decide on rejection criteria	Reject H_0 if $\lvert z\rvert > 1.96$ 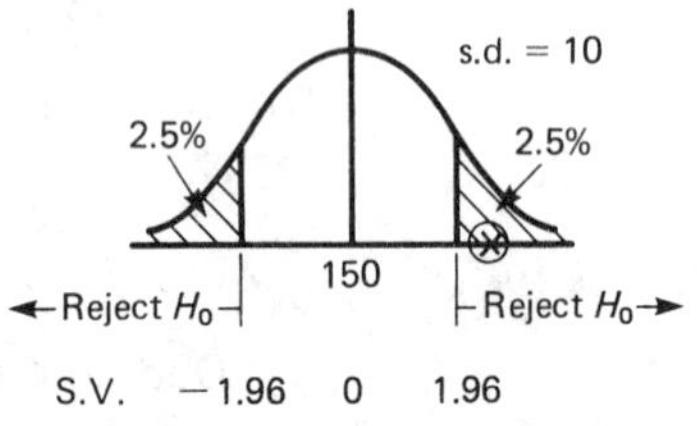
(5) Calculate the value of the test statistic	where $z = \dfrac{x-\mu}{\sigma}$ $= \dfrac{172-150}{10}$ $= 2.2$
(6) Make conclusion	As $\lvert z\rvert > 1.96$, we reject H_0 and conclude that there is significant evidence, at the 5% level, to suggest that the sample value does not come from a population with mean 150.

NOTE: Once z has been calculated, its position can be noted on the diagram thus, ⊗, indicating whether it lies in the rejection region or not. In Example 9.1, $z = 2.2$ so ⊗ is placed to the right of the standardised value 1.96.

Example 9.2 Test at the 1% level whether the single sample value 54 has been drawn from a normal population with mean 65 and variance 30, or whether the mean is less than 65.

Solution 9.2 Let the population mean be μ and the population variance be σ^2.

$$H_0: \quad \mu = 65$$

$$H_1: \quad \mu < 65 \quad \text{(one-tailed test)}$$

Now under H_0, $X \sim \mathrm{N}(\mu, \sigma^2)$ with $\mu = 65$, $\sigma = \sqrt{30}$.

We perform a one-tailed test at the 1% level, and reject H_0 if $z < -2.326$, where

$$z = \frac{x-\mu}{\sigma} = \frac{54-65}{\sqrt{30}} = -2.01$$

s.d. = $\sqrt{30}$; 1%; 65; ← Reject H_0; S.V. −2.326 0

Conclusion: as $z > -2.326$, we do not reject H_0 and conclude at the 1% level, that the sample value could have been drawn from a population with mean 65.

Example 9.3 If 100 seeds are planted, and 83 seeds germinate, use the normal approximation to the binomial distribution to test the manufacturer's claim of a 90% germination rate. Use a 5% level of significance.

Solution 9.3 Let X be the r.v. 'the number of seeds that germinate'. Then we have a binomial situation, and $X \sim \mathrm{Bin}(n, p)$ with $n = 100$.

$$H_0: \quad p = 0.9 \quad \text{(the germination rate is 90\%)}$$

$$H_1: \quad p < 0.9 \quad \text{(the germination rate is less than 90\%)}$$

(We have chosen a one-tailed test, as this seems more appropriate to the situation.)

Under H_0, $X \sim \mathrm{Bin}(n, p)$ with $n = 100, p = 0.9$.

Now, as n is large, we use the normal approximation to the binomial distribution

so $X \sim \mathrm{N}(np, npq)$ where $np = (100)(0.9) = 90$

i.e. $X \sim \mathrm{N}(90, 9)$ $\qquad npq = (100)(0.9)(0.1) = 9$

Perform a one-tailed test, at the 5% level, and reject H_0 if $z < -1.645$, where

$$z = \frac{x-np}{\sqrt{npq}} = \frac{83-90}{3} = -2.33$$

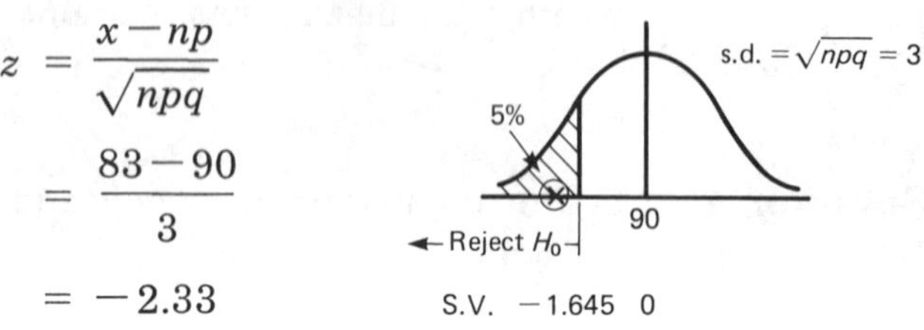

Conclusion: As $z < -1.645$ we reject H_0 and conclude that there is significant evidence, at the 5% level, to suggest that the manufacturer's claim is false.

NOTE: on the continuous scale, 83 lies between 82.5 and 83.5, so we should really apply a continuity correction. In this case we would try 83.5. If this value is in the critical region, then obviously 82.5 is also.

With the continuity correction,

$$z = \frac{83.5-90}{3} = -2.17$$

and the same conclusion is reached.

Exercise 9a

1. Test whether the sample value could have been drawn from the normal population indicated in the null hypothesis. Test at (*a*) the 5% level, (*b*) the 1% level.

	Sample value	Hypotheses	Population variance (σ^2)
(i)	104	H_0: $\mu = 96$, H_1: $\mu \neq 96$	16
(ii)	90	H_0: $\mu = 96$, H_1: $\mu < 96$	16
(iii)	132	H_0: $\mu = 120$, H_1: $\mu > 120$	32
(iv)	106	H_0: $\mu = 120$, H_1: $\mu \neq 120$	32
(v)	22	H_0: $\mu = 18$, H_1: $\mu \neq 18$	1.2
(vi)	16	H_0: $\mu = 18$, H_1: $\mu < 18$	1.2

2. A coin is tossed 64 times. Test at the 5% level of significance whether the coin is fair, or whether it is biased in favour of showing heads, if (*a*) 38 heads occur, (*b*) 42 heads occur.

3. A manufacturer claims that 8 out of 10 dogs prefer his brand to any other. In a random sample of 120 dogs, it was found that 88 ate that brand. Test at the 5% level whether you would accept the manufacturer's claim.

4. In a survey it was found that 3 out of 10 people supported a particular political party. A month later the party representative claimed that the popularity of the party had increased. Would you accept that the number who supported the party was still 3 out of 10 if a further survey revealed that 38 people in a random sample of 100 supported the party. Test at the 3% level.

5. A gardener sows 150 'Special' cabbage seeds and knows that the germination rate is 75%. (*a*) By using a suitable approximation find the probability that (i) more than 122 seeds germinate, (ii) less than 106 seeds germinate. (*b*) The gardener also sows 120 'Everyday' cabbage seeds and finds that 81 germinate. Test whether the 'Everyday' seeds have a germination rate less than 75%. Test at the 4% level.

TEST 2 — TESTING A MEAN

We may wish to investigate whether a random sample of size n, with mean $\bar{x}$, could have been drawn from a normal population with mean μ.

Case 1 — Population variance σ^2 known

Under the null hypothesis that the population mean is μ, taking samples of size n,

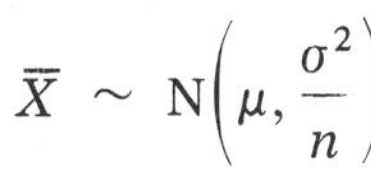

$$\bar{X} \sim \mathrm{N}\left(\mu, \frac{\sigma^2}{n}\right)$$

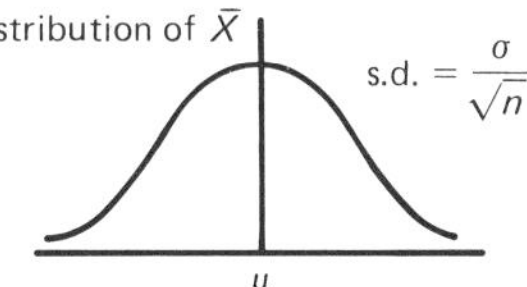

If X is distributed normally then this holds for all sample sizes, but if X does not follow a normal distribution then n must be large (Central Limit Theorem).

Reminders: the distribution of $\bar{X}$ is known as the sampling distribution of means; the standard deviation of this distribution ($\sigma/\sqrt{n}$) is known as the standard error of the mean (see p. 399).

Now, we want to investigate whether there is a significant difference between the sample mean and the population mean given by the null hypothesis.

Standardising, we have $Z = \dfrac{\bar{X} - \mu}{\sigma/\sqrt{n}}$ where $Z \sim \mathrm{N}(0, 1)$.

> We use the test statistic $Z = \dfrac{\bar{X} - \mu}{\sigma/\sqrt{n}}$ which is distributed as $\mathrm{N}(0, 1)$ under the null hypothesis H_0 that the true population mean is μ.

Example 9.4 The lengths of metal bars produced by a particular machine are normally distributed with mean length 420 cm and standard deviation 12 cm. The machine is serviced, after which a sample of 100 bars gives a mean length of 423 cm. Is there evidence, at the 5% level, of a change in the mean length of the bars produced by the machine, assuming that the standard deviation remains the same?

Solution 9.4 Let X be the r.v. 'the length, in cm, of a metal bar'. Let the population mean be μ and the population variance be σ^2.

Reminders:

(1) State H_0 and H_1; decide whether the test is one-tailed or two-tailed	H_0: $\mu = 420$ cm (there is no change in the population mean μ) H_1: $\mu \neq 420$ cm (there is a change in the population mean μ) (Two-tailed test)
(2) Consider the distribution given by H_0	Consider the sampling distribution of means under H_0, $\bar{X} \sim \mathrm{N}\left(\mu, \dfrac{\sigma^2}{n}\right)$ with $\sigma = 12$, $n = 100$
(3) Decide on the level of the test	We perform a two-tailed test at the 5% level
(4) Decide on rejection criteria	Reject H_0 if $\lvert z \rvert > 1.96$ s.d. $= \dfrac{\sigma}{\sqrt{n}}$ 2.5% 2.5% 420 ← Reject H_0 Reject H_0 → S.V. −1.96 0 1.96 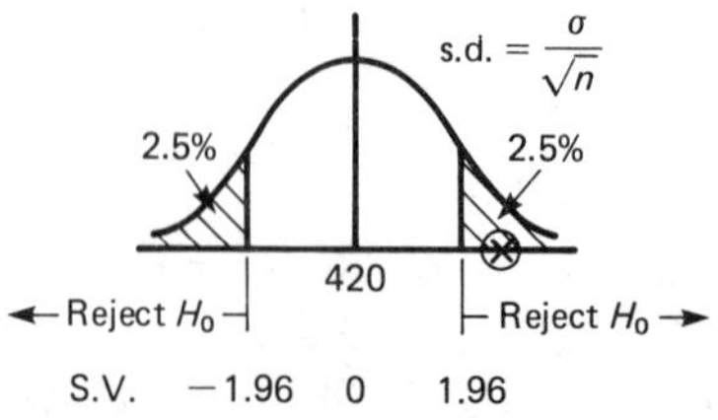
(5) Calculate the value of the test statistic	where $z = \dfrac{\bar{x} - \mu}{\sigma/\sqrt{n}}$ $= \dfrac{423 - 420}{12/\sqrt{100}}$ $= 2.5$
(6) Make conclusion	As $\lvert z \rvert > 1.96$. we reject H_0 and conclude that there is sufficient evidence, at the 5% level, of a change in the mean length of the bars produced by the machine

Example 9.5 Experience has shown that the scores obtained in a particular test are normally distributed with mean score 70 and variance 36. When the test is taken by a random sample of 36 students, the mean score is 68.5. Is there sufficient evidence, at the 3% level, that these students have not performed as well as expected?

Solution 9.5 Let X be the r.v. 'the score of a student'. Let the population mean be μ and the population variance be σ^2.

H_0: $\mu = 70$ (the population mean μ is 70 and the students have not under-achieved)

H_1: $\mu < 70$ (the population mean is less than 70 and the students have not done as well as expected)

Consider the sampling distribution of means where, under H_0

$$\bar{X} \sim \mathrm{N}\left(\mu, \frac{\sigma^2}{n}\right) \quad \text{with} \quad \mu = 70, \quad \sigma^2 = 36, \quad n = 36$$

We perform a one-tailed test at the 3% level.

Now, if a is the critical value of the test statistic, then

$$\Phi(a) = 0.03$$
$$a = -1.881$$

So, we reject H_0 if $z < -1.881$, where

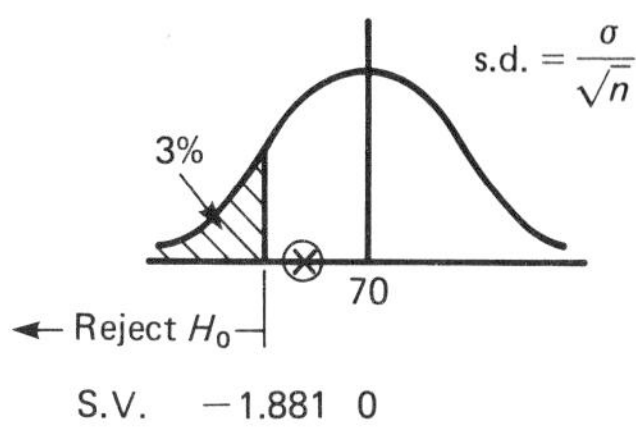

$$z = \frac{\bar{x} - \mu}{\sigma/\sqrt{n}}$$
$$= \frac{68.5 - 70}{6/\sqrt{36}}$$
$$= -1.5$$

Conclusion: as $z > -1.881$ we do not reject H_0 and conclude that at the 3% level the students have not under-achieved.

Example 9.6 It is claimed that the masses of components produced at a particular workshop are normally distributed with a mean mass of 6 g and a standard deviation of 0.8 g. If this claim is accepted, at the 5% level, on the basis of the mean mass obtained from a random sample of 50 components, between what values must the mean mass of the 50 components in the sample lie?

Solution 9.6 Let X be the r.v. 'the mass, in g, of a component'. Let the population mean be μ and the population variance be σ^2.

H_0: $\mu = 6$ g

H_1: $\mu \neq 6$ g (two-tailed test)

Consider the sampling distribution of means where, under H_0

$$\bar{X} \sim \mathrm{N}\left(\mu, \frac{\sigma^2}{n}\right) \quad \text{with} \quad \sigma = 0.8 \quad \text{and} \quad n = 50$$

If the test is performed at the 5% level then H_0 is accepted if $|z| < 1.96$, where

$$z = \frac{\bar{x} - \mu}{\sigma/\sqrt{n}}$$

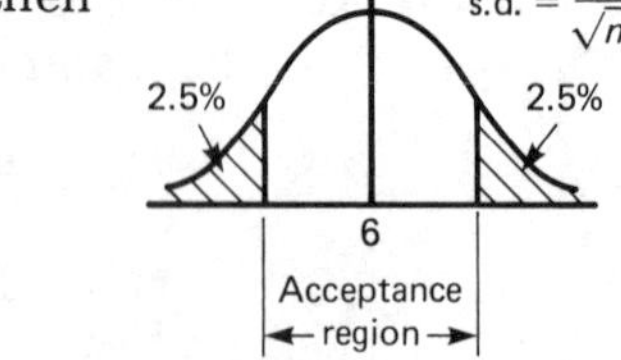

Now, as H_0 is accepted,

$$-1.96 < z < 1.96$$

i.e.
$$-1.96 < \frac{\bar{x} - 6}{0.8/\sqrt{50}} < 1.96$$

$$-1.96\left(\frac{0.8}{\sqrt{50}}\right) < \bar{x} - 6 < 1.96\left(\frac{0.8}{\sqrt{50}}\right)$$

$$6 - 1.96\left(\frac{0.8}{\sqrt{50}}\right) < \bar{x} < 6 + 1.96\left(\frac{0.8}{\sqrt{50}}\right)$$

$$5.78 < \bar{x} < 6.22$$

Therefore the mean mass of the 50 components must lie in the range $5.78\text{ g} < \bar{x} < 6.22\text{ g}$.

Example 9.7 A machine produces elastic bands with breaking tension normally distributed with mean 45.00 N and s.d. 4.36 N. On a certain day a sample of 50 was tested and found to have a mean breaking tension of 43.46 N. Test at the 5% level of significance whether this indicates a change in the mean, explaining what is meant by '5% level of significance'.

Find a 95% confidence interval for the population mean based on the sample mean assuming an unchanged s.d.

If the s.d. has changed to σ, find the least value of σ for a 95% confidence interval for the population mean to contain 45.00 N.

(SUJB)

Solution 9.7 Let X be the r.v. 'the breaking tension, in N, of an elastic band'. Let the population mean be μ and the population variance be σ_1^2.

H_0: $\mu = 45.00$ (there is no change in the mean)

H_1: $\mu \neq 45.00$ (there is a change in the mean)

Consider the sampling distribution of means where, under H_0

$$\bar{X} \sim \mathrm{N}\left(\mu, \frac{\sigma_1^2}{n}\right) \quad \text{with} \quad \mu = 45.00\text{ N},$$

$$\sigma_1 = 4.36\text{ N}, \quad n = 50$$

We use a two-tailed test at the 5% level and reject H_0 if $|z| > 1.96$, where

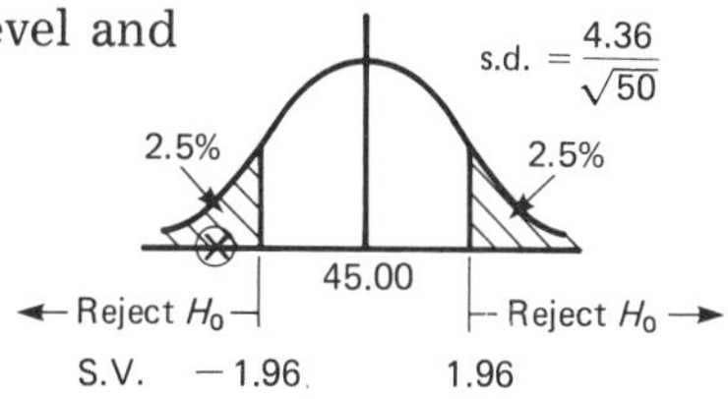

$$z = \frac{\bar{x}-\mu}{\sigma_1/\sqrt{n}} = \frac{43.46-45.00}{4.36/\sqrt{50}} = -2.498$$

s.d. $= \frac{4.36}{\sqrt{50}}$
2.5%
2.5%
45.00
← Reject H_0
Reject H_0 →
S.V. −1.96 1.96

Conclusion: As $|z| > 1.96$, we reject H_0 and conclude that there is evidence, at the 5% level, of a change in the mean.

As the level of the test chosen is the 5% level of significance, the probability that we have wrongly rejected H_0 is 0.05.

A 95% confidence interval for the population mean is given by

$$\bar{x} \pm 1.96\frac{\sigma_1}{\sqrt{n}} = 43.46 \pm 1.96\frac{4.36}{\sqrt{50}} = 43.46 \pm 1.209 = (42.25, 44.67)$$

So the 95% confidence interval for the mean breaking tension is (42.25 N, 44.67 N).

NOTE: as expected, the value 45.00 is not in this interval.

If the standard deviation has changed to σ, then the least value of σ for the 95% confidence interval to contain 45.00 is such that

$$43.46 + 1.96\frac{\sigma}{\sqrt{50}} = 45.00$$

i.e.

$$\sigma = \frac{(45.00-43.46)}{1.96}\sqrt{50} = 5.56 \quad (2\,\text{d.p.})$$

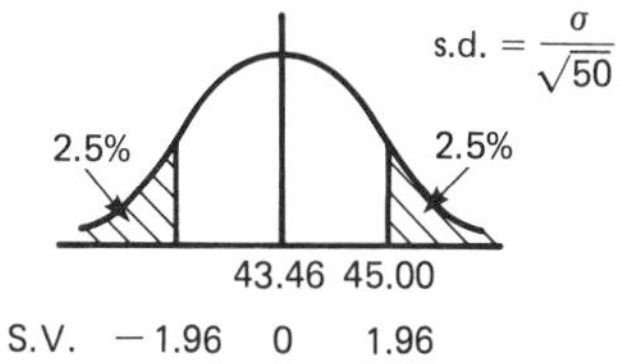

s.d. $= \frac{\sigma}{\sqrt{50}}$
2.5%
2.5%
43.46 45.00
S.V. −1.96 0 1.96

So the least value of σ must be 5.56 N.

Example 9.8 Describe, referring to your projects if you wish, the steps used in carrying out a significance test.

Over a long period it has been found that the breaking strains of cables produced by a factory are normally distributed with mean 6000 N and standard deviation 150 N. Find, to 3 decimal places, the probability that a cable chosen at random from the production will have a breaking strain of more than 6200 N.

A modification is introduced into the production process which only affects the value of the mean breaking strain. Six cables, chosen at random from the modified process, are tested and found to have a mean breaking of 5920 N.

(a) Test, at the 5% significance level, whether the sample evidence is sufficient to conclude that the mean breaking strain of the cables is actually less than 6000 N.

(b) Find, to 3 significant figures, the value C for which we can state with 90% confidence that the mean breaking strain of the cables exceeds C N. (L)

Solution 9.8 For steps used in carrying out a significance test, see p. 461.

Let X be the r.v. 'the breaking strain, in N, of a cable'.
$X \sim N(6000, 150^2)$

$$\begin{aligned} P(X > 6200) &= P\left(Z > \frac{6200-6000}{150}\right) \\ &= P(Z > 1.333) \\ &= 0.091 \quad \text{(3 d.p.)} \end{aligned}$$

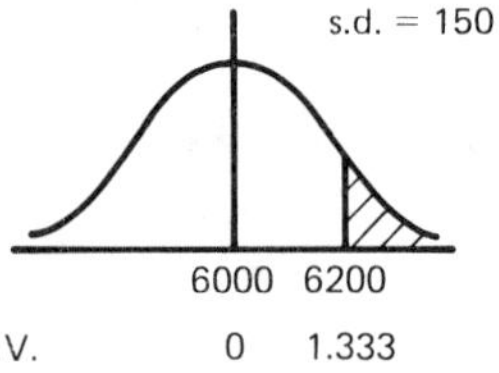

s.d. = 150

6000 6200

S.V. 0 1.333

The probability that a cable chosen at random will have a breaking strength of more than 6200 N is 0.091 (3 d.p.).

A sample of six cables is tested, giving $\bar{x} = 5920$ N. Let μ be the population mean.

(a) $H_0: \mu = 6000$ N

$H_1: \mu < 6000$ N

Consider the sampling distribution of means, where under H_0,

$$\bar{X} \sim N\left(\mu, \frac{\sigma^2}{n}\right) \text{ with } \mu = 6000\text{ N}, \ \sigma = 150\text{ N}, \ n = 6$$

so $$\bar{X} \sim N\left(6000, \frac{150^2}{6}\right)$$

We perform a 1-tailed test, at 5% level and reject H_0 if $z < -1.645$ where

$$\begin{aligned} z &= \frac{\bar{x}-\mu}{\sigma/\sqrt{n}} \\ &= \frac{5920-6000}{150/\sqrt{6}} \\ &= -1.306 \end{aligned}$$

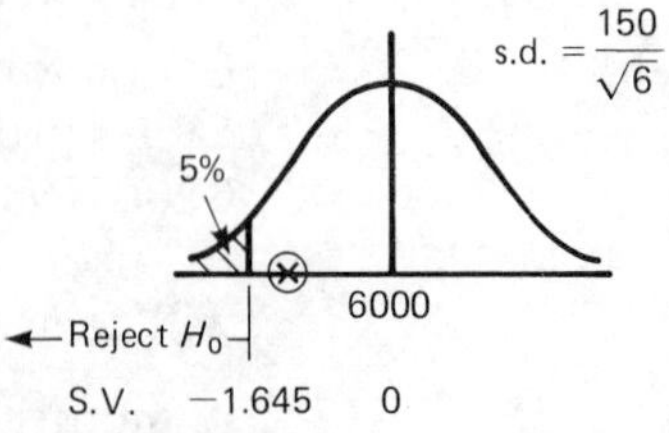

s.d. = $\frac{150}{\sqrt{6}}$

5%

6000

← Reject H_0

S.V. −1.645 0

Since $z > -1.645$ we do not reject H_0 and conclude that the mean breaking strength is not less than 6000 N.

(b) We require a one-sided (not symmetric) confidence interval such that

$$P(C < \mu < \infty) = 0.9$$

This is given by

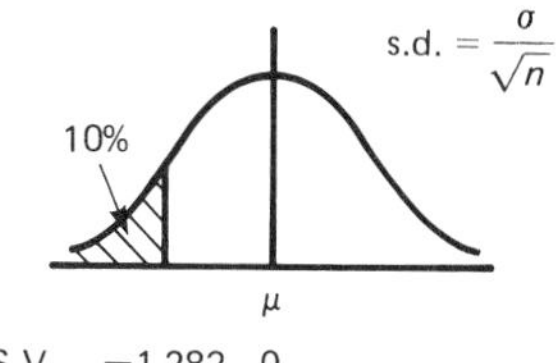

$$\left(\bar{x} - 1.282\frac{\sigma}{\sqrt{n}},\ \infty\right)$$

so

$$\begin{aligned} C &= \bar{x} - 1.282\frac{\sigma}{\sqrt{n}} \\ &= 5920 - 1.282\left(\frac{150}{\sqrt{6}}\right) \\ &= 5840 \quad (3\text{ S.F.}) \end{aligned}$$

Therefore we can state, with 90% confidence, that the mean breaking strength of the cables exceeds 5840 N.

Exercise 9b

1. For each of the following, a random sample of size n is taken from a normal distribution with mean μ and variance σ^2. The sample mean is $\bar{x}$. Test the hypotheses stated, at the level of significance indicated.

	n	$\bar{x}$	σ	Hypotheses	Level of significance
(a)	30	15.2	3	H_0: $\mu = 15.8$, H_1: $\mu \neq 15.8$	5%
(b)	10	27	1.2	H_0: $\mu = 26.3$, H_1: $\mu > 26.3$	5%
(c)	49	125	4.2	H_0: $\mu = 123.5$, H_1: $\mu > 123.5$	1%
(d)	100	4.35	0.18	H_0: $\mu = 4.40$, H_1: $\mu < 4.40$	2%

2. The masses of components produced by a certain machine are normally distributed with mean 15.4 g and standard deviation 2.3 g. The setting on the machine is altered, following which a random sample of 81 components is found to have a mean mass of 15.0 g. Does this provide evidence, at the 5% level, of a reduction in the mean mass of components produced by this machine? Assume that the standard deviation is not altered.

3. A variable with known variance of 32 is thought to have a mean of 55. A random sample of 81 independent observations of the variable gives a mean of 56.2. Is there sufficient evidence that the mean is not 55 (*a*) at the 10% level, (*b*) at the 5% level, (*c*) at the 1% level?

4. A manufacturer claims that his cassettes, advertised as having a playing time of 90 minutes, actually have a mean playing time of 92 minutes, with standard deviation 1.8 minutes. 36 tapes are selected at random and tested. The investigator rejects the manufacturer's claim, at the 5% level, saying that the mean playing time of the tapes is less than 92 minutes. What can be said about the value of the sample mean obtained for this decision to be taken?

5. Mass-produced washers have thicknesses which are normally distributed with mean 3 mm and standard deviation 0.2 mm.
(*a*) Find, correct to three decimal places, the probability that the mean thickness of a random sample of 4 washers will lie between 2.9 mm and 3.1 mm.

(*b*) During a check on the manufacturing process a random sample of 25 washers is taken from production and the mean thickness $\bar{x}$ mm is calculated. Find the interval in which the value of $\bar{x}$ must lie in order that the hypothesis that the production mean thickness is 3 mm will not be rejected when the significance level is 5 per cent. (JMB)

6. Describe briefly how the Central Limit Theorem may be demonstrated.

The distance driven by a long distance lorry driver in a week is a normally distributed variable having mean 1130 km and standard deviation 106 km. Find, to 3 decimal places, the probability that in a given week he will drive less than 1000 km. Find, to 3 decimal places the probability that in 20 weeks his average distance driven per week is more than 1200 km.

New driving regulations are introduced and, in the first 20 weeks after their introduction, he drives a total of 21 900 km. Assuming that the standard deviation of the weekly distances he drives is unchanged, test, at the 10% level of significance, whether his mean weekly driving distance has been reduced. State clearly your null and alternative hypotheses. (L)

7. A machine packs flour into bags. A random sample of eleven filled bags was taken and the masses of the bags to the nearest 0.1 g were: 1506.8, 1506.6, 1506.7, 1507.2, 1506.9, 1506.8, 1506.6, 1507.0, 1507.5, 1506.3, 1506.4. Obtain the mean and the variance of this sample showing your working clearly. Filled bags are supposed to have a mass of 1506.5 g. Assuming that the mass of a bag has normal distribution with variance 0.16 g test whether the sample provides significant evidence at the 5% level that the machine produces overweight bags. Give the 99% confidence interval for the mass of a filled bag. (C)

8. A sample of size 25 is taken from the distribution of X where $X \sim N(\mu, 4)$. The sample mean $\bar{x}$ is 10.72. At what level test would we reject the null hypothesis that $\mu = 10$ in favour of the alternative hypothesis
(*a*) $\mu > 10$, (*b*) $\mu \neq 10$?

9. Explain, briefly, the roles of a null hypothesis and a level of significance in a project which you have undertaken.

Records of the diameters of spherical ball bearings produced on a certain machine indicate that the diameters are normally distributed with mean 0.824 cm and standard deviation 0.046 cm. Two hundred samples, each consisting of 100 ball bearings, are chosen. Calculate the expected number of the 200 samples having a mean diameter less than 0.823 cm.

On a certain day it was suspected that the machine was malfunctioning. It may be assumed that if the machine is malfunctioning it will change the mean of the diameters without changing their standard deviation. On that day a random sample of 100 ball bearings had a mean diameter of 0.834 cm. Determine a 98% confidence interval for the mean diameter of the ball bearings being produced that day.

Hence state whether or not you would conclude that the machine is malfunctioning on that day given that the significance level is 2%. (L)

10. X_1 and X_2 are independent random variables with means μ_1 and μ_2, variances σ_1^2 and σ_2^2 respectively. Give the mean and variance of $X_1 - X_2$. If $Y = \lambda X_1$, where λ is a constant, give the mean and variance of Y.

The random variable $\bar{X}_1$ denotes the mean of a random sample of size n_1 from the second of the above distributions, show how to obtain the mean and variance of the distribution of $\lambda_1\bar{X}_1 + \lambda_2\bar{X}_2$ from the results you have stated, λ_1 and λ_2 being constants.

The yield of a certain crop per plot of standard area is normally distributed with mean 253.0 kg and variance 67.1 kg^2. A new fertiliser is applied to 10 randomly selected plots, and their mean yield is found to be 257.8 kg. Is there any evidence of significant improvement in the yield? Assume the new fertiliser does not affect the variance of the yields. (O)

11. The masses of loaves from a certain bakery are normally distributed with mean 500 g and standard deviation 20 g.
(*a*) Determine what percentage of the output would fall below 475 g and what percentage would be above 530 g.
(*b*) The bakery produces 1000 loaves daily at a cost of 8 p per loaf and can sell all those above 475 g for 20p each but is not allowed to sell the rest. Calculate the expected daily profit.
(*c*) A sample of 25 loaves yielded a mean mass of 490 g. Does this provide evidence of a reduced population mean? Use the 5% level of significance and state whether the test is one-tailed or two. (SUJB)

12. Illustrate the role of the null hypothesis with reference, if possible, to one of your projects making sure that you state the alternative hypothesis and the level of significance used. Explain how you decided whether to use a one-tail or a two-tail test.

Research workers measured the body lengths, in mm, of 10 specimens of fish spawn of a certain species off the coast of Eastern Scotland and found these lengths to be

12.5 10.2 11.1 9.6 12.1
10.7 11.4 14.7 10.4 9.3

Obtain unbiased estimates for the mean and variance of the lengths of all such fish spawn off Eastern Scotland.

Research shows that, for a very large number of specimens of spawn of this species off the coast of Wales, the mean body length is 10.2 mm. Assuming that the variance of the lengths of spawn off Eastern Scctland is 2.56, perform a significance test at the 5% level to decide whether the mean body length of fish spawn off the coast of Eastern Scotland is larger than that of fish spawn off the coast of Wales. (L)

13. Give an example, from your projects if you wish, of the steps used in carrying out a test of significance.

Climbing rope produced by a manufacturer is known to be such that one-metre lengths have breaking strengths that are normally distributed with mean 170.2 kg and standard deviation 10.5 kg. Find, to 3 decimal places, the probability that
(*a*) a one-metre length of rope chosen at random from those produced by the manufacturer will have a breaking strength of 175 kg to the nearest kg.
(*b*) a random sample of 50 one-metre lengths will have a mean breaking strength of more than 172.4 kg.

A new component material is added to the ropes being produced. The manufacturer believes that this will increase the mean breaking strength without changing the standard deviation. A random sample of 50 one-metre lengths of the new rope is found to have a mean breaking strength of 172.4 kg. Perform a significance test at the 5% level to decide whether this result provides sufficient evidence to confirm the manufacturer's belief that the mean breaking strength is increased. State clearly the null and alternative hypotheses which you are using. (L)

14. (*a*) Write down the mean and the variance of the distribution of the means of all possible samples of size n taken from an infinite population having mean μ and variance σ^2.

Describe the form of this distribution of sample means when
(i) n is large,
(ii) the distribution of the population is normal.

Explain briefly how you acquired empirical evidence for the Central Limit Theorem.

(*b*) The standard deviation of all the till receipts of a supermarket during 1984 was £4.25.
(i) Given that the mean of a random sample of 100 of the till receipts is £18.50, obtain an approximate 95% confidence interval for the mean of all the till receipts during 1984.
(ii) Find the size of sample that should be taken so that the management can be 95% confident that the sample mean will not differ from the true mean by more than 50p.
(iii) The mean of all the till receipts of the supermarket during 1983 was £19.40. Using a 5% significance level, investigate whether the sample in (i) above provides sufficient evidence to conclude that the mean of all the 1984 till receipts is different from that in 1983. (L)

Testing a Mean. Case 2 — Population variance σ^2 unknown, sample size large ($n \geq 30$, say)

Again, we have the sampling distribution of means, where

$$\bar{X} \sim \mathrm{N}\left(\mu, \frac{\sigma^2}{n}\right)$$

Now, since σ^2 is unknown, we use an estimator $\hat{\sigma}^2$ for it.

We have $\hat{\sigma}^2 = \dfrac{nS^2}{n-1}$ where S^2 is the r.v. 'the sample variance'. But since n is large, $\dfrac{n}{n-1} \approx 1$, so $\hat{\sigma}^2 = \dfrac{nS^2}{n-1} \approx S^2$.

We use the test statistic $Z = \dfrac{\bar{X} - \mu}{\hat{\sigma}/\sqrt{n}}$ which is distributed as N(0, 1) under the null hypothesis H_0 that the true population mean is μ with $\hat{\sigma} \approx S$.

Example 9.9 A normal distribution is thought to have a mean of 50. A random sample of 100 gave a mean of 52.6 and a standard deviation of 14.5. Is there evidence that the population mean has increased (**a**) at the 5% level, (**b**) at the 1% level?

Solution 9.9 Let the population mean be μ and the population variance be σ^2.

The sample mean $\bar{x}$ is 52.6 and the sample standard deviation s is 14.5.

H_0: $\mu = 50$ (there is no change in the population mean μ)

H_1: $\mu > 50$ (there is an increase in the population mean μ)

Consider the sampling distribution of means.

Under H_0 $\qquad \bar{X} \sim \mathrm{N}\left(\mu, \dfrac{\sigma^2}{n}\right)$

Now, σ^2 is unknown, so since n is large we use $\hat{\sigma}^2 = \dfrac{ns^2}{n-1} \approx s^2$

i.e. $\bar{X} \sim \mathrm{N}\left(\mu, \dfrac{\hat{\sigma}^2}{n}\right)$, with $\hat{\sigma} \approx 14.5$, $\mu = 50$, $n = 100$

(**a**) We use a one-tailed test at the 5% level and reject H_0 if $z > 1.645$, where

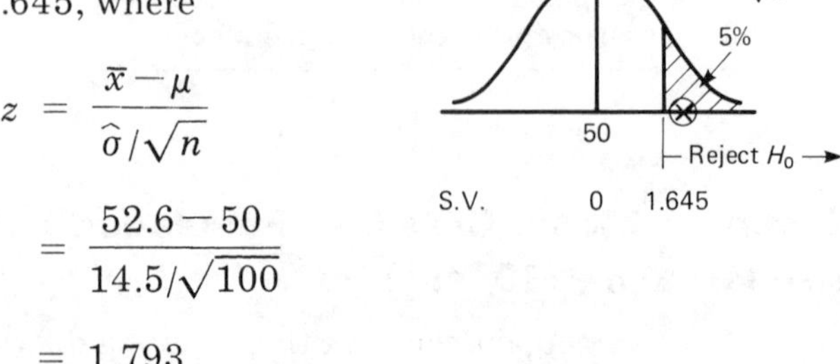

$$z = \frac{\bar{x} - \mu}{\hat{\sigma}/\sqrt{n}} = \frac{52.6 - 50}{14.5/\sqrt{100}} = 1.793$$

Conclusion: As $z > 1.645$, we reject H_0 and conclude that there is evidence, at the 5% level, that the population mean has increased.

(**b**) We use a one-tailed test at the 1% level and reject H_0 if $z > 2.326$, where

$$z = \frac{\bar{x}-\mu}{\hat{\sigma}/\sqrt{n}}$$

$$= 1.793 \quad \text{(as before)}$$

Conclusion: As $z < 2.326$, we do not reject H_0 and conclude that there is not sufficient evidence, at the 1% level, that the population mean has increased.

NOTE: the value of the test statistic z is significant at the 5% level, but not at the 1% level, giving some, but not strong, evidence that the mean has increased.

If we do not approximate $\hat{\sigma}$ and use $\hat{\sigma}^2 = \frac{100}{99}(14.5)^2$, then $\hat{\sigma} = 14.57 \ldots$ and $z = 1.784 \ldots$ giving the same conclusions. There is little difference in the z value when n is large so we usually use $\hat{\sigma} \approx s$.

Example 9.10 A manufacturer claims that the average life of his electric light bulbs is 2000 hours. A random sample of 64 bulbs is tested and the life x in hours recorded. The results obtained are as follows: $\Sigma x = 127\,808$, $\Sigma(x-\bar{x})^2 = 9694.6$. Is there sufficient evidence, at the 2% level, that the manufacturer is over-estimating the length of life of his light bulbs? Assume that the distribution of the length of life of light bulbs is normal.

Solution 9.10 Sample readings:

$$\Sigma x = 127\,808, \quad \Sigma(x-\bar{x})^2 = 9694.6, \quad n = 64$$

$$\bar{x} = \frac{\Sigma x}{n} = \frac{127\,808}{64} = 1997 \qquad \text{and} \qquad s^2 = \frac{\Sigma(x-\bar{x})^2}{n} = \frac{9694.6}{64} = 151.48$$

The sample mean $\bar{x} = 1997$ and the sample variance $s^2 = 151.48$ and standard deviation $s = 12.31$.

Significance test: Let X be the r.v. 'the life, in hours, of a light bulb'. Let the population mean be μ and the population variance be σ^2.

H_0: $\mu = 2000$ (the manufacturer is not over-estimating the length of life)

H_1: $\mu < 2000$ (the manufacturer is over-estimating the length of life)

Consider the sampling distribution of means.

Under H_0, $$\overline{X} \sim \mathrm{N}\left(\mu, \frac{\sigma^2}{n}\right)$$

Now, as σ^2 is unknown, and we have a large sample, we estimate for it, using $\hat{\sigma}^2 = \dfrac{ns^2}{n-1} \approx s^2$

and so $$\overline{X} \sim \mathrm{N}\left(\mu, \frac{\hat{\sigma}^2}{n}\right) \text{ with } \hat{\sigma} \approx 12.31,\ \mu = 2000,\ n = 64$$

We use a one-tailed test at the 2% level, and reject H_0 if $z < -2.054$, where

$$z = \frac{\bar{x}-\mu}{\hat{\sigma}/\sqrt{n}} = \frac{1997-2000}{12.31/\sqrt{64}} = -1.95$$

Conclusion: As $z > -2.054$, we do not reject H_0 and conclude that there is not sufficient evidence, at the 2% level, that the manufacturer is over-estimating the length of life of the light bulbs.

NOTE: If we do not approximate for $\hat{\sigma}$, then $\hat{\sigma}^2 = \dfrac{64}{63}(151.48)$, $\hat{\sigma} = 12.40\ldots$ and $z = -1.93$ giving the same conclusion.

Exercise 9c

In this exercise use $\hat{\sigma} \approx s$ unless otherwise stated.

1. For each of the following, a random sample of size n is taken from a normal distribution with mean μ and variance σ^2. The sample mean is $\bar{x}$. Test the hypotheses stated, at the level of significance indicated.

	n	$\bar{x}$	$\Sigma(x-\bar{x})^2$	Hypotheses	Level
(*a*)	65	100	842.4	H_0: $\mu = 99.2$, H_1: $\mu \neq 99.2$	5%
(*b*)	65	100	842.4	H_0: $\mu = 99.2$, H_1: $\mu > 99.2$	5%
(*c*)	80	85.3	2508.8	H_0: $\mu = 86.2$, H_1: $\mu < 86.2$	10%
(*d*)	100	6.85	36	H_0: $\mu = 7$, H_1: $\mu \neq 7$	1%

2. A sample of 40 observations from a normal distribution gave $\Sigma x = 24$ and $\Sigma x^2 = 596$. Test, at the 5% level, whether the mean of the distribution is zero. Perform a two-tailed test.

3. A random sample of 75 eleven-year-olds performed a particular task. Denoting the time taken by $(15 + y)$ minutes, the results are summarised as follows: $\Sigma y = 90$, $\Sigma(y-\bar{y})^2 = 2025$. Test whether there is sufficient evidence, at the 4% level, to suggest that the mean time to perform the task is greater than 15 minutes.

   Determine a symmetric 96% confidence interval for the mean time, based on the sample observations.

4. Explain, briefly, the roles of a null hypothesis, an alternative hypothesis and a level of significance in a statistical test, referring to your projects where possible.

A shopkeeper complains that the average weight of chocolate bars of a certain type that he is buying from a wholesaler is less than the stated value of 8.50 g. The shopkeeper weighed 100 bars from a large delivery and found that their weights had a mean of 8.36 g and a standard deviation of 0.72 g. Using a 5% significance level, determine whether or not the shopkeeper is justified in his complaint. State clearly the null and alternative hypotheses that you are using, and express your conclusion in words.

Obtain, to 2 decimal places, the limits of a 98% confidence interval for the mean weight of the chocolate bars in the shopkeeper's delivery. (L)

5. An electronic device is advertised as being able to retain information stored in it 'for 70 to 90 hours' after power has been switched off. In experiments carried out to test this claim, the retention time in hours, X, was measured on 250 occasions, and the data obtained is summarised by $\Sigma(x-76)=683$ and $\Sigma(x-76)^2=26\,132$. The population mean and variance of X are denoted by μ and σ^2 respectively.
(*a*) Show that, correct to one decimal place, an unbiased estimate of σ^2 is 97.5.
(*b*) Test the hypothesis that $\mu=80$ against the alternative hypothesis that $\mu<80$, using a 5% significance level.
(*c*) Calculate a symmetric 95% confidence interval for μ. (C)

6. At an early stage in analysing the marks scored by the large number of candidates in an examination paper, the Examination Board takes a random sample of 250 candidates and finds that the marks, x, of these candidates give $\Sigma x=11\,872$ and $\Sigma x^2=646\,193$. Calculate a 90% confidence interval for the population mean, μ, for this paper.

Using the figures obtained in this sample, the null hypothesis $\mu=49.5$ is tested against the alternative hypothesis $\mu<49.5$ at the α% significance level. Determine the set of values of α for which the null hypothesis is rejected in favour of the alternative hypothesis.

It is subsequently found that the population mean and standard deviation for the paper are 45.292 and 18.761 respectively. Find the probability of a random sample of size 250 giving a sample mean at least as high as the one found in the sample above. (C)

Testing a Mean. Case 3 — Population variance σ^2 unknown, sample size small ($n<30$, say)

Consider $X \sim \mathrm{N}(\mu, \sigma^2)$

Since σ^2 is unknown, we use $\hat{\sigma}^2=\dfrac{nS^2}{n-1}$ where S^2 is the r.v. 'the sample variance'.

Since n is small, $\dfrac{n}{n-1}$ is not approximately equal to 1 and we cannot use $\hat{\sigma}^2 \approx S^2$.

Now $\dfrac{\hat{\sigma}}{\sqrt{n}}=\dfrac{\sqrt{n}\,S}{\sqrt{n}\sqrt{n-1}}=\dfrac{S}{\sqrt{n-1}}$

The test statistic $\dfrac{\overline{X}-\mu}{\sigma/\sqrt{n}}$ becomes

$$\frac{\overline{X}-\mu}{\hat{\sigma}/\sqrt{n}}=\frac{\overline{X}-\mu}{S/\sqrt{n-1}}$$

and the distribution of the test statistic changes from $\mathrm{N}(0,1)$ to $t(n-1)$.

We use the test statistic $T = \dfrac{\overline{X}-\mu}{S/\sqrt{n-1}}$ which is distributed as $t(n-1)$ under the null hypothesis H_0 that the true population mean is μ.

Example 9.11 Five readings of the resistance, in ohms, of a piece of wire gave the following results

$$1.51,\ 1.49,\ 1.54,\ 1.52,\ 1.54$$

If the wire were pure silver, its resistance would be 1.50 ohms. If the wire were impure, the resistance would be increased. Test, at the 5% level, the hypothesis that the wire is pure silver.

Solution 9.11 Let X be the r.v. 'the resistance, in ohms, of a piece of wire'. We assume X is normally distributed.

H_0: $\mu = 1.50$ ohms (the population mean μ is 1.50 ohms and the wire is pure silver)

H_1: $\mu > 1.50$ ohms (the population mean μ is greater than 1.50 ohms and the wire is not pure silver)

(One-tailed test)

Under H_0, the test statistic is $T = \dfrac{\overline{X}-\mu}{S/\sqrt{n-1}}$, where $T \sim t(n-1)$.

Now $n-1 = 4$, so $T \sim t(4)$.

We use a one-tailed test at the 5% level.

The critical value for t is found from row $\nu = 4$, column $Q = 5\%$, giving $t = 2.132$, so we reject H_0 if $t_{\text{test}} > 2.132$.

Considering the sample readings, we have

$$\bar{x} = \frac{\Sigma x}{n} = \frac{7.6}{5} = 1.52 \qquad \text{and} \qquad s^2 = \frac{\Sigma(x-\bar{x})^2}{n} = \frac{0.0018}{5} = 0.000\,36$$

$$\underline{\bar{x} = 1.52} \qquad \underline{s = 0.019}$$

Now $$t_{test} = \frac{\bar{x}-\mu}{s/\sqrt{n-1}} = \frac{1.52-1.50}{0.019/\sqrt{4}} = 2.105$$

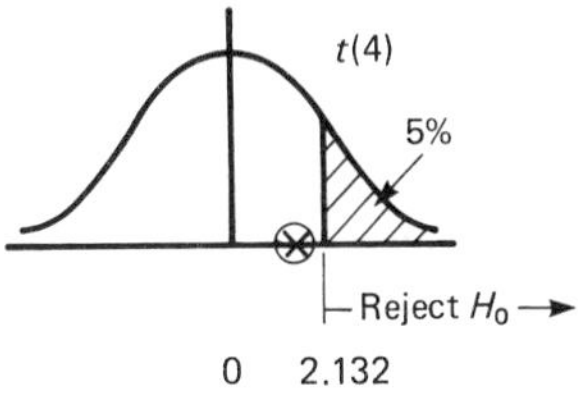

t(4)
5%
Reject H_0 →
0 2.132

Conclusion: As $t_{test} < 2.132$, we do not reject H_0 and conclude that there is not sufficient evidence, at the 5% level, that the wire is impure.

Example 9.12 A random sample of eight observations of a normal variable gave $\bar{x} = 4.65$, $\Sigma(x-\bar{x})^2 = 0.74$. Test, at the 2% level, whether the mean of the distribution is 4.3.

Solution 9.12 Now $$s^2 = \frac{\Sigma(x-\bar{x})^2}{n} = \frac{0.74}{8} = 0.0925$$

$$s = 0.304, \quad \bar{x} = 4.65$$

$$H_0: \quad \mu = 4.3$$

$$H_1: \quad \mu \neq 4.3$$

Under H_0, the test statistic is $T = \dfrac{\bar{X}-\mu}{S/\sqrt{n-1}}$, where $T \sim t(n-1)$.

Now, $n = 8$, therefore $T \sim t(7)$.

We perform a two-tailed test, at the 2% level.

The critical value for t is found from row $\nu = 7$, column $2Q = 2\%$, giving $t = 2.998$.

So we reject H_0 if $|t_{test}| > 2.998$, where

$$t_{test} = \frac{\bar{x}-\mu}{s/\sqrt{n-1}} = \frac{4.65-4.3}{0.304/\sqrt{7}} = 3.05$$

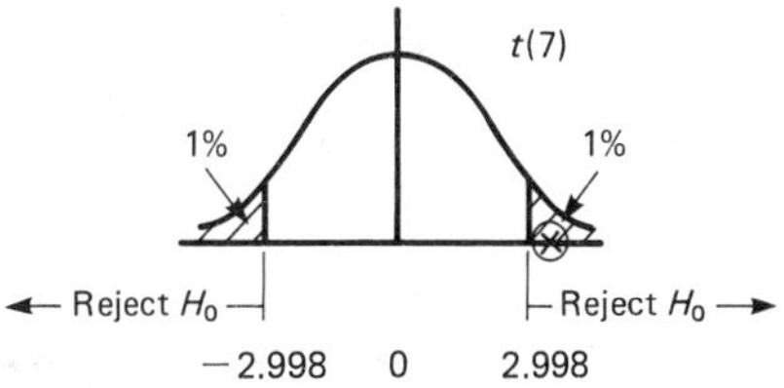

t(7)
1% 1%
← Reject H_0 Reject H_0 →
−2.998 0 2.998

Conclusion: As $|t_{test}| > 2.998$, we reject H_0 at the 2% level and conclude that the mean of the distribution is not 4.3.

Exercise 9d

1. For each of the following, a random sample of size n is taken from a normal distribution with mean μ and variance σ^2. The sample mean is $\bar{x}$. Test the hypotheses stated, at the level of significance indicated.

	n	$\bar{x}$	$\Sigma(x-\bar{x})^2$	Hypotheses	Level
(a)	12	24.9	12.3	H_0: $\mu = 24.1$, H_1: $\mu > 24.1$	$2\frac{1}{2}\%$
(b)	17	35.6	1471.8	H_0: $\mu = 40$, H_1: $\mu \neq 40$	5%
(c)	6	1505.8	50.8	H_0: $\mu = 1503$, H_1: $\mu > 1503$	5%
(d)	10	129.8	97.6	H_0: $\mu = 133.0$, H_1: $\mu < 133.0$	1%

2. A machine is supposed to produce steel pins of length 2 cm. A sample of 10 pins was taken and their lengths measured in cm. The following results were obtained:

   1.98, 1.96, 1.99, 2.00, 2.01, 1.95, 1.97, 1.96, 1.97, 1.99

   Assuming that the lengths are normally distributed, test, at the 1% level of significance, whether the machine is in good working order.

3. An athlete finds that his times for running the 100 m race follow a normal distribution with mean 10.6 seconds. He trains intensively for a week and then runs 100 m on each of 5 consecutive days. His times (measured in seconds) were 10.7, 10.65, 10.75, 10.8, 10.6. Is there evidence, at the 5% level, that the training has improved his times?

4. 'Family' packs of bacon slices are sold in 1.5 kg packs. A sample of 12 packs was selected at random and the masses, measured in kg, noted. The following results were obtained $\Sigma x = 17.81$, $\Sigma x^2 = 26.4357$.

   Assuming that the masses of the packs follow a normal distribution, with variance σ^2, test at the 1% level whether the packs are significantly underweight (*a*) if σ^2 is unknown, (*b*) if $\sigma^2 = 0.0003$.

5. It is thought that a certain Normal population has a mean of 1.6. A sample of 10 gives $\bar{x} = 1.49$ and $s = 0.3$. Does this provide evidence, at the 5% level, that the population mean is less than 1.6?

6. A marmalade manufacturer produces thousands of jars of marmalade each week. The mass of marmalade in a jar is an observation from a normal distribution having mean 455 g and standard deviation 0.8 g. Determine the probability that a randomly chosen jar contains less than 454 g.

   Following a slight adjustment to the filling machine, a random sample of 10 jars is found to contain the following masses (in g) of marmalade:

   454.8, 453.8, 455.0, 454.4, 455.4, 454.4, 454.4, 455.0, 455.0, 453.6

   (*a*) Assuming that the variance of the distribution is unaltered by the adjustment, test, at the 5% significance level, the hypothesis that there has been no change in the mean of the distribution.
   (*b*) Assuming that the variance of the distribution may have altered, obtain an unbiased estimate of the new variance and, using this estimate, test, at the 5% significance level, the hypothesis that there has been no change in the mean of the distribution. (C Further Maths)

7. A random sample of 8 women yielded the following cholesterol levels:

   3.1 2.8 1.5 1.7 2.4 1.9 3.3 1.6

   It is required to test whether the sample could be drawn from a population whose mean cholesterol level is 3.1.
   (*a*) Assuming that the sample is drawn from a normal distribution give two reasons why a t-test is appropriate.
   (*b*) Perform the test, stating your null and alternative hypotheses. What conclusions do you draw?
   (*c*) Calculate a 90% symmetric confidence interval for the mean cholesterol level in the population. (SUJB)

TEST 3 — TESTING THE DIFFERENCE BETWEEN MEANS

Consider two unpaired, independent samples of sizes n_1 and n_2 such that

$$X_1 \sim \mathrm{N}(\mu_1, \sigma_1^2) \quad \text{and} \quad X_2 \sim \mathrm{N}(\mu_2, \sigma_2^2)$$

Then
$$\bar{X}_1 - \bar{X}_2 \sim \mathrm{N}\left(\mu_1 - \mu_2, \frac{\sigma_1^2}{n_1} + \frac{\sigma_2^2}{n_2}\right)$$

This distribution is known as the **sampling distribution of the difference between means.**

The following may be used to test whether there is a significant difference between means.

(1) If σ_1^2, σ_2^2 are *known*

we use the test statistic

$$Z = \frac{\bar{X}_1 - \bar{X}_2 - (\mu_1 - \mu_2)}{\sqrt{\dfrac{\sigma_1^2}{n_1} + \dfrac{\sigma_2^2}{n_2}}}$$

which is distributed as N(0, 1).

(2) If there is a *known common population variance* such that $\sigma_1^2 = \sigma_2^2 = \sigma^2$, then

$$\bar{X}_1 - \bar{X}_2 \sim \mathrm{N}\left(\mu_1 - \mu_2, \sigma^2\left(\frac{1}{n_1} + \frac{1}{n_2}\right)\right)$$

we use the test statistic

$$Z = \frac{\bar{X}_1 - \bar{X}_2 - (\mu_1 - \mu_2)}{\sigma\sqrt{\dfrac{1}{n_1} + \dfrac{1}{n_2}}} \quad \text{where} \quad Z \sim \mathrm{N}(0, 1)$$

(3) If there is an *unknown common population variance* σ^2 then we use an estimate $\hat{\sigma}^2$ for it, where

$$\hat{\sigma}^2 = \frac{n_1 s_1^2 + n_2 s_2^2}{n_1 + n_2 - 2} \quad \text{where} \quad s_1^2, s_2^2 \text{ are the sample variances}$$

For small samples we use the test statistic

$$T = \frac{\bar{X}_1 - \bar{X}_2 - (\mu_1 - \mu_2)}{\hat{\sigma}\sqrt{\dfrac{1}{n_1} + \dfrac{1}{n_2}}} \quad \text{where} \quad T \sim t(n_1 + n_2 - 2)$$

For large samples we use the test statistic

$$Z = \frac{\bar{X}_1 - \bar{X}_2 - (\mu_1 - \mu_2)}{\hat{\sigma}\sqrt{\dfrac{1}{n_1} + \dfrac{1}{n_2}}} \quad \text{where} \quad Z \sim \mathrm{N}(0, 1)$$

Example 9.13 A random sample of size 100 is taken from a normal population with variance $\sigma_1^2 = 40$. The sample mean $\bar{x}_1$ is 38.3. Another random sample, of size 80, is taken from a normal population with variance $\sigma_2^2 = 30$. The sample mean $\bar{x}_2$ is 40.1. Test, at the 5% level, whether there is a significant difference in the population means μ_1 and μ_2.

Solution 9.13 Sample 1: $n_1 = 100, \quad \bar{x}_1 = 38.3, \quad \sigma_1^2 = 40,$

population mean $= \mu_1$

Sample 2: $n_2 = 80, \quad \bar{x}_2 = 40.1, \quad \sigma_2^2 = 30,$

population mean $= \mu_2$

H_0: $\mu_1 = \mu_2$ (there is no difference between the means)

H_1: $\mu_1 \neq \mu_2$ (there is a difference)

We consider the sampling distribution of the difference between means where under H_0,

$$\bar{X}_1 - \bar{X}_2 \sim N\left(\mu_1 - \mu_2, \frac{\sigma_1^2}{n_1} + \frac{\sigma_2^2}{n_2}\right)$$

We use a two-tailed test, at the 5% level and reject H_0 if $|z| > 1.96$, where

$$z = \frac{\bar{x}_1 - \bar{x}_2 - (\mu_1 - \mu_2)}{\sqrt{\frac{\sigma_1^2}{n_1} + \frac{\sigma_2^2}{n_2}}}$$

$$= \frac{38.3 - 40.1 - (0)}{\sqrt{\frac{40}{100} + \frac{30}{80}}}$$

$$= -2.04$$

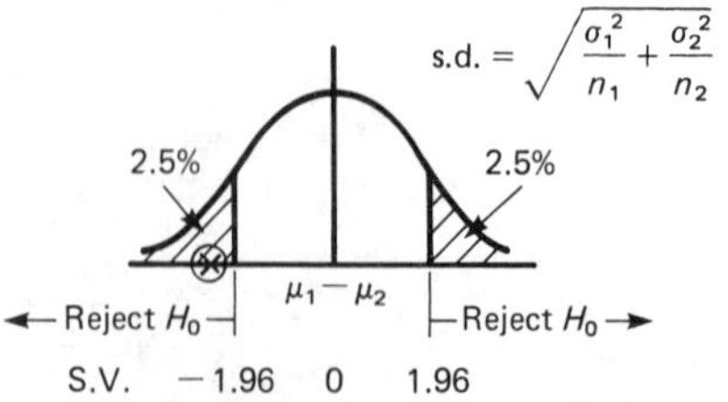

Conclusion: As $|z| > 1.96$, we reject H_0 and conclude that there is evidence, at the 5% level, of a difference in population means.

Example 9.14 The same test was given to a group of 100 scouts and to a group of 144 guides. The mean score for the scouts was 27.53 and the mean score for the guides was 26.81. Assuming a common population standard deviation of 3.48, test, using a 5% level of significance, whether the scouts' performance in the test was better than that of the guides. Assume that the scores are normally distributed.

Solution 9.14 Let X be the r.v. 'a scout's score'.

Scouts: $\bar{x} = 27.53$, $n_1 = 100$, population mean $= \mu_1$

Let Y be the r.v. 'a guide's score'.

Guides: $\bar{y} = 26.81$, $n_2 = 144$, population mean $= \mu_2$

Common population standard deviation $\sigma = 3.48$.

H_0: $\mu_1 = \mu_2$ (there is no difference in the performances)

H_1: $\mu_1 > \mu_2$ (the performance of the scouts was better)

Consider the sampling distribution of the difference between means, where under H_0,

$$\bar{X} - \bar{Y} \sim \mathrm{N}\left(\mu_1 - \mu_2, \sigma^2\left(\frac{1}{n_1} + \frac{1}{n_2}\right)\right)$$

We use a one-tailed test, at the 5% level, and reject H_0 if $z > 1.645$, where

$$z = \frac{(\bar{x} - \bar{y}) - (\mu_1 - \mu_2)}{\sigma\sqrt{\dfrac{1}{n_1} + \dfrac{1}{n_2}}}$$

$$= \frac{27.53 - 26.81 - (0)}{3.48\sqrt{\dfrac{1}{100} + \dfrac{1}{144}}}$$

$$= 1.589$$

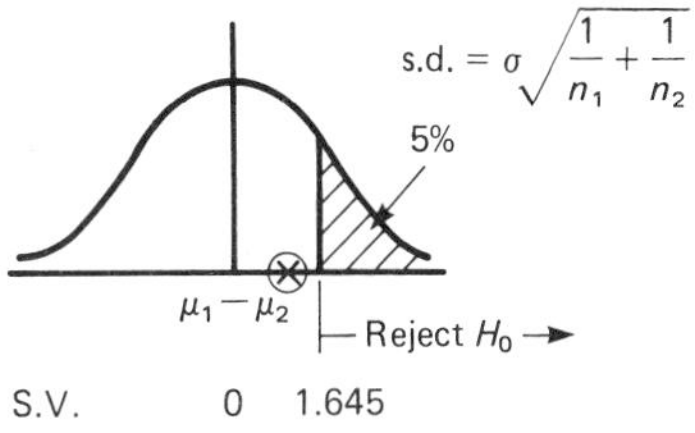

Conclusion: As $z < 1.645$, we do not reject H_0 and conclude that there is not sufficient evidence, at the 5% level, to show that the performance of the scouts in the test was better than that of the guides.

Example 9.15 A certain political group maintains that girls reach a higher standard in single-sex classes than in mixed classes. To test this hypothesis 140 girls of similar ability are split into two groups, with 68 attending classes containing only girls and 72 attending classes with boys. All the classes follow the same syllabus and after a specified time the girls are given a test. The test results are summarised thus:

Girls in the mixed classes: $\Sigma x = 7920$, $\Sigma x^2 = 879\,912$

Girls in single-sex classes: $\Sigma y = 7820$, $\Sigma y^2 = 904\,808$

Treating both samples as large samples from normal distributions having the same variance, obtain a 2-sample pooled estimate of the common population variance. Test whether the results provide significant evidence, at the 1% level, that girls reach a higher standard in single-sex classes.

Solution 9.15 *Mixed classes*:

Let X be the r.v. 'score of a girl in a mixed class'.

$\Sigma x = 7920, \quad \Sigma x^2 = 879\,912, \quad n_1 = 72,$
population mean μ_1

Therefore $$\bar{x} = \frac{\Sigma x}{n_1} = \frac{7920}{72} = 110 \quad \text{and} \quad s_1^2 = \frac{\Sigma x^2}{n_1} - \bar{x}^2 = \frac{879\,912}{72} - 110^2 = 121$$

Single-sex classes:

Let Y be the r.v. 'the score of a girl in a single-sex class'.

$\Sigma y = 7820, \quad \Sigma y^2 = 904\,808, \quad n_2 = 68,$
population mean μ_2

Therefore $$\bar{y} = \frac{\Sigma y}{n_2} = \frac{7820}{68} = 115 \quad \text{and} \quad s_2^2 = \frac{\Sigma y^2}{n_2} - \bar{y}^2 = \frac{904\,808}{68} - 115^2 = 81$$

The pooled 2-sample estimate of the common population variance is given by $\hat{\sigma}^2$, where

$$\hat{\sigma}^2 = \frac{n_1 s_1^2 + n_2 s_2^2}{n_1 + n_2 - 2}$$

So $$\hat{\sigma}^2 = \frac{72(121) + 68(81)}{72 + 68 - 2} = 103.04$$

Therefore $\hat{\sigma}^2 = 103.04$ and $\hat{\sigma} = 10.15$ (2 d.p.).

Significance test:

H_0: $\mu_1 = \mu_2$ (there is no difference in the test scores)

H_1: $\mu_1 < \mu_2$ (the girls in the single-sex classes reach a higher standard)

We consider the sampling distribution of the difference between means, where under H_0

$$\bar{X} - \bar{Y} \sim \mathrm{N}\left(\mu_1 - \mu_2, \hat{\sigma}^2\left(\frac{1}{n_1} + \frac{1}{n_2}\right)\right)$$

We use a one-tailed test at the 1% level, and reject H_0 if $z < -2.326$,

where
$$z = \frac{\bar{x}-\bar{y}-(\mu_1-\mu_2)}{\hat{\sigma}\sqrt{\frac{1}{n_1}+\frac{1}{n_2}}}$$
$$= \frac{110-115-(0)}{10.15\sqrt{\frac{1}{72}+\frac{1}{68}}}$$
$$= -2.913$$

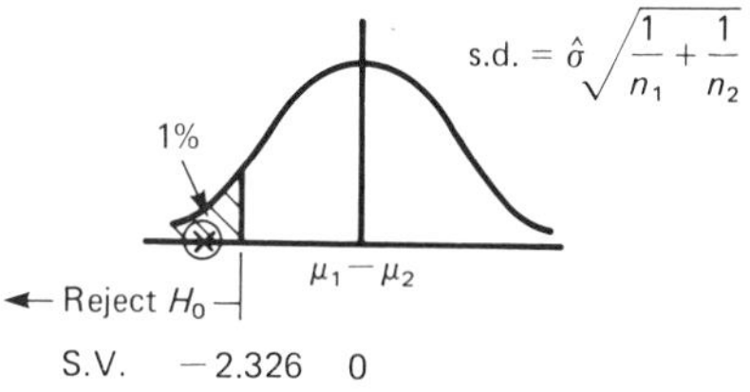

s.d. $= \hat{\sigma}\sqrt{\frac{1}{n_1}+\frac{1}{n_2}}$

1%

$\mu_1-\mu_2$

← Reject H_0

S.V. -2.326 0

Conclusion: As $z < -2.326$ we reject H_0 and conclude that there is evidence at the 1% level to suggest that girls in single-sex classes reach a higher standard.

Example 9.16 Two statistics teachers, Mr Chalk and Mr Talk argue about their abilities at golf. Mr Chalk claims that with a number 7 iron he can hit the ball, on average, at least 10 m further than Mr Talk. Denoting the distance Mr Chalk hits the ball by $(100+c)$ metres, the following results were obtained: $n_1 = 40$, $\Sigma c = 80$, $\Sigma(c-\bar{c})^2 = 1132$.

Denoting the distance Mr Talk hits the ball by $(100+t)$ metres, the following results were obtained: $n_2 = 35$, $\Sigma t = -175$,
$$\Sigma(t-\bar{t})^2 = 1197$$

If the distances for both teachers are normally distributed with a common variance, show that an unbiased estimate of this common variance is 31.90.

Test whether there is any evidence, at the 1% level, to support Mr Chalk's claim.

Solution 9.16 Mr Chalk:

Let X_1 be the r.v. 'the distance, in m, for Mr Chalk'.

$$\text{Distance} = (100+c) \text{ metres}$$
$$\Sigma c = 80, \quad \Sigma(c-\bar{c})^2 = 1132, \quad n_1 = 40,$$
sample mean $\bar{x}_1$, population mean μ_1

Now $$\bar{x}_1 = 100+\bar{c} = 100+\frac{80}{40} = 102\text{ m}$$

Mr Talk:

Let X_2 be the r.v. 'the distance, in m, for Mr Talk'.

$$\text{Distance} = (100+t) \text{ metres}$$
$$\Sigma t = -175, \quad \Sigma(t-\bar{t})^2 = 1197, \quad n_2 = 35,$$
sample mean $\bar{x}_2$, population mean μ_2

Now $$\bar{x}_2 = 100+\bar{t} = 100+\frac{-175}{35} = 95\text{ m}$$

The unbiased estimate of the population variance is $\hat{\sigma}^2$ where

$$\hat{\sigma}^2 = \frac{\Sigma(c-\bar{c})^2 + \Sigma(t-\bar{t})^2}{n_1+n_2-2}$$

$$= \frac{1132+1197}{40+35-2}$$

$$= \frac{2329}{73}$$

$$= 31.90$$

The unbiased estimate of the common population variance is 31.90.

Mr Chalk claims that he can hit the ball *at least* 10 m further than Mr Talk. Therefore the alternative hypothesis is that Mr Chalk hits the ball *less than* 10 m, and a one-tailed test is performed.

H_0: $\mu_1 - \mu_2 = 10$ (Mr Chalk hits the ball 10 m further than Mr Talk)

H_1: $\mu_1 - \mu_2 < 10$ (Mr Chalk hits the ball less than 10 m further than Mr Talk)

We consider the sampling distribution of the difference between means where under H_0,

$$\bar{X}_1 - \bar{X}_2 \sim \mathrm{N}\left(\mu_1-\mu_2, \hat{\sigma}^2\left(\frac{1}{n_1}+\frac{1}{n_2}\right)\right)$$

We perform a one-tailed test, at the 1% level, and reject H_0 if $z < -2.326$ where

$$z = \frac{\bar{x}_1 - \bar{x}_2 - (\mu_1-\mu_2)}{\hat{\sigma}\sqrt{\frac{1}{n_1}+\frac{1}{n_2}}}$$

$$= \frac{102-95-(10)}{5.648\sqrt{\frac{1}{40}+\frac{1}{35}}}$$

$$= -2.29$$

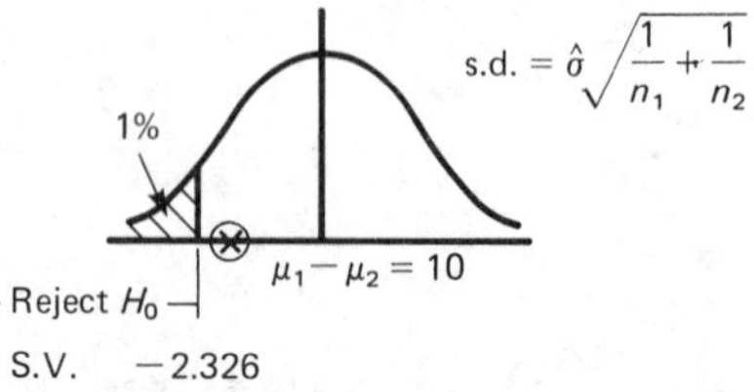

Conclusion: As $z > -2.326$, we do not reject H_0 and conclude that there is not sufficient evidence, at the 1% level, to reject Mr Chalk's claim that he hits the ball, on average, at least 10 m further than Mr Talk.

Example 9.17 The heights (measured to the nearest cm) of a random sample of six policemen from a certain force in Wales were found to be

$$176,\ 180,\ 179,\ 181,\ 183,\ 179$$

The heights (measured to the nearest cm) of a random sample of eleven policemen from a certain force in Scotland gave the following data:

$$\Sigma y = 1991, \quad \Sigma(y-\bar{y})^2 = 54$$

Test at the 5% level, the hypothesis that Welsh policemen are shorter than Scottish policemen. Assume that the heights of policemen in both forces are normally distributed and have a common population variance.

Solution 9.17 *Welsh policemen*:

Let X be the r.v. 'the height, in cm, of a Welsh policeman'.

$$\Sigma x = 1078, \quad \Sigma x^2 = 193\,708, \quad n_1 = 6.$$

population mean $= \mu_1$

$$\text{So } \bar{x} = \frac{\Sigma x}{n_1} = \frac{1078}{6} = 179.67\text{ cm} \quad (2\text{ d.p.})$$

$$s_1^2 = \frac{\Sigma x^2}{n_1} - \left(\frac{\Sigma x}{n_1}\right)^2 = \frac{193\,708}{6} - \left(\frac{1078}{6}\right)^2 = 4.556 \quad (3\text{ d.p.})$$

Scottish policemen:

Let Y be the r.v. 'the height, in cm, of a Scottish policeman'.

$$\Sigma y = 1991, \quad \Sigma(y-\bar{y})^2 = 54,$$

$$n_2 = 11, \quad \text{population mean} = \mu_2$$

So

$$\bar{y} = \frac{\Sigma y}{n_2} = \frac{1991}{11} = 181\text{ cm}$$

$$n_2 s_2^2 = \Sigma(y-\bar{y})^2 = 54$$

Let the unbiased estimate of the common population variance be $\hat{\sigma}^2$.

We have

$$\hat{\sigma}^2 = \frac{n_1 s_1^2 + n_2 s_2^2}{n_1 + n_2 - 2} = \frac{6(4.556) + 54}{6 + 11 - 2} = 5.422 \qquad \text{(3 d.p.)}$$

so $\hat{\sigma} = 2.329\text{ cm}$ (3 d.p.)

Significance test:

H_0: $\mu_1 = \mu_2$ (there is no difference in the mean heights)

H_1: $\mu_1 < \mu_2$ (the Welsh policemen are shorter)

The test statistic is $$T = \frac{\bar{X} - \bar{Y} - (\mu_1 - \mu_2)}{\hat{\sigma}\sqrt{\dfrac{1}{n_1} + \dfrac{1}{n_2}}}$$

where under H_0 $T \sim t(n_1 + n_2 - 2)$

Now $n_1 + n_2 - 2 = 15$, so $T \sim t(15)$.

We use a one-tailed test at the 5% level.

The critical value for t is found from row $\nu = 15$, column $Q = 5\%$, giving $t = -1.753$.

So we reject H_0 if $t_{test} < -1.753$, where

$$t_{test} = \frac{\bar{x} - \bar{y} - (\mu_1 - \mu_2)}{\hat{\sigma}\sqrt{\dfrac{1}{n_1} + \dfrac{1}{n_2}}}$$

$$= \frac{(179.67 - 181) - (0)}{2.329\sqrt{\dfrac{1}{6} + \dfrac{1}{11}}}$$

$$= -1.13 \qquad \text{(2 d.p.)}$$

Conclusion: As $t_{test} > -1.753$, we do not reject H_0 and conclude that there is not sufficient evidence, at the 5% level, to suggest that Welsh policemen are shorter than Scottish policemen.

Exercise 9e

1. For each of the following sets of data, perform a test to decide whether there is a significant difference between the means, μ_1 and μ_2, of the normal populations from which the samples are drawn.

	n_1	Σx	σ_1^2	n_2	Σy	σ_2^2	Hypotheses	Level
(a)	100	4250	30	80	3544	35	H_0: $\mu_1 = \mu_2$ H_1: $\mu_1 \neq \mu_2$	5%
(b)	20	95	2.3	25	135	2.5	H_0: $\mu_1 = \mu_2$ H_1: $\mu_1 < \mu_2$	2%
(c)	50	1545	6.5	50	1480	7.1	H_0: $\mu_1 = \mu_2$ H_1: $\mu_1 > \mu_2$	1%
	n_1	Σx	n_2	Σy	Common population standard deviation (σ)		Hypotheses	Level
(d)	50	2480	40	1908	4.5		H_0: $\mu_1 = \mu_2$ H_1: $\mu_1 \neq \mu_2$	2%
(e)	100	12 730	100	12 410	10.9		H_0: $\mu_1 = \mu_2$ H_1: $\mu_1 > \mu_2$	5%
(f)	30	192	45	315	1.25		H_0: $\mu_1 = \mu_2$ H_1: $\mu_1 < \mu_2$	1%
(g)	200	18 470	300	27 663	0.86		H_0: $\mu_1 = \mu_2$ H_1: $\mu_1 \neq \mu_2$	10%
	n_1	Σx	$\Sigma(x-\bar{x})^2$	n_2	Σy	$\Sigma(y-\bar{y})^2$	Hypotheses	Level
(h)	40	2128	810	50	2580	772	H_0: $\mu_1 = \mu_2$ H_1: $\mu_1 > \mu_2$	5%
(i)	80	6824	2508	100	8740	3969	H_0: $\mu_1 = \mu_2$ H_1: $\mu_1 \neq \mu_2$	2%
(j)	65	5369	8886	80	4672	5026	H_0: $\mu_1 - \mu_2 = 20$ H_1: $\mu_1 - \mu_2 > 20$	1%
(k)	6	171	83	7	164.5	112	H_0: $\mu_1 = \mu_2$ H_1: $\mu_1 > \mu_2$	5%
(l)	5	678.5	562.3	7	971.6	308.6	H_0: $\mu_1 = \mu_2$ H_1: $\mu_1 \neq \mu_2$	5%
(m)	8	238.4	296	10	206	145	H_0: $\mu_1 - \mu_2 = 4$ H_1: $\mu_1 - \mu_2 > 4$	1%
(n)	12	116.16	45.1	18	156.96	72	H_0: $\mu_1 = \mu_2$ H_1: $\mu_1 \neq \mu_2$	10%

2. An investigation was carried out to assess the effects of adding certain vitamins to the diet. 64 two-week old rats were given a vitamin supplement in their diet for a period of one month, after which time their masses were noted. A control group of 36 rats of the same age were fed on an ordinary diet and their masses were also noted after one month. The results are summarised in the table.

	Number in sample	Mean mass	Standard deviation
Rats with vitamin supplement	64	89.6 g	12.96 g
Rats without vitamin supplement	36	83.5 g	11.41 g

Treating the samples as large samples from normal distributions with the same variance, test whether the results provide evidence, at the 5% level, that rats given the vitamin supplement have a greater mass, at age six weeks, than those not given the vitamin supplement.

3. (*a*) In one county in England, a random sample of 225 12-year old boys and 250 12-year-old girls was given an arithmetic test. The average mark for the boys was

57 with a standard deviation of 12, whilst the average for the girls was 60 with a standard deviation of 15.

Assuming that the distributions are normal, does this provide evidence at the 2% level that 12-year-old girls are superior to 12-year-old boys at arithmetic?

(*b*) An IQ test which had been standardised giving a mean of 100 and a standard deviation of 12 was given to a random sample of 50 children in one area. The average mark obtained was 105.

Does this provide evidence, at the 5% level, that children from this area are generally more intelligent? (SUJB)

4. The mean height of 50 male students of a college who took an active part in athletic activities was 178 cm with a standard deviation of 5 cm, while 50 male students who showed no interest in such activities had a mean height of 176 cm with a standard deviation of 7 cm. Test the hypothesis that male students who take an active part in athletic activities have the same mean height as the other male students.

If both samples had been of size n, instead of 50, find the least value of n which would ensure that the observed difference of 2 cm in the mean height would be significant at the 1% level. (Assume that the samples continue to have the same means and standard deviations.) (C)

5. Mr Mean notes the time, in minutes, that it takes him to drive to work in the mornings. The results are $n_1 = 8$, $\Sigma x_1 = 120$, $\Sigma x_1^2 = 1827$.
(*a*) Show that the value of the unbiased estimate of the population variance obtained from this sample is 10.29.
(*b*) Assuming that the times are normally distributed, find a 98% confidence interval for the average journey time, and explain what it means.

For his return journey in the rush hour, Mr Mean notes that for $n_2 = 10$, $\Sigma x_2 = 230$, $\Sigma x_2^2 = 5436$.

He maintains that, on average, it takes him at least 10 minutes longer to drive home.
(*c*) Using the results from the two samples, find an unbiased estimate of the common population variance.
(*d*) Assuming that the times of all journeys are normally distributed, use the two-sample t-test at the 5% level to test Mr Mean's claim.

6. Random samples of fourth-year pupils at two schools are given the same mathematics test. The results are summarised thus:

School A: $n_1 = 20$, $\bar{x} = 43$, $\Sigma(x-\bar{x})^2 = 1296$

School B: $n_2 = 17$, $\bar{y} = 36$, $\Sigma(y-\bar{y})^2 = 1388$

Assuming that the distributions of marks are normal with a common population variance, and treating the samples as large, test at the 2% level whether there is a significant difference in the mathematical ability of the fourth-year pupils at the two schools.

7. A random sample of 27 individuals from the population of young men aged 18 and of high intelligence have foot lengths (in cm, to the nearest cm) as summarised below.

Foot length (in cm)	24	25	26	27	28	29	30
Number with this foot length	1	2	3	9	6	5	1

Obtain the sample mean and show that the unbiased estimate of the population variance, based on this sample, is 2.00. Obtain a 96% confidence interval for the mean foot length of this type of person.

A random sample of 48 individuals from the population of young men aged 18 and of moderate intelligence have foot lengths summarised by $\bar{x} = 26.6$, $\Sigma(x_i-\bar{x})^2 = 123.20$. A complex genetic theory suggests that persons of high intelligence have a greater foot length than do those of moderate intelligence. The two samples described above may be assumed to have been drawn at random from independent normal distributions having a common variance. Obtain an unbiased two-sample estimate of this common variance. Treating the samples as large samples, test this genetic theory, using a significance test at the 1% significance level and stating clearly the hypotheses under comparison. (C)

8. If the mean of n numbers, $x_1, x_2, \ldots, x_n$ is $\bar{x}$, prove that their variance is

$$\frac{1}{n}\sum_{i=1}^{n}(x_i-m)^2-(\bar{x}-m)^2$$

for any constant m.

A garage wished to estimate the average time spent in servicing and repairing cars during a certain month. A sample of 100 cars yielded:

$\Sigma x_i = 325.5$, $\Sigma x_i^2 = 1076$, x_i being the time spent, in hours, on the ith car.

Assuming that the measurements are from a normal population, give 95% confidence limits for the population mean.

Could the restriction of the population being normal be dropped? A sample of 25 cars from the following month yielded a mean repair time of 3.55 hours. Is this evidence of an increase in population mean over the previous month? (SUJB)

9. Mr Brown and Mr Green work at the same office and live next door to each other. Each day they leave for work together but travel by different routes. Mr Brown maintains that his route is quicker, on average, by at least 4 minutes. Both men time their journeys in minutes over a period of 10 weeks. The results obtained were:

Mr Brown: $n_1 = 50$, $\bar{x}_1 = 21$, $s_1^2 = 10.24$

Mr Green: $n_2 = 50$, $\bar{x}_2 = 24$, $s_2^2 = 7.84$

Assuming that the times are normally distributed and that they have a common population variance, test at the 5% level whether Mr Brown's claim can be accepted.

10. Hischi and Taschi are two makes of video tapes. They are both advertised as having a recording time of 3 hours. A sample of 49 Hischi tapes was tested and, denoting the actual recording time by $(180+h)$ minutes, the following results were obtained:

$\Sigma h = -147$, $\Sigma(h-\bar{h})^2 = 12\,720$

A sample of 81 Taschi tapes was also tested. Denoting the actual recording time by $(180+t)$ minutes, the results obtained were

$\Sigma t = 324$, $\Sigma(t-\bar{t})^2 = 33\,488$

If the recording times for the two makes are normally distributed and have a common variance, show that the unbiased estimate of this common variance is 361. Test whether there is significant evidence, at the 5% level, of a difference in the mean recording times. Is the difference significant at the 4% level?

11. The lengths (in millimetres) of nine screws selected at random from a large consignment are found to be 7.99, 8.01, 8.00, 8.02, 8.03, 7.99. 8.00, 8.01, 8.01. Calculate unbiased estimates of the population mean and variance. Assuming a normal distribution with variance 0.0001, test, at the 5% level, the hypothesis that the population mean is 8.00 against the alternative hypothesis that the population mean is not 8.00.

From a second large consignment, sixteen screws are selected at random and their mean length (in millimetres) is found to be 7.992. Assuming a normal distribution with variance 0.0001, test, at the 5% level, the hypothesis that this population has the same mean as the first population, against the alternative hypothesis that this population has a smaller mean than the first population. (C)

12. A random sample of size n_1 is taken from a population P_1 whose mean is μ_1 and variance σ_1^2 and a random sample of size n_2 is taken from population P_2 with mean μ_2 and variance σ_2^2. Under what circumstances is it valid to test the hypothesis $\mu_1 - \mu_2 = 0$ using a two-sample t-test?
A machine fills bags of sugar and a random sample of 20 bags selected from a week's production yielded a mean weight of 499.1 g with standard deviation 0.63 g. A week later a sample of 25 bags yielded a mean weight of 500.2 g with standard deviation 0.48 g. Assuming that your stated conditions are satisfied perform a test to determine whether the mean has increased significantly during the second week. Test whether the mean during the second week could be 500 g. (Use a 5% significance level for both tests.) (SUJB)

13. A large number of tomato plants are grown under controlled conditions. Half of the plants, chosen at random, are treated with a new fertilizer, and the other half of the plants are treated with a standard fertilizer. Random samples of 100 plants are selected from each half, and records are kept of the total crop mass of each plant. For those treated with the new fertilizer, the crop masses (in suitable units) are summarized by the figures $\Sigma x = 1030.0$, $\Sigma x^2 = 11\,045.59$. Obtain an unbiased estimate of the population variance, and, treating the sample as a large sample from a normal distribution, obtain a symmetric 96% confidence interval for the mean crop mass.
The corresponding figures for those plants treated with the standard fertilizer are $\Sigma y = 990.0$, $\Sigma y^2 = 10\,079.19$. Treating the sample as a large sample from a normal distribution, and assuming that

the population variances of both distributions are equal, obtain a two-sample pooled estimate of the common population variance.

Assuming that it is impossible for the new fertilizer to be less efficacious than the old fertilizer and assuming that both distributions are normal, test whether the results provide significant evidence (at the 3% level) that the new fertilizer is associated with a greater mean crop mass, stating clearly your null and alternative hypotheses. (C)

14. From a large population of students, 120 males and 160 females are chosen at random. Their heights x in metres are summarised in the table below. The males and females may be treated as random samples from two independent populations.

	Number	Σx	Σx^2
Males	120	198	327
Females	160	248	385

(a) Find the sample means and variances.
(b) Assuming that in both populations the heights are normally distributed with these means and variances, find the probability that a randomly-chosen female will be taller than a randomly-chosen male.
(c) Assuming only that height is distributed normally in both populations, test the hypothesis that the mean height of the population of male students exceeds the mean height of the population of female students by less than 0.08 metres. (O & C)

15. A large group of sunflowers is growing in the shady side of a garden. A random sample of 36 of these sunflowers is measured. The sample mean height is found to be 2.86 m, and the sample standard deviation is found to be 0.60 m. Treating the sample as a large sample and assuming the heights to be normally distributed, give a symmetric 99% confidence interval for the mean height of the sunflowers in the shady side of the garden.

A second group of sunflowers is growing in the sunny side of the garden. A random sample of 26 of these sunflowers is measured. The sample mean height is found to be 3.29 m and the sample standard deviation is found to be 0.90 m. Treating the samples as large samples from normal distributions having the same variance but possibly different means, obtain a pooled estimate of the variance and test whether the results provide significant evidence (at the 5% level) that the sunny-side sunflowers grow taller, on average, than the shady-side sunflowers. (C)

TEST 4 — TESTING A PROPORTION

We may wish to test whether a random sample of size n, with proportion of 'successes' p_s could have been drawn from a population with proportion of 'successes' p.

The sampling distribution of proportions gives

$P_s \sim N\left(p, \frac{pq}{n}\right)$ where $q = 1-p$ and n is large

(see p. 407)

s.d. $= \sqrt{\frac{pq}{n}}$

p

The test statistic used is

$$Z = \frac{P_s - p}{\sqrt{\frac{pq}{n}}}$$

which is distributed as N(0, 1) under the null hypothesis H_0 that the proportion of 'successes' in the population is p.

Example 9.18 The manufacturer of 'Chummy Morsels' claims that 8 out of 10 dogs choose his product rather than that produced by a rival firm. In a random sample of 200 dogs, 152 chose 'Chummy Morsels', and the rest chose the rival brand. Comment on the manufacturer's claim.

Solution 9.18 From the sample: $p_s = \dfrac{152}{200} = 0.76, \quad n = 200$

Let p be the population proportion of dogs who prefer 'Chummy Morsels'.

H_0: $p = 0.8$ (80% of dogs prefer 'Chummy Morsels' and the manufacturer's claim is correct)

H_1: $p < 0.8$ (less than 80% of dogs prefer 'Chummy Morsels' and the manufacturer's claim is not correct)

Consider the sampling distribution of proportions, where under H_0,

$$P_s \sim \mathrm{N}\left(p, \frac{pq}{n}\right) \quad \text{where} \quad q = 1-p$$

Use a one-tailed test at the 5% level. We will reject H_0 if $z < -1.645$ where

$$z = \frac{p_s - p}{\sqrt{\dfrac{pq}{n}}}$$

$$= \frac{0.76-0.80}{\sqrt{\dfrac{(0.8)(0.2)}{200}}}$$

$$= -1.414$$

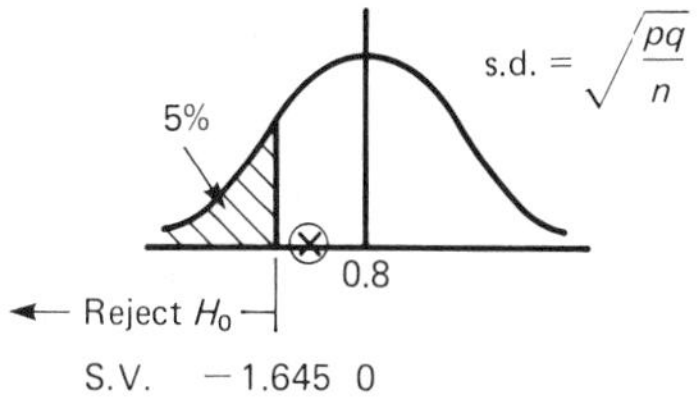

Conclusion: As $z > -1.645$, we do not reject H_0 and conclude that there is not sufficient evidence, at the 5% level, to refute the manufacturer's claim.

NOTE: as we are dealing with proportions, we should use the *continuity correction* of $\pm\dfrac{1}{2n}$. However, *if n is large*, this makes very little difference to the calculation of z.

With the continuity correction,

$$z = \frac{\left(0.76 - \dfrac{1}{400}\right) - 0.8}{\sqrt{\dfrac{(0.8)(0.2)}{200}}}$$

$$= -1.503$$

and the conclusion is the same as before.

NOTE: an alternative approach to this type of problem was introduced in Test 1, on p. 463. The method is shown again below.

Alternative Solution 9.18 Let X be the r.v. 'the number of dogs who prefer Chummy Morsels'. Then $X \sim \text{Bin}(n, p)$.

H_0: $p = 0.8$ (80% prefer Chummy Morsels)

H_1: $p < 0.8$ (less than 80% prefer Chummy Morsels)

Under H_0

$$X \sim \text{Bin}(n, p) \quad \text{with} \quad n = 200 \quad \text{and} \quad p = 0.8$$

Now, using the normal approximation to the binomial distribution,

$$X \sim \text{N}(np, npq) \quad \text{where} \quad q = 1-p$$

$$X \sim \text{N}(np, npq) \quad \text{with} \quad np = (200)(0.8) = 160$$

$$npq = (200)(0.8)(0.2) = 32$$

We use a one-tailed test, at the 5% level and reject H_0 if $z < -1.645$ where

$$z = \frac{x-np}{\sqrt{npq}} = \frac{152-160}{\sqrt{32}} = -1.414$$

s.d. = $\sqrt{32}$

160

← Reject H_0

S.V. −1.645 0

Conclusion: As $z > -1.645$, we do not reject H_0 and conclude that there is not sufficient evidence at the 5% level to refute the manufacturer's claim.

If the continuity correction is used,

$$z = \frac{151.5-160}{\sqrt{32}}$$

$$= -1.503 \quad \text{and the conclusion is the same}$$

Problems of this type can be tackled by either method. The calculations performed correspond exactly.

Example 9.19 A large college claims that it admits equal numbers of men and women. A random sample of 500 students at the college gave 267 males. Is there any evidence, at the 5% level, that the college population is not evenly divided into males and females?

Solution 9.19 From the sample: $p_s = \dfrac{267}{500} = 0.534, \quad n = 500$

Let p be the proportion of males in the population.

H_0: $p = 0.5$ (there are equal numbers of males and females)

H_1: $p \neq 0.5$ (the college population is not evenly divided into males and females)

Consider the sampling distribution of proportions, where

$$P_s \sim N\left(p, \frac{pq}{n}\right) \qquad \text{where} \quad q = 1-p$$

We use a two-tailed test, at the 5% level, and reject H_0 if $|z| > 1.96$ where

$$z = \frac{p_s - p}{\sqrt{\frac{pq}{n}}}$$

$$= \frac{0.534-0.5}{\sqrt{\frac{(0.5)(0.5)}{500}}}$$

$$= 1.52$$

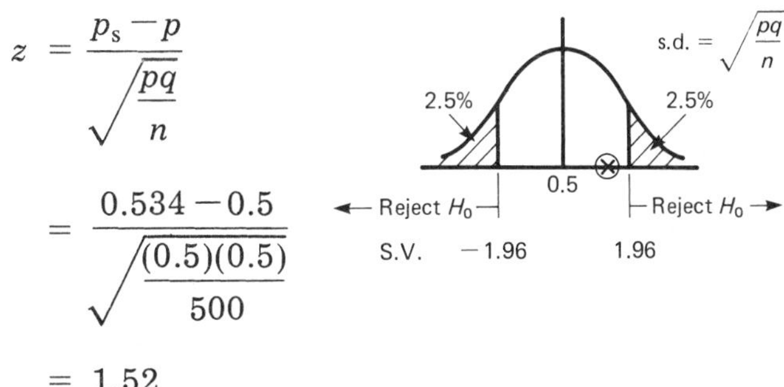

Conclusion: As $z < 1.96$, we do not reject H_0 and conclude that, at the 5% level, there is not sufficient evidence to refute the claim that the population is evenly divided into males and females.

Exercise 9f

1. For each of the following sets of data, carry out a significance test for the hypotheses stated.

	Number in sample	Number of 'successes'	Hypotheses	Level
(*a*)	50	45	H_0: $p = 0.8$, H_1: $p > 0.8$	5%
(*b*)	60	42	H_0: $p = 0.55$, H_1: $p \neq 0.55$	2%
(*c*)	120	21	H_0: $p = \frac{1}{4}$, H_1: $p \neq \frac{1}{4}$	5%
(*d*)	300	213	H_0: $p = 0.65$, H_1: $p \neq 0.65$	1%
(*e*)	90	56	H_0: $p = 0.76$, H_1: $p < 0.76$	1%

2. A theory predicts that the probability of an event is 0.4. The theory is tested experimentally and in 400 independent trials the event occurred 140 times. Is the number of occurrences significantly less than that predicted by the theory. Test at the 1% level.

3. It is thought that the proportion of defective items produced by a particular machine is 0.1. A random sample of 100 items is inspected and found to contain 15 defective items. Does this provide evidence, at the 5% level, that the machine is producing more defective items than expected?

4. A coin is tossed 100 times and 38 heads are obtained. Is there evidence, at the 2% level, that the coin is biased in favour of tails?

5. A government report states that one-third of teenagers in Great Britain belong to a youth organisation. A survey, conducted among a random sample of 1000 teenagers from a certain city, revealed that 370 belonged to a youth organisation. Does this provide significant evidence, at the 2% level, that the proportion of teenagers who belong to a youth organisation is greater in this city than the national average?

   Based on the results of this sample, calculate a 95% confidence interval for the proportion of teenagers in this city who belong to a youth organisation.

6. The probability that an oyster larva will develop in unpolluted water is 0.9, while in polluted water this probability is less than 0.9. Given that 20 oyster larvae are placed in unpolluted water, find the probabilities, each to two decimal places, that the number that will develop is
   (*a*) at least 17,
   (*b*) exactly 17.

   An oyster breeder put 20 larvae in a sample of water and observed that only 16 of them developed. Use a 10% significance level to determine whether the breeder would be justified in concluding that the water is polluted. (JMB)

7. A fruit farm grows 'Golden Delicious' apples, and it can be assumed that the distribution of the masses of the apples is described by a normal probability function. The apples are graded by mass (x g) into three grades: 'small' for $x < 80$, 'medium' for $80 < x < 100$, 'large' for $x > 100$. In 1979, 20% of the apples were graded as 'small' and 54% as 'medium'. Estimate, to one decimal place, the mean and standard deviation of the masses of the apples produced on the farm in that year. Estimate also what proportion of the apples had masses exceeding 105 g.

   When he begins to harvest his 1980 crop the grower picks out a sample of 100 apples at random and finds that only 9 are 'small'. Find, on the hypothesis that the proportion of 'small' apples in the whole crop is the same as in 1979, the probability of getting 9 or fewer 'small' apples in such a sample. Would he be justified in concluding, on the evidence of this sample, that there has been a reduction (for 1980 as compared with 1979) in the percentage of 'small' apples in the crop as a whole? (SMP)

8. A factory produces large numbers of sweets in a variety of colours. Automatic machines select the sweets at random and pack them in boxes of 20. A random sample of 100 boxes was chosen, the contents of each box examined and the number of black sweets in each box recorded. The results obtained are summarised in the following table.

No. of black sweets	0	1	2	3	4	5	6 or more
No. of boxes	11	29	27	22	7	4	0

   (*a*) Find an unbiased estimate for the proportion p of sweets produced which are black, and, to three significant figures, an estimate of its standard error.

   (*b*) Using a distributional approximation and a 5 per cent significance level, test the null hypothesis $p = 0.1$ against the alternative hypothesis $p \neq 0.1$. State your conclusion.

   (*c*) Given that $p = 0.1$, use tables to find, to the nearest integer, the expected frequencies corresponding to the observed frequencies tabulated above. (JMB)

9. In a public opinion poll, 1000 randomly chosen electors were asked whether they would vote for the 'Purple Party' at the next election and 357 replied 'Yes'. Find a 95% confidence interval for the proportion p of the population who would answer 'Yes' to the same question.

   Twenty similar polls are taken and the 95% confidence interval is determined for each poll. State the expected number of these intervals which will enclose the true value of p.

   The leader of the 'Purple Party' believes that the true value of p is 0.4. Test, at the 8% level, whether he is overestimating his support. (C)

10. In an investigation into ownership of calculators, 200 randomly chosen school students were interviewed, and 143 of them owned a calculator. Using the evidence of this sample, test, at the 5% level of significance, the hypothesis that the proportion of school students owning a calculator is 75% against the alternative hypothesis that the proportion is less than 75%. (C)p

TEST 5 — TESTING THE DIFFERENCE BETWEEN PROPORTIONS

Consider two random samples, sizes n_1 and n_2, with proportion of 'successes' p_{s_1} and p_{s_2}. If the population proportions are p_1 and p_2, and the sample sizes are large,

then
$$P_{s_1} \sim N\left(p_1, \frac{p_1 q_1}{n_1}\right) \quad \text{with} \quad q_1 = 1 - p_1$$

$$P_{s_2} \sim N\left(p_2, \frac{p_2 q_2}{n_2}\right) \quad \text{with} \quad q_2 = 1 - p_2$$

So
$$P_{s_1} - P_{s_2} \sim N\left(p_1 - p_2, \frac{p_1 q_1}{n_1} + \frac{p_2 q_2}{n_2}\right)$$

The distribution is known as the *sampling distribution of the difference between proportions.*

We usually wish to test whether the samples have been drawn from populations which have a **common proportion** p.

In this case

$$P_{s_1} - P_{s_2} \sim N\left(0, pq\left(\frac{1}{n_1} + \frac{1}{n_2}\right)\right) \quad \text{where} \quad q = 1 - p$$

Case 1 — If p is known

We use the test statistic

$$Z = \frac{P_{s_1} - P_{s_2} - (0)}{\sqrt{pq\left(\frac{1}{n_1} + \frac{1}{n_2}\right)}}$$

This is distributed as $N(0, 1)$ under the null hypothesis H_0 that the common population proportion is p.

Case 2 — If p is unknown

We use an estimate $\hat{p}$ for it, where

$$\hat{p} = \frac{n_1 p_{s_1} + n_2 p_{s_2}}{n_1 + n_2}$$

We use the test statistic

$$Z = \frac{P_{s_1} - P_{s_2} - (0)}{\sqrt{\hat{p}\hat{q}\left(\frac{1}{n_1} + \frac{1}{n_2}\right)}} \quad \text{with} \quad \hat{q} = 1 - \hat{p}$$

This is distributed as $N(0, 1)$ under H_0.

Example 9.20 Two companies 'Consumer Opinion' and 'People's Choice' conduct research in a large city before an election. 'Consumer Opinion' finds that in a random sample of 500 people, 325 said that they would vote for Mr A. The 'People's Choice' finds that in a random sample of 300 people, 201 said that they would vote for Mr A. Test, at the 5% level, whether there is a significant difference between the two proportions.

Solution 9.20 Consumer Opinion: $p_{s_1} = \frac{325}{500} = 0.65, \quad n_1 = 500$

Population proportion $= p_1$

People's Choice: $p_{s_2} = \frac{201}{300} = 0.67, \quad n_2 = 300$

Population proportion $= p_2$

Let the true population proportion voting for Mr A be p.

H_0: $p_1 = p_2 = p$ (there is no significant difference between the proportions)

H_1: $p_1 \neq p_2$ (there is a significant difference between the proportions)

Consider the sampling distribution of the difference between proportions

where $$P_{s_1} - P_{s_2} \sim \mathrm{N}\left(0, pq\left(\frac{1}{n_1} + \frac{1}{n_2}\right)\right)$$

Now, p is unknown, so we use

$$\hat{p} = \frac{n_1 p_{s_1} + n_2 p_{s_2}}{n_1 + n_2}$$

$$= \frac{325 + 201}{500 + 300}$$

$$= 0.6575$$

and $$\hat{q} = 1 - \hat{p} = 0.3425$$

We use a two-tailed test, at the 5% level, and reject H_0 if $|z| > 1.96$, where

$$z = \frac{p_{s_1} - p_{s_2} - 0}{\sqrt{\hat{p}\hat{q}\left(\frac{1}{n_1} + \frac{1}{n_2}\right)}}$$

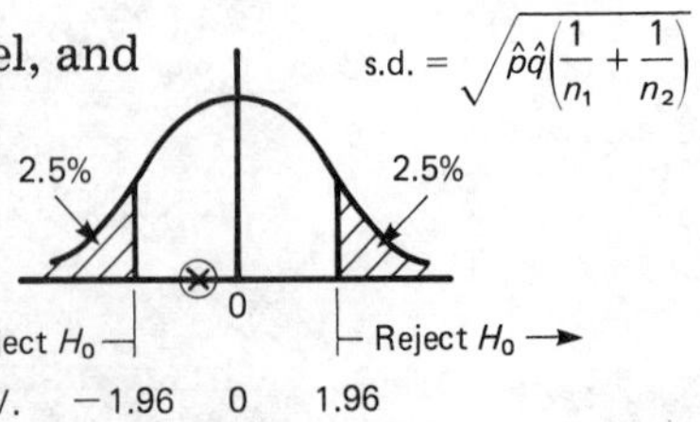

$$= \frac{0.65 - 0.67}{\sqrt{(0.6575)(0.3425)\left(\frac{1}{500} + \frac{1}{300}\right)}}$$

$$= -0.577$$

Conclusion: As $|z| < 1.96$, we do not reject H_0 and conclude that there is no evidence at the 5% level of a significant difference between the proportions obtained by the two companies.

Example 9.21 The manufacturer of Stay-White toothpaste maintains that by the addition of a certain chemical he can guarantee that (over a given period of time) people will have less dental decay if they clean their teeth regularly with Stay-White, than if they clean their teeth with ordinary toothpaste.

To test this theory, free tubes of Stay-White were given to 150 children at a school dental clinic, and at the same time, free tubes of ordinary toothpaste were given to a further 100 children at the clinic.

All the children guaranteed to clean their teeth regularly for 6 months and then return for a check-up.

At the end of 6 months it was found that 118 children using Stay-White had no dental decay, whereas 72 children using the ordinary toothpaste had no dental decay.

Dentists at the clinic believe that there is no real difference between the proportions of children who had no dental decay over the 6 month period. Use the sample results to test, at the 10% level, the dentists' theory.

Assuming that the proportions of children with no further dental decay are the same, give an approximate 98% confidence interval for this common proportion.

Solution 9.21 Stay-White toothpaste: $p_{s_1} = \frac{118}{150} = 0.787, \quad n_1 = 150,$

population proportion $= p_1$

Ordinary toothpaste: $p_{s_2} = \frac{72}{100} = 0.72, \quad n_2 = 100,$

population proportion $= p_2$

The dentists' theory is that the proportions are the same.

Let the common population proportion be p. Now, as we do not know p, we use

$$\hat{p} = \frac{n_1 p_{s_1} + n_2 p_{s_2}}{n_1 + n_2} = \frac{118 + 72}{250} = 0.76$$

and $$\hat{q} = 1 - \hat{p} = 0.24$$

H_0: $p_1 = p_2 = p$ (the proportions are the same)

H_1: $p_1 > p_2$ (the proportion with no dental decay is greater if Stay-White is used)

Consider the sampling distribution of the difference between proportions, where

$$P_{s_1} - P_{s_2} \sim \text{N}\left(0, \hat{p}\hat{q}\left(\frac{1}{n_1} + \frac{1}{n_2}\right)\right)$$

We use a one-tailed test at the 10% level, and reject H_0 if $z > 1.282$ where

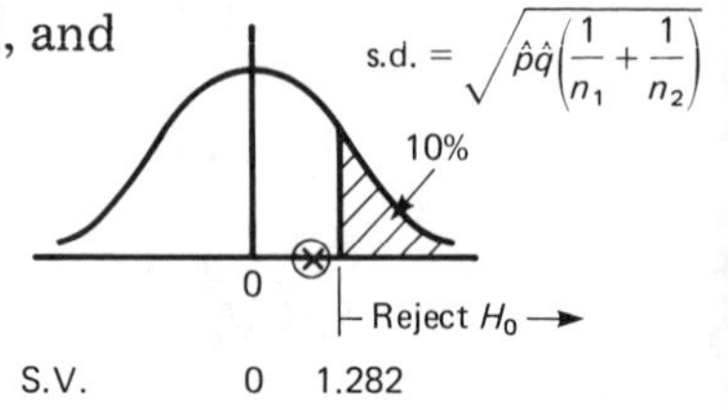

$$z = \frac{p_{s_1} - p_{s_2} - 0}{\sqrt{\hat{p}\hat{q}\left(\frac{1}{n_1} + \frac{1}{n_2}\right)}} = \frac{0.787 - 0.72}{\sqrt{(0.76)(0.24)\left(\frac{1}{150} + \frac{1}{100}\right)}} = 1.21 \text{ (3 S.F.)}$$

Conclusion: As $z < 1.282$ we do not reject H_0 and conclude that there is no evidence, at the 10% level, to suggest that people have fewer occurrences of dental decay if they use Stay-White toothpaste.

A 98% confidence interval for the common population proportion p of people who have no further dental decay after 6 months is given by

$$\hat{p} \pm 2.326\sqrt{\frac{\hat{p}\hat{q}}{N}} \quad \text{where} \quad \hat{p} = 0.76, \quad \hat{q} = 0.24, \quad \text{and} \quad N = 250$$

The 98% C.I. is

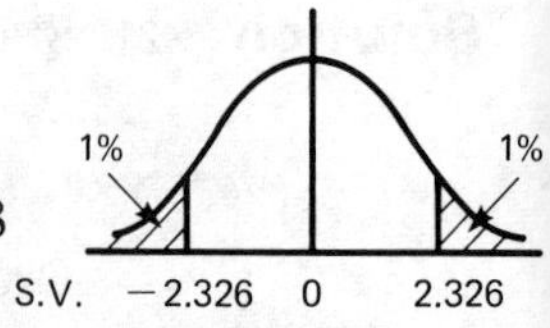

$$0.76 \pm 2.326\sqrt{\frac{(0.76)(0.24)}{250}} = 0.76 \pm 0.0628 = (0.697, 0.823)$$

The 98% confidence interval for p is $(0.697, 0.823)$.

Exercise 9g

1. For each of the following sets of data, test the hypothesis that there is a common proportion, p.

	Sample I		Sample II			
	Number in sample	Number of 'successes'	Number in sample	Number of 'successes'	Hypotheses	Level
(*a*)	150	125	200	176	H_0: $p_1 = p_2 = p$ H_1: $p_1 < p_2$	5%
(*b*)	1000	542	900	427	H_0: $p_1 = p_2 = p$ H_1: $p_1 \neq p_2$	1%
(*c*)	100	63	120	80	H_0: $p_1 = p_2 = p$ H_1: $p_1 \neq p_2$	10%
(*d*)	80	40	60	23	H_0: $p_1 = p_2 = p$ H_1: $p_1 > p_2$	5%

2. A shipment of Golden Delicious apples was tested for bruising. A random sample of 1000 apples was found to contain 30 which were bruised. A second random sample of 2000 apples contained 78 bruised apples. Show that the unbiased estimate for the proportion of bruised apples in a shipment is 3.6%. Find whether the proportions obtained from the two samples differ significantly at the 5% level.

3. See Example 8.22 (p. 451). Answer the end part of that question using a significance test for the difference between proportions. Use a 5% level of significance.

4. A region has a large population, 60% of whom have surnames beginning with a letter between A and M in the alphabet. In a random sample of 400 people from a town in the region it is found that 260 have surnames beginning with a letter between A and M. Test whether this result indicates any significant difference between the town and the region. Give full details of your test, stating any assumptions made and the hypotheses under test.
   In a sample of 300 from another town in the region, 200 have surnames beginning with a letter between A and M. Test whether the results for the two towns indicate a significant difference between them. Give full details of your test. (C)

5. A certain country in a fairy tale is populated by elves and fairies. A random sample of 100 fairies were each asked the question 'Do you believe in people': 72 fairies replied 'Yes' and the remainder replied 'No'. Give an approximate symmetric 98% confidence interval for the proportion of fairies that say they believe in people.
   A random sample of 62 elves were each asked the same question: 54 elves replied 'Yes' and the remainder replied 'No'. A question of interest is whether the proportion of fairies that say they believe in people differs from the proportion of elves that say they believe in people. Assuming that these proportions are equal, obtain an unbiased estimate of the common proportion. Using this estimate, test the question of interest at the 10% significance level, stating clearly your null and alternative hypotheses. (C)

6. A farmer has an orchard containing large numbers of two varieties (A and B) of apple trees. One year the farmer selected at random one tree of each variety, and kept a careful count of the fates of all the apples from these two trees. Some apples fell from the trees before picking time; some were eaten by insects; some eaten apples fell and some apples remained uneaten and on the trees until picking time. The farmer's results are given below.

	Variety A			Variety B	
	Fallen	On tree		Fallen	On tree
Eaten	150	50	Eaten	90	50
Uneaten	40	160	Uneaten	20	240

   You may assume, when answering the questions below, that the fate of an individual apple was independent of the fate of all other apples.

(*a*) Before any apple had fallen or been eaten, the farmer selected at random a variety *A* apple and stated that it would not fall before picking time. Estimate the probability that he was correct.
(*b*) At picking time the farmer accidentally trod on a fallen apple. Assuming that this apple was equally likely to have been any one of the fallen apples, estimate the probability that it was of variety *A*.
(*c*) Give an approximate symmetric 95% confidence interval for p_A, the proportion of variety *A* apples remaining on the tree and uneaten until picking time.
(*d*) The proportion of variety *B* apples remaining on the tree and uneaten until picking time is p_B. Determine whether there is evidence at the 0.1% significance level of a difference between p_A and p_B. (C)

7. Assuming that the mean and variance of a random variable X having a binomial distribution with parameters n and p are np and $np(1-p)$ respectively, prove the mean and variance of a proportion based on a sample of size n are p and $p(1-p)/n$ respectively, where p is the true proportion. Of a random sample of 50 shoppers in a certain city store 13 stated that they lived more than 10 miles from the city centre. Of a random sample of 50 shoppers from another store in the same city 9 lived more than 10 miles from the city centre. Stating your null and alternative hypotheses and using a significance level of 5%
(*a*) test that the true proportion in both stores could be 0.15;
(*b*) show that the two samples do not offer evidence of a difference in proportions between the two stores. (SUJB)

8. An organisation interviews a randomly chosen sample of 1000 adults from the population of the United Kingdom, and 517 of those interviewed claim to support the Conservative party. A second organisation independently interviews a random sample of 2000 adults, of whom 983 claim to support the Conservative party.
(*a*) Verify that the results of the two organisations do not differ significantly at the 5% level.
(*b*) Obtain a symmetric 99% confidence interval (based on the combined results) for the proportion of the population who claim to support the Conservative party. (C)

9. Countries *A* and *B* contain large numbers of elm trees, many of which are affected by Dutch elm disease. It is found that in country *A*, out of a random sample of 100 elm trees, 67 are affected by the disease and that in country *B*, out of a random sample of 150 elm trees, 93 are affected by the disease.

Tree experts have a theory that the proportions of affected trees in the two countries are the same, although there is a possibility that, since the disease affected the trees in country *A* before those in country *B*, the proportion of trees affected in country *A* may be greater than the proportion affected in country *B*. Using the sample results, test, at the 10% significance level, the theory of the experts. For the test that you perform, state clearly the hypotheses under comparison.

Assuming that the proportions of affected trees in the two countries are the same, give an approximate symmetric 98% confidence interval for this common proportion. (C)

10. According to some recent accident statistics, a random sample of 800 car drivers injured in road accidents comprised of 250 who were wearing seat belts and 550 who were not wearing seat belts. Determine a symmetric 99% confidence interval for the proportion of injured drivers wearing seat belts.
The injuries of the car drivers were described as either slight or serious. Of those wearing seat belts, 50 were seriously injured; of those not wearing seat belts, 150 were seriously injured. Determine an unbiased estimate of the overall proportion of serious injuries amongst the injured drivers. Test, at the 5% significance level, the hypothesis that the proportion of serious injuries is greater amongst those injured drivers not wearing seat belts than amongst those injured drivers wearing seat belts. (C)

11. It is known that in a large population there is a proportion p with a certain attribute. A random sample of size n is taken and it is found that x of them have the attribute. If $\hat{p} = x/n$ show that the mean and variance of $\hat{p}$ are p and $p(1-p)/n$ respectively. (You may assume any result relating to a binomial distribution.). What is the approximate distribution of $\hat{p}$ when n is large?

An ambulance station claims that at least 30% of its calls are life-threatening emergencies. To check this a random sample of 150 of its records were examined and, of these, only 38 were found to be life-threatening emergencies. Test the claim using a 1% significance level.

At a neighbouring station a random sample of 150 records showed that 50 were life-threatening emergencies. Test whether there is a difference between percentage rates in the two stations. (SUJB)

12. A drug research company has produced two compounds A and B for reducing blood pressure. They are administered to two sets of volunteers. One group of 90 was treated with A and 59 responded with lower blood pressure. The other group of 80 was treated with B and 51 responded with lower blood pressure.
(*a*) Find an approximate 95% confidence interval for the population proportion for which A is effective. In what way is your interval approximate?
(*b*) Test (at the 5% significance level) if there is any difference between the effectiveness of the two drugs. (SUJB)

SUMMARY — SIGNIFICANCE TESTING, USING NORMAL AND *t*-DISTRIBUTIONS

Type of test	Test statistic		
Single sample value	$Z = \dfrac{X-\mu}{\sigma}$		
Binomial situation	$Z = \dfrac{X-np}{\sqrt{npq}}$		
Means	σ known	σ unknown	
		Large n	Small n
	$Z = \dfrac{\overline{X}-\mu}{\sigma/\sqrt{n}}$	$Z = \dfrac{\overline{X}-\mu}{\sigma/\sqrt{n}}$ where $\widehat{\sigma}^2 = \dfrac{nS^2}{n-1} \approx S^2$	$T_{(n-1)} = \dfrac{\overline{X}-\mu}{S/\sqrt{n-1}}$
Difference between means	Unequal population variances σ_1, σ_2, known		
	$Z = \dfrac{\overline{X}_1-\overline{X}_2-(\mu_1-\mu_2)}{\sqrt{\dfrac{\sigma_1^2}{n_1}+\dfrac{\sigma_2^2}{n_2}}}$		
	Equal population variance σ^2		
	σ known	σ unknown	
		Large samples	Small samples
	$Z = \dfrac{\overline{X}_1-\overline{X}_2-(\mu_1-\mu_2)}{\sigma\sqrt{\dfrac{1}{n_1}+\dfrac{1}{n_2}}}$	$Z = \dfrac{\overline{X}_1-\overline{X}_2-(\mu_1-\mu_2)}{\widehat{\sigma}\sqrt{\dfrac{1}{n_1}+\dfrac{1}{n_2}}}$	$T_{(n_1+n_2-2)} = \dfrac{\overline{X}_1-\overline{X}_2-(\mu_1-\mu_2)}{\widehat{\sigma}\sqrt{\dfrac{1}{n_1}+\dfrac{1}{n_2}}}$
		where $\widehat{\sigma}^2 = \dfrac{n_1S_1^2+n_2S_2^2}{n_1+n_2-2}$	
Proportions	p known		
	$Z = \dfrac{P_{s_1}-p}{\sqrt{\dfrac{pq}{n}}}$ (large n)		
Difference between proportions	Equal population proportion p		
	p known	p unknown	
	$Z = \dfrac{P_{s_1}-P_{s_2}}{\sqrt{pq\left(\dfrac{1}{n_1}+\dfrac{1}{n_2}\right)}}$	$Z = \dfrac{P_{s_1}-P_{s_2}}{\sqrt{\widehat{p}\widehat{q}\left(\dfrac{1}{n_1}+\dfrac{1}{n_2}\right)}}$ where $\widehat{p} = \dfrac{n_1P_{s_1}+n_2P_{s_2}}{n_1+n_2}$	

Miscellaneous Exercise 9h

1. The heights of men can be assumed to be normally distributed with standard deviation 0.11 m.

   In 1928 the mean height of men in a certain city was 1.72 m. In a survey in 1978 the mean height of a random sample of 16 men from the same city was 1.77 m. On the hypothesis that the population mean height has not changed, calculate the probability of obtaining a sample mean height greater than that measured.

   In another survey in 1978 the mean height of a random sample of 32 men from a second city was 1.73 m. Assuming that the population mean heights are the same in the two cities, calculate the probability that a difference in sample mean heights greater than that measured would be obtained. (MEI)

2. Jack says 'Boys are better than girls at Watology'. Jill says 'Not true, girls are better'. Assume ability at Watology can be measured by a test with a maximum score of 100 and that scores are approximately normally distributed. Explain how to investigate Jack and Jill's assertions by describing how to conduct an experiment in which the measurements made are the Watology test scores of boys and girls.

   Assume your experiment has been done and the following scores obtained:

Boys, x	92	80	76	79	84	80	87	88	81	91
Girls, y	94	86	78	77	85	83	96	88	82	90

   Test if there is any difference in the ability of boys and girls at Watology. (The following sums may be of use to some candidates: $\Sigma x = 838$, $\Sigma y = 859$, $\Sigma x^2 = 70\,492$, $\Sigma y^2 = 74\,143$.) (O)

3. An expert golfer wishes to discover whether the average distances travelled by two different brands of golf ball differ significantly. He tests each ball by hitting it with his driver and measuring the distance X (in metres) that it travels. The distribution of X may be assumed to be normal.

   His results for a random sample of nine 'Farfly' golf balls were $\bar{x} = 214$ and $\Sigma(x-\bar{x})^2 = 2048$. Making the assumption that the population variance is equal to the sample variance, obtain a 95% symmetric confidence interval for the mean of X for 'Farfly' golf balls.

   His results for a random sample of sixteen 'Gofar' golf balls were $\bar{x} = 224$ and $\Sigma(x-\bar{x})^2 = 2460$. Assuming that the variance of X is the same for both types of golf ball, obtain a pooled (two-sample) estimate of this variance and, making the assumption that the true variance is equal to this estimate, test at the 5% level whether his results for 'Gofar' golf balls differ significantly from those for 'Farfly' golf balls. (C)

4. Mr Smith and Mr Jones are neighbours who work at the same office. Mr Smith drives to work in his old car, and each day records the time (x minutes) his journey takes. After 250 journeys his observations are summarised by $\Sigma x = 6250$, $\Sigma x^2 = 158\,491$. Regarding his observations as constituting a large random sample, give a symmetric 97% confidence interval for his average journey time.

   Mr Jones drives to work in his new car, and his average time over a random sample of 50 journeys is found to be 21 minutes. Mr Jones claims that if he leaves home 3 minutes after Mr Smith he will, on average, arrive at work before him. Assuming that Mr Smith and Mr Jones take different routes to work, that their journey times have standard deviation 3 minutes and that the samples may be treated as being large samples, test whether Mr Jones' claim may be accepted at the 2% significance level. (C)

5. Let p denote the probability of obtaining a head when a certain coin is tossed.
   (*a*) If $p = 0.4$, find the probability of obtaining at least 3 heads in 10 independent tosses of the coin.
   (*b*) If $p = 0.6$, find the probability of obtaining exactly 12 heads in 20 independent tosses of the coin.
   (*c*) Write down an appropriate null hypothesis and an appropriate alternative hypothesis for testing whether the coin is unbiased.

   To carry out this test 20 independent tosses of the coin are made and the number of heads that occurs is observed. Given that 15 heads occurred, carry out the test, assuming a 5 per cent significance level. Write down a statement of the conclusion you draw about the value of p for this coin. (JMB)

6. (*a*) After a survey a market research company asserted that 75% of T.V. viewers watched a certain programme. Another company interviewed 75 viewers and found that 51 watched the programme and 24 did not. Does this provide evidence at the 5% level of significance that the first company's figure of 75% was incorrect?
(*b*) Samples of leaves were collected from two oak trees *A* and *B*. The number of galls was counted on each leaf and the mean and standard deviation of the number of galls per leaf were calculated with the following results:

Tree	*A*	*B*
Sample size	60	80
Mean	11.4	10.7
S.D.	2.6	3.1

Assuming normal distributions, do the data provide evidence at the 5% significance level of different population means for the two trees? (SUJB)

7. An investigation was conducted into the dust content in the flue gases of two types of solid-fuel boilers. Thirteen boilers of type *A* and nine boilers of type *B* were used under identical fuelling and extraction conditions. Over a similar period, the following quantities (Table A), in grams, of dust were deposited in similar traps inserted in each of the twenty-two flues.
Assuming that these independent samples came from normal populations with the same variance
(*a*) use a two sample *t*-test at the 5% level of significance to determine whether there is any difference between the two samples as regards the mean dust deposit.
(*b*) test at the 5% level of significance whether there is any difference between the two samples as regards the mean dust deposit, where this time you should also assume that the population variances are both known to be 196.0.
Explain the apparent contradiction in your test results. (AEB 1980)

8. Explain what is meant by a random sample.
Random samples of size n_1 and n_2 are taken, one from each of two normal distributions with means μ_1, μ_2 and variances σ_1^2, σ_2^2 respectively. The sample means are $\bar{x}_1$ and $\bar{x}_2$ respectively. Write down expressions for the population mean and variance of $\bar{x}_1-\bar{x}_2$.
Given that $\sigma_1^2 = 0.04$, $\sigma_2^2 = 0.05$ and $n_1 = n_2 = 100$, write down a symmetrical two-sided 99% confidence interval for $\mu_1-\mu_2$ in terms of $\bar{x}_1-\bar{x}_2$.
If in fact $\mu_1 = 3.06$ and $\mu_2 = 3.00$, determine the probability that the hypothesis $\mu_1 = \mu_2$ would not be rejected using a two-tailed test with significance level 1%. State how this probability would be affected if the values of the population means were $\mu_1 = 3.00$, $\mu_2 = 3.06$.
Determine whether or not the hypothesis $\mu_1 = \mu_2$ should be rejected at the 1% level of significance in each of the cases when
(*a*) $\bar{x}_1 = 3.07$, $\bar{x}_2 = 2.99$,
(*b*) $\bar{x}_1 = 3.07$, $\bar{x}_2 = 3.12$. (JMB)

9. The length *X* of a certain component made by a machine is specified by the manufacturer to be 10 cm. *X* may be considered to be a random variable distributed normally with mean 10 cm and standard deviation 0.05 cm. All components are tested and are acceptable if they lie between 9.95 cm and 10.03 cm. Those less than 9.95 cm are rejected at a loss of 40 p each to the manufacturer; those between 10.03 cm and 10.05 cm can be shortened at a loss of 20 p and those greater than 10.05 cm can be shortened resulting in a loss of 25 p. Calculate the probabilities that if a component is tested the loss $L = 0, 20, 25, 40$ pence and hence calculate the expected value of *L*.
In order to test the accuracy of the machine a random sample of 25 components is measured and found to have a mean length of 10.014 cm. Is this sufficient evidence at the 5% level of significance to indicate that the mean is greater than 10 cm? If a further random sample of 25 yielded a mean of 10.012 cm, by pooling the two samples determine whether your conclusion about the mean alters. (SUJB)

10. Blocks of wood used for flooring are cut by machine. Their lengths are normally distributed with mean 230 mm and standard deviation 2 mm, while their widths are normally distributed with mean 80 mm and standard deviation 1.5 mm; the two measurements are independent. Calculate the probabilities
(*a*) that a block selected at random will lie within the tolerance limits 226.5 mm to 233 mm in length,
(*b*) that a block selected at random will lie within the tolerance limits 77 mm to 82 mm for width,
(*c*) that a block selected at random will satisfy both tolerances,

(*d*) that a block selected at random will be within the tolerance limits for width but not for length.

The setting on the machine which cuts the blocks to length is to be changed so that, while the standard deviation remains unchanged, 95% of the blocks will be no longer than 232.7 mm. Calculate the new mean length. After this resetting a block is produced that is only 224.6 mm long. Does this suggest that the machine is not correctly set? (SUJB)

11. HUM and WOW are two makes of cassette tapes both having a nominal playing time of 5400 seconds. Each tape of a random sample of 64 HUM tapes was timed using standard equipment. Denoting the actual playing time of a HUM tape by $(5400+h)$ seconds, the results obtained are summarised by $\Sigma h = 2624$, $\Sigma(h-\bar{h})^2 = 22\,743$. Obtain an unbiased estimate of the variance of the playing time of HUM tapes and obtain a symmetric 96% confidence interval for the average playing time of HUM tapes.
A random sample of 36 WOW tapes was also tested. Denoting the actual playing time of a WOW tape by $(5400+w)$ seconds, the results obtained are summarised by $\Sigma w = 1800$, $\Sigma(w-\bar{w})^2 = 9009$. Assuming that the two types of tape have playing time distributions with equal variances, show that the unbiased two-sample estimate of this common variance is 324. Treating both samples as large samples, test whether there is significant evidence, at the 3% level, of a difference in the means of the two playing time distributions. State your hypotheses clearly. (C)

12. The lifetime, T, in hours of a certain type of electric lamp is a random variable with distribution

$$\begin{aligned} f(t) &= A\,e^{-t/1200}, \quad 0 \leqslant t < \infty \\ &= 0, \quad t < 0 \end{aligned}$$

Find the value of A and show that the mean and standard deviation of T are both 1200 hours.

To test the reliability of the production a random sample of 40 bulbs was tested and found to have a mean life of 1020 hours. Does this indicate at the 5% level of significance that the batch from which the sample was taken was sub-standard? (SUJB)

Table A

Dust deposit—Type A boilers (g)	73.1, 56.4, 82.1, 67.2, 78.7, 75.1, 48.0, 53.3, 55.5, 61.5, 60.6, 55.2, 63.1
Dust deposit—Type B boilers (g)	53.0, 39.3, 55.8, 58.8, 41.2, 66.6, 46.0, 56.4, 58.9

TESTS INVOLVING THE BINOMIAL DISTRIBUTION

We have already seen that if $X \sim \text{Bin}(n, p)$, where *n is large*, we can test whether a value comes from a particular distribution by using the normal approximation (see p. 463). We now consider the case when *n is not large*.

The significance test can be designed in a similar manner to those involving the normal distribution. When dealing with a normal distribution, which is continuous, we look to see whether a particular *point* lies in the critical region or not, but when dealing with a binomial distribution, which is discrete, we look to see whether a particular *rectangle* lies in the critical region or not.

Consider $X \sim \text{Bin}(8, p)$, where $p = 0.4$, and suppose we wish to test, at the 5% level, whether a single sample value $x = 7$ comes from this distribution or from a distribution with a higher value of p.

We make the hypotheses: $H_0: p = 0.4$

$H_1: p > 0.4$

Under H_0, $X \sim \text{Bin}(8, 0.4)$ and $P(X = x) = {}^8C_x(0.6)^{8-x}(0.4)^x$, $x = 0, 1, \ldots, 8$.

The diagram shows the probability distribution of X.

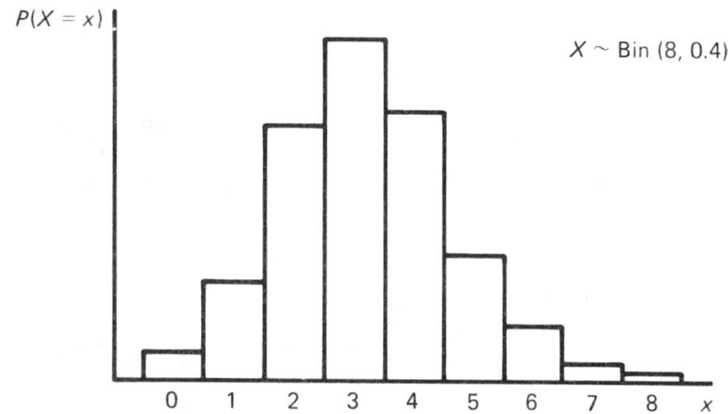

We use a 1-tailed test and look at the right-hand tail of the distribution.

We want to draw the boundary line for the critical region so that 5% of the area lies to the right of the boundary.

We find, from tables or calculations, that

$$P(X \geqslant 5) = 0.1737 \quad (> 0.05)$$

$$P(X \geqslant 6) = 0.0498 \quad (< 0.05)$$

So the boundary line must be drawn slightly to the left of the rectangle for $x = 6$.

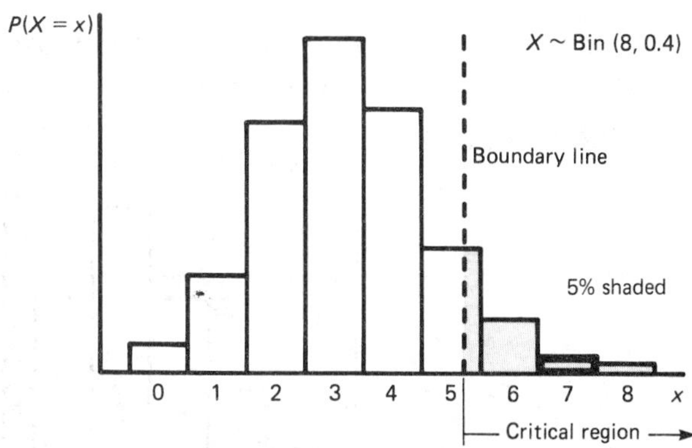

Now we wish to test the value $x = 7$ and will reject H_0 if the rectangle representing $x = 7$ lies *wholly* in the critical region.

From the diagram we see that this is the case, and conclude that $x = 7$ is unlikely to occur in the distribution Bin(8, 0.4).

Example 9.22 A coin is tossed 6 times. Test, at the 5% level, whether the coin is biased towards heads if **(a)** 6 heads are obtained, **(b)** 5 heads are obtained.

Solution 9.22 Let X be the r.v. 'the number of heads when the coin is tossed 6 times', and let p be the probability that the coin shows heads.

$H_0: p = 0.5$ (the coin is fair)

$H_1: p > 0.5$ (the coin is biased so that it is more likely to show heads)

Under H_0, $X \sim \text{Bin}(6, 0.5)$

and
$$P(X = x) = {}^6C_x(0.5)^{6-x}(0.5)^x \qquad x = 0, 1, \ldots, 6$$
$$= {}^6C_x(0.5)^6$$

From tables or calculations

$$P(X \geqslant 5) = 0.109\,375 \quad (> 0.05)$$
$$P(X = 6) = 0.015\,625 \quad (< 0.05)$$

so the boundary line for the critical region will be drawn as shown in the diagram, to give an area of 5% in the critical region.

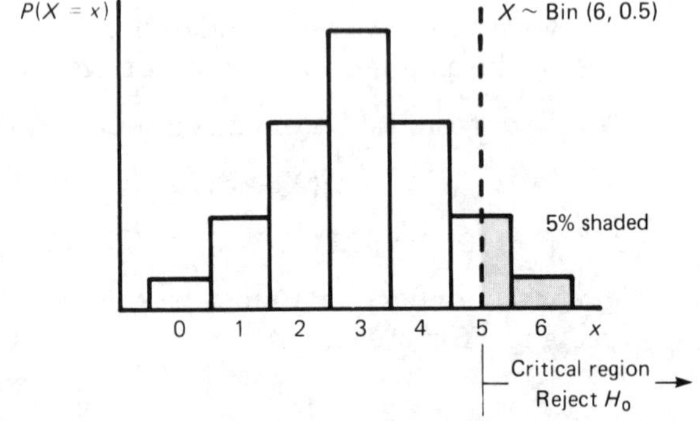

Using a 1-tailed test, at the 5% level, we reject H_0 if our observation lies *wholly* in the critical region.

(a) 6 heads are obtained:

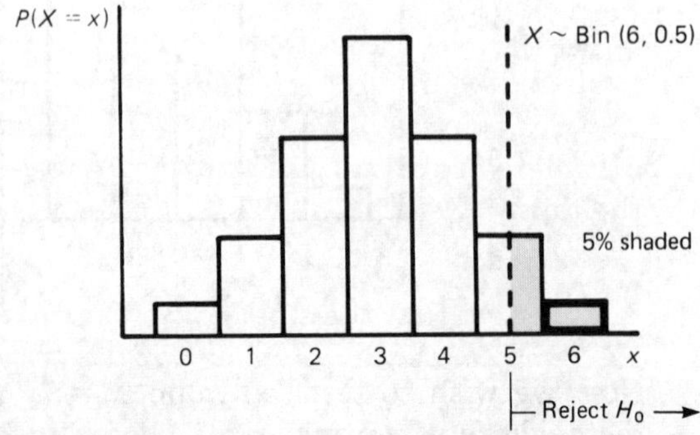

We see that the rectangle for $x = 6$ lies wholly in the critical region, and conclude that there is evidence, at the 5% level, to suggest that the coin is biased towards heads if 6 heads are obtained in 6 tosses.

(b) 5 heads are obtained:

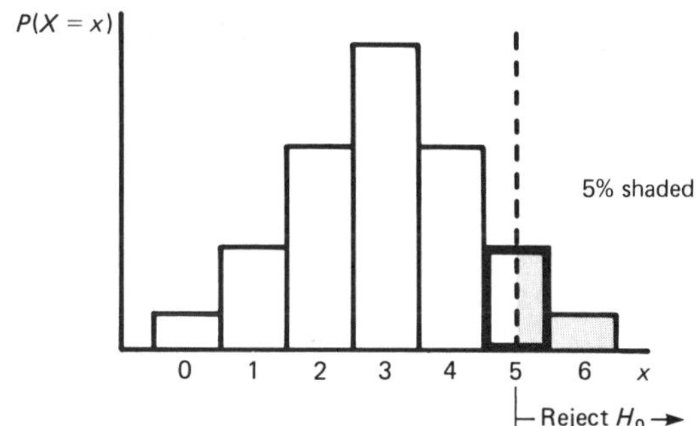

The rectangle for $x = 5$ does not lie wholly in the critical region, so we do not reject H_0 and conclude that there is no evidence, at the 5% level, to suggest that the coin is biased towards heads if 5 heads are obtained in 6 tosses.

Example 9.23 The discrete r.v. X is distributed binomially with $n = 10$. If a single observation x is taken from the distribution, test, at the 8% level, the hypothesis that $p = 0.45$ against the alternative hypothesis $p \neq 0.45$ when **(a)** $x = 7$, **(b)** $x = 1$.

Solution 9.23

$$H_0{:}p = 0.45$$

$$H_1{:}p \neq 0.45 \qquad \text{(2-tailed test)}$$

Under H_0, $\qquad X \sim \text{Bin}(10, 0.45)$

Since the test is 2-tailed, we need to consider both tails of the probability distribution and find boundary lines such that 4% of the area is in each tail. Note that it is not necessary to draw the complete probability distribution, since we are interested only in the tails.

(a) We test first the single observation $x = 7$. We are interested in the position of the boundary of the critical region in the right-hand tail, and need to know whether the rectangle for $x = 7$ lies wholly to the right of the boundary, i.e. wholly in the critical region.

Now if $x = 7$ does lie wholly in the critical region we would have $P(X \geqslant 7) < 0.04$.

From tables, $P(X \geqslant 7) = 0.102 > 0.04$, so the rectangle for $x = 7$ does not lie wholly in the critical region. We do not reject H_0, and conclude, at the 8% level, that $p = 0.45$.

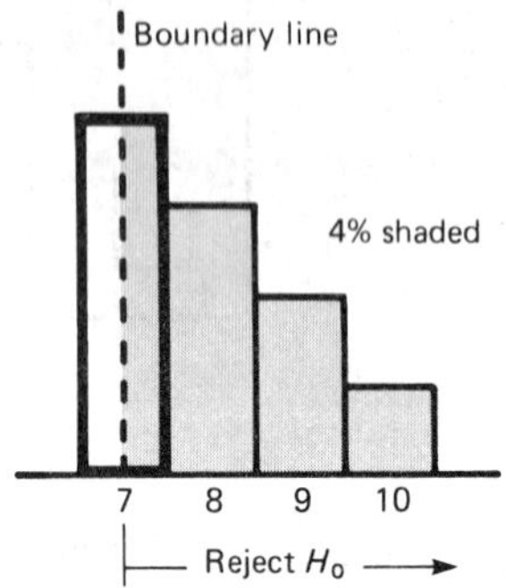

NOTE: Since $P(X \geqslant 8) = 0.0274$ the boundary line comes within the rectangle for $x = 7$.

(b) We now test the single observation $x = 1$. This time we are interested in the position of the boundary of the critical region in the left-hand tail. Now the rectangle for $x = 1$ will lie wholly in the critical region if $P(X \leqslant 1) < 0.04$.

From tables, $P(X \leqslant 1) = 0.0233 < 0.04$, indicating that the rectangle for $x = 1$ does lie wholly in the critical region. Therefore we reject H_0 and conclude, at the 8% level, that there is evidence that $p \neq 0.45$.

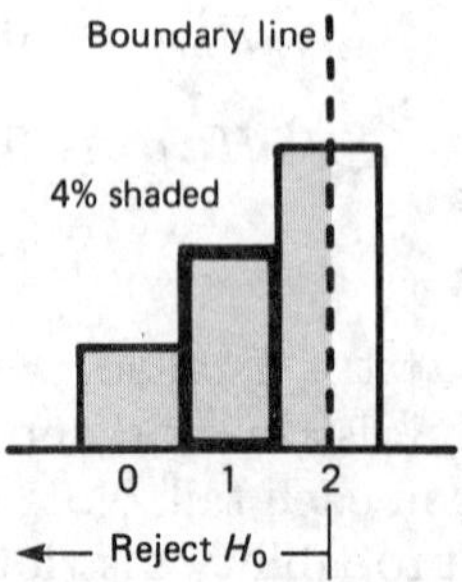

NOTE: Since $P(X \leqslant 2) = 0.0996$, the boundary line comes within the rectangle for $x = 2$.

Example 9.24 State the conditions under which the binomial distribution may be used. Illustrate your answer by referring to a specific example preferably from a project.

Records kept in a hospital show that 3 out of every 10 casualties who come to the casualty department have to wait more than half an hour before receiving medical attention. Find, to 3 decimal places, the probability that of the first 8 casualties who come to that casualty department **(a)** none, **(b)** more than two will have to wait more than half an hour before receiving medical attention. Find also the most probable number of the 8 casualties that will have to wait more than half an hour.

The hospital decided to increase the staff of the department by one member and it was then found that of the next 20 casualties 2 had to wait more than half an hour for medical attention. Test **(c)** at the 2% level, **(d)** at the 5% level whether the new staffing has decreased the number of casualties who have to wait more than half an hour for medical attention. (L)

Solution 9.24 For the first part see p. 209.

Let X be the r.v. 'the number of casualties who wait more than half an hour', and let p be the probability that a casualty has to wait more than half an hour.

Then $X \sim \text{Bin}(n, p)$ with $n = 8,\ p = 0.3$

and $P(X = x) = {}^8C_x(0.7)^{8-x}(0.3)^x, \quad x = 0, 1, 2, \ldots, 8$

(a) $P(X = 0) = (0.7)^8 = \underline{0.058}$ (3 d.p.)

(b)
$$\begin{aligned} P(X > 2) &= 1 - P(X \leqslant 2) \\ &= 1 - 0.5518 \quad \text{(from tables)} \\ &= \underline{0.488} \quad \text{(3 d.p.)} \end{aligned}$$

$$\begin{aligned} P(X = 0) &= 0.058 \\ P(X = 1) &= 8(0.7)^7(0.3) = 0.1977 \\ P(X = 2) &= 28(0.7)^6(0.3)^2 = 0.2965 \\ P(X = 3) &= 56(0.7)^5(0.3)^3 = 0.2541 \end{aligned}$$

and so on . . .

$P(X = 2) > P(X = 1)$ and $P(X = 2) > P(X = 3)$ and so the most probable number of casualties that will have to wait more than half an hour is 2.

Now let X be the r.v. 'the number of casualties in 20 who wait more than half an hour'.

Then $$X \sim \text{Bin}(20, p)$$

$H_0: p = 0.3$ (there is no change in the waiting pattern)

$H_1: p < 0.3$ (there is a decrease in the number who wait more than half an hour)

Under H_0, $X \sim \text{Bin}(20, 0.3)$.

(c) To test the significance of $x = 2$, at the 2% level, we perform a 1-tailed test, and reject H_0 if $P(X \leqslant 2) < 0.02$ (indicating that the rectangle for $x = 2$ lies wholly in the critical region).

Now, from tables, $P(X \leqslant 2) = 0.0355 > 0.02$, so we do not reject H_0 and conclude that there is no evidence at the 2% level to suggest a decrease in the number of casualties who wait more than half an hour.

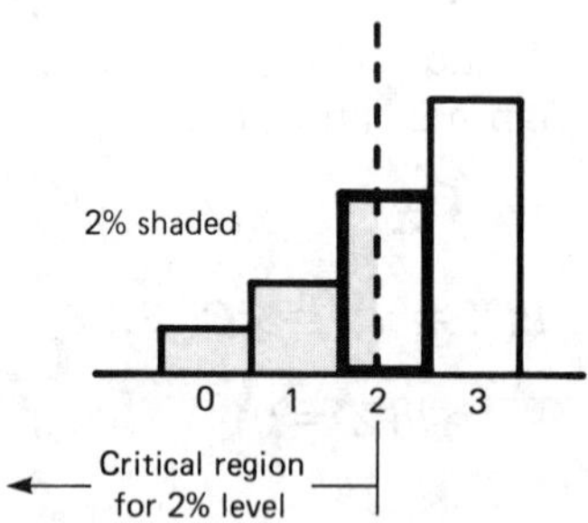

(d) At the 5% level, reject H_0 if $P(X \leqslant 2) < 0.05$.

Now $P(X \leqslant 2) = 0.0355 < 0.05$, so we reject H_0 and conclude that there is sufficient evidence, at the 5% level, to suggest a decrease in the number of casualties who wait more than half an hour.

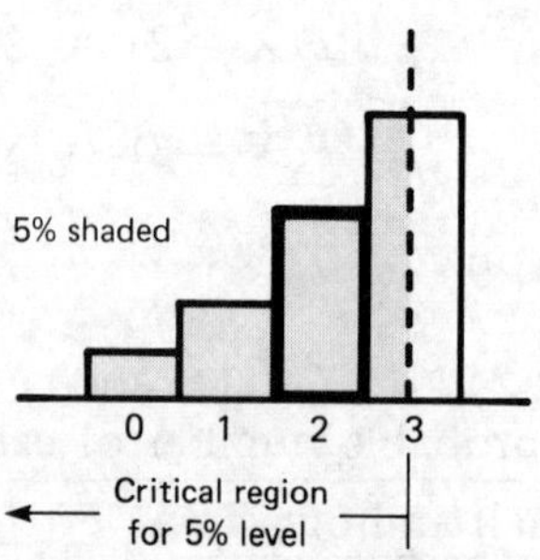

SUMMARY — TESTING A BINOMIAL PROPORTION, *n* NOT LARGE

$X \sim \text{Bin}(n, p)$ Level of test: $\alpha\%$ Test the single value $x = r$	
1-tailed test	
$H_0: p = p_0$ $H_1: p > p_0$	Reject H_0 if $P(X \geqslant r) < \dfrac{\alpha}{100}$
$H_0: p = p_0$ $H_1: p < p_0$	Reject H_0 if $P(X \leqslant r) < \dfrac{\alpha}{100}$
2-tailed test	
$H_0: p = p_0$ $H_1: p \neq p_0$	Reject H_0 if $P(X \leqslant r) < \dfrac{\frac{1}{2}\alpha}{100}$ or if $P(X \geqslant r) < \dfrac{\frac{1}{2}\alpha}{100}$

Exercise 9i

1. For each of the following, a single observation x is taken from a binomial distribution where $X \sim \text{Bin}(n, p)$. Test the hypotheses at the level of significance stated.

	x	n	Hypotheses	Level of significance
(*a*)	6	8	$H_0: p = 0.45,\ H_1: p > 0.45$	5%
(*b*)	1	10	$H_0: p = 0.45,\ H_1: p < 0.45$	5%
(*c*)	9	15	$H_0: p = 0.35,\ H_1: p > 0.35$	5%
(*d*)	9	15	$H_0: p = 0.35,\ H_1: p \neq 0.35$	5%
(*e*)	2	9	$H_0: p = 0.45,\ H_1: p < 0.45$	5%
(*f*)	16	20	$H_0: p = 0.45,\ H_1: p > 0.45$	1%
(*g*)	5	7	$H_0: p = 0.4,\ H_1: p > 0.4$	10%
(*h*)	2	20	$H_0: p = 0.3,\ H_1: p < 0.3$	1%

2. A die is thrown 15 times and it shows a six on twelve occasions. Is the die biased in favour of showing a six? Test at the 1% level.

3. The probability that a certain type of seed germinates is 0.7. The seeds undergo a new treatment, and when a packet of 10 seeds is tested 9 germinate. Is this evidence, at the 5% level, of an increase in the germination rate?

4. In a test of 10 true-false questions a student gets 8 correct. The student claims she was not guessing. Test this claim at the 5% level.

TESTS INVOLVING THE POISSON DISTRIBUTION

As with the test involving the binomial distribution we again look to see whether a particular rectangle lies wholly in the critical region or not.

Consider $X \sim \text{Po}(\lambda)$ and suppose we wish to test, at the 5% level, whether the single sample value $x = 14$ comes from the distribution where $\lambda = 8.5$ or from a distribution with a higher value of λ.

We make the hypotheses: $H_0: \lambda = 8.5$

$H_1: \lambda > 8.5$

We use a 1-tailed test and look at the right hand tail of the distribution.

We want to draw the boundary line for the critical region so that 5% of the area lies to the right of the boundary.

If $X \sim \text{Po}(8.5)$, $P(X \geqslant 14) = 0.0514 > 5\%$ and $P(X \geqslant 15) = 0.0274 < 5\%$, so the boundary line must be drawn as shown.

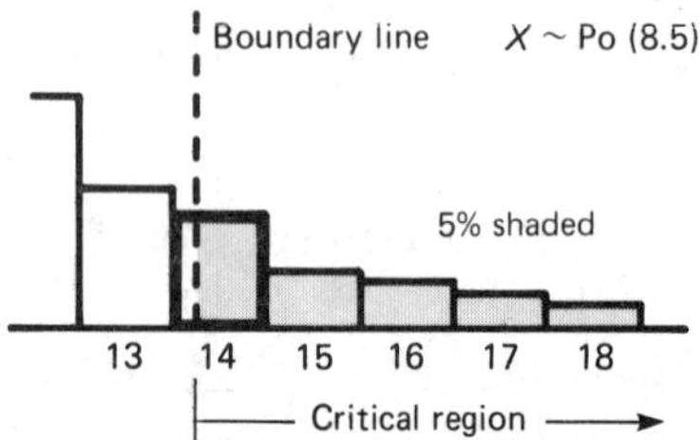

NOTE: Since $P(X \geqslant 15) = 0.0274$, the boundary line comes within the rectangle for $x = 14$.

The rectangle representing $P(X = 14)$ does not lie wholly in the critical region and we would therefore accept H_0 and conclude that $x = 14$ does come from the distribution $\text{Po}(8.5)$.

Example 9.25 The number of misprints on the front page of the *Daily Informer* is found to have a Poisson distribution with mean 6.5. A new proof-reader is employed and shortly afterwards the front page was found to have 12 misprints. The editor says that the mean number of misprints has increased. Test this claim at the 5% level.

Solution 9.25 Let X be the r.v. 'the number of misprints on the front page'.

Then $X \sim \text{Po}(\lambda)$

$H_0: \lambda = 6.5$ (the mean is unchanged)

$H_1: \lambda > 6.5$ (the mean has increased)

We test at the 5% level and will reject H_0 if $P(X \geqslant 12) < 0.05$, indicating that the rectangle for $P(X = 12)$ lies wholly within the critical region.

Now, from tables or by calculation, we find that $P(X \geqslant 12) = 0.0339 < 0.05$, so we reject H_0 and conclude that there is evidence, at the 5% level, to suggest that the mean has increased.

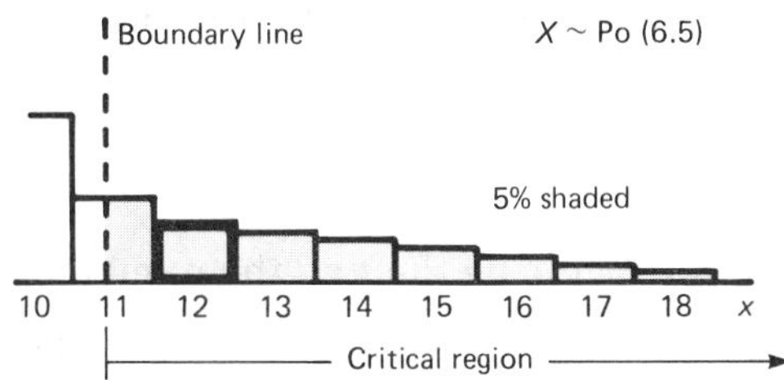

NOTE: Since $P(X \geqslant 11) = 0.0668$, the boundary line is drawn within the rectangle for $P(X = 11)$.

Example 9.26 Consider $X \sim \text{Po}(\lambda)$ and $H_0: \lambda = 6.5$. If $x = 2$, test, at the 5% level, (a) $H_1: \lambda \neq 6.5$, (b) $H_1: \lambda < 6.5$.

Solution 9.26 (a)

$$H_0: \lambda = 6.5$$
$$H_1: \lambda \neq 6.5$$

We perform a 2-tailed test, at the 5% level, and reject H_0 if $P(X \leqslant 2) < 0.025$, indicating that the rectangle for $P(X = 2)$ lies wholly in the critical region.

We find, from tables or calculation, that $P(X \leqslant 2) = 0.043 > 0.025$, therefore we do not reject H_0 and conclude that $\lambda = 6.5$.

(b)

$$H_0: \lambda = 6.5$$
$$H_1: \lambda < 6.5$$

We perform a 1-tailed test, at the 5% level, and reject H_0 if $P(X \leqslant 2) < 0.05$.

Now, since $P(X \leqslant 2) = 0.043 < 0.05$, we reject H_0 and conclude that the mean is less than 6.5.

Example 9.27 State conditions under which the Poisson distribution is a suitable model to use in statistical work. Describe briefly how a Poisson distribution was used, or could have been used, in a project.

(a) The number, X, of breakdowns per day of the lifts in a large block of flats has a Poisson distribution with mean 0.2. Find, to 3 decimal places, the probability that on a particular day
  (*i*) there will be at least one breakdown,
  (*ii*) there will be at most two breakdowns.

(b) Find, to 3 decimal places, the probability that, during a 20 day period, there will be no lift breakdowns.

(c) The maintenance contract for the lifts is given to a new company. With this company it is found that there are 2 breakdowns over a period of 30 days. Perform a significance test at the 5% level to decide whether or not the number of breakdowns has decreased. (L)

Solution 9.27 **(a)** Let X be the r.v. 'the number of breakdowns per day'.

Then $$X \sim \text{Po}(0.2).$$

(*i*)
$$\begin{aligned} P(X \geqslant 1) &= 1 - P(X = 0) \\ &= 1 - e^{-0.2} \\ &= \underline{0.181 \quad (3 \text{ d.p.})} \end{aligned}$$

(*ii*)
$$\begin{aligned} P(X \leqslant 2) &= P(X = 0) + P(X = 1) + P(X = 2) \\ &= e^{-0.2}\left(1 + 0.2 + \frac{0.2^2}{2}\right) \\ &= 1.22e^{-0.2} \\ &= \underline{0.999 \quad (3 \text{ d.p.})} \end{aligned}$$

(b) In 1 day we 'expect' 0.2 breakdowns, so in 20 days we 'expect' $20 \times 0.2 = 4$ breakdowns.

Let Y be the r.v. 'the number of breakdowns in 20 days'.

Then

$$Y \sim \text{Po}(4) \quad \text{and} \quad \underline{P(Y = 0) = e^{-4} = 0.018 \quad (3 \text{ d.p.})}$$

NOTE: We could consider

$$\begin{aligned} P(\text{no breakdowns in 20 days}) &= (P(X = 0))^{20} \\ &= (e^{-0.2})^{20} \\ &= e^{-4} \qquad \text{as before.} \end{aligned}$$

(c) In 30 days we 'expect' $30 \times 0.2 = 6$ breakdowns.

Let B be the r.v. 'the number of breakdowns in 30 days'.

Then $$B \sim \text{Po}(\lambda) \quad \text{where} \quad \lambda = 6.$$

Now there are 2 breakdowns in 30 days and we wish to test whether there has been a decrease in the average number of breakdowns.

$H_0: \lambda = 6$ (there is no change)

$H_1: \lambda < 6$ (the average number of breakdowns has decreased)

We perform a 1-tailed test, at the 5% level, and will reject H_0 if $P(B \leqslant 2) < 0.05$.

Now, from tables, $P(B \leqslant 2) = 0.062 > 0.05$. Therefore we do not reject H_0 and conclude that there is no evidence, at the 5% level, to suggest that the average number of breakdowns has decreased.

SUMMARY — TESTING A POISSON MEAN

$X \sim \text{Po}(\lambda)$ Level of test: $\alpha\%$ Test the single value $x = r$	
1-tailed test	
$H_0: \lambda = \lambda_0$ $H_1: \lambda > \lambda_0$	Reject H_0 if $P(X \geqslant r) < \dfrac{\alpha}{100}$
$H_0: \lambda = \lambda_0$ $H_1: \lambda < \lambda_0$	Reject H_0 if $P(X \leqslant r) < \dfrac{\alpha}{100}$
2-tailed test	
$H_0: \lambda = \lambda_0$ $H_1: \lambda \neq \lambda_0$	Reject H_0 if $P(X \geqslant r) < \dfrac{\frac{1}{2}\alpha}{100}$ or if $P(X \leqslant r) < \dfrac{\frac{1}{2}\alpha}{100}$

Exercise 9j

1. For each of the following a single observation x is taken from a Poisson distribution, where $X \sim \text{Po}(\lambda)$. Test the hypotheses at the level of significance stated.

	x	Hypotheses	Level of significance
(a)	11	$H_0: \lambda = 7,\ H_1: \lambda > 7$	5%
(b)	12	$H_0: \lambda = 7,\ H_1: \lambda \neq 7$	5%
(c)	4	$H_0: \lambda = 10,\ H_1: \lambda < 10$	1%
(d)	18	$H_0: \lambda = 10,\ H_1: \lambda > 10$	5%
(e)	2	$H_0: \lambda = 6.5,\ H_1: \lambda \neq 6.5$	5%
(f)	2	$H_0: \lambda = 6.5,\ H_1: \lambda > 6.5$	5%

2. The number of white corpuscles on a slide has Poisson distribution with mean 3.5. After certain treatment another sample was taken and the number of white corpuscles was found to be 8. Test, at the 5% level, whether the mean has increased.

3. The number of breakdowns in a computer is known to follow a Poisson distribution with a mean of 4.5 per month. A new computer is installed and in the first month there are 2 breakdowns. Test, at the 5% level, the claim that the mean has decreased.

4. The number of telephone calls to an office follows a Poisson distribution with a mean number of 6 per hour on a week-day.
(*a*) On Monday there were 5 calls between 10.00 and 10.30. Test, at the 5% level, whether the mean has increased.
(*b*) On Wednesday there were 3 calls between 11.00 and 12.30. Test, at the 5% level, whether the mean has decreased.

5. The number of flaws per 100 m of fabric is known to follow a Poisson distribution with mean 2. A 200 m length of fabric is tested and found to have 7 flaws. Test at the 5% level, whether the mean has increased.

6. Describe, briefly, the experimental evidence which you obtained in order to illustrate the Poisson distribution. State carefully any assumptions which you made.

The number X of emergency telephone calls to a gas board office in t minutes at weekends is known to follow a Poisson distribution with mean $\frac{1}{90}t$. Given that the telephone in that office is unmanned for 10 minutes, calculate, to 2 significant figures, the probability that there will be at least 2 emergency telephone calls to the office during that time.

Find, to the nearest minute, the length of time that the telephone can be left unmanned for there to be a probability of 0.9 that no emergency telephone call is made to the office during the period the telephone is unmanned.

During a week of very cold weather it was found that there had been 10 emergency telephone calls to the office in the first 12 hours of the weekend. Using the tables provided, or otherwise, determine whether the increase in the average number of emergency telephone calls to that office is significant at the 5% level. (L)

7. Explain briefly, referring to your projects if possible, the role of the null hypothesis and of the alternative hypothesis in a test of significance.

Over a long period, John has found that the bus taking him to school arrives late on average 9 times per month. In the month following the start of new summer schedules, John finds that his bus arrives late 13 times. Assuming that the number of times the bus is late has a Poisson distribution, test, at the 5% level of significance, whether the new schedules have in fact increased the number of times on which the bus is late. State clearly your null and alternative hypotheses. (L)P

TYPE I AND TYPE II ERRORS

When conducting a significance test we reach one of four possible conclusions. These are summarised in the table below.

	True situation	Our conclusion		
(1)	H_0 is true	Accept H_0	Correct decision	
(2)	H_0 is true	Reject H_0	Wrong decision	Type I error
(3)	H_0 is false	Accept H_0	Wrong decision	Type II error
(4)	H_0 is false	Reject H_0	Correct decision	

We say that

(a) Type I error is made if we reject H_0 when it is true.

(b) Type II error is made if we accept H_0 when it is false.

We write

(a) $P(\text{Type I error}) = P(\text{rejecting } H_0 \mid H_0 \text{ is true})$

(b) $P(\text{Type II error}) = P(\text{accepting } H_0 \mid H_1 \text{ is true})$

NOTE: when considering Type II errors we must state a definite value of the parameter in the alternative hypothesis H_1.

Example 9.28 Define Type I and Type II errors in testing hypotheses.

A box is known to contain either (H_0) ten white counters and 90 black counters or (H_1) 50 white counters and 50 black counters. In order to test hypothesis H_0 against hypothesis H_1, four counters are drawn at random from the box, without replacement. If all four counters are black, H_0 is accepted. Otherwise is it rejected.

Find the size of the Type I and Type II errors for this test.

(AEB)

Solution 9.28 For the first part, see preceding paragraph.

H_0: The bag contains 10 white and 90 black counters

H_1: The bag contains 50 white and 50 black counters

H_0 is accepted if all four counters, drawn without replacement, are black.

$$\begin{aligned} P(\text{Type I error}) &= P(\text{rejecting } H_0 \mid H_0 \text{ is true}) \\ &= P(\text{at least 1 white} \mid \text{there are 10 white and 90 black}) \end{aligned}$$

Now, if there are 10 white and 90 black

$$\begin{aligned} P(\text{drawing 4 black}) &= \left(\frac{90}{100}\right)\left(\frac{89}{99}\right)\left(\frac{88}{98}\right)\left(\frac{87}{97}\right) \\ &= 0.652 \end{aligned}$$

$$\begin{aligned} P(\text{drawing at least 1 white}) &= 1-0.652 \\ &= 0.348 \end{aligned}$$

Therefore $P(\text{Type I error}) = 0.348$.

$$\begin{aligned} P(\text{Type II error}) &= P(\text{accepting } H_0 \mid H_1 \text{ is true}) \\ &= P(\text{all 4 are black} \mid \text{there are 50 white and 50 black}) \end{aligned}$$

If there are 50 white and 50 black

$$\begin{aligned} P(\text{drawing 4 black}) &= \left(\frac{50}{100}\right)\left(\frac{49}{99}\right)\left(\frac{48}{98}\right)\left(\frac{47}{97}\right) \\ &= 0.059 \end{aligned}$$

Therefore $P(\text{Type II error}) = 0.059$.

Example 9.29 A man claims that he can throw a six with a fair die five times out of six on the average. Calculate the probability that he will throw four or more sixes in six throws (*i*) if his claim is justified (*ii*) if he can throw a six, on the average, only once in six throws.

To test the claim, he is invited to throw the die six times, his claim being accepted if he throws at least four sixes.

Find the probability that the test will **(a)** accept the man's claim when hypothesis (*ii*) is true, or **(b)** reject the claim when it is justified, that is, when hypothesis (*i*) is true. (AEB 1974)

Solution 9.29 (*i*) *If his claim is justified*

$$P(\text{throws a six}) = \frac{5}{6}$$

So, in six throws, let X be the r.v. 'the number of sixes obtained'.

Then $\quad X \sim \text{Bin}(n, p) \quad \text{with} \quad n = 6, \quad p = \frac{5}{6}$

Now $\quad P(X = x) = {}^6C_x\left(\frac{1}{6}\right)^{6-x}\left(\frac{5}{6}\right)^x \quad x = 0, 1, \ldots, 6$

$$\begin{aligned} P(X \geqslant 4) &= P(X = 4) + P(X = 5) + P(X = 6) \\ &= 15\left(\frac{1}{6}\right)^2\left(\frac{5}{6}\right)^4 + 6\left(\frac{1}{6}\right)\left(\frac{5}{6}\right)^5 + \left(\frac{5}{6}\right)^6 \\ &= \left(\frac{5}{6}\right)^4\left(\frac{1}{6}\right)^2(15 + 30 + 25) \\ &= 0.938 \quad (3\text{ d.p.}) \end{aligned}$$

Therefore $P(X \geqslant 4) = 0.938$ (3 d.p.).

(*ii*) If $P(\text{throws a six}) = \frac{1}{6}$.

Then $\quad X \sim \text{Bin}(n, p) \quad \text{with} \quad n = 6, \quad p = \frac{1}{6}$

$$P(X = x) = {}^6C_x\left(\frac{5}{6}\right)^{6-x}\left(\frac{1}{6}\right)^x \quad x = 0, 1, 2, \ldots, 6$$

$$\begin{aligned} P(X \geqslant 4) &= P(X = 4) + P(X = 5) + P(X = 6) \\ &= 15\left(\frac{5}{6}\right)^2\left(\frac{1}{6}\right)^4 + 6\left(\frac{5}{6}\right)\left(\frac{1}{6}\right)^5 + \left(\frac{1}{6}\right)^6 \\ &= \left(\frac{1}{6}\right)^6(375 + 30 + 1) \\ &= 0.0087 \quad (2\text{ S.F.}) \end{aligned}$$

Therefore $P(X \geqslant 4) = 0.0087$ (2 S.F.).

H_0: The man can throw a six five times out of six

H_1: The man can throw a six, on the average, only once in six throws

H_0 is accepted if the man throws at least four sixes in six throws.

(a) $$P(\text{accepting } H_0 \mid H_1 \text{ is true}) = P\left(X \geqslant 4 \mid p = \frac{1}{6}\right)$$
$$= 0.0087$$

Therefore the probability of accepting the man's claim, when hypothesis (*ii*) is true = *P*(Type II error) = 0.0087 (2 S.F.).

(b) $$P(\text{rejecting } H_0 \mid H_0 \text{ is true}) = P\left(X < 4 \mid p = \frac{5}{6}\right)$$
$$= 1 - 0.938$$
$$= 0.062 \quad (3 \text{ d.p.})$$

Therefore the probability of rejecting the man's claim when it is justified (probability of a Type I error) is 0.062 (3 d.p.).

Example 9.30 Dating of archaeological specimens is a difficult task. It is known that specimens emit a certain type of radioactive particle; the number of particles emitted in n minutes having a Poisson distribution with parameter $n\lambda$, where the value of λ depends upon the age of the specimen.

Two hypotheses concerning the age of one particular specimen are put forward:

H_A: specimen is 7000 years old (in which case $\lambda = 1.0$)

H_B: specimen is 15 000 years old (in which case $\lambda = 4.0$)

It is decided to count the number, X, of radioactive particles emitted in n minutes and

accept H_A (and reject H_B) if $X \leqslant 1$

and accept H_B (and reject H_A) if $X \geqslant 2$

If $n = 1$ what is **(a)** the probability of rejecting H_A when H_A is in fact true, **(b)** the probability of rejecting H_B when H_B is in fact true?

If the probability of rejecting H_B when H_B is in fact true is to be less than 0.001, show that the minimum number of complete minutes for which counting should be recorded is three. What is the corresponding probability of rejecting H_A when H_A is in fact true?

(AEB 1980)

Solution 9.30

H_A: specimen is 7000 years old ($\lambda = 1.0$)

H_B: specimen is 15 000 years old ($\lambda = 4.0$)

Let X be the r.v. 'the number of particles emitted in n minutes'. Then $X \sim \text{Po}(n\lambda)$.

We accept H_A (and reject H_B) if $X \leqslant 1$ and accept H_B (and reject H_A) if $X \geqslant 2$.

If $n = 1$, then $X \sim \text{Po}(\lambda)$.

(a) $P(\text{rejecting } H_A \mid H_A \text{ is true}) = P(X \geqslant 2 \mid X \sim \text{Po}(1.0))$

Now, if $X \sim \text{Po}(1.0)$,

$$P(X = x) = e^{-1.0}\frac{(1.0)^x}{x!} \qquad x = 0, 1, \ldots$$

So
$$\begin{aligned} P(X \geqslant 2) &= 1 - P(X = 0) - P(X = 1) \\ &= 1 - e^{-1} - e^{-1} \\ &= 1 - 2e^{-1} \\ &= 1 - 0.736 \\ &= 0.264 \end{aligned}$$

Therefore the probability of rejecting H_A when H_A is true is 0.264 (3 d.p.).

(b) $P(\text{rejecting } H_B \mid H_B \text{ is true}) = P(X \leqslant 1 \mid X \sim \text{Po}(4.0))$

If $X \sim \text{Po}(4.0)$, then

$$P(X = x) = e^{-4.0}\frac{(4.0)^x}{x!} \qquad x = 0, 1, \ldots$$

So
$$\begin{aligned} P(X \leqslant 1) &= P(X = 0) + P(X = 1) \\ &= e^{-4} + e^{-4}4 \\ &= 5e^{-4} \\ &= 0.092 \end{aligned}$$

Therefore the probability of rejecting H_B when H_B is true is 0.092 (3 d.p.).

Now under H_A: $\quad X \sim \text{Po}((1.0)n)$

under H_B: $\quad X \sim \text{Po}((4.0)n)$

If $P(\text{rejecting } H_B \mid H_B \text{ is true}) \leqslant 0.001$ then

$P(X \leqslant 1 \mid X \sim \text{Po}(4n)) \leqslant 0.001$.

If $X \sim \text{Po}(4n)$ then

$$P(X = x) = e^{-4n}\frac{(4n)^x}{x!} \qquad x = 0, 1, \ldots$$

We have
$$\begin{aligned} P(X \leqslant 1) &= P(X = 0) + P(X = 1) \\ &= e^{-4n} + e^{-4n}4n \\ &= e^{-4n}(1 + 4n) \end{aligned}$$

So, we require to find n such that

$$e^{-4n}(1+4n) \leqslant 0.001$$

By, trial, when $n=1$ $e^{-4}(5) = 0.092 > 0.001$

when $n=2$ $e^{-8}(9) = 0.003\,02 > 0.001$

when $n=3$ $e^{-12}(13) = 0.000\,08 < 0.001$

Therefore, the minimum number of complete minutes is 3.

In this case, when $n=3$, $X \sim \text{Po}(3\lambda)$.

So $P(\text{rejecting } H_A \mid H_A \text{ is true}) = P(X \geqslant 2 \mid X \sim \text{Po}(3(1.0)))$

If $X \sim \text{Po}(3)$ then

$$P(X=x) = e^{-3}\frac{(3)^x}{x!} \qquad x = 0, 1, \ldots$$

So

$$\begin{aligned} P(X \geqslant 2) &= 1 - P(X=0) - P(X=1) \\ &= 1 - e^{-3} - e^{-3}3 \\ &= 1 - 4e^{-3} \\ &= 0.801 \quad (3 \text{ d.p.}) \end{aligned}$$

Therefore the probability of rejecting H_A when H_A is true is 0.801.

Example 9.31 Two hypotheses concerning the probability density function of a random variable X are

$$H_0: \quad f(x) = \begin{cases} \frac{1}{4}(x+1) & 0 < x < 2 \\ 0 & \text{otherwise} \end{cases}$$

$$H_1: \quad f(x) = \begin{cases} \frac{1}{4}x^3 & 0 < x < 2 \\ 0 & \text{otherwise} \end{cases}$$

Sketch the p.d.f. in each case.

The following test procedure is decided upon. A single observation of X is made and if X is less than a particular value k, where $0 < k < 2$, then H_0 is accepted, otherwise H_1 is accepted.

(a) Find k if $P(\text{Type I error}) = 0.1$.

(b) With this value of k, find $P(\text{Type II error})$.

Solution 9.31

Under H_0

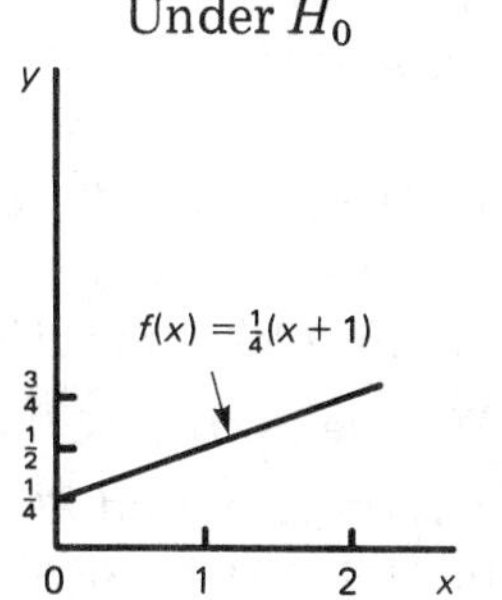

Under H_1

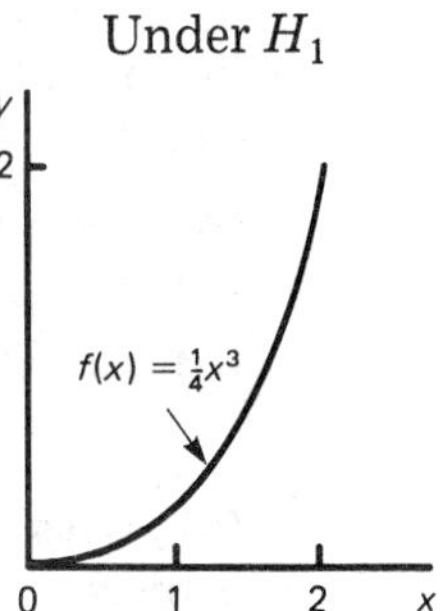

(a) If $X < k$, we accept H_0. If $X \geqslant k$, we accept H_1.

We need to find k such that $P(\text{Type I error}) = 0.1$,

i.e. $$P(\text{accept } H_1 \mid H_0 \text{ is true}) = 0.1$$

At this stage it is important to rewrite the statements as follows:

$$P(X \geqslant k \mid f(x) = \tfrac{1}{4}(x+1)) = 0.1$$

Under H_0, $$P(X \geqslant k) = \int_k^2 \tfrac{1}{4}(x+1)\,dx$$

So $$\int_k^2 \tfrac{1}{4}(x+1)\,dx = 0.1$$

$$[\tfrac{1}{2}x^2 + x]_k^2 = 0.4$$

$$2 + 2 - \tfrac{1}{2}k^2 - k = 0.4$$

$$k^2 + 2k - 7.2 = 0$$

$$(k+1)^2 = 8.2$$

$$k + 1 = \pm 2.86$$

Therefore $k = 1.86$ since $0 < k < 2$.

(b) Now, when $k = 1.86$

$$\begin{aligned} P(\text{Type II error}) &= P(\text{accept } H_0 \mid H_1 \text{ is true}) \\ &= P(X < 1.86 \mid f(x) = \tfrac{1}{4}x^3) \\ &= \int_0^{1.86} \tfrac{1}{4}x^3\,dx \\ &= \left[\frac{x^4}{16}\right]_0^{1.86} \\ &= 0.748 \end{aligned}$$

Therefore, when $k = 1.86$, $P(\text{Type II error}) = 0.748$.

Example 9.32 To test whether a coin is fair, the following decision rule is adopted. Toss the coin 120 times; if the number of heads is between 50 and 70 inclusive, accept the hypothesis that the coin is fair, otherwise reject it.

(a) Find the probability of rejecting the hypothesis when it is correct.

(b) How should the decision rule be modified if

$$P(\text{Type I error}) < 0.01?$$

(c) With the original decision rule, find $P(\text{Type II error})$ if the coin is biased and the probability that a head is obtained is in fact 0.6.

Solution 9.32 Let X be the r.v. 'the number of heads obtained'.
Then $X \sim \text{Bin}(n, p)$ with $n = 120$.

Now, as n is large, $X \sim \text{N}(np, npq)$ approximately, where $q = 1 - p$

$$H_0: \text{ The coin is fair } \quad (p = \tfrac{1}{2})$$

$$H_1: \text{ The coin is biased } \quad (p \neq \tfrac{1}{2})$$

Under H_0,

$$np = (120)\left(\frac{1}{2}\right) = 60 \quad \text{and} \quad npq = (120)\left(\frac{1}{2}\right)\left(\frac{1}{2}\right) = 30$$

So $\qquad X \sim \text{N}(60, 30)$

(a) Under H_0,

$$P(50 \leqslant X \leqslant 70) \rightarrow P(49.5 < X < 70.5) \qquad \text{(continuity correction)}$$

$$= P\left(\frac{49.5-60)}{\sqrt{30}} < \frac{X-60}{\sqrt{30}} < \frac{70.5-60}{\sqrt{30}}\right)$$

$$= P(-1.917 < Z < 1.917)$$

$$= 0.9446$$

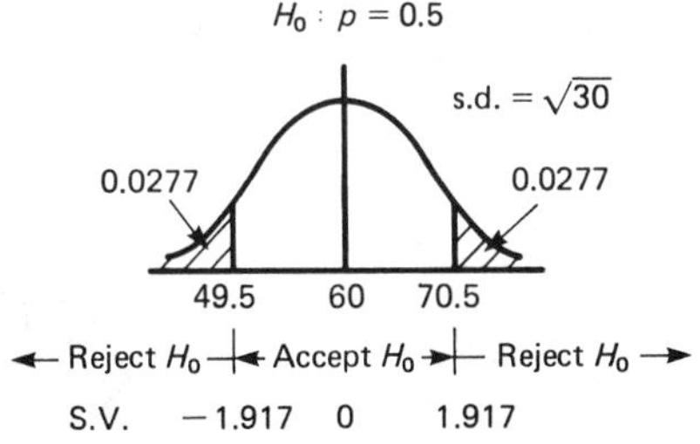

So $\qquad P(\text{accepting } H_0 \mid H_0 \text{ is true}) = 0.9446$

$$P(\text{rejecting } H_0 \mid H_0 \text{ is true}) = 1 - 0.9446 = 0.0554.$$

(b) If $P(\text{Type I error}) < 0.01$ then we need to find a value a such that $\Phi(a) = 0.995$. Now, from tables, $a = 2.575$.

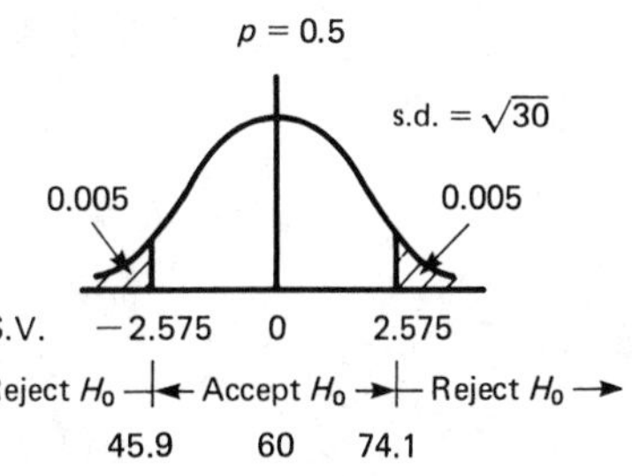

So, if $X \sim \text{N}(60, 30)$, then the value of X corresponding to the standardised value of 2.575 is given by

$$60 + 2.575\sqrt{30} = 74.1 \quad (3 \text{ S.F.})$$

The value corresponding to the standardised value of -2.575 is

$$60 - 2.575\sqrt{30} = 45.9 \quad (3 \text{ S.F.})$$

Therefore, the decision rule becomes:

Accept the hypothesis that the coin is fair if the number of heads lies between 46 and 74 inclusive, otherwise reject it.

(c) H_0: coin is fair $(p = 0.5)$

H_1: coin is biased $(p = 0.6)$

We accept H_0 if the number of heads lies between 50 and 70 inclusive.

$$\text{Now} \quad P(\text{Type II error}) = P(\text{accepting } H_0 \mid H_1 \text{ is true})$$
$$= P(49.5 < X < 70.5 \mid p = 0.6)$$

Now, if $p = 0.6$, $np = (120)(0.6) = 72$

$npq = (120)(0.6)(0.4) = 28.8$

So $X \sim \text{N}(72, 28.8)$.

$$P(49.5 < X < 70.5) = P\left(\frac{49.5-72}{\sqrt{28.8}} < \frac{X-72}{\sqrt{28.8}} < \frac{70.5-72}{\sqrt{28.8}}\right)$$
$$= P(-4.193 < Z < -0.2795)$$
$$= 0.390 \quad (3\text{ S.F.})$$

Therefore, $P(\text{Type II error}) = 0.390$ (3 S.F.).

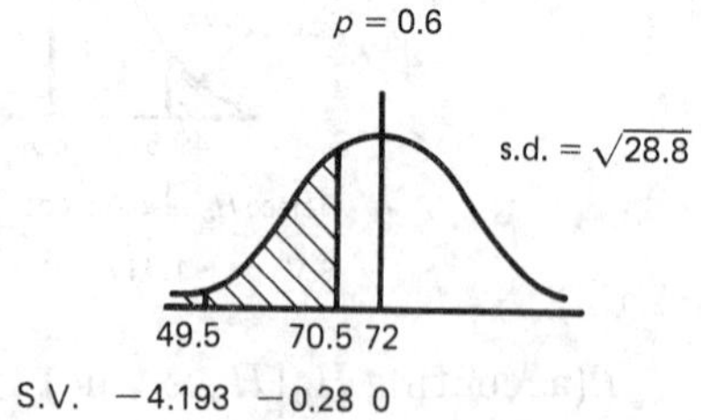

We see that, with the given decision rule, there is a fairly high probability that the coin will be accepted as fair when in fact $p = 0.6$.

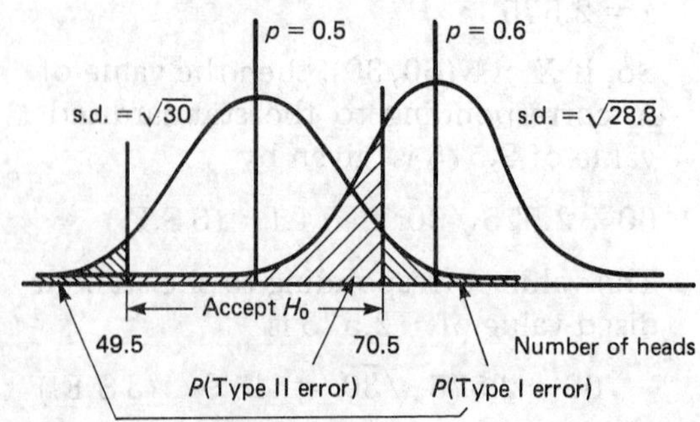

Example 9.33 A sample of size 100 is taken from a normal population with unknown mean μ and known variance 36. An investigator wishes to test the hypotheses H_0: $\mu = 65$, H_1: $\mu > 65$. He decides on the following criteria:

$$\text{accept } H_0 \text{ if the sample mean } \bar{x} \leqslant 66.5$$

$$\text{reject } H_0 \text{ if } \bar{x} > 66.5$$

Find the probability that he makes a Type I error.

If he uses as alternative hypothesis H_1: $\mu = 67.9$, find the probability that he makes a Type II error.

On which critical value should he decide for the sample mean if he wants $P(\text{Type I error}) = P(\text{Type II error})$?

Solution 9.33 Under H_0,

$$\bar{X} \sim \mathrm{N}\left(\mu, \frac{\sigma^2}{n}\right) \quad \text{with} \quad \mu = 65, \quad \sigma = 6, \quad n = 100$$

He rejects H_0 if $\bar{x} > 66.5$.

Now

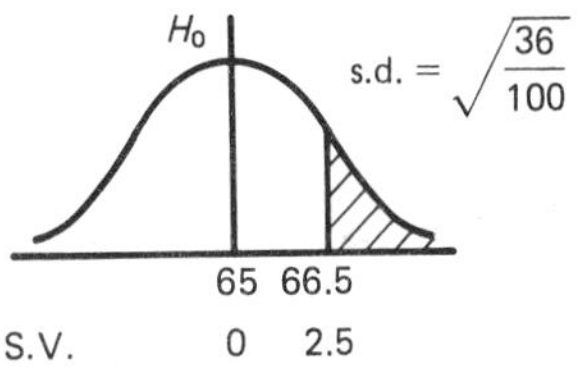

$$\begin{aligned} P(\bar{X} > 66.5) &= P\left(\frac{\bar{X}-65}{6/10} > \frac{66.5-65}{6/10}\right) \\ &= P(Z > 2.5) \\ &= 0.006\,21 \end{aligned}$$

Therefore, the probability that he rejects H_0, when in fact H_0 is true (Type I error) is 0.006 21.

If H_1: $\mu = 67.9$, then under H_1

$$\bar{X} \sim \mathrm{N}\left(67.9, \frac{36}{100}\right)$$

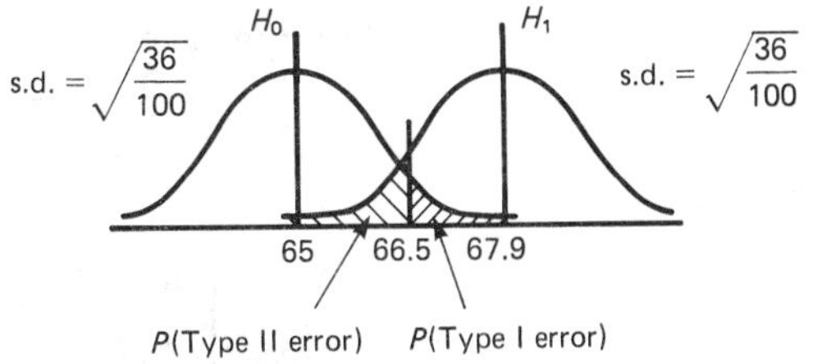

$$\begin{aligned} P(\text{Type II error}) &= P(\text{accept } H_0 \mid H_1 \text{ is true}) \\ &= P(\bar{X} \leqslant 66.5 \mid \mu = 67.9) \\ &= P\left(\frac{\bar{X}-67.9}{6/10} \leqslant \frac{66.5-67.9}{6/10}\right) \end{aligned}$$

$$= P(Z \leqslant -2.333)$$
$$= 0.009\,82$$

Therefore, $P(\text{Type II error}) = 0.009\,82$.

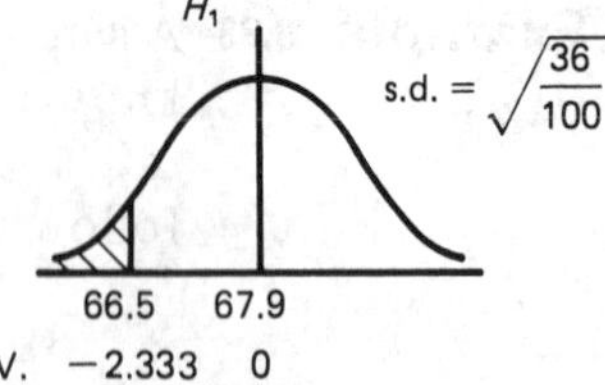

If he wants $P(\text{Type I error}) = P(\text{Type II error})$ then the critical value of $\bar{x}$ should be fixed so that

$$P(\bar{X} > \bar{x} \mid H_0 \text{ is true}) = P(\bar{X} \leqslant \bar{x} \mid H_1 \text{ is true})$$

As the variances of the distributions given by H_0 and H_1 are equal, we see, by symmetry, that the value of $\bar{x}$ lies mid-way between 65 and 67.9.

Therefore, he should take as critical value, $\bar{x} = \frac{1}{2}(65 + 67.9) = 66.45$.

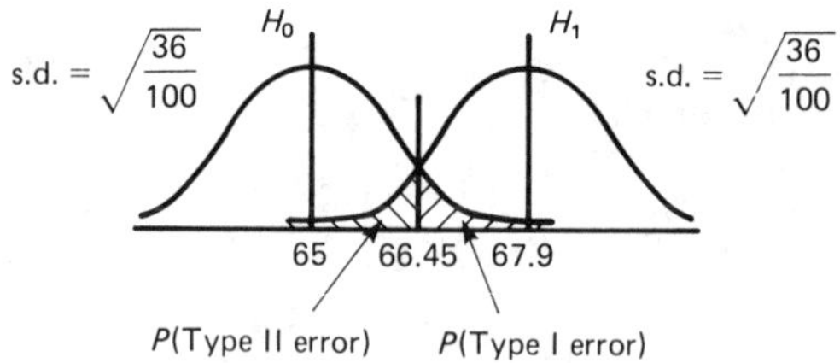

Example 9.34 The ingredients for concrete are mixed together to obtain a mean breaking strength of 2000 newtons. If the mean breaking strength drops below 1800 newtons then the composition must be changed. The distribution of the breaking strength is normal with standard deviation 200 newtons.

Samples are taken in order to investigate the hypotheses:

$$H_0: \quad \mu = 2000 \text{ newtons}$$
$$H_1: \quad \mu = 1800 \text{ newtons}$$

How many samples must be tested so that

$$P(\text{Type I error}) = \alpha = 0.05$$

and

$$P(\text{Type II error}) = \beta = 0.1?$$

Solution 9.34

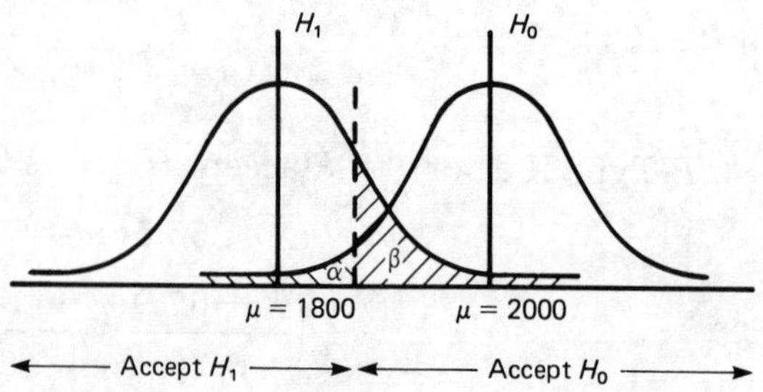

Under H_0, $\quad X \sim N(2000, 200^2)$

So, for a random sample of size n

$$\bar{X} \sim N\left(2000, \frac{200^2}{n}\right)$$

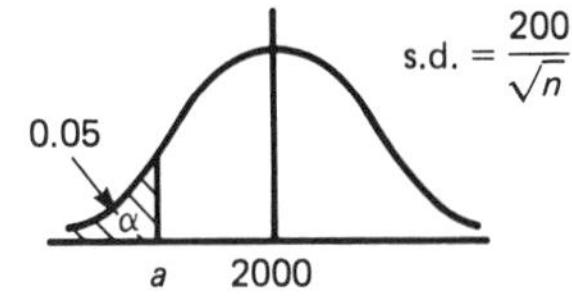

Now a corresponds to a standardised value of -1.645,

i.e. $$a = 2000 - 1.645\left(\frac{200}{\sqrt{n}}\right) \qquad \text{(i)}$$

Under H_1, $\quad X \sim N(1800, 200^2)$

So, for a random sample of size n

$$\bar{X} \sim N\left(1800, \frac{200^2}{n}\right)$$

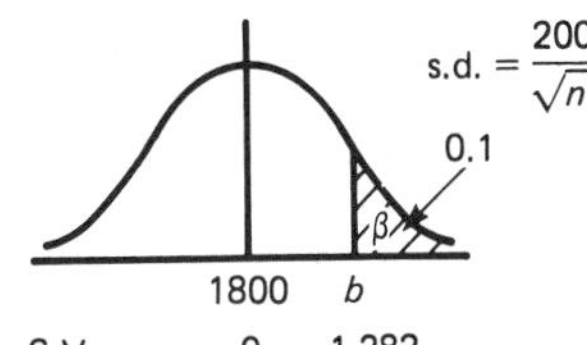

Now b corresponds to a standardised value of 1.282,

i.e. $$b = 1800 + 1.282\left(\frac{200}{\sqrt{n}}\right) \qquad \text{(ii)}$$

Now, if we find a value for n such that $\alpha = 0.05$ and $\beta = 0.1$

then $$a = b$$

Equating (i) and (ii)

$$2000 - 1.645\left(\frac{200}{\sqrt{n}}\right) = 1800 + 1.282\left(\frac{200}{\sqrt{n}}\right)$$

$$200 = \left(\frac{200}{\sqrt{n}}\right)(1.282 + 1.645)$$

$$\sqrt{n} = 2.927$$

$$n = 8.57$$

So the estimated number of samples which need to be tested is 9.

Exercise 9k

1. Two separate tests are proposed to determine whether a coin is biased or unbiased. These are:

   Test 1 — Toss the coin 4 times, and conclude that it is biased if all 4 tosses give the same result, and unbiased otherwise.

   Test 2 — Toss the coin 7 times, and conclude that it is biased if at least 6 of the tosses give the same result, and unbiased otherwise.

(*a*) Suppose that the coin is unbiased. Show that each test has the same probability of giving a wrong conclusion.
(*b*) Suppose that the coin is such that the probability of a head in any toss is $\frac{2}{3}$. Determine which test is more likely to give the conclusion that the coin is biased. (MEI)

2. Two alternative hypotheses concerning the probability density function of a random variable are

$$H_0: \quad f(x) = 2x \qquad 0<x<1$$
$$= 0 \qquad \text{otherwise}$$
$$H_1: \quad f(x) = 2(1-x) \qquad 0<x<1$$
$$= 0 \qquad \text{otherwise}$$

Give a sketch of the probability density function for each case.

The following test procedure is decided upon. A single observation of X is made and if X exceeds a particular value a, where $0<a<1$, then H_0 is accepted, otherwise H_1 is accepted. Find the value of a if the probability of accepting H_1 given that H_0 is true is $\frac{1}{9}$. With this value of a, find the probability of accepting H_0 given that H_1 is true. (C)

3. A manufacturer makes two grades of squash ball — 'slow' and 'fast'. Slow balls have a 'bounce' (measured under standard conditions) which is known to be a normal variable with mean 10 cm and standard deviation 2 cm. The 'bounce' of fast balls is a normal variable with mean 15 cm and standard deviation 2 cm. A box of balls is unlabelled so that it is not known whether they are all slow or all fast. Devise a test, based on a single observation of the bounce of one ball such that the probability of deciding that the box contains fast balls when in fact it contains slow balls, i.e. the Type I error, is equal to the Type II error.

Devise a test, based on an observation of the mean bounce of a sample of 4 balls from the box such that the Type I error is 0.05 and state the magnitude of the Type II error for this test. (C)

4. Two hypotheses concerning the probability density function of a random variable are

$$H_0: \quad f(x) = \begin{cases} 1 & 1<x<2 \\ 0 & \text{otherwise} \end{cases}$$

$$H_1: \quad f(x) = \begin{cases} 3x^2/7 & 1<x<2 \\ 0 & \text{otherwise} \end{cases}$$

Give a sketch of the probability density function for each case.

The following test procedure is decided upon:

A single observation of X is made and if X is less than a particular value a, where $1<a<2$, then H_0 is accepted; otherwise H_1 is accepted.

Find a such that, when H_0 is true, the test procedure leads, with probability 0.1, to the acceptance of H_1. With this value of a, find the probability that, when H_1 is true, the test procedure leads to the acceptance of H_0. (C)

5. One of two dice is loaded so that there is a probability of 0.2 of throwing a six with it, nothing being known about the other scores. The other die is fair. A person is given one of these dice (which is just as likely to be the fair as the biased one), together with the above information and is asked to discover which die it is. He decides to throw the die 10 times; if there are two or more sixes he will assert that the die is biased, otherwise he will assert that it is fair. Calculate the probability of his asserting that the die is (*a*) biased when it is, in fact, fair; (*b*) fair when it is, in fact, biased. What is the probability that his choice will be incorrect?

If, instead, he decided to throw the die 240 times and will assert that the die is biased if there are N or more sixes, use the normal approximation to the binomial distribution to estimate N if the probability of his asserting that it is fair when it is biased is to be 0.2. (SUJB)

6. An automated engineering process for manufacturing components includes an automatic screening of the output to reject defective components. The process gives on average 5% of defectives. The probability that the screening stage identifies correctly a defective component is 98% but there is also a probability of 6% that a component which is not defective is rejected at the screening stage. What is the proportion of all components which is rejected and what is the proportion of all components passed from the screening stage that is still defective? (MEI)

7. In order to examine a six-sided die for bias, one face is marked, the die is tossed a pre-determined number of times, and the number of times the marked face is uppermost is recorded.

(a) If this occurred r times in n tosses, explain how you would decide if this provided significant evidence of bias. Do not consider any approximate methods in this part.
(b) Would you consider it likely to be biased if the marked face came up once in 30 tosses?
(c) Would you consider it likely to be biased if the marked face came up 39 times in 180 tosses? (O)

8. Flour is packed in bags. The combined mass, X grams, of a full bag and its contents is a normally distributed random variable with mean μ grams and standard deviation 5 grams. When the packing machine is working correctly $\mu = 136$, but when the packing machine is working incorrectly $\mu = 130$. Show that the probabilities of a randomly chosen bag having a combined mass of less than 131.5 grams when the machine is working (a) correctly, (b) incorrectly, are approximately 0.2 and 0.6 respectively.

   When X is less than 131.5 the bag is underweight. Using the approximate probability 0.2, determine the probability that, when the machine is working correctly, in a random sample of five bags there are precisely k bags which are underweight, for $k = 3$, $k = 4$ and $k = 5$.

   The machine is presumed to be working incorrectly if the number of underweight bags found in a random sample of five bags is equal to or greater than r. Determine the minimum value of r which gives a probability less than 0.01 of presuming the machine to be working incorrectly when it is working correctly. (C)

9. One suggested test for deciding whether a coin is fair or not is to toss it four times and call it 'biased' if four heads or four tails are obtained. A second suggested way is to toss it seven times and call it biased if six or seven heads, or six or seven tails, are obtained. Show that both these tests would be equally likely to conclude wrongly that a fair coin was biased.

   Which of these two suggested tests would be better for correctly judging as biased a coin whose probability of coming down heads was 2/3?

   Are any of the above results statistically significant? (SMP)

10. A fair coin is tossed 100 times. Use a normal approximation to determine the probability of obtaining (a) more than 57 heads, (b) more than 58 heads.

    It is desired to construct a significance test to choose between the following two hypotheses concerning the possible bias of a coin:

    H_0: the probability that the coin falls heads is 0.5

    H_1: the probability that the coin falls heads is 0.6

    The coin is to be tossed 100 times and the number of heads, X, recorded. Construct a significance test based upon the observed value of X such that the probability of accepting H_1 when H_0 is true is as close as possible to 0.05. For this test calculate the probability of accepting H_0 when H_1 is true. (C)

11. Two alternative hypotheses for the probability density function of a random variable X are given below.

    H_0: $f(x) = a+\frac{1}{4}x \quad -1 \leqslant x \leqslant 1$
    $\quad = 0 \quad$ otherwise

    H_1: $f(x) = b-\frac{1}{4}x \quad -1 \leqslant x \leqslant 1$,
    $\quad = 0 \quad$ otherwise.

    Design a test, based on a single observation of X such that the probability of wrongly accepting H_0 is 0.05.

    Design also a test, based on a single observation of X, such that the probability of wrongly accepting H_0 is twice the probability of wrongly accepting H_1. (C)

12. You are provided with a coin which may be biased. In order to test this you are allowed to toss it 12 times and count the number, r, of heads and to use the value of r to decide. If the coin is really fair you wish to have at least a 95% chance of saying so. For what values of r should you say that the coin is fair?

    If you adopt your procedure with a coin which is actually biased two to one in favour of heads, what is the probability that you decide the coin is biased? (O)

13. Random samples of 400 seeds are taken from a large batch. For this batch the probability of a randomly chosen seed germinating is α. The r.v. X is defined as the number of germinating seeds in a sample. Use an appropriate normal approximation to determine the values of
    (a) $P(X \leqslant 340 \mid \alpha = 0.9)$
    (b) $P(X \geqslant 340 \mid \alpha = 0.8)$

The seed assessor knows that the value of α is either 0.8 or 0.9. Suppose that, in fact out of the 400 seeds in a particular sample a total of x germinate. The assessor decides that the value of α is 0.8 if $Z = P(X \geqslant x \mid \alpha = 0.8) - P(X \leqslant x \mid \alpha = 0.9)$ is positive. Otherwise he decides that α is 0.9. Determine the assessor's decision for each of the cases $x = 330$, $x = 340$, $x = 350$. (C)

10

THE χ^2 TEST

We now investigate the use of the chi-squared distribution in significance testing. It has a very complicated p.d.f. which has been included only for completeness. The information required for the test will be obtained from tables.

THE CHI-SQUARED DISTRIBUTION

Consider the r.v. X with p.d.f. $f(x)$ where

$$f(x) = K_\nu(\tfrac{1}{2}x)^{\frac{1}{2}\nu-1}\,\mathrm{e}^{-\frac{1}{2}x} \qquad x > 0$$

X has one parameter, ν and the constant K_ν depends on this parameter.

If X is distributed in this way, we write

$$X \sim \chi^2(\nu)$$

NOTE: χ^2 is pronounced 'kye-squared'.

The shape of the distribution for various values of ν is shown:

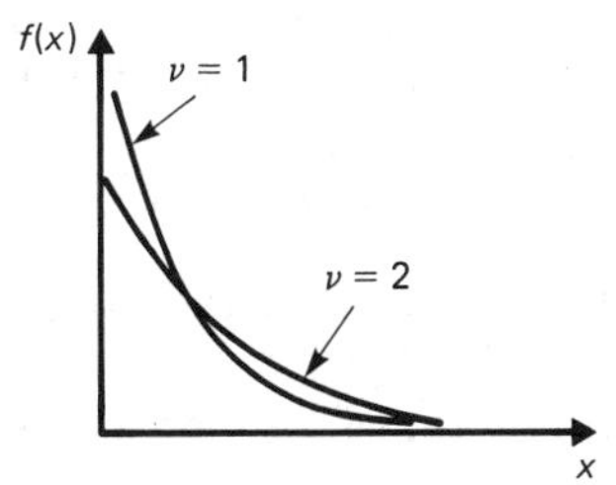

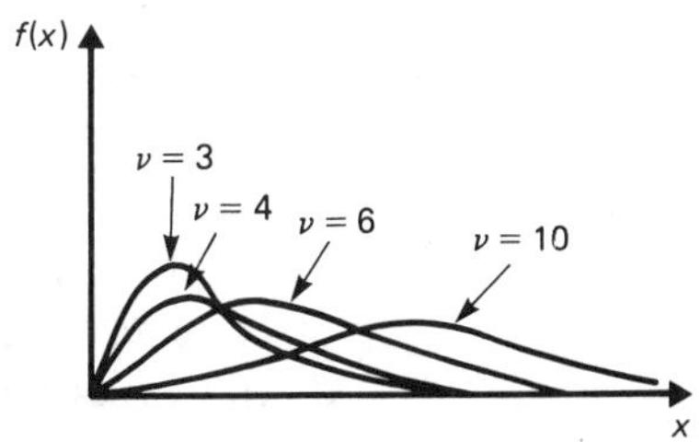

The probability that the value of X is greater than a particular value χ_p^2 is given by the tail area, shaded in the diagram.

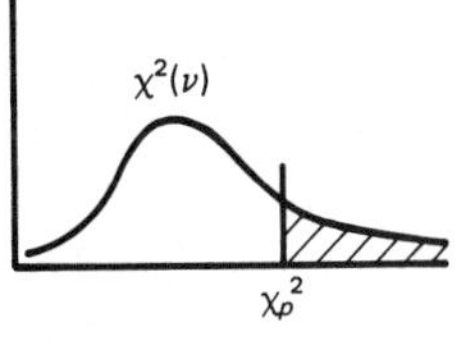

So $$P(X > \chi_p^2) = \int_{\chi_p^2}^{\infty} K_\nu(\tfrac{1}{2}x)^{\frac{1}{2}\nu-1}\mathrm{e}^{-\frac{1}{2}x}\,\mathrm{d}x$$

However, this integral is *very difficult* to evaluate, so we refer to χ^2 tables.

In significance tests we are usually concerned with the values of χ_p^2 such that

$$P(X > \chi_p^2) = 0.05 \quad \text{(5\% of the area is in the tail)}$$
$$P(X > \chi_p^2) = 0.01 \quad \text{(1\% of the area is in the tail)}$$

These values are summarised on p. 637; the first few lines are reproduced below:

ν	$\chi^2_{5\%}$	$\chi^2_{1\%}$
1	3.84	6.64
2	5.99	9.21
3	7.82	11.34
4	9.49	13.28

For example, if $\nu = 4$,

$$\chi^2_{5\%} = 9.49 \qquad \chi^2_{1\%} = 13.28$$
$$P(X > 9.49) = 0.05 \qquad P(X > 13.28) = 0.01$$

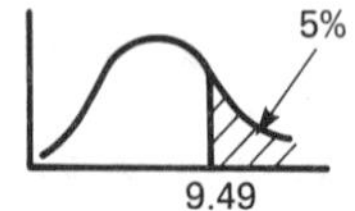

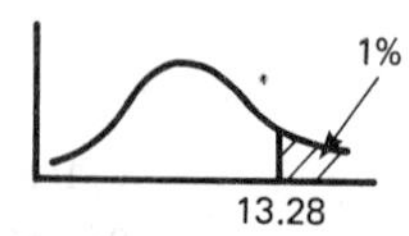

THE χ^2 TEST

Consider an experiment or situation which results in n observed frequencies, written O_i, $i = 1, 2, \ldots, n$.

Say we wish to make a hypothesis about the distribution; we could then calculate the frequencies expected under this hypothesis, written E_i, $i = 1, 2, \ldots, n$. We now decide how well the 'observed' data fits the 'expected' data, and consider whether it is likely that the differences can be attributed to chance.

Now, the comparison between the observed frequencies (O_i) and the expected frequencies (E_i), for $i = 1, 2, \ldots, n$ (that is, for n pairs of values, or classes) is made by considering the statistic

$$\sum_{i=1}^{n} \frac{(O_i - E_i)^2}{E_i}$$

For high values of O_i and E_i *this statistic approximates to the chi-squared distribution.*

We define

$$\chi^2_{\text{calc}} = \sum_{i=1}^{n} \frac{(O_i - E_i)^2}{E_i}$$

Now

if $\chi^2_{\text{calc}} = 0$ then there is *exact* agreement between the observed and the expected data,

if $\chi^2_{\text{calc}} > 0$ then O_i and E_i do not agree exactly and, for a given value of ν, the larger the value of χ^2_{calc} the greater the discrepancy.

For a test performed at the 5% level:

if $\chi^2_{\text{calc}} \geqslant \chi^2_{5\%}$ then we consider the discrepancy to be too large and reject the null hypothesis,

if $\chi^2_{\text{calc}} < \chi^2_{5\%}$ we do not reject the null hypothesis.

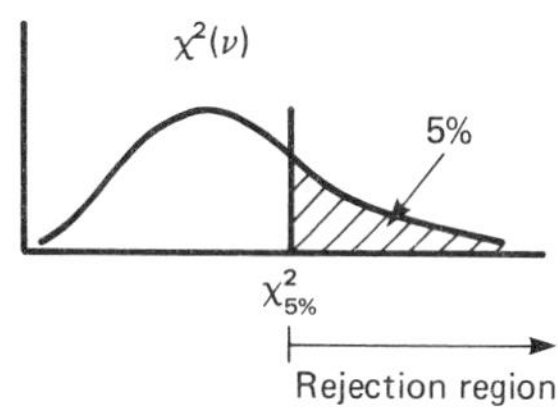

NOTE: **(a)** When using the chi-squared distribution we are approximating from a discrete to a continuous distribution.
(*i*) The approximation is not valid if the expected frequency is less than 5. This problem can be overcome by combining two or more classes with small frequencies to form a class sufficiently large.
(*ii*) If $\nu = 1$, it is advisable to use Yates' continuity correction. In this case

$$\chi^2_{\text{calc}} = \sum_{i=1}^{n} \frac{(|O_i - E_i| - 0.5)^2}{E_i}$$

(b) When the value of χ^2_{calc} is *very small*, it is wise to query the reliability of the observed data and to question whether they have been 'fiddled'.

Degrees of freedom

The parameter ν is known as the *number of degrees of freedom*. Now the number of degrees of freedom associated with a statistic is given by

ν = number of independent variables involved in calculating the statistic

This can be found by considering

ν = number of classes − number of restrictions

When considering the statistic $\sum_{i=1}^{n} \frac{(O_i - E_i)^2}{E_i}$ the number of classes is the number of pairs of values, i.e. there are n classes. The number of restrictions involved depends on the null hypothesis. We consider several cases in the following examples.

UNIFORM DISTRIBUTION

Example 10.1 The table shows the number of employees absent for a single day during a particular period of time:

Day	Monday	Tuesday	Wednesday	Thursday	Friday	Total
Number of absentees	121	87	87	91	114	500

(a) Find the frequencies expected under the hypothesis that the number of absentees is independent of the day of the week.

(b) Test, at the 5% level, whether the differences in the observed and expected data are significant.

Solution 10.1 (a) If the number of absentees is independent of the day of the week, then we would expect the total of 500 to be spread uniformly throughout the week, so that the expected number of absentees for any day is 100.

Expected frequencies:

Day	Monday	Tuesday	Wednesday	Thursday	Friday	Total
Number of absentees	100	100	100	100	100	500

(b) *Degrees of freedom*, ν

Now ν = number of independent variables

= number of classes − number of restrictions

There are 5 classes and, since the total expected and observed frequencies each have to be 500, there is one restriction.

Therefore $\nu = 5-1 = 4$

The χ^2 test is carried out as follows:

Reminders:	Significance test
(1) Make the null hypothesis (H_0)	H_0: the number of absentees is independent of the day of the week
(2) Work out the number of degrees of freedom, ν	ν = number of classes − number of restrictions $= 5 - 1$ $= 4$
(3) Decide on the level of the test	Test at the 5% level
(4) Decide on rejection criterion	and reject H_0 if $\chi^2_{calc} > \chi^2_{5\%}(4)$, i.e. if $\chi^2_{calc} > 9.49$ $\chi^2(4)$ 5% 9.49 Reject H_0
(5) Calculate χ^2_{calc}	where $\chi^2_{calc} = \sum_{i=1}^{5} \frac{(O_i - E_i)^2}{E_i}$ (see table below) Therefore $\chi^2_{calc} = 10.56$
(6) Make conclusion	As $\chi^2_{calc} > 9.49$ we reject H_0 and conclude that the number of absentees is not independent of the day of the week.

O_i	E_i	$\lvert O_i - E_i \rvert$	$(O_i - E_i)^2$	$\frac{(O_i - E_i)^2}{E_i}$
121	100	21	441	4.41
87	100	13	169	1.69
87	100	13	169	1.69
91	100	9	81	0.81
114	100	14	196	1.96
$\Sigma O_i = 500$	$\Sigma E_i = 500$			10.56

NOTE: **(a)** The test does not indicate what the relationship might be between number of absentees and the day of the week. However, a look at the observed frequencies suggests a tendency towards a greater number of absentees on Mondays and Fridays.

(b) When working out the table we are not concerned whether $O_i - E_i$ is positive or negative, as we require $(O_i - E_i)^2$. Therefore we find $\lvert O_i - E_i \rvert$.

Example 10.2 An ordinary die is thrown 120 times and each time the number on the uppermost face is noted. The results are as follows:

Number on die	1	2	3	4	5	6	Total
Frequency	14	16	24	22	24	20	120

Perform a χ^2 test, at the 5% level, to investigate whether the die is fair.

Solution 10.2 H_0: The die is fair.

Now $\Sigma O_i = \Sigma E_i$, therefore there is one restriction, namely that the totals agree.

$$\begin{aligned} \nu &= \text{number of classes} - \text{number of restrictions} \\ &= 6 - 1 \\ &= 5 \end{aligned}$$

Therefore $\nu = 5$ and we consider the $\chi^2(5)$ distribution.

We will test at the 5% level and reject H_0 if $\chi^2_{\text{calc}} > \chi^2_{5\%}(5)$ i.e. if $\chi^2_{\text{calc}} > 11.07$

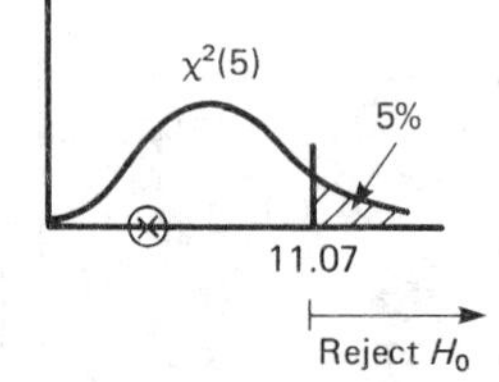

where $\quad \chi^2_{\text{calc}} = \displaystyle\sum_{i=1}^{6} \frac{(O_i - E_i)^2}{E_i}$

Now, under H_0 (that the die is fair) we would expect each number to occur the same number of times,

so $\quad E_i = 20 \quad \text{for} \quad i = 1, 2, \ldots, 6$

O_i	E_i	$\lvert O_i - E_i \rvert$	$(O_i - E_i)^2$	$\dfrac{(O_i - E_i)^2}{E_i}$
14	20	6	36	1.8
16	20	4	16	0.8
24	20	4	16	0.8
22	20	2	4	0.2
24	20	4	16	0.8
20	20	0	0	0
$\Sigma O_i = 120$	$\Sigma E_i = 120$			4.4

Therefore $\chi^2_{\text{calc}} = 4.4$.

As $\chi^2_{\text{calc}} < 11.07$ we do not reject H_0 and conclude that the differences between the observed and expected frequencies are not significant at the 5% level and the die is fair.

DISTRIBUTION IN A GIVEN RATIO

Example 10.3 According to genetic theory the number of colour-strains pink, white and blue in a certain flower should appear in the ratio 3 : 2 : 5. For 100 plants, the results were as follows:

Colour	Pink	White	Blue	Total
Number of plants	24	14	62	100

Are the differences between the observed and expected frequencies significant, at the 1% level?

Solution 10.3 H_0: the colours pink, white and blue occur in the ratio 3 : 2 : 5.

Now $\Sigma O_i = \Sigma E_i = 100$, so there is one restriction, that the totals agree.

So ν = number of classes − number of restrictions
$= 3 - 1$
$= 2$

Therefore $\nu = 2$ and we consider the $\chi^2(2)$ distribution.

We test at the 1% level and reject H_0 if $\chi^2_{\text{calc}} > \chi^2_{1\%}$, i.e. if $\chi^2_{\text{calc}} > 9.21$ where

$$\chi^2_{\text{calc}} = \sum_{i=1}^{3} \frac{(O_i - E_i)^2}{E_i}$$

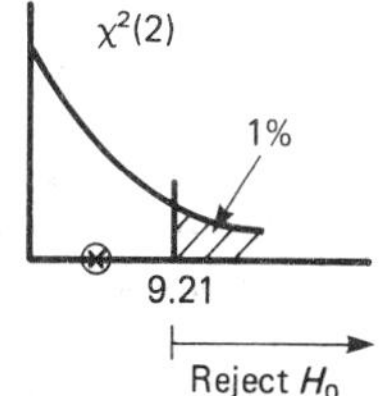

Now under H_0 we expect the colours pink, white and blue to appear in the ratio 3 : 2 : 5, so the expected frequencies are

$$\frac{3}{10}(100) : \frac{2}{10}(100) : \frac{5}{10}(100) = 30 : 20 : 50$$

O_i	E_i	$\lvert O_i - E_i \rvert$	$(O_i - E_i)^2$	$\frac{(O_i - E_i)^2}{E_i}$
24	30	6	36	1.2
14	20	6	36	1.8
62	50	12	144	2.88
$\Sigma O_i = 100$	$\Sigma E_i = 100$			5.88

As $\chi^2_{\text{calc}} < 9.21$, we do not reject H_0 and conclude that the differences in observed and expected frequencies are not significant at the 1% level.

Exercise 10a

1. A tetrahedral die is thrown 120 times and the number on which it lands is noted.

Number	1	2	3	4	
Frequency	35	32	25	28	Total 120

Test, at the 5% level whether the die is fair.

2. From a list of 500 digits, the occurrence of each digit is noted.

Digit	0	1	2	3	4	5	6	7	8	9
Frequency	40	58	49	53	38	56	61	53	60	32

Test, at the 1% level, whether the sequence is a random sample from a uniform distribution.

3. The outcomes, A, B and C, of a certain experiment are thought to occur in the ratio 1 : 2 : 1. The experiment is performed 200 times and the observed frequencies of A, B and C are 36, 115 and 49 respectively. Is the difference in the observed and expected results significant? Test at the 5% level.

4. According to genetic theory the number of colour strains red, yellow, blue and white in a certain flower should appear in proportions 4 : 12 : 5 : 4. Observed frequencies of red, yellow, blue and white strains amongst 800 plants were 110, 410, 150, 130 respectively. Are these differences from the expected frequencies significant at the 5% level? If the number of plants had been 1600 and the observed frequencies 220, 820, 300, 260, would the difference have been significant at the 5% level? (C Additional)

5. It is thought that each of the 8 outcomes of an experiment is equally likely to occur. When the experiment is performed 400 times, the observed frequencies are 45, 42, 55, 53, 40, 62, 47 and 56. Perform a test at the 1% level to investigate the validity of the theory.

6. In a particular subject students are set multiple choice questions each of which contain 5 alternatives A, B, C, D and E. A teacher suggests that when students do not know the correct answer they are twice as likely to choose one of B, C or D than to choose A or E. For 160 questions where it was known that the student answered without knowing the correct answer, A, B, C, D, E were chosen 23, 45, 36, 43 and 13 times respectively. Is there evidence, at the 5% level, to support the teacher's theory?

7. For a given set of data the observed and expected frequencies are shown:

Result	1	2	3	4	5
Observed frequency	30	31	42	40	57
Expected frequency	38	45	36	36	45

Are the differences between the observed and expected frequencies significant at the 1% level?

'GOODNESS OF FIT' TESTS

We now illustrate the use of the chi-squared test to investigate whether an observed distribution fits a well-known distribution such as the binomial, Poisson or normal. The test is often referred to as a 'goodness of fit' test.

BINOMIAL DISTRIBUTION, p known

Example 10.4 Four coins are thrown 160 times, and the distribution of the number of heads is observed to be

x (number of heads)	0	1	2	3	4
f (frequency)	5	35	67	41	12

Find the expected frequencies if the coins are unbiased. Compare the observed and expected frequencies and apply the χ^2 test. Is there any evidence that the coins are biased? (AEB 1974)

Solution 10.4 Let X be the r.v. 'the number of heads obtained when four coins are thrown'. Then if the coins are unbiased $X \sim \text{Bin}(4, \frac{1}{2})$

and $P(X = x) = {}^4C_x\left(\frac{1}{2}\right)^{4-x}\left(\frac{1}{2}\right)^x = {}^4C_x\left(\frac{1}{2}\right)^4 \quad x = 0, 1, \ldots, 4$

	x (number of heads)	Expected frequencies $[160P(X = x)]$
$P(X=0) = \left(\frac{1}{2}\right)^4 = \frac{1}{16}$	0	10
$P(X=1) = 4\left(\frac{1}{2}\right)^4 = \frac{4}{16}$	1	40
$P(X=2) = 6\left(\frac{1}{2}\right)^4 = \frac{6}{16}$	2	60
$P(X=3) = 4\left(\frac{1}{2}\right)^4 = \frac{4}{16}$	3	40
$P(X=4) = \left(\frac{1}{2}\right)^4 = \frac{1}{16}$	4	10
		Total 160

χ^2 test:

H_0: the coins are not biased and $P(\text{head}) = \frac{1}{2}$

Degrees of freedom: number of classes $= 5$
number of restrictions $= 1$
(totals must agree)

Therefore $\nu = 5 - 1 = 4$, and we consider the $\chi^2(4)$ distribution.

We will test at the 5% level and reject H_0 if $\chi^2_{\text{calc}} > \chi^2_{5\%}(4)$
i.e. if $\chi^2_{\text{calc}} > 9.49$.

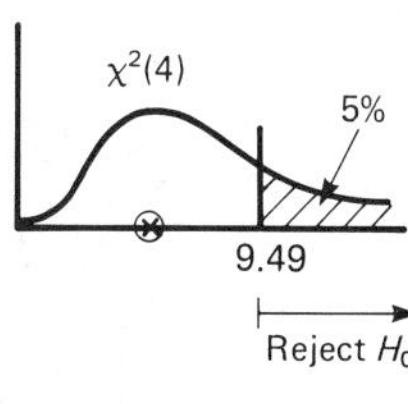

Now $$\chi^2_{\text{calc}} = \sum_{i=1}^{5} \frac{(O_i - E_i)^2}{E_i}$$

O_i	E_i	$\lvert O_i - E_i \rvert$	$(O_i - E_i)^2$	$\frac{(O_i - E_i)^2}{E_i}$
5	10	5	25	2.5
35	40	5	25	0.625
67	60	7	49	0.817
41	40	1	1	0.025
12	10	2	4	0.4
$\Sigma O_i = 160$	$\Sigma E_i = 160$			4.367

Therefore $\chi^2_{calc} = 4.367$ (3 d.p.).

As $\chi^2_{calc} < 9.49$, we do not reject H_0 and conclude that there is no evidence that the coins are biased.

BINOMIAL DISTRIBUTION, *p* unknown

Example 10.5 Samples of size 5 are selected regularly from a production line and tested. During one week 500 samples are taken and the number of defective items in each sample recorded.

Number of defectives, x	0	1	2	3	4	5
Frequency, f	170	180	120	20	8	2

(a) Find the frequencies of the number of defectives per sample given by the binomial distribution having the same mean and total as the observed distribution.

(b) Test whether the observed distribution follows a binomial pattern.

Solution 10.5 Now

$$\bar{x} = \frac{\Sigma fx}{\Sigma f} = 1.044$$

and $n = 5$, $p = P(\text{defective item})$.

We require $np = \bar{x}$

i.e. $$5p = 1.044$$

So $$p = 0.2088$$

Let X be the r.v. 'the number of defectives in a sample'.

Then $X \sim \text{Bin}(n, p)$ with $n = 5, \quad p = 0.2088,$

$$q = 1 - p = 0.7912$$

We have

$$P(X = x) = {}^5C_x (0.7912)^{5-x}(0.2088)^x \quad x = 0, 1, \ldots, 5$$

The expected frequencies can be found by calculating $500P(X = x)$ for $x = 0, 1, \ldots, 5$.

These are shown in the table. Frequencies have been rounded to the nearest integer.

Number of defectives	0	1	2	3	4	5
Expected frequency	155	205	108	28	4	0

(b) *Perform a* χ^2 test.

H_0: the distribution follows a binomial pattern

Now, we note that the expected frequencies for two of the classes are less than 5. So we combine the last three classes to read '3 or more defectives'. Therefore the classes are 0, 1, 2, 3 or more.

Degrees of freedom: The number of classes = 4

The number of restrictions = 2
(totals agree, means agree)

Therefore $\nu = 4-2 = 2$ and we consider the $\chi^2(2)$ distribution.

We test at the 5% level and reject H_0 if $\chi^2_{\text{calc}} > \chi^2_{5\%}$,

i.e. if $\chi^2_{\text{calc}} > 5.99$.

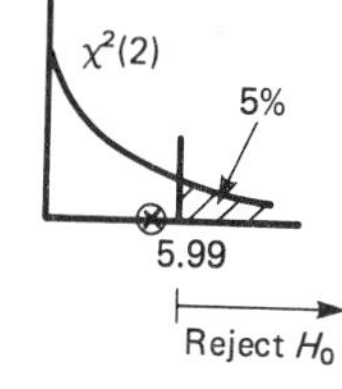

Now $$\chi^2_{\text{calc}} = \sum_{i=1}^{4} \frac{(O_i - E_i)^2}{E_i}$$

O_i	E_i	$\lvert O_i - E_i \rvert$	$(O_i - E_i)^2$	$\frac{(O_i - E_i)^2}{E_i}$
170	155	15	225	1.452
180	205	25	625	3.049
120	108	12	144	1.333
30	32	2	4	0.125
$\Sigma O_i = 500$	$\Sigma E_i = 500$			5.959

Therefore $\chi^2_{\text{calc}} = 5.959$.

As $\chi^2_{\text{calc}} < 5.99$ we do not reject H_0 and conclude that the distribution follows a binomial pattern.

POISSON DISTRIBUTION

Example 10.6 Analysis of the goals scored per match by a certain football team gave the following results:

No. goals per match (x)	0	1	2	3	4	5	6	7
No. of matches (f)	14	18	29	18	10	7	3	1

Calculate the mean of the above distribution and the frequencies (each correct to 1 decimal place) associated with a Poisson distribution having the same mean. Perform a χ^2 goodness of fit test to determine whether or not the above distribution can be reasonably modelled by this Poisson distribution. (SUJB)

Solution 10.6 Now

$$\bar{x} = \frac{\Sigma fx}{\Sigma f} = \frac{230}{100} = 2.3$$

Consider the r.v. X where $X \sim \text{Po}(2.3)$, X is 'the number of goals per match'.

Then

$$P(X=x) = \frac{e^{-2.3}(2.3)^x}{x!} \qquad x = 0, 1, 2, \ldots$$

and the expected frequencies are given by $100P(X=x)$.

These have been calculated and are shown in the table.

Number of goals per match	0	1	2	3	4	5	6	7	8 or more
Expected frequency (1 d.p.)	10.0	23.1	26.5	20.3	11.7	5.4	2.1	0.7	0.2

χ^2 *test*:

H_0: the distribution is Poisson

As the χ^2 test is not valid for expected frequencies less than 5, we combine the end categories into '5 or more goals'

Degrees of freedom: number of classes $= 6$

number of restrictions $= 2$
(totals agree, means agree)

Therefore $\nu = 6 - 2 = 4$ and we consider the $\chi^2(4)$ distribution.

We wish to test at the 5% level and reject H_0 if $\chi^2_{\text{calc}} > \chi^2_{5\%}(4)$,

i.e. if $\chi^2_{\text{calc}} > 9.49$,

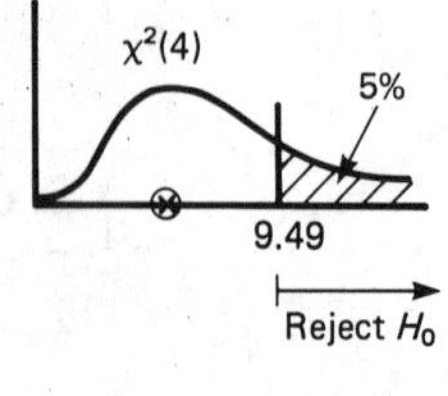

where

$$\chi^2_{\text{calc}} = \sum_{i=1}^{6} \frac{(O_i - E_i)^2}{E_i}$$

O_i	E_i	$\lvert O_i - E_i \rvert$	$(O_i - E_i)^2$	$\frac{(O_i - E_i)^2}{E_i}$
14	10.0	4.0	16	1.6
18	23.1	5.1	26.01	1.126
29	26.5	2.5	6.25	0.236
18	20.3	2.3	5.29	0.261
10	11.7	1.7	2.89	0.247
11	8.4	2.6	6.76	0.805
$\Sigma O_i = 100$	$\Sigma E_i = 100$			4.275

Therefore $\chi^2_{\text{calc}} = 4.275$.

As $\chi^2_{calc} < 9.49$, we do not reject H_0 and conclude that the distribution can be reasonably modelled by the Poisson distribution having the same mean.

NORMAL DISTRIBUTION, mean and variance known

Example 10.7 For a period of six months 100 similar hamsters were given a new type of feedstuff. The gains in mass are recorded in the table below:

Gain in mass (g) x	Observed frequency f
$-\infty < x \leqslant -10$	3
$-10 < x \leqslant -5$	6
$-5 < x \leqslant 0$	9
$0 < x \leqslant 5$	15
$5 < x \leqslant 10$	24
$10 < x \leqslant 15$	16
$15 < x \leqslant 20$	14
$20 < x \leqslant 25$	8
$25 < x \leqslant 30$	3
$30 < x \leqslant \infty$	2

It is thought that these data follow a normal distribution, with mean 10 and variance 100. Use the χ^2 distribution at the 5% level of significance to test this hypothesis.

Describe briefly how you would modify this test if the mean and variance were unknown. (AEB)

Solution 10.7 Let X be the r.v. 'the gain in mass', then $X \sim \mathrm{N}(10, 100)$. We calculate $P(a < X \leqslant b) = p$ from the normal distribution tables (p. 634) and work out the expected frequencies using $E_i = 100p$.

Interval $(a < x \leqslant b)$	Upper class boundary (u.c.b.)	Standardised u.c.b. (z)	$P(Z \leqslant z)$	$P(a < X \leqslant b)$	$E_i = 100p$
$-\infty < x \leqslant -10$	-10	-2	0.0228	0.0228	2.28
$-10 < x \leqslant -5$	-5	-1.5	0.0668	0.044	4.4
$-5 < x \leqslant 0$	0	-1	0.1587	0.0919	9.19
$0 < x \leqslant 5$	5	-0.5	0.3085	0.1498	14.98
$5 < x \leqslant 10$	10	0	0.5000	0.1915	19.15
$10 < x \leqslant 15$	15	0.5	0.6915	0.1915	19.15
$15 < x \leqslant 20$	20	1	0.8413	0.1498	14.98
$20 < x \leqslant 25$	25	1.5	0.9332	0.0919	9.19
$25 < x \leqslant 30$	30	2	0.9772	0.044	4.4
$30 < x \leqslant \infty$	∞	∞	1	0.0228	2.28
					$\Sigma E_i = 100$

χ^2 test:

H_0: the distribution is normal with mean 10 and variance 100

We note that the expected frequencies given by H_0 are such that the first two classes contain less than 5, similarly the last two classes. So these are combined to give two classes instead of 4.

Degrees of freedom: number of classes = 8

number of restrictions = 1
(totals agree)

Therefore $\nu = 8-1 = 7$ and the $\chi^2(7)$ distribution is considered.

We test at the 5% level and reject H_0 if
$\chi^2_{calc} > \chi^2_{5\%}(7)$,
i.e. if $\chi^2_{calc} > 14.07$,

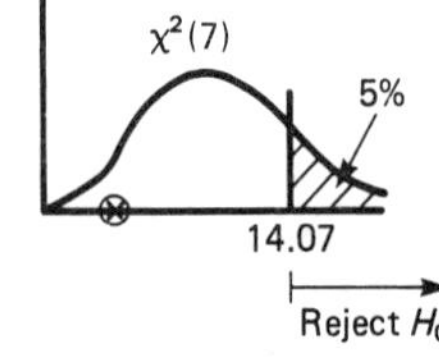

where $$\chi^2_{calc} = \sum_{i=1}^{8} \frac{(O_i - E_i)^2}{E_i}$$

O_i	E_i	$\lvert O_i - E_i \rvert$	$(O_i - E_i)^2$	$\frac{(O_i - E_i)^2}{E_i}$
9	6.68	2.32	5.3824	0.806
9	9.19	0.19	0.0361	0.004
15	14.98	0.02	0.0004	0.000
24	19.15	4.85	23.5225	1.228
16	19.15	3.15	9.9225	0.518
14	14.98	0.98	0.9604	0.064
8	9.19	1.19	1.4161	0.154
5	6.68	1.68	2.8224	0.423
$\Sigma O_i = 100$	$\Sigma E_i = 100$			3.197

Therefore $\chi^2_{calc} = 3.197$.

As $\chi^2_{calc} < 14.07$, we do not reject H_0 and conclude that the data follows a normal distribution with mean 10 and variance 100.

NORMAL DISTRIBUTION, mean and variance unknown

If the mean and variance are not given for the normal distribution then these have to be *estimated* from the observed data. The expected frequencies are then calculated using these estimates.

This alters the number of degrees of freedom, for if estimates of the mean and the variance are used, then

number of restrictions = 3 (totals agree, means agree, standard deviations agree)

For Example 10.7, $\nu = 8-3 = 5$ and the $\chi^2(5)$ distribution would be considered.

Exercise 10b

1. Perform a χ^2 test to investigate whether the following data is drawn from a binomial distribution with $p = 0.3$. Use a 5% level of significance.

x	0	1	2	3	4	5
f	12	39	27	15	4	3

2. Using the data of Question 1, Exercise 4d (p. 225) carry out a χ^2 test, at the 5% level of significance whether the observed results are sufficiently close to the theoretical results to support an assumption of a binomial distribution with mean as calculated from the observed data.

3. Repeat the procedure in Question 2 for the data given in Exercise 4e (p. 228) (*a*) question 4 (*b*) question 6.

4. Two dice were thrown 216 times, and the number of sixes at each throw were counted. The results were:

No. of sixes	0	1	2	
Frequency	130	76	10	Total 216

Test the hypothesis that the distribution is binomial with the parameter $p = \frac{1}{6}$. Explain how the test would be modified if the hypothesis to be tested is that the distribution is binomial with the parameter p unknown. (Do not carry out the test.) (O)

5. A six-sided die with faces numbered as usual from 1 to 6 was thrown 5 times and the number of sixes was recorded. The experiment was repeated 200 times, with the following results:

x	0	1	2	3	4	5
Frequency	66	82	40	10	2	0

On this evidence, would you consider the die to be biased? Fit a suitable distribution to the data and test and comment on the goodness of fit. (MEI)

6. Under what circumstances would you expect a variate, X, to have a binomial distribution? What is the mean of X if it has a binomial distribution with parameters n and p?

A new fly spray is applied to 50 samples each of 5 flies and the number of living flies counted after one hour. The results were as follows:

Number living	0	1	2	3	4	5
Frequency	7	20	12	9	1	1

Calculate the mean number of living flies per sample and hence an estimate for p, the probability of a fly surviving the spray. Using your estimate calculate the expected frequencies (each correct to one place of decimals) corresponding to a binomial distribution and perform a χ^2 goodness-of-fit test using a 5% significance level. (SUJB)

7. Table A overleaf gives the distribution for the number of heavy rainstorms reported by 330 weather stations in the United States of America over a one year period.

(*a*) Find the expected frequencies of rainstorms given by the Poisson distribution having the same mean and total as the observed distribution.
(*b*) Use the χ^2 distribution to test the adequacy of the Poisson distribution as a model for these data. (AEB 1977)

8. Use the χ^2 distribution to test the adequacy of the Poisson distribution as a model for the data given in Example 4.31 (p. 256).

9. The numbers of cars passing a checkpoint during 100 intervals each of time 5 minutes, were noted:

Number of cars	0	1	2	3	4	5	6 or more
Frequency	5	23	23	25	14	10	0

Fit a Poisson distribution to these data and test the goodness of fit.

10. Write a short account of the χ^2 test of goodness of fit, giving some indication of its shortcomings.

During the weaving of cloth the thread sometimes breaks. 147 lengths of thread of equal length were observed during weaving and the table records the number of these threads for which the indicated number of breaks occurred.

Number of breaks per thread	0	1	2	3	4	5
Number of threads	48	46	30	12	9	2

Fit a Poisson distribution to the data and examine whether the deviation between theory and experiment is significant. (MEI)

11. The table below gives the distribution of the number of hits by flying bombs in 450 equally sized areas in South London during World War II.

Number of hits (x)	0	1	2	3	4	5	6 or more
Frequency (f)	180	173	69	20	6	2	0

(*a*) Find the expected frequencies of hits given by a Poisson distribution having the same mean and total as the observed distribution.
(*b*) Use the χ^2 distribution and a 10% level of significance to test the adequacy of the Poisson distribution as a model for these data. (AEB 1980)

See Exercise 10e, Question 5, (p. 556) for values of $\chi^2_{10\%}(\nu)$.

12. The following data (Table B) gives the heights in cm of 100 male students:
(*a*) Test, at the 5% level, whether the data follows a normal distribution with mean 173.5 cm and standard deviation 7 cm.
(*b*) Find the expected frequencies for a normal distribution having the same mean and variance as the data given, and test the goodness of fit, using a 5% level of significance.

13. During observations on a patch of white dead nettles it was noticed that the numbers of flowers visited by bees during 100 5-minute intervals were as follows:

Number of flowers visited/5-minute interval	Frequency
0–5	4
6–11	8
12–17	13
18–23	16
24–29	22
30–35	17
36–41	10
42–47	6
48–53	3
54–59	1

(*a*) Calculate the mean and variance for the data.
(*b*) Find the expected frequencies for a normal distribution with the same mean and variance.
(*c*) Test, at the 5% level of significance, how well the observed data fits this normal distribution.

Table A

Number of rainstorms (x)	0	1	2	3	4	5	more than 5
Number of stations (f) reporting x rainstorms	102	114	74	28	10	2	0

Table B

Height (cm)	155–160	161–166	167–172	173–178	179–184	185–190
Frequency	5	17	38	25	9	6

USE OF χ^2 TESTS IN CONTINGENCY TABLES

Sometimes situations arise when individuals are classified according to two sets of attributes. We may then wish to investigate whether the two sets of attributes are independent, or whether there is evidence of an association between them.

2 × 2 Contingency tables (2 rows and 2 columns)

Example 10.8 A driving school examined the results of 100 candidates who were taking their driving test for the first time. They found that of the 40 men, 28 passed and out of the 60 women, 34 passed. Do these results indicate, at the 5% level of significance, a relationship between the sex of a candidate and the ability to pass first time?

Solution 10.8 The results can be shown in a table, known as a 2×2 (read '2 by 2') contingency table:

		Results of first-time candidates		
		Pass	Fail	Totals
Sex	Male	28	12	40
	Female	34	26	60
	Totals	62	38	100

H_0: there is no relationship between the sex of a candidate and the ability to pass first time; the attributes are independent

To calculate the expected frequencies:

$$P(\text{candidate is male}) = \frac{40}{100}$$

$$P(\text{candidate passes first time}) = \frac{62}{100}$$

Under H_0, the events are independent. Therefore

$$P(\text{candidate passes first time and is male}) = \left(\frac{40}{100}\right)\left(\frac{62}{100}\right)$$

$$\text{Expected number who pass and are male} = 100\left(\frac{40}{100}\right)\left(\frac{62}{100}\right)$$

$$= \frac{(40)(62)}{100} = 24.8$$

NOTE: Expected frequency $= \dfrac{\text{(row total)(column total)}}{\text{grand total}}$.

We could work through this procedure to give the other expected frequencies, but this is unnecessary, as the other frequencies can be found by using the fact that the sub-totals and totals must agree with those in the observed data:

Expected frequencies:

		Results of first-time candidates		
		Pass	Fail	Totals
Sex	Male	**24.8**	15.2	40
	Female	37.2	22.8	60
	Totals	62	38	100

Degrees of freedom: number of independent variables = 1
(once one expected frequency is known, the others are determined by agreement of totals).

Therefore $\nu = 1$ and we consider the $\chi^2(1)$ distribution.

NOTE: As $\nu = 1$, we use Yates' continuity correction when calculating χ^2_{calc}.

We test at the 5% level and reject H_0 if $\chi^2_{calc} > \chi^2_{5\%}(1)$,
i.e. if $\chi^2_{calc} > 3.84$,

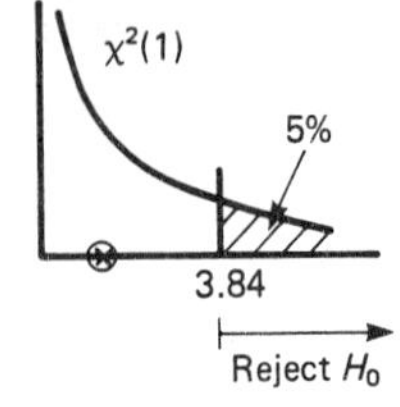

where $$\chi^2_{calc} = \sum_{i=1}^{4} \frac{(|O_i - E_i| - 0.5)^2}{E_i}$$

(using Yates' continuity correction)

O_i	E_i	$\|O_i - E_i\|$	$(\|O_i - E_i\| - 0.5)$	$(\|O_i - E_i\| - 0.5)^2$	$\frac{(\|O_i - E_i\| - 0.5)^2}{E_i}$
28	24.8	3.2	2.7	7.29	0.294
12	15.2	3.2	2.7	7.29	0.480
34	37.2	3.2	2.7	7.29	0.196
26	22.8	3.2	2.7	7.29	0.320
$\Sigma O_i = 100$	$\Sigma E_i = 100$				1.290

Therefore $\chi^2_{calc} = 1.29$.

As $\chi^2_{calc} < 3.84$, we do not reject H_0 and conclude that these results do not indicate a relationship between the sex of the candidate and the ability to pass first time.

Exercise 10c

1. In an investigation into eye colour and left or right handedness the following results were obtained:

		Handedness	
		Left	Right
Eye colour	Blue	15	85
	Brown	20	80

Is there evidence, at the 5% level, of an association between eye colour and left or right handedness?

2. An investigation into colourblindness and the sex of a person gave the following results:

		Colourblindness	
		Colourblind	Not colourblind
Sex	Male	36	964
	Female	19	981

Is there evidence, at the 5% level, of an association between the sex of a person and whether or not they are colourblind?

3. Consider the following 2 × 2 contingency tables, and for each test whether A and B are independent. Use a 5% level of significance.

(*a*)

	A_1	A_2
B_1	25	60
B_2	32	83

(*b*)

	A_1	A_2
B_1	32	58
B_2	62	48

(c)

	A_1	A_2
B_1	43	17
B_2	82	58

(d)

	A_1	A_2
B_1	7	18
B_2	48	47

4. In a 2×2 contingency table, the observed frequencies are as shown:

	A	B	Totals
Group I	a	b	c
Group II	d	e	f
Totals	g	h	k

Show that $\sum_{i=1}^{4} \frac{(O_i - E_i)^2}{E_i} = \frac{k(ae-bd)^2}{cfgh}$

(do not use the continuity correction).

5. In an examination 37 out of 47 boys passed and 27 out of 41 girls passed. By considering a suitable 2×2 contingency table, test whether boys and girls differ in their ability in this subject.

6. The results obtained by 200 students in chemistry and biology are shown in the table. Test, at the 5% level, whether the performances in both subjects are related.

		Chemistry	
		Pass	Fail
Biology	Pass	102	45
	Fail	21	32

$h \times k$ Contingency tables (h rows and k columns)

Example 10.9 In the principality of Viewmania a survey of 200 families known to be regular television viewers was undertaken. They were asked which of the three television channels they watched most during an average week. A summary of their replies is given in the following table, together with the region in which they lived.

		Region			
		North	East	South	West
Channel watched most	CCB1	29	16	42	23
	CCB2	6	11	26	7
	VIT	15	3	12	10

Find the expected frequencies on the hypothesis that there is no association between the channel watched most and the region.

Use the χ^2 distribution and a 5% level of significance to test the above hypothesis.

By considering the contribution to the value of your test statistic from each cell and the relative sizes of the observed and expected frequencies in each cell, indicate the main source of the association, if any exists. (AEB 1980)

Solution 10.9 H_0: there is no association between the channel watched most and the region.

The observed frequencies are first totalled, and then the expected frequencies under H_0 are calculated from

$$\text{Expected frequency} = \frac{\text{(row total)(column total)}}{\text{grand total}}$$

Observed data:

	North	East	South	West	Totals
CCB1	29	16	42	23	110
CCB2	6	11	26	7	50
VIT	15	3	12	10	40
Totals	50	30	80	40	200

This is a 3×4 contingency table

Expected data:

Expected frequency for northern viewers of

$$\text{CCB1} = \frac{(110)(50)}{200} = 27.5$$

This process is continued for the expected frequencies shown in heavy type. The remaining frequencies are found by ensuring that totals and sub-totals agree.

	North	East	South	West	Totals
CCB1	**27.5**	**16.5**	**44**	22	110
CCB2	**12.5**	**7.5**	**20**	10	50
VIT	10	6	16	8	40
Totals	50	30	80	40	200

Degrees of freedom: Once 6 expected frequencies have been found, the others are known automatically (by agreement of totals).

So ν = number of independent variables = 6, and we consider the $\chi^2(6)$ distribution.

We test at the 5% level and reject H_0 if $\chi^2_{\text{calc}} > \chi^2_{5\%}(6)$,

i.e. if $\chi^2_{\text{calc}} > 12.59$,

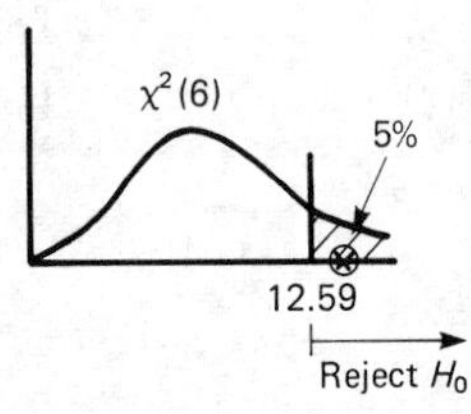

where $$\chi^2_{\text{calc}} = \sum_{i=1}^{12} \frac{(O_i - E_i)^2}{E_i}$$

O_i	E_i	$O_i - E_i$	$(O_i - E_i)^2$	$\frac{(O_i - E_i)^2}{E_i}$
29	27.5	1.5	2.25	0.082
16	16.5	0.5	0.25	0.015
42	44	2	4	0.091
23	22	1	1	0.045
6	12.5	6.5	42.25	3.38
11	7.5	3.5	12.25	1.633
26	20	6	36	1.8
7	10	3	9	0.9
15	10	5	25	2.5
3	6	3	9	1.5
12	16	4	16	1
10	8	2	4	0.5
$\Sigma O_i = 200$	$\Sigma E_i = 200$			13.446

Therefore $\chi^2_{\text{calc}} = 13.446$.

As $\chi^2_{\text{calc}} > 12.59$, we reject H_0 and conclude that there is an association between the channel watched most and the region.

The largest pair of contributions to χ^2_{calc} are the northern viewers of CCB2 and VIT, indicating that these viewers watch less CCB2 and more VIT than might be expected; the reverse effect is seen to some extent in the eastern and southern regions.

Number of degrees of freedom for an $h \times k$ contingency table

In general, if there are h rows, then once $h-1$ expected frequencies in a row have been calculated, the last value in the row is known (agreement of totals). Similarly, if there are k columns, once $k-1$ expected frequencies in a column have been calculated, the last value in the column is known.

Therefore, number of independent variables $= (h-1)(k-1)$.

So, for $h \times k$ contingency table, $\nu = (h-1)(k-1)$.

Exercise 10d

1. Consider the following contingency tables and for each one, test at the 5% level whether A and B are independent.

(a)

	B_1	B_2	B_3
A_1	16	19	15
A_2	26	14	10

(b)

	B_1	B_2	B_3	B_4
A_1	20	42	28	12
A_2	8	50	24	6
A_3	38	42	94	26

(c)

	B_1	B_2
A_1	20	9
A_2	31	14
A_3	16	12
A_4	23	15

(d)

	B_1	B_2	B_3
A_1	10	26	34
A_2	8	29	43
A_3	27	43	50
A_4	9	14	27

2. A thousand households are taken at random and divided into three groups *A*, *B* and *C*, according to the total weekly income. The following table shows the numbers in each group having a colour television receiver, a black and white receiver, or no television at all.

	A	*B*	*C*
Colour television	56	51	93
Black and white	118	207	375
None	26	42	32

Calculate the expected frequencies if there is no association between total income and television ownership.

Apply a test to find whether the observed frequencies suggest that there is such an association. (AEB 1974)

3. The following table shows the numbers of students passed and failed by three examiners *A*, *B* and *C*.

	Examiners			
	A	*B*	*C*	Totals
Pass	51	48	58	157
Fail	4	14	7	25
Totals	55	62	65	182

Test the hypothesis that the three examiners fail equal proportions of students by applying χ^2 tests with and without Yates' correction. Comment on the results. (AEB 1976)

4. At St. Trinian's College for Young Ladies there are 1000 pupils. Of these 75 have represented the College at both hockey and netball, 10 have represented the College at hockey but do not play netball, 35 have represented the College at netball but do not play hockey, and 100 do not play games at all. In all 100 girls have represented the College at hockey, and 150 at netball. The number who do not play hockey is 200 and the number who do not play netball is 125.

Arrange the above data in the form of a 3×3 contingency table, and state how many pupils play both hockey and netball but have not represented the College in either.

Apply the χ^2 test to your 3×3 table, and state the hypothesis which it tests. (AEB 1976)

5. The following are data on 150 chickens, divided into two groups according to breed, and into three groups according to yield of eggs:

	Yield		
	High	Medium	Low
Rhode Island Red	46	29	28
Leghorn	27	14	6

Are these data consistent with the hypothesis that the yield is not affected by the type of breed?

6. In a small survey 350 car owners from four districts P, Q, R, S were found to have cars in price ranges A, B, C, D, the frequencies of the prices being as shown in the table.

		P	Q	R	S
Price of car	A	9	10	12	19
	B	13	20	18	29
	C	24	29	12	25
	D	34	41	18	37

Find the expected frequencies on the hypothesis that there is no association between the district and the price of the car.

Use the χ^2 distribution to test this hypothesis. (AEB 1975)

SUMMARY — χ^2 TEST AND DEGREES OF FREEDOM

$$\chi^2_{\text{calc}} = \sum_{i=1}^{n} \frac{(O_i - E_i)^2}{E_i}$$

For $\nu = 1$, using Yates' continuity correction,

$$\chi^2_{\text{calc}} = \sum_{i=1}^{n} \frac{(|O_i - E_i| - 0.5)^2}{E_i}$$

Degree of freedom (ν):

Uniform distribution and distributions in a given ratio	n classes 1 restriction (totals agree) $n-1$ independent variables	$\nu = n-1$
Binomial distribution	(a) *p known* n classes 1 restriction (totals agree) $n-1$ independent variables (b) *p unknown* n classes 2 restrictions (totals agree, means agree) $n-2$ independent variables	$\nu = n-1$ $\nu = n-2$
Poisson distribution	n classes 2 restrictions (totals agree, means agree) $n-2$ independent variables	$\nu = n-2$
Normal distribution	(a) *mean and variance known* n classes 1 restriction (totals agree) $n-1$ independent variables (b) *mean and variance unknown* n classes 3 restrictions (totals agree, means agree, variances agree) $n-3$ independent variables	$\nu = n-1$ $\nu = n-3$
2×2 contingency table	4 classes 1 independent variable	$\nu = 1$
$h \times k$ contingency	$h \times k$ classes $(h-1)(k-1)$ independent variables	$\nu = (h-1)(k-1)$

Miscellaneous Exercise 10e

1. A random sample of 100 housewives were asked by a market research team whether or not they used Sudsey Soap. 58 said yes and 42 said no. In a second random sample of 80 housewives, 62 said yes and 18 said no. By considering a suitable 2×2 contingency table, test whether these two samples are consistent with each other. (O & C)

2. Two fair dice are thrown 432 times. Find the expected frequencies of the scores 2, 3, 4, . . . , 12.

   Two players, A and B are each given two dice and told to throw them 432 times, recording the results. The frequencies reported are given in Table A below.

   Is there any evidence that either pair of dice is biased? What can be said about B's alleged results? (AEB 1976)

3. Over a period of 50 weeks the numbers of road accidents reported to a police station are shown in the table below.

No. of accidents	0	1	2	3
No. of weeks	23	13	10	4

   Find the mean number of accidents per week.

   Use this means, a 5% level of significance, and your table of χ^2 to test the hypothesis that these data are a random sample from a population with a Poisson distribution. (O & C)

4. Table B below shows the girths of one type of fir tree in a plantation of 480 trees set alongside the distribution that would be expected if the distribution were normal. Use the χ^2 test, with a 5% significance level, to determine whether the observed distribution can be taken as normal, with the same mean and standard deviation as the observed distribution. (C Additional)

5. Smallwoods Ltd. run a weekly football pools competition. One part of this involves a fixed-odds contest where the entrant has to forecast correctly the result of each of five given matches. In the event of a fully correct forecast the entrant is paid out at odds of 100 to 1. During the last two years Miss Fortune has entered this fixed-odds contest 80 times. The table below summarises her results.

Number of matches correctly forecast per entry (x)	0	1	2	3	4	5
Number of entries with x correct forecasts (f)	8	19	25	22	5	1

   (a) Find the frequencies of the number of matches correctly forecast per entry given by a binomial distribution having the same mean and total as the observed distribution.
   (b) Use the χ^2 distribution and a 10% level of significance to test the adequacy of the binomial distribution as a model for these data.
   (c) On the evidence before you, and assuming that the point of entering is to win money, would you advise Miss Fortune to continue with this competition and why? (AEB 1981)

   (*NOTE*: $\chi^2_{10\%}(1) = 2.71$, $\chi^2_{10\%}(2) = 4.61$, $\chi^2_{10\%}(3) = 6.25$, $\chi^2_{10\%}(4) = 7.78$, $\chi^2_{10\%}(5) = 9.24$.)

6. The table summarises the incidence of cerebral tumours in 141 neurosurgical patients.

		Type of tumour		
		Benign	Malignant	Others
Site of tumour	Frontal lobes	23	9	6
	Temporal lobes	21	4	3
	Elsewhere	34	24	17

Table A

Scores	2	3	4	5	6	7	8	9	10	11	12
A's frequency	18	33	28	54	62	65	66	42	30	27	7
B's frequency	14	22	34	51	58	73	63	45	38	25	9

Table B

Girth of trees (in metres)	0.6–0.8	0.8–1.0	1.0–1.2	1.2–1.4	1.4–1.6
No. of trees (observed)	37	137	152	119	35
No. of trees (expected)	34	136	162	115	33

Find the expected frequencies on the hypothesis that there is no association between the type and site of a tumour. Use the χ^2 distribution to test this hypothesis. (AEB 1977)

7. Explain how to calculate the degrees of freedom for the χ^2 statistic in (*a*) a goodness-of-fit test, (*b*) a test of no association of the two factors in an $n \times k$ contingency table.

An ecologist collected organisms of a particular species from three beaches and counted the number of females in each sample (the remainder were males).

Beach	1	2	3
No. of females	44	86	110
Total no. in sample	100	200	200

Test if the proportion of females differed significantly between the beaches.

Find the percentage of females at each beach and comment on the results. (O)

8. A factory operates four production lines. Maintenance records show that the daily number of stoppages due to mechanical failure were as shown in Table C below (it is possible for a production line to break down more than once on the same day). You may assume that $\Sigma f = 1400$, $\Sigma fx = 1036$.
(*a*) Use a χ^2 distribution and a 1% significance level to determine whether the Poisson distribution is an adequate model for the data.
(*b*) The maintenance engineer claims that breakdowns occur at random and that the mean rate has remained constant throughout the period. State, giving a reason, whether your answer to (*a*) is consistent with this claim.
(*c*) Of the 1036 breakdowns which occurred 230 were on production line *A*, 303 on *B*, 270 on *C* and 233 on *D*. Test at the 5% significance level whether these data are consistent with breakdowns occurring at an equal rate on each production line. (AEB 1988)

9. The number of accidents in a large factory over a period of one month is recorded in 7 hourly periods in Table D below. Display these data in a suitable diagram and comment on them.

Test the hypothesis that accidents are equally likely to happen at any time of the day. Comment on your conclusion in relation to the diagram drawn. (SUJB)

10. (*a*) As part of a statistics project, students observed five private cars passing a college and counted the number which were carrying the driver only, with no passengers. This was repeated 80 times. The results of a particular student were as follows:

Number of cars with driver only	Number of times observed
0	0
1	3
2	12
3	27
4	26
5	12

Use the χ^2 distribution and a 5% significance level to test whether the binomial distribution provides an adequate model for the data.
(*b*) In a further part of the project the students counted the number of cars passing the college in 130 intervals each of length 5 seconds. Table E overleaf shows the results obtained by the same student together with the expected numbers if a Poisson distribution, with the same mean as the observed data, is fitted.

Use the χ^2 distribution and a 5% significance level to test whether the Poisson distribution provides an adequate model for the data.

Table C

Number of stoppages, x	0	1	2	3	4	5	6 or more
Number of days, f	728	447	138	48	26	13	0

Table D

Period	09.00–10.00	10.00–11.00	11.00–12.00	13.00–14.00	14.00–15.00	15.00–16.00	16.00–17.00
No. of accidents	12	17	15	21	19	26	32

(*c*) The teacher suspected that this student had not observed the data but invented them. Explain why the teacher was suspicious and comment on the strength of the evidence supporting her suspicions. (AEB 1987)

11. One formula for the χ^2 statistic is

$$\chi^2 = \sum \frac{(f_o - f_e)^2}{f_e}$$

where f_o is the observed frequency, f_e is the expected frequency and the summation is over the number of groups. Show that the formula may also be written as

$$\chi^2 = \left(\sum \frac{f_o^2}{f_e}\right) - N$$

where N is the total number of observations.

(*a*) Ballpoint pens come off a production line and are packed into batches of 100. It is believed that the number of defective pens in each batch follows a Poisson distribution with mean 2.8. 100 batches of pens were examined and the observed frequencies of the number of defective pens in each batch found to be those in the table below. Test whether the suggested Poisson model fits these data.

Number of defective pens	0	1	2	3	4	5	6+
Frequency	5	19	25	20	16	7	8

(*b*) To find whether there is any association between a person's eye colour and his or her skin's susceptibility to sunburn, a random sample of 180 people was taken and the data in the table below obtained. Test whether there is significant evidence of association.

Eye colour	Susceptibility to sunburn			Total
	High	Medium	Low	
Blue	19	27	4	50
Brown	1	13	16	30
Grey-green	27	48	25	100

(O)

12. Over a long period of time, a research team monitored the number of car accidents which occurred in a particular county. Each accident was classified as being trivial (minor damage and no personal injuries), serious (damage to vehicles and passengers, but no deaths) or fatal (damage to vehicles and loss of life). The colour of the car which, in the opinion of the research team, caused the accident was also recorded, together with the day of the week on which the accident occurred. The following data were collected.

Colour	Trivial	Serious	Fatal
White	50	25	16
Black	35	39	18
Green	28	23	13
Red	25	17	11
Yellow	17	20	16
Blue	24	33	10

Analyse these data for evidence of association between the colour of the car and the type of accident.

State the condition which sometimes necessitates the amalgamation of rows or columns in contingency tables. Explain why amalgamation might not be appropriate for this table.

The following table summarises the data relating to the day of the week on which the accident occurred.

Day	Number of accidents
Monday	60
Tuesday	54
Wednesday	48
Thursday	53
Friday	53
Saturday	75
Sunday	77

Investigate the hypothesis that these data are a random sample from a uniform distribution. (AEB 1987)

Table E

Number of cars passing a point in a 5 second interval	0	1	2	3	4	5	6	7 or more
Number of intervals observed	28	40	32	19	7	3	1	0
Number of intervals expected	25.85	41.75	33.72	18.16	7.33	2.37	0.64	0.18

11

REGRESSION AND CORRELATION

SCATTER DIAGRAM

Sometimes we wish to investigate the results of a statistical enquiry or experiment by *comparing two sets of data*, x and y, for example

x	y
The weight at the end of a spring	The length of the spring
Pupil's mark in French	Pupil's mark in German
The diameter of the stem of a plant	The average length of leaf of the plant
The age of a plant	The quantity of fruit produced by a plant

Consider the set of points (x_1, y_1), (x_2, y_2), ..., (x_n, y_n). If the values of y are plotted against the values of x, then a **scatter diagram** is obtained.

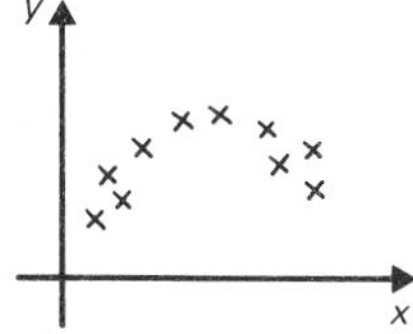

REGRESSION FUNCTION

We then look for a relationship $y = f(x)$, where the function f is to be determined, i.e. *given* the points only we have to 'work backwards' or '*regress*' to the original function f. Hence this function is called the **regression function.**

LINEAR CORRELATION AND REGRESSION LINES

We shall consider only the simplest type of function where $y = f(x)$ is a *straight line*. If all the points in the scatter diagram seem to lie near a straight line, we say that there is **linear correlation** between x and y.

We try to *estimate* fairly accurately the position of this line, and having done so we call it a **regression line.**

(**a**) If y tends to increase as x increases, then there is **positive linear correlation.**

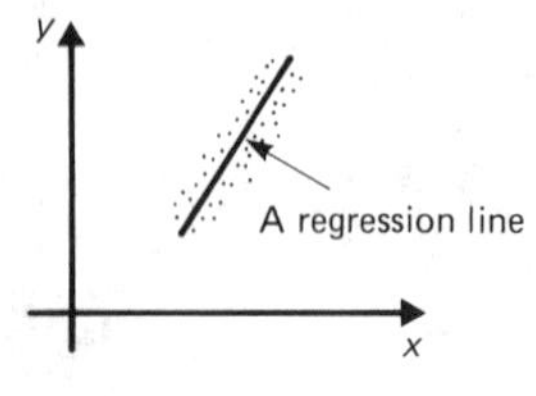

(**b**) If y tends to decrease as x increases, then there is **negative linear correlation.**

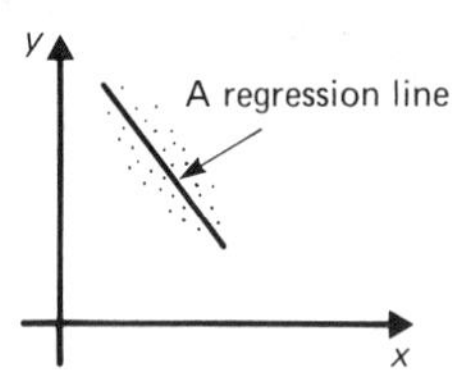

(**c**) If there is no relationship between x and y, then there is *no correlation.*

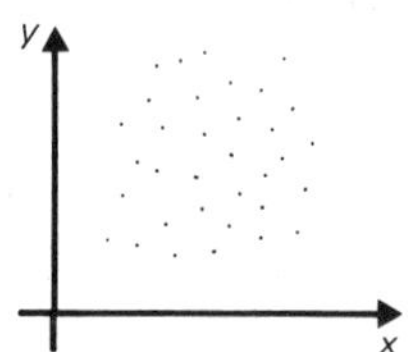

NOTE: common sense is needed when interpreting scatter diagrams; for instance we might find that there is an increase in the number of bank robberies and an increase in the number of health food shops over the last 5 years in a certain town—however, it would be foolish to look for a relationship between the variables.

There are many ways of obtaining regression lines for different purposes—two methods are indicated below.

METHOD I — DRAWING A REGRESSION LINE 'BY EYE'

(**a**) If there is very little scatter.

First calculate the co-ordinates of the point $(\bar{x}, \bar{y})$ where $\bar{x} = \dfrac{\Sigma x_i}{n}$ and $\bar{y} = \dfrac{\Sigma y_i}{n}$.

Then draw a line of good fit, ensuring that it passes through $(\bar{x}, \bar{y})$.

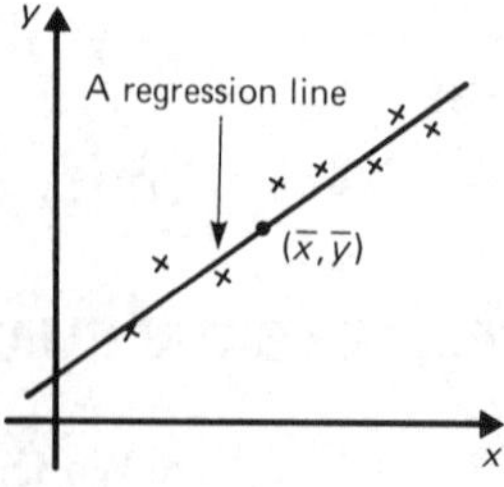

(**b**) If there is a fair degree of scatter.

In this case we can distinguish *two regression lines*:

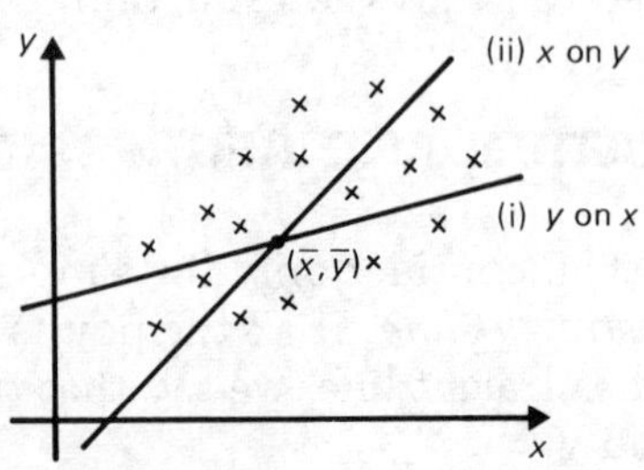

(i) *A line of regression of y on x*. This can be used to estimate y, given a value of x.
(ii) *A line of regression of x on y*. This can be used to estimate x, given a value of y.

Method for drawing these two regression lines by eye

(i) *A line of regression of y on x* — we assume the values of x to be accurate and draw a regression line as follows:
  (a) Find the mean $M(\bar{x}, \bar{y})$ of the distribution.
  (b) Through M draw a line parallel to the y axis. This divides the points into two groups.
  (c) Find the mean M_L of the points on the left.
  (d) Find the mean M_R of the points on the right.
  (e) Draw a line of best fit through M, M_L and M_R.

(ii) *A line of regression of x on y* — we assume the values of y to be accurate and draw a regression line as follows:
  (a) Find the mean $M(\bar{x}, \bar{y})$ of the distribution.
  (b) Through M draw a line parallel to the x axis.
  (c) Find the mean M_A of the points above.
  (d) Find the mean M_B of the points below.
  (e) Draw the line of best fit through M, M_A and M_B.

Example 11.1 The following table gives the test results for 10 children.

Child	*A*	*B*	*C*	*D*	*E*	*F*	*G*	*H*	*I*	*J*
Arithmetic mark, x	1	8	15	18	23	28	33	39	45	45
English mark, y	3	14	8	20	19	17	36	26	14	29

(a) (*i*) Draw a scatter diagram, and by finding the means of certain points draw a regression line y on x.
(*ii*) Estimate an English mark for a child who missed the English test, but who had 20 in the arithmetic test.

(b) (*i*) On the scatter diagram draw a regression line x on y.
(*ii*) Estimate an arithmetic mark for a child who was absent for the arithmetic test, but who had 30 in the English test.

(c) Would you use one of these lines to estimate an English mark for a child who had 60 in the arithmetic test?

Solution 11.1 (a) (*i*) $\bar{x} = \dfrac{\Sigma x}{10} = \dfrac{255}{10} = 25.5, \quad \bar{y} = \dfrac{\Sigma y}{10} = \dfrac{186}{10} = 18.6$

So we plot the point $M(25.5, 18.6)$ and ensure that the line passes through it.

For a regression line y on x, draw a line through M parallel to the y axis.

For the points on the left		For the points on the right	
x	y	x	y
1	3	28	17
8	14	33	36
15	8	39	26
18	20	45	14
23	19	45	29
$\Sigma x = 65$	$\Sigma y = 64$	$\Sigma x = 190$	$\Sigma y = 122$
So $\bar{x}_L = \frac{65}{5} = 13,$		$\bar{x}_R = \frac{190}{5} = 38,$	
$\bar{y}_L = \frac{64}{5} = 12.8$		$\bar{y}_R = \frac{122}{5} = 24.4$	
We plot M_L (13, 12.8)		We plot M_R (38, 24.4)	

Now draw a line of good fit through M, M_L and M_R. This is a regression line y on x.

(*ii*) If a child had 20 in the arithmetic test, since x is given, we use the line y on x to estimate the English mark. From the line, the estimated mark for English is 16.

(b) (*i*) *x on y*: Draw the line through M parallel to the x axis.

For the points above		For the points below	
x	y	x	y
18	20	1	3
23	19	8	14
33	36	15	8
39	26	28	17
45	29	45	14
$\Sigma x = 158$	$\Sigma y = 130$	$\Sigma x = 97$	$\Sigma y = 56$
$\bar{x}_A = \frac{158}{5} = 31.6,$		$\bar{x}_B = \frac{97}{5} = 19.4,$	
$\bar{y}_A = \frac{130}{5} = 26$		$\bar{y}_B = \frac{56}{5} = 11.2$	
We plot M_A (31.6, 26)		We plot M_B (19.4, 11.2)	

Now draw a line of good fit through M, M_A, M_B. This is a regression line x on y.

(*ii*) If a child had 30 in the English test we use the line x on y, as y is given. From this line, the estimated arithmetic mark is 35.

(c) The mark of 60 in the arithmetic test is outside the range of the data. We could use the regression line y on x as drawn on the scatter diagram to give an estimated English mark of 35, but this result should be used with caution. As a general rule, keep within the range of the data.

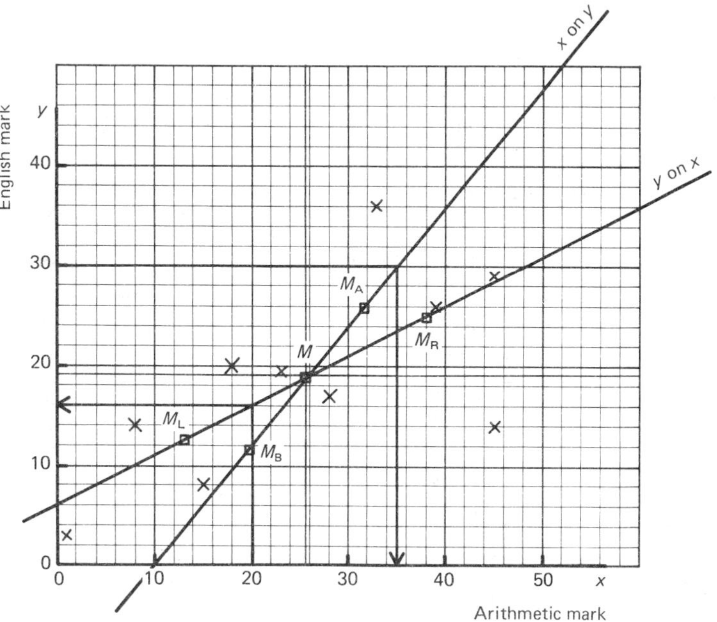

Scatter diagram to show English and Arithmetic marks for 10 pupils

NOTE: it may appear strange to have two regression lines, but it does matter which is considered. Suppose that there is a positive correlation between the height and mass of males. A result of this might be that the average mass of all males of height 1.93 m (6 ft 4 inches) is 85.7 kg ($13\frac{1}{2}$ stone).

So, if you were given a height of 6 ft 4 inches you would guess $13\frac{1}{2}$ stone for the mass.

But, if you were given a mass of $13\frac{1}{2}$ stone, would you guess 6 ft 4 inches for the height? If you would not, then the two regression lines are different.

NOTE: if we had a set of data such as

x	5	10	15	20	25
y	20	21	23	24	23

then it is obvious that the value of x has been *controlled*. In this case we would use a regression line of y on x to estimate y, given x, but *not* a regression line x on y to estimate x, given y.

Exercise 11a

1. For the following sets of data, draw scatter diagrams and comment on the correlation. Draw regression lines y on x and x on y.

(*a*) Use these 11 pairs of data:

x	3	7	9	11	14	14	15	21	22	23	26
y	5	12	5	12	10	17	23	16	10	20	25

(*b*) Use these 13 pairs of data:

x	1	5	5	5	6	7.5	7.5
y	85	82	85	89	78	66	77

x	7.5	10	11	12.5	14	14.5
y	81	70	74	65	69	63

(*c*) Use these 10 pairs of data:

x	0.6	1	2	2.5	2.8
y	5	10	15	10	2.5

x	3.6	4	4	4	5
y	7.5	2.5	5	15	10

2. Values of two variables x and y obtained from a survey are recorded below.

x	1	2	3	4	5	6	7	8
y	81	73	53	55	43	29	15	3

Represent these data on a scatter diagram and draw in the line of best fit. Obtain the equation of the line of best fit in the form $y = mx + c$ and estimate the value of y when $x = 5.5$. (SUJB)

3. Four identical money boxes contain different numbers of a particular type of coin and no coins of other types. From the information on the combined weights, which is given below, it is desired to estimate the weight of a box and the mean weight of a coin.

Number of coins in box	x	10	20	30	40
Combined weight of coins and box	y	312	509	682	865

(*a*) Plot these data on a scatter diagram, labelling the axes clearly. State whether the data display strong positive, strong negative, or near zero correlation (or otherwise).
(*b*) State the co-ordinates of one point through which the line of regression of y upon x must pass.
(*c*) Draw on your diagram, by eye, this regression line.
(*d*) Estimate, from your regression line, (*i*) the weight of an empty box, (*ii*) the mean weight of a single coin. (C)

4. Table A gives the rainfall, in cm, for the first nine months of a year at two weather stations. Calculate the mean monthly rainfall over this period at each station and plot the information given in the table on a scatter diagram, drawing a line of best fit.

Find the equation of this line and use it to predict the rainfall at B in a month when 2.5 cm of rain fell at A. (C Additional)

Table A

	Jan.	Feb.	Mar.	Apr.	May	June	July	Aug.	Sept.
A	5.2	4.8	6.1	5.0	3.2	2.9	4.4	4.0	3.1
B	4.6	4.2	5.4	4.4	2.9	2.8	3.9	3.6	3.0

METHOD II — CALCULATING THE EQUATIONS OF THE LEAST SQUARES REGRESSION LINES

(a) The least squares regression line y on x

Let the equation of the least squares regression line y on x be $y = ax + b$.

Consider the set of points (x_i, y_i), where $i = 1, 2, \dots, n$.

We find the values of a and b such that Σm_i^2 is a *minimum*, where m_i is as shown in the diagram.

The lengths $P_1Q_1, P_2Q_2, \dots, P_iQ_i$ are called *residuals*.

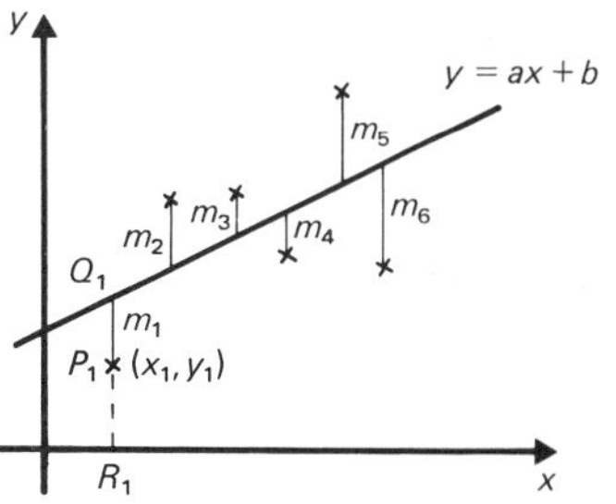

$$\text{Now}\qquad m_1 = Q_1P_1 = Q_1R_1 - P_1R_1 = (ax_1 + b) - y_1$$
$$m_1^2 = (ax_1 + b - y_1)^2$$

$$\text{So}\qquad \Sigma m_i^2 = \Sigma(ax_i + b - y_i)^2 \quad i = 1, 2, \dots, n$$

Σm_i^2 is the *sum of the squares of the residuals* and if we can find values of a and b such that Σm_i^2 is a *minimum*, then the line $y = ax + b$ is called the *least squares* regression line of y on x.

If we allow b to vary, keeping all other quantities constant, we can obtain the value of b for which Σm_i^2 is a minimum by setting $\dfrac{\mathrm{d}\Sigma m_i^2}{\mathrm{d}b} = 0$.

$$\text{Now}\qquad \frac{\mathrm{d}\Sigma m_i^2}{\mathrm{d}b} = \frac{\mathrm{d}\Sigma(ax_i + b - y_i)^2}{\mathrm{d}b} = \Sigma 2(ax_i + b - y_i)$$
$$= 2(a\Sigma x_i + nb - \Sigma y_i)$$

$$\text{So}\qquad \frac{\mathrm{d}\Sigma m_i^2}{\mathrm{d}b} = 0 \quad \text{when} \quad \Sigma y_i = a\Sigma x_i + nb \qquad \text{(i)}$$

Similarly, if we allow a to vary, keeping all the other quantities constant,

$$\frac{\mathrm{d}\Sigma m_i^2}{\mathrm{d}a} = \frac{\mathrm{d}\Sigma(ax_i + b - y_i)^2}{\mathrm{d}a} = \Sigma 2(ax_i + b - y_i)x_i$$
$$= 2(a\Sigma x_i^2 + b\Sigma x_i - \Sigma x_iy_i)$$

$$\text{So}\qquad \frac{\mathrm{d}\Sigma m_i^2}{\mathrm{d}a} = 0 \quad \text{when} \quad \Sigma x_iy_i = a\Sigma x_i^2 + b\Sigma x_i \qquad \text{(ii)}$$

If the least squares regression line y on x is $y = ax + b$, the values of a and b are found by solving the simultaneous equations

$$\Sigma y = a\Sigma x + nb$$
$$\Sigma xy = a\Sigma x^2 + b\Sigma x$$

These equations are called the **normal equations for y on x.**

Example 11.2 Show that the least squares regression line y on x passes through the mean of the data.

Solution 11.2 Consider the regression line $y = ax + b$.

To calculate a and b the following equation is used:

$$\Sigma y = a\Sigma x + nb$$

Divide through by n

$$\frac{\Sigma y}{n} = a\frac{\Sigma x}{n} + b$$

so

$$\bar{y} = a\bar{x} + b$$

Hence, the point $(\bar{x}, \bar{y})$ lies on the regression line $y = ax + b$.

(b) The least squares regression line x on y

Let the equation of the least squares regression line x on y be $x = cy + d$.

We find the values c and d such that Σn_i^2 is a *minimum*.

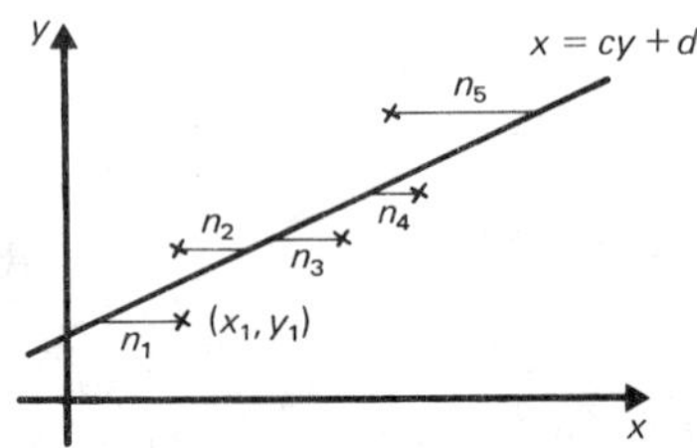

It can be shown that

if the least squares regression line x on y is $x = cy + d$, the values of c and d are found by solving the simultaneous equations

$$\Sigma x = c\Sigma y + nd$$

$$\Sigma xy = c\Sigma y^2 + d\Sigma y$$

These equations are called the **normal equations for x on y.**

It can be shown that the point $(\bar{x}, \bar{y})$ lies on the line $x = cy + d$.

NOTE: in general, these lines will *not* coincide with those obtained by the earlier methods described.

Example 11.3 Obtain the normal equations for the least squares regression line y on x for the following data:

x	1	2	4	6	7	8	10
y	10	14	12	13	15	12	13

Hence find the equation of the least squares regression line y on x.

Solution 11.3

x	y	x^2	y^2	xy
1	10	1	100	10
2	14	4	196	28
4	12	16	144	48
6	13	36	169	78
7	15	49	225	105
8	12	64	144	96
10	13	100	169	130
$\Sigma x = 38$	$\Sigma y = 89$	$\Sigma x^2 = 270$	$\Sigma y^2 = 1147$	$\Sigma xy = 495$

For these data $n = 7$. The normal equations for y on x are

$$\Sigma y = a\Sigma x + nb, \quad \text{so} \quad 89 = 38a + 7b \qquad \text{(i)}$$

$$\Sigma xy = a\Sigma x^2 + b\Sigma x \qquad 495 = 270a + 38b \qquad \text{(ii)}$$

Solving equations (i) and (ii) gives $a = 0.186$, $b = 11.7$.

The least squares regression line y on x is

$$\underline{y = 0.186x + 11.7}$$

Exercise 11b

1. Calculate the equation of the least squares regression line x on y for the data given in Example 11.3.

2. Calculate the equation of the least squares regression line (a) y on x, (b) x on y for the data given in (i) Exercise 11a, Question 1(a) (p. 564), (ii) Example 11.1 (p. 561).

COVARIANCE

The covariance of $(x_1, y_1), (x_2, y_2) \ldots, (x_n, y_n)$ is defined as

$$s_{xy} = \frac{1}{n}\Sigma(x-\bar{x})(y-\bar{y})$$

Now $$s_{xy} = \frac{1}{n}\Sigma(x-\bar{x})(y-\bar{y}) = \frac{\Sigma xy}{n} - \bar{x}\frac{\Sigma y}{n} - \bar{y}\frac{\Sigma x}{n} + \frac{n\bar{x}\bar{y}}{n}$$

$$= \frac{\Sigma xy}{n} - \bar{x}\bar{y} - \bar{y}\bar{x} + \bar{x}\bar{y} = \frac{\Sigma xy}{n} - \bar{x}\bar{y}$$

so $$s_{xy} = \frac{\Sigma xy}{n} - \bar{x}\bar{y} = \frac{\Sigma xy}{n} - \left(\frac{\Sigma x}{n}\right)\left(\frac{\Sigma y}{n}\right)$$

NOTE: this is simply an extension of the formula for the variance, since

$$s_{xx} = \frac{1}{n}\Sigma(x-\bar{x})(x-\bar{x}) = s_x^2$$

$$s_{yy} = \frac{1}{n}\Sigma(y-\bar{y})(y-\bar{y}) = s_y^2$$

In the following we shall use the alternative forms of the formulae:

$$s_x^2 = \frac{\Sigma x^2}{n} - \bar{x}^2, \quad s_y^2 = \frac{\Sigma y^2}{n} - \bar{y}^2$$ (see p. 49)

For the least squares regression line y on x, let $y = ax + b$.

ALTERNATIVE METHOD FOR CALCULATING THE EQUATIONS OF LEAST SQUARES REGRESSION LINES

Using regression coefficients

The normal equations are

$$\Sigma y = a\Sigma x + nb \qquad \text{(i)}$$

$$\Sigma xy = a\Sigma x^2 + b\Sigma x \qquad \text{(ii)}$$

Multiplying (i) by Σx,

$$\Sigma x \Sigma y = a(\Sigma x)^2 + nb\Sigma x$$

Multiplying (ii) by n,

$$n\Sigma xy = an\Sigma x^2 + nb\Sigma x$$

Subtracting and rearranging

$$a = \frac{n\Sigma xy - \Sigma x\Sigma y}{n\Sigma x^2 - (\Sigma x)^2}$$

$$= \frac{\dfrac{\Sigma xy}{n} - \left(\dfrac{\Sigma x}{n}\right)\left(\dfrac{\Sigma y}{n}\right)}{\dfrac{\Sigma x^2}{n} - \left(\dfrac{\Sigma x}{n}\right)^2}$$ (dividing both the numerator and denominator by n^2)

$$= \frac{s_{xy}}{s_x^2}$$

a is known as the coefficient of regression of y on x, where

$$a = \frac{s_{xy}}{s_x^{\,2}}$$

a is the gradient of the least squares regression line y on x.

Now, since the regression line passes through $(\bar{x}, \bar{y})$, its equation must be of the form $y - \bar{y} = a(x - \bar{x})$,

The equation of the least squares regression line y on x is

$$y - \bar{y} = \frac{s_{xy}}{s_x^{\,2}}(x - \bar{x})$$

Similarly, for the least squares regression line x on y, $x = cy + d$, it can be shown that

$c = \dfrac{s_{xy}}{s_y^{\,2}}$ where c is the coefficient of regression of x on y

The equation of the regression line x on y is

$$x - \bar{x} = \frac{s_{xy}}{s_y^{\,2}}(y - \bar{y})$$

MINIMUM SUM OF SQUARES OF RESIDUALS

The equation of the least squares regression line y on x is

$$y - \bar{y} = \frac{s_{xy}}{s_x^{\,2}}(x - \bar{x}).$$

The point Q_i lies on this line and its x-coordinate is x_i. So its y-coordinate is y_Q where

$$y_Q - \bar{y} = \frac{s_{xy}}{s_x^{\,2}}(x_i - \bar{x})$$

So

$$\begin{aligned} m_i &= y_i - y_Q \\ &= y_i - \left[\bar{y} + \frac{s_{xy}}{s_x^{\,2}}(x_i - \bar{x})\right] \end{aligned}$$

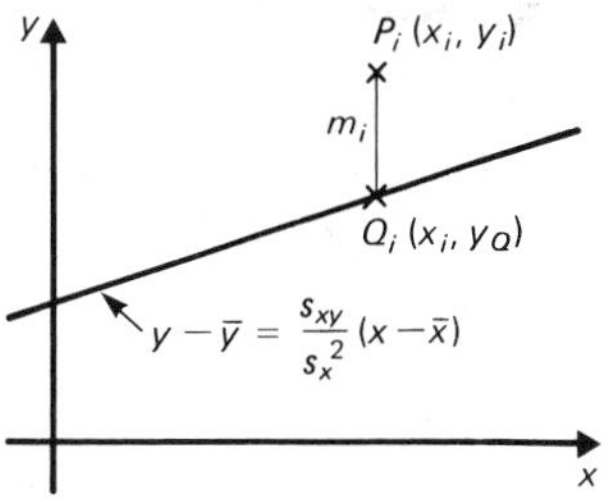

We shall denote the minimum value of the sum of the squares of residuals by $\Sigma m_i^2{}_{(\min)}$ where

$$\begin{aligned}
\Sigma m_{i(\min)}^2 &= \Sigma\left[(y_i-\bar{y})-\frac{s_{xy}}{s_x^2}(x_i-\bar{x})\right]^2\\
&= \Sigma(y_i-\bar{y})^2-2\frac{s_{xy}}{s_x^2}\Sigma(x_i-\bar{x})(y_i-\bar{y})+\frac{s_{xy}^2}{(s_x^2)^2}\Sigma(x_i-\bar{x})^2\\
&= ns_y^2-2\frac{s_{xy}}{s_x^2}ns_{xy}+\frac{s_{xy}^2}{(s_x^2)^2}ns_x^2\\
&= ns_y^2-2n\frac{s_{xy}^2}{s_x^2}+n\frac{s_{xy}^2}{s_x^2}\\
&= ns_y^2-n\frac{s_{xy}^2}{s_x^2}
\end{aligned}$$

The minimum sum of squares of residuals for y on x is

$$n\left(s_y^2-\frac{s_{xy}^2}{s_x^2}\right)$$

Similarly, it can be shown that

The minimum sum of squares of residuals for x on y is

$$n\left(s_x^2-\frac{s_{xy}^2}{s_y^2}\right)$$

Example 11.4 Draw a scatter diagram for the following data. Calculate the equations of the lines of regression **(a)** y on x, **(b)** x on y, and draw these on the diagram.

Find also the minimum sum of squares of residuals **(c)** for y on x, **(d)** for x on y.

x	1	2	4	6	7	8	10
y	10	14	12	13	15	12	13

Solution 11.4 This is the same data as in Example 11.3, so we refer to the table on p. 567.

We have $\bar{x} = \dfrac{\Sigma x}{n} = \dfrac{38}{7} = 5.4$ (1 d.p.),

$$\bar{y} = \frac{\Sigma y}{n} = \frac{89}{7} = 12.7 \quad \text{(1 d.p.)}$$

On the scatter diagram, plot $M(5.4, 12.7)$.

Now $$s_{xy} = \frac{\Sigma xy}{n} - \overline{x}\overline{y} = \frac{495}{7} - \left(\frac{38}{7}\right)\left(\frac{89}{7}\right) = 1.694$$

$$s_x^{\,2} = \frac{\Sigma x^2}{n} - \overline{x}^2 = \frac{270}{7} - \left(\frac{38}{7}\right)^2 = 9.102$$

$$s_y^{\,2} = \frac{\Sigma y^2}{n} - \overline{y}^2 = \frac{1147}{7} - \left(\frac{89}{7}\right)^2 = 2.204$$

(a) Equation of least squares regression line y on x

$$y - \overline{y} = \frac{s_{xy}}{s_x^{\,2}}(x - \overline{x})$$

so $$y - \frac{89}{7} = \frac{1.694}{9.102}\left(x - \frac{38}{7}\right)$$

Rearranging, $\underline{y = 0.186x + 11.7}$ (as before)

Draw this on the scatter diagram, by plotting $M(5.4, 12.7)$ and two other points say $(0, 11.7)$ and $(1, 11.886)$.

(b) Equations of least squares regression line x on y

$$x - \overline{x} = \frac{s_{xy}}{s_y^{\,2}}(y - \overline{y})$$

so $$x - \frac{38}{7} = \frac{1.694}{2.204}\left(y - \frac{89}{7}\right)$$

Rearranging, $\underline{x = 0.769y - 4.34}$

Draw this on the scatter diagram by plotting $M(5.4, 12.7)$ and two other points, say $(0, 5.64)$ and $(1, 6.94)$.

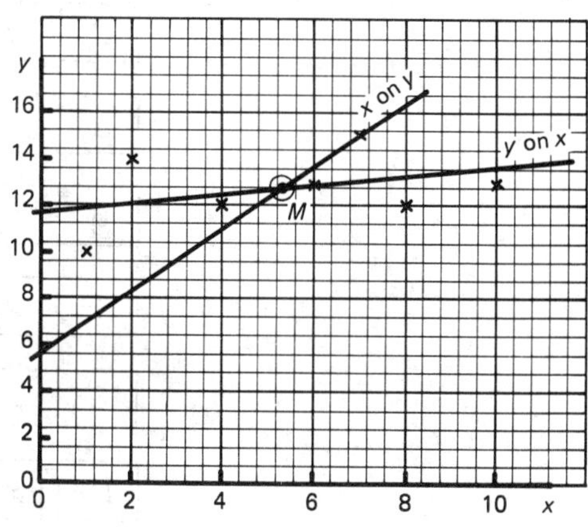

(c) For y on x

$$\Sigma m_{i\,(\text{min})}^2 = n\left(s_y^2 - \frac{s_{xy}^2}{s_x^2}\right)$$

$$= 7\left(2.204 - \frac{1.694^2}{9.102}\right)$$

$$= 13.2 \quad (3\ \text{S.F.})$$

The minimum sum of squares of residuals for y on x is 13.2 (3 S.F.).

(d) For x on y

$$\Sigma n_{i\,(\text{min})}^2 = n\left(s_x^2 - \frac{s_{xy}^2}{s_y^2}\right)$$

$$= 7\left(9.102 - \frac{1.694^2}{2.204}\right)$$

$$= 54.6 \quad (3\ \text{S.F.})$$

The minimum sum of squares of residuals for x on y is 54.6 (3 S.F.).

CALCULATOR NOTE: working in SD mode, the values of $\bar{x}$, $\bar{y}$, s_x and s_y can be obtained directly and used in the calculations. The value of s_{xy} will need to be calculated separately.

If your calculator has LR mode (linear regression) then the regression line y on x can be found directly. The following calculations were done on a Casio 100C or 115N:

For the data in Example 11.4 on p. 570.

x	1	2	4	6	7	8	10
y	10	14	12	13	15	12	13

set the calculator to LR (linear regression) by pressing

[MODE] [2].

Then [SHIFT] [KAC] [1] [$x_D y_D$] [10] [DATA]

[2] [$x_D y_D$] [14] [DATA]

[4] [$x_D y_D$] [12] [DATA]

[6] [$x_D y_D$] [13] [DATA]

[7] [$x_D y_D$] [15] [DATA]

[8] [$x_D y_D$] [12] [DATA]

[10] [$x_D y_D$] [13] [DATA]

(Try to use both hands, the left hand for the numbers and the right hand for the [$x_D y_D$] [DATA] keys.)

If the regression line y on x is written in the form

$$y = A + Bx$$

then [SHIFT] [7] gives [A] $= 11.704\,035\,87$

[SHIFT] [8] gives [B] $= 0.186\,098\,654$

so that the regression line y on x is $\underline{y = 0.186x + 11.7}$ (as before).

You also have access to the following information:

[K_{out}] [1]	gives	[Σx^2]	$= 270$
[K_{out}] [2]	gives	[Σx]	$= 38$
[K_{out}] [3]	gives	[n]	$= 7$
[K_{out}] [4]	gives	[Σy^2]	$= 1147$
[K_{out}] [5]	gives	[Σy]	$= 89$
[K_{out}] [6]	gives	[Σxy]	$= 495$
[SHIFT] [1]	gives	[$\bar{x}$]	$= 5.428\,571\,429$
[SHIFT] [2] [x^2]	gives	$s_x{}^2$	$= 9.102\,040\,816$
[SHIFT] [4]	gives	[$\bar{y}$]	$= 12.714\,285\,71$
[SHIFT] [5] [x^2]	gives	$s_y{}^2$	$= 2.204\,081\,634$

SPECIAL NOTE: you also have access to r, the product-moment correlation coefficient (see p. 579) and this is obtained by pressing [SHIFT] [9] which gives [r] $= 0.378\,180\,198$.

To clear the LR mode, press [MODE] [0].

Exercise 11c

In the following questions, check your answers using your calculator in LR mode if possible.

1. Calculate (i) the covariance, (ii) the equations of the two least squares regression lines for the following data. Plot the scatter diagrams and draw in the regression lines. Find also the minimum sum of squares of residuals (iii) for y on x, (iv) for x on y.

(*a*)

x	1	2	3	4	4	5	6	8	8	9
y	6	5	7	4	5	4.8	3	1	6	4.6

(*b*)

x	20	20.2	21.4	21.6	22.8	23.4	24.6
y	5	6	4.9	12	8	13.5	12.5

(*c*)

x	2	6.5	6.5	11.5	14	16.5
y	5	7.5	4.5	10	12.5	15.5

(*d*)

x	11	12	12	14	15	16	18	19
y	65	63	64	65	63	62	60	61

2. Calculate the equation of the regression line of y on x for the following distribution:

x	25	30	35	40	45	50
y	78	70	65	58	48	42

Is it possible to calculate from the equation you have just found (*a*) an estimate for the value of x when $y = 54$? (*b*) an estimate for the value of y when $x = 37$? In each case, if the answer is 'Yes', calculate the estimate. If the answer is 'No', say why not. (SUJB)

3. The following data show, in convenient units, the yield (y) of a chemical reaction run at various different temperatures (x):

Temperature (x)	110	120	130	140	150	160	170
Yield (y)	2.1	4.3	3.1	3.4	2.9	5.5	3.3

(*a*) Plot the data. Comment on whether it appears that the usual simple linear regression model is appropriate.
(*b*) Assuming that such a model is appropriate, estimate the regression line of yield on temperature.
(*c*) Plot your estimated line on your graph, and indicate clearly on your graph the distances, the sum of whose squares is minimised by the linear regression procedure. (MEI)

4. To test the effect of a new drug twelve patients were examined before the drug was administered and given an initial score (I) depending on the severity of various symptoms. After taking the drug they were examined again and given a final score (F). A decrease in score represented an improvement. The scores for the twelve patients are given in Table A.

Calculate the equation of the line of regression of F on I.

On the average what improvement would you expect for a patient whose initial score was 30? (MEI)

5. A straight line regression equation is fitted by the least squares method to the n points (x_r, y_r), $r = 1, 2, \ldots, n$. For the regression equation $y = ax + b$, show in a sketch the distances whose sum of squares is minimised, and mark clearly which axis records the dependent variable and which axis records the independent (controlled) variable.

In a chemical reaction it is known that the amount, A grams, of a certain compound produced is a linear function of the temperature T°C. Eight trial runs of this reaction are performed, two at each of four different temperatures. The observed values of A are subject to error. The results are shown in the table.

T	10	15	20	25
A	10 12	15 12	18 16	16 20

Draw a scatter diagram for these data.
Calculate $\overline{A}$ and $\overline{T}$.

Table A

Patient		1	2	3	4	5	6	7	8	9	10	11	12
Score	Initial (I)	61	23	8	14	42	34	32	31	41	25	20	50
	Final (F)	49	12	3	4	28	27	20	20	34	15	16	40

Obtain the equation of the regression line of A on T giving the coefficients to 2 decimal places.

Draw this line on your scatter diagram.

Use the regression equation to obtain an estimate of the mean value of A when $T = 20$, and explain why this estimate is preferable to averaging the two observed values of A when $T = 20$.

Estimate the mean increase in A for a one degree increase in temperature.

State any reservations you would have about estimating the mean value of A when $T = 0$. (L)

6. In an attempt to increase the yield (kg/h) of an industrial process a technician varies the percentage of a certain additive used, while keeping all other conditions as constant as possible. The results are shown below.

Yield, y	% additive, x
127.6	2.5
130.2	3.0
132.7	3.5
133.6	4.0
133.9	4.5
133.8	5.0
133.3	5.5
131.9	6.0

You may assume that $\Sigma x = 34$, $\Sigma y = 1057$, $\Sigma xy = 4504.55$, $\Sigma x^2 = 155$.

(*a*) Draw a scatter diagram of the data.
(*b*) Calculate the equation of the regression line of yield on percentage additive and draw it on the scatter diagram.

The technician now varies the temperature (°C) while keeping other conditions as constant as possible and obtains the following results.

Yield, y	Temperature, t
127.6	70
128.7	75
130.4	80
131.2	85
133.6	90

He calculates (correctly) that the regression line is $y = 107.1 + 0.29t$.
(*c*) Draw a scatter diagram of these data together with the regression line.
(*d*) The technician reports as follows, 'The regression coefficient of yield on percentage additive is larger than that of yield on temperature, hence the most effective way of increasing the yield is to make the percentage additive as large as possible, within reason.'

Criticise the report and make your own recommendations on how to achieve the maximum yield. (AEB 1988)

7. Referring to your projects if possible, explain clearly the purpose of obtaining a linear regression equation, and describe what use was, or could be, made of this equation.

A large field used for growing potatoes was divided into 6 equal plots, and each plot was treated with a different concentration of a certain fertiliser. At harvest time the yield from each plot was recorded, and the results are given in the table, with potato yield ($Y\,\text{kg m}^{-2}$) and fertilizer concentration ($C\,\text{g l}^{-1}$).

Concentration, C	$\frac{1}{2}$	1	2	3	4	6
Yield, Y	10	16	26	36	50	72

Draw a scatter diagram for these data, and mark on your diagram the point representing the mean of the data.

Find the equation of a suitable regression line from which the yield to be expected for a concentration of $5\,\text{g l}^{-1}$ can be predicted, and give the value of this expected yield. Sketch the regression line on your scatter diagram.

Calculate the sum of squares of the residuals and explain what this value represents with regard to your regression line.

[If required, you may assume in your working that $\Sigma C^2 = 66.25$, $\Sigma CY = 813$, $\Sigma Y^2 = 10\,012$.] (L)

8. In an experiment the temperature of a metal rod was raised from 300 K. The extensions E mm of the rod at selected temperatures T K are shown in the table.

T	E
300	0
350	0.38
400	0.80
450	1.22
500	1.60
550	2.00
600	2.42
650	2.80
700	3.18

Draw a scatter diagram of the data and mark on your diagram the point representing the means of T and E.

Find the equation of the regression line of E on T and draw this line on your diagram. Estimate the extension of the rod at 430 K. (L)P

THE PRODUCT-MOMENT CORRELATION COEFFICIENT

The least squares regression line y on x is

$$y-\bar{y} = \frac{s_{xy}}{s_x^2}(x-\bar{x})$$

and the least squares regression line x on y is

$$x-\bar{x} = \frac{s_{xy}}{s_y^2}(y-\bar{y})$$

We 'standardise' these equations as follows:

$$\frac{y-\bar{y}}{s_y} = \frac{s_{xy}}{s_x s_y}\frac{(x-\bar{x})}{s_x}$$

$$\frac{x-\bar{x}}{s_x} = \frac{s_{xy}}{s_x s_y}\frac{(y-\bar{y})}{s_y}$$

Now take new axes, with origin $(\bar{x}, \bar{y})$ and the X axis graduated in units of s_x, the Y axis graduated in units of s_y.

$$Y = \frac{y-\bar{y}}{s_y} \quad \text{and} \quad X = \frac{x-\bar{x}}{s_x}$$

Now the regression lines can be written

$$Y = rX \quad \text{where} \quad r = \frac{s_{xy}}{s_x s_y}$$

and $$X = rY$$

The diagram from Example 11.4 would change from diagram (*a*) to diagram (*b*):

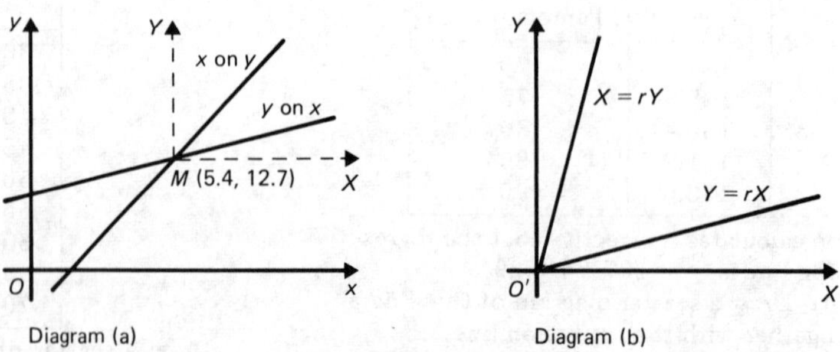

Diagram (a) Diagram (b)

where the new origin is at the point (5.4, 12.7) and

$$r = \frac{s_{xy}}{s_x s_y} = \frac{1.694}{\sqrt{9.102}\sqrt{2.204}} = 0.38 \quad \text{(2 S.F.)}$$

Note that if $Y = rX$ makes an angle θ with the X-axis, then $r = \tan\theta$, so that $X = rY$ makes *the same angle* θ with the Y-axis.

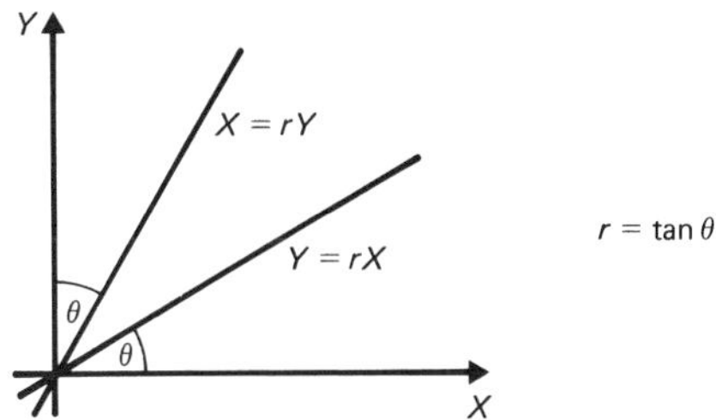

Some examples of regression lines together with the corresponding lines $Y = rX$ and $X = rY$ are shown below.

Generally, the more correlated the variables are, the closer are the two regression lines.

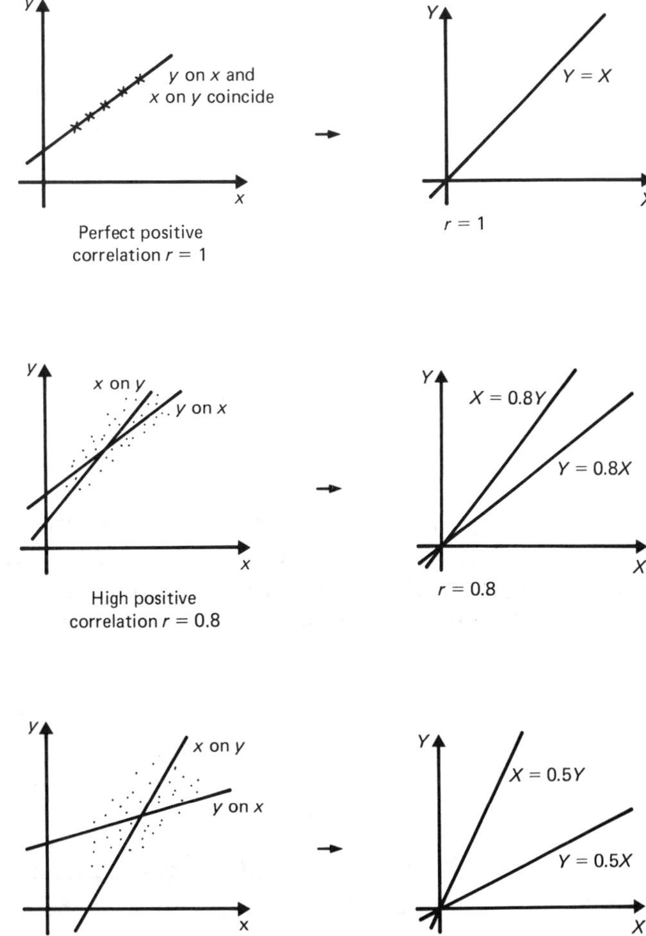

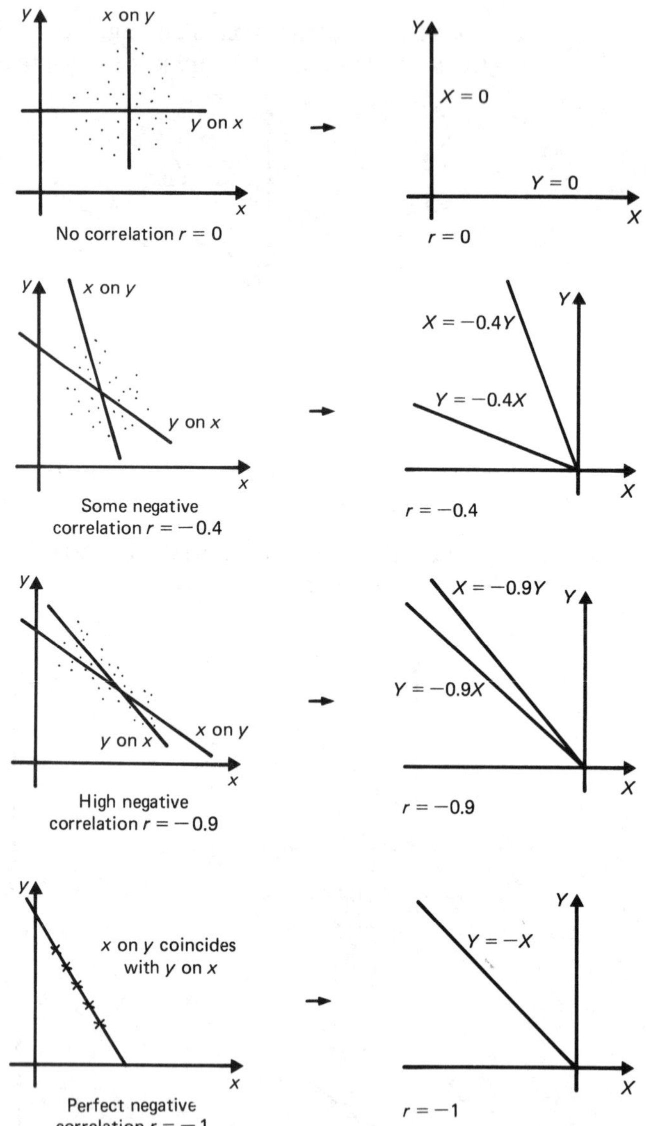

Thus r is **a measure of the degree of scatter.** It is independent of the units in which the data are measured. Note that it is a measure of *linear correlations only*, so that even if there were a perfect quadratic relation between the variables we should still have $r \approx 0$.

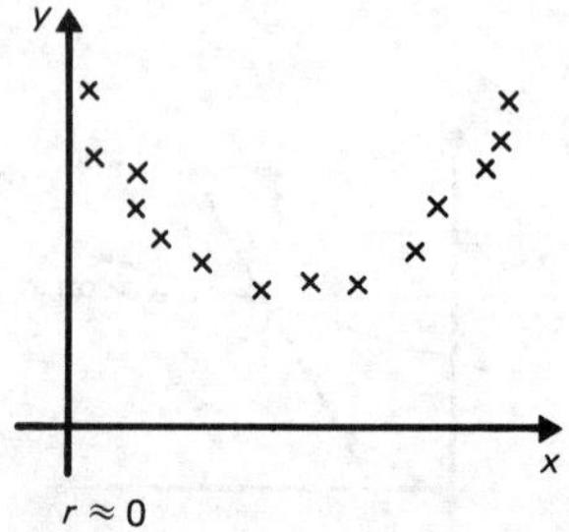

r is known as the **product-moment correlation coefficient.**

The product-moment correlation coefficient, r, is given by

$$r = \frac{s_{xy}}{s_x s_y}$$

ALTERNATIVE METHOD FOR FINDING THE MINIMUM SUM OF SQUARES OF RESIDUALS

The minimum sum of squares of residuals can now be written in a slightly simpler form, using the product-moment correlation coefficient, r.

For y on x

$$\Sigma m_{i(\min)}^2 = n\left(s_y{}^2 - \frac{s_{xy}{}^2}{s_x{}^2}\right)$$

$$= n(s_y{}^2 - r^2 s_y{}^2) \qquad \text{since} \qquad r = \frac{s_{xy}}{s_x s_y}$$

$$= n(1-r^2)s_y{}^2$$

For y on x, the minimum sum of squares of residuals is $n(1-r^2)s_y{}^2$.

For x on y

$$\Sigma n_{i(\min)}^2 = n\left(s_x{}^2 - \frac{s_{xy}{}^2}{s_y{}^2}\right)$$

$$= n(s_x{}^2 - r^2 s_x{}^2) \qquad \text{since} \qquad r = \frac{s_{xy}}{s_x s_y}$$

$$= n(1-r^2)s_x{}^2$$

For x on y, the minimum sum of squares of residuals is $n(1-r^2)s_x{}^2$.

Example 11.5 For the following data, find the product-moment correlation coefficient. Find also the minumum sum of squares of residuals for y on x.

x	20	30	40	46	54	60	80	88	92
y	54	60	54	62	68	80	66	80	100

Solution 11.5

x	y	x^2	y^2	xy
20	54	400	2916	1080
30	60	900	3600	1800
40	54	1600	2916	2160
46	62	2116	3844	2852
54	68	2916	4624	3672
60	80	3600	6400	4800
80	66	6400	4356	5280
88	80	7744	6400	7040
92	100	8464	10 000	9200
$\Sigma x = 510$	$\Sigma y = 624$	$\Sigma x^2 = 34\,140$	$\Sigma y^2 = 45\,056$	$\Sigma xy = 37\,884$

For these data, $n = 9$.

$$s_{xy} = \frac{\Sigma xy}{n} - \overline{xy} = \frac{37\,884}{9} - \left(\frac{510}{9}\right)\left(\frac{624}{9}\right) = 280.4445$$

$$s_x{}^2 = \frac{\Sigma x^2}{n} - \bar{x}^2 = \frac{34\,140}{9} - \left(\frac{510}{9}\right)^2 = 582.2222$$

$$s_y{}^2 = \frac{\Sigma y^2}{n} - \bar{y}^2 = \frac{45\,056}{9} - \left(\frac{624}{9}\right)^2 = 199.1111$$

So $$r = \frac{s_{xy}}{s_x s_y} = \frac{280.4445}{\sqrt{(582.2222)(199.1111)}} = 0.8237$$

Therefore the product-moment correlation coefficient is 0.82 (2 d.p.), indicating a high positive correlation.

Now, for y on x

$$\begin{aligned}\Sigma m_i{}^2{}_{(\text{min})} &= n(1-r^2)s_y{}^2 \\ &= 9(1-0.823^2)(199.1111) \\ &= 576 \qquad (3 \text{ S.F.})\end{aligned}$$

The minimum sum of squares of residuals for y on x is 576 (3 S.F.).

CALCULATOR NOTE: if your calculator has pre-programmed SD and/or LR modes then make use of them whenever possible.

In LR mode, r can be obtained directly as follows:

Set LR mode by pressing [MODE] [2].

[SHIFT] [KAC] [20] [$x_D y_D$] [54] [DATA]
[30] [$x_D y_D$] [60] [DATA]
[40] [$x_D y_D$] [54] [DATA]
[46] [$x_D y_D$] [62] [DATA]
[54] [$x_D y_D$] [68] [DATA]
[60] [$x_D y_D$] [80] [DATA]
[80] [$x_D y_D$] [66] [DATA]
[88] [$x_D y_D$] [80] [DATA]
[92] [$x_D y_D$] [100] [DATA]

[SHIFT] [9] gives [r] = 0.823 67 . . . , as before.

To obtain s_y^2 press [SHIFT] [5] [x^2] and

s_x^2 by pressing [SHIFT] [2] [x^2]

So to calculate $n(1-r^2)s_y^2$, where $n = 9$, press

[1] [−] [SHIFT] [9] [x^2] [=]

[×] [SHIFT] [5] [x^2] [×] [9] [=]

which gives $\Sigma m_{i\,(\min)}^2 = 576$ (3 S.F.), as before.

RELATIONSHIP BETWEEN REGRESSION COEFFICIENTS AND *r*

For the regression line y on x

$$y = ax + b \qquad \text{where} \qquad a = \frac{s_{xy}}{s_x^2}$$

and for the regression line x on y

$$x = cy + d \qquad \text{where} \qquad c = \frac{s_{xy}}{s_y^2}.$$

where a and c are the regression coefficients.

Now
$$ac = \frac{s_{xy}}{s_x^2}\frac{s_{xy}}{s_y^2}$$
$$= \left(\frac{s_{xy}}{s_x s_y}\right)^2$$
$$= r^2$$

Now, either a and c are both positive

or a and c are both negative,

so $r^2 = ac$ and $r = +\sqrt{ac}$ if a, c are positive
$r = -\sqrt{ac}$ if a, c are negative

Example 11.6 For the data given in Example 11.4, find r, the product-moment correlation coefficient.

x	1	2	4	6	7	8	10
y	10	14	12	13	15	12	13

Solution 11.6 *Method 1*
From p. 571 the least squares regression line y on x is

$$y = 0.186x + 11.7 \qquad (a = 0.186)$$

and x on y is

$$x = 0.769y - 4.34 \qquad (c = 0.769)$$

So
$$r = +\sqrt{ac}$$
$$= \sqrt{(0.186)(0.769)}$$
$$= 0.378 \qquad \text{(3 d.p.)}$$

$r = 0.378$, indicating that there is some positive correlation.

Method 2

We have shown that $s_{xy} = 1.694$, $s_x^2 = 9.102$, $s_y^2 = 2.204$ (p. 571).

So
$$r = \frac{s_{xy}}{s_x s_y} = \frac{1.694}{\sqrt{(9.102)(2.204)}} = 0.378 \quad \text{(3 d.p.)}$$

Example 11.7 The moisture content, M, in grams of water per 100 grams of dried solids, of core samples of mud from an estuary was measured at depth D metres. The results are shown in the table:

Depth (D)	0	5	10	15	20	25	30	35
Moisture content (M)	90	82	56	42	30	21	21	18

(a) On graph paper, draw a scatter diagram for these data.

(b) Obtain, to 3 decimal places, the product-moment correlation coefficient. Without performing a significance test, interpret the meaning of your result.

(c) Find the equation of the regression line of M on D, giving the coefficients to 2 decimal places.

(d) Find, to 2 decimal places, the minimum sum of squares of the residuals and explain using words and a diagram what this number represents.

(e) From your equation estimate, to 2 decimal places, the decrease in M when D increases by 1. (L)

Solution 11.7 (a) **Scatter diagram to show moisture content, M, and depth, D.**

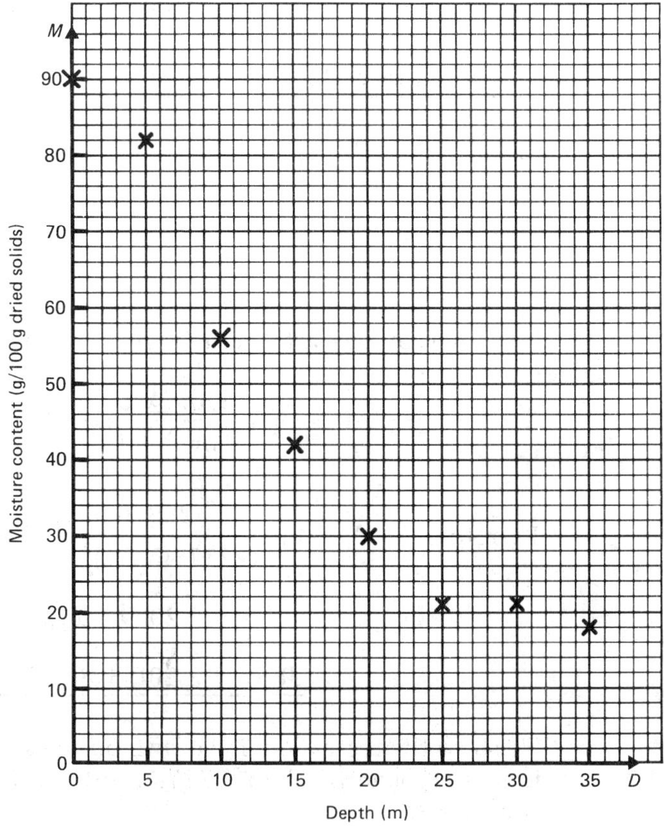

Method 1 — using calculator in SD mode.

(b)

D	0	5	10	15	20	25	30	35
M	90	82	56	42	30	21	21	18

$$r = \frac{s_{DM}}{s_D s_M} \quad \text{where} \quad s_{DM} = \frac{\Sigma DM}{8} - \overline{D}\,\overline{M}$$

Now $\Sigma DM = (0)(90)+(5)(82)+\ldots+(35)(18) = 3985$

From calculator,

$$\overline{D} = 17.5, \qquad s_D = 11.456\,439$$

$$\overline{M} = 45, \qquad s_M = 26.528\,287$$

So
$$s_{DM} = \frac{3985}{8} - (17.5)(45)$$

$$= -289.375$$

Therefore
$$r = \frac{-289.375}{(11.45\ldots)(26.52\ldots)}$$

$$= \underline{-0.952 \qquad (3\text{ d.p.})}$$

<u>This is almost perfect negative correlation.</u>

(c) Equation of least squares regression line M on D is

$$M - \overline{M} = \frac{s_{MD}}{s_D{}^2}(D - \overline{D})$$

Therefore
$$M - 45 = \frac{-289.375}{(11.45\ldots)^2}(D - 17.5)$$

$$= -2.20(D - 17.5)$$

$$\underline{M = -2.20D + 83.58 \qquad (2\text{ d.p.})}$$

We show the regression line drawn on the scatter diagram. Note that it goes through $(\overline{D}, \overline{M})$ i.e. (17.5, 45) and the intercept on the M-axis is 83.58.

The least squares regression line of M on D: $\underline{M = 83.58 - 2.20D}$.

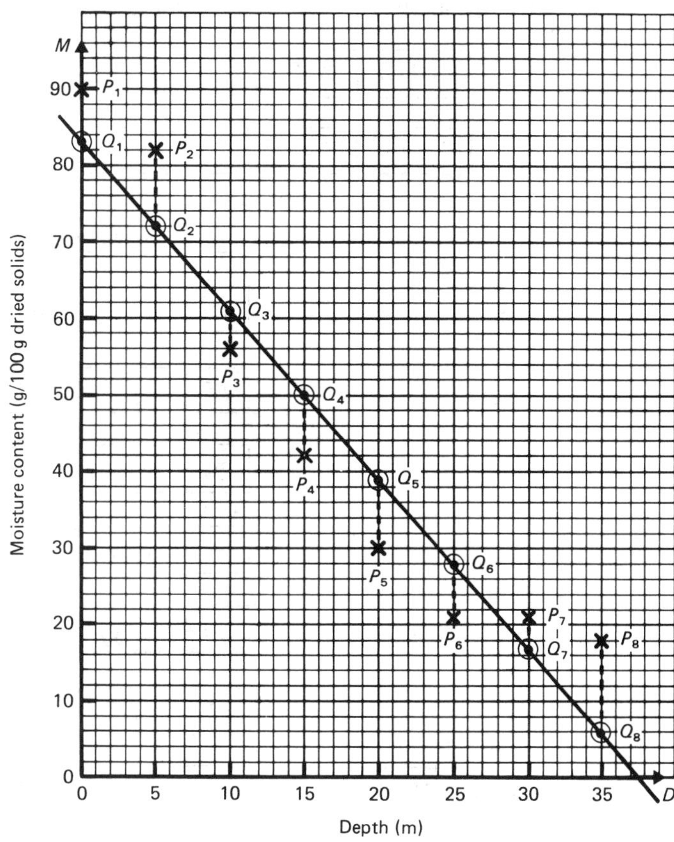

(d) The lengths $P_1Q_1, P_2Q_2, \ldots, P_8Q_8$ are called the residuals. The sum of the squares of the residuals is given by

$$\Sigma m_i^2 = P_1Q_1{}^2 + \ldots + P_8Q_8{}^2$$

and the minimum value of this sum is given by

$$\begin{aligned}\Sigma m_{i\,(\min)}^2 &= n(1 - r^2)s_M{}^2 \\ &= 8(1-(-0.952\ldots)^2)\,(26.5\ldots)^2 \\ &= \underline{525.98 \qquad (2\text{ d.p.})}\end{aligned}$$

(e) The gradient of the regression line is -2.20 so that when D increases by 1, $\underline{M \text{ decreases by } 2.20.}$

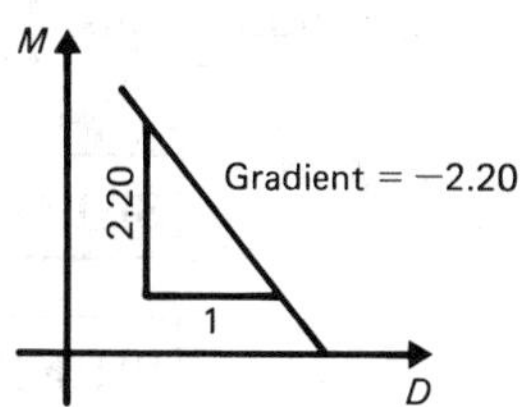

Method 2 — using calculator in LR mode.

(b) We shall use x for depth D and y for moisture content M.

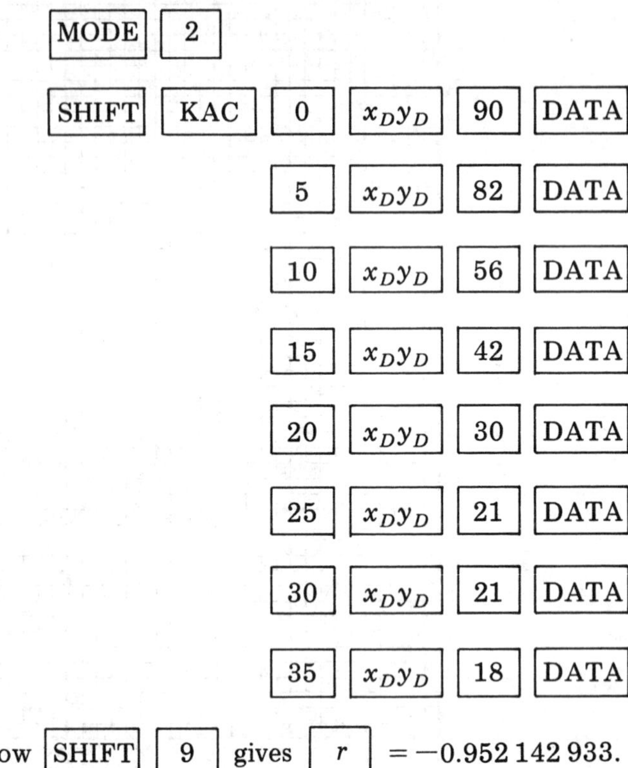

Now [SHIFT] [9] gives [r] $= -0.952\,142\,933$.

So the product moment correlation coefficient $= -0.952$ (3 d.p.).

This is almost perfect negative correlation.

(c) The least squares regression line M on D is given by $M = A + BD$.

where [SHIFT] [7] gives [A] $= 83.583$
$= 83.58$ (2 d.p.)

and [SHIFT] [8] gives [B] $= -2.204\,161\,905$
$= -2.20$ (2 d.p.)

So the regression line is $M = 83.58 - 2.20D$ (2 d.p.).

NOTE: [SHIFT] [1] gives [$\bar{D}$] $= 17.5$

[SHIFT] [4] gives [$\bar{M}$] $= 45$.

(d) $$\Sigma m_{i\,(\min)}^2 = n(1-r^2)s_M{}^2$$

Now $n = 8$ and from calculator

SHIFT 5 x^2 gives $s_M{}^2 = 703.75$

From (b) SHIFT 9 gives r = $-0.952\,142\,933$

so to calculate $n(1-r^2)s_M{}^2$:

1 − SHIFT 9 x^2 =

× SHIFT 5 x^2 × 8 =

So $\Sigma m_{i\,(\min)}^2 = 525.98$ (2 d.p.).

(e) As in *Method 1*.

Exercise 11d

1. Calculate the product-moment correlation coefficient for the sets of data given in Exercise 11c, Question 1, and comment on your answers.

2. If the equations of the least squares regression lines are

$$y = 0.648x + 2.64 \quad (y \text{ on } x) \quad \text{and}$$
$$x = 0.917y - 1.91 \quad (x \text{ on } y)$$

find the product-moment correlation coefficient for the data.

3. For a given set of data $\Sigma x = 680$, $\Sigma y = 996$, $\Sigma x^2 = 20\,154$, $\Sigma y^2 = 34\,670$, $\Sigma xy = 24\,844$, $n = 30$. Find the product-moment correlation coefficient and the equations of the two least squares regression lines.

4. For a given set of data the equations of the least squares regression lines are

$$y = -0.219x + 20.8 \quad (y \text{ on } x) \quad \text{and}$$
$$x = -0.785y + 16.2 \quad (x \text{ on } y)$$

Find the product-moment correlation coefficient for the data.

5. For a given set of data $\Sigma x = 21$, $\Sigma y = 33$, $\Sigma x^2 = 91$, $\Sigma y^2 = 205$, $\Sigma xy = 128$, $n = 6$. Find the product-moment correlation coefficient for the data. Find also the minimum sum of squares of residuals for y on x.

USING A METHOD OF CODING

When the values of x and y are very large or very small we need to avoid exceeding the capacity of the calculator. The least squares calculations can be better done by a change of origin and scaling, that is, using a method of coding.

For the data $(x_1, y_1), (x_2, y_2), \ldots, (x_n, y_n)$ suppose we use the coding

$$X = \frac{x-a}{b} \quad \text{and} \quad Y = \frac{y-c}{d}$$

NOTE: do not confuse the scaling constants a and c used here with the regression coefficients a and c.

Now, rearranging we have

$$x_i = a + bX_i \quad \text{and} \quad y_i = c + dY_i \quad \text{for} \quad i = 1, 2, \ldots, n$$

We have already seen (pp. 44, 63) that

$$\bar{x} = a + b\bar{X}, \qquad \bar{y} = c + d\bar{Y}$$

and

$$s_x = bs_X, \qquad s_y = ds_Y$$

For the covariance

$$s_{xy} = \frac{\Sigma(x_i - \bar{x})(y_i - \bar{y})}{n}$$

$$= \frac{1}{n}\Sigma[a + bX_i - (a + b\bar{X})][c + dY_i - (c + d\bar{Y})]$$

$$= \frac{1}{n}\Sigma b(X_i - \bar{X})d(Y_i - \bar{Y})$$

So $\quad s_{xy} = bds_{XY}$

For the product-moment correlation coefficient

$$r_{XY} = \frac{s_{XY}}{s_X s_Y}$$

$$= \frac{\frac{1}{bd}s_{xy}}{\frac{1}{b}s_x \cdot \frac{1}{d}s_y}$$

$$= \frac{s_{xy}}{s_x s_y}$$

$$= r_{xy}$$

So $\quad r_{XY} = r_{xy},$

i.e. the product-moment correlation coefficient remains *unchanged*. This is because r is a measure of the degree of scatter and this is unchanged by a change of origin and scaling.

Example 11.8 For the following data, use a method of coding to find (a) the covariance, (b) the product-moment correlation coefficient, (c) the least squares regression lines y on x and x on y.

x	1000	1012	1009	1007	1010	1015	1010	1011
y	235	240	245	250	255	260	265	270

Solution 11.8 We use the codings

$$X = x - 1000, \quad Y = \frac{y-250}{5}$$

So, referring to the results on p. 588 with $a = 1000$, $b = 1$, $c = 250$, $d = 5$, we have $s_x = s_X$, $s_y = 5s_Y$ and $s_{xy} = 5s_{XY}$.

X	Y	X^2	Y^2	XY
0	−3	0	9	0
12	−2	144	4	−24
9	−1	81	1	−9
7	0	49	0	0
10	1	100	1	10
15	2	225	4	30
10	3	100	9	30
11	4	121	16	44
$\Sigma X = 74$	$\Sigma Y = 4$	$\Sigma X^2 = 820$	$\Sigma Y^2 = 44$	$\Sigma XY = 81$

(a) $$s_{XY} = \frac{\Sigma XY}{n} - \overline{XY} = \frac{81}{8} - \left(\frac{74}{8}\right)\left(\frac{4}{8}\right) = 5.5$$

Therefore
$$s_{xy} = 5s_{XY} = 5(5.5) = 27.5$$

The covariance s_{xy} is 27.5.

(b) Now

$$s_X^2 = \frac{\Sigma X^2}{n} - \overline{X}^2 = \frac{820}{8} - \left(\frac{74}{8}\right)^2 = 16.9375$$

$$s_Y^2 = \frac{\Sigma Y^2}{n} - \overline{Y}^2 = \frac{44}{8} - \left(\frac{4}{8}\right)^2 = 5.25$$

Therefore

$$r_{XY} = \frac{s_{XY}}{s_X s_Y} = \frac{5.5}{\sqrt{(16.9375)(5.25)}} = 0.58 \quad (2\,\text{d.p.})$$

So $\quad r_{xy} = r_{XY} = 0.58 \quad (2\,\text{d.p.})$

The product-moment correlation coefficient is 0.58 (2 d.p.).

(c) The equation of the least squares regression line Y on X is

$$Y-\bar{Y} = \frac{s_{XY}}{s_X^{\,2}}(X-\bar{X})$$

i.e.
$$Y-\frac{4}{8} = \frac{5.5}{16.9375}\left(X-\frac{74}{8}\right)$$

so
$$Y = 0.3247X-2.5037$$

Now, since $Y=\dfrac{y-250}{5}$ and $X = x-1000$, this equation may be written

$$\frac{y-250}{5} = 0.3247\,(x-1000)-2.5037$$

$$y = 1.6235x-1386.0185 \quad \text{(least squares regression line } y \text{ on } x\text{)}$$

The equation of the least squares regression line X on Y is

$$X-\bar{X} = \frac{s_{XY}}{s_Y^{\,2}}(Y-\bar{Y})$$

$$X-\frac{74}{8} = \frac{5.5}{5.25}\left(Y-\frac{4}{8}\right)$$

i.e.
$$X = 1.048Y+8.726$$

This equation may be written

$$x-1000 = 1.048\left(\frac{y-250}{5}\right)+8.726$$

$$x = 0.2096y+956.326 \quad \text{(least squares regression line } x \text{ on } y\text{)}$$

Exercise 11e

For the following sets of data, use appropriate methods of coding to calculate (*a*) the covariance, (*b*) the product-moment correlation coefficient, (*c*) the least squares lines of regression of y on x and x on y.

1.

x	1701	1722	1717	1718	1703	1701
y	45.1	45.8	45.6	45.3	45.1	45.1

2.

x	981.2	981.3	981.9	981.6	981.5
y	55.6	55.2	54.8	56.1	55.5

3.

x	0.001 57	0.001 56	0.001 49	0.001 65
y	100.4	100.7	100.0	100.4

COEFFICIENTS OF RANK CORRELATION

For the data $(x_1, y_1), \ldots, (x_n, y_n)$ the product-moment correlation coefficient is $\dfrac{s_{xy}}{s_x s_y}$.

Now suppose that, instead of using precise values of the variables, or when such information is not available, we *rank* the numbers in order of size using the numbers $1, 2, \ldots, n$.

A correlation coefficient can be determined on the basis of the *ranks*. There are two useful rank correlation coefficients:

Spearman's Coefficient of Rank Correlation, r_S

Kendall's Coefficient of Rank Correlation, r_k.

SPEARMAN'S COEFFICIENT OF RANK CORRELATION r_S

Suppose that

$X_1, X_2, \ldots, X_n$ are the *ranks* of $x_1, x_2, \ldots, x_n$

$Y_1, Y_2, \ldots, Y_n$ are the *ranks* of $y_1, y_2, \ldots, y_n$

Then

$X_1, X_2, \ldots, X_n$ are the numbers $1, 2, \ldots, n$ in some order

and $Y_1, Y_2, \ldots, Y_n$ are the numbers $1, 2, \ldots, n$ in some order

Consider the rank difference $d_1, d_2, \ldots, d_n$ given by

$$d_1 = X_1 - Y_1, \quad d_2 = X_2 - Y_2, \quad \ldots, \quad d_n = X_n - Y_n$$

so that

$$\begin{aligned} \Sigma d_i^2 &= \Sigma (X_i - Y_i)^2 \quad i = 1, 2, \ldots, n \\ &= \Sigma X_i^2 + \Sigma Y_i^2 - 2\Sigma X_i Y_i \\ &= 2\left(\frac{1}{6}\right) n(n+1)(2n+1) - 2\Sigma X_i Y_i \end{aligned}$$

since $1^2 + 2^2 + \ldots + n^2 = \frac{1}{6}n(n+1)(2n+1)$.

So, dropping the subscript, we have

$$\Sigma XY = \frac{1}{6}n(n+1)(2n+1) - \frac{1}{2}\Sigma d^2 \qquad \text{(i)}$$

Now, if we substitute for the x's and y's in the original data their corresponding *ranks* in the formula for r, the product-moment correlation coefficient, we obtain an *approximation to* r. This approximation is called *Spearman's coefficient of rank correlation*, r_S.

We write

$$r_S = \frac{s_{XY}}{s_X s_Y} \qquad \text{where} \qquad s_{XY} = \frac{\Sigma XY}{n} - \left(\frac{\Sigma X}{n}\right)\left(\frac{\Sigma Y}{n}\right)$$

Substituting from (i) and using

$$\Sigma X = \Sigma Y = 1+2+\ldots+n = \tfrac{1}{2}n(n+1)$$

$$\begin{aligned} s_{XY} &= \frac{1}{6}(n+1)(2n+1) - \frac{1}{2n}\Sigma d^2 - \frac{1}{4}(n+1)^2 \\ &= \frac{[2(n+1)(2n+1)-3(n+1)^2]}{12} - \frac{1}{2n}\Sigma d^2 \\ &= \frac{(n+1)(n-1)}{12} - \frac{1}{2n}\Sigma d^2 \\ &= \frac{(n^2-1)}{12} - \frac{1}{2n}\Sigma d^2 \end{aligned} \qquad \text{(ii)}$$

Now

$$s_X^2 = \frac{\Sigma X^2}{n} - \left(\frac{\Sigma X}{n}\right)^2 = \frac{(n+1)(2n+1)}{6} - \frac{(n+1)^2}{4} = \frac{(n^2-1)}{12} \qquad \text{(iii)}$$

Similarly

$$s_Y^2 = \frac{(n^2-1)}{12}$$

$$\text{so } s_X s_Y = \frac{(n^2-1)}{12}$$

Therefore from (ii) and (iii)

$$r_S = \frac{s_{XY}}{s_X s_Y} = \frac{\dfrac{(n^2-1)}{12} - \dfrac{1}{2n}\Sigma d^2}{\dfrac{(n^2-1)}{12}} = 1 - \frac{6\Sigma d^2}{n(n^2-1)}$$

Spearman's coefficient of rank correlation, r_S, is given by

$$r_S = 1 - \frac{6\Sigma d^2}{n(n^2-1)}$$

It is much easier to calculate r_S than to calculate r, the product-moment correlation coefficient, as there is far less working involved. However, in general, r is a more accurate measure of correlation.

Method of ranking

Suppose we have the masses, x, (in kg) of five men

66, 68, 65, 69, 70

Arranged in ascending order of magnitude, these are 65, 66, 68, 69, 70, so we assign the ranks as follows:

x	66	68	65	69	70
Rank x	2	3	1	4	5

If we have *two or more equal values* we proceed as follows:

x	66	68	65	68	70
Rank x	2	3.5	1	3.5	5

Here, the 3rd and the 4th places represent the same mass (68 kg), so we assign the average rank 3.5 to both these places.

Similarly for the eight values:

x	66	65	66	67	66	64	68	68
Rank x	4	2	4	6	4	1	7.5	7.5

Here the 3rd, 4th and 5th places represent the same mass (66 kg) so we assign the average rank 4 to these places; also the 7th and the 8th places represent the same mass (68 kg) so we assign the average rank 7.5 to both these places.

NOTE: if there are more than just a few equal values, then this method is not appropriate.

Example 11.9 Two competitors rank the eight photographs in a competition as follows:

Photograph	A	B	C	D	E	F	G	H
1st competitor	2	5	3	6	1	4	7	8
2nd competitor	4	3	2	6	1	8	5	7

Calculate Spearman's coefficient of rank correlation for the data.

Solution 11.9 In this example, the data has been ranked already.

Let $d = \text{rank}(x) - \text{rank}(y)$.

Photograph	Rank(x)	Rank(y)	$\lvert d \rvert$	d^2
A	2	4	2	4
B	5	3	2	4
C	3	2	1	1
D	6	6	0	0
E	1	1	0	0
F	4	8	4	16
G	7	5	2	4
H	8	7	1	1
				$\Sigma d^2 = 30$

$$r_S = 1 - \frac{6\Sigma d^2}{n(n^2-1)} \quad \text{where} \quad n = 8$$

$$= 1 - \frac{6(30)}{8(64-1)}$$

$$= 0.64 \quad (2\,\text{d.p.})$$

Spearman's coefficient of rank correlation for the data is 0.64, indicating some positive correlation between the competitors.

Example 11.10 The marks of 10 pupils in French and German tests are as follows.

French, x	12	8	16	12	7	10	12	16	12	9
German, y	6	5	7	7	4	6	8	13	10	10

Calculate Spearman's coefficient of rank correlation.

Solution 11.10 Let $d = \text{rank}(x) - \text{rank}(y)$.

French, x	German, y	Rank(x)	Rank(y)	$\lvert d \rvert$	d^2
12	6	6.5	3.5	3	9
8	5	2	2	0	0
16	7	9.5	5.5	4	16
12	7	6.5	5.5	1	1
7	4	1	1	0	0
10	6	4	3.5	0.5	0.25
12	8	6.5	7	0.5	0.25
16	13	9.5	10	0.5	0.25
12	10	6.5	8.5	2	4
9	10	3	8.5	5.5	30.25
					$\Sigma d^2 = 61$

$$\begin{aligned} r_S &= 1 - \frac{6\Sigma d^2}{n(n^2-1)} \\ &= 1 - \frac{6(61)}{10(100-1)} \\ &= 0.63 \quad (2\text{ d.p.}) \end{aligned}$$

Spearman's coefficient of rank correlation is 0.63, indicating some positive correlation between the marks in the two tests.

Exercise 11f

1. The table shows the marks awarded to six children in a competition. Calculate a coefficient of rank correlation for the data:

Child	A	B	C	D	E	F
Judge 1	6.8	7.3	8.1	9.8	7.1	9.2
Judge 2	7.8	9.4	7.9	9.6	8.9	6.9

2. The following table shows the marks of eight pupils in biology and chemistry. Rank the results and find the value of Spearman's coefficient of rank correlation.

Biology, x	65	65	70	75	75	80	85	85
Chemistry, y	50	55	58	55	65	58	61	65

3. Mr and Mrs Brown and their son John all drive the family car. Before ordering a new car they decide to list in order their preferences for five optional extras independently. The rank order of their choices is as shown:

Optional extra	Mr Brown	Mrs Brown	John
Heated rear window	1st	2nd	3rd
Anti-rust treatment	2nd	4th	2nd
Headrests	3rd	1st	1st
Inertia-reel seat belts	4th	5th	5th
Radio	5th	3rd	4th

(a) Calculate coefficients of rank correlation between each pair of members of the Brown family. (b) A salesman offered to supply three of these extras free with the new car. The family agreed to choose those three which were ranked highest by the two members who agreed most. Which three did they choose, and in what order. (L Additional)

4. Two adjudicators at a Music Competition award marks to ten Pianists as follows:

	Pianist									
	A	B	C	D	E	F	G	H	I	J
Adjudicator I	78	66	73	73	84	66	89	84	67	77
Adjudicator II	81	68	81	75	80	67	85	83	66	78

Calculate a coefficient of rank correlation for these data. Name the method you have used and describe briefly, without proof, the principle on which it is based. (SUJB Additional)

5. Calculate Spearman's coefficient of rank correlation for the set of data given in Exercise 11c, Question 1 (p. 574). Compare these with the product-moment correlation coefficients found in Exercise 11d, Question 1 (p. 587).

6. In a skating competition one judge awards the same mark to all 4 competitors. Show that the coefficient of rank correlation (Spearman's) is 0.5, irrespective of the marks awarded to the competitors by the other judge.

7. Each of the variables x and y takes the values $1, 2, \dots, n$ but not necessarily in the same order as each other. Prove that the covariance of x and y is

$$\operatorname{cov}(x, y) = \frac{n^2-1}{12} - \frac{1}{2n}\sum_{i=1}^{n}(x_i - y_i)^2$$

Hence show that Spearman's coefficient of rank correlation between x and y may be written as

$$1 - \frac{6}{n(n^2-1)}\sum_{i=1}^{n}(x_i - y_i)^2$$

Seven army recruits ($A, B, \dots, G$) were given two separate aptitude tests. Their orders of merit in each test were

Order of merit	1st	2nd	3rd	4th	5th	6th	7th
1st test	G	F	A	D	B	C	E
2nd test	D	F	E	B	G	C	A

Find Spearman's coefficient of rank correlation between the two orders and comment briefly on the correlation obtained. (O & C)

8. Sketch two scatter diagrams illustrating the following situations:
(a) two variables having a large, negative correlation;
(b) two variables having a small, positive correlation.

The mean rainfall per day and the mean number of hours of sunshine per day observed at a weather station are given below.

Month	Rainfall (mm)	Sunshine (hours)
January	1.26	1.1
February	1.25	2.7
March	0.65	4.5
April	2.10	5.1
May	2.45	5.5
June	2.17	7.6
July	2.84	5.2
August	1.74	5.7
September	2.57	4.8
October	1.65	2.9
November	1.47	2.8
December	1.94	1.8

Calculate, correct to two decimal places, the rank correlation coefficient between rainfall and hours of sunshine.

What is the rank correlation coefficient between rainfall and minutes of sunshine? (SUJB Additional)

9. In a study of population density in eight suburbs of a town the statistics shown in the table were obtained. The population density is denoted by p, and the distance of the suburb from the centre of the town by d.

Suburb	A	B	C	D	E	F	G	H
p (persons/hectare)	55	11	68	38	46	43	21	25
d (km)	0.7	3.8	1.7	2.6	1.5	2.6	3.4	1.9

(*a*) Plot p against d on a scatter diagram. (*b*) Calculate and mark on the diagram the mean of the array. (*c*) Calculate a coefficient of rank correlation between p and d, stating the system of ranking adopted for both quantities. (*d*) State what conclusions can be drawn from your answers to (*a*) and (*c*) concerning the general trend of the results. (*e*) Giving a reason for your answer, state which suburb in your opinion fits the general trend least well. (L Additional)

10.

Candidate	A	B	C	D	E	F
English	38	62	56	42	59	48
History	64	84	84	60	73	69

The table shows the original marks of six candidates in two examinations. Calculate a coefficient of rank correlation and comment on the value of your result.

The History papers are re-marked and one of the six candidates is awarded five additional marks. Given that the other marks, and the coefficient of rank correlation, are unchanged, state, with reasons, which candidate received the extra marks. (C Additional)

11. (*a*) X and Y were judges at a beauty contest in which there were 10 competitors. Their rankings are shown below.

Competitor	A	B	C	D	E	F	G	H	I	J
X	4	9	2	5=	3	10	5=	7	8	1
Y	6	10	2	8	1	9	7	4	5	3

Calculate a coefficient of rank correlation between these two sets of ranks and comment briefly on your result.

(*b*) Illustrate by means of two scatter diagrams rank correlation coefficients of 0 and -1 between two variables X and Y. (C Additional)

12. (*a*) Sketch scatter diagrams which illustrate (i) positive linear correlation, (ii) negative linear correlation, (iii) no correlation, between two variables X and Y.

(*b*) A doctor asked ten of his patients, who were smokers, how many years they had smoked. In addition, for each patient, he gave a grade between 0 and 100 indicating the extent of their lung damage. The following table shows the results:

Patient	A	B	C	D	E	F	G	H	I	J
Number of years smoking	15	22	25	28	31	33	36	39	42	48
Lung damage grade	30	50	55	30	57	35	60	72	70	75

Calculate a coefficient of rank correlation between the number of years of smoking and the extent of lung damage. Comment on the figure which you obtain. (C Additional)

SIGNIFICANCE OF SPEARMAN'S RANK CORRELATION COEFFICIENT

In order to test the significance of the calculated value of r_S, it is necessary to calculate the probability of obtaining a given value of Σd^2.

We look at the distribution of Σd^2 in the following situation: These are the rankings of four samples of sparkling wine by two wine-tasters, Enrico and Claude.

Wine	A	B	C	D
Enrico's ranking	1	2	3	4
Claude's ranking	2	1	4	3

If we leave the first row in its natural ranking order, 1, 2, 3, 4, then the second row *could* be ranked in 4! different ways, assuming that there are no equal ranks. These 24 arrangements are shown here, with the corresponding values of Σd^2.

	1 2 3 4	1 2 3 4	1 2 3 4	1 2 3 4	1 2 3 4	1 2 3 4
	1 2 3 4	1 2 4 3	1 3 4 2	1 3 2 4	1 4 3 2	1 4 2 3
Σd^2	0	2	6	2	8	6
	($r_S = 1$)					

	1 2 3 4	1 2 3 4	1 2 3 4	1 2 3 4	1 2 3 4	1 2 3 4
	2 1 3 4	2 1 4 3	2 3 1 4	2 3 4 1	2 4 1 3	2 4 3 1
Σd^2	2	4	6	12	10	14
					($r_S = 0$)	

	1 2 3 4	1 2 3 4	1 2 3 4	1 2 3 4	1 2 3 4	1 2 3 4
	3 1 2 4	3 1 4 2	3 2 4 1	3 2 1 4	3 4 1 2	3 4 2 1
Σd^2	6	10	14	8	16	18
		($r_S = 0$)				

	1 2 3 4	1 2 3 4	1 2 3 4	1 2 3 4	1 2 3 4	1 2 3 4
	4 1 2 3	4 1 3 2	4 2 3 1	4 2 1 3	4 3 1 2	4 3 2 1
Σd^2	12	14	18	14	18	20
						($r_S = -1$)

Arranging Σd^2 in the form of a frequency distribution:

Σd^2	0	2	4	6	8	10	12	14	16	18	20
Frequency	1	3	1	4	2	2	2	4	1	3	1

The distribution is symmetrical about $\Sigma d^2 = 10$.

Note the following results:

Using $r_S = 1 - \dfrac{6\Sigma d^2}{n(n^2-1)}$, $n = 4$

$\Sigma d^2 = 0$	$\Sigma d^2 = 10$	$\Sigma d^2 = 20$
$r_S = 1 - \dfrac{6(0)}{4(15)}$ $= 1$	$r_S = 1 - \dfrac{6(10)}{4(15)}$ $= 0$	$r_S = 1 - \dfrac{6(20)}{4(15)}$ $= -1$
Perfect position correlation — the rankings agree exactly.	No correlation	Perfect negative correlation — one rank is the exact reverse of the other.

The frequency distribution of Σd^2.

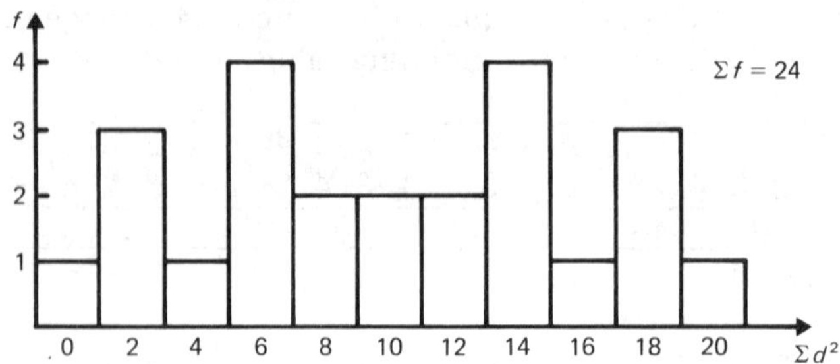

We can now use this bar chart to find probabilities associated with various values of Σd^2.

(a)

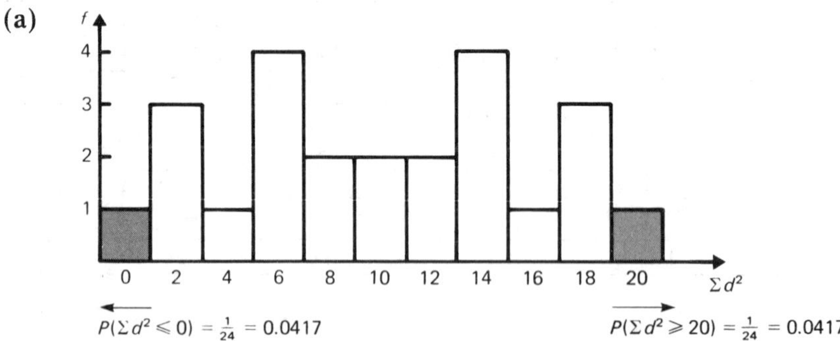

(b)

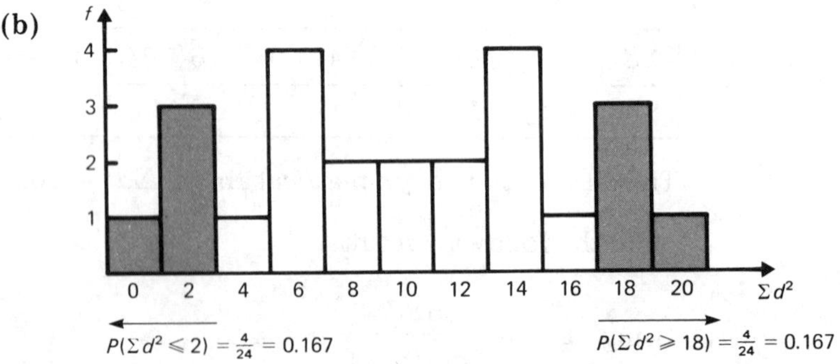

(c)

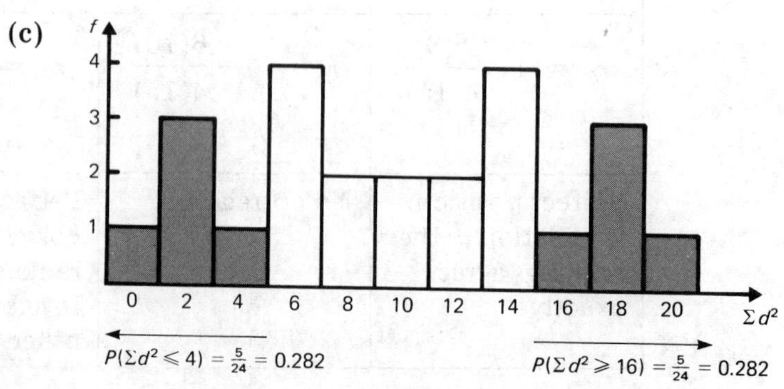

(d)

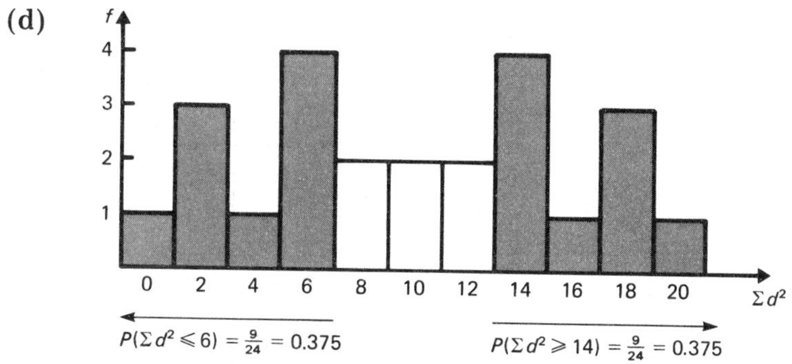

(e)

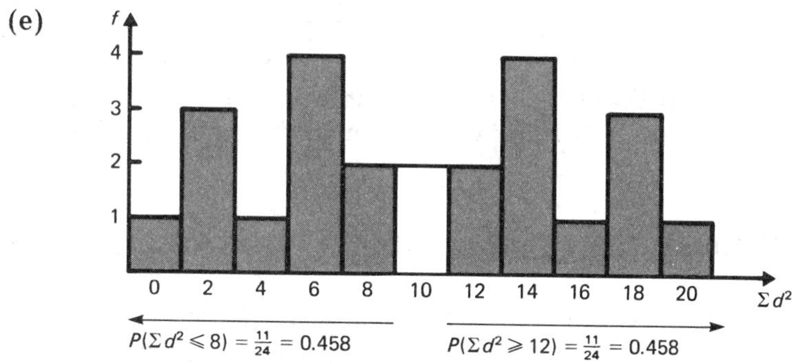

If we consider one of these diagrams, (e) say,

$$P(\Sigma d^2 \leqslant 8) = 0.458 \quad \text{and} \quad P(\Sigma d^2 \geqslant 12) = 0.458$$

Also, if $\Sigma d^2 \leqslant 8$, with $n = 4$,

$$r_S \geqslant 1 - \frac{6(8)}{4(15)} \quad \text{so} \quad r_S \geqslant 0.2$$

So $\underline{P(\Sigma d^2 \leqslant 8) = P(r_S \geqslant 0.2) = 0.458.}$

Also if $\Sigma d^2 \geqslant 12$, with $n = 4$,

$$r_S \leqslant 1 - \frac{6(12)}{4(15)} \quad \text{so} \quad r_S \leqslant -0.2$$

So $\underline{P(\Sigma d^2 \geqslant 12) = P(r_S \leqslant -0.2) = 0.458.}$

Putting together the results from all the diagrams we have

	Probability	
$\Sigma d^2 \leqslant 0$, $r_S \geqslant 1$	0.0417	$\Sigma d^2 \geqslant 20$, $r_S \leqslant -1$
$\Sigma d^2 \leqslant 2$, $r_S \geqslant 0.8$	0.167	$\Sigma d^2 \geqslant 18$, $r_S \leqslant -0.8$
$\Sigma d^2 \leqslant 4$, $r_S \geqslant 0.6$	0.208	$\Sigma d^2 \geqslant 16$, $r_S \leqslant -0.6$
$\Sigma d^2 \leqslant 6$, $r_S \geqslant 0.4$	0.375	$\Sigma d^2 \geqslant 14$, $r_S \leqslant -0.4$
$\Sigma d^2 \leqslant 8$, $r_S \geqslant 0.2$	0.458	$\Sigma d^2 \geqslant 12$, $r_S \leqslant -0.2$

The results for Σd^2 are also summarised thus:

$n = 4$ MAX $\Sigma d^2 = 20$		
Σd^2		P
$\leqslant$	$\geqslant$	
8	12	.458
6	14	.375
4	16	.208
2	18	.167
0	20	.0417

This is an extract from a larger table, given in the Appendix on p. 638 which gives probabilities for Σd^2 for values of n in the range $4 \leqslant n \leqslant 10$. We will refer to this table as Table A.

SIGNIFICANCE TEST FOR r_S, USING PROBABILITIES OF Σd^2

When we test r_S for significance, a suitable null hypothesis is $H_0 : \rho = 0$, where ρ is the true *population* correlation coefficient and $\rho = 0$ indicates that there is no predictable correlation between the rankings. The method is illustrated as follows:

Example 11.11 For 8 pairs of rankings, $\Sigma d^2 = 28$, giving Spearman's coefficient of rank correlation $r_S = 0.667$ (3 d.p.).

Does this value indicate **(a)** a correlation significantly different from zero, at the 10% level, **(b)** a significant positive correlation at the 1% level?

Solution 11.11 (a) $H_0 : \rho = 0$ (there is no correlation)

$H_1 : \rho \neq 0$ (there is some correlation different from zero)

We refer to Table A (p. 639).

Use a 2-tailed test at the 10% level and reject H_0 if $P(\Sigma d^2 \leqslant 28) < 0.05$.

From Table A:

$n = 8$		
Σd^2		P
$\leqslant$	$\geqslant$	
28	140	0.0415

This indicates that $P(\Sigma d^2 \leqslant 28) = 0.0415 < 0.05$, so we reject H_0 and conclude that there is evidence at the 10% level of a correlation different from zero.

(b) $H_0: \rho = 0$ (there is no correlation)

$H_0: \rho > 0$ (there is some positive correlation)

Use a 1-tailed test at 1% level and reject H_0 if $P(\Sigma d^2 \leqslant 28) < 0.01$.

Now, from Table A, $P(\Sigma d^2 \leqslant 28) = 0.0415 > 0.01$ so we do not reject H_0 and conclude that there is no evidence at the 1% level of a positive correlation.

Example 11.12 For 9 pairs of rankings it is found that $\Sigma d^2 = 214$, giving $r_S = -0.783$. Does this provide evidence, at the 1% level, of a negative correlation?

Solution 11.12 $H_0: \rho = 0$ (there is no correlation)

$H_1: \rho < 0$ (there is some negative correlation)

Use a 1-tailed test at 1% level and reject H_0 if $P(\Sigma d^2 \geqslant 214) < 0.01$.

Now, from Table A, $P(\Sigma d^2 \geqslant 214) = 0.0086 < 0.01$, so we reject H_0 and conclude that there is evidence, at the 1% level, of negative correlation.

Example 11.13 An expert on porcelain is asked to place 7 china bowls in date order of manufacture assigning the rank 1 to the oldest bowl. The actual dates of manufacture and the order given by the expert are shown.

Bowl	*A*	*B*	*C*	*D*	*E*	*F*	*G*
Date of manufacture	1920	1857	1710	1896	1810	1690	1780
Order given by expert	7	3	4	6	2	1	5

Find, to 3 decimal places, the Spearman rank correlation coefficient between the order of manufacture and the order given by the expert.

Refer to one of the tables of critical values provided to comment on the significance of your result. State clearly the null hypothesis which is being tested. (L)P

Solution 11.13

	Bowl							
	A	B	C	D	E	F	G	
Rank (x)	7	5	2	6	4	1	3	
Rank (y)	7	3	4	6	2	1	5	
$\lvert d \rvert$	0	2	2	0	2	0	2	
Σd^2	0	4	4	0	4	0	4	$\Sigma d^2 = 16,\ n = 7$

Now
$$r_S = 1 - \frac{6\Sigma d^2}{n(n^2-1)}$$
$$= 1 - \frac{6(16)}{7(48)}$$
$$= \underline{0.714 \quad (3 \text{ d.p.})}$$

$H_0: \rho = 0$ (no evidence of correlation)

$H_1: \rho > 0$ (some positive correlation)

Use a 1-tailed test, at the 5% level, and reject H_0 if $P(\Sigma d^2 \leqslant 16) < 0.05$.

From Table A, $P(\Sigma d^2 \leqslant 16) = 0.044 < 0.05$, so we reject H_0 and conclude that there is evidence, at the 5% level, of agreement between the order given by the expert and the actual dates of manufacture.

Exercise 11g

In each of the following questions use Table A to test the hypotheses.

	n	Σd^2	Hypotheses	Level of significance
1.	9	212	$H_0: \rho = 0,\ H_1: \rho < 0$	1%
2.	8	30	$H_0: \rho = 0,\ H_1: \rho > 0$	5%
3.	8	30	$H_0: \rho = 0,\ H_1: \rho \neq 0$	5%
4.	10	78	$H_0: \rho = 0,\ H_1: \rho > 0$	5%
5.	10	252	$H_0: \rho = 0,\ H_1: \rho < 0$	10%
6.	10	274	$H_0: \rho = 0,\ H_1: \rho \neq 0$	5%
7.	7	18	$H_0: \rho = 0,\ H_1: \rho \neq 0$	10%
8.	7	106	$H_0: \rho = 0,\ H_1: \rho < 0$	1%
9.	7	14	$H_0: \rho = 0,\ H_1: \rho \neq 0$	5%

SIGNIFICANCE TEST FOR r_S, USING CRITICAL VALUES

Instead of working with Σd^2 it is much easier to refer to Table B which gives critical values of r_S. This is printed below and in the Appendix on p. 639.

Critical values of the Spearman rank correlation coefficient

n	Significance level (one-tailed test)	
	0.05	0.01
4	1.000	
5	0.900	1.000
6	0.829	0.943
7	0.714	0.893
8	0.643	0.833
9	0.600	0.783
10	0.564	0.746
12	0.506	0.712
14	0.456	0.645
16	0.425	0.601
18	0.399	0.564
20	0.377	0.534
22	0.359	0.508
24	0.343	0.485
26	0.329	0.465
28	0.317	0.448
30	0.306	0.432

TABLE B

Example 11.14 Using Table B, for $n = 8$ and $r_S = 0.667$, test the following hypotheses:

(a) $H_0: \rho = 0$, $H_1: \rho \neq 0$ (10% level of significance)

(b) $H_0: \rho = 0$, $H_1: \rho > 0$ (1% level of significance)

Solution 11.14 (a) Using a 2-tailed test, at the 10% level, and considering Table B, with $n = 8$, significance 0.05 (because test is 2-tailed) we reject H_0 if $r_S \geqslant 0.643$.

Now $r_S = 0.667$, so we reject H_0 and conclude that there is evidence at the 10% level of a correlation different from zero.

(b) Using a 1-tailed test, at 1% level, and considering Table B, with $n = 8$, significance 0.01, we reject H_0 if $r_S \geqslant 0.833$.

Now $r_S = 0.667 < 0.833$, so we do not reject H_0 and conclude that there is no evidence, at the 1% level, of positive correlation.

Example 11.15 For 9 pairs of values, r_S is found to be -0.765. Test, at the 1% level, whether there is evidence of a negative correlation.

Solution 11.15 $H_0: \rho = 0$ (there is no correlation)

$H_1: \rho < 0$ (there is negative correlation)

Using Table B, with $n = 9$, significance level 0.01, we reject H_0 if $r_S \leqslant -0.783$.

Now $r_S = -0.765 > -0.783$, so we do not reject H_0 and conclude that there is no evidence, at the 1% level, of a negative correlation.

Exercise 11h

In each of the following, use Table B to comment on the significance of the value for r_S.

	n	r_S	Hypotheses	Level of significance
1.	7	0.900	$H_0: \rho = 0,\ H_1: \rho \neq 0$	2%
2.	14	0.500	$H_0: \rho = 0,\ H_1: \rho > 0$	1%
3.	28	0.324	$H_0: \rho = 0,\ H_1: \rho \neq 0$	10%
4.	28	0.324	$H_0: \rho = 0,\ H_1: \rho > 0$	1%
5.	16	−0.420	$H_0: \rho = 0,\ H_1: \rho < 0$	5%
6.	12	−0.689	$H_0: \rho = 0,\ H_1: \rho \neq 0$	10%
7.	12	0.689	$H_0: \rho = 0,\ H_1: \rho > 0$	1%
8.	10	−0.738	$H_0: \rho = 0,\ H_1: \rho > 0$	1%

Exercise 11i

In the following questions, either Table A or Table B may be used.

1. Find r_S and comment on the significance of the result. X and Y have been ranked.

(a) X 1 2 3 4 5 6
Y 1 2 3 4 5 6

(b) X 1 2 3 4 5 6
Y 6 5 4 3 2 1

(c) X 1 2 3 4 5 6
Y 3 5 1 4 6 2

(d) X 1 2 3 4 5 6
Y 2 1 3 5 4 6

(e) X 1 2 3 4 5 6 7 8 9 10
Y 1 2 3 4 5 6 7 8 9 10

(f) X 1 2 3 4 5 6 7 8 9 10
Y 10 9 8 7 6 5 4 3 2 1

(g) X 1 2 3 4 5 6 7 8 9 10
Y 3 1 6 4 5 8 10 9 2 7

(h) X 1 2 3 4 5 6 7 8 9 10
Y 9 8 10 7 6 5 2 4 3 1

2. Calculate r_S for the following data and comment on the significance of the results.

(a) (20, 13), (47, 29), (50, 33), (33, 20), (57, 32), (44, 23), (38, 25), (25, 19).

(b) (4.8, 81), (6.2, 79), (8.4, 86), (4.1, 63), (7.5, 90), (5.1, 87).

KENDALL'S COEFFICIENT OF RANK CORRELATION r_k

To calculate Kendall's coefficient of rank correlation the data is ranked and then written so that one set is in rank order, for example

Rank (x)	1	2	3	4	5
Rank (y)	3	1	4	2	5

We investigate the agreement between the rankings by calculating a 'score', S in the following way. We denote the y values by y_1, y_2, y_3, y_4, y_5.

Rank (x)	1	2	3	4	5
Rank (y)	3 y_1	1 y_2	4 y_3	2 y_4	5 y_5

Start at y_1 and work *from left to right.*

Add 1 to the 'score' S for each y-value *greater* than y_1.

Subtract 1 from the 'score' S for each y-value *less* than y_1.

Then repeat the process, starting in turn from y_2, y_3 and y_4 and always working from left to right.

2 values greater than y_1 ↘ 2 values less than y_1 ↙

Score to the right of $y_1 = +2 \quad -2 = 0$

3 values greater than y_2 ↘ 0 values less than y_2 ↙

Score to the right of $y_2 = +3 \quad -0 = +3$

1 value greater than y_3 ↘ 1 value less than y_3 ↙

Score to the right of $y_3 = +1 \quad -1 = 0$

1 value greater than y_4 ↘ 0 values less than y_4 ↙

Score to the right of $y_4 = +1 \quad +1 = 0$

Therefore the total 'score' S

$= 0+3+0+1 = 4.$

Now, Kendall's coefficient of rank correlation is defined as

$$r_k = \frac{S}{\frac{1}{2}n(n-1)}$$

So, when $n = 5$, $S = 4$

$$r_k = \frac{4}{\frac{1}{2}(5)(4)} = 0.4$$

Now, when there is perfect agreement, Kendall's coefficient r_k should take the value $+1$. We check that this is the case. Consider

	A	B	C	D	E	F	G	H	I
Rank (x)	1	2	3	4	5	6	7	8	9
Rank (y)	1	2	3	4	5	6	7	8	9
	y_1	y_2	y_3	y_4	y_5	y_6	y_7	y_8	y_9

$$r_k = \frac{S}{\frac{1}{2}n(n-1)}$$

$$= \frac{36}{\frac{1}{2}(9)(8)}$$

$$= 1 \qquad \text{as expected.}$$

y_i	Score to the right of y_i
y_1	+8
y_2	+7
y_3	+6
y_4	+5
y_5	+4
y_6	+3
y_7	+2
y_8	+1
	Maximum score = 36

NOTE: In general, when there are n pairs of rankings, the maximum score S is given by

$$S = 1+2+3+\ldots+(n-1) = \tfrac{1}{2}n(n-1)$$

(sum of an A.P.)

Example 11.16 Nine applicants are interviewed for a teaching post by the headteacher and the head of department. Each ranked the applicants in order of merit as follows:

Applicant	A	B	C	D	E	F	G	H	J
Headteacher	2	1	5	4	3	7	8	6	9
Head of Department	3	1	2	4	5	9	8	7	6

Investigate the extent of the agreement between the rankings of the two interviewers.

Solution 11.16 We will consider Kendall's rank correlation coefficient, and must first put one set in rank order and allocate letters y_1 to y_9 to the other set:

Headteacher (x)	1	2	3	4	5	6	7	8	9
Head of Department (y)	1	3	5	4	2	7	9	8	6
	y_1	y_2	y_3	y_4	y_5	y_6	y_7	y_8	y_9

Calculating the total 'score' S,

$$\text{Now} \qquad r_k = \frac{S}{\frac{1}{2}n(n-1)} = \frac{20}{\frac{1}{2}(9)(8)} = 0.556$$

y_i	Score to the right of y_i
y_1	$+8 = +8$
y_2	$+6-1 = +5$
y_3	$+4-2 = +2$
y_4	$+4-1 = +3$
y_5	$+4 = +4$
y_6	$+2-1 = +1$
y_7	$-2 = -2$
y_8	$-1 = -1$
	Total score $S = 20$

Therefore there is some agreement between the rankings of the two interviewers. We now investigate the level of significance of this value, and will complete the solution after the next section, in Solution 11.17.

SIGNIFICANCE OF KENDALL'S RANK CORRELATION COEFFICIENT

In order to test the significance of the calculated value of r_k, it is necessary to calculate the probability of obtaining a given value of S. We look at the distribution of S in the same way that we looked at the distribution of Σd^2 on p. 597.

For $n = 4$, all the possible ranks are listed below, showing the value of S in each case.

	1 2 3 4	1 2 3 4	1 2 3 4	1 2 3 4	1 2 3 4	1 2 3 4
	1 2 3 4	1 2 4 3	1 3 4 2	1 3 2 4	1 4 3 2	1 4 2 3
S	6	4	2	4	0	2
	$(r_k = 1)$				$(r_k = 0)$	

	1 2 3 4	1 2 3 4	1 2 3 4	1 2 3 4	1 2 3 4	1 2 3 4
	2 1 3 4	2 1 4 3	2 3 1 4	2 3 4 1	2 4 1 3	2 4 3 1
S	4	2	2	0	0	−2
				$(r_k = 0)$	$(r_k = 0)$	

	1 2 3 4	1 2 3 4	1 2 3 4	1 2 3 4	1 2 3 4	1 2 3 4
	3 1 2 4	3 1 4 2	3 2 4 1	3 2 1 4	3 4 1 2	3 4 2 1
S	2	0	−2	0	−2	−4
		$(r_k = 0)$		$(r_k = 0)$		

	1 2 3 4	1 2 3 4	1 2 3 4	1 2 3 4	1 2 3 4	1 2 3 4
	4 1 2 3	4 1 3 2	4 2 3 1	4 2 1 3	4 3 1 2	4 3 2 1
S	0	−2	−4	−2	−4	−6
						$(r_k = -1)$

Arranging S in a frequency distribution:

S	−6	−4	−2	0	2	4	6
Frequency	1	3	5	6	5	3	1

The distribution is symmetrical about $S = 0$.

Note the following results:

Using $$r_k = \frac{S}{\frac{1}{2}n(n-1)} \quad \text{with } n = 4$$

$S = 6$	$S = 0$	$S = -6$
$r_k = 1$	$r_k = 0$	$r_k = -1$
Perfect positive correlation — the rankings agree exactly	No correlation	Perfect negative correlation — one rank is the exact reverse of the other

The frequency distribution of S

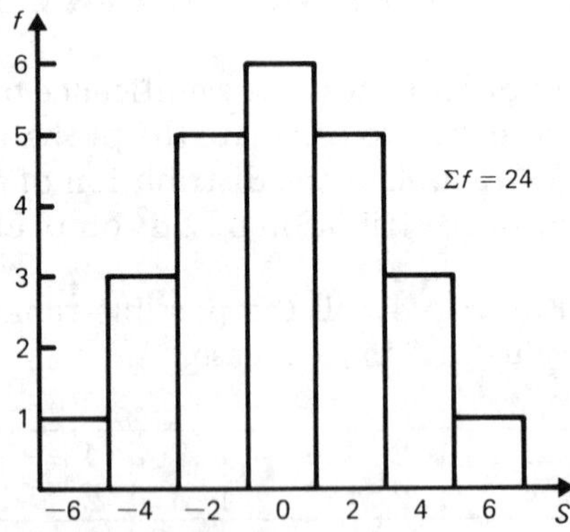

We can now use this diagram to find the probabilities associated with various values of S.

(a)

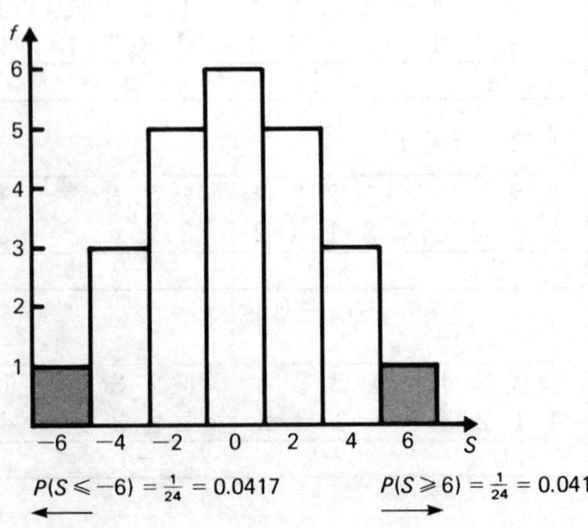

(b)

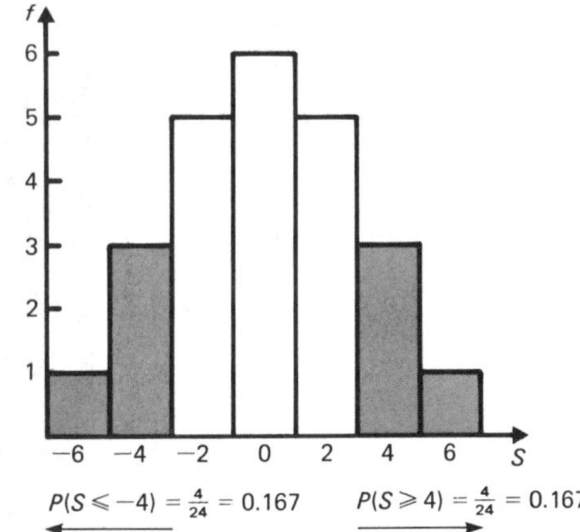
f
6
5
4
3
2
1
−6 −4 −2 0 2 4 6
S
P(S ⩽ −4) = 4/24 = 0.167
P(S ⩾ 4) = 4/24 = 0.167

(c)

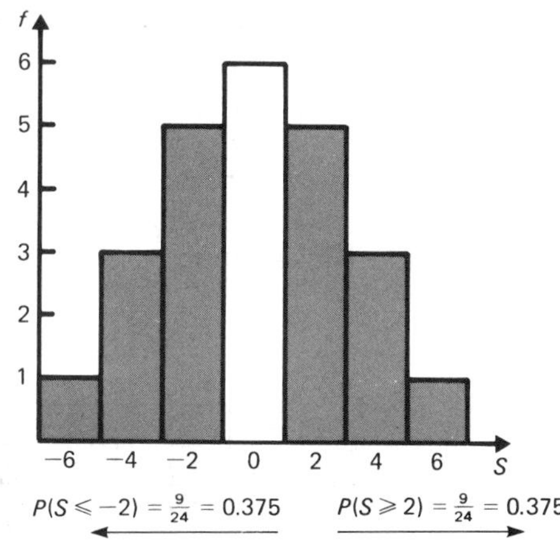
f
6
5
4
3
2
1
−6 −4 −2 0 2 4 6
S
P(S ⩽ −2) = 9/24 = 0.375
P(S ⩾ 2) = 9/24 = 0.375

(d)

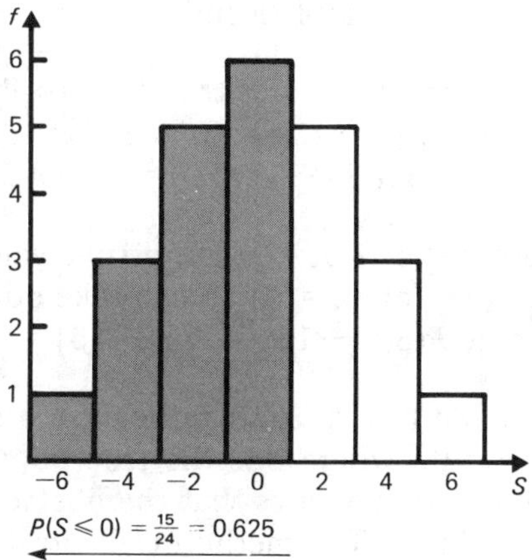
f
6
5
4
3
2
1
−6 −4 −2 0 2 4 6
S
P(S ⩽ 0) = 15/24 = 0.625

(e)

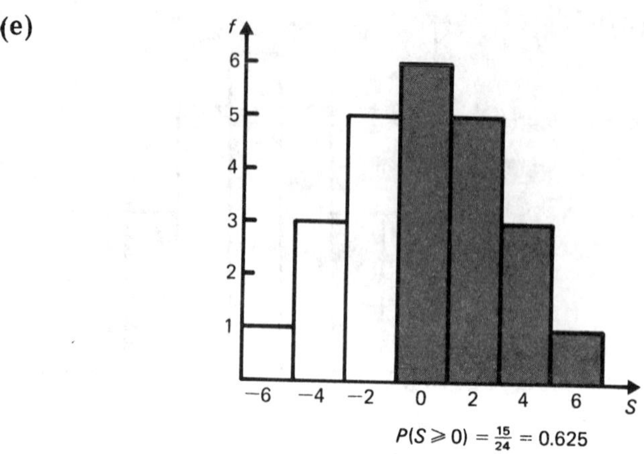

Putting together the results from all the diagrams:

$n = 4$	Probability	
$S \leqslant -6$	0.0417	$S \geqslant 6$
$S \leqslant -4$	0.167	$S \geqslant 4$
$S \leqslant -2$	0.375	$S \geqslant 2$
$S \leqslant 0$	0.625	$S \geqslant 0$

The results are also summarised thus:

S	$n = 4$ Max $S = 6$
0	.625
2	.375
4	.167
6	.0417

This is an extract from a larger table, given in the Appendix on p. 639 which gives probabilities for S for n in the range $4 \leqslant n \leqslant 10$. We will refer to this as Table C.

Note that the table is in two parts:

When $n = 4, 5, 8$ or 9 then the possible values of S are *even* numbers. For example, for $n = 8$, S can take even values from -28 to 28 inclusive and $P(S \leqslant -20) = P(S \geqslant 20) = 0.0071$.

When $n = 6, 7$ or 10 the possible values of S are *odd* numbers. For example, for $n = 10$, S can take odd values from -45 to 45 inclusive and $P(S \leqslant -15) = P(S \geqslant 15) = 0.108$.

When we test r_k for significance, a suitable null hypothesis is $H_0: \rho = 0$, where ρ is the true population correlation coefficient and $\rho = 0$ indicates that there is no predictable correlation between the rankings. The method is illustrated as follows:

Example 11.17 Considering the data given in 11.16, perform a significance test to determine the extent of agreement between the headteacher and head of department when ranking nine applicants for a teaching post.

Solution 11.17 We found that $S = 20$, $n = 9$ and $r_k = 0.556$, and we want to test r_k for significance.

So $H_0: \rho = 0$ (no correlation)

$H_1: \rho \neq 0$ (there is some positive correlation)

Use a 1-tailed test at the 5% level and reject H_0 if $P(S \geqslant 20) < 0.05$. Now, from Table C,

S	$n = 9$ Max $S = 36$
20	0.0223

indicates that $P(S \geqslant 20) = 0.0223 < 0.05$, so we reject H_0 and conclude that there is evidence, at the 5% level, of a positive correlation between the rankings of the headteacher and head of department.

Example 11.18 Calculate r_k for the following data and comment on the result.

(6.9, 89), (5.8, 73), (4.8, 81), (6.2, 79), (8.4, 86),
(4.1, 63), (7.5, 90), (5.1, 87), (9.9, 96), (4.3, 72).

Solution 11.18

x	6.9	5.8	4.8	6.2	8.4	4.1	7.5	5.1	9.9	4.3
y	89	73	81	79	86	63	90	87	96	72
Rank (x)	7	5	3	6	9	1	8	4	10	2
Rank (y)	8	3	5	4	6	1	9	7	10	2

Now re-arrange the pairs so that the x-values are in rank order.

Rank (x)	1	2	3	4	5	6	7	8	9	10
Rank (y)	1 y_1	2 y_2	5 y_3	7 y_4	3 y_5	4 y_6	8 y_7	9 y_8	6 y_9	10 y_{10}

y_i	Score to the right of y_i
y_1	$9 = 9$
y_2	$8 = 8$
y_3	$5 - 2 = 3$
y_4	$3 - 3 = 0$
y_5	$5 = 5$
y_6	$4 = 4$
y_7	$2 - 1 = 1$
y_8	$1 - 1 = 0$
y_9	$1 = 1$
	$S = 31$

Now $r_k = \dfrac{S}{\frac{1}{2}n(n-1)}$

$= \dfrac{31}{\frac{1}{2}(10)(9)}$

$= 0.689$ (3 d.p.)

$H_0: \rho = 0$ (no correlation)

$H_1: \rho > 0$ (evidence of positive correlation)

Use a 1-tailed test, at the 1% level, and reject H_0 if $P(S \geqslant 31) < 0.01$.

From Table C, $n = 10$, $P(S \geqslant 31) = 0.0023 < 0.01$, so reject H_0 and conclude that there is evidence, at the 1% level, of positive correlation.

Example 11.19 When calculating r_k, with 9 pairs of data, it is found that $S = -24$. Test, at the 1% level, the hypotheses: $H_0: \rho = 0$, $H_1: \rho \neq 0$.

Solution 11.19 Use a 2-tailed test, at the 1% level, and reject H_0 if $P(S \leqslant -24) < 0.005$.

Now, from Table C,
$n = 9$, $P(S \leqslant -24) = P(S \geqslant 24) = 0.0063 > 0.005$. Therefore we do not reject H_0 and conclude that, at the 1% level, there is no evidence of correlation.

Exercise 11j

In each of questions 1 to 7, use Table C to test the hypotheses, at the level of significance indicated.

	n	S	Hypotheses	Level of significance
1.	7	11	$H_0: \rho = 0$, $H_1: \rho > 0$	5%
2.	5	6	$H_0: \rho = 0$, $H_1: \rho > 0$	5%
3.	9	−22	$H_0: \rho = 0$, $H_1: \rho < 0$	1%
4.	10	23	$H_0: \rho = 0$, $H_1: \rho \neq 0$	5%
5.	10	−23	$H_0: \rho = 0$, $H_1: \rho < 0$	1%
6.	4	4	$H_0: \rho = 0$, $H_1: \rho > 0$	5%
7.	8	16	$H_0: \rho = 0$, $H_1: \rho \neq 0$	5%

8. Calculate r_K for number 1, Exercise 11i and comment on the significance of the result.

9. Calculate r_K for number 2 of Exercise 11i and comment on the significance of the result.

10. Three judges in a bouncing baby competition rank the babies as shown.

Baby	*A*	*B*	*C*	*D*	*E*	*F*	*G*	*H*
Judge 1	6	5	4	7	1	2	3	8
Judge 2	4	3	1	6	7	8	2	5
Judge 3	4	5	6	7	1	3	2	8

Calculate r_K for each pair of judges and comment on the significance of your results.

11. These were the marks obtained by 8 pupils in Mathematics and Physics.

Mathematics	67	42	85	51	39	97	81	70
Physics	70	59	71	38	55	62	80	76

Calculate r_K and comment on the significance of the result.

MISCELLANEOUS WORKED EXAMPLES

Example 11.20 (a) The marks of eight candidates in English and Mathematics are:

Candidate	1	2	3	4	5	6	7	8
English (x)	50	58	35	86	76	43	40	60
Mathematics (y)	65	72	54	82	32	74	40	53

Rank the results and hence find a rank correlation coefficient between the two sets of marks.

(b) Using the data in part (a), obtain the product-moment correlation coefficient. To assist in the lengthy calculation, you may use the information $s_x = 16.67$. (SUJB)

Solution 11.20 (a)

Candidate	English, x	Maths, y	Rank(x)	Rank(y)	$\lvert d \rvert$	d^2
1	50	65	4	5	1	1
2	58	72	5	6	1	1
3	35	54	1	4	3	9
4	86	82	8	8	0	0
5	76	32	7	1	6	36
6	43	74	3	7	4	16
7	40	40	2	2	0	0
8	60	53	6	3	3	9
						$\Sigma d^2 = 72$

$$r_S = 1 - \frac{6\Sigma d^2}{n(n^2-1)}$$

$$= 1 - \frac{6(72)}{8(64-1)}$$

$$= 0.17 \quad (2\text{ d.p.})$$

Spearman's coefficient of rank correlation is 0.17 (2 d.p.).

(b)

x	y	y^2	xy
50	65	4225	3250
58	72	5184	4176
35	54	2916	1890
86	82	6724	7052
76	32	1024	2432
43	74	5476	3182
40	40	1600	1600
60	53	2809	3180
$\Sigma x = 448$	$\Sigma y = 472$	$\Sigma y^2 = 29\,958$	$\Sigma xy = 26\,762$

$$\bar{x} = \frac{\Sigma x}{n} = \frac{448}{8} = 56; \quad \bar{y} = \frac{\Sigma y}{n} = \frac{472}{8} = 59$$

$$\text{Covariance } s_{xy} = \frac{\Sigma xy}{n} - \bar{x}\bar{y} = \frac{26\,762}{8} - (56)(59) = 41.25$$

$$s_y{}^2 = \frac{\Sigma y^2}{n} - \bar{y}^2 = \frac{29\,958}{8} - (59)^2 = 263.75; \quad s_y = 16.24$$

$$s_x = 16.67 \quad \text{(given)}$$

Therefore

$$r = \frac{s_{xy}}{s_x s_y} = \frac{41.25}{(16.24)(16.67)} = 0.15 \quad (2\,\text{d.p.})$$

The product-moment correlation coefficient is 0.15 (2 d.p.).

Example 11.21 It is suspected that two quantities Q and W are related according to the formula $Q = aW^b$, where a and b are constants. Observations on Q and W were made and the results were as follows:

W	13	16	20	25	32	40	50	60
Q	71	40	50	32	24	31	25	16

Plot a scatter diagram of $\log_{10}Q$ against $\log_{10}W$ and estimate the equation of the regression line of $\log Q$ on $\log W$, using the means of certain points, or otherwise. Use your results to estimate values for a and b.

State how you would obtain the product moment correlation coefficient between $\log Q$ and $\log W$. (SUJB)

Solution 11.21

$x = \log_{10}W$	$y = \log_{10}Q$
1.114	1.851
1.204	1.602
1.301	1.699
1.398	1.505
1.505	1.380
1.602	1.491
1.699	1.398
1.778	1.204
$\Sigma x = 11.601$	$\Sigma y = 12.13$

$$\bar{x} = \frac{\Sigma x}{n} = \frac{11.601}{8} = 1.45$$

$$\bar{y} = \frac{\Sigma y}{n} = \frac{12.13}{8} = 1.52$$

We plot $M(\bar{x}, \bar{y})$ and draw a line parallel to the y axis.

For the points on the left

x	y
1.114	1.851
1.204	1.602
1.301	1.699
1.398	1.505
$\Sigma x = 5.017$	$\Sigma y = 6.657$

$$\bar{x}_L = \frac{5.017}{4} = 1.25$$

$$\bar{y}_L = \frac{6.657}{4} = 1.66$$

We plot M_L (1.25, 1.66)

For the points on the right

x	y
1.505	1.380
1.602	1.491
1.699	1.398
1.778	1.204
$\Sigma x = 6.584$	$\Sigma y = 5.473$

$$\bar{x}_R = \frac{6.584}{4} = 1.65$$

$$\bar{y}_R = \frac{5.473}{4} = 1.37$$

We plot M_R (1.65, 1.37)

The line of best fit is drawn through M, M_L and M_R, ensuring that the line passes through M. This is a regression line y on x.

Using these points, we estimate the gradient to be

$$\frac{\bar{y}_L - \bar{y}_R}{\bar{x}_L - \bar{x}_R} = \frac{1.66 - 1.37}{1.25 - 1.65} = -0.73 \quad (2\,\text{d.p.})$$

Now $Q = aW^b$ so

$$\log_{10}Q = \log_{10}a + b\log_{10}W$$

i.e.

$$y = \log_{10}a + bx$$

The gradient, $b = -0.73$.

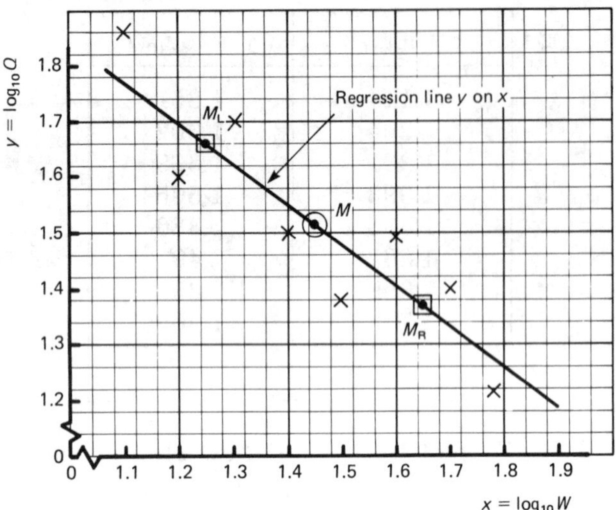

Scatter diagram of $\log_{10}Q$ against $\log_{10}W$

To find a, use the fact that $(\bar{x}, \bar{y})$ lies on the line.

Then $\quad \bar{y} = \log_{10}a + b\bar{x}$, so $\quad 1.52 = \log_{10}a + (-0.73)(1.45)$

$\log_{10}a = 2.5785$ and $a = 10^{2.5785} = 380$ (2 S.F.)

We estimate the values of a and b to be 380 and -0.73 respectively.

The product-moment correlation coefficient would be found by calculating

$$r = \frac{s_{xy}}{s_x s_y}$$

Example 11.22 The body and heart masses of fourteen 10-month-old male mice are tabulated below:

Body mass (x) (grams)	27	30	37	38	32	36	32	32	38	42	36	44	33	38
Heart mass (y) (milligrams)	118	136	156	150	140	155	157	114	144	159	149	170	131	160

(a) Draw a scatter diagram of these data.

(b) Calculate the equation of the regression line of y on x and draw this line on the scatter diagram.

(c) Calculate the product-moment coefficient of correlation.

(AEB)

Solution 11.22 (a)

x	y	xy	x^2	y^2
27	118	3186	729	13 924
30	136	4080	900	18 496
37	156	5772	1369	24 336
38	150	5700	1444	22 500
32	140	4480	1024	19 600
36	155	5580	1296	24 025
32	157	5024	1024	24 649
32	114	3648	1024	12 996
38	144	5472	1444	20 736
42	159	6678	1764	25 281
36	149	5364	1296	22 201
44	170	7480	1936	28 900
33	131	4323	1089	17 161
38	160	6080	1444	25 600
$\Sigma x = 495$	$\Sigma y = 2039$	$\Sigma xy = 72\,867$	$\Sigma x^2 = 17\,783$	$\Sigma y^2 = 300\,405$

$$\bar{x} = \frac{\Sigma x}{n} = \frac{495}{14} = 35.36 \quad \text{(2 d.p.)}$$

$$\bar{y} = \frac{\Sigma y}{n} = \frac{2039}{14} = 145.64 \quad \text{(2 d.p.)}$$

(b) The least squares regression line of y on x is given by

$$y - \bar{y} = \frac{s_{xy}}{s_x^{\,2}}(x - \bar{x})$$

where

$$s_{xy} = \frac{\Sigma xy}{n} - \bar{x}\bar{y} = \frac{72\,867}{14} - \left(\frac{495}{14}\right)\left(\frac{2039}{14}\right) = 55.27 \quad \text{(2 d.p.)}$$

and

$$s_x^{\,2} = \frac{\Sigma x^2}{n} - \bar{x}^2 = \frac{17\,783}{14} - \left(\frac{495}{14}\right)^2 = 20.09 \quad \text{(2 d.p.)}$$

Therefore the coefficient of regression $= \dfrac{s_{xy}}{s_x^{\,2}} = \dfrac{55.27}{20.09} = 2.75$

So the equation of the least squares regression line y on x is

$$y - 145.64 = 2.75(x - 35.36)$$

To draw this on the scatter diagram, first plot $(\bar{x}, \bar{y})$. Then find two further points, e.g.

$$\text{when} \quad x = 40, \quad y = 145.64 + 2.75(40 - 35.36) = 158.4$$

$$\text{when} \quad x = 30, \quad y = 145.64 + 2.75(30 - 35.36) = 130.9$$

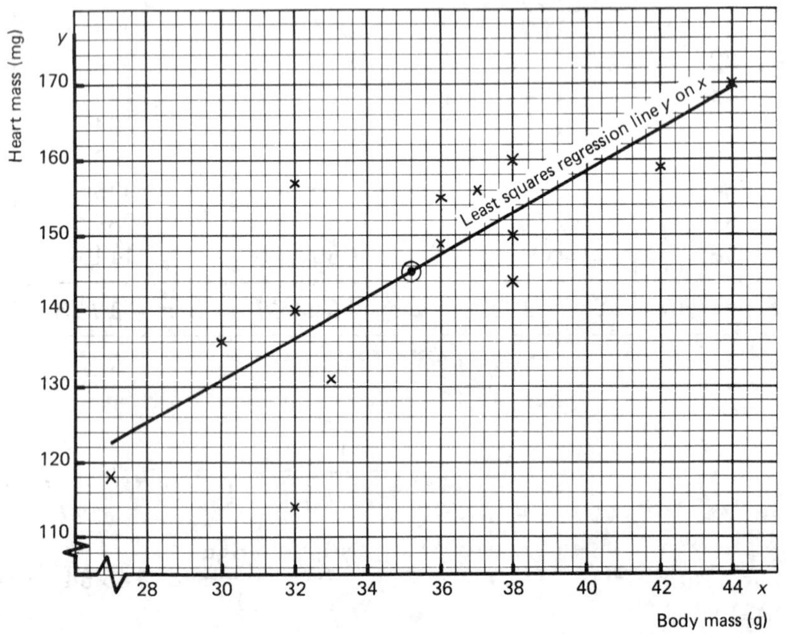

Scatter diagram to show body and heart masses of 14 mice

(c) The product-moment correlation coefficient r is given by

$$r = \frac{s_{xy}}{s_x s_y}$$

Now $\quad s_y{}^2 = \dfrac{\Sigma y^2}{n} - \bar{y}^2 = \dfrac{300\,405}{14} - \left(\dfrac{2039}{14}\right)^2 = 245.66$

Therefore

$$r = \frac{s_{xy}}{s_x s_y} = \frac{55.27}{\sqrt{(20.09)(245.66)}} = 0.79 \quad (2\text{ d.p.})$$

The product-moment correlation coefficient is 0.79 (2 d.p.).

Example 11.23 The positions in a league of 8 hockey clubs at the end of a season are shown in the table.

Shown also are the average attendances (in hundreds) at home matches during that season.

Club	A	B	C	D	E	F	G	H
Position	1	2	3	4	5	6	7	8
Average attendance	30	32	12	19	27	18	15	25

Calculate a coefficient of rank correlation between position in the league and average home attendance.

Refer to the appropriate table of critical values provided in the formulae booklet to comment on the significance of your result, stating clearly the null hypothesis being tested. (L)P

Solution 11.23 Either r_S or r_k could be calculated. We show the working for both.

Spearman

Club (x)	A	B	C	D	E	F	G	H
Position (x)	1	2	3	4	5	6	7	8
Attendance rank (y)	2	1	8	5	3	6	7	4
$\|d\|$	1	1	5	1	2	0	0	4
d^2	1	1	25	1	4	0	0	16

We have $\Sigma d^2 = 48$.

Now
$$r_S = 1 - \frac{6\Sigma d^2}{n(n^2-1)}$$
$$= 1 - \frac{6(48)}{8(63)}$$
$$= 0.4286 \quad (4 \text{ d.p.})$$

Significance test:

$H_0 : \rho = 0$ (no correlation between the two ranks)

$H_1 : \rho > 0$ (some positive correlation)

NOTE: We would expect a positive correlation between league position and average attendance.

Use a 1-tailed test, at the 5% level, and reject H_0 if $P(\Sigma d^2 \leqslant 48) < 0.05$.

From Table A, with $n = 8$, $P(\Sigma d^2 \leqslant 48) = 0.15 > 0.05$ so we do not reject H_0 and conclude that there is no evidence, at the 5% level, of positive correlation between the two sets of ranks.

Kendall

Position (x)	1	2	3	4	5	6	7	8
Attendance rank (y)	2 y_1	1 y_2	8 y_3	5 y_4	3 y_5	6 y_6	7 y_7	4 y_8

y_i	Score to the right of y_i
y_1	$6-1=5$
y_2	$6=6$
y_3	$-5=-5$
y_4	$2-2=0$
y_5	$3=3$
y_6	$1-1=0$
y_7	$-1=-1$
	$S=8$

$$r_k = \frac{S}{\frac{1}{2}n(n-1)} = \frac{8}{\frac{1}{2}(8)(7)} = 0.286 \quad \text{(3 d.p.)}$$

Significance test:

$$H_0: \rho = 0 \qquad \text{(no correlation)}$$

$$H_1: \rho > 0 \qquad \text{(some positive correlation)}$$

Use a 1-tailed test, at the 5% level, and reject H_0 if $P(S \geqslant 8) < 0.05$.

From Table C, with $n = 8$, $P(S \geqslant 8) = 0.199 > 0.05$, so we do not reject H_0 and conclude that there is no evidence, at the 5% level, of positive correlation between the two sets of ranks.

NOTE: The significance tests gave the same conclusion.

SUMMARY — REGRESSION AND CORRELATION

Least squares regression lines	
y on x If equation of line is $y = ax + b$ $\Sigma y = a\Sigma x + nb$ $\Sigma xy = a\Sigma x^2 + b\Sigma x$	*x on y* If equation of line is $x = cy + d$ $\Sigma x = c\Sigma y + nd$ $\Sigma xy = c\Sigma y^2 + d\Sigma y$

$$s_{xy} = \frac{1}{n}\Sigma(x-\bar{x})(y-\bar{y}) = \frac{\Sigma xy}{n} - \overline{xy} \quad \text{(covariance)}$$

$$s_x^2 = \frac{1}{n}\Sigma(x-\bar{x})^2 = \frac{\Sigma x^2}{n} - \bar{x}^2;$$

$$s_y^2 = \frac{1}{n}\Sigma(y-\bar{y})^2 = \frac{\Sigma y^2}{n} - \bar{y}^2$$

<table>
<tr><td colspan="2">Equation of least squares regression line can be written:</td></tr>
<tr><td>y on x
$y-\bar{y} = \dfrac{s_{xy}}{s_x^2}(x-\bar{x})$</td><td>x on y
$x-\bar{x} = \dfrac{s_{xy}}{s_y^2}(y-\bar{y})$</td></tr>
<tr><td colspan="2">Minimum sum of squares of residuals</td></tr>
<tr><td>y on x
$n(1-r^2)s_y^2$</td><td>x on y
$n(1-r^2)s_x^2$</td></tr>
<tr><td colspan="2">Product-moment correlation coefficient, r
$r = \dfrac{s_{xy}}{s_x s_y}$</td></tr>
<tr><td colspan="2">In terms of the regression coefficients: $r^2 = ac$
where $a = \dfrac{s_{xy}}{s_x^2}$ (regression coefficient of y on x)
$c = \dfrac{s_{xy}}{s_y^2}$ (regression coefficient of x on y)</td></tr>
<tr><td colspan="2">Spearman's coefficient of rank correlation, r_S
$r_S = 1-\dfrac{6\Sigma d^2}{n(n^2-1)}$</td></tr>
<tr><td colspan="2">Kendall's coefficient of rank correlation, r_k
$r_k = \dfrac{S}{\frac{1}{2}n(n-1)}$</td></tr>
</table>

Miscellaneous Exercise 11k

In questions involving regression lines assume, unless stated otherwise, that the least squares regression lines are required.

1. 12 students were given a prognostic test at the beginning of a course and their scores X_i in the test were compared with their scores Y_i obtained in an examination at the end of the course ($i = 1, 2, \ldots, 12$). The results were as follows:

X_i	1	2	2	4	5	5	6	7	8	8	9	9
Y_i	3	4	5	5	4	8	6	6	6	7	8	10

Find the equation of the regression line of Y on X and determine the correlation coefficient between X and Y. (SUJB)

2. The heights h, in cm, and weights W, in kg, of 10 people are measured. It is found that $\Sigma h = 1710$, $\Sigma W = 760$, $\Sigma h^2 = 293\,162$, $\Sigma hW = 130\,628$ and $\Sigma W^2 = 59\,390$.

   Calculate the correlation coefficient between the values of h and W.

   What is the equation of the regression line of W on h? (O & C)

3. Ten boys compete in throwing a cricket ball, and the following table shows the height of each boy (x cm) to the nearest cm and the distance (y m) to which he can throw the ball.

Boy	A	B	C	D	E	F	G	H	I	J
x	122	124	133	138	144	156	158	161	164	168
y	41	38	52	56	29	54	59	61	63	67

Find the equations of the regression lines of y on x, and of x on y. No diagram is needed. Calculate also the coefficient of correlation.

Estimate the distance to which a cricket ball can be thrown by a boy 150 cm in height. (AEB)

4. Sketch scatter diagrams for which
(*a*) the product moment correlation coefficient is -1,
(*b*) Spearman's correlation coefficient is $+1$, but the product moment correlation coefficient is *less than* 1.

Five independent observations of the random variables X and Y were:

X	0	1	4	3	2
Y	11	8	5	4	7

Find
(*c*) the sample product moment correlation coefficient,
(*d*) Spearman's correlation coefficient. (O & C)

5. The state of Tempora demands that every household in the country shall have a reliable clock; inspectors are being introduced throughout the country to implement the policy. The Chief Inspector has the following data on the population size of towns, where Inspection Units have been set up, and the number of man-hours spent on inspection.

Population (thousands)	3	4	5	9	13	15	18	20	21	22
Man-hours (thousands)	8	11	13	18	24	26	31	32	34	33

(*a*) Calculate the regression line for predicting the number of man-hours from the population size (note that the mean value of each variate is a whole number).
(*b*) Predict the manpower required (in man-hours) for a new Inspection Unit to be installed in a town with a population of 17 000. (O)

6. In a certain heathland region there is a large number of alder trees where the ground is marshy but very few where the ground is dry. The number x of alder trees and the ground moisture content y are found in each of 10 equal areas (which have been chosen to cover the range of x in all such areas). The following is a summary of the results of the survey:

$$\Sigma x = 500, \quad \Sigma y = 300,$$
$$\Sigma x^2 = 27\,818, \quad \Sigma xy = 16\,837,$$
$$\Sigma y^2 = 10\,462$$

Find the equation of the regression line of y on x.

Estimate the ground moisture content in an area equal to one of the chosen areas which contains 60 alder trees. (O & C)

7. (*a*) The following marks were awarded by 2 judges at a music competition:

	Judge 1	Judge 2
Child 1	10	9
Child 2	5	6
Child 3	8	10
Child 4	7	5
Child 5	9	8

Calculate a coefficient of rank correlation.

(*b*) Determine, by calculation, the equation of the regression line of x on y based on the following information about 8 children:

Child	1	2	3	4	5	6	7	8
Arithmetic mark (x)	45	33	27	23	18	14	8	0
English mark (y)	31	33	18	20	19	9	13	1

(SUJB)

8. The following data (Table A) represent the lengths (x) and breadths (y) of 12 cuckoos' eggs measured in millimetres.

Draw a scatter diagram for the data.

Obtain the least squares regression lines of y on x and plot this on the scatter diagram. (JMB)

9. $(X_i, Y_i), i = 1, 2, \ldots, n$ is a sample from a bivariate population. The least-square regression lines of Y on X and X on Y are calculated. Why would you not expect the two lines to coincide? Under what circumstances would they coincide?

In the table, Y_i is the mass (in grammes) of potassium bromide which will dissolve in 100 grammes of water at a temperature of X_i °C.

Table A

x	22.3	23.6	24.2	22.6	22.3	22.3	22.1	23.3	22.2	22.2	21.8	23.2
y	16.5	17.1	17.3	17.0	16.8	16.4	17.2	16.8	16.7	16.2	16.6	16.4

X	10	20	30	40	50
Y	61	64	70	73	78

Find the equation of the regression line of Y on X.

Find, also, the product-moment correlation coefficient between X and Y. (SUJB)

10. (*a*) The 1973 and 1980 catalogue prices (in pence) of five British postage stamps are as follows:

1973 price	50	45	65	25	15
1980 price	500	350	600	500	120

(i) Plot these results on a scatter diagram.
(ii) Write down the coordinates of one point through which the regression line of the 1980 price on the 1973 price must pass.
(iii) Fit, by eye, the regression line of the 1980 price on the 1973 price.
(iv) Denoting the 1980 price by y and the 1973 price by x, write down the equation of your fitted regression line in the form $y = ax + b$, giving the constants a and b to one decimal place. Use this equation to determine the value of y when x has the value 20.

(*b*) (i) On a certain island there are large numbers of each of two clans, the Fatties and the Thinnies. Two random samples of 50 adult males are taken, one sample from each clan. Each of the clansmen is weighed and measured. For each clan, the value of the correlation coefficient between the heights and weights of the clansmen is found to be near $+1$. However, for the combined sample of 100 adult males, the value of the correlation coefficient is found to be near -1. Show, by a sketch of the scatter diagram for the combined sample, how this could arise.

(ii) A large sample survey of three-person families is conducted. The value of X, the greatest amount earned by any one member of the family, and the value of Y, the total amount earned by the entire family, are both recorded. Would you expect the value of the correlation coefficient between X and Y to be near $+1$, near -1 or near 0? Justify your answer. (C)

11. Explain clearly what is meant by the statistical term 'correlation'.

Vegboost Industries, a small chemical firm specializing in garden fertilizers set up an experiment to study the relationship between a new fertilizer compound and the yield from tomato plants. Eight similar plants were selected and treated regularly throughout their life with x grams of fertilizer diluted in a standard volume of water. The yield y, in kilograms, of good tomatoes was measured for each plant. The following table summarizes the results.

Plant	*A*	*B*	*C*	*D*	*E*	*F*	*G*	*H*
Amount of fertilizer x (g)	1.2	1.8	3.1	4.9	5.7	7.1	8.6	9.8
Yield y (kg)	4.5	5.9	7.0	7.8	7.2	6.8	4.5	2.7

(*a*) Calculate the product-moment correlation coefficient for these data.
(*b*) Calculate Spearman's rank correlation coefficient for these data.
(*c*) Is there any evidence of a relationship between these variables? Justify your answer. (No formal test is required.) (AEB 1978)

12. State the effect on the product-moment correlation coefficient between two variables x and y of (*a*) changing the origin for x and (*b*) changing the units of x.

Table B below gives the daily output of the substance creatinine from the body of each of ten nutrition students together with the student's body mass.

Draw a scatter diagram for the data.

Calculate, correct to two decimal places, the product-moment correlation coefficient.

Comment on any relationship which is indicated by the scatter diagram and the correlation coefficient. (JMB)

13. In a regression calculation for five pairs of observations one pair of values was lost when the data were filed. For the regression of y on x the equation was calculated as

$$y = 2x - 0.1$$

Table B

Output of creatinine (grammes)	1.32	1.54	1.45	1.06	2.13	1.00	0.90	2.00	2.70	0.75
Body mass (kilogrammes)	55	48	55	53	74	44	49	68	78	51

The four recorded pairs of values are

x	0.1	0.2	0.4	0.3
y	0.1	0.3	0.7	0.4

Find the missing pair of values, using the following data for the four pairs above: $\Sigma x = 1$, $\Sigma x^2 = 0.3$, $\Sigma xy = 0.47$, $\Sigma y = 1.5$. (MEI)

14. (*a*) On two separate occasions, ranks 1, 2, 3 are assigned at random to three objects A, B, C. Obtain the probability distribution of a coefficient of rank correlation between the pair of rankings.
(*b*) Five sacks of coal, A, B, C, D and E have different weights, with A being heavier than B, B being heavier than C, and so on. A weight lifter ranks the sacks (heaviest first) in the order A, D, B, E, C. Calculate a coefficient of rank correlation between the weight lifter's ranking and the true ranking of the weights of the sacks. (C)

15. In an investigation into prediction using the stars and planets, a celebrated astrologist Horace Cope predicted the ages at which thirteen young people would first marry. The complete data, of predicted and actual ages at first marriage, are now available and are summarised in Table C below.
(*a*) Draw a scatter diagram of these data.
(*b*) Calculate the equation of the regression line of y on x and draw this line on the scatter diagram.
(*c*) Comment upon the results obtained, particularly in view of the data for person G. What further action would you suggest? (AEB 1981)

16. Explain how you used, or could have used, a correlation coefficient to analyse the results of an experiment. State briefly when it is appropriate to use a rank correlation coefficient rather than a product-moment correlation coefficient.
Seven rock samples taken from a particular locality were analysed. The percentages, C and M, of two oxides contained in each sample were recorded. The results are shown in the table.

Sample	C	M
1	0.60	1.06
2	0.42	0.72
3	0.51	0.94
4	0.56	1.04
5	0.31	0.84
6	1.04	1.16
7	0.80	1.24

Given that

$$\Sigma CM = 4.459 \qquad \Sigma C^2 = 2.9278$$
$$\Sigma M^2 = 7.196,$$

find, to 3 decimal places, the product-moment correlation coefficient of the percentages of the two oxides. Calculate also, to 3 decimal places, a rank correlation coefficient.
Using the tables provided state any conclusions which you draw from the value of your rank correlation coefficient. State clearly the null hypothesis being tested. (L)

17. Giving an example from your projects if you wish, describe conditions under which you would use a rank correlation coefficient as a measure of association.
In a ski-jumping contest each competitor made 2 jumps. The orders of merit for the 10 competitors who completed both jumps are shown in the table.

Ski-jumper	A	B	C	D	E	F	G	H	I	J
First jump	2	9	7	4	10	8	6	5	1	3
Second jump	4	10	5	1	8	9	2	7	3	6

(*a*) Calculate, to 2 decimal places, a rank correlation coefficient for the performances of the ski-jumpers in the two jumps.
(*b*) Using a 5% level of significance and quoting from the tables of critical values provided, interpret your result. State clearly your null and alternative hypotheses. (L)P

Table C

Person	A	B	C	D	E	F	G	H	I	J	K	L	M
Predicted age x (years)	24	30	28	36	20	22	31	28	21	29	40	25	27
Actual age y (years)	23	31	28	35	20	25	45	30	22	27	40	27	26

18. In Table D below x is the average weekly household income in £ and y the infant mortality per 1000 live births in 11 regions of the UK in 1985.
It is hypothesised that a high value of x will be associated with a low value of y. Explain why it would not be appropriate to use the product moment correlation coefficient to investigate this. Calculate a rank correlation coefficient and test its significance. The values below give the probabilities of exceeding the given values of r_S and r_k calculated from 10 and 11 pairs of uncorrelated variables.

One-sided test probability		5%	2.5%	1.0%
Two-sided test probability		10%	5.0%	2.0%
$n = 10$	r_S	0.5636	0.6485	0.7455
$n = 11$	r_S	0.5364	0.6182	0.7091
$n = 10$	r_k	0.4667	0.5111	0.6000
$n = 11$	r_k	0.4182	0.4909	0.5636

It appears that region A is exceptional. What would your findings be if this region were omitted from the analysis?
(SUJB)

19. Define a ranking scale and give an example to illustrate your definition. Explain how you would rank values of equal magnitude.

At the end of the academic year students on a particular course are given examinations in Sociology (S). Social Administration (SA) and Quantitative Methods (QM). The final grade awarded to each student is based on the total of the marks scored on the three papers. Table E shows the marks obtained by a sample of ten students who sat the three papers.

The following matrix of Spearman rank correlation coefficients was obtained for this sample of ten students.

	S	SA	QM	Total mark
S	1	0.24	−0.01	0.78
SA		1	x	0.77
QM			1	y
Total mark				1

Find the values of x and y.

It has been decided that in future students should only be required to sit two papers. Use these data to decide which two examinations should be used. Give a reason for your choice. (AEB 1987)

20. (a) Explain briefly, referring to your project work if you wish, the conditions under which you would measure association using a rank correlation coefficient rather than a product moment correlation coefficient.

Table D

Region	A	B	C	D	E	F	G	H	I	J	K
x	170.4	183.2	172.9	187.1	203.2	204.8	208.8	248.0	198.3	187.1	179.1
y	8.4	9.4	10.3	10.5	8.3	9.4	8.5	9.0	9.4	9.8	9.6

Table E

Student	Sociology (S)	Social Administration (SA)	Quantitative Methods (QM)
1	66	48	44
2	50	46	48
3	44	46	47
4	58	72	64
5	64	68	54
6	26	64	55
7	74	65	59
8	67	42	48
9	36	40	56
10	48	55	48

At an agricultural show 10 Shetland sheep were ranked by a qualified judge and by a trainee judge. Their rankings are shown in the table.

Qualified judge	1	2	3	4	5	6	7	8	9	10
Trainee judge	1	2	5	6	7	8	10	4	3	9

Calculate a rank correlation coefficient for these data.

Using one of the tables provided and a 10% significance level, state your conclusions as to whether there is some degree of agreement between the two sets of ranks.

(*b*) The variables H and T are known to be linearly related. Fifty pairs of experimental observations of the two variables gave the following results:

$$\Sigma H = 83.4, \quad \Sigma T = 402.0,$$
$$\Sigma HT = 680.2, \quad \Sigma H^2 = 384.6,$$
$$\Sigma T^2 = 3238.2.$$

Obtain the regression equation from which one can estimate H when T has the value 7.8, and give, to 1 decimal place, the value of this estimate. (L)

21. A company is to replace its fleet of cars. Eight possible models are considered and the transport manager is asked to rank them, from 1 to 8, in order of preference. A saleswoman is asked to use each type of car for a week and grade them according to their suitability for the job (A — very suitable to E — unsuitable). The price is also recorded.

Model	Transport manager's ranking	Saleswoman's grade	Price (£10's)
S	5	B	611
T	1	$B+$	811
U	7	$D-$	591
V	2	C	792
W	8	$B+$	520
X	6	D	573
Y	4	$C+$	683
Z	3	$A-$	716

(*a*) Calculate Spearman's rank correlation coefficient between

(i) price and transport manager's rankings,

(ii) price and saleswoman's grades.

(*b*) Based on the results of (*a*) state, giving a reason, whether it would be necessary to use all three different methods of assessing the cars.

(*c*) A new employee is asked to collect further data and to do some calculations. He produces the following results.

The correlation coefficient between

(i) price and boot capacity is 1.2,

(ii) maximum speed and fuel consumption in miles per gallon is −0.7,

(iii) price and engine capacity is −0.9.

For each of his results say, giving a reason, whether you think it is reasonable.

(*d*) Suggest two sets of circumstances where Spearman's rank correlation coefficient would be preferred to the product moment correlation coefficient as a measure of association. (AEB 1988)

22. A sample of n pairs (x_i, y_i), $i = 1, 2, \ldots, n$, is drawn from a bivariate population (X, Y) and a rank correlation coefficient, r, calculated.

(*a*) What range of values is it possible for r to have?

(*b*) What information about the sample does r indicate?

(*c*) What can be concluded about the sample points when $r = 1$? Can the same be said about the population from which the sample is drawn? Explain your answer.

Table F below gives the average share rate and average mortgage rate calculated on the first day of the months shown for the years 1976 to 1985.

Plot a scatter diagram and comment on its implication for r. Indicate on your diagram which point appears to be an outlier (i.e. one that is far from the trend line).

Calculate a rank correlation coefficient between share rate and mortgage rate.

Under the null hypothesis that the population $r = 0$ against the alternative $r \neq 0$ the following are critical values of Spear-

Table F

Year	1976	1977	1978	1979	1980	1981	1982	1983	1984	1985
Month	Nov	Nov	Dec	Dec	Jan	Nov	Dec	Jul	Dec	Apr
Share %	7.8	6.0	8.0	10.5	10.5	9.8	6.2	7.2	6.8	8.2
Mortgage %	12.2	9.5	11.8	11.8	15.0	15.0	10.0	11.2	11.5	14.0

man's and Kendall's coefficients for a sample size of 10.

Significance level	Spearman's r_S	Kendall's r_K
10%	0.564	0.467
5%	0.648	0.511
2%	0.745	0.600
1%	0.794	0.644

Comment on the significance of your calculated value of r. (SUJB)

23. Explain briefly what is measured by the product moment correlation coefficient.

The manager of a large office supervises 15 clerical assistants, each using a word-processor. Because of the pressure of work, the assistants did not all receive the same amount of training in the use of their word-processors. In order to make an assessment of the need for training the manager monitored their work during a given week, recording the number of pieces of work correctly produced without any errors (x_1), the number produced containing errors (x_2) together with the number of days training received (y). The results are summarised in Table G below.

(*a*) Given that $\Sigma x_1^2 = 11\,513$, $\Sigma y^2 = 728$ and $\Sigma x_1 y = 2676$, show that the product moment correlation coefficient between y and x_1 is 0.491.
(*b*) Without using a comment of the form 'The correlation between x_1 and y is not very strong', suggest how the manager might have attempted to interpret this value as part of the assessment.

The manager then decided to investigate whether there is any association between the number of days training and a perceived measure of accuracy based on the difference between x_1 and x_2. Consequently a new variable $z = x_1 - x_2$ was created.
(*c*) Plot a scatter diagram of y against z.

Explain why the manager should not correlate y and z using the product moment correlation coefficient.
(*d*) Explain why z^2 might be a better variable to correlate with y using the product moment correlation coefficient. Evaluate the correlation coefficient between y and z^2 and explain why the manager might be pleased with the value obtained. Suggest how this new variable would present the manager with a practical problem. (AEB 1988)

24. The experimental data below were obtained by measuring the horizontal distance y cm, rolled by an object released from the point P on a plane inclined at $\theta°$ to the horizontal, as shown in the diagram.

P
y cm
θ°

Distance y	Angle θ
44	8.0
132	25.0
152	31.5
87	17.5
104	20.0
91	10.5
142	28.5
76	14.5

$\Sigma y = 828$, $\Sigma y\theta = 18\,147$,
$\Sigma\theta = 155.5$, $\Sigma\theta^2 = 3520.25$.

Table G

Number correct (x_1)	Number incorrect (x_2)	Number of days training (y)
35	23	10
26	25	2
33	24	7
22	28	5
40	27	11
31	21	8
22	32	9
20	21	3
24	33	8
23	23	2
30	23	6
22	26	4
30	26	5
26	22	3
23	34	11

(*a*) Illustrate the data by a scatter diagram.
(*b*) Calculate the equation of the regression line of distance on angle and draw this line on the scatter diagram.
(*c*) It later emerged that one of the points was obtained using a different object.
  (i) Suggest which point this was.
  (ii) Draw by eye a line of best fit on the scatter diagram ignoring the point apparently obtained with the different object.
  (iii) Use the line drawn by eye to estimate the distance the original object would roll if released at an angle of **(a)** 12°, **(b)** 40°.

Discuss the uncertainty of each of these estimates. (AEB 1987)

25. Table H below gives the average cost per hundredweight of zinc manufactures imported into the UK during each of the years 1873 to 1882.
(*a*) Plot the data on graph paper, by coding with (year − 1872) as the x variable and (cost − 100) as the y variable.
(*b*) Given that $\Sigma y = 270$ and $\Sigma xy = 1057$, show that the gradient of the equation of the least squares regression line of y on x is -5.2 (to 2 significant figures). Calculate the equation of this line and plot it on your graph.
(*c*) Use your equation to predict the cost of zinc manufactures imported in 1883. Comment on your prediction.
(*Source:* Statistical Abstract for the United Kingdom 1871 to 1885.) (O)

26. A purchasing manager, of a London-based company, believes that the time in transit of goods sent by road depends upon the distance between the supplier and the company. In an attempt to measure this dependence, twelve packages, sent from different parts of the country, have their transit times (y, days) accurately recorded, together with the distance (x, miles) of the supplier from the company. The results are summarised as follows:

$$\Sigma x = 1800, \quad \Sigma y = 36.0,$$
$$\Sigma xy = 6438.6, \quad \Sigma x^2 = 336\,296,$$
$$\Sigma y^2 = 126.34.$$

Obtain the least squares straight line regression equation of y on x.

Explain the significance of the regression coefficient.

Predict the transit time of a package sent from a supplier 200 miles away from the company.

Give two reasons why you would not use the equation to predict transit time for a package sent from a supplier 1500 miles away.

Calculate the product moment correlation coefficient between x and y.

Explain why the value you have obtained supports the purchasing manager's attempt to establish a regression equation of y on x. (AEB 1987)

Table H

Year	1873	1874	1875	1876	1877	1878	1879	1880	1881	1882
Cost (p)	147	147	144	140	129	119	112	116	107	109

APPENDIX 1

RANDOM NUMBERS

65 23	68 00	77 82	58 14	10 85	11 85	57 11	73 74	45 25	50 46
09 56	76 51	04 73	94 30	16 74	69 59	04 38	83 98	30 20	87 85
55 99	98 60	01 33	06 93	85 13	23 17	25 51	92 04	52 31	38 70
72 82	45 44	09 53	04 83	03 83	98 41	67 41	01 38	66 83	11 99
04 21	28 72	73 25	02 74	35 81	78 49	52 67	61 40	60 50	47 50
87 01	80 59	89 36	41 59	60 27	64 89	47 45	18 21	69 84	76 06
31 62	46 53	84 40	56 31	74 76	52 23	72 95	96 06	56 83	85 22
29 81	57 94	35 91	90 70	94 24	19 35	50 22	23 72	87 34	83 15
39 98	74 22	77 19	12 81	29 42	04 50	62 34	36 81	43 07	97 92
56 14	80 10	76 52	38 54	84 13	99 90	22 55	41 04	72 37	89 33
29 56	62 74	12 67	09 35	89 33	04 28	44 75	01 57	87 45	52 21
93 32	57 38	39 36	87 42	72 55	73 97	98 36	57 41	76 09	11 68
95 69	51 54	43 19	20 49	57 25	90 55	26 20	70 98	43 73	56 45
65 71	32 43	64 67	22 55	65 65	48 86	10 88	20 12	40 18	49.25
90 27	33 43	97 84	20 57	49 91	41 20	17 64	29 60	66 87	55 97
90 29	42 45	61 34	30 13	30 39	21 52	59 28	64 98	08 76	09 27
99 74	06 29	20 55	72 70	11 43	95 82	75 37	90 24	77 43	63 21
87 87	56 91	16 97	51 50	61 36	96 47	76 68	49 11	50 56	51 06
46 24	17 74	97 37	39 03	54 83	34 00	74 61	77 51	43 63	15 67
66 79	81 43	40 92	84 72	88 32	83 24	67 01	41 34	70 19	26 93
36 42	94 58	83 30	92 39	18 40	03 00	12 90	32 37	91 65	48 15
07 66	25 08	99 27	69 48	85 32	16 46	19 31	85 02	86 36	22 96
93 10	05 72	18 26	36 67	68 48	31 69	68 58	93 49	45 86	99 29
49 50	63 99	26 71	47 94	32 71	72 91	34 18	74 06	32 14	40 80
20 75	58 89	39 04	42 73	37 93	11 07	28 77	91 36	60 47	82 62
02 40	62 09	00 71	09 37	80 44	50 37	32 70	20 38	71 86	75 34
59 87	21 38	29 78	72 67	42 83	65 21	54 79	66 42	47 86	31 15
48 08	99 66	43 38	28 13	50 25	47 93	11 15	07 84	28 30	19 07
54 26	86 75	44 15	20 39	20 03	58 54	80 29	62 53	06 97	71 51
35 35	58 45	23 58	63 66	09 62	80 92	14 55	81 41	21 48	87 34
73 84	90 49	01 21	90 29	57 06	68 73	51 10	51 95	63 08	57 99
34 64	78 00	92 59	67 74	58 48	92 09	42 20	40 37	63 80	58 93
68 56	87 47	63 06	24 71	41 98	79 06	07 18	58 29	16 49	67 37
72 47	05 42	88 07	27 55	58 74	82 08	42 28	26 48	25 32	00 31
44 44	96 75	89 57	12 60	42 38	77 36	45 69	21 68	32 70	04 96
28 11	57 47	61 57	89 88	62 18	93 67	57 32	96 72	21 17	13 54
87 22	38 88	91 99	16 08	17 76	27 47	52 14	98 86	35 68	23 85
44 93	14 59	67 40	24 10	11 63	40 47	07 56	14 22	62 74	93 39
81 84	37 25	90 43	56 62	94 58	49 03	84 22	57 22	47 98	86 37
09 75	35 21	04 47	54 08	98 44	08 16	44 86	69 71	20 52	64 94
77 65	05 04	22 18	20 10	81 87	05 69	43 70	96 76	42 05	21 10
19 06	51 61	34 03	61 55	98 58	83 50	01 48	99 85	08 67	15 91
52 91	87 07	19 62	32 28	04 91	42 48	65 24	86 09	87 68	55 51
52 47	25 14	93 91	75 51	49 26	49 41	20 83	30 30	43 22	69 08
52 67	87 40	63 41	91 86	10 47	80 70	56 87	25 86	89 94	21 42
65 25	71 73	78 60	50 62	91 04	95 97	64 16	71 31	32 80	19 61
29 97	56 42	56 90	16 75	74 95	99 26	01 63	25 16	54 18	54 46
15 25	03 68	92 45	53 00	06 29	46 43	46 66	27 12	85 05	22 44
82 08	65 67	64 13	51 14	38 28	24 30	39 62	20 35	23 90	57 36
81 35	03 25	87 24	83 59	04 67	51 52	26 21	69 75	87 28	61 50

Each digit in this table is an independent sample from a population where each of the digits 0 to 9 has a probability of occurrence of 0.1. It should be noted that these digits have been computer generated, and are therefore ‘pseudo’ random numbers.

CUMULATIVE BINOMIAL PROBABILITIES

The tabulated value is $P(X \leqslant r)$ where $X \sim \text{Bin}(n, p)$

	p =	0.05	0.10	0.15	0.20	0.25	0.30	0.35	0.40	0.45	0.50
$n = 2$	$r = 0$	0.9025	0.8100	0.7225	0.6400	0.5625	0.4900	0.4225	0.3600	0.3025	0.2500
	1	0.9975	0.9900	0.9775	0.9600	0.9375	0.9100	0.8775	0.8400	0.7975	0.7500
	2	1.0000	1.0000	1.0000	1.0000	1.0000	1.0000	1.0000	1.0000	1.0000	1.0000
$n = 3$	$r = 0$	0.8574	0.7290	0.6141	0.5120	0.4219	0.3430	0.2746	0.2160	0.1664	0.1250
	1	0.9928	0.9720	0.9393	0.8960	0.8438	0.7840	0.7183	0.6480	0.5748	0.5000
	2	0.9999	0.9990	0.9966	0.9920	0.9844	0.9730	0.9571	0.9360	0.9089	0.8750
	3	1.0000	1.0000	1.0000	1.0000	1.0000	1.0000	1.0000	1.0000	1.0000	1.0000
$n = 4$	$r = 0$	0.8145	0.6561	0.5220	0.4096	0.3164	0.2401	0.1785	0.1296	0.0915	0.0625
	1	0.9860	0.9477	0.8905	0.8192	0.7383	0.6517	0.5630	0.4752	0.3910	0.3125
	2	0.9995	0.9963	0.9880	0.9728	0.9492	0.9163	0.8735	0.8208	0.7585	0.6875
	3	1.0000	0.9999	0.9995	0.9984	0.9961	0.9919	0.9850	0.9744	0.9590	0.9375
	4		1.0000	1.0000	1.0000	1.0000	1.0000	1.0000	1.0000	1.0000	1.0000
$n = 5$	$r = 0$	0.7738	0.5905	0.4437	0.3277	0.2373	0.1681	0.1160	0.0778	0.0503	0.0313
	1	0.9774	0.9185	0.8352	0.7373	0.6328	0.5282	0.4284	0.3370	0.2562	0.1875
	2	0.9988	0.9914	0.9734	0.9421	0.8965	0.8369	0.7648	0.6826	0.5931	0.5000
	3	1.0000	0.9995	0.9978	0.9933	0.9844	0.9692	0.9460	0.9130	0.8688	0.8125
	4		1.0000	0.9999	0.9997	0.9990	0.9976	0.9947	0.9898	0.9815	0.9688
	5			1.0000	1.0000	1.0000	1.0000	1.0000	1.0000	1.0000	1.0000
$n = 6$	$r = 0$	0.7351	0.5314	0.3771	0.2621	0.1780	0.1176	0.0754	0.0467	0.0277	0.0156
	1	0.9672	0.8857	0.7765	0.6554	0.5339	0.4202	0.3191	0.2333	0.1636	0.1094
	2	0.9978	0.9842	0.9527	0.9011	0.8306	0.7443	0.6471	0.5443	0.4415	0.3438
	3	0.9999	0.9987	0.9941	0.9830	0.9624	0.9295	0.8826	0.8208	0.7447	0.6563
	4	1.0000	0.9999	0.9996	0.9984	0.9954	0.9891	0.9777	0.9590	0.9308	0.8906
	5		1.0000	1.0000	0.9999	0.9998	0.9993	0.9982	0.9959	0.9917	0.9844
	6				1.0000	1.0000	1.0000	1.0000	1.0000	1.0000	1.0000
$n = 7$	$r = 0$	0.6983	0.4783	0.3206	0.2097	0.1335	0.0824	0.0490	0.0280	0.0152	0.0078
	1	0.9556	0.8503	0.7166	0.5767	0.4449	0.3294	0.2338	0.1586	0.1024	0.0625
	2	0.9962	0.9743	0.9262	0.8520	0.7564	0.6471	0.5323	0.4199	0.3164	0.2266
	3	0.9998	0.9973	0.9879	0.9667	0.9294	0.8740	0.8002	0.7102	0.6083	0.5000
	4	1.0000	0.9998	0.9988	0.9953	0.9871	0.9712	0.9444	0.9037	0.8471	0.7734
	5		1.0000	0.9999	0.9996	0.9987	0.9962	0.9910	0.9812	0.9643	0.9375
	6			1.0000	1.0000	0.9999	0.9998	0.9994	0.9984	0.9963	0.9922
	7					1.0000	1.0000	1.0000	1.0000	1.0000	1.0000
$n = 8$	$r = 0$	0.6634	0.4305	0.2725	0.1678	0.1001	0.0576	0.0319	0.0168	0.0084	0.0039
	1	0.9428	0.8131	0.6572	0.5033	0.3671	0.2553	0.1691	0.1064	0.0632	0.0352
	2	0.9942	0.9619	0.8948	0.7969	0.6785	0.5518	0.4278	0.3154	0.2201	0.1445
	3	0.9996	0.9950	0.9786	0.9437	0.8862	0.8059	0.7064	0.5941	0.4770	0.3633
	4	1.0000	0.9996	0.9971	0.9896	0.9727	0.9420	0.8939	0.8263	0.7396	0.6367
	5		1.0000	0.9998	0.9988	0.9958	0.9887	0.9747	0.9502	0.9115	0.8555
	6			1.0000	0.9999	0.9996	0.9987	0.9964	0.9915	0.9819	0.9648
	7				1.0000	1.0000	0.9999	0.9998	0.9993	0.9983	0.9961
	8						1.0000	1.0000	1.0000	1.0000	1.0000

CUMULATIVE BINOMIAL PROBABILITIES

The tabulated value is $P(X \leqslant r)$ where $X \sim \text{Bin}(n, p)$

$p =$		0.05	0.10	0.15	0.20	0.25	0.30	0.35	0.40	0.45	0.50
$n = 9$	$r = 0$	0.6302	0.3874	0.2316	0.1342	0.0751	0.0404	0.0207	0.0101	0.0046	0.0020
	1	0.9288	0.7748	0.5995	0.4362	0.3003	0.1960	0.1211	0.0705	0.0385	0.0195
	2	0.9916	0.9470	0.8591	0.7382	0.6007	0.4628	0.3373	0.2318	0.1495	0.0898
	3	0.9994	0.9917	0.9661	0.9144	0.8343	0.7297	0.6089	0.4826	0.3614	0.2539
	4	1.0000	0.9991	0.9944	0.9804	0.9511	0.9012	0.8283	0.7334	0.6214	0.5000
	5		0.9999	0.9994	0.9969	0.9900	0.9747	0.9464	0.9006	0.8342	0.7461
	6		1.0000	1.0000	0.9997	0.9987	0.9957	0.9888	0.9750	0.9502	0.9102
	7				1.0000	0.9999	0.9996	0.9986	0.9962	0.9909	0.9805
	8					1.0000	1.0000	0.9999	0.9997	0.9992	0.9980
	9							1.0000	1.0000	1.0000	1.0000
$n = 10$	$r = 0$	0.5987	0.3487	0.1969	0.1074	0.0563	0.0282	0.0135	0.0060	0.0025	0.0010
	1	0.9139	0.7361	0.5443	0.3758	0.2440	0.1493	0.0860	0.0464	0.0233	0.0107
	2	0.9885	0.9298	0.8202	0.6778	0.5256	0.3828	0.2616	0.1673	0.0996	0.0547
	3	0.9990	0.9872	0.9500	0.8791	0.7759	0.6496	0.5138	0.3823	0.2660	0.1719
	4	0.9999	0.9984	0.9901	0.9672	0.9219	0.8497	0.7515	0.6331	0.5044	0.3770
	5	1.0000	0.9999	0.9986	0.9936	0.9803	0.9527	0.9051	0.8338	0.7384	0.6230
	6		1.0000	0.9999	0.9991	0.9965	0.9894	0.9740	0.9452	0.8980	0.8281
	7			1.0000	0.9999	0.9996	0.9984	0.9952	0.9877	0.9726	0.9453
	8				1.0000	1.0000	0.9999	0.9995	0.9983	0.9955	0.9893
	9						1.0000	1.0000	0.9999	0.9997	0.9990
	10								1.0000	1.0000	1.0000
$n = 15$	$r = 0$	0.4633	0.2059	0.0874	0.0352	0.0134	0.0047	0.0016	0.0005	0.0001	0.0000
	1	0.8290	0.5490	0.3186	0.1671	0.0802	0.0353	0.0142	0.0052	0.0017	0.0005
	2	0.9638	0.8159	0.6042	0.3980	0.2361	0.1268	0.0617	0.0271	0.0107	0.0037
	3	0.9945	0.9444	0.8227	0.6482	0.4613	0.2969	0.1727	0.0905	0.0424	0.0176
	4	0.9994	0.9873	0.9383	0.8358	0.6865	0.5155	0.3519	0.2173	0.1204	0.0592
	5	0.9999	0.9978	0.9832	0.9389	0.8516	0.7216	0.5643	0.4032	0.2608	0.1509
	6	1.0000	0.9997	0.9964	0.9819	0.9434	0.8689	0.7548	0.6098	0.4522	0.3036
	7		1.0000	0.9994	0.9958	0.9827	0.9500	0.8868	0.7869	0.6535	0.5000
	8			0.9999	0.9992	0.9958	0.9848	0.9578	0.9050	0.8182	0.6964
	9			1.0000	0.9999	0.9992	0.9963	0.9876	0.9662	0.9231	0.8491
	10				1.0000	0.9999	0.9993	0.9972	0.9907	0.9745	0.9408
	11					1.0000	0.9999	0.9995	0.9981	0.9937	0.9824
	12						1.0000	0.9999	0.9997	0.9989	0.9963
	13							1.0000	1.0000	0.9999	0.9995
	14									1.0000	1.0000
$n = 20$	$r = 0$	0.3585	0.1216	0.0388	0.0115	0.0032	0.0008	0.0002	0.0000	0.0000	0.0000
	1	0.7358	0.3917	0.1756	0.0692	0.0243	0.0076	0.0021	0.0005	0.0001	0.0000
	2	0.9245	0.6769	0.4049	0.2061	0.0913	0.0355	0.0121	0.0036	0.0009	0.0002
	3	0.9841	0.8670	0.6477	0.4114	0.2252	0.1071	0.0444	0.0160	0.0049	0.0013
	4	0.9974	0.9568	0.8298	0.6296	0.4148	0.2375	0.1182	0.0510	0.0189	0.0059
	5	0.9997	0.9887	0.9327	0.8042	0.6172	0.4164	0.2454	0.1256	0.0553	0.0207
	6	1.0000	0.9976	0.9781	0.9133	0.7858	0.6080	0.4166	0.2500	0.1299	0.0577
	7		0.9996	0.9941	0.9679	0.8982	0.7723	0.6010	0.4159	0.2520	0.1316
	8		0.9999	0.9987	0.9900	0.9591	0.8867	0.7624	0.5956	0.4143	0.2517
	9		1.0000	0.9998	0.9974	0.9861	0.9520	0.8782	0.7553	0.5914	0.4119
	10			1.0000	0.9994	0.9961	0.9829	0.9468	0.8725	0.7507	0.5881
	11				0.9999	0.9991	0.9949	0.9804	0.9435	0.8692	0.7483
	12				1.0000	0.9998	0.9987	0.9940	0.9790	0.9420	0.8684
	13					1.0000	0.9997	0.9985	0.9935	0.9786	0.9423
	14						1.0000	0.9997	0.9984	0.9936	0.9793
	15							1.0000	0.9997	0.9985	0.9941
	16								1.0000	0.9997	0.9987
	17									1.0000	0.9998
	18										1.0000

CUMULATIVE POISSON PROBABILITIES

The tabulated value is $P(X \leqslant r)$ where $X \sim \text{Po}(\lambda)$

$\lambda =$	0.2	0.4	0.5	0.6	0.8	1.0	1.2	1.4	1.5
$r = 0$	0.8187	0.6703	0.6065	0.5488	0.4493	0.3679	0.3012	0.2466	0.2231
1	0.9825	0.9384	0.9098	0.8781	0.8088	0.7358	0.6626	0.5918	0.5578
2	0.9989	0.9921	0.9856	0.9769	0.9526	0.9197	0.8795	0.8335	0.8088
3	0.9999	0.9992	0.9982	0.9966	0.9909	0.9810	0.9662	0.9463	0.9344
4	1.0000	0.9999	0.9998	0.9996	0.9986	0.9963	0.9923	0.9857	0.9814
5		1.0000	1.0000	1.0000	0.9998	0.9994	0.9985	0.9968	0.9955
6					1.0000	0.9999	0.9997	0.9994	0.9991
7						1.0000	1.0000	0.9999	0.9998
8								1.0000	1.0000

$\lambda =$	1.6	1.8	2.0	2.2	2.4	2.5	2.6	2.8	3.0
$r = 0$	0.2019	0.1653	0.1353	0.1108	0.0907	0.0821	0.0743	0.0608	0.0498
1	0.5249	0.4628	0.4060	0.3546	0.3084	0.2873	0.2674	0.2311	0.1991
2	0.7834	0.7306	0.6767	0.6227	0.5697	0.5438	0.5184	0.4695	0.4232
3	0.9212	0.8913	0.8571	0.8194	0.7787	0.7576	0.7360	0.6919	0.6472
4	0.9763	0.9636	0.9473	0.9275	0.9041	0.8912	0.8774	0.8477	0.8153
5	0.9940	0.9896	0.9834	0.9751	0.9643	0.9580	0.9510	0.9349	0.9161
6	0.9987	0.9974	0.9955	0.9925	0.9884	0.9858	0.9828	0.9756	0.9665
7	0.9997	0.9994	0.9989	0.9980	0.9967	0.9958	0.9947	0.9919	0.9881
8	1.0000	0.9999	0.9998	0.9995	0.9991	0.9989	0.9985	0.9976	0.9962
9		1.0000	1.0000	0.9999	0.9998	0.9997	0.9996	0.9993	0.9989
10				1.0000	1.0000	0.9999	0.9999	0.9998	0.9997
11						1.0000	1.0000	1.0000	0.9999
12									1.0000

$\lambda =$	3.2	3.4	3.5	3.6	3.8	4.0	4.5	5.0	5.5
$r = 0$	0.0408	0.0334	0.0302	0.0273	0.0224	0.0183	0.0111	0.0067	0.0041
1	0.1712	0.1468	0.1359	0.1257	0.1074	0.0916	0.0611	0.0404	0.0266
2	0.3799	0.3397	0.3208	0.3027	0.2689	0.2381	0.1736	0.1247	0.0884
3	0.6025	0.5584	0.5366	0.5152	0.4735	0.4335	0.3423	0.2650	0.2017
4	0.7806	0.7442	0.7254	0.7064	0.6678	0.6288	0.5321	0.4405	0.3575
5	0.8946	0.8705	0.8576	0.8441	0.8156	0.7851	0.7029	0.6160	0.5289
6	0.9554	0.9421	0.9347	0.9267	0.9091	0.8893	0.8311	0.7622	0.6860
7	0.9832	0.9769	0.9733	0.9692	0.9599	0.9489	0.9134	0.8666	0.8095
8	0.9943	0.9917	0.9901	0.9883	0.9840	0.9786	0.9597	0.9319	0.8944
9	0.9982	0.9973	0.9967	0.9960	0.9942	0.9919	0.9829	0.9682	0.9462
10	0.9995	0.9992	0.9990	0.9987	0.9981	0.9972	0.9933	0.9863	0.9747
11	0.9999	0.9998	0.9997	0.9996	0.9994	0.9991	0.9976	0.9945	0.9890
12	1.0000	0.9999	0.9999	0.9999	0.9998	0.9997	0.9992	0.9980	0.9955
13		1.0000	1.0000	1.0000	1.0000	0.9999	0.9997	0.9993	0.9983
14						1.0000	0.9999	0.9998	0.9994
15							1.0000	0.9999	0.9998
16								1.0000	0.9999
17									1.0000
18									

CUMULATIVE POISSON PROBABILITIES

The tabulated value is $P(X \leqslant r)$ where $X \sim \text{Po}(\lambda)$

$\lambda =$	6.0	6.5	7.0	7.5	8.0	8.5	9.0	9.5	10.0
$r = 0$	0.0025	0.0015	0.0009	0.0006	0.0003	0.0002	0.0001	0.0001	0.0000
1	0.0174	0.0113	0.0073	0.0047	0.0030	0.0019	0.0012	0.0008	0.0005
2	0.0620	0.0430	0.0296	0.0203	0.0138	0.0093	0.0062	0.0042	0.0028
3	0.1512	0.1118	0.0818	0.0591	0.0424	0.0301	0.0212	0.0149	0.0103
4	0.2851	0.2237	0.1730	0.1321	0.0996	0.0744	0.0550	0.0403	0.0293
5	0.4457	0.3690	0.3007	0.2414	0.1912	0.1496	0.1157	0.0885	0.0671
6	0.6063	0.5265	0.4497	0.3782	0.3134	0.2562	0.2068	0.1649	0.1301
7	0.7440	0.6728	0.5987	0.5246	0.4530	0.3856	0.3239	0.2687	0.2202
8	0.8472	0.7916	0.7291	0.6620	0.5925	0.5231	0.4557	0.3918	0.3328
9	0.9161	0.8774	0.8305	0.7764	0.7166	0.6530	0.5874	0.5218	0.4579
10	0.9574	0.9332	0.9015	0.8622	0.8159	0.7634	0.7060	0.6453	0.5830
11	0.9799	0.9661	0.9467	0.9208	0.8881	0.8487	0.8030	0.7520	0.6968
12	0.9912	0.9840	0.9730	0.9573	0.9362	0.9091	0.8758	0.8364	0.7916
13	0.9964	0.9929	0.9872	0.9784	0.9658	0.9486	0.9261	0.8981	0.8645
14	0.9986	0.9970	0.9943	0.9897	0.9827	0.9726	0.9585	0.9400	0.9165
15	0.9995	0.9988	0.9976	0.9954	0.9918	0.9862	0.9780	0.9665	0.9513
16	0.9998	0.9996	0.9990	0.9980	0.9963	0.9934	0.9889	0.9823	0.9730
17	0.9999	0.9998	0.9996	0.9992	0.9984	0.9970	0.9947	0.9911	0.9857
18	1.0000	0.9999	0.9999	0.9997	0.9993	0.9987	0.9976	0.9957	0.9928
19		1.0000	1.0000	0.9999	0.9997	0.9995	0.9989	0.9980	0.9965
20				1.0000	0.9999	0.9998	0.9996	0.9991	0.9984
21					1.0000	0.9999	0.9998	0.9996	0.9993
22						1.0000	0.9999	0.9999	0.9997
23							1.0000	0.9999	0.9999
24								1.0000	1.0000

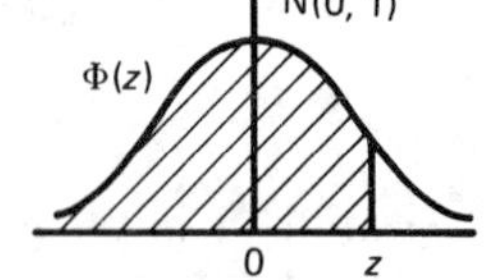

THE DISTRIBUTION FUNCTION $\Phi(z)$ OF THE NORMAL DISTRIBUTION N(0, 1)

z	0	1	2	3	4	5	6	7	8	9	ADD								
											1	2	3	4	5	6	7	8	9
0.0	.5000	.5040	.5080	.5120	.5160	.5199	.5239	.5279	.5319	.5359	4	8	12	16	20	24	28	32	36
0.1	.5398	.5438	.5478	.5517	.5557	.5596	.5636	.5675	.5714	.5753	4	8	12	16	20	24	28	32	36
0.2	.5793	.5832	.5871	.5910	.5948	.5987	.6026	.6064	.6103	.6141	4	8	12	15	19	23	27	31	35
0.3	.6179	.6217	.6255	.6293	.6331	.6368	.6406	.6443	.6480	.6517	4	7	11	15	19	22	26	30	34
0.4	.6554	.6591	.6628	.6664	.6700	.6736	.6772	.6808	.6844	.6879	4	7	11	14	18	22	25	29	32
0.5	.6915	.6950	.6985	.7019	.7054	.7088	.7123	.7157	.7190	.7224	3	7	10	14	17	20	24	27	31
0.6	.7257	.7291	.7324	.7357	.7389	.7422	.7454	.7486	.7517	.7549	3	7	10	13	16	19	23	26	29
0.7	.7580	.7611	.7642	.7673	.7704	.7734	.7764	.7794	.7823	.7852	3	6	9	12	15	18	21	24	27
0.8	.7881	.7910	.7939	.7967	.7995	.8023	.8051	.8078	.8106	.8133	3	5	8	11	14	16	19	22	25
0.9	.8159	.8186	.8212	.8238	.8264	.8289	.8315	.8340	.8365	.8389	3	5	8	10	13	15	18	20	23
1.0	.8413	.8438	.8461	.8485	.8508	.8531	.8554	.8577	.8599	.8621	2	5	7	9	12	14	16	19	21
1.1	.8643	.8665	.8686	.8708	.8729	.8749	.8770	.8790	.8810	.8830	2	4	6	8	10	12	14	16	18
1.2	.8849	.8869	.8888	.8907	.8925	.8944	.8962	.8980	.8997	.9015	2	4	6	7	9	11	13	15	17
1.3	.9032	.9049	.9066	.9082	.9099	.9115	.9131	.9147	.9162	.9177	2	3	5	6	8	10	11	13	14
1.4	.9192	.9207	.9222	.9236	.9251	.9265	.9279	.9292	.9306	.9319	1	3	4	6	7	8	10	11	13
1.5	.9332	.9345	.9357	.9370	.9382	.9394	.9406	.9418	.9429	.9441	1	2	4	5	6	7	8	10	11
1.6	.9452	.9463	.9474	.9484	.9495	.9505	.9515	.9525	.9535	.9545	1	2	3	4	5	6	7	8	9
1.7	.9554	.9564	.9573	.9582	.9591	.9599	.9608	.9616	.9625	.9633	1	2	3	4	4	5	6	7	8
1.8	.9641	.9649	.9656	.9664	.9671	.9678	.9686	.9693	.9699	.9706	1	1	2	3	4	4	5	6	6
1.9	.9713	.9719	.9726	.9732	.9738	.9744	.9750	.9756	.9761	.9767	1	1	2	2	3	4	4	5	5
2.0	.9772	.9778	.9783	.9788	.9793	.9798	.9803	.9808	.9812	.9817	0	1	1	2	2	3	3	4	4
2.1	.9821	.9826	.9830	.9834	.9838	.9842	.9846	.9850	.9854	.9857	0	1	1	2	2	2	3	3	4
2.2	.9861	.9864	.9868	.9871	.9875	.9878	.9881	.9884	.9887	.9890	0	1	1	1	2	2	2	3	3
2.3	.9893	.9896	.9898								0	1	1	1	1	2	2	2	2
				.9901	.99036	.99061	.99086				3	5	8	10	13	15	18	20	23
								.99111	.99134	.99158	2	5	7	9	12	14	16	18	21
2.4	.99180	.99202	.99224	.99245	.99266						2	4	6	8	11	13	15	17	19
						.99286	.99305	.99324	.99343	.99361	2	4	6	7	9	11	13	15	17
2.5	.99379	.99396	.99413	.99430	.99446	.99461	.99477	.99492	.99506	.99520	2	3	5	6	8	9	11	12	14
2.6	.99534	.99547	.99560	.99573	.99585	.99598	.99609	.99621	.99632	.99643	1	2	3	5	6	7	8	9	10
2.7	.99653	.99664	.99674	.99683	.99693	.99702	.99711	.99720	.99728	.99736	1	2	3	4	5	6	7	8	9
2.8	.99744	.99752	.99760	.99767	.99774	.99781	.99788	.99795	.99801	.99807	1	1	2	3	4	4	5	6	6
2.9	.99813	.99819	.99825	.99831	.99836	.99841	.99846	.99851	.99856	.99861	0	1	1	2	2	3	3	4	4
3.0	.99865	.99869	.99874	.99878	.99882	.99886	.99889	.99893	.99896	.99900	0	1	1	2	2	2	3	3	4
3.1	$.9^{3}032$	$.9^{3}065$	$.9^{3}096$								3	6	9	13	16	19	22	25	28
				$.9^{3}126$	$.9^{3}155$	$.9^{3}184$	$.9^{3}211$				3	6	8	11	14	17	20	22	25
								$.9^{3}238$	$.9^{3}264$	$.9^{3}289$	2	5	7	10	12	15	17	20	22
3.2	$.9^{3}313$	$.9^{3}336$	$.9^{3}359$	$.9^{3}381$	$.9^{3}402$						2	4	7	9	11	13	15	18	20
						$.9^{3}423$	$.9^{3}443$	$.9^{3}462$	$.9^{3}481$	$.9^{3}499$	2	4	6	8	9	11	13	15	17
3.3	$.9^{3}517$	$.9^{3}534$	$.9^{3}550$	$.9^{3}566$	$.9^{3}581$						2	3	5	6	8	10	11	13	14
						$.9^{3}596$	$.9^{3}610$	$.9^{3}624$	$.9^{3}638$	$.9^{3}651$	1	3	4	5	7	8	9	10	12
3.4	$.9^{3}663$	$.9^{3}675$	$.9^{3}687$	$.9^{3}698$	$.9^{3}709$	$.9^{3}720$	$.9^{3}730$	$.9^{3}740$	$.9^{3}749$	$.9^{3}758$	1	2	3	4	5	6	7	8	9
3.5	$.9^{3}767$	$.9^{3}776$	$.9^{3}784$	$.9^{3}792$	$.9^{3}800$	$.9^{3}807$	$.9^{3}815$	$.9^{3}822$	$.9^{3}828$	$.9^{3}835$	1	1	2	3	4	4	5	6	7
3.6	$.9^{3}841$	$.9^{3}847$	$.9^{3}853$	$.9^{3}858$	$.9^{3}864$	$.9^{3}869$	$.9^{3}874$	$.9^{3}879$	$.9^{3}883$	$.9^{3}888$	0	1	1	2	2	3	3	4	5
3.7	$.9^{3}892$	$.9^{3}896$	$.9^{3}90$	$.9^{4}04$	$.9^{4}08$	$.9^{4}12$	$.9^{4}15$	$.9^{4}18$	$.9^{4}22$	$.9^{4}250$									
3.8	$.9^{4}28$	$.9^{4}31$	$.9^{4}33$	$.9^{4}36$	$.9^{4}38$	$.9^{4}41$	$.9^{4}43$	$.9^{4}46$	$.9^{4}48$	$.9^{4}500$									
3.9	$.9^{4}52$	$.9^{4}54$	$.9^{4}56$	$.9^{4}58$	$.9^{4}59$	$.9^{4}61$	$.9^{4}63$	$.9^{4}64$	$.9^{4}66$	$.9^{4}670$									

For negative values of z use $\Phi(z) = 1 - \Phi(-z)$

UPPER QUANTILES $z_{[P]}$ OF THE NORMAL DISTRIBUTION N(0, 1)

P	Q	z	P	Q	z	P	Q	z	P	Q	z	Q	z
.50	.50	0.000	.85	.15	1.036	.975	.025	1.960	.990	.010	2.326	$.0^{3}4$	3.353
.55	.45	0.126	.86	.14	1.080	.976	.024	1.977	.991	.009	2.366	$.0^{3}3$	3.432
.60	.40	0.253	.87	.13	1.126	.977	.023	1.995	.992	.008	2.409	$.0^{3}2$	3.540
.65	.35	0.385	.88	.12	1.175	.978	.022	2.014	.993	.007	2.457	$.0^{3}1$	3.719
.70	.30	0.524	.89	.11	1.227	.979	.021	2.034	.994	.006	2.512	$.0^{4}5$	3.891
.75	.25	0.674	.90	.10	1.282	.980	.020	2.054	.995	.005	2.576	$.0^{4}1$	4.265
.76	.24	0.706	.91	.09	1.341	.981	.019	2.075	.996	.004	2.652	$.0^{5}5$	4.417
.77	.23	0.739	.92	.08	1.405	.982	.018	2.097	.997	.003	2.748	$.0^{5}1$	4.753
.78	.22	0.772	.93	.07	1.476	.983	.017	2.120	.998	.002	2.878	$.0^{6}5$	4.892
.79	.21	0.806	.94	.06	1.555	.984	.016	2.144	.999	.001	3.090	$.0^{6}1$	5.199
.80	.20	0.842	.950	.050	1.645	.985	.015	2.170	.9991	$.0^{3}9$	3.121	$.0^{7}5$	5.327
.81	.19	0.878	.955	.045	1.695	.986	.014	2.197	.9992	$.0^{3}8$	3.156	$.0^{7}1$	5.612
.82	.18	0.915	.960	.040	1.751	.987	.013	2.226	.9993	$.0^{3}7$	3.195	$.0^{8}5$	5.731
.83	.17	0.954	.965	.035	1.812	.988	.012	2.257	.9994	$.0^{3}6$	3.239	$.0^{8}1$	5.998
.84	.16	0.994	.970	.030	1.881	.989	.011	2.290	.9995	$.0^{3}5$	3.291	$.0^{9}5$	6.109

Reproduced, by permission, from Miller and Powell, *The Cambridge Elementary Mathematical Tables* (Cambridge University Press).

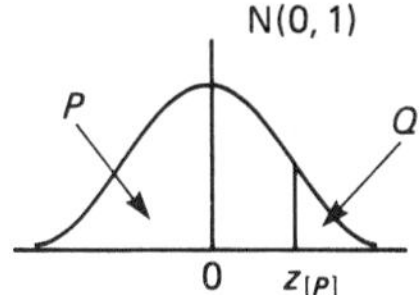

UPPER QUANTILES $t_{[P]}$ OF t-DISTRIBUTIONS $t(\nu)$

P	.75	.90	.95	.975	.99	.995	.9975	.999	.9995
Q	.25	.10	.05	.025	.01	.005	.0025	.001	.0005
$2Q$	.50	.20	.10	.050	.02	.010	.0050	.002	.0010
$\nu = 1$	1.000	3.078	6.314	12.71	31.82	63.66	127.3	318.3	636.6
2	0.816	1.886	2.920	4.303	6.965	9.925	14.09	22.33	31.60
3	0.765	1.638	2.353	3.182	4.541	5.841	7.453	10.21	12.92
4	0.741	1.533	2.132	2.776	3.747	4.604	5.598	7.173	8.610
5	0.727	1.476	2.015	2.571	3.365	4.032	4.773	5.893	6.869
6	0.718	1.440	1.943	2.447	3.143	3.707	4.317	5.208	5.959
7	0.711	1.415	1.895	2.365	2.998	3.499	4.029	4.785	5.408
8	0.706	1.397	1.860	2.306	2.896	3.355	3.833	4.501	5.041
9	0.703	1.383	1.833	2.262	2.821	3.250	3.690	4.297	4.781
10	0.700	1.372	1.812	2.228	2.764	3.169	3.581	4.144	4.587
11	0.697	1.363	1.796	2.201	2.718	3.106	3.497	4.025	4.437
12	0.695	1.356	1.782	2.179	2.681	3.055	3.428	3.930	4.318
13	0.694	1.350	1.771	2.160	2.650	3.012	3.372	3.852	4.221
14	0.692	1.345	1.761	2.145	2.624	2.977	3.326	3.787	4.140
15	0.691	1.341	1.753	2.131	2.602	2.947	3.286	3.733	4.073
16	0.690	1.337	1.746	2.120	2.583	2.921	3.252	3.686	4.015
17	0.689	1.333	1.740	2.110	2.567	2.898	3.222	3.646	3.965
18	0.688	1.330	1.734	2.101	2.552	2.878	3.197	3.610	3.922
19	0.688	1.328	1.729	2.093	2.539	2.861	3.174	3.579	3.883
20	0.687	1.325	1.725	2.086	2.528	2.845	3.153	3.552	3.850
21	0.686	1.323	1.721	2.080	2.518	2.831	3.135	3.527	3.819
22	0.686	1.321	1.717	2.074	2.508	2.819	3.119	3.505	3.792
23	0.685	1.319	1.714	2.069	2.500	2.807	3.104	3.485	3.767
24	0.685	1.318	1.711	2.064	2.492	2.797	3.091	3.467	3.745
25	0.684	1.316	1.708	2.060	2.485	2.787	3.078	3.450	3.725
26	0.684	1.315	1.706	2.056	2.479	2.779	3.067	3.435	3.707
27	0.684	1.314	1.703	2.052	2.473	2.771	3.057	3.421	3.690
28	0.683	1.313	1.701	2.048	2.467	2.763	3.047	3.408	3.674
29	0.683	1.311	1.699	2.045	2.462	2.756	3.038	3.396	3.659
30	0.683	1.310	1.697	2.042	2.457	2.750	3.030	3.385	3.646
40	0.681	1.303	1.684	2.021	2.423	2.704	2.971	3.307	3.551
60	0.679	1.296	1.671	2.000	2.390	2.660	2.915	3.232	3.460
120	0.677	1.289	1.658	1.980	2.358	2.617	2.860	3.160	3.373
∞	0.674	1.282	1.645	1.960	2.326	2.576	2.807	3.090	3.291

The figure shows the form of the distribution for $\nu = 2$; the shaded area represents the tail probability Q. For large ν the distributions approximate to the normal distribution N(0, 1), shown by the broken line.

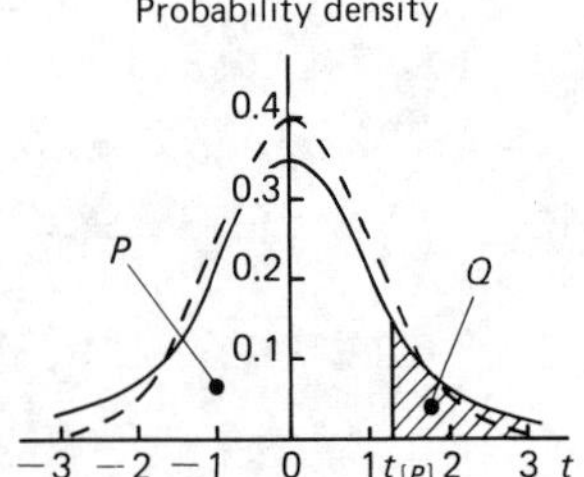

CHI-SQUARED TABLES showing

$\chi^2_{5\%}$ where $P(X > \chi^2_{5\%}) = 0.05$
$\chi^2_{1\%}$ where $P(X > \chi^2_{1\%}) = 0.01$

Degrees of freedom	$\chi^2_{5\%}$	$\chi^2_{1\%}$
1	3.84	6.64
2	5.99	9.21
3	7.82	11.34
4	9.49	13.28
5	11.07	15.09
6	12.59	16.81
7	14.07	18.48
8	15.51	20.09
9	16.92	21.67
10	18.31	23.21
11	19.68	24.73
12	21.03	26.22
13	22.36	27.69
14	23.68	29.14
15	25.00	30.58
20	31.41	37.57
30	43.77	50.89
40	55.76	63.69

NOTE: Extended tables can be found in J. White, A. Yeats and G. Skipworth, *Tables for Statisticians*, 3rd edition, (Stanley Thornes Publishers Ltd).

TABLE A

Table of probabilities associated with Σd^2 in Spearman's rank correlation coefficient, r_S.
Probability that Σd^2 exceeds, or is less than, certain values, for $4 \leqslant n \leqslant 10$.

$n=4$ MAX. $\Sigma d^2=20$			$n=5$ MAX. $\Sigma d^2=40$			$n=6$ MAX. $\Sigma d^2=70$			$n=7$ MAX. $\Sigma d^2=112$			$n=8$ MAX. $\Sigma d^2=168$			$n=9$ MAX. $\Sigma d^2=240$			$n=10$ MAX. $\Sigma d^2=330$		
Σd^2		P	Σd^2		P	Σd^2		P	Σd^2		P	Σd^2		P	Σd^2		P	Σd^2		P
$\leqslant$	$\geqslant$		$\leqslant$	$\geqslant$		$\leqslant$	$\geqslant$		$\leqslant$	$\geqslant$		$\leqslant$	$\geqslant$		$\leqslant$	$\geqslant$		$\leqslant$	$\geqslant$	
8	12	.458	18	22	.475	28	42	.357	54	58	.482	78	90	.441	112	128	.440	160	170	.473
6	14	.375	16	24	.392	26	44	.329	52	60	.453	74	94	.397	108	132	.405	154	176	.433
4	16	.208	14	26	.342	24	46	.282	50	62	.420	68	100	.332	104	136	.371	148	182	.393
2	18	.167	12	28	.258	22	48	.249	48	64	.391	64	104	.291	100	140	.339	140	190	.341
0	20	.0417	10	30	.225	20	50	.210	46	66	.356	60	108	.250	96	144	.307	132	198	.292
			8	32	.175	18	52	.178	44	68	.331	56	112	.214	92	148	.276	124	206	.246
			6	34	.117	16	54	.149	42	70	.297	54	114	.195	88	152	.247	116	214	.204
			4	36	.0667	14	56	.121	40	72	.278	52	116	.180	84	156	.218	114	216	.194
			2	38	.0417	12	58	.0875	38	74	.249	48	120	.150	80	160	.193	108	222	.165
			0	40	.0083	10	60	.0681	36	76	.222	46	122	.134	76	164	.168	100	230	.132
						8	62	.0514	34	78	.198	44	124	.122	72	168	.146	92	238	.102
						6	64	.0292	32	80	.177	42	126	.108	68	172	.125	90	240	.0956
						4	66	.0167	30	82	.151	40	128	.0983	64	176	.106	84	246	.0774
						2	68	.0083	28	84	.133	38	130	.0855	62	178	.0969	78	252	.0616
						0	70	.0014	26	86	.118	36	132	.0756	56	184	.0738	74	256	.0524
									24	88	.100	34	134	.0661	50	190	.0540	72	258	.0481
									22	90	.0833	32	136	.0575	48	192	.0484	68	262	.0403
									20	92	.0694	30	138	.0481	40	200	.0294	62	268	.0302
									18	94	.0548	28	140	.0415	34	206	.0184	56	274	.0219
									16	96	.0440	24	144	.0288	28	212	.0107	50	280	.0153
									14	98	.0331	20	148	.0184	26	214	.0086	44	286	.0101
									12	100	.0240	16	152	.0109	22	218	.0054	42	288	.0087
									10	102	.0171	14	154	.0077	20	220	.0041	36	294	.0053
									8	104	.0119	12	156	.0054	16	224	.0023	30	300	.0029
									6	106	.0062	10	158	.0036	14	226	.0015	22	308	.0011
									4	108	.0034	6	162	.0011	12	228	.0010	16	314	.0004
									2	110	.0014	4	164	.0006	10	230	.0007	10	320	.0001

TABLE B

Table of critical values of the Spearman's rank correlation coefficient.

n	Significance level (one-tailed test)	
	0.05	0.01
4	1.000	
5	0.900	1.000
6	0.829	0.943
7	0.714	0.893
8	0.643	0.833
9	0.600	0.783
10	0.564	0.746
12	0.506	0.712
14	0.456	0.645
16	0.425	0.601
18	0.399	0.564
20	0.377	0.534
22	0.359	0.508
24	0.343	0.485
26	0.329	0.465
28	0.317	0.448
30	0.306	0.432

TABLE C

Table of probabilities associated with S in Kendall's rank correlation coefficient, r_k. Probability that S is equal to, or greater than, certain values, for $4 \leqslant n \leqslant 10$.

S	$n = 4$ MAX. S $= 6$	$n = 5$ MAX. S $= 10$	$n = 8$ MAX. S $= 28$	$n = 9$ MAX. S $= 36$	$n = 6$ MAX. S. $= 15$	$n = 7$ MAX. S $= 21$	$n = 10$ MAX. S $= 45$	S
0	.625	.592	.548	.540	.500	.500	.500	1
2	.375	.408	.452	.460	.360	.386	.431	3
4	.167	.242	.360	.381	.235	.281	.364	5
6	.0417	.117	.274	.306	.136	.191	.300	7
8		.0417	.199	.238	.0681	.119	.242	9
10		.0083	.138	.179	.0278	.0681	.190	11
12			.0894	.130	.0083	.0345	.146	13
14			.0543	.0901	.0014	.0151	.108	15
16			.0305	.0597		.0054	.0779	17
18			.0156	.0376		.0014	.0542	19
20			.0071	.0223		.0002	.0363	21
22			.0028	.0124			.0233	23
24			.0009	.0063			.0143	25
26			.0002	.0029			.0083	27
28				.0012			.0046	29
30				.0004			.0023	31
32				.0001			.0011	33
							.0005	35
							.0002	37
							.0001	39

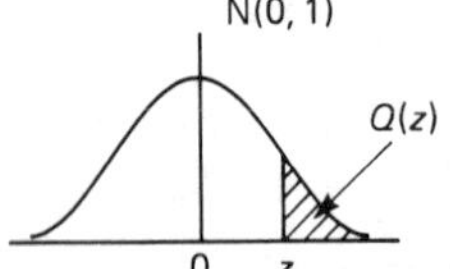

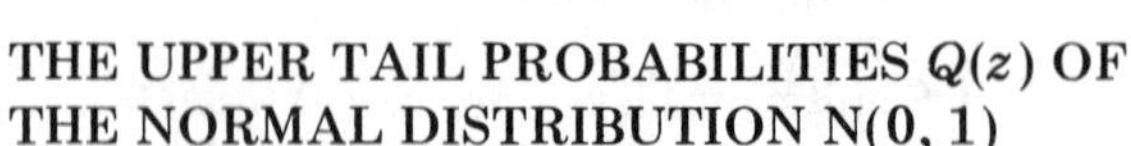

THE UPPER TAIL PROBABILITIES $Q(z)$ OF THE NORMAL DISTRIBUTION N(0, 1)

z	0	1	2	3	4	5	6	7	8	9	1	2	3	4	5 SUBTRACT	6	7	8	9
0.0	.5000	.4960	.4920	.4880	.4840	.4801	.4761	.4721	.4681	.4641	4	8	12	16	20	24	28	32	36
0.1	.4602	.4562	.4522	.4483	.4443	.4404	.4364	.4325	.4286	.4247	4	8	12	16	20	24	28	32	36
0.2	.4207	.4168	.4129	.4090	.4052	.4013	.3974	.3936	.3897	.3859	4	8	12	15	19	23	27	31	35
0.3	.3821	.3783	.3745	.3707	.3669	.3632	.3594	.3557	.3520	.3483	4	7	11	15	19	22	26	30	34
0.4	.3446	.3409	.3372	.3336	.3300	.3264	.3228	.3192	.3156	.3121	4	7	11	14	18	22	25	29	32
0.5	.3085	.3050	.3015	.2981	.2946	.2912	.2877	.2843	.2810	.2776	3	7	10	14	17	20	24	27	31
0.6	.2743	.2709	.2676	.2643	.2611	.2578	.2546	.2514	.2483	.2451	3	7	10	13	16	19	23	26	29
0.7	.2420	.2389	.2358	.2327	.2296	.2266	.2236	.2206	.2177	.2148	3	6	9	12	15	18	21	24	27
0.8	.2119	.2090	.2061	.2033	.2005	.1977	.1949	.1922	.1894	.1867	3	5	8	11	14	16	19	22	25
0.9	.1841	.1814	.1788	.1762	.1736	.1711	.1685	.1660	.1635	.1611	3	5	8	10	13	15	18	20	23
1.0	.1587	.1562	.1539	.1515	.1492	.1469	.1446	.1423	.1401	.1379	2	5	7	9	12	14	16	19	21
1.1	.1357	.1335	.1314	.1292	.1271	.1251	.1230	.1210	.1190	.1170	2	4	6	8	10	12	14	16	18
1.2	.1151	.1131	.1112	.1093	.1075	.1056	.1038	.1020	.1003	.0985	2	4	6	7	9	11	13	15	17
1.3	.0968	.0951	.0934	.0918	.0901	.0885	.0869	.0853	.0838	.0823	2	3	5	6	8	10	11	13	14
1.4	.0808	.0793	.0778	.0764	.0749	.0735	.0721	.0708	.0694	.0681	1	3	4	6	7	8	10	11	13
1.5	.0668	.0655	.0643	.0630	.0618	.0606	.0594	.0582	.0571	.0559	1	2	4	5	6	7	8	10	11
1.6	.0548	.0537	.0526	.0516	.0505	.0495	.0485	.0475	.0465	.0455	1	2	3	4	5	6	7	8	9
1.7	.0446	.0436	.0427	.0418	.0409	.0401	.0392	.0384	.0375	.0367	1	2	3	4	4	5	6	7	8
1.8	.0359	.0351	.0344	.0336	.0329	.0322	.0314	.0307	.0301	.0294	1	1	2	3	4	4	5	6	6
1.9	.0287	.0281	.0274	.0268	.0262	.0256	.0250	.0244	.0239	.0233	1	1	2	2	3	4	4	5	5
2.0	.0228	.0222	.0217	.0212	.0207	.0202	.0197	.0192	.0188	.0183	0	1	1	2	2	3	3	4	4
2.1	.0179	.0174	.0170	.0166	.0162	.0158	.0154	.0150	.0146	.0143	0	1	1	2	2	2	3	3	4
2.2	.0139	.0136	.0132	.0129	.0125	.0122	.0119	.0116	.0113	.0110	0	1	1	1	2	2	2	3	3
2.3	.0107	.0104	.0102								0	1	1	1	1	2	2	2	2
				$.0^{2}990$	$.0^{2}964$	$.0^{2}939$	$.0^{2}914$				3	5	8	10	13	15	18	20	23
								$.0^{2}889$	$.0^{2}866$	$.0^{2}842$	2	5	7	9	12	14	16	18	21
2.4	$.0^{2}820$	$.0^{2}798$	$.0^{2}776$	$.0^{2}755$	$.0^{2}734$						2	4	6	8	11	13	15	17	19
						$.0^{2}714$	$.0^{2}695$	$.0^{2}676$	$.0^{2}657$	$.0^{2}639$	2	4	6	7	9	11	13	15	17
2.5	$.0^{2}621$	$.0^{2}604$	$.0^{2}587$	$.0^{2}570$	$.0^{2}554$	$.0^{2}539$	$.0^{2}523$	$.0^{2}508$	$.0^{2}494$	$.0^{2}480$	2	3	5	6	8	9	11	12	14
2.6	$.0^{2}466$	$.0^{2}453$	$.0^{2}440$	$.0^{2}427$	$.0^{2}415$	$.0^{2}402$	$.0^{2}391$	$.0^{2}379$	$.0^{2}368$	$.0^{2}357$	1	2	3	5	6	7	8	9	10
2.7	$.0^{2}347$	$.0^{2}336$	$.0^{2}326$	$.0^{2}317$	$.0^{2}307$	$.0^{2}298$	$.0^{2}289$	$.0^{2}280$	$.0^{2}272$	$.0^{2}264$	1	2	3	4	5	6	7	8	9
2.8	$.0^{2}256$	$.0^{2}248$	$.0^{2}240$	$.0^{2}233$	$.0^{2}226$	$.0^{2}219$	$.0^{2}212$	$.0^{2}205$	$.0^{2}199$	$.0^{2}193$	1	1	2	3	4	4	5	6	6
2.9	$.0^{2}187$	$.0^{2}181$	$.0^{2}175$	$.0^{2}169$	$.0^{2}164$	$.0^{2}159$	$.0^{2}154$	$.0^{2}149$	$.0^{2}144$	$.0^{2}139$	0	1	1	2	2	3	3	4	4
3.0	$.0^{2}135$	$.0^{2}131$	$.0^{2}126$	$.0^{2}122$	$.0^{2}118$	$.0^{2}114$	$.0^{2}111$	$.0^{2}107$	$.0^{2}104$	$.0^{2}100$	0	1	1	2	2	2	3	3	4
3.1	$.0^{3}968$	$.0^{3}935$	$.0^{3}904$								3	6	9	13	16	19	22	25	28
				$.0^{3}874$	$.0^{3}845$	$.0^{3}816$	$.0^{3}789$				3	6	8	11	14	17	20	22	25
								$.0^{3}762$	$.0^{3}736$	$.0^{3}711$	2	5	7	10	12	15	17	20	22
3.2	$.0^{3}687$	$.0^{3}664$	$.0^{3}641$	$.0^{3}619$	$.0^{3}598$						2	4	7	9	11	13	15	18	20
						$.0^{3}577$	$.0^{3}557$	$.0^{3}538$	$.0^{3}519$	$.0^{3}501$	2	4	6	8	9	11	13	15	17
3.3	$.0^{3}483$	$.0^{3}466$	$.0^{3}450$	$.0^{3}434$	$.0^{3}419$						2	3	5	6	8	10	11	13	14
						$.0^{3}404$	$.0^{3}390$	$.0^{3}376$	$.0^{3}362$	$.0^{3}349$	1	3	4	5	7	8	9	10	12
3.4	$.0^{3}337$	$.0^{3}325$	$.0^{3}313$	$.0^{3}302$	$.0^{3}291$	$.0^{3}280$	$.0^{3}270$	$.0^{3}260$	$.0^{3}251$	$.0^{3}242$	1	2	3	4	5	6	7	8	9
3.5	$.0^{3}233$	$.0^{3}224$	$.0^{3}216$	$.0^{3}208$	$.0^{3}200$	$.0^{3}193$	$.0^{3}185$	$.0^{3}178$	$.0^{3}172$	$.0^{3}165$	1	1	2	3	4	4	5	6	7
3.6	$.0^{3}159$	$.0^{3}153$	$.0^{3}147$	$.0^{3}142$	$.0^{3}136$	$.0^{3}131$	$.0^{3}126$	$.0^{3}121$	$.0^{3}117$	$.0^{3}112$	0	1	1	2	2	3	3	4	5
3.7	$.0^{3}108$	$.0^{3}104$	$.0^{3}100$	$.0^{4}96$	$.0^{4}92$	$.0^{4}88$	$.0^{4}85$	$.0^{4}82$	$.0^{4}78$	$.0^{4}75$									
3.8	$.0^{4}72$	$.0^{4}69$	$.0^{4}67$	$.0^{4}64$	$.0^{4}62$	$.0^{4}59$	$.0^{4}57$	$.0^{4}54$	$.0^{4}52$	$.0^{4}50$									
3.9	$.0^{4}48$	$.0^{4}46$	$.0^{4}44$	$.0^{4}42$	$.0^{4}41$	$.0^{4}39$	$.0^{4}37$	$.0^{4}36$	$.0^{4}34$	$.0^{4}33$									

For negative values of z, use

$$Q(z) = 1 - Q(-z)$$

APPENDIX 2

USE OF THE STANDARD NORMAL TABLES USING $Q(z)$ (see p.640)

Only positive values of z are printed in the tables, so for negative values of z the symmetrical properties of the curve are used:

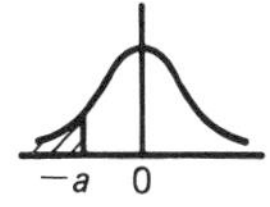

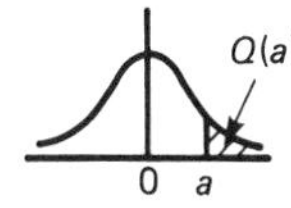

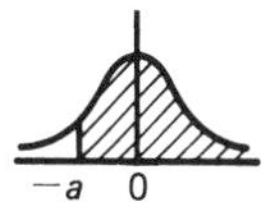

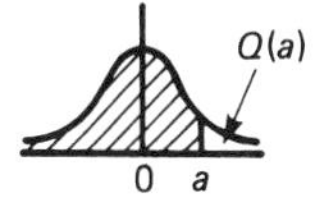

$$P(Z < -a) = P(Z > a) = Q(a)$$

$$P(Z > -a) = 1 - P(Z < -a) = 1 - Q(a)$$

NOTE: We have $Q(-a) = 1 - Q(a)$.

Example 6.2A If $Z \sim N(0, 1)$ find from tables (a) $P(Z > 1.377)$, (b) $P(Z < 1.377)$, (c) $P(Z < -1.377)$, (d) $P(Z > -1.377)$.

Solution 6.2A (a)

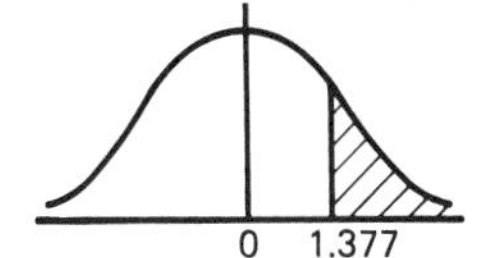

$$P(Z > 1.377) = Q(1.377) = 0.0842$$

(b)

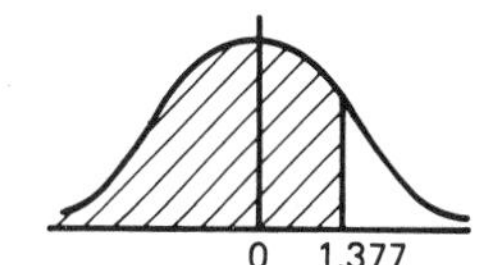

$$P(Z < 1.377) = 1 - Q(1.377) = 1 - 0.0842 = 0.9158$$

(c)

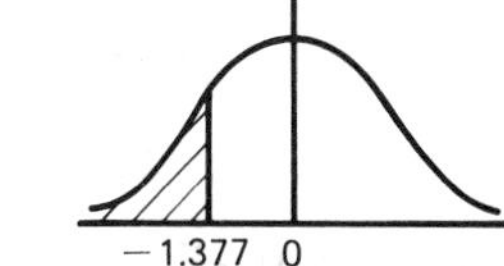

$$P(Z < -1.377) = P(Z > 1.377) = Q(1.377) = 0.0842$$

(d)

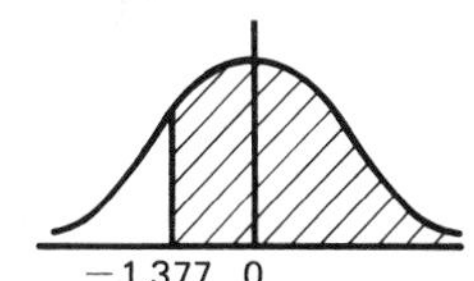

$$P(Z > -1.377) = 1 - Q(1.377) = 1 - 0.0842 = 0.9158$$

Example 6.3A If $Z \sim N(0, 1)$, find
(a) $P(0.345 < Z < 1.751)$, (b) $P(-2.696 < Z < 1.865)$,
(c) $P(-1.4 < Z < -0.6)$, (d) $P(|Z| < 1.433)$,
(e) $P(Z > 0.863$ or $Z < -1.527)$.

Solution 6.3A (a)

$$P(0.345 < Z < 1.751) = Q(0.345) - Q(1.751) = 0.3650 - 0.0400 = 0.3250$$

So $P(0.345 < Z < 1.751) = 0.3250$.

(b)
$$\begin{aligned}P(-2.696 < Z < 1.865) &= Q(-2.696) - Q(1.865)\\ &= 1 - Q(2.696) - Q(1.865)\\ &= 1 - 0.003\,50 - 0.0310\\ &= 0.9655\end{aligned}$$

<u>So $P(-2.696 < Z < 1.865) = 0.9655$.</u>

−2.696 0 1.865

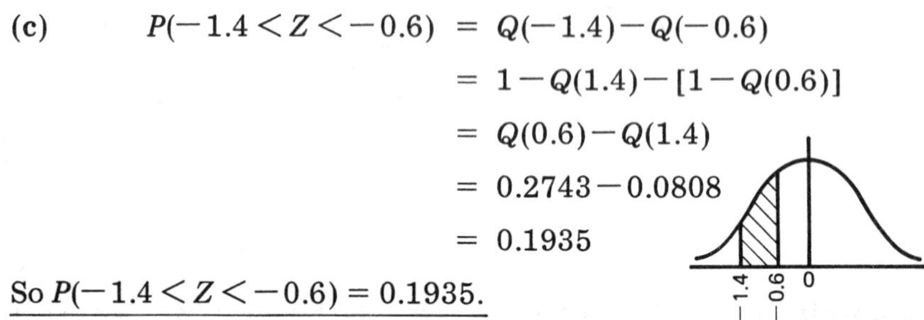

(c)
$$\begin{aligned}P(-1.4 < Z < -0.6) &= Q(-1.4) - Q(-0.6)\\ &= 1 - Q(1.4) - [1 - Q(0.6)]\\ &= Q(0.6) - Q(1.4)\\ &= 0.2743 - 0.0808\\ &= 0.1935\end{aligned}$$

<u>So $P(-1.4 < Z < -0.6) = 0.1935$.</u>

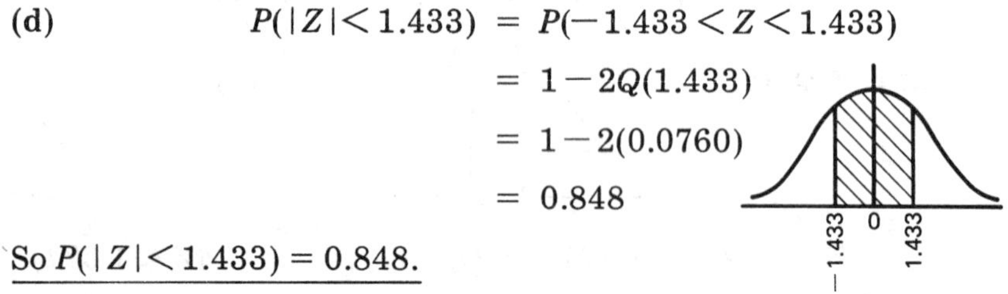

(d)
$$\begin{aligned}P(|Z| < 1.433) &= P(-1.433 < Z < 1.433)\\ &= 1 - 2Q(1.433)\\ &= 1 - 2(0.0760)\\ &= 0.848\end{aligned}$$

<u>So $P(|Z| < 1.433) = 0.848$.</u>

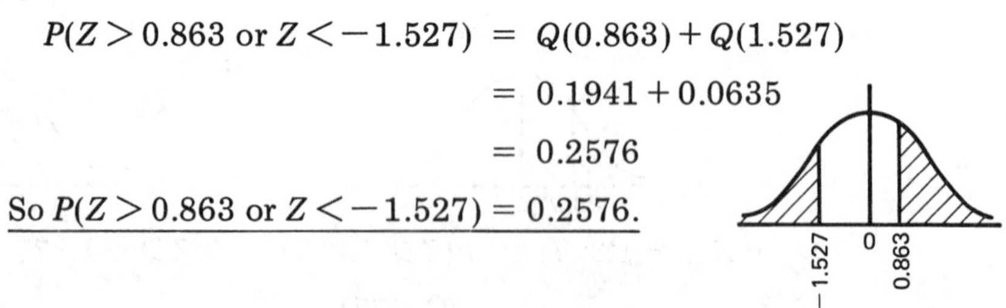

(e)
$$\begin{aligned}P(Z > 0.863 \text{ or } Z < -1.527) &= Q(0.863) + Q(1.527)\\ &= 0.1941 + 0.0635\\ &= 0.2576\end{aligned}$$

<u>So $P(Z > 0.863 \text{ or } Z < -1.527) = 0.2576$.</u>

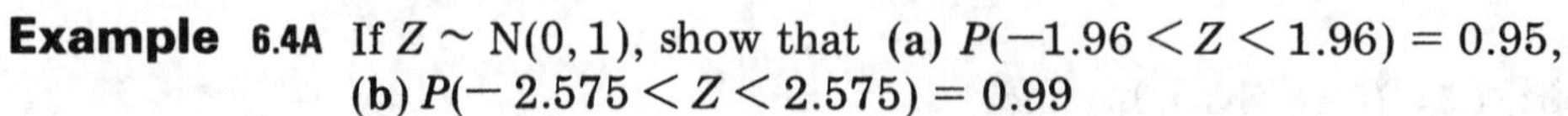

Example 6.4A If $Z \sim N(0, 1)$, show that (a) $P(-1.96 < Z < 1.96) = 0.95$, (b) $P(-2.575 < Z < 2.575) = 0.99$

Solution 6.4A (a)
$$\begin{aligned}P(-1.96 < Z < 1.96) &= 1 - 2Q(1.96)\\ &= 1 - 2(0.025)\\ &= 0.95\end{aligned}$$

95%
2.5% 2.5%
−1.96 0 1.96

<u>Therefore $P(-1.96 < Z < 1.96) = 0.95$</u>

NOTE: this is an important result:

The central 95% of the distribution lies between ±1.96.

(b) $P(-2.575 < Z < 2.575) = 1 - 2Q(2.575)$

$= 1 - 2(0.005)$

$= 1 - 0.01$

$= 0.99$

Therefore $P(-2.575 < Z < 2.575) = 0.99.$

The central 99% of the distribution lies between ± 2.575.

Exercise 6a (page 337)

Example 6.5A If $Z \sim N(0, 1)$, find the value of a if (a) $P(Z > a) = 0.3802$, (b) $P(Z > a) = 0.7818$, (c) $P(Z < a) = 0.0793$, (d) $P(Z < a) = 0.9693$, (e) $P(|Z| < a) = 0.9$.

Solution 6.5A (a) $P(Z > a) = 0.3802$.

i.e. $Q(a) = 0.3802$

so from tables

$a = 0.305$

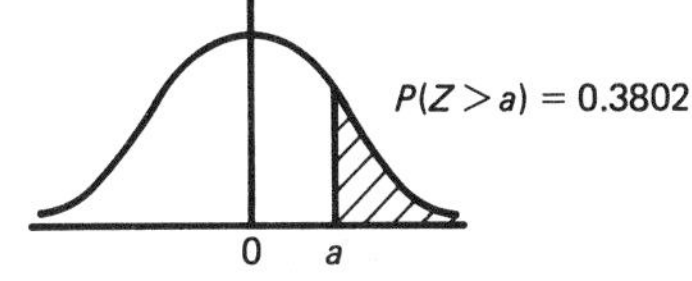

(b) $P(Z > a) = 0.7818$.

Now, since the probability is greater than 0.5, a must be negative.

Now $Q(a) = 1 - Q(-a) = 0.7818$

$Q(-a) = 1 - 0.7818$

$= 0.2182$

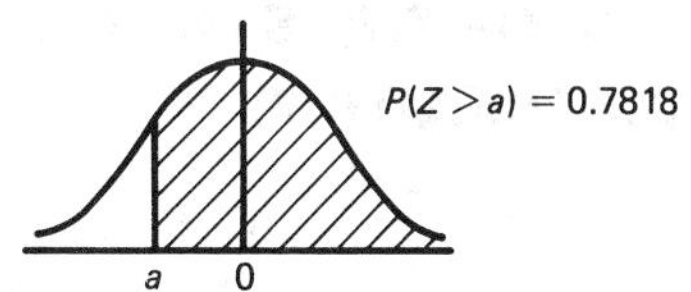

But from tables

$$Q(0.778) = 0.2182$$

therefore

$$-a = 0.778$$

so $a = -0.778$

(c) $P(Z < a) = 0.0793$.

From the diagram it is obvious that a must be negative

By symmetry

$$Q(-a) = 0.0793$$

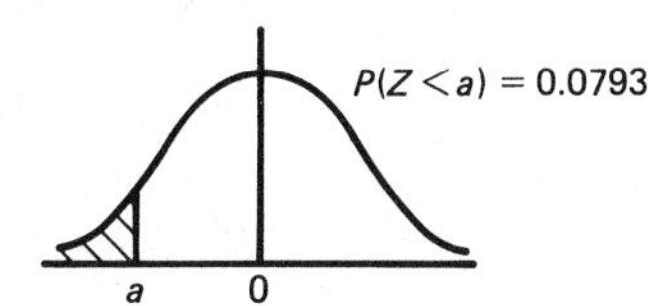

From tables

$$-a = 1.41$$

therefore $a = -1.41$

(d) $P(Z < a) = 0.9693$.

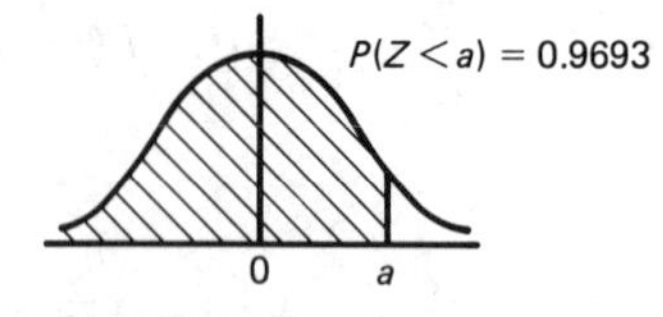

Now $\quad 1 - Q(a) = 0.9693$

so $\quad Q(a) = 0.0307$

from tables $\quad \underline{a = 1.87}$

(e) $P(|Z| < a) = 0.9$,

i.e. $\quad P(-a < Z < a) = 0.9$

From symmetry

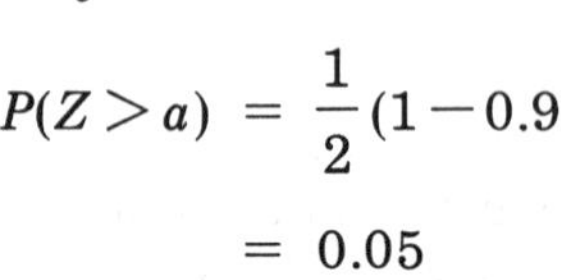

$$P(Z > a) = \frac{1}{2}(1 - 0.9)$$

$$= 0.05$$

i.e. $\quad Q(a) = 0.05$

From tables

$$Q(1.645) = 0.05$$

so $\quad \underline{a = 1.645}$

Exercise 6b (page 338)

USE OF THE STANDARD NORMAL TABLES FOR ANY NORMAL DISTRIBUTION

We now show how the tables for the standard normal distribution can be adapted for use with any random variable X where $X \sim \mathrm{N}(\mu, \sigma^2)$.

Example 6.6A The r.v. $X \sim \mathrm{N}(300, 25)$. Find (a) $P(X > 305)$, (b) $P(X < 291)$, (c) $P(X < 312)$, (d) $P(X > 286)$.

Solution 6.6A (a) $P(X > 305)$.

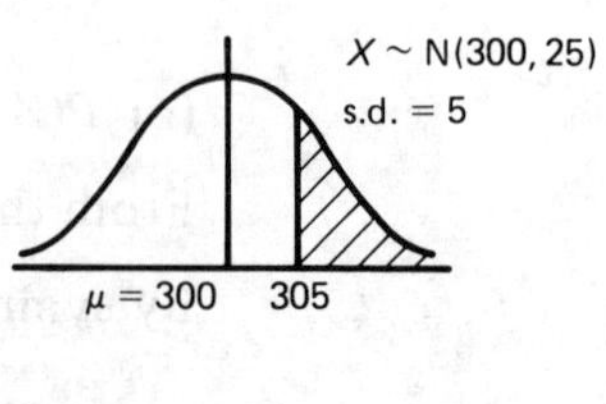

First we have to standardise the random variable X by subtracting the mean, 300, and dividing by the standard deviation, (s.d.), 5, so that $Z = \dfrac{X - 300}{5}$.

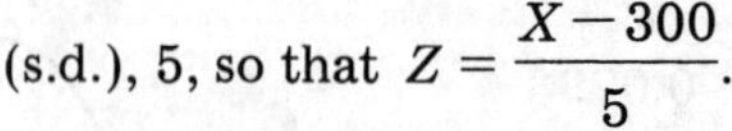

We also use the following properties of inequalities:

$$X > 305 \Rightarrow X - 300 > 305 - 300 \Rightarrow \frac{X - 300}{5} > \frac{305 - 300}{5}$$

So $P(X > 305) = P\left(\frac{X-300}{5} > \frac{305-300}{5}\right)$

$= P(Z > 1)$

$= Q(1)$

$= 0.1587$

<u>Therefore $P(X > 305) = 0.1587$.</u>

Standard normal curve

$Z \sim N(0, 1)$

s.d. = 1

0 1

NOTE: if the two curves had been drawn to scale, the curve for X would have been much more spread out and not as steep as the curve for Z. However, for convenience of drawing, we use the same sketch.

Often, again for convenience, we draw one sketch and write the values of the standardised variable underneath the x values. We use the abbreviation S.V. for 'standardised variable'.

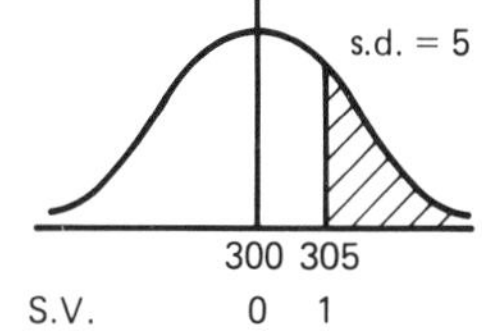

(b) $P(X < 291) = P\left(\frac{X-300}{5} < \frac{291-300}{5}\right)$

$= P(Z < -1.8)$

$= Q(1.8)$

$= 0.0359$

<u>Therefore $P(X < 291) = 0.0359$.</u>

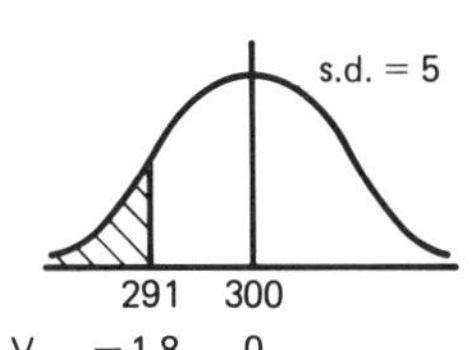

(c) $P(X < 312) = P\left(\frac{X-300}{5} < \frac{312-300}{5}\right)$

$= P(Z < 2.4)$

$= 1 - Q(2.4)$

$= 1 - 0.0082$

$= 0.9918$

<u>Therefore $P(X < 312) = 0.9918$.</u>

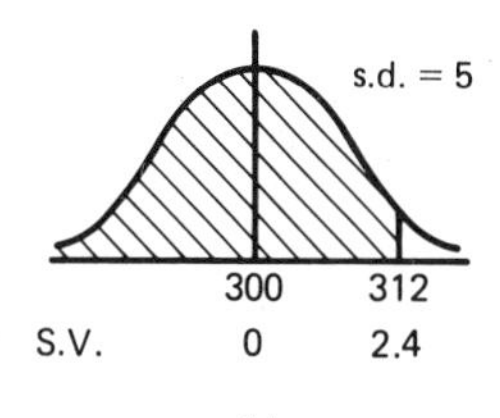

(d) $P(X > 286) = P\left(\frac{X-300}{5} > \frac{286-300}{5}\right)$

$= P(Z > -2.8)$

$= 1 - Q(2.8)$

$= 1 - 0.002\,56$

$= 0.997\,44$

<u>Therefore $P(X > 286) = 0.997\,44$.</u>

s.d. = 5

286 300

S.V. −2.8 0

Example 6.7A The r.v. X is such that $X \sim N(50, 8)$. Find **(a)** $P(48 < X < 54)$, **(b)** $P(52 < X < 55)$, **(c)** $P(46 < X < 49)$, **(d)** $P(|X-50| < \sqrt{8})$.

Solution 6.7A Standardise X so that $Z = \dfrac{X-50}{\sqrt{8}}$.

(a)
$$\begin{aligned} P(48 < X < 54) &= P\left(\frac{48-50}{\sqrt{8}} < \frac{X-50}{\sqrt{8}} < \frac{54-50}{\sqrt{8}}\right) \\ &= P(-0.707 < Z < 1.414) \\ &= 1-[Q(0.707)+Q(1.414)] \\ &= 1-(0.2399+0.0787) \\ &= 1-0.3186 \\ &= 0.6814 \end{aligned}$$

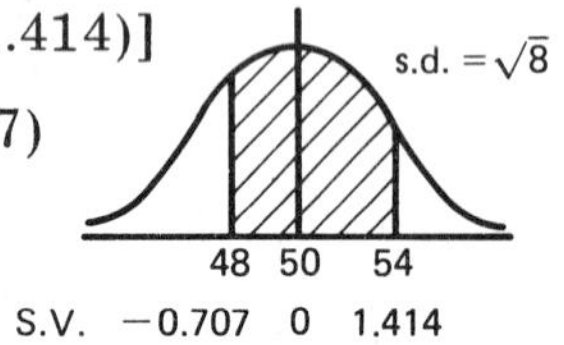

s.d. $= \sqrt{8}$
48 50 54
S.V. −0.707 0 1.414

Therefore $P(48 < X < 54) = 0.6814.$

(b)
$$\begin{aligned} P(52 < X < 55) &= P\left(\frac{52-50}{\sqrt{8}} < \frac{X-50}{\sqrt{8}} < \frac{55-50}{\sqrt{8}}\right) \\ &= P(0.707 < Z < 1.768) \\ &= Q(0.707)-Q(1.768) \\ &= 0.2399-0.0385 \\ &= 0.2014 \end{aligned}$$

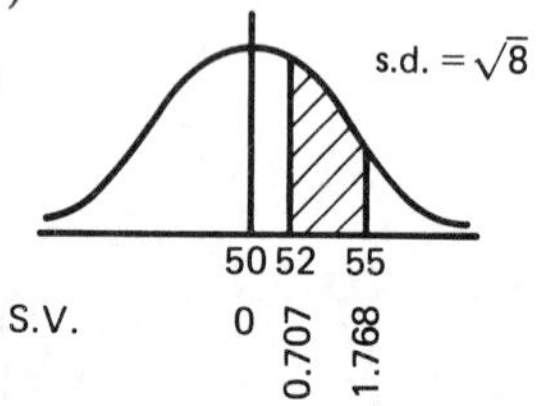

s.d. $= \sqrt{8}$
50 52 55
S.V. 0 0.707 1.768

Therefore $P(52 < X < 55) = 0.2014.$

(c)
$$\begin{aligned} P(46 < X < 49) &= P\left(\frac{46-50}{\sqrt{8}} < \frac{X-50}{\sqrt{8}} < \frac{49-50}{\sqrt{8}}\right) \\ &= P(-1.414 < Z < -0.354) \\ &= Q(0.354)-Q(1.414) \\ &= 0.3617-0.0787 \\ &= 0.283 \end{aligned}$$

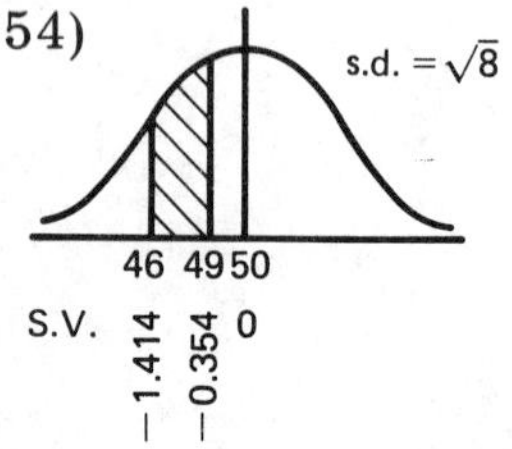

s.d. $= \sqrt{8}$
46 49 50
S.V. −1.414 −0.354 0

Therefore $P(46 < X < 49) = 0.283.$

(d)
$$\begin{aligned} P(|X-50| < \sqrt{8}) &= P(-\sqrt{8} < X-50 < \sqrt{8}) \\ &= P\left(-1 < \frac{X-50}{\sqrt{8}} < 1\right) \\ &= P(-1 < Z < 1) \\ &= 1-2Q(1) \\ &= 1-2(0.1587) \\ &= 0.6826 \end{aligned}$$

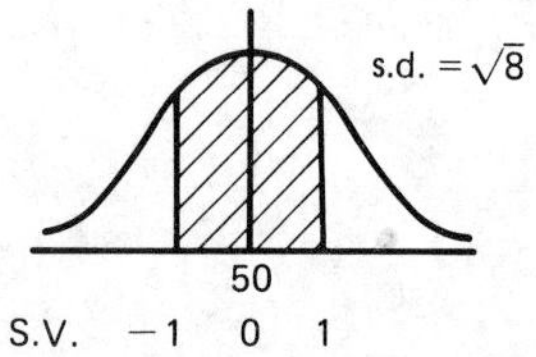

s.d. $= \sqrt{8}$
50
S.V. −1 0 1

Therefore $P(|X-50| < \sqrt{8}) = 0.6826.$

Example 6.8A The time taken by a milkman to deliver milk to the High Street is normally distributed with mean 12 minutes and standard deviation 2 minutes. He delivers milk every day. Estimate the number of days during the year when he takes **(a)** longer than 17 minutes, **(b)** less than 10 minutes, **(c)** between 9 and 13 minutes.

Solution 6.8A Let X be the r.v. 'the time taken to deliver the milk to the High Street'. Then $X \sim N(12, 2^2)$.

We standardise X so that $Z = \dfrac{X-12}{2}$.

$$\text{(a)}\quad P(X>17) = P\left(\frac{X-12}{2} > \frac{17-12}{2}\right) = P(Z>2.5) = Q(2.5) = 0.006\,21$$

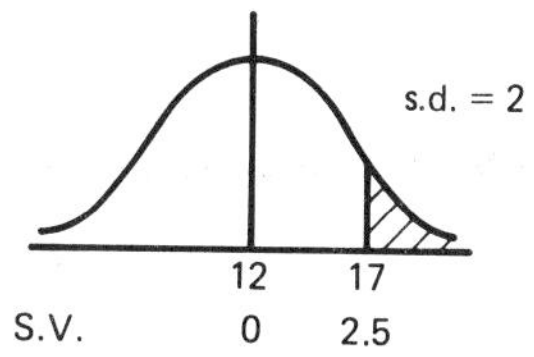

s.d. = 2
12 17
S.V. 0 2.5

The number of days when he takes longer than 17 minutes

$$= 365(0.006\,21) = 2.27 \approx 2$$

<u>Therefore on approximately 2 days in the year he takes longer than 17 minutes.</u>

$$\text{(b)}\quad P(X<10) = P\left(\frac{X-12}{2} < \frac{10-12}{2}\right) = P(Z<-1) = Q(1) = 0.1587$$

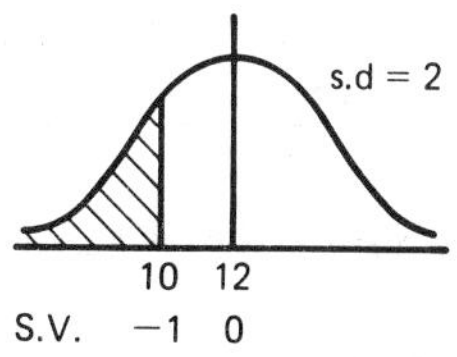

s.d = 2
10 12
S.V. −1 0

The number of days when he takes less than 10 minutes

$$= 365(0.1587) = 57.9 \approx 58$$

<u>Therefore on approximately 58 days in the year he takes less than 10 minutes.</u>

(c) $P(9<X<13) = P\left(\frac{9-12}{2}<\frac{X-12}{2}<\frac{13-12}{2}\right)$

$= P(-1.5<Z<0.5)$

$= 1-Q(1.5)-Q(0.5)$

$= 1-0.068-0.3085$

$= 0.6247$

s.d. = 2

9 12 13

S.V. −1.5 0 0.5

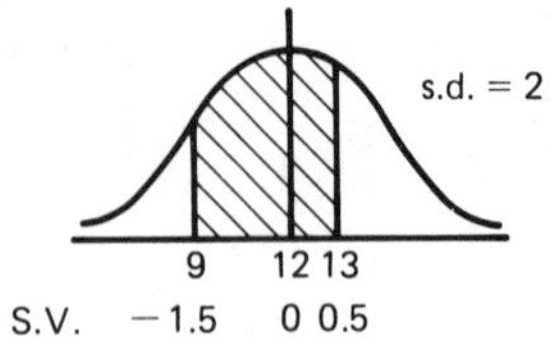

The number of days when he takes between 9 and 13 minutes

$= 365(0.6247)$

$= 228$ days

Therefore on 288 days he takes between 9 and 13 minutes.

Exercise 6c (page 342)

De-standardising

Sometimes it is necessary to find a value X which corresponds to the standardised value Z. We use $Z=\frac{X-\mu}{\sigma}$, so that $X=\mu+\sigma Z$.

Example 6.9A If $X \sim N(50, 6.8)$, find the value of X which corresponds to a standardised value of (a) -1.2, (b) 0.6.

Solution 6.9A Now $X=\mu+\sigma Z$, where $\mu=50$ and $\sigma=\sqrt{6.8}$, so that $X=50+\sqrt{6.8}Z$.

(a) when $z=-1.2$,

$x = 50+\sqrt{6.8}(-1.2)$

$= 46.87$ (2 d.p.)

(b) when $z=0.6$,

$x = 50+\sqrt{6.8}(0.6)$

$= 51.56$ (2 d.p.)

Exercise 6d (page 343)

Example 6.10A If $X \sim N(100, 36)$ and $P(X>a)=0.1093$, find the value of a.

Solution 6.10A As $P(X>a)$ is less than 0.5, a must be greater than the mean, 100.

Now $P(X>a) = 0.1093$

so $P\left(\frac{X-100}{6}>\frac{a-100}{6}\right) = 0.1093$

i.e. $P\left(Z>\frac{a-100}{6}\right) = 0.1093$

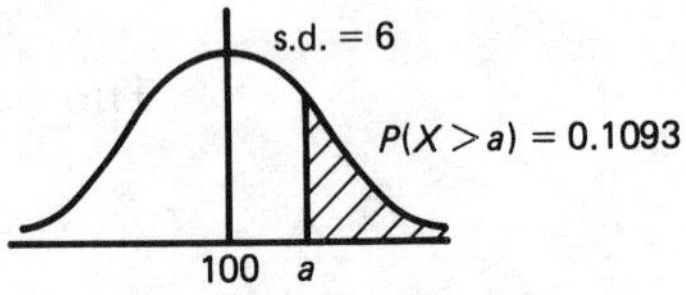

s.d. = 6

$P(X>a) = 0.1093$

100 a

We have $Q\left(\frac{a-100}{6}\right) = 0.1093$

But from tables,

$$Q(1.23) = 0.1093$$

Therefore $$\frac{a-100}{6} = 1.23$$

$$a = 100+6(1.23) = 107.38$$

Therefore, if $P(X > a) = 0.1093$, then $a = 107.38$.

Example 6.11A If $X \sim N(24, 9)$ and $P(X > a) = 0.974$, find the value of a.

Solution 6.11A As $P(X > a)$ is greater than 0.5, a must be less than the mean 24.

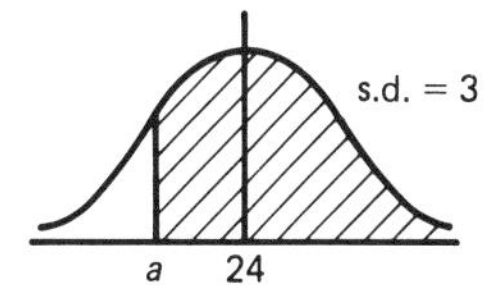

Now $$P(X > a) = 0.974$$

so $$P\left(\frac{X-24}{3} > \frac{a-24}{3}\right) = 0.974$$

i.e. $$P\left(Z > \frac{a-24}{3}\right) = 0.974$$

Now $\dfrac{a-24}{3}$ must be negative and

$$Q\left(\frac{a-24}{3}\right) = 1-Q\left[-\left(\frac{a-24}{3}\right)\right]$$

$$= 1-Q\left(\frac{24-a}{3}\right)$$

so $$1-Q\left(\frac{24-a}{3}\right) = 0.974$$

$$Q\left(\frac{24-a}{3}\right) = 0.026$$

But, from the tables, $Q(1.943) = 0.026$

Therefore $$\left(\frac{24-a}{3}\right) = 1.943$$

$$a = 24-(3)(1.943)$$

$$= 18.171$$

Therefore, if $P(X > a) = 0.974$, then $a = 18.171$.

Example 6.12A If $X \sim N(70, 25)$, find the value of a such that $P(|X-70| < a) = 0.8$. Hence find the limits within which the central 80% of the distribution lies.

Solution 6.12A $P(|X-70|<a) = 0.8$

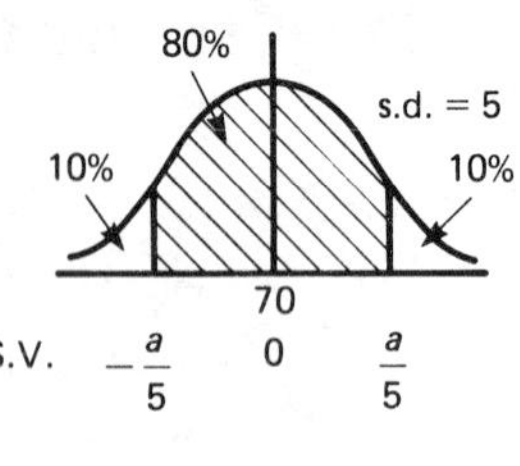

Therefore

$$P(-a<X-70<a) = 0.8$$

$$P\left(-\frac{a}{5}<\frac{X-70}{5}<\frac{a}{5}\right) = 0.8$$

$$P\left(-\frac{a}{5}<Z<\frac{a}{5}\right) = 0.8$$

Now, by symmetry

$$P\left(Z>\frac{a}{5}\right) = \tfrac{1}{2}(1-0.8)$$

$$= 0.1$$

and from tables, $Q(1.282) = 0.1$

Therefore $\dfrac{a}{5} = 1.282$

$$a = 6.41$$

So $P(-6.41<X-70<6.41) = 0.8$

or $P(63.59<X<76.41) = 0.8$

The central 80% of the distribution lies between 63.59 and 76.41.

Exercise 6e (page 345)

PROBLEMS THAT INVOLVE FINDING THE VALUE OF μ OR σ OR BOTH

Example 6.13A The lengths of certain items follow a normal distribution with mean μ cm and standard deviation 6 cm. It is known that 4.78% of the items have a length greater than 82 cm. Find the value of the mean μ.

Solution 6.13A Let X be the r.v. 'the length of an item in cm'.

$X \sim N(\mu, 36)$ and $P(X>82) = 0.0478$.

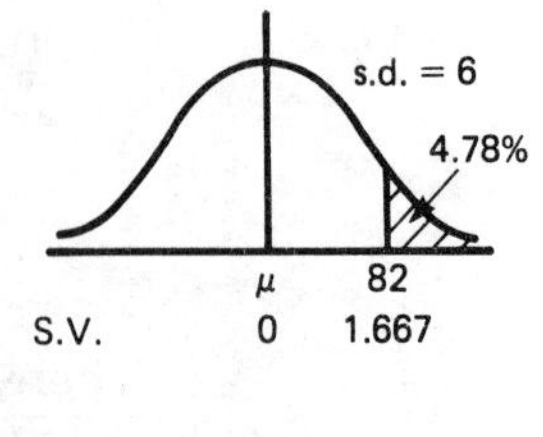

Now $P(X>82) = P\left(\frac{X-\mu}{6}>\frac{82-\mu}{6}\right)$

$$= P\left(Z>\frac{82-\mu}{6}\right)$$

$$= Q\left(\frac{82-\mu}{6}\right)$$

so $Q\left(\dfrac{82-\mu}{6}\right) = 0.0478$

But from tables

$$Q(1.667) = 0.0478$$

so
$$\frac{82-\mu}{6} = 1.667$$

$$82-\mu = 10.002$$

$$\mu = 72 \quad (2\,\text{S.F.})$$

The mean of the distribution is 72 cm.

Example 6.14A $X \sim N(100, \sigma^2)$ and $P(X < 106) = 0.8849$. Find the standard deviation, σ.

Solution 6.14A

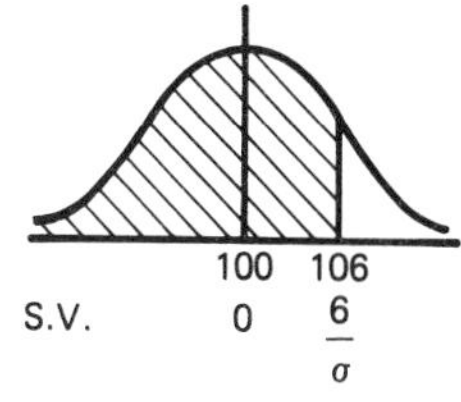

$$P(X < 106) = 0.8849$$

$$P\left(\frac{X-100}{\sigma} < \frac{106-100}{\sigma}\right) = 0.8849$$

$$P\left(Z < \frac{6}{\sigma}\right) = 0.8849$$

$$P\left(Z > \frac{6}{\sigma}\right) = 1-0.8849$$

$$= 0.1151$$

$$Q\left(\frac{6}{\sigma}\right) = 0.1151$$

But from tables

$$Q(1.2) = 0.1151$$

Therefore
$$\frac{6}{\sigma} = 1.2$$

$$\sigma = \frac{6}{1.2}$$

$$= 5$$

The standard deviation of the distribution is 5.

Example 6.15A The masses of articles produced in a particular workshop are normally distributed with mean μ and standard deviation σ. 5% of the articles have a mass greater than 85 g and 10% have a mass less than 25 g. Find the values of μ and σ, and find the ranges symmetrical about the mean, within which 75% of the masses lie.

Solution 6.15A Let X be the r.v. 'the mass, in g, of an article'. Then $X \sim N(\mu, \sigma^2)$ where μ and σ are unknown.

Now $P(X > 85) = 0.05$

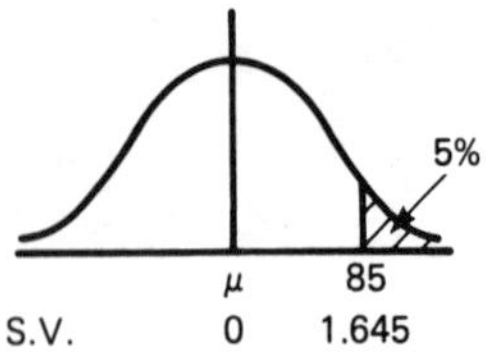

i.e. $P\left(\frac{X-\mu}{\sigma} > \frac{85-\mu}{\sigma}\right) = 0.05$

$$P\left(Z > \frac{85-\mu}{\sigma}\right) = 0.05$$

$$Q\left(\frac{85-\mu}{\sigma}\right) = 0.05$$

5%
μ 85
S.V. 0 1.645

But from tables

$$Q(1.645) = 0.05$$

Therefore $\frac{85-\mu}{\sigma} = 1.645$

$$85-\mu = 1.645\sigma \qquad \text{(i)}$$

Also $P(X < 25) = 0.10$

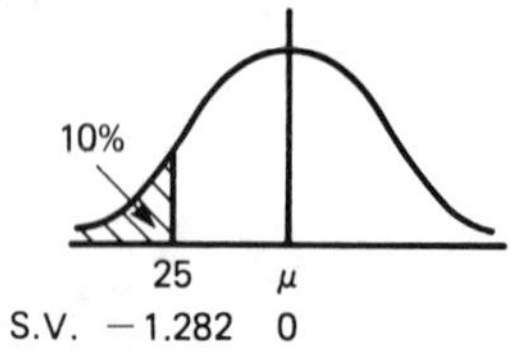

$$P\left(\frac{X-\mu}{\sigma} < \frac{25-\mu}{\sigma}\right) = 0.10$$

$$P\left(Z < \frac{25-\mu}{\sigma}\right) = 0.10$$

10%
25 μ
S.V. −1.282 0

But $\frac{25-\mu}{\sigma}$ is negative, and by symmetry,

$$Q\left[-\left(\frac{25-\mu}{\sigma}\right)\right] = 0.10$$

From tables $Q(1.282) = 0.10$

Therefore $-\left(\frac{25-\mu}{\sigma}\right) = 1.282$

i.e. $\mu - 25 = 1.282\sigma \qquad \text{(ii)}$

Adding (i) and (ii) we have

$$60 = 2.927\sigma$$

$$\sigma = 20.5 \quad (3\text{ S.F.})$$

Substituting for σ in (ii)

$$\mu = 25 + (1.282)(20.5)$$

$$= 51.3 \quad (3\text{ S.F.})$$

Therefore the distribution has mean mass 51.3 g and standard deviation 20.5 g.

Now consider values a and b such that

$$P(a \leqslant X \leqslant b) = 0.75$$

and a and b are symmetrical about the mean.

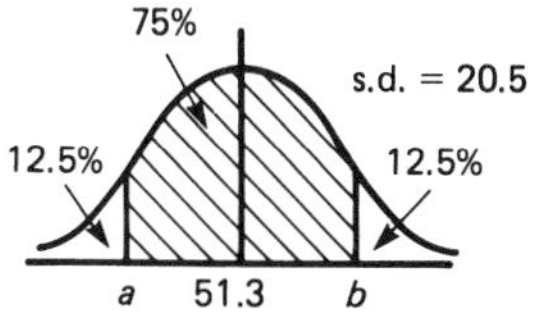

75%
s.d. = 20.5
12.5%
12.5%
a 51.3 b

Now $$P(X > b) = 0.125$$

$$P\left(\frac{X-51.3}{20.5} > \frac{b-51.3}{20.5}\right) = 0.125$$

so $$Q\left(\frac{b-51.3}{20.5}\right) = 0.125$$

but from tables $Q(1.15) = 0.125$

therefore $$\frac{b-51.3}{20.5} = 1.15$$

$$b = 51.3+(20.5)(1.15) = 74.9 \quad \text{(3 S.F.)}$$

From symmetry $$a = 51.3-(20.5)(1.15) = 27.7 \quad \text{(3 S.F.)}$$

Therefore, the central 75% of the distribution lies between the limits 27.7 g and 74.9 g.

Exercise 6f (page 349)

ANSWERS

CHAPTER 1

Exercise 1a (page 10)

1. 10, 9, 11, 13, 14, 19, 13, 9, 3
2. 4, 6, 7, 13, 10, 5, 5

For Questions 3–9, heights of rectangles are in the proportions indicated:

3. 6, 6, 7, 4, 3
4. 3, 20, 30, 42, 12, 18, 3
5. 2, 6, 12, 14, 8, 4, 1
6. 62, 40, 88, 100, 112, 60, 12, 3
7. 11, 18, 22, 24, 28, 24, 16
8. 2, 6, 7, 5, 2
9. 12, 30, 65, 48, 20, 10

Exercise 1b (page 14)

1. $66^\circ, 60^\circ, 45^\circ, 75^\circ, 30^\circ, 84^\circ$
2. $27.4^\circ, 56.7^\circ, 115.1^\circ, 160.8^\circ$
3. (i) 660 km^2 (ii) 4°, (iii) 2700 km^2
4. Radii in the ratio 7.7 : 6.7
5. $66^\circ, 156^\circ, 24^\circ, 42^\circ, 72^\circ$; 5.5 cm, 6 cm; 50°
6. Radii in the ratio 20.3 : 22.5 : 26.8
7. $67.5^\circ, 157.5^\circ, 135^\circ$; £6; $72^\circ, 115.2^\circ, 172.8^\circ$

Exercise 1c (page 16)

Ratio of standard frequencies

3. 6 : 14 : 22 : 10 : 6 : 3
4. 87.5 : 470 : 535 : 280 : 59

Exercise 1d (page 21)

1. (b) (i) Approx 74%
   (ii) approx 50 marks
2. (a) Cumulative frequencies 2, 4, 7, 13, 25, 41, 47, 50
   (b) 24 (c) 26 (d) 23
   (e) 2, 2, 3, 6, 12, 16, 6, 3
3. (a) Cumulative frequencies 3, 5, 12, 30, 48, 51, 52
   (b) 21 (c) 14 (d) 62 kg
4. (b) 84% (c) 6.5
   (d) 1, 1, 3, 5, 9, 19, 5, 3, 2, 1

Exercise 1e (page 24)

1. (a) 9 (b) 207 (c) 1896
   (d) 0.55
2. 4
3. (a) 7 (b) 14 (c) 17
   (d) 5.4

Exercise 1f (page 28)

1. Approx 33 mins
2. 61.5 g
3. 51 km/h
4. 0.56 cm
5. 59 marks
6. 687.25 hrs

Exercise 1g (page 32)

1. (i) (a) 61 (b) 52 (c) 73
   (ii) (a) 8 (b) 7 (c) 10
2. (a) 6 (b) 3 marks
3. (i) 46 (ii) 29, 9%
4. 6.6, 4, 9.25
5. (i) 153 mm (ii) 15 mm
   (iii) 11%
6. (i) 96.5 mins (ii) 5 mins
   (iii) 61 approx
7. 437.5, 413, 455
8. (a) 46 (b) 24
   (c) 30
9. (i) 57 (ii) 71.5 (iii) 32%
10. (a) 135 cm (c) 176, 162, 170, 14
11. (a) 179.8 (b) 4.4 cm
    (c) 285 approx

Exercise 1h (page 37)

1. (a) 27 (b) 412, 435
   (c) 4, 6 (d) none
2. 4
3. (a) 103 (b) 7 (c) 32; 101
4. Approx 45
5. (a) $659.5 \leqslant x < 669.5$
   (b) 4, 9, 21, 45, 60, 70, 77, 80
   (c) (i) 697.5 (ii) 723.5
   (d) point of inflexion

Exercise 1i (page 39)

1. (i) 9.7 (ii) 154.8 (iii) 51.375
   (iv) $1775\frac{5}{7}$ (v) 0.908 (3 S.F.)
2. 21
3. 19
4. 8
5. 7

Exercise 1j (page 42)

1. (i) 4 (ii) 29.54 (iii) 122.82
   (iv) 18.625 (v) 109.4 (1 d.p.)
2. 12; 4, 3
3. (b) 15, 7; 1, 2
4. (a) $7\frac{5}{28}$ (b) bimodal 8, 10
   (c) median 8
5. 49.3
6. 58.95
7. 6, 2; 15
8. 45 (2 S.F.)

Exercise 1k (page 45)

1. 35
2. $232\frac{5}{6}$
3. 75.6 (3 S.F.)
4. (a) 33.3 (3 S.F.) (b) 28.9
   (c) 7.1
5. $46\frac{2}{3}$
6. 31.7 (3 S.F.)

Answers are given to 3 S.F. where appropriate in Exercises 1l to 1q

Exercise 1l (page 50)

1. (a) 5, 2 (b) 8.5, 1.80
   (c) 18.8, 6.46 (d) $10\frac{5}{6}$, 4.10
   (e) 3.42, 1.91 (f) 205, 3.16
2. 3.74
3. 29, 5.9
4. 5.10
5. 5
6. 6, 4
7. 10.5, 5.77
8. $\frac{1}{2}(n+1), \sqrt{\dfrac{n^2-1}{12}}$
9. (a) 121, 6.19 (b) 14, 1703.8
   (c) 1716, 3.59 (d) 1026, 58770
10. $5, \sqrt{7.5}; 5, 11$

Exercise 1m (page 52)

Answers as for Exercise 1l

Exercise 1n (page 53)

1. Each multiplied by 3
3. (a) 6, 2.14 (b) 516, 2.14
   (c) 78, 27.8
4. (a) 4, 4; 7, 4; 40, 400; 43, 400
   (b) $d_i = 10a_i + 3$
5. 17, 8

Exercise 1o (page 55)

1. (i) $a = \frac{3}{4}$, $b = 22$ (ii) 70 (iii) 76
2. (i) 38, 8.99 (3 S.F.) (ii) 34, 77
3. $a = 0.8$, $b = -5$; 6.25
4. (i) $\mu + c, \sigma$ (ii) $k\mu, k\sigma$;
   $a = \frac{5}{6}$, $b = 22$
5. (a) $f(x) = 2x + 3$ (b) $5, 12\frac{1}{3}$
   (c) $13, 49\frac{1}{3}$ (d) 26 (e) $64\frac{4}{7}$
6. (a) 2 (b) 200 (c) 2.02
   (d) −4, −1, 2, 5, 8, 11, 14

Exercise 1p (page 59)

1. (a) 7.6, 3.14 (b) 30.4, 6.76
   (c) 13.65, 3.02
2. 15.6, 7.66
3. 12.1, 1.83
4. 25.9, 1.99

Exercise 1q (page 62)

1. 69.3, 1.7
2. 115.8 (4 S.F.), 7.58
3. (a) $\frac{11}{3}$, 1.23 (b) 7.85, 3.07
   (c) 31, 4.36 (d) 34.1, 13.4
   (e) 16.04, 7.01 (f) 10, 1.44
4. 28.15, 3.84
5. 159
6. (a) 294.55, 28.15, 3.84
   (b) 10, 7450, 4 (c) 500, 5450, 450
   (d) 159, 5.3, 2.47 (e) 12, 300, 5

Exercise 1r (page 65)

1. (a) 313.76, 5.19 (b) 42.6, 13.2
   (c) 1954, 348.4 (d) 17.1, 8.18
   (e) 1.02, 0.507 (f) 321.9, 68.1
2. 61.5, 18.6
3. 277.2, 95.7; 254.4, 92.8
4. 71.2, 3.82

Exercise 1s (page 72)

1. (a) 180.5 (b) 175.5;
   second part (a) 187 (b) 189.5
2. (i) 2, 0, 1, 0, 4, 4, 8, 5, 8, 14, 12, 5, 0, 0, 1
   (ii) 34.4 (3 S.F.), 7.88 (3 S.F.)
   (iii) 74%
3. 2 min 38 secs, 1 min 54 secs,
   2 min 18 secs, 1 min 25.5 secs,
   2 min 59 secs
4. 86.6, 44.1, $N = 188$
5. (i) 4, 8 (ii) 6; mean = 7, n = ±6
6. 16, 6 (i) 5.86 (3 S.F.) (ii) 15, 7
7. 3−4 (i) 3.25 (ii) 2.2
   (iii) approx 41%
8. (b) Mid-point is representative of
   interval (i) 11 (ii) 101
9. 6 hr 14 min, 13 min; 6 hr 18 min, 16 min
10. 63; intervals 30-, 40-, . . . frequencies
    1, 3, 7, 14, 22, 32, 35, 32, 25, 16, 8, 3, 2;
    63.05, 11.3 (3 S.F.): normal (see
    chapter 6)
11. (i) 100.7, 14.5 (3 S.F.) (ii) approx 73
12. (a) 5, 6, 4.07 (3 S.F.) (b) $\dfrac{n}{m+n}$

13. (a) (i) $\mu+k, \sigma$ (ii) $p\mu, p\sigma$; $3\mu+5, 3\sigma$; (b) $a = 1.6, b = 10$
14. 0, 1; better in algebra
15. 44.5, 51.75, 64, 40.5; $a = 0.89$ (2 S.F.), $b = 1$; yes
16. (i) 17 − 22; 19.2 (ii) 20.2 (iii) 6 (iv) 20.6 (3 S.F.), 8.19 (3 S.F.)
17. 34.9, 32.7, 186.5, 13.7; 61%
18. (a) 51.5 (b) 52 (c) 50 or 54 (d) 51 (e) 6 (f) 57.5 (g) 109
19. 1, 3, 8.1, 0.03; 814, 300; $a = -7.2$, $b = 2$
20. (a) 0.785 (3 S.F.) (b) 4.44 (3 S.F.)
21. (b) (i) 7 (ii) 1; $x = 5, y = 9$
22. 6.49, 1.71 (3 S.F.), 7
23. (i) 17, 4 (approx) (ii) 17.85, 5.57 (3 S.F.)
24. (a) 20.1, 5.7 (b) 0.46 (c) 18.4
25. 35 years 1 month, 11 years 3 months (a) Approx 33 years 10 months (b) Approx 17 years 10 months (c) Approx 65.6%
26. Taking mark intervals $0 \leqslant \text{mark} < 10$, etc. (c) 40.4, 15.4; $a = 24$ (2 S.F.), $b = 0.65$ (2 S.F.)
27. £195.45, £14.12
28. (b) 5.21 (3 S.F.), 2.70 (3 S.F.)
29. 36.7 mm, 15.5 mm, $\frac{6}{25}$
30. 11.87, 0.80
31. (c) 5, 1 (d) 4.9, 1.15; 4.86, 2.84

CHAPTER 2

Exercise 2a (page 82)

1. (a) $\frac{1}{3}$ (b) 1 (c) $\frac{2}{3}$
2. (a) $\frac{1}{52}$ (b) $\frac{7}{26}$ (c) $\frac{10}{13}$
3. (a) $\frac{1}{4}$ (b) $\frac{1}{4}$ (c) $\frac{1}{10}$ (d) $\frac{9}{19}$
4. (a) $\frac{3}{8}$ (b) $\frac{5}{8}$ (c) $\frac{3}{4}$ (d) $\frac{3}{4}$ (e) 1
5. (a) $\frac{1}{18}$ (b) $\frac{1}{6}$ (c) $\frac{1}{6}$ (d) $\frac{1}{3}$ (e) $\frac{3}{4}$
6. $\frac{4}{15}$
7. (a) $\frac{3}{10}$ (b) $\frac{3}{4}$
8. (a) $\frac{1}{2}$ (b) $\frac{1}{2}$ (c) $\frac{1}{4}$ (d) $\frac{3}{8}$ (e) $\frac{1}{2}$
9. (a) $\frac{1}{12}$ (b) 0 (c) $\frac{1}{4}$
10. (a) $\frac{1}{12}$ (b) $\frac{1}{12}$ (c) $\frac{1}{4}$ (d) $\frac{1}{8}$
11. (a) $\frac{1}{36}$ (b) $\frac{1}{12}$ (c) 0 (d) $\frac{1}{3}$; $t = 6$ or 12

Exercise 2b (page 88)

1. (a) $\frac{1}{2}$ (b) $\frac{1}{2}$ (c) $\frac{5}{6}$
2. $\frac{11}{30}$
3. (a) $\frac{4}{17}$ (b) $\frac{4}{51}$ (c) $\frac{5}{17}$ (d) $\frac{5}{17}$
4. $\frac{1}{2}$
5. $\frac{3}{5}$
6. 0.4
7. 0.7
8. (a) $\frac{7}{36}$ (b) $\frac{1}{6}$ (c) $\frac{5}{18}$ (d) $\frac{1}{12}$
9. $\frac{3}{4}$
10. (a) $\frac{11}{36}$ (b) $\frac{11}{36}$ (c) $\frac{5}{9}$

Exercise 2c (page 94)

1. $\frac{1}{2}$
2. (a) $\frac{3}{10}$ (b) $\frac{1}{10}$
3. (a) $\frac{3}{10}$ (b) $\frac{1}{16}$ (c) $\frac{1}{6}$
4. (a) $\frac{1}{3}$ (b) $\frac{2}{15}$ (c) $\frac{8}{15}$
5. (a) $\frac{1}{17}$ (b) $\frac{13}{204}$ (c) $\frac{13}{51}$
6. (a) $\frac{4}{15}$ (b) $\frac{8}{15}$ (c) $\frac{2}{3}$
7. (a) $\frac{1}{6}$ (b) $\frac{1}{5}$ (c) $\frac{1}{3}$
8. (a) $\frac{1}{3}$ (b) 0
9. (a) $\frac{1}{8}$ (b) $\frac{1}{2}$
10. (a) $\frac{7}{10}$ (b) No
11. Yes
12. $\frac{1}{2}$
15. (a) $\frac{3}{16}$ (b) $\frac{1}{4}$ (c) $\frac{1}{16}$ (d) $\frac{3}{8}$

Exercise 2d (page 98)

1. $\frac{1}{9}$
2. (a) $\frac{1}{2704}$ (b) $\frac{1}{16}$ (c) $\frac{1}{2}$ (d) $\frac{25}{169}$
3. $\frac{1}{4}$
4. (a) $\frac{1}{3}$ (b) $\frac{1}{2}$ (c) $\frac{1}{6}$
5. (a) 0.0025 (b) 0.095
6. (a) $\frac{1}{10}$ (b) $\frac{3}{10}$ (c) $\frac{9}{20}$
7. (a) 0.15 (b) 0.65; No
8. (a) $\frac{1}{4}$ (b) $\frac{1}{6}$
9. (a) $\frac{1}{6}$ (b) Not independent
10. (a) $\frac{1}{4}$ (b) $\frac{7}{12}$
11. $\frac{7}{16}$

Exercise 2e (page 99)

1. 0.4
2. (a) 0.24 (b) 0.42
3. (a) $\frac{5}{21}$ (b) $\frac{2}{3}$ (c) $\frac{5}{12}$
4. $\frac{9}{14}$

5. (a) $\frac{3}{4}$ (b) $\frac{1}{4}$
6. 0.008%; 0.625

Exercise 2f (page 101)

1. $\frac{31}{90}$
2. (a) $\frac{1}{28}$ (b) 0
   (c) $\frac{1}{14}$; A and B, A and C, $\frac{3}{14}$
3. $\frac{3}{8}$
4. (a) 0.02 (b) 0.45
5. (a) 0.5 (b) 0.35 (c) 0.375
   (d) 0.4
6. (i) 0.02 (ii) 0.78 (iii) 0.76
   (iv) $\frac{1}{30}$
7. (a) $\frac{1}{25}$ (b) $\frac{106}{125}$ (c) $\frac{14}{19}$
8. (i) 0.4 (ii) 0.5 (iii) 0.52
9. $\frac{1}{25}$, $\frac{16}{25}$; $\frac{1}{16}$, $\frac{3}{8}$; (a) 0.04
   (b) 0.6225 (c) 0.1825

Exercise 2g (page 107)

1. (a) $\frac{1}{64}$ (b) $\frac{169}{10200}$
2. $\frac{25}{72}$
3. $\frac{5}{16}$
4. (a) $\frac{2}{11}$ (b) $\frac{1}{2}$ (c) $\frac{43}{66}$
   (d) $\frac{2}{55}$ (e) $\frac{12}{55}$
5. (b) (i) $\frac{1}{6}$ (ii) 0
   (c) (i) H and R (ii) H and S
6. (a) (i) $\frac{4}{55}$ (ii) $\frac{14}{55}$ (b) 0.3
7. (i) $\frac{1}{216}$ (ii) $\frac{1}{36}$ (iii) $\frac{125}{216}$ (iv) $\frac{1}{36}$
8. (b) (i) $\frac{55}{108}$ (ii) $\frac{1}{12}$
9. $\frac{3}{64}$

Exercise 2h (page 115)

1. (b) $\frac{8}{25}$ (d) $\frac{28}{75}$
2. (a) (i) $\frac{7}{92}$ (ii) $\frac{34}{69}$ (b) $\frac{49}{253}$
3. (a) $\frac{1}{27}$ (b) $\frac{20}{27}$
4. (a) (i) $\frac{8}{27}$ (ii) $\frac{4}{9}$ (iii) $\frac{7}{27}$
   (b) (i) $\frac{5}{21}$ (ii) $\frac{15}{28}$ (iii) $\frac{19}{84}$
5. (a) $\frac{12}{49}$ (b) $\frac{20}{49}$
6. (a) $\frac{23}{63}$ (b) $\frac{65}{98}$
7. (a) $\frac{1}{2}$ (b) $\frac{3}{10}$ (c) $\frac{3}{5}$
8. (a) $\frac{5}{14}$ (b) $\frac{17}{42}$
9. (a) 0.34 (b) 0.063
   (c) 0.19 (d) 0.97; 3 white

Exercise 2i (page 119)

1. (a) $\frac{7}{18}$ (b) (i) $\frac{5}{8}$ (ii) $\frac{8}{25}$
2. (a) $\frac{5}{12}$ (b) $\frac{3}{5}$
3. (a) 0.66 (b) $\frac{9}{17}$
4. (a) 0.024 (b) 0.452
   (c) 0.496 (3 S.F.)
5. $\frac{36}{95}$, $\frac{5}{38}$, $\frac{7}{190}$, $\frac{43}{95}$, 26-64 age group
6. $\frac{23}{45}$, $\frac{18}{23}$
7. (a) $\frac{5}{8}$ (b) $\frac{133}{200}$ (c) $\frac{11}{25}$
   (d) $\frac{42}{47}$
8. (a) $\frac{3}{10}$ (b) $\frac{4}{15}$ (c) $\frac{3}{10}$
   (d) $\frac{1}{3}$ (e) $\frac{5}{16}$
9. (b) (i) $\frac{1}{10}$ (ii) $\frac{1}{5}$ (iii) $\frac{3}{5}$
   (d) (i) $\frac{1}{3}$ (ii) $\frac{4}{7}$
10. (ii) $\frac{6}{13}$
11. Machine 1

Exercise 2j (page 123)

1. (a) 0.763 (3 S.F.) (b) 14
2. (a) 5 (b) 6
3. $\frac{1}{2}$, 6
4. 0.999 (3 S.F.)
5. $\frac{5}{11}$
6. 1:8
7. $\frac{1}{2}$; (a) $\frac{1}{6}$ (b) $\frac{25}{216}$
   (c) $\frac{625}{7776}$; $\frac{6}{11}$

Exercise 2k (page 130)

1. 9!; $\frac{1}{72}$
2. (a) 6! (b) $\frac{1}{3}$
3. (a) 4!9! (b) $\frac{54}{55}$
4. $\frac{9}{11}$
5. (a) 8! (b) $\frac{1}{28}$
6. $\frac{1}{126}$
7. (a) $\frac{12!}{(2!)^4}$ (b) $\frac{1}{66}$

Exercise 2l (page 138)

1. $\frac{28}{153}$
2. $\frac{49}{143}$
3. $\frac{60}{143}$
4. (a) 210 (b) $\frac{2}{15}$ (c) $\frac{1}{30}$
5. (i) $\frac{1}{14}$ (ii) $\frac{3}{7}$ (iii) $\frac{1}{30}$
6. (i) 65 268 (ii) 4263
7. 510
8. $\frac{37}{42}$
9. 4608
10. (a) 1260 (b) 2520
11. (a) 420 (b) Boys 252, Girls 462
    (c) 120 (d) $\frac{44}{133}$
12. (a) 2.5×10^{-7} (b) 3 193 344
13. (a) $\frac{2}{5}$ (b) $\frac{1}{5}$
14. 130
15. (a) 360 (b) 6 (d) 12
    (e) 1170
16. (a) 64 (b) 18 (c) $\frac{21}{32}$
17. (a) 9! (b) $\frac{7}{36}$ (c) 1260
    (d) $\frac{5}{9}$
18. (a) 75 (c) $\frac{181}{456}$ (d) (i) 6! (ii) 72

Exercise 2m (page 158)

1. (a) (i) $\frac{5}{18}$ (ii) $\frac{1}{3}$ (b) $\frac{12}{35}$ (c) 294
2. (a) $\beta+\frac{1}{5}(\alpha-\beta)(\alpha+4\beta)$
3. 0.005 99, 0.987 (3 S.F.)
4. (a) $\frac{1}{2}$ (b) $\frac{1}{3}$ (c) $\frac{1}{6}$
   (d) $\frac{3}{4}$ (e) $\frac{1}{15}$
5. (a) 6720 (b) 8106; $\frac{9}{80}$
6. $\frac{1}{5}, \frac{1}{5}, \frac{3}{7}, \frac{1}{4}$; (a) No (b) No
7. (a) $\frac{64}{729}$ (b) $\frac{28}{243}$ (c) $\frac{28}{243}$
   (d) $\frac{604}{729}$
8. (a) (i) $\frac{1}{4}$ (ii) $\frac{1}{4}$ (iii) $\frac{1}{17}$ (iv) $\frac{15}{34}$ (b) $\frac{13}{30}$
   (c) (i) $\frac{95}{253}$ (ii) $\frac{285}{506}$
9. 0.59; (i) 0.352 (ii) 0.4576
   (iii) 0.480 64
10. (a) $\frac{1}{22}$ (b) $\frac{41}{55}$ (c) $\frac{3}{11}$
    (d) $\frac{3}{44}$
11. (a) $\frac{1}{4}$ (b) $\frac{1}{4}$ (c) $\frac{1}{16}$
    (d) $\frac{1}{4}$ (e) $\frac{3}{4}$
12. (a) $\frac{1}{36}$ (b) $\frac{5}{12}$; $\frac{73}{648}, \frac{25}{81}$
13. 0.624
14. $\frac{5}{24}; \frac{1}{24}; \frac{5}{6}; \frac{1}{12}$; (a) No (b) No
15. (a) (i) $\frac{25}{49}$ (ii) $\frac{24}{49}$ (iii) $\frac{39}{49}$
    (b) 0.013 824 (c) $\frac{1}{2}, \frac{49}{288}$
16. (a) $\frac{1}{27}$ (b) $\frac{2}{9}$ (c) $\frac{4}{9}$
    (d) $\frac{8}{27}$ (e) $\frac{43}{144}$ (f) $\frac{65}{72}$
    (g) $\frac{64}{195}$
17. (i) 0.36 (ii) 0.6875
18. (a) $\frac{2}{5}$ (b) $\frac{2}{15}$
19. (a) (i) $\frac{1}{1050}$ (ii) $\frac{4}{35}$ (b) (i) $\frac{1}{25}$
    (ii) $\frac{12}{25}$; $\frac{24}{625}$
20. (a) 0.12 (b) 0.184 (c) 0.32
    (d) 0.25
21. (a) $\frac{1}{8}$ (b) $\frac{5}{32}$; $P(D)=0.0325$,
    $P(C\cap D)=0.025$, $P(C|D)=\frac{10}{13}$
22. (a) 0.875, $\frac{19}{30}$ (b) $\frac{22}{47}$
23. (a) $\frac{1}{2}$ (b) $\frac{1}{4}$ (c) $\frac{3}{10}$
    (d) $\frac{19}{24}$; No, no
24. (a) $\frac{25}{216}:\frac{27}{216}$ (b) 0.5177, 0.4914
    (c) 0.6651, 0.6186
25. (a) $\frac{5}{33}$ (b) $\frac{5}{33}$ (c) $\frac{1}{792}$
    (d) $\frac{41}{132}$
26. (a) $\frac{2}{261}$ (b) $\frac{16}{609}, \frac{308}{435}, \frac{204}{1015}$
27. (a) $\frac{2}{15}$ (b) $\frac{1}{2}$
28. (a) $\frac{6}{323}$ (b) $\frac{135}{323}$ (c) $\frac{1}{5}$
    (d) $\frac{1}{5}$ (i) Yes, no (ii) No, yes
29. 70 (a) 55 (b) 30 (c) 65
    (d) $\frac{2}{7}$ (e) $\frac{1}{7}$ (f) $\frac{1}{7}$
30. (a) $\frac{6}{13}$ (b) 0.0481 (3 S.F.)
31. (a) (i) $\frac{4}{17}$ (ii) $\frac{1}{17}$ (iii) $\frac{5}{17}$
    (b) (i) $\frac{1}{4}$ (ii) $\frac{1}{13}$ (iii) $\frac{4}{13}$ (c) $\frac{17}{108}$
32. (a) $\frac{4}{13}$ (b) $\frac{2}{39}$ (c) $\frac{7}{78}$
    (d) $\frac{1}{6}$ (e) $\frac{4}{7}$
33. (a) $\frac{3}{64}$ (b) $\frac{243}{256}$ (c) $\frac{18}{25}$
    (d) $\frac{2}{25}$ (e) $\frac{58}{35}$
34. $(p+p^2+p^3)/(1+p^2+p^3)$
35. $1-4p+6p^2$
36. $\frac{1}{7}, \frac{324}{343}$, 0.617
37. (a) $1/3^7$ (b) $16/3^3$ (c) $593/3^7$
    (d) $784/3^8$
38. (a) (i) 120 (ii) 12 600
    (b) $\frac{4}{35}, \frac{18}{35}, \frac{12}{35}, \frac{1}{35}; \frac{11}{24}$
39. 0.336, 0.452, 0.188, 0.024; 0.9
40. (a) $\frac{59}{663}$ (b) 0.0546
41. (a) ${}^{13}C_3\times{}^{13}C_4\times{}^{26}C_6/{}^{52}C_{13}$
    (b) 0.355 (3 d.p.) (c) 0.920 (3 d.p.)
42. (a) $\frac{135}{323}$ (b) $\frac{70}{323}$ (c) $\frac{3}{8}$
    (d) $\frac{27}{128}$
43. (a) (i) $\frac{1}{2}$ (ii) $\frac{1}{2}$ (iii) $\frac{6}{7}$ (iv) $\frac{1}{8}$, No, no
    (b) (i) $\frac{5}{21}$ (ii) $\frac{3}{14}$
44. (a) 48 (c) 63, 32
45. (a) 0.45, not
    (b) (i) 0.33 (ii) $\frac{7}{11}$
46. (a) (i) $\frac{9}{22}$ (ii) $\frac{6}{11}$ (iii) $\frac{2}{11}$
    (iv) $\frac{4}{7}$
    (b) (i) 0.0303 (ii) 0.450 (iii) 0.0348
    (c) (i) 0.36 (ii) 0.848
47. (a) 0.7, 0.68 (b) 0.28
    (c) 0.656 25
48. (a) (i) $\frac{1}{6}$ (ii) $\frac{1}{12}$ (iii) $\frac{2}{3}$
    (b) $\frac{7}{12}$
49. (a) 0.88, 0.05
    (b) (i) 0.346 (ii) 0.476
50. $\frac{15}{44}$, no, 0.1
51. (a) 0.096 (ii) 0.156; $\frac{5}{13}$
52. (a) (ii) 0.43, 0.67 (iii) $\frac{15}{26}$
    (b) (i) $\frac{36}{415}$ (ii) $\frac{11}{13}$
53. (a) $\frac{180}{1001}$ (b) $(1-p-p'+2pp')^n$
    (c) $\frac{5}{9}$
54. (a) 0.071, 0.929 (b) 0.600, 0.572

CHAPTER 3

Exercise 3a (page 170)

1. (a)

x	0	1	2
$P(X=x)$	$\frac{1}{4}$	$\frac{1}{2}$	$\frac{1}{4}$

(b) $P(X=x)=\dfrac{x-1}{36}$, $x=2,\ldots,7$

$P(X=x)=\dfrac{13-x}{36}$, $x=8,\ldots,12$

(c)

x	0	1	2
$P(X=x)$	$\frac{9}{16}$	$\frac{3}{8}$	$\frac{1}{16}$

(d) $P(X=x)=0.1,\ x=0, 1, \ldots, 9$

(e)

x	0	1	2	3
$P(X=x)$	$\frac{1}{8}$	$\frac{3}{8}$	$\frac{3}{8}$	$\frac{1}{8}$

(f) $P(X=0)=\frac{1}{6},\ P(X=x)=\dfrac{6-x}{18},$

$x=1, 2, \ldots, 5$

2. $\frac{1}{39}$
3. $\frac{1}{6}$
4. (a)

x	0	1	2	3
$P(X=x)$	0.216	0.432	0.288	0.064

(b) 0.648

5.

x	0	1	2	3
$P(X=x)$	$\frac{1}{27}$	$\frac{2}{9}$	$\frac{4}{9}$	$\frac{8}{27}$

6. $\frac{1}{5}$

Exercise 3b (page 177)

1. $2\frac{1}{4}$
2. 7
3. (a) 0.3 (b) 2.9
4. 1
5. 0.5
6. $\frac{12}{11}$
7. 0.75p
8.

x	10	20
$P(X=x)$	0.4	0.6

9. (a) 0.3 (b) 0.2
10.

x	4	6	8	9	11	14
$P(X=x)$	0.16	0.32	0.16	0.16	0.16	0.04

; loss of £1.20

11. (i) £$\frac{3}{8}(7+x)$ (a) 5 (b) Loss of £3.75
12. $\frac{12}{11}$
13. $\frac{11}{3}$
14. (a) 24

(c)

x	0	1	2	3	4
$P(X=x)$	$\frac{3}{8}$	$\frac{1}{3}$	$\frac{1}{4}$	0	$\frac{1}{24}$

(d) 1

15. (a) 0.2 (b) 2.08
16. (a) $\frac{1}{21}$ (b) $\frac{2}{7}$ (c) $\frac{4}{9}$ (d) $\frac{5}{42}$ (e) 1 (f) $\frac{4}{3}$, 0
17. 2
18. $\frac{2}{3}, \frac{2}{3}, \dfrac{b}{6}$ or $\dfrac{2b}{b+w}$

Exercise 3c (page 182)

1. (a) 2.3 (b) 5.9
3. (a) 3.5 (b) $15\frac{1}{6}$ (c) 14.5
4. (a) $\frac{24}{11}$ (b) $\frac{61}{11}$ (c) $\frac{50}{11}$ (d) $16\frac{9}{11}$
5. (a) 3.5 (b) 14 (c) 5.5 (d) 84 (e) 95
6. (a) 2 (b) 3 or -3

Exercise 3d (page 188)

1. (a) 6 (b) 22
2. (a) 5 (b) 2.5 (c) 10 (d) 10
3. (a) 4.2 (b) $7\frac{1}{3}$ (c) 3.67 (3 S.F.)
4. (a) $\frac{1}{10}$ (b) $3\frac{1}{2}$ (c) $15\frac{7}{30}$ (d) $2\frac{59}{60}$ (e) $47\frac{11}{15}$
5. (a) $1\frac{2}{3}$ (b) $3\frac{1}{3}$ (c) $\frac{5}{9}$
6. (a) 21.34
7.

x	0	1	2
$P(X=x)$	$\frac{1}{7}$	$\frac{4}{7}$	$\frac{2}{7}$

(a) $\frac{8}{7}$ (b) $\frac{12}{7}$ (c) $\frac{20}{49}$ (d) $\frac{180}{49}$

8. (a) -0.7 (b) 3.5 (c) 3.01
9. $P(X=x)=\dfrac{10-x}{45},\ x=1, 2, \ldots, 9$

$3\frac{2}{3}$, 2.21 (2 d.p.), 1;

$P(X=x)=\left(\dfrac{4}{5}\right)^{x-1}\left(\dfrac{1}{5}\right),\ x=1, 2, \ldots$

10. (a) $\frac{1}{12}$ (b) 0 (c) 6 (d) 2.45 (2 d.p.)
11. (a) 0.04 (b) 5 (c) 4 (d) 7 (e) 16

Exercise 3e (page 191)

1. (a)

x	0	1	2
$F(x)$	$\frac{25}{36}$	$\frac{35}{36}$	1

(b)

x	1	2	3	4	5	6
$F(x)$	$\frac{11}{36}$	$\frac{5}{9}$	$\frac{3}{4}$	$\frac{8}{9}$	$\frac{35}{36}$	1

(c)

x	0	1	2	3
$F(x)$	$\frac{1}{8}$	$\frac{1}{2}$	$\frac{7}{8}$	1

2.

y	0.1	0.2	0.3	0.4	0.5
$P(Y \leqslant y)$	0.05	0.3	0.6	0.75	1

3. (a) 0.41 (b) 0.87 (c) 0.46 (d) 0.13 (e) 2.58
4.

x	3	4	5	6	7
$P(X=x)$	0.01	0.22	0.41	0.22	0.14

; 0.9724

5. (a) $\frac{4}{9}$ (b) $\frac{1}{3}$

(c) $P(X=x)=\dfrac{2x-1}{9},\ x=1, 2, 3$

(d) $\frac{17}{9}$

6. (a) $\frac{1}{3}$ (b) $\frac{2}{3}$

(c) $P(X=x)=\frac{1}{3},\ x=1, 2, 3$

(d) 0.816 (3 S.F.)

7. (b)

x	1	2	3	4
$P(X=x)$	$\frac{1}{4}$	$\frac{1}{2}$	$\frac{15}{64}$	$\frac{1}{64}$

(c) $2\frac{1}{64}$, 0.547 (3 S.F.) (d) $\frac{1}{4}$

Exercise 3f (page 203)

1. (a) 1.3, 1, 1.01, 0.8

(b)

$x+y$	0	1	2	3	4	5
$P(X+Y=x+y)$	0.12	0.14	0.32	0.2	0.18	0.04

(e)

$x-y$	−2	−1	0	1	2	3
$P(X-Y=x-y)$	0.12	0.14	0.32	0.2	0.18	0.04

2. (a) 26 (b) 15 (c) 17 (d) 59 (e) 59
3. (a) $\frac{1}{2}$ (b) $\frac{5}{12}$ (c) $2\frac{2}{3}$
4. (a) 0 or 12 or −12 (b) 294
5. (a) 1 (b) −1 (c) 34 (d) 14 (e) 14 (f) 30
6. (a) (i) 7 (ii) $\frac{35}{6}$ (b) (i) 0 (ii) $\frac{35}{6}$
7. (a) 1.2, 0.36 (b) 0.09 (c) 2.4, 0.72 (d) 0.3 (e) 2.4, 1.44
8. (a) 2.6, 0.24 (b) 5.2, 0.48 (c) 7.8, 0.72
9. $29\frac{1}{6}$
10. (a) 0.1 (b) 3 (c) 1 (d) 0.2 (e) 12 (f) 3

Exercise 3g (page 205)

1. 6.25
2. $2\frac{1}{2}$, $\frac{10}{21}$
3. (a) $\frac{1}{36}$ (b) $\frac{5}{36}$ (c) $\frac{11}{36}$; $-\frac{1}{36}$, 7
4.

x	6	7	8	9	10
$P(X=x)$	$\frac{1}{12}$	$\frac{1}{3}$	$\frac{1}{3}$	$\frac{2}{9}$	$\frac{1}{36}$

, 0.975 (3 S.F.), 0.640 (3 S.F.)

5. (a) (i) $\frac{1}{3}$ (ii) $\frac{1}{12}$ (b) £$\frac{1}{12}$ (c) $1\frac{1}{6}$ (d) $\left(\frac{1}{6}\right)^{r-1}\left(\frac{1}{3}\right)$ (e) $\frac{2}{5}$
6. $\frac{35}{18}$; (a) $\frac{1}{2}$ (b) $\frac{1}{12}$ (c) $\left(\frac{1}{6}\right)^{r-1}\left(\frac{1}{2}\right)$; $\frac{3}{5}$, £1.50
7. (a) $\frac{16}{81}, \frac{32}{81}, \frac{24}{81}, \frac{8}{81}, \frac{1}{81}$ (b) −50 p
8. (a) 1, $\frac{4}{5}$ (b) $\frac{3}{5}, \frac{6}{25}$ (c) 11.2, 7.28

t	0	1	2	3	4
$P(T=t)$	$\frac{2}{15}$	$\frac{9}{25}$	$\frac{8}{25}$	$\frac{11}{75}$	$\frac{1}{25}$

9. $\frac{1}{15}, \frac{2}{3}, \frac{34}{45}, \frac{1}{75}, \frac{4}{3}, \frac{68}{45}$
10. $P(X=x)=\frac{1}{6}$, $x=1,2,3,4,5$; $P(X=6)=0$, $P(X=x)=\frac{1}{36}$, $x=7,8,\ldots,12$; $4\frac{1}{12}$, $\frac{6}{17}$
11.

x	2	3	4	5	6	7	8	9
$P(X=x)$	$\frac{1}{36}$	$\frac{1}{12}$	$\frac{1}{6}$	$\frac{1}{6}$	$\frac{1}{6}$	$\frac{1}{6}$	$\frac{5}{36}$	$\frac{1}{12}$

; $5\frac{5}{6}$, 0.001 37 (3 S.F.)

12.

x	−3	0	3	6
$P(X=x)$	p^3	$3p^2(1-p)$	$3p(1-p)^2$	$(1-p)^3$

; 450p, 30p

13.

y	0	1	2	3	4
$P(Y=y)$	0.09	0.24	0.34	0.24	0.09

z	0	1	2	3	4
$P(Z=z)$	0.447	0.232	0.222	0.072	0.027

, 1; 1.2

14. (a) $\frac{1}{13}$ (b) 2, $\frac{12}{13}$
15. (a) 0.01 (b) 3.54, 0.4684 (c) 14.7, 11.71
16. (a) $\frac{4}{5}$ (b) −0.24 p (c) 3.34 p^2 (2 d.p.)
17. 0.1, 1
18. (a) (i) $\frac{1}{64}$ (ii) $\frac{9}{64}$ (iii) $\frac{27}{64}$ (iv) $\frac{27}{64}$
19. 2, 1

y	−4	−2	0	2	4
$P(Y=y)$	$\frac{1}{16}$	$\frac{1}{4}$	$\frac{3}{8}$	$\frac{1}{4}$	$\frac{1}{16}$

; 4, 3

20. (i) $\frac{1}{8}, \frac{5}{24}$ (ii) 2.78 (3 S.F.) (iii) 0.260 (3 S.F.)
21. $\frac{1}{2}(n+1)$, $\frac{1}{12}(n^2-1)$, $\frac{1}{2}$; (a) $\frac{2}{105}$ (b) 16
22. (a)

x	1	2	4	5
$P(X=x)$	$\frac{1}{12}$	$\frac{5}{12}$	$\frac{1}{3}$	$\frac{1}{6}$

y	2	3	4	5	6	7
$P(Y=y)$	$\frac{1}{144}$	$\frac{5}{72}$	$\frac{25}{144}$	$\frac{1}{18}$	$\frac{11}{36}$	$\frac{5}{36}$

y	8	9	10
$P(Y=y)$	$\frac{1}{9}$	$\frac{1}{9}$	$\frac{1}{36}$

; $6\frac{1}{6}$, $\frac{251}{72}$

23. $\frac{1}{28}$, 3.5, 1.25, 12, 20

CHAPTER 4

Answers are given to 3 S.F. where applicable.

Exercise 4a (page 214)

1. (a) 0.0823 (b) 0.680
2. (a) 0.209 (b) 0.0168 (c) 0.008 52
3. (a) 0.531 (b) 0.000 055 (c) 0.984
4. 0.002 00
5. 0.891
6. 0.5
7. (a) 0.0808 (b) 0.428
8. 0.0819
9. (a) 0.329 (b) 0.461
10. 0.0962
12. 4
13. 68
14. 5
15. 9
16. (a) 0.0563 (b) 0.000 416
17. (a) 0.267 (b) 0.000 144

Exercise 4b (page 217)

1. 2.5, 1.5
2. 0.844
3. 8, 1.30
4. (a) 0.2 (b) 0.005 51
5. (a) 0.25 (b) 2.5 (c) 0.282
6. 0.1, 0.23 (2 d.p.)
7. (ii) (a) 0.68 (2 S.F.) (b) 8, 1.6
8. $\frac{1}{5}, \frac{3}{4}$

Exercise 4c (page 221)

1. (a) (i) 0.9830 (ii) 0.0170 (iii) 0.0015
   (b) (i) 0.1596 (ii) 0.2660 (iii) 0.5044 (iv) 0.9004
   (c) (i) 0.0037 (ii) 0.0037 (iii) 0.2916
   (d) (i) 0.5551 (ii) 0.0706 (iii) 0.9294 (iv) 0.3114
2.

x	0	1	2	3
$P(X=x)$	0.0467	0.1866	0.311	0.2765

x	4	5	6
$P(X=x)$	0.1382	0.0369	0.0041

3.

x	0	1	2	3
$P(X=x)$	0.0053	0.0487	0.1812	0.3364

x	4	5
$P(X=x)$	0.3124	0.116

Exercise 4d (page 225)

1. (a) 1
2. 0.946
3. (a) 3 (b) 3 (c) 0.633
4. (a) 2 (b) 0.994
5. 0.922
6. (a) 3 (b) 0.826 (c) 0.406

Exercise 4e (page 228)

1. (a) 1.2 (b) 0.4
   (c) 0.216, 0.432, 0.288, 0.064
   (d) 39, 78, 52, 11
2. 0.06; 293, 94, 12, 1, 0, 0
3. 1; 0.894 (a) 5 (b) 0.2
4. 5, 22, 37, 28, 8
5. 0, 0, 3, 13, 30, 36, 18
6. 16.5, 42.4, 45.4, 25.9, 8.3, 1.4, 0.1

Exercise 4f (page 233)

1. 0.0243
2. (a) (i) 0.201 (ii) 0.00637 (b) 2
   (c) 5, 2 (d) 14
3. (a) 4.8, 0.98 (2 d.p.) (c) 0.737
   (d) 0.388
4. 1, 0.336, 20
5. $s^3 + 3sd^2$
6. (a) 0.940 (b) 0.0432 (c) 0.0167
7. (a) $(1-p)^8(36p^2+8p+1)$;
   $(1-p)^5 + 5p(1-p)^8(1+4p)$
   (b) 0.678, 0.630, 0.0547, 0.0605
8. ${}^nC_r(1-p)^{n-r}p^r$ (a) 0.1296
   (b) 0.1792;

x	0	1	2
$P(X=x)$	$\frac{1}{4}$	$\frac{1}{2}$	$\frac{1}{4}$

   ;

   0.4816
9. (a) $\frac{2}{3}$ (b) 0.0424
10. (a) 0.4 (b) (i) 0.4516 (ii) 1.8

Exercise 4g (page 241)

1. (b) (i) $\frac{5}{36}$ (ii) $\frac{5}{216}$ (c) $\frac{1}{5}$
2. 2
3. (a) 0.128
   (b) $P(X=r) = (0.8)^{r-1}(0.2)$ Geometric
   (c) 0.512; 10, 40, 0.0768
4. $\frac{5}{4}, \frac{7}{4}, 2\frac{13}{16}$, 0.000 26, $\frac{1}{4}$
5. (a) (i) $\frac{1}{6}$ (ii) $\frac{25}{216}$ (iii) $\frac{125}{216}$
   (iv) 1 (v) 6
   (c) 17
6. 0.0047, December 22
7. (a) 0.504 (b) 0.432 (c) 0.5904
   (d) 44
8. $\frac{q}{p}$

Exercise 4h (page 245)

NOTE: Answers are given to 3 S.F. but *all* the numbers are retained in the calculator when addition of probabilities is required.

1. (a) 0.0302 (b) 0.106
   (c) 0.185 (d) 0.216 (e) 0.321
   (f) 0.463
2. (a) 0.007 81 (b) 0.000 452
   (c) 0.731 (d) 0.109
3. (a) 0.0907 (b) 0.308 (c) 0.570
   (d) 0.779
4. (a) 1.6 (b) 0.976
5. (a) 0.607, 0.303, 0.0758, 0.0126, 0.001 58
   (b) 0.0608, 0.170, 0.238, 0.222, 0.156
   (c) 0.0273, 0.0984, 0.177, 0.212, 0.191
6. (a) 2 (b) 0.271
7. 0.433

Exercise 4i (page 247)

1. (a) 0.513 (b) 0.004 23 (c) 0.0302
2. (a) 0.143 (b) 0.762 (c) 0.670
3. (a) 0.0821 (b) 0.242 (c) 0.759
   (d) 0.0486 (e) 0.125
4. (a) 0.0821 (b) 0.109 (c) 0.265
   (d) 0.0631
5. (a) 0.567 (b) 0.184
6. (a) 1.2 (b) 0.879 (c) 0.570

7. (a) 0.607 (b) 0.185
8. (a) 0.0408 (b) 0.219 (c) 0.0463
(d) 0.145

Exercise 4j (page 251)

1. (i) 0.0476, 0.0498
(ii) 0.225, 0.224 (iii) 0.171, 0.168
2. (a) 0.879 (b) 0.001 50
3. (a) 0.287 (b) 0.191
4. (a) (i) 0.368 (ii) 0.184 (iii) 0.0190
(b) 0.677
5. (a) (i) 0.195 (ii) 0.0916
(b) 0.075
6. 0.647, 0.185
7. 0.121

Exercise 4k (page 255)

1. (i) 0.165, 0.298, 0.268, 0.161, 0.0723, 0.0260
(ii) 0.0743, 0.1931, 0.2510, 0.2176, 0.1414, 0.0736
(iii) 0.0111, 0.05, 0.113, 0.169, 0.190, 0.171
(iv) 0.0224, 0.0850, 0.162, 0.205, 0.194, 0.148

Exercise 4l (page 257)

1. (a) 44, 44, 22, 8, 2
(b) 90, 72, 29, 8, 1, 0
2. 2, 7, 15, 20, 20, 16, 10, 6, 3, 1, 0, 0, 0; 71; 23 (78, 26 if do not round figures)
3. 0.5, 0.481; 31, 16, 4, 1, 0
4. 95, 137, 98, 47, 17, 5, 1; Approx 58

Exercise 4m (page 261)

1. 0.121
2. (a) 0.189 (b) 0.308 (c) 0.184
3. (a) 0.323 (b) 0.0119
4. (a) 0.301 (b) 0.080 (c) 0.251

Exercise 4n (page 268)

1. 0.407, 0.366, 0.165, 0.0629; 0.816, 0.0518
2. 3, 18.5%
3. m; (a) 0.983 (b) 0.184
(c) 0.199
4. (a) 0.788 (b) 0.002 93
5. (a) 0.368 (b) 0.264 (c) 3.16
(d) 0.199
6. (a) 22 (b) 19; 39
7. (a) 0.100 (b) 0.0702
8. (a) 0.600 (b) 0.0741
9. (a) 0.0902 (b) 0.0613; 4
10. (a) 0.647 (b) 6
11. (a) 0.185 (b) 4 (c) 2.68
(d) 6
12. (a) (i) 0.238 (ii) 0.841 (b) 0.083
13. (a) 0.677 (b) 0.017; 1498
14. (a) 0.082 (b) 0.242; 6.15
15. 0.371, £60.37
16. (a) 0.135 (b) 0.323; 0.81
17. (d) 0.387 (e) 0.929 (f) 0.893
(g) 0.205 (h) 0.816; 0.0290
18. (a) (ii) 1.5 (b) 0.577 (c) 0.0249
19. (c) $e^{-\lambda}\dfrac{\lambda^3}{6}$ (d) $1-e^{-\lambda}\left(1+\lambda+\dfrac{\lambda^2}{2}\right)$;
0.013, 0.014, 0.182

CHAPTER 5

Exercise 5a (page 275)

1. (a) $\frac{3}{8}$ (c) $\frac{7}{8}$ (d) $\frac{13}{32}$
2. (a) $\frac{1}{5}$ (c) 0.74
3. (a) $\frac{1}{4}$ (c) 0.66
4. (a) $\frac{3}{56}$ (c) $\frac{19}{56}, \frac{37}{56}$
5. $c = 1$, $k = 4$
6. (a) $\frac{1}{4}$ (c) $\frac{1}{4}$ (d) $\frac{5}{16}$
(e) 0.3475
7. (a) $\frac{1}{8}$ (c) $\frac{19}{24}$ (d) $\frac{1}{6}$

Exercise 5b (page 280)

1. (a) $\frac{3}{4}$ (b) $\frac{3}{5}$ (d) 6.45
2. (a) 1 (b) 1.2
3. (a) 2 (b) 3 (c) 6 (d) 1
4. (a) $\frac{1}{4}$ (b) 1.6 (c) 4.8
(d) $-\frac{11}{15}$
5. (a) $2\frac{1}{24}$ (b) $8\frac{5}{24}$ (c) 4.86 (3 S.F.)
6. (a) $\frac{1}{3}$ (b) 2 (e) $4\frac{5}{6}$
7. 6 m
8. (a) $\frac{2}{75}$ (b) $\frac{70}{9}$
(c) 0.48; Money bond
9. 2, 0.124 (3 S.F.)
10. 2.5, 0.803 (3 d.p.), 0.456 (3 d.p.)
11. (a) 2.875 kg (b) £4.75, $\frac{3}{16}$
12. (i) 0.4 (ii) 2.6 (iii) 1.5

Exercise 5c (page 287)

1. (a) $\frac{3}{2}$ (b) $\frac{12}{5}$ (c) $\frac{3}{20}$
(d) 0.387 (3 S.F.)
2. (a) $\frac{1}{2}$ (b) $\frac{7}{3}$ (c) $\frac{25}{12}$
(d) 1.44 (3 S.F.)
3. (a) $\frac{11}{6}$ (b) $\frac{11}{3}$ (c) $\frac{11}{36}$
(d) 0.553 (3 S.F.)
4. (a) $\frac{17}{14}$ (b) $\frac{62}{35}$ (c) $\frac{291}{980}$
(d) 0.545 (3 S.F.)
5. (a) $\frac{4}{5}$ (b) $\frac{2}{3}$ (c) $\frac{2}{75}$
(d) 0.163 (3 S.F.)
6. (a) $\frac{43}{24}$ (b) $\frac{97}{24}$ (c) $\frac{479}{576}$
(d) 0.912 (3 S.F.)

7. (a) $\frac{5}{18}$ (b) $\frac{214}{405}$ (c) $\frac{731}{1620}$
(d) 0.672 (3 S.F.)

8. (a) $\frac{3}{64}$ (b) 3, $\frac{3}{5}$ (c) $\frac{7}{64}$

9. (a) 1 (b) 1 (c) $\frac{1}{6}$ (d) $\frac{19}{32}$
(e) 1

Exercise 5d (page 294)

1. (a) $F(x) = \begin{cases} \dfrac{x^3}{8} & 0 \leqslant x \leqslant 2 \\ 1 & x \geqslant 2 \end{cases}$

(b) 1.59 (3 S.F.)

2. (a) $F(x) = \begin{cases} \dfrac{1}{5}(x+2) & -2 \leqslant x \leqslant 3 \\ 1 & x \geqslant 3 \end{cases}$

(b) 0.5

3. (a) $F(x) = \begin{cases} \dfrac{1}{8}(8x - x^2 - 7) & 1 \leqslant x \leqslant 3 \\ 1 & x \geqslant 3 \end{cases}$

4. (a) $F(x) = \begin{cases} \dfrac{x}{56}(x^2 + 6x + 12) & 0 \leqslant x \leqslant 2 \\ 1 & x \geqslant 2 \end{cases}$

5. (a) $F(x) = \begin{cases} x^4 & 0 \leqslant x \leqslant 1 \\ 1 & x \geqslant 1 \end{cases}$

(b) 0.841 (3 S.F.)

6. (a) $F(x) = \begin{cases} \dfrac{x}{4} & 0 \leqslant x \leqslant 2 \\ \dfrac{1}{4}(x^2 - 3x + 4) & 2 \leqslant x \leqslant 3 \\ 1 & x \geqslant 3 \end{cases}$

7. (a) $F(x) = \begin{cases} \dfrac{1}{24}(x+2)^3 & -2 \leqslant x \leqslant 0 \\ \dfrac{1}{3} + \dfrac{x}{2} & 0 \leqslant x \leqslant \frac{4}{3} \\ 1 & x \geqslant \frac{4}{3} \end{cases}$

(b) $\frac{1}{3}$

8. (a) 1.5 (b) 0.75

(c) $F(x) = \begin{cases} \dfrac{x}{3} & 0 \leqslant x \leqslant 3 \\ 1 & x \geqslant 3 \end{cases}$

(d) 0.4 (e) 0.2

9. (a) $\frac{3}{7}$ (b) 0.272 (3 S.F.)

(c) $F(x) = \begin{cases} \dfrac{1}{7}(x^3 - 1) & 1 \leqslant x \leqslant 2 \\ 1 & x \geqslant 2 \end{cases}$

(d) 1.65 (3 S.F.)

10. (a) $\frac{1}{3}$, $\frac{1}{3}$

(b) $F(x) = \begin{cases} \dfrac{x^2}{6} - \dfrac{2x}{3} + \dfrac{2}{3} & 2 \leqslant x \leqslant 3 \\ \dfrac{x}{3} - \dfrac{5}{6} & 3 \leqslant x \leqslant 5 \\ 2x - \dfrac{x^2}{6} - 5 & 5 \leqslant x \leqslant 6 \\ 1 & x \geqslant 6 \end{cases}$

(c) $\frac{1}{3}$ (d) $\frac{1}{24}$ (e) 4 (f) $\frac{5}{6}$

11. (a) 0.455, 3 (b) 3.64, 4.95

(c) $F(x) = \begin{cases} \dfrac{1}{\ln 9} \ln x & 1 \leqslant x \leqslant 9 \\ 1 & x \geqslant 9 \end{cases}$

12. $\frac{3}{4}$, $\frac{19}{80}$

$F(x) = \begin{cases} \frac{3}{4}x - \frac{1}{16}x^3 & 0 \leqslant x \leqslant 2 \\ 1 & x \geqslant 2 \end{cases}$

0.007

13. $F(x) = \begin{cases} \frac{1}{4}x & 0 \leqslant x \leqslant 1 \\ \frac{1}{5} + \dfrac{x^4}{20} & 1 \leqslant x \leqslant 2 \\ 1 & x \geqslant 2 \end{cases}$

1.565, 0.821

14. (a) 5 (b) $\frac{1}{6}$
(c) $\frac{5}{252}$; 543 tonnes

Exercise 5e (page 301)

1. (a) $\frac{1}{3}$ (b) $f(x) = \begin{cases} \dfrac{2}{3} & 0 \leqslant x < 1 \\ \dfrac{1}{3} & 1 < x \leqslant 2 \\ 0 & \text{otherwise} \end{cases}$

(c) $\frac{5}{6}$ (d) $\dfrac{\sqrt{11}}{6}$ (e) 0.608 (3 S.F.)

2. (a) $f(x) = \begin{cases} \dfrac{x-1}{6} & 1 \leqslant x \leqslant 3 \\ \dfrac{7-x}{12} & 3 \leqslant x \leqslant 7 \\ 0 & \text{otherwise} \end{cases}$

(b) $\frac{11}{3}$, $\frac{14}{9}$ (c) 3.54 (3 S.F.)
(d) 0.595

3. (a) $\frac{1}{3}$ (b) $f(x)=\begin{cases}0 & x<-1\\ \alpha & -1\leqslant x<0\\ 2\alpha & 0\leqslant x<1\\ 0 & x\geqslant 1\end{cases}$
(c) $\frac{1}{6}$ (d) 0.553 (e) $\frac{11}{18}$
4. (a) 2 (b) $f(x)=\begin{cases}2 & 0\leqslant x\leqslant 0.5\\ 0 & \text{otherwise}\end{cases}$
(c) 0.25 (d) 0.144
5. (a) $-\frac{1}{16}$, 1 (b) $f(x)=\begin{cases}\dfrac{x}{8} & 0\leqslant x\leqslant 4\\ 0 & \text{otherwise}\end{cases}$
6. 1, $\frac{5}{6}$, $\frac{1}{4}$

Exercise 5f (page 306)

1. (a) $f(x)=\begin{cases}\dfrac{1}{3} & 3\leqslant x\leqslant 6\\ 0 & \text{otherwise}\end{cases}$
(b) 4.5 (c) 0.75 (d) $\frac{1}{3}$
2. (a) $\frac{1}{3}$ (b) 0.5 (c) -3.5
(d) 0.75
3. (a) 5 (b) 0.325 (c) 3 (d) $\frac{4}{3}$
4. $f(a)=\frac{1}{6}a^{-1/2}$, $1\leqslant a\leqslant 16$; 7, 19.2
5. $f(a)=\dfrac{1}{4\sqrt{\pi a}}$, $\pi\leqslant a\leqslant 9\pi$; $\dfrac{13\pi}{3}$, $\dfrac{244\pi^2}{45}$
6. (i) 0.25 (ii) 0.845
7. $\frac{26}{3}$, $\frac{964}{45}$
8. 3π, $3\pi^2$, 0.63 (2 d.p.)
9. $\dfrac{3f}{2}\leqslant v\leqslant 2f$, $f(1+\ln 2)$, $\dfrac{5f}{3}$
10. (b) $\frac{125}{36}$

Exercise 5g (page 315)

1. (a) 0.0821 (b) 0.2 (c) 0.632
(d) 0.2 (e) 0.139 (f) 0
2. (a) 2000 hrs (b) (i) 0.287 (ii) 0.593
(c) 0.465 (d) 0.0515
3. (a) 6.93 (b) $0.1e^{-0.1x}$
(c) 10, 100 (d) 0.24 (2 S.F.)
4. 0.1386, £26.30, 0.225
5. (a) $e^{-0.4t}$ (b) $1-e^{-0.4t}$ (c) 2.5
(d) 1.73; 0.135
6. $a=92.2$, $A=0.0108$
(a) 0.114 (b) 0.338 (c) 0.202
7. (i) 0.62 (ii) 0.38
8. 2.895
9. $\dfrac{1}{\lambda}$, $e^{-2\lambda t}$, $e^{-2\lambda t}$, $\dfrac{1-e^{-\lambda(1-k)t}}{1-e^{-\lambda t}}$
10. 2, $\frac{1}{2}$, $\frac{1}{4}$, $1-e^{-2x}$, 0.368

Exercise 5h (page 327)

1. (a) 2.4 (b) 20, $\frac{1}{3}$, 0.178 (3 S.F.)
2. (i) $-\frac{3}{16}$ (ii) $\frac{19}{80}$ (iv) 2
3. (b) $\sqrt{\dfrac{3}{5}}$
4. 4, $\frac{8}{15}$, $\frac{11}{225}$, 0.541 (3 S.F.)
5. 8, $\frac{1}{9}$, 39
6. (a) $\frac{2}{3}$ (b) $f(x)=\begin{cases}\dfrac{2x}{3} & 0\leqslant x\leqslant 1\\ 1-\dfrac{x}{3} & 1\leqslant x\leqslant 3;\\ 0 & \text{otherwise}\end{cases}$
$\frac{4}{3}$, $\frac{7}{18}$ (c) 1.27 (3 S.F.), 0.875
7. $\lambda=\frac{1}{3}$, $2\frac{7}{12}$, $\frac{79}{96}$
8. $a=2$, $k=\frac{3}{4}$, 0.2
9. 0.6, 0.2, 0.166
10. (c) $F(x)=\begin{cases}-\dfrac{1}{12x^3} & x\leqslant -1\\ \dfrac{1}{2}+\dfrac{x}{2}-\dfrac{x^3}{12} & -1\leqslant x\leqslant 1\\ 1-\dfrac{1}{12x^3} & x\geqslant 1\end{cases}$
(d) 0, $\frac{11}{15}$
11. (b) $\dfrac{\theta}{\theta+3}$, $\dfrac{\theta(\theta+1)}{(\theta+3)(\theta+4)}$
(c) $\dfrac{3\theta}{(\theta+3)^2(\theta+4)}$ (d) 0.2
12. (a) 0.991 (b) 0.983 (c) 0.28
(d) 0.0017 (e) £15.40
13. (i) $b=\dfrac{a^2}{2}$, $\dfrac{2}{3a}$, $\dfrac{2}{9a^2}$ (ii) $c=a$
14. $\frac{4}{5}$, $\frac{4}{25}$, £8
15. (a) 2.1, 1.29 (b) 1, $\frac{1}{2}$
16. $\frac{1}{4}$, 2, $\frac{8}{3}$, 20
17. 0.181, 0.0498; 11.6 miles (3 S.F.)
18. (a) 98p (b) 83p

CHAPTER 6

Exercise 6a (page 337)

1. (a) 0.1911 (b) 0.8089
(c) 0.1911 (d) 0.8089
2. (a) 0.0359 (b) 0.2578
(c) 0.99973 (d) 0.9131
(e) 0.004 94 (f) 0.991 11
(g) 0.9686 (h) 0.2343
(i) 0.0312 (j) 0.9484
(k) 0.9803 (l) 0.002 01

3. (a) 0.05 (b) 0.05
(c) 0.0999 (d) 0.025
(e) 0.005 (f) 0.01
(g) 0.0025 (h) 0.975
4. (a) 0.1709 (b) 0.548 07
(c) 0.3639 (d) 0.4582
(e) 0.4798 (f) 0.997 92
(g) 0.033 68 (h) 0.9082
(i) 0.2729 (j) 0.030 61
(k) 0.925 (l) 0.4508
(m) 0.9 (n) 0.02

Exercise 6b (page 338)

1. (a) 3.03 (b) 2.326/7/8/9
(c) 1.96 (d) 0.849
(e) 0.047/8 (f) −0.501/2
(g) −0.885 (h) −2.272/3/4
2. (a) −3.432 (b) −1.887
(c) −0.454 (d) 0.015
(e) 0.796 (f) 1.231/2
(g) 0.953 (h) 1.94
3. (a) 0.91 (b) 1.66
(c) 0.674 (d) 2.05
4. 0.674, −0.674, 0.524
5. (a) 1.645 (b) 1.96
(c) 2.054/5 (d) 2.326
(e) 2.575 (f) 2.808/9
6. (a) 1.282 (b) 2.054/5
(c) 2.17 (d) 2.575

Exercise 6c (page 342)

1. (a) 0.0548 (b) 0.0107
(c) 0.8849 (d) 0.9713
(e) 0.6554 (f) 0.9918
(g) 0.4602 (h) 0.0808
2. (a) 0.0106 (b) 0.273
(c) 0.5971 (d) 0.2168
(e) 0.9857 (f) 0.997 02
3. (a) 0.3015 (b) 0.0105
(c) 0.9079 (d) 0.2533
(e) 0.2097 (f) 0.0323
(g) 0.5231
4. (a) 0.1587 (b) 0.8413
(c) 0.6915 (d) 0.3085
(e) 0.9332
5. (a) 0.8634 (b) 0.2413
(c) 0.1388 (d) 0.6826
(e) 0.2565
6. (a) 0.8014 (b) 0.085
(c) 0.2714 (d) 0.4028
(e) 0.188 62
7. (a) 0.5923 (b) 0.4208
(c) 0.9544
8. (a) 0.0668 (b) 0.4013
(c) 0.1747
9. (a) 0.7054 (b) 0.3228
(c) 0.0618 (d) 0.8962
(e) 0.1818 (f) 0.4621
10. (a) 0.0478 (b) 8.17×10^{-4}
11. (a) 735 (b) 646
(c) 546 (d) 740

Exercise 6d (page 343)

(i) (a) 51.55 (b) 63.55
(ii) (a) 117.44 (b) 126.752
(iii) (a) 70.00 (b) 90.58
(iv) (a) 49.66 (b) 67.60
(v) (a) $\mu - 2.05\sigma$ (b) $\mu + 0.86\sigma$
(vi) (a) $a - 2.05\sqrt{b}$ (b) $a + 0.86\sqrt{b}$
(vii) (a) $-1.05a$ (b) $1.86a$
(viii) (a) 34.65 (b) 55.02

Exercise 6e (page 345)

1. (i) 63.655 (ii) 67.37
(iii) 55.09 (iv) 56.69 or 56.695
2. (a) 37.572 (b) 50.012
(c) 38.244 (d) 55.608
3. 9.87; $70.13 < X < 89.87$
4. (i) 9.2 (ii) 18.608
(iii) 15.68 (iv) 17.92
(v) (384.32, 415.68)
5. (a) 0.6247 (b) 629.52 g (c) 3
6. (a) 290 (b) 78 (c) 27
7. (i) 1.645 (ii) 2.575
(iii) 1.96 (iv) 2.808
8. (458.92, 546.52)
9. 8, 1.158, (6.10, 9.90)

Exercise 6f (page 349)

1. 10.7
2. 4
3. 30
4. 35.5
5. 52.73, 11.96
6. 100.8, 5.71
7. 50, 6.12
8. 39.5, 5.32
9. 53.87, 16.48
10. 0.7725
11. 0.203
12. (a) 92.7% (b) 1.32
(c) 1.7%
13. 2080, 236
14. (a) 9.1% (b) 99.69
(c) 0.4 mm
15. 4.46
16. 4.299

Exercise 6g (page 353)

1. (a) 37.8% (b) (125.5, 194.5)
(c) 0.405
2. (a) 7, 3.5 (b) 0.075
3. 979.27, 17.27, 133
4. 5.2007, 0.003 46; 0.0269; 0.002 61, 1.4%
5. 24.97, 53.03
6. 0.0038, 230.65, 1.29
7. 433.7 hrs
8. 137, 149.5
9. 0.30 (2 d.p.), 0.26 (2 d.p.); steeper
10. (a) 1.2 (b) 53.6
(c) 54.2; 0.066
11. 0.4013, 0.0031
12. 0.159, 0.775, 0.067, 2.7, £37.56

Exercise 6h (page 361)

1. (a) $P(2.5 < X < 9.5)$
(b) $P(3.5 < X < 8.5)$
(c) $P(10.5 < X < 24.5)$
(d) $P(1.5 < X < 7.5)$
(e) $P(X > 54.5)$ (f) $P(X > 75.5)$
(g) $P(45.5 < X < 66.5)$
(h) $P(X < 108.5)$ (i) $P(X < 45.5)$
(j) $P(55.5 < X < 56.5)$
(k) $P(400.5 < X < 560.5)$
(l) $P(66.5 < X < 67.5)$
(m) $P(X > 59.5)$
(n) $P(99.5 < X < 100.5)$
(o) $P(33.5 < X < 42.5)$
(p) $P(6.5 < X < 7.5)$ (q) $P(X > 508.5)$
(r) $P(X < 6.5)$
(s) $P(26.5 < X < 28.5)$
(t) $P(52.5 < X < 53.5)$
2. (a) 0.9474 (b) 0.6325
(c) 0.5914 (d) 0.0269
(e) 0.2106
3. (a) 0.0154 (b) 0.8145
(c) 0.02
4. 0.1127
5. (a) 0.3729 (b) 0.9501 (c) 0.1039
(d) 0.929
6. $\frac{1}{4}$, $\frac{3}{4}$, 27; 6.75, 2.25, 0.8413
7. 20, 16, 0.004 36
8. ${}^nC_r(1-p)^{n-r}p^r$, np, $np(1-p)$
(a) 0.2304
(b) 0.922 24; 0.8531 (incl.), 0.7946 (not incl.)
9. (a) (i) 0.0432 (ii) 0.1845
(iii) 0.7723 (b) at least 9
10. (a) 61.7 (b) 0.075 (c) 163.5
(d) 134.3 (e) 702.2
11. (a) 0.0566 (b) 0.2171 (c) 0.4708
(d) 0.1432
12. (a) 285 (b) 43
13. 0.6886

Exercise 6i (page 365)

1. (a) 0.6201 (b) 0.39
(c) 0.5406
2. (a) 0.3998 (b) 0.2004
(c) 0.3361 (d) 0.0637
3. (a) 0.313 (b) 0.5078
(c) 0.8335 (d) 0.1101
4. (a) 0.2614 (b) 0.2343
(c) 0.0558
5. 0.8901
6. 0.6887; 4
7. (a) 0.4574 (b) 0.173 (c) 0.8312
8. (a) 0.4594 (b) 0.5363
9. (a) (i) 0.999 767 (ii) 0.000 177
(iii) 0.924 41 (b) 0.009 44
10. 86

Exercise 6j (page 369)

1. 0.5455 (a) 0.5462 (b) 0.3983
2. 0.1036 (a) 0.098 812
(b) 0.1061
3. 0.1360 (a) 0.1381
(b) 0.0936
4. 0.063 03 (a) 0.0579
(b) 0.0655
5. (a) 0.061 84 (b) 0.0651
6. 0.2192, 0.2075

Exercise 6k (page 370)

1. (a) 0.55 (b) 0.18
2. (a) 0.649 (b) 0.965 (c) 0.371
3. ${}^{10}C_1(0.96)^9(0.04)$ (a) 0.20
(b) 0.77
4. (a) 0.2025 (b) 0.410 (c) 0.0238
5. ${}^nC_r(0.211)^r(0.789)^{n-r}$; (a) 0.306
(b) 21 (c) 0.203
6. (a) 0.6247 (b) 93.32% (c) 0.7852
7. (a) 0.3154
(b) 0.3068; worse; 0.5245
8. 0.1796, 3500
9. (a) 0.194 (b) 0.933 (c) 0.986
10. (a) (i) 0.1353 (ii) 0.3233
(b) 0.250
11. (a) 2.04×10^{-10}
(b) 0.004 34; $x = 73$
12. $\dfrac{2000!}{(2000-N)!N!}\left(\dfrac{1}{30}\right)^N\left(\dfrac{29}{30}\right)^{2000-N}$;
86 lines; 2×47 lines > 86; No
13. (a) (i) 0.0525 (ii) 0.358 75
(b) (i) 0.143 (ii) 0.145
14. (a) 0.315 (b) 0.5644
15. (a) 0.199 (b) 0.353
(c) $e^{-3} - e^{-6}$
(d) $3e^{-6}(1-e^{-3})$; 0.047 31, 0.007 066; 0.870
16. 0.859 (c) 0.204 (d) 0.034
17. 0.043
18. (a) (i) 0.5987 (ii) 0.149
(b) (i) 0.0294 (ii) 0.751
(c) (i) 0.5987 (ii) 0.9772
19. (a) 0.215
20. 0.360, 0.734
21. (a) 0.927 (b) 0.0102; 0.297

CHAPTER 7

Exercise 7a (page 377)

1. (a) 0.6554 (b) 0.7698
(c) 0.3446 (d) 0.8301
2. (a) 0.0359 (b) 0.269 64
(c) 0.6554 (d) 0.2743
(e) 0.9918
3. (a) 0.001 35 (b) 0.0228
(c) 0.0913
4. (a) 0.9044 (b) 0.9522
(c) 0.6826

5. (i) 6.68% (ii) 6.1, $\sqrt{0.13}$
(iii) 4.81% (iv) £74 (v) $7\frac{1}{2}$p
6. (a) 0.0478 (b) 0.0668
(d) 0.9324
7. 0.12, 0.0583, 1.98%

Exercise 7b (page 382)

1. (a) 0.0228 (b) 0.8621
(c) 0.9638
2. 0.6915
3. (a) 0.1728 (b) 0.6127
(c) 0.5
4. 0.0561
5. (a) 0.0289 (b) 0.0200
(c) 0.6252
6. 0.5402
7. (a) 0.1247 (b) 0.6957
8. 0.1103, 0.753
9. 0.9043
10. 0.0651
11. 0.2575
12. 9.6, 0.522; (a) 0.0177
(b) 0.2218
13. (a) (94.4, 105.6) (b) 92.55%
(c) 22.14%
14. (a) 0.0787 (b) 3.019×10^{-6}
15. (a) 0.6298 (b) 0.1056

Exercise 7c (page 389)

1. (a) 0.0745 (b) 0.9736
(c) 0.9386 (d) 0.0271
2. (a) 0.8131 (b) 0.0478
(c) 0.1078 (d) 0.0306
(e) 0.995 53 (f) 0.2762
3. (a) 6, 2 (b) 0.2074
(c) 0.7601 (d) 0.5143
4. 0.8681
5. (a) 0.990 39 (b) 0.9772
(c) 0.7373
6. (a) 0.1587 (b) 0.0127
7. (a) 0.244 (b) 0.659
(c) 0.409

Exercise 7d (page 392)

1. (a) 0.60 (b) 0.20 (c) 0.95
(d) 0.5
2. (a) 0.051 (b) 0.001 55 (c) 0.9782
3. 1000, 172.4, 3000, 298.6, 0.16, 0.02
4. (a) 0.0888 (b) 0.6611
5. 0.0625, 0.2574, 0.5, 0.7123, 7
6. $Y \sim N(a_1\mu_1 + a_2\mu_2, a_1{}^2\sigma_1{}^2 + a_2{}^2\sigma_2{}^2)$, 0.84
7. (a) 0.8413 (b) 0.5
(c) 0.4207; 0.9938
8. 12 kg, 57.0 g, 3.97%, 765 g
9. (a) 0.3446
(b) 0.6915; 0.003 29, 0.304
10. 0.9192, 08 13, 0.999 912, 08 10, No
11. 0.8603, 0.1574, 0.3909
12. (a) 0.106 (b) 438.2 ml
(c) 0.800 (d) 0.961
(e) 0.244 (f) 388.6 ml
13. $N(\mu_1 + \mu_2 - \mu_3, 3\sigma^2)$
(a) 0.1657 (b) 108p
(c) 0.4148
14. (a) 0.0139 (b) 0.1587
(c) 0.9332
15. (a) 0.159 (c) 0.584

Exercise 7e (page 398)

1. (a) (i) 0.5, 0.45 (ii) 1.5, 1.05
(iii) 0.6, 9.24
2. 0, $\frac{7}{6}$;

$\bar{x}$	-2	-1.5	-1	-0.5	0	0.5	1	1.5	2
$P(\bar{X} = \bar{x})$	$\frac{1}{144}$	$\frac{6}{144}$	$\frac{17}{144}$	$\frac{30}{144}$	$\frac{36}{144}$	$\frac{30}{144}$	$\frac{17}{144}$	$\frac{6}{144}$	$\frac{1}{144}$

;
0, $\frac{7}{18}$
3. 4.75, 8.1875; 4.75, 4.09 (3 S.F.)
4. 5, 7.5;

Mean	2.5	4	4.5	5.5	6	7.5
f	2	2	2	2	2	2

;
5; 2.5
5. 0.84, 1.68
6. (a) 24.5 (b) 2.57; 2.35, 7, 6

Exercise 7f (page 401)

1. (a) 0.0401 (b) 0.3891
2. 0.3206
3. (a) 0.0668 (b) 0.9893
(c) 0.1974
4. 0.0228
5. (a) 0.0401 (b) 0.7571
(c) 0.2660
6. (a) 12 (b) 25
7. (a) 0.2399 (b) 0.0787
(c) 0.0127; $n > 108$
8. 62
9. (a) N(10, 3.2) (b) N(50, 3.2)
(c) N(−10, 3.2) (d) N(210, 48)
(e) N(80, 27.2)
10. 0.009 61
11. 44
12. (a) $2\mu, \sqrt{2}\sigma$ (b) $0, \sqrt{2}\sigma$
(c) $\mu, \dfrac{\sigma}{\sqrt{2}}$; 0.7078, 0.9213
13. 0.0968, 0.0828, 0.000 907, 0.2295
14. N(960, 21.2)
15. 0.0983
16. (a) 0.1457
(b) Distribution of $V^{1/3}$ different, prob. $< 10^{-4}$
(c) 0.1210
17. 0.332, 0.0587, 0.009
18. (a) (i) 0.7881 (ii) 0.673
(b) 0.0749 (c) 0.0548

Exercise 7g (page 405)

1. 1580, 1200^2
2. 3.21, 0.265 (3 S.F.), 0.001 44
3. (a) 0.034 (b) 0.8194
4. (a) 3.85 (b) 62.34
   (c) 1.7
5. (a) 0.9145 (b) 0.7081
   (c) 0.6226
6. 50
7. 60
8. 35
9. 42
10. 45
11. 20 500, 1768
12. 0.9212
13. (b) 8.86, 7.82 (c) 0.331
14. 0.25, 0.0228

Exercise 7h (page 409)

1. (a) 0.0745 (b) 0.003 67
2. (a) 0.005 68 (b) 0.527 02
   (c) 0.0958
3. (a) 0.000 215 (b) 0.5229
   (c) 0.0367
4. (a) 0.3085 (b) 0.0970
5. 0.7181
6. (a) 0.0648 (b) 0.0851
   (c) 0.3068
7. 0.22

Exercise 7i (page 417)

Some answers depend on the random numbers used and on the method of allocation. These are possible answers.

10. (a) 1, 1, 1, 0, 3 (b) 4
11. 33.134, 34.193, 28.712
12. (a) 3, 5 (b) 1, 5
    (c) 1007.2, 1016.8
13. 1.52, 1.48
14. 3.3, 1.41
17. (a) 3 (b) 6.1826
18. (a) 5.36, 5.53

CHAPTER 8

Exercise 8a (page 424)

1. a, c, e
2. e
4. $\frac{2}{3}, \frac{4Z}{3}$, Smaller variance
5. $6\frac{1}{3}$
6. $a = 0.24$, $b = 0.28$; 8.8, Unbiased, Minimum variance

Exercise 8b (page 427)

1. 48.875, 6.98 (2 d.p.)
2. 51.5, 241.1
3. 1.69 (2 d.p.), 8×10^{-6} (1 S.F.)
4. 15, 43.14 (2 d.p.)
5. 15, 43.14 (2 d.p.)
6. 10, 3.11 (2 d.p.)
7. 9.71 (2 d.p.), 621.12 (2 d.p.)
8. 57.78 (2 d.p.), 6496.15 (2 d.p.)
9. 46.9, 242.46 (2 d.p.)
10. 10.96, 17.35 (2 d.p.)
11. 22.79 (2 d.p.), 1.81 (2 d.p.)
12. 236, 7.58
13. 3.17, 1.5611

Exercise 8c (page 432)

Answers are given to 3 S.F. where applicable

1. 0.663, 0.002 21
2. 9.88, 0.796
3. 3.69, 1.33
4. 9.19, 10.0
5. 2.27, 10.2
6. 5.46, 0.0481
7. 30.15, 11.9
8. 2.39, 0.0275
9. 0.348
10. 0.838
11. $\frac{2}{3}$
12. 0.307

Exercise 8d (page 439)

1. (a) (139.494, 140.506)
   (b) (139.399, 140.601)
2. (a) (74.026, 77.974)
   (b) (73.396, 78.604)
   (c) (72.91, 79.09)
3. (a) (747.516, 748.484)
   (b) (747.424, 748.576)
   (c) (747.316, 748.684)
4. (a) (79.209, 84.791)
   (b) (78.91, 85.09)
5. (a) (68.123, 69.877)
   (b) (67.848, 70.152)
6. (1011, 1114)
7. (10.821, 14.079)
8. 10.82, 1.70 (3 S.F.), 11.19, 0.646 (3 S.F.), 10.968, (10,821, 11.115)
9. 85.2, 15.45 (2 d.p.), 85.01, 2.01, 85.08 (2 d.p.), (84.628, 85.540)
10. 25.3, 3.63, (24.85, 25.75)
11. 91.32 (2 d.p.), 7.42 (2 d.p.), (90.5, 92.2)
12. 194, 176.41 (2 d.p.), (173.48, 214.52)
13. 9.71, (172.3, 173.3)
14. (b) 38.1, 1080.39 (c) 38.1 ± 4.56
    (d) 28.5
15. (b) $\frac{2}{3}$, 0.471 (c) $\frac{5}{9}$
    (d) 0.9 ± 0.173
16. 3.71

Exercise 8e (page 444)

1. $(-3.707, 3.707)$
2. (a) $(-1.943, 1.943)$
   (b) $(-2.447, 2.447)$
   (c) $(-3.143, 3.143)$
   (d) $(-4.317, 4.317)$
3. (a) $(-3.499, 3.499)$
   (b) $(-3.055, 3.055)$
   (c) $(-2.947, 2.947)$
   (d) $(-2.921, 2.921)$
4. 0.945
5. 0.045
6. 0.075
7. 1.86
8. 2.179

Exercise 8f (page 447)

1. (a) (177.74, 181.59)
   (b) (177.21, 182.12)
   (c) (175.82, 183.52)
2. (3.77, 4.51)
3. (a) (0.285, 0.335)
   (b) (0.275, 0.345)
4. (14.98, 15.78)
5. (a) (8.08, 9.12) (b) (8.01, 9.19)
6. (4.70, 5.56)
7. (32.08, 33.22)

Exercise 8g (page 454)

1. (a) (0.323, 0.517)
   (b) (0.696, 0.904)
   (c) (0.222, 0.418)
   (d) (0.529, 0.791)
   (e) (0.146, 0.254)
   (f) (0.693, 0.847)
   (g) (0.469, 0.531)
2. (0.622, 0.738)
3. (a) (0.293, 0.427)
   (b) (0.273, 0.447)
4. (0.156, 0.344)
5. (0.510, 0.574); Yes
6. 267.2 (1 d.p.), 227.9 (1 d.p.), (0.256, 0.410)
7. (a) 3, (2.04, 3.96)
   (b) 30%, (25.2, 34.8)
8. $\dfrac{mn}{n_0}$, (0.002 41, 0.007 59), (13 175, 41 493)
9. 10 000, (7236, 16 181)
10. $\text{Bin}(n\theta, n\theta(1-\theta))$, n large; (0.826, 0.945)

Exercise 8h (page 456)

1. (a) (92.32, 99.68)
   (b) (0.351, 0.369), 5277
2. 6.6 mm, 3.5 mm

Mesh size	0 to 1	>1 to 2	>2 to 4
Additional diamonds	1	1	4

Mesh size	>4 to 6	>6 to 8	>8 to 12
Additional diamonds	2	6	6

   6.30, 9.93
3. 14.01, 0.04, (13.92, 14.10); 0.40
4. (0.123, 0.392), (170.84, 178.16), (165.57, 186.83), (£488, £531)

CHAPTER 9

Exercise 9a (page 464)

1. (i) $z = 2$ (a) No (b) Yes
   (ii) $z = -1.5$ (a) Yes (b) Yes
   (iii) $z = 2.12$ (a) No (b) Yes
   (iv) $z = -2.475$ (a) No (b) Yes
   (v) $z = 3.645$ (a) No (b) No
   (vi) $z = -1.826$ (a) No (b) Yes

In questions 2–5 the continuity correction has been omitted.

2. (a) $z = 1.5$, Fair (b) $z = 2.5$, Biased
3. $z = -1.826$, Reject claim
4. $z = 1.746$, Accept
5. (a) (i) 0.0298 (ii) 0.0934
   (b) $z = -1.897$, Yes, less than 75%

Exercise 9b (page 471)

1. (a) $z = -1.095$, Accept H_0
   (b) $z = 1.845$, Reject
   (c) $z = 2.5$, Reject
   (d) $z = -2.778$, Reject
2. $z = -1.565$, No
3. $z = 1.909$ (a) Yes (b) No
   (c) No
4. $\bar{x} < 91.51$ min
5. (a) 0.683
   (b) $2.9216 < \bar{x} < 3.0784$
6. 0.1101, 0.001 58, Reduced
7. $z = 2.487$, Yes; 1506.8 ± 0.311
8. (a) $\alpha \geqslant 3.59\%$ (b) $\alpha \geqslant 7.18\%$
9. Approx. 83, (0.823, 0.845), No
10. $\mu_1 - \mu_2$, $\sigma_1^2 + \sigma_2^2$; $\lambda_1\mu_1 + \lambda_2\mu_2$, $\dfrac{\lambda_1^2\sigma_1^2}{n_1} + \dfrac{\lambda_2^2\sigma_2^2}{n_2}$
    $z = 1.853$, Yes at the 5% level
11. (a) 10.6%, 6.7% (b) £98.80
    (c) $z = -2.5$, One tailed, Yes
12. 11.2, 2.54, Reject at the 5% level
13. 0.0341, 0.069, Do not reject
14. (b) (i) (£$17\frac{2}{3}$, £$19\frac{1}{3}$)
    (ii) 279 (iii) Yes

Exercise 9c (page 476)

1. (a) $z = 1.792$, Accept
   (b) $z = 1.792$, Reject
   (c) $z = -1.437$, Reject
   (d) $z = -2.5$, Accept
2. $z = 0.995$, Accept
3. $z = 2.00$, Yes; 16.2 ± 1.232
4. Justified; (8.19, 8.53)
5. (b) Reject H_0 (c) (77.50, 79.96)
6. (45.6, 49.4), $\alpha > 4$, 0.0321

Exercise 9d (page 480)

1. (a) $t = 2.622$, Reject
   (b) $t = -1.892$, Accept
   (c) $t = 2.152$, Reject
   (d) $t = -3.073$, Reject
2. $t = -3.601$, Not in good working order
3. $t = 2.828$, Yes
4. (a) $t = -3.54$, Yes
   (b) $z = -3.2$, Yes
5. $t = 1.1$, No
6. 0.1056;
   (a) $z = -1.660$, Do not reject
   (b) 0.324; $t = -2.33$, Reject
7. (b) H_0 rejected at better than 2% significance level, population mean unlikely to be 3.1
   (c) (1.81, 2.76)

Exercise 9e (page 491)

1. (a) $z = -2.096$, S
   (b) $z = -1.402$, NS
   (c) $z = 2.493$, S
   (d) $z = 1.99$, NS
   (e) $z = 2.076$, S
   (f) $z = -2.036$, NS
   (g) $z = 1.783$, S
   (h) $z = 1.779$, S
   (i) $z = -2.321$, (only just) NS
   (j) $z = 2.55$, S
   (k) $t = 2.135$, S
   (l) $t = -0.567$, NS
   (m) $t = 2.088$, NS
   (n) $t = 1.260$, NS
2. $z = 2.33$, Yes
3. (a) $|z| = 2.385$, Yes
   (b) $z = 2.946$, Yes
4. $z = 1.627$, Accept; least $n = 124$
5. (b) 15 ± 3.4 (c) 13.625
   (d) $t = -1.42$, Accept claim
6. $z = 2.423$, S
7. (26.77, 27.89); 2.4, $z = 1.97$, NS, Those of high intelligence do not have greater foot length
8. (3.175, 3.335), Yes, n large, $z = -3.248$, Yes
9. $z = -1.646$, Just S, Reject Mr Brown's claim
10. $|z| = 2.036$, S at 5%, NS at 4%
11. 8.0067, 0.000 175, $z = 2.00$, S, Reject population mean is 8.00, $z = 3.52$, S, Second population has smaller mean than first
12. $|t| = 6.496$, Yes; $t = 2.041$, Yes
13. 4.41, (9.87, 10.73), 3.61, $z = 1.49$, NS
14. (a) 1.65, 0.0025; 1.55, 0.003 75
    (b) 0.1038 (c) $z = 2.911$, Reject
15. (2.602, 3.118), 0.567, $z = -2.219$, S, Flowers on sunny side grow taller

Exercise 9f (page 495)

Continuity corrections have been omitted.

1. (a) $z = 1.768$, S, Reject
   (b) $z = 2.335$, S, Reject
   (c) $z = -1.897$, NS, Accept
   (d) $z = 2.179$, NS, Accept
   (e) $z = -3.060$, S, Reject
2. $z = -2.04$, No
3. $z = 1.667$, Yes
4. $z = -2.4$, Yes
5. $z = 2.46$, Yes, (0.340, 0.400)
6. (a) 0.87 (b) 0.19, $z = -1.49$, Yes
7. 91.3, 13.5; 0.1571; 0.003 (without continuity correction), $z = -2.75$, Yes
8. (a) 0.0985, 0.0666
   (b) $z = -0.0223$, Do not reject
   (c) 12, 27, 29, 19, 9, 3, 1
9. (0.3273, 0.3867), 19, $z = -2.775$ Reject H_0, Yes
10. $z = -1.143$, Do not reject

Exercise 9g (page 501)

1. (a) $z = -1.245$, NS, Accept
   (b) $z = 2.941$, S, Reject
   (c) $z = -0.568$, NS, Accept
   (d) $z = 1.373$, NS, Accept
2. $z = -1.247$, NS
3. $|z| = 1.35$, NS
4. $z = 2.04$, S, $z = -0.45$, NS (5%)
5. (0.616, 0.824), $\frac{126}{162}$, $z = 2.25$, S, Difference in proportions
6. (a) $\frac{21}{40}$ (b) $\frac{19}{30}$ (c) 0.4 ± 0.048
   (d) $z = -5.657$, Yes
7. (a) $z_1 = 2.178$, Reject; $z_2 = 0.594$, Accept (b) $z = 0.966$, NS
8. (a) $z = 1.317$, NS
   (b) (0.476, 0.524)
9. $z = 0.807$, Proportions same, (0.569, 0.711)
10. (0.270, 0.355), 0.25, $z = 2.20$, Yes, Greater among those not wearing seat belts
11. $z = -1.247$, Accept claim
    $z = -1.521$, No at 5% level
12. (a) $0.557 < p < 0.754$ (b) No

Exercise 9h (page 504)

1. 0.0345, 0.1174
2. $\hat{\sigma}^2 = 34.583$, $|t| = 0.798$, NS
3. (204.1, 223.9), 196, $|z| = 1.714$, NS
4. (24.59, 25.41), $|z| = 2.15$, Accept Mr Jones' claim
5. (a) 0.833 (b) 0.180
   (c) H_0: $p = 0.5$, H_1: $p \neq 0.5$; $z = 2.236$, Biased, (0.560, 0.940), $p > 0.5$
6. (a) $z = -1.4$, NS, no
   (b) $|z| = 1.405$, NS
7. (a) $t = 2.52$, evidence that means are different
   (b) $z = 1.80$, No evidence that means are different. Assumption $\sigma_1^2 = \sigma_2^2 = 196.0$ is clearly suspect; a value of 100 would be more consistent with the data
8. $\bar{x}_1 - \bar{x}_2 \pm 0.077\,25$, 0.7174, Same
   (a) $z = 2.667$, Reject
   (b) $z = -1.667$, Accept
9. 0.567, 0.1156, 0.1587, 0.1587; 12.6 pence; $z = 1.4$, Not sufficient evidence, $z = 1.838$, Sufficient evidence
10. (a) 0.8931 (b) 0.8859
    (c) 0.7912 (d) 0.0947;
    $\mu = 229.41$, $z = -2.405$, Yes, Not correctly set
11. 361, 5441 ± 4.88, $|z| = 2.4$, S, Yes
12. 1/1200, $z = -0.949$, NS, No

Exercise 9i (page 513)

1. (a) N.S. (b) S. (c) S.
   (d) N.S. (e) N.S. (f) S.
   (g) S. (h) N.S.
2. N.S.
3. No
4. N.S. (She could have been guessing)

Exercise 9j (page 518)

1. (a) N.S. (b) N.S. (c) N.S.
   (d) S. (e) N.S. (f) S.
2. S.
3. N.S.
4. (a) N.S. (b) S.
5. N.S.
6. 0.0057, 9 mins, N.S.
7. No

Exercise 9k (page 529)

1. (a) 0.125, 0.125
   (b) 0.2099, 0.2702, Test 2
2. $\frac{1}{3}$, $\frac{4}{9}$
3. Accept as slow if 'bounce' < 12.5, Accept as slow if 'mean bounce' < 11.64; 0.0004
4. 1.9, 0.837
5. 0.515, 0.376, 0.4455, 43
6. 0.106, 0.02
7. (a) H_0: $p = \frac{1}{6}$, H_1: $p \neq \frac{1}{6}$;
   (b) $z = -1.9595\ldots$, Just NS at 5% level, No (c) $z = 1.8$, NS, no
8. 0.000 32, 0.006 72, 0.057 92; 4
9. See question 1
10. (a) 0.0668
    (b) 0.0446; If $X > 58$ accept H_1, If $X \leqslant 58$ accept H_0; 0.380
11. Accept H_0 if $X > 0.817$; Accept H_0 if $X > -0.255$
12. $3 \leqslant r \leqslant 9$; 0.182
13. (a) 0.000 577
    (b) 0.007 38; 0.8, 0.8, 0.9

CHAPTER 10

NOTE: There will be variation in answers, depending on the degree of approximation used at various stages in the working.

Exercise 10a (page 540)

	χ^2 calc	ν	Decision
1.	1.93	3	accept fair
2.	18.16	9	accept
3.	6.19	2	reject H_0, yes
4.	4.95	3	no
	9.90	3	yes
5.	8.24	7	accept
6.	4.15	4	yes
7.	10.68	4	NS

Exercise 10b (page 547)

1. $X \sim \text{Bin}(5, 0.3)$, $E_i = 17, 36, 31, 13, 3, 0$ (combine last three classes), $\nu = 3$, $\chi^2_{\text{calc}} = 4.49$, accept
2. $X \sim \text{Bin}(3, 0.4)$, $E_i = 39, 78, 52, 11$, $\nu = 2$, $\chi^2_{\text{calc}} = 26.9$, No
3. (a) $X \sim \text{Bin}(4, 0.53)$, $E_i = 5, 22, 37, 28, 8$, $\nu = 3$, $\chi^2_{\text{calc}} = 1.23$, Yes
   (b) $X \sim \text{Bin}(6, 0.3)$, $E_i = 17, 42, 45, 26, 10$ (with last three classes combined), $\nu = 3$, $\chi^2_{\text{calc}} = 11.8$, No
4. $X \sim \text{Bin}(2, \frac{1}{6})$, $E_i = 150, 60, 6$, $\nu = 2$, $\chi^2_{\text{calc}} = 9.6$, Reject; Use $\bar{x} = 0.444$, $p = 0.222$, Find E_i, $\nu = 1$

5. $X \sim \text{Bin}(5, \frac{1}{6})$, $E_i = 80.5, 80.5, 32, 7$ (with last three classes combined), $\nu = 3$, $\chi^2_{\text{calc}} = 8.21$, yes, biased; $\bar{x} = 1$, $p = 0.2$, $X \sim \text{Bin}(5, 0.2)$
$E_i = 66, 82, 41, 11$ (with last three classes combined), $\nu = 2$, χ^2_{calc} is very small, too good a fit, query data
6. np, 1.6, 0.32, $E_i = 7.3, 17.1, 16.1, 7.5, 1.8, 0.2$ (combine last three classes), $\nu = 2$, $\chi^2_{\text{calc}} = 1.79$, Good fit
7. (a) $\bar{x} = 1.2$, $E_i = 99, 119, 72, 29, 9, 2$ (combine last two classes),
(b) $\chi^2_{\text{calc}} = 0.48$, $\nu = 3$, Very good fit
8. $E_i = 53, 55, 29, 10, 3$ (combine last two classes), $\nu = 2$, $\chi^2_{\text{calc}} = 3.01$, Good fit
9. $\bar{x} = 2.5$, $E_i = 8, 21, 26, 21, 13, 11$ (combining end classes), $\nu = 4$, $\chi^2_{\text{calc}} = 2.59$, Good
10. $\bar{x} = 1.28$ (2 d.p.), $E_i = 41, 52, 34, 14, 6$ (combining end classes), $\nu = 3$, $\chi^2_{\text{calc}} = 6.81$, NS
11. (a) $\bar{x} = 0.9$, $E_i = 183, 165, 74, 22, 6$ (combining end classes)
(b) $\nu = 3$, $\chi^2_{\text{calc}} = 1.62$, Adequate
12. (a) $E_i = 3, 13, 28, 32, 18, 6$ (combine first two classes), $\nu = 4$, $\chi^2_{\text{calc}} = 11.9$, S, Reject normal
(b) $\bar{x} = 171.54$, $s = 7.11$ (2 d.p.), $E_i = 6, 18, 32, 28, 13, 3$ (combine last two classes), $\nu = 2$, $\chi^2_{\text{calc}} = 1.73$, Accept normal, Good fit
13. $\bar{x} = 25.9$, $s = 11.8$ (1 d.p.), $E_i = 4, 7, 13, 18, 20, 17, 12, 6, 3, 1$ (combine first two classes and last three classes), $\nu = 4$, $\chi^2_{\text{calc}} = 0.95$, Very good fit

Exercise 10c (page 550)

1. $E_i = 17.5, 82.5, 17.5, 82.5$, $\nu = 1$, $\chi^2_{\text{calc}} = 0.58$, no
2. $E_i = 27.5, 972.5, 27.5, 972.5$, $\nu = 1$, $\chi^2_{\text{calc}} = 4.79$, yes
3. (a) $E_i = 24.225, 60.775, 32.775, 82.225$, $\nu = 1$, $\chi^2_{\text{calc}} = 7.59 \times 10^{-3}$, Independent, query whether agreement too close
(b) $E_i = 42.3, 47.7, 51.7, 58.3$, $\nu = 1$, $\chi^2_{\text{calc}} = 7.79$, Not independent
(c) $E_i = 37.5, 22.5, 87.5, 52.5$, $\nu = 1$, $\chi^2_{\text{calc}} = 2.54$, Independent
(d) $E_i = 11.5, 13.5, 43.5, 51.5$, $\nu = 1$, $\chi^2_{\text{calc}} = 3.18$, Independent
5. $E_i = 34.2, 29.8, 12.8, 11.2$, $\nu = 1$, $\chi^2_{\text{calc}} = 1.24$, NS
6. $E_i = 90.405, 56.595, 32.595, 20.405$ $\nu = 1$, $\chi^2_{\text{calc}} = 13.3$, Related

Exercise 10d (page 553)

NOTE: Minor adjustments need to be made when approximating, so that totals agree.

	E_i	ν	χ^2_{calc}	Conclusion
1.	(a) 21, 16.5, 12.5, 21, 16.5, 12.5	2	4.14	Independent
	(b) 17.3, 35.0 38.2, 11.5 14.9, 30.3, 32.9, 9.9, 33.8, 68.7, 74.9, 22.6	6	40.1	Not independent
	(c) 18.6, 10.4, 28.9, 16.1, 18, 10, 24.5, 13.5	3	1.60	Independent
	(d) 11.8, 24.5, 33.7, 13.5, 28, 38.5, 20.3, 42, 57.7, 8.4, 17.5, 24.1	6	7.53	Independent
2.	40, 60, 100, 140, 210, 350, 20, 30, 50	4	26.6	Yes
3.	40, 60, 100, 140, 210, 350, 20, 30, 50	2	6.57 (without c.c.)	Reject hypothesis
	47.4, 53.5, 56.1, 7.6, 8.5, 8.9	6	5.13 (with c.c.)	Accept hypothesis
4.	(645); 25, 145, 30, 87.5, 507.5, 105, 12.5, 72.5, 15	4	694	Performance in both sports not independent
5.	50.1, 29.5, 23.4, 22.9, 13.5, 10.6	2	4.00	Yes
6.	11.4, 14.3, 8.6, 15.7, 18.3, 22.9, 13.7, 25.1, 20.6, 25.7, 15.4, 28.3, 29.7, 37.1, 22.3, 40.9	9	12.0	Accept

Exercise 10e (page 556)

1. $E_i = 66.7, 33.3, 53.3, 26.7$, $\nu = 1$, $\chi^2_{\text{calc}} = 6.81$, No
2. A: $E_i = 12, 24, 36, 48, 60, 72, 60, 48, 36, 24, 12$, $\nu = 10$, $\chi^2_{\text{calc}} = 14.5$, Not biased
B: E_i as in A, $\chi^2_{\text{calc}} = 2.12$, Unbiased, but query whether data have been fiddled
3. $\bar{x} = 0.9$, $E_i = 21, 18, 8, 2, 1$ (combine last three classes), $\nu = 1$, $\chi^2_{\text{calc}} = 1.80$, Yes, Consistent

4. $\nu = 2$, $\chi^2_{calc} = 1.15$, Yes, Normal
5. (a) $\bar{x} = 2$, $p = 0.4$, $E_i = 6, 21, 28, 18, 6, 1$ (combine last two classes),
   (b) $\chi^2_{calc} = 2.21$, $\nu = 3$, Yes, Binomial adequate
6. $E_i = 21.0, 10.0, 7.0, 15.5, 7.5, 5, 41.5$ $19.5, 14$, $\nu = 4$, $\chi^2_{calc} = 7.86$, Accept hypothesis
7. $E_i = 48, 52, 96, 104, 96, 104$, $\nu = 2$, $\chi^2_{calc} = 6.57$, Yes, Proportions different, 1:44%, 2:42%, 3:55%, Beach 3 contributed to the high value of χ^2_{calc}
8. (a) $\lambda = 0.74$; combine $\geqslant 5$,
   $E_i = 667.96, 494.29, 182.89, 45.11, 8.35, 1.4$, $\nu = 4$, $\chi^2_{calc} = 13.8$,
   Not adequate
   (c) $\nu = 3$, $\chi^2_{calc} = 13.8$, not consistent
9. $\chi^2 = 13.77$, $\nu = 6$, Reject uniform distribution
10. (a) Combine first 3 classes, $\nu = 2$
    $E_i = 15.24, 25.76, 27.37, 11.63$,
    $\chi^2_{calc} = 0.144$; N.S., Yes
    (b) Combine $\geqslant 4$, $\nu = 3$, $\chi^2_{calc} = 0.404$
    $E_i = 25.85, 41.75, 33.72, 18.16, 10.52$, N.S., Yes
    (c) Both values of χ^2_{calc} are very small
11. (a) $E_i = 6.081, 17.026, 23.837, 22.248, 15.573, 8.721, 6.514$
    $\nu = 6$ (totals agree),
    $\chi^2_{calc} = 1.39$, Yes
    (b) $\chi^2_{calc} = 24.6$, $\nu = 4$, Yes
12. $\nu = 10$, $\chi^2_{calc} \simeq 16$
    No association at 5% level
    $\nu = 6$, $\chi^2_{calc} = 13.2$
    Reject hypothesis

CHAPTER 11

Exercise 11a (page 564)

NOTE: Answers may vary, depending on the final approximation used and the number of figures retained in the calculator during working.

1. (a) Positive correlation
   $(\bar{x}, \bar{y}) = (15, 14.1)$
   (b) Negative correlation,
   $(\bar{x}, \bar{y}) = (8.2, 75.7)$
   (c) No correlation, $(\bar{x}, \bar{y}) = (2.95, 8.25)$
2. $y = -10.75x + 93, 34$
3. (a) Strong positive (b) (25, 592)
   (d) (i) 130 (ii) 18.5
4. 4.3, 3.87; $y = 0.8125x + 0.376$; 2.4

Exercise 11b (page 567)

1. $x = 0.769y - 4.34$
2. (i) (a) $y = 0.64x + 4.50$
   (b) $x = 0.75y + 4.42$
   (ii) (a) $y = 0.43x + 7.57$
   (b) $x = 1.04y + 6.18$

Exercise 11c (page 574)

1. (a) (i) -1.96
   (ii) $y = -0.297x + 6.125$; $x = -0.757y + 8.51$
   (iii) 20.083 (iv) 51.170
   (b) (i) 4.18
   (ii) $y = 1.709x - 28.76$; $x = 0.348y + 18.92$
   (iii) 34.129 (iv) 6.944
   (c) (i) 18.5
   (ii) $y = 0.75x + 2.02$; $x = 1.19y - 1.37$
   (iii) 9.924 (iv) 15.601
   (d) (i) -3.92
   (ii) $y = -0.524x + 70.58$; $x = -1.37y + 100.8$
   (iii) 6.434 (iv) 16.84
2. $y + 1.45x = 114.4$;
   (a) No, x is controlled
   (b) $y = 61$ (2 S.F.)
3. (b) $y = 0.0207x + 0.614$
4. $F = 0.901 - 6.33$, $F = 20.8$
5. $\bar{A} = 14.875$, $\bar{T} = 17.5$,
   $A = 6.3 + 0.49T$, 16.1, 0.49
6. (b) $y = 127.15 + 1.17x$
7. $\bar{C} = 2.75$, $\bar{Y} = 35$,
   $Y = 3.976 + 11.28C$, 5.22
8. $\bar{T} = 500$, $\bar{E} = 1.6$,
   $E = -2.4 + 0.008T$, 1.04

Exercise 11d (page 587)

1. (a) -0.47, Some negative correlation
   (b) 0.77, High positive
   (c) 0.95, Very high positive
   (d) -0.84, High negative
2. 0.77
3. 0.82; $y = 0.48x + 22.36$,
   $x = 1.42y - 24.3$
4. -0.415
5. 0.616, 14.57

Exercise 11e (page 590)

1. (a) 2.17 (b) 0.895
   (c) $y = 0.028x - 2.37$,
   $x = 28.8y + 406.6$
2. (a) -0.038 (b) -0.359
   (c) $y = -0.63x + 677.0$;
   $x = -0.204y + 992.8$
3. (a) 7.1875×10^{-6} (b) 0.51
   (c) $y = 2233x + 96.9$;
   $x = 0.000\,116y - 0.0102$

Exercise 11f (page 594)

1. 0.26
2. 0.75
3. (a) 0.3, 0.5, 0.7
   (b) Mrs Brown and John
   1) Headrests 2) Heated rear window
   3) Anti-rust treatment

4. 0.86
5. (a) -0.52 (b) 0.82 (c) 0.9
(d) -0.9; Quite good agreement
6. Assume second judge gives no tied ranks
7. -0.036, Very little negative correlation
8. 0.60, Same
9. (b) (38.375, 2.275) (c) -0.84
(d) High negative correlation
10. 0.84, E
11. (a) 0.75, High positive correlation
12. (b) 0.84, High positive correlation

Exercise 11g (page 602)

N.S. 1, 3, 4, 7, 9,
S. 2, 5, 6, 8,

Exercise 11h (page 604)

N.S. 2, 4, 5, 7, 8,
S. 1, 3, 6,

Exercise 11i (page 604)

1. (a) 1 (b) -1 (c) 0.028
(d) 0.886 (e) 1 (f) -1
(g) 0.479 (h) -0.927
2. (a) 0.952 (S.) (b) 0.6 (N.S.)

Exercise 11j (page 612)

N.S. 1, 2, 3, 5, 6, 7,
S. 4,
8. (a) 1 (b) -1 (c) 0.067
(d) 0.733 (e) 1 (f) -1
(g) 0.378 (h) -0.822
9. (a) 0.857 (S.) (b) 0.467 (N.S.)
10. Judges 1, 2; $r_k = 0$, N.S.
Judges 1, 3; $r_k = 0.714$, S. 1% level
Judges 2, 3; $r_k = -0.143$, N.S.
11. 0.429, N.S.

Exercise 11k (page 623)

1. $Y = 0.56X + 2.9$; 0.79
2. 0.60, $W = 0.89h - 76$
3. $y = 0.53x - 25.4$, $x = 1.01y + 94.4$;
0.73, 53.7
4. (a) Straight line, negative slope
(b) Monotonically increasing curve
(c) -0.92 (d) -0.9
5. (a) $y = 1.33x + 5.7$ (b) 28 330
6. $y = 0.65x - 2.59$, 36.5
7. (a) 0.6 (b) $x = 1.23y - 1.17$
8. $y = 0.23x + 11.5$
9. $Y = 0.43X + 56.3$; 0.995
10. (a) (ii) (40, 414) (iv) 7, 135; 275
(b) (ii) Near $+1$
11. (a) -0.37 (b) -0.26
(c) Very little linear negative correlation
12. 0.91, High positive
13. 0.3, 0.6
14. (a) $P(X = -1) = P(X = 1) = \frac{1}{6}$;
$P(X = -0.5) = P(X = 0.5) = \frac{1}{3}$;
(b) 0.5
15. (b) $y = 1.03x + 0.53$
16. $r = 0.825$, $r_S = 0.929$, S at 1% level
17. (a) $r_k = 0.511$, $r_S = 0.660$
(b) r_k significant at 5% level
r_S significant at 5% level
18. $r_S = -0.3341$, $r_k = -0.2545$,
N.S. 5%;
$r_S = -0.6939$, $r_k = -0.5111$,
S. $2\frac{1}{2}$%
19. 0.48, 0.36
20. (a) $r_S = 0.527$, S.
(b) $H = 1.579T - 11.03$, $H = 1.3$
21. (a) (i) -0.976 (ii) -0.292
22. $r_S = 0.845$, $r_k = 0.71$ or 0.76
Both highly significant indicating an association.
23. (d) 0.962
24. (b) $y = 4.12\theta + 22.3$
(c) (i) (10.5, 91)
(iii) (a) Approx 64 cm
(b) Approx 196 cm
25. (b) $y = -5.2x + 55.5$ (c) 98.3p
26. $y = 0.0157x + 0.65$,
3 days 19 hours, 0.9419

INDEX